Large Networks
and Graph Limits

American Mathematical Society

Colloquium Publications
Volume 60

Large Networks and Graph Limits

László Lovász

American Mathematical Society
Providence, Rhode Island

Editorial Board

Lawrence C. Evans

Yuri Manin

Peter Sarnak (Chair)

2010 *Mathematics Subject Classification.* Primary 58J35, 58D17, 58B25, 19L64, 81R60, 19K56, 22E67, 32L25, 46L80, 17B69.

For additional information and updates on this book, visit
www.ams.org/bookpages/coll-60

ISBN-13: 978-0-8218-9085-1

Copying and reprinting. Individual readers of this publication, and nonprofit libraries acting for them, are permitted to make fair use of the material, such as to copy a chapter for use in teaching or research. Permission is granted to quote brief passages from this publication in reviews, provided the customary acknowledgment of the source is given.

Republication, systematic copying, or multiple reproduction of any material in this publication is permitted only under license from the American Mathematical Society. Requests for such permission should be addressed to the Acquisitions Department, American Mathematical Society, 201 Charles Street, Providence, Rhode Island 02904-2294 USA. Requests can also be made by e-mail to reprint-permission@ams.org.

© 2012 by the author. All rights reserved.
Printed in the United States of America.

∞ The paper used in this book is acid-free and falls within the guidelines
established to ensure permanence and durability.
Visit the AMS home page at http://www.ams.org/

10 9 8 7 6 5 4 3 2 1   17 16 15 14 13 12

To Kati
as all my books

Contents

Preface	xi
Part 1. Large graphs: an informal introduction	1
Chapter 1. Very large networks	3
1.1. Huge networks everywhere	3
1.2. What to ask about them?	4
1.3. How to obtain information about them?	5
1.4. How to model them?	8
1.5. How to approximate them?	11
1.6. How to run algorithms on them?	18
1.7. Bounded degree graphs	22
Chapter 2. Large graphs in mathematics and physics	25
2.1. Extremal graph theory	25
2.2. Statistical physics	32
Part 2. The algebra of graph homomorphisms	35
Chapter 3. Notation and terminology	37
3.1. Basic notation	37
3.2. Graph theory	38
3.3. Operations on graphs	39
Chapter 4. Graph parameters and connection matrices	41
4.1. Graph parameters and graph properties	41
4.2. Connection matrices	42
4.3. Finite connection rank	45
Chapter 5. Graph homomorphisms	55
5.1. Existence of homomorphisms	55
5.2. Homomorphism numbers	56
5.3. What hom functions can express	62
5.4. Homomorphism and isomorphism	68
5.5. Independence of homomorphism functions	72
5.6. Characterizing homomorphism numbers	75
5.7. The structure of the homomorphism set	79
Chapter 6. Graph algebras and homomorphism functions	83
6.1. Algebras of quantum graphs	83
6.2. Reflection positivity	88

6.3.	Contractors and connectors	94
6.4.	Algebras for homomorphism functions	101
6.5.	Computing parameters with finite connection rank	106
6.6.	The polynomial method	108

Part 3. Limits of dense graph sequences — 113

Chapter 7. Kernels and graphons — 115
- 7.1. Kernels, graphons and stepfunctions — 115
- 7.2. Generalizing homomorphisms — 116
- 7.3. Weak isomorphism I — 121
- 7.4. Sums and products — 122
- 7.5. Kernel operators — 124

Chapter 8. The cut distance — 127
- 8.1. The cut distance of graphs — 127
- 8.2. Cut norm and cut distance of kernels — 131
- 8.3. Weak and L_1-topologies — 138

Chapter 9. Szemerédi partitions — 141
- 9.1. Regularity Lemma for graphs — 141
- 9.2. Regularity Lemma for kernels — 144
- 9.3. Compactness of the graphon space — 149
- 9.4. Fractional and integral overlays — 151
- 9.5. Uniqueness of regularity partitions — 154

Chapter 10. Sampling — 157
- 10.1. W-random graphs — 157
- 10.2. Sample concentration — 158
- 10.3. Estimating the distance by sampling — 160
- 10.4. The distance of a sample from the original — 164
- 10.5. Counting Lemma — 167
- 10.6. Inverse Counting Lemma — 169
- 10.7. Weak isomorphism II — 170

Chapter 11. Convergence of dense graph sequences — 173
- 11.1. Sampling, homomorphism densities and cut distance — 173
- 11.2. Random graphs as limit objects — 174
- 11.3. The limit graphon — 180
- 11.4. Proving convergence — 185
- 11.5. Many disguises of graph limits — 193
- 11.6. Convergence of spectra — 194
- 11.7. Convergence in norm — 196
- 11.8. First applications — 197

Chapter 12. Convergence from the right — 201
- 12.1. Homomorphisms to the right and multicuts — 201
- 12.2. The overlay functional — 205
- 12.3. Right-convergent graphon sequences — 207
- 12.4. Right-convergent graph sequences — 211

Chapter 13.	On the structure of graphons	217
13.1.	The general form of a graphon	217
13.2.	Weak isomorphism III	220
13.3.	Pure kernels	222
13.4.	The topology of a graphon	225
13.5.	Symmetries of graphons	234
Chapter 14.	The space of graphons	239
14.1.	Norms defined by graphs	239
14.2.	Other norms on the kernel space	242
14.3.	Closures of graph properties	247
14.4.	Graphon varieties	250
14.5.	Random graphons	256
14.6.	Exponential random graph models	259
Chapter 15.	Algorithms for large graphs and graphons	263
15.1.	Parameter estimation	263
15.2.	Distinguishing graph properties	266
15.3.	Property testing	268
15.4.	Computable structures	276
Chapter 16.	Extremal theory of dense graphs	281
16.1.	Nonnegativity of quantum graphs and reflection positivity	281
16.2.	Variational calculus of graphons	283
16.3.	Densities of complete graphs	285
16.4.	The classical theory of extremal graphs	293
16.5.	Local vs. global optima	294
16.6.	Deciding inequalities between subgraph densities	299
16.7.	Which graphs are extremal?	307
Chapter 17.	Multigraphs and decorated graphs	317
17.1.	Compact decorated graphs	318
17.2.	Multigraphs with unbounded edge multiplicities	325

Part 4. Limits of bounded degree graphs — 327

Chapter 18.	Graphings	329
18.1.	Borel graphs	329
18.2.	Measure preserving graphs	332
18.3.	Random rooted graphs	338
18.4.	Subgraph densities in graphings	344
18.5.	Local equivalence	346
18.6.	Graphings and groups	349
Chapter 19.	Convergence of bounded degree graphs	351
19.1.	Local convergence and limit	351
19.2.	Local-global convergence	360
Chapter 20.	Right convergence of bounded degree graphs	367
20.1.	Random homomorphisms to the right	367
20.2.	Convergence from the right	375

Chapter 21. On the structure of graphings	383
21.1. Hyperfiniteness	383
21.2. Homogeneous decomposition	393
Chapter 22. Algorithms for bounded degree graphs	397
22.1. Estimable parameters	397
22.2. Testable properties	402
22.3. Computable structures	405
Part 5. Extensions: a brief survey	**413**
Chapter 23. Other combinatorial structures	415
23.1. Sparse (but not very sparse) graphs	415
23.2. Edge-coloring models	416
23.3. Hypergraphs	421
23.4. Categories	425
23.5. And more...	429
Appendix A. Appendix	433
A.1. Möbius functions	433
A.2. The Tutte polynomial	434
A.3. Some background in probability and measure theory	436
A.4. Moments and the moment problem	441
A.5. Ultraproduct and ultralimit	444
A.6. Vapnik–Chervonenkis dimension	445
A.7. Nonnegative polynomials	446
A.8. Categories	447
Bibliography	451
Author Index	465
Subject Index	469
Notation Index	473

Preface

Within a couple of months in 2003, in the Theory Group of Microsoft Research in Redmond, Washington, three questions were asked by three colleagues. Michael Freedman, who was working on some very interesting ideas to design a quantum computer based on methods of algebraic topology, wanted to know which graph parameters (functions on finite graphs) can be represented as partition functions of models from statistical physics. Jennifer Chayes, who was studying internet models, asked whether there was a notion of "limit distribution" for sequences of graphs (rather than for sequences of numbers). Vera T. Sós, a visitor from Budapest interested in the phenomenon of quasirandomness and its connections to the Regularity Lemma, suggested to generalize results about quasirandom graphs to multitype quasirandom graphs. It turned out that these questions were very closely related, and the ideas which we developed for the answers have motivated much of my research for the next years.

Jennifer's question recalled some old results of mine characterizing graphs through homomorphism numbers, and another paper with Paul Erdős and Joel Spencer in which we studied normalized versions of homomorphism numbers and their limits. Using homomorphism numbers, Mike Freedman, Lex Schrijver and I found the answer to Mike's question in a few months. The method of solution, the use of graph algebras, provided a tool to answer Vera's. With Christian Borgs, Jennifer Chayes, Lex Schrijver, Vera Sós, Balázs Szegedy, and Kati Vesztergombi, we started to work out an algebraic theory of graph homomorphisms and an analytic theory of convergence of graph sequences and their limits. This book will try to give an account of where we stand.

Finding unexpected connections between the three questions above was stimulating and interesting, but soon we discovered that these methods and results are connected to many other studies in many branches of mathematics. A couple of years earlier Itai Benjamini and Oded Schramm had defined convergence of graph sequences with bounded degree, and constructed limit objects for them (our main interest was, at least initially, the convergence theory of dense graphs). Similar ideas were raised even earlier by David Aldous. The limit theories of dense and bounded-degree graphs have lead to many analogous questions and results, and each of them is better understood thanks to the other.

Statistical physics deals with very large graphs and their local and global properties, and it turned out to be extremely fruitful to have two statistical physicists (Jennifer and Christian) on the (informal) team along with graph theorists. This put the burden to understand the other person's goals and approaches on all of us, but at the end it was the key to many of the results.

Another important connection that was soon discovered was the theory of property testing in computer science, initiated by Goldreich, Goldwasser and Ron several years earlier. This can be viewed as statistics done on graphs rather than on numbers, and probability and statistics became a major tool for us.

One of the most important application areas of these results is extremal graph theory. A fundamental tool in the extremal theory of dense graphs is Szemerédi's Regularity Lemma, and this lemma turned out to be crucial for us as well. Graph limit theory, we hope, repaid some of this debt, by providing the shortest and most general formulation of the Regularity Lemma ("compactness of the graphon space"). Perhaps the most exciting consequence of the new theory is that it allows the precise formulation of, and often the exact answer to, some very general questions concerning algorithms on large graphs and extremal graph theory. Independently and about the same time as we did, Razborov developed the closely related theory of flag algebras, which has lead to the solution of several long-standing open problems in extremal graph theory.

Speaking about limits means, of course, analysis, and for some of us graph theorists, it meant hard work learning the necessary analytical tools (mostly measure theory and functional analysis, but even a bit of differential equations). Involving analysis has advantages even for some of the results that can be stated and proved purely graph-theoretically: many definitions and proofs are shorter, more transparent in the analytic language. Of course, combinatorial difficulties don't just disappear: sometimes they are replaced by analytic difficulties. Several of these are of a technical nature: Are the sets we consider Lebesgue/Borel measurable? In a definition involving an infimum, is it attained? Often this is not really relevant for the development of the theory. Quite often, on the other hand, measurability carries combinatorial meaning, which makes this relationship truly exciting.

There were some interesting connections with algebra too. Balázs Szegedy solved a problem that arose as a dual to the characterization of homomorphism functions, and through his proof he established, among others, a deep connection with the representation theory of algebras. This connection was later further developed by Schrijver and others. Another one of these generalizations has lead to a combinatorial theory of categories, which, apart from some sporadic results, has not been studied before. The limit theory of bounded degree graphs also found very strong connections to algebra: finitely generated infinite groups yield, through their Cayley graphs, infinite bounded degree graphs, and representing these as limits of finite graphs has been studied in group theory (under the name of sofic groups) earlier.

These connections with very different parts of mathematics made it quite difficult to write this book in a readable form. One way out could have been to focus on graph theory, not to talk about issues whose motivation comes from outside graph theory, and sketch or omit proofs that rely on substantial mathematical tools from other parts. I felt that such an approach would hide what I found the most exciting feature of this theory, namely its rich connections with other parts of mathematics (classical and non-classical). So I decided to explain as many of these connections as I could fit in the book; the reader will probably skip several parts if he/she does not like them or does not have the appropriate background, but perhaps the flavor of these parts can be remembered.

The book has five main parts. First, an informal introduction to the mathematical challenges provided by large networks. We ask the "general questions" mentioned above, and try to give an informal answer, using relatively elementary mathematics, and motivating the need for those more advanced methods that are developed in the rest of the book.

The second part contains an algebraic treatment of homomorphism functions and other graph parameters. The two main algebraic constructions (connection matrices and graph algebras) will play an important role later as well, but they also shed some light on the seemingly completely heterogeneous set of "graph parameters".

In the third part, which is the longest and perhaps most complete within its own scope, the theory of convergent sequences of dense graphs is developed, and applications to extremal graph theory and graph algorithms are given.

The fourth part contains an analogous theory of convergent sequences of graphs with bounded degree. This theory is more difficult and less well developed than the dense case, but it has even more important applications, not only because most networks arising in real life applications have low density, but also because of connections with the theory of finitely generated groups. Research on this topic has been perhaps the most active during the last months of my work, so the topic was a "moving target", and it was here where I had the hardest time drawing the line where to stop with understanding and explaining new results.

The fifth part deals with extensions. One could try to develop a limit theory for almost any kind of finite structures. Making a somewhat arbitrary selection, we only discuss extensions to edge-coloring models and categories, and say a few words about hypergraphs, to much less depth than graphs are discussed in parts III and IV.

I included an Appendix about several diverse topics that are standard mathematics, but due to the broad nature of the connections of this material in mathematics, few readers would be familiar with all of them.

One of the factors that contributed to the (perhaps too large) size of this book was that I tried to work out many examples of graph parameters, graph sequences, limit objects, etc. Some of these may be trivial for some of the readers, others may be tough, depending on one's background. Since this is the first monograph on the subject, I felt that such examples would help the reader to digest this quite diverse material.

In addition, I included quite a few exercises. It is a good trick to squeeze a lot of material into a book through this, but (honestly) I did try to find exercises about which I expected that, say, a graduate student of mathematics could solve them with not too much effort.

Acknowledgements. I am very grateful to my coauthors of those papers that form the basis of this book: Christian Borgs, Jennifer Chayes, Michael Freedman, Lex Schrijver, Vera Sós, Balázs Szegedy, and Kati Vesztergombi, for sharing their ideas, knowledge, and enthusiasm during our joint work, and for their advice and extremely useful criticism in connection with this book. The creative atmosphere and collaborative spirit at Microsoft Research made the successful start of this research project possible. It was a pleasure to do the last finishing touches on the book in Redmond again. The author acknowledges the support of ERC Grant No. 227701,

OTKA grant No. CNK 77780, the hospitality of the Institute for Advanced Study in Princeton, and in particular of Avi Wigderson, while writing most of this book.

My wife Kati Vesztergombi has not only contributed to the content, but has provided invaluable professional, technical and personal help all the time.

Many other colleagues have very unselfishly offered their expertise and advice during various phases of our research and while writing this book. I am particularly grateful to Miklós Abért, Noga Alon, Endre Csóka, Gábor Elek, Guus Regts, Svante Janson, Dávid Kunszenti-Kovács, Gábor Lippner, Russell Lyons, Jarik Nešetřil, Yuval Peres, Oleg Pikhurko, the late Oded Schramm, Miki Simonovits, Vera Sós, Kevin Walker, and Dominic Welsh. Without their interest, encouragement and help, I would not have been able to finish my work.

Part 1

Large graphs: an informal introduction

CHAPTER 1

Very large networks

1.1. Huge networks everywhere

In the last decade it became apparent that a large number of the most interesting structures and phenomena of the world can be described by networks: often the system consists of discrete, well separable elements, with connections (or interactions) between certain pairs of them. To understand the behavior of the whole system, one has to study the behavior of the individual elements as well as the structure of the underlying network. Let us see some examples.

- Among very large networks, probably the best known and the most studied is the *internet*. Moreover, the internet (as the physical underlying network) gives rise to many other networks: the network of hyperlinks (web, logical internet), internet based social networks, distributed data bases, etc. The size of the internet is growing fast: currently the number of web pages may be 30 billion ($3 \cdot 10^{10}$) or more, and the number of interconnected devices is probably more than a billion. The graph theoretic structure of the internet determines, to a large degree, how communication protocols should be designed, how likely certain parts get jammed, how fast computer viruses spread etc.

- Social networks are basic objects of many studies in the area of sociology, history, epidemiology and economics. They are not necessarily formally established, like Facebook and other internet networks: The largest social network is the acquaintance graph of all living people, with about 7 billion nodes. The structure of this acquaintance graph determines, among others, how fast news, inventions, religions, diseases spread over the world, now and during history.

- Biology contributes ecological networks, networks of interactions between proteins, and the human brain, just to mention a few. The human brain, a network of neurons, is really large for its mass, having about a hundred billion nodes. One of the greatest challenges is, of course, to understand ourselves.

- Statistical physics studies the interactions between large numbers of discrete particles, where the underlying structure is often described by a graph. For example, a crystal can be thought of as a graph whose nodes are the atoms and whose edges represent chemical bonds. A perfect crystal is a rather boring graph, but impurities and imperfections create interesting graph-theoretical digressions. 10 gram of a diamond has about 5×10^{23} nodes. The structure of a crystal influences important macroscopic properties like whether the material is magnetizable, or how it melts.

- Some of the largest networks in engineering occur in chip design. There can be more than a billion transistors on a chip nowadays. Even though these networks are man-made and carefully designed, many of their properties, like the exact time they will need to perform some computation, are difficult to determine from their design, due to their huge size.

- To be pretentious, we can say that the whole universe is a single (really huge, possibly infinite) network, where the nodes are events (interactions between elementary particles), and the edges are the particles themselves. This is a network with perhaps 10^{80} nodes. It is an ongoing debate in physics how much additional structure the universe has, but perhaps understanding the graph-theoretical structure of this graph can help with understanding the global structure of the universe.

These huge networks pose exciting challenges for the mathematician. Graph Theory (the mathematical theory of networks) has been one of the fastest developing areas of mathematics in the last decades; with the appearance of the Internet, however, it faces fairly novel, unconventional problems. In traditional graph theoretical problems the whole graph is exactly given, and we are looking for relationships between its parameters or efficient algorithms for computing its parameters. On the other hand, very large networks (like the Internet) are never completely known, in most cases they are not even well defined. Data about them can be collected only by indirect means like random local sampling or by monitoring the behavior of various global processes.

Dense networks (in which a node is adjacent to a positive percent of other nodes) and very sparse networks (in which a node has a bounded number of neighbors) show a very different behavior. From a practical point of view, sparse networks are more important, but at present we have more complete theoretical results for dense networks. In this introduction, most of the discussion will focus on dense graphs; we will survey the additional challenges posed by sparse networks in Section 1.7.

1.2. What to ask about them?

Think of a really large graph, say the internet, and try to answer the following four simple questions about it.

Question 1. *Does the graph have an odd or even number of nodes?*

This is a very basic property of a graph in the classical setting. For example, it is one of the first theorems or exercises in a graph theory course that every graph with an odd number of nodes must have a node with even degree.

But for the internet, this question is clearly nonsense. Not only does the number of nodes change all the time, with devices going online and offline, but even if we fix a specific time like 12:00am today, it is not well-defined: there will be computers just in the process of booting up, breaking down etc.

Question 2. *What is the average degree of nodes?*

This, on the other hand, is a meaningful question. Of course, the average degree can only be determined with a certain error, and it will change as technology or the social composition of users change; but at a given time, a good approximation can be sought (I am not speaking now about how to find it).

Question 3. *Is the graph connected?*

To this question, the answer is almost certainly no: somewhere in the world there will be a faulty router with some unhappy users on the wrong side of it. But this is not the interesting way to interpret the question: we should consider the internet "disconnected" if, say, an earthquake combined with a sunflare severs all connections between the Old and New worlds. So we want to ignore small components that are negligible in comparison with the whole graph, and consider the graph "disconnected" only if it decomposes into two parts which are commeasurable with the whole. On the other hand, we may want to allow that the two large parts be connected by a very few edges, and still consider the graph "disconnected".

Question 4. *Where is the largest cut in the graph?*

(This means to find the partition of the nodes into two classes so as to maximize the number of edges connecting the two classes.) This example shows that even if the question is meaningful, it is not clear in what form can we expect the answer. We can ask for the fraction of edges contained in the largest cut (depending on the model, this can be determined relatively easily, with an error that is small with high probability, although it is not easy to prove that the algorithm works). But suppose we want to "compute" the largest cut itself; how to return the result, i.e., how to specify the largest cut (or even an approximate version of it)? We cannot just list all nodes and tell on which side do they belong: this would be too much time and memory space. Is there a better way to answer the question?

1.3. How to obtain information about them?

If we face a large network (think of the internet) the first challenge is to obtain information about it. Often, we don't even know the number of nodes.

1.3.1. Sampling. Properties of very large graphs can be studied by randomly sampling small subgraphs. The theory of this, sort of a statistics where we work with graphs instead of numbers, is called *property testing* in computer science. Initiated by Goldreich, Goldwasser and Ron [1998], this theory emerged in the last 15-20 years, and will be one of the main areas of applications of the methods developed in this book.

In the case of dense graphs G, it is simple to describe a reasonably realistic sampling process: we select independently a fixed number k of random nodes, and determine the edges between them, to get a random induced subgraph (Figure 1.1). We have to assume, of course, that we have methods to select a uniformly distributed random node of the graph, and to determine whether two nodes are adjacent. We'll call this *subgraph sampling*. For each graph F on $[k] = \{1, 2, \ldots, k\}$, there is a certain probability of seeing F when k nodes are sampled, which we denote by $\sigma_{G,k}(F)$. So every graph G defines a probability distribution $\sigma_{G,k}$ on all graphs with k nodes. It turns out that this sample contains enough information to determine many properties and parameters of the graph, with some error of course. This error can be made arbitrarily small with high probability if we choose the sample size k sufficiently large, depending on the error bound (and only on the error bound, not on the graph!).

One may try to strengthen this and allow this sampling process to be repeated a bounded number of times. This would not give anything new, however: sampling k nodes r times gives less information than sampling kr nodes once. For clarity, it is sometimes better to describe algorithms saying that we repeat a certain sampling process, but this could always be replaced by taking a single sample (larger, but still of bounded size).

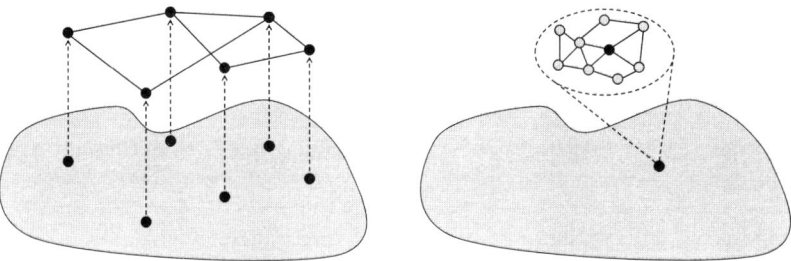

FIGURE 1.1. Sampling from a dense graph and from a graph with bounded degree.

1.3.2. Global observables. Instead of taking a random subset of the nodes (sampling) and studying the subgraph induced by them, we can take a random partition of the nodes into a small number of classes, and study the "quotient", the small graph obtained by merging the classes of the partition. (This will have very large edge multiplicities, which we have to normalize appropriately.) These quotients carry information about global measurements (like the number of stable sets, the maximum cut, various quantities in statistical physics, etc.). The remarkable fact is that under the right conditions, these "global" observables carry the same information as "local" sampling (see Sections 12.3 and 20.2.)

Another source of information about a very large network is the observation of the behavior of various processes on the graph, through a longer time interval. The observation can be global (measurement of some global parameter), or local (at one node, or a few neighboring nodes). Observing heat propagation through a material is an example of the first kind of approach; web crawlers can be considered as examples of the second, and in a sense so is our observation of the universe. There are some sporadic results about the local observation of simple random processes (Benjamini and Lovász [2002], Benjamini, Kozma, Lovász, Romik, and Tardos [2006], but a general theory of such local observation of global processes has not emerged yet.

1.3.3. Left and right homomorphisms. In theoretical studies, it is often more convenient to talk about homomorphisms (adjacency-preserving maps) between graphs, instead of looking at randomly chosen induced subgraphs. For two finite simple graphs F and G, let $\hom(F,G)$ denote the number of homomorphisms of F into G (adjacency-preserving maps from $V(F)$ to $V(G)$). We often normalize these homomorphism numbers, to get *homomorphism densities*:

$$(1.1) \qquad t(F,G) = \frac{\hom(F,G)}{\mathsf{v}(G)^{\mathsf{v}(F)}}.$$

This number is the probability that a random map of $V(F)$ into $V(G)$ preserves adjacency. (We denote by $V(F)$ and $E(F)$ the sets of nodes and edges of the graph F, respectively, and their cardinalities by $\mathsf{v}(F) = |V(F)|$ and $\mathsf{e}(F) = |E(F)|$.)

Homomorphisms will be basic tools throughout the book. We introduce them in Chapter 5 (where we survey some of our knowledge about them), but use them all the time thereafter. There will be different versions like homomorphisms into weighted graphs, which play an important role in statistical physics (we will return to them at the end of the Introduction).

Homomorphism densities can be expressed in terms of the distribution of samples, and vice versa (at least asymptotically, as the size of G tends to infinity). For example, let us consider the homomorphism density of the quadrilateral C_4 in a large graph G. If we map the four nodes of C_4 into G, if may be that the images are different, and so the image is a quadrilateral. It could also happen that the images of two nodes coincide. This cannot happen to adjacent nodes, because the image of an edge must be an edge; but it may happen to two opposite nodes. In this case, the image is a "V" (path with two edges). Or it can happen that both pairs of opposite nodes have the same image, and the image is a single edge. If we know the numbers of edges, V's and quadrilaterals in G, then we can compute the number of homomorphisms of the quadrilateral into G. (Warning: the same quadrilateral in G can be the image in 8 different ways, the same V can be the image in 4 ways, and the same edge, in 2 ways!). The numbers of quadrilaterals, V's and edges can be estimated by sampling. In fact, the last two will not matter much for very large graphs G, since a random map of 4 elements into $\mathsf{v}(G)$ elements will be one-to-one with high probability.

So homomorphism densities and sampling distributions carry the same information, why bother to introduce both? Using homomorphisms has several advantages (and some disadvantages).

- Homomorphism numbers are better behaved algebraically, and they have been used before to study various algebraic questions concerning direct product of graphs, like cancellation laws (see Section 5.4.2). Furthermore, a lot is known about other issues concerning homomorphisms: existence, structure, etc.

- When looking at a (large) graph G, we may try to study its local structure by counting homomorphisms from various "small" graphs F into G; we can also study its global structure by counting homomorphisms from G into various small graphs H. The first type of information is closely related (in many cases, equivalent) to sampling, while the second is related to global observables. This way homomorphisms are pointing at a certain duality between sampling and global observation. We can sum up our framework for studying large graphs in the following formula:

$$F \longrightarrow G \longrightarrow H.$$

 We will informally talk about "left-homomorphisms" and "right-homomorphisms" to refer to these two kind of mappings.

- We will characterize which distributions come from sampling k nodes from a (large) graph G, and we will characterize homomorphism densities as well. It turns out that a characterization of sample distributions is simpler and more natural, but putting it in another way, the characterization of homomorphism densities is more surprising, and therefore has more interesting applications.

- Using homomorphisms leads us to looking at things through the spectacles of category theory, and this point of view is very fruitful. For example, sometimes one can simply "turn arrows around", and get new results almost for free. We will say more about this generalization to categories near the end of the book, in Section 23.4.

1.4. How to model them?

1.4.1. Random graphs. We celebrated the 50th birthday of random graphs recently: The simplest random graph model was developed by Erdős and Rényi [1959] and Gilbert [1959]. Given a positive integer n and a real number $0 \leq p \leq 1$, we generate a random graph $\mathbb{G}(n,p)$ by taking n nodes, say $[n] = \{1,\ldots,n\}$, and connecting any two of them with probability p, making an independent decision about each pair.

There are alternate models, which are essentially equivalent from the point of view of many properties. Two of these were introduced in the early papers by Erdős–Rényi [1959, 1960]: We could fix the number of edges m, and then choose a random m-element subset of the set of pairs in $[n]$, uniformly from all such subsets. This random graph $\mathbb{G}(n,m)$ has very similar properties to $\mathbb{G}(n,p)$ when $m = p\binom{n}{2}$. Another model, closer to some of the more recent developments, is *evolving random graphs*, where edges are added one by one, always choosing uniformly from the set of unconnected pairs. Stopping this process after m steps, we get $\mathbb{G}(n,m)$.

Random graphs have many interesting, often surprising properties, and a huge literature, see Bollobás [2001], Janson, Łuczak and Ruczinski [2000], or Alon and Spencer [2000].

One conventional wisdom about random graphs with a given edge density is that they are all alike. Their basic parameters, like chromatic number, maximum clique, triangle density, spectra etc. are highly concentrated. This fact will be an important motivation when defining the right measure of global similarity of two graphs in Chapter 8.

Many generalizations of this random graph model have been studied. For example, we can consider a "template" for the random graph in the form of a weighted graph H on q nodes, with a weight $\alpha_i > 0$ associated with each node, and a weight $0 \leq \beta_{ij} \leq 1$ associated with each edge ij. We assume that the nodeweights sum to 1. We may also assume that H is complete with a loop at every node, since the missing edges can be added with weight 0. A *multitype random graph* $\mathbb{G}(n; H)$ with template H is generated as follows: We take $[n] = \{1,\ldots,n\}$ as its node set, where we think of n as a much larger number than q. We partition $[n]$ into q sets $V_1,\ldots,V_q$, by putting node u in V_i with probability α_i, and connecting each pair $u \in V_i$ and $v \in V_j$ with probability β_{ij} (all these decisions are made independently).

While multitype random graphs are too close to the original Erdős–Rényi model to be useful as, say, internet models, they play an extremely important role by serving as simple objects approximating arbitrarily large graphs (see Section 1.5.2 and Chapter 9).

More generally, one could have different probabilities assigned to different edges (this was suggested by Erdős and Rényi in their second paper [1960] already). The construction we will use a lot, namely constructing a random graph from a symmetric measurable function $[0,1]^2 \to [0,1]$ is a related idea, discovered independently several times, first time, probably, by Diaconis and Freedman [1981]. These random graphs, which we call W-random, will be discussed in Section 10.1 and will play an important role throughout the book.

1.4.2. Quasirandom graphs. Deterministic objects that look and behave like randomly generated ones are important in various branches of science. For example, pseudorandom number generators are basic algorithms in computer science, with many applications in Monte-Carlo algorithms, computer security and elsewhere. Exact definitions are usually difficult to give, and they vary according to need. It is very remarkable that in graph theory it is possible to give a very robust definition of quasirandom graphs, where many related or even seemingly quite different formalizations of properties of random graphs capture the same notion. We know that random graphs have a variety of quite strict properties (with high probability); it turns out that for several of these basic properties, the exceptional graphs are the same. In other words, any of these properties implies the others, regardless of any stochastic consideration.

A measure of quasirandomness of a graph was introduced by Thomason [1987]; the theory of quasirandom graph sequences, which has been an important example for convergent graph sequences central to this book, was developed by Chung, Graham and Wilson [1989].

To make this idea precise, we consider a sequence of graphs (G_n) with $\mathsf{v}(G_n) \to \infty$. For simplicity of notation, assume that $\mathsf{v}(G_n) = n$. Let $0 < p < 1$ be a real number. Consider the following properties of the sequence of graphs:

(QR1) Almost all degrees are asymptotically pn and almost all codegrees (numbers of common neighbors of two nodes) are asymptotically $p^2 n$.

(QR2) For every fixed graph F, the number of homomorphisms of F into G_n is asymptotically $p^{\mathsf{e}(F)} n^{\mathsf{v}(F)}$.

(QR3) The number of edges is asymptotically $pn^2/2$ and the number of 4-cycles is asymptotically $p^4 n^4/8$. (We have to divide by 2 and 8, because we are counting unlabeled copies rather than homomorphisms.)

(QR4) The number of edges induced by any set of $n/2$ nodes is asymptotically $pn^2/8$.

(QR5) For any two disjoint sets X, Y of nodes, the number of edges between X and Y is $p|X||Y| + o(n^2)$.

All these properties hold with probability 1 if $G_n = \mathbb{G}(n,p)$. However, more is true: if a graph sequence satisfies either one of them, then it satisfies all. Perhaps the most surprising fact along these lines is the equivalence of the second and third: prescribing the right asymptotic number of copies in G_n for just two small graphs (the edge and the 4-cycle) forces every other simple graph to have (asymptotically) the right number of copies.

Such graph sequences are called *quasirandom*. The five properties above are only a sampler; there are many other properties of random graphs that are also equivalent to these (Chung, Graham and Wilson [1989], Simonovits and Sós [1991, 1997, 2003]).

Many interesting deterministic graph sequences are quasirandom. We mention an important example from number theory:

EXAMPLE 1.1 (**Paley graphs**). Let q be any prime congruent 1 modulo 4, and let us define a graph on $\{0, \ldots, q-1\}$ by connecting i and j if and only if $i-j$ is a quadratic residue modulo q. We construct this graph for every such prime, and order them in a sequence.

This graph sequence is quasirandom with density $1/2$. (To verify the first property above is perhaps the easiest; see Exercise 1.2). This example also illustrates how some of the equivalent conditions above may be much easier to verify than others: in this case, the third would be as easy as the verification of the first, but the second and fourth would be quite difficult: How would you count the number of copies of, say, the Petersen graph in a Paley graph? How would you count the number of those differences that are quadratic residues between, say, square-free integers in $\{0, \ldots, q-1\}$? When posed directly, these questions sound formidable; but the equivalence of the above conditions provides answers to them. ♦

We should emphasize that in this setting, quasirandomness is a property of a sequence of graphs, not of a single graph. Of course, one could introduce a measure of deviation from the "ideal" quasirandomness in each of the conditions (QR1)–(QR5), and prove explicit relationships between them. Since our interest is the limit theory, we will not go in this direction.

Sometimes we need to consider quasirandom bipartite graphs, which can be defined, *mutatis mutandis*, by any of the properties above. More generally, just as there are multitype random graphs, there are also *multitype quasirandom graph sequences*. Similarly as for random graphs, a multitype quasirandom graph sequence (G_n) is defined by a "template" weighted graph H on q nodes, with a nodeweights $\alpha_i > 0$ and edgeweights β_{ij}. The sequence is multitype quasirandom with template H, if the node set $V(G_n)$ can be partitioned into q sets $V_1, \ldots, V_q$ such that $|V_i| \sim \alpha_i \mathsf{v}(G_n)$, the subgraphs $G_n[V_i]$ induced by V_i form a quasirandom sequence for every $i \in [q]$, and the bipartite subgraphs $G_n[V_i, V_j]$ between V_i and V_j form a quasirandom bipartite graph sequence for each pair $i \neq j \in [q]$.

The same remark applies as for multitype random graphs: they play an extremely important role by serving as simple objects approximating arbitrarily large graphs. The equivalence of conditions (Q1)–(Q5) can be generalized appropriately (with a larger, but finite set of graphs in (Q3) instead of just 2), as it will be discussed in Section 16.7.1.

The main topic of the book, the theory of convergent graph sequences, can be considered as a further, rather far-reaching generalization of quasirandom sequences.

1.4.3. Randomly growing graphs. Random graph models on a fixed set of nodes, discussed above, fail to reproduce important properties of real-life networks. For example, the degrees of Erdős–Rényi random graphs follow a binomial distribution, and so they are asymptotically normal if the edge probability p is a constant, and asymptotically Poisson if the expected degree is constant (i.e., $p = p(n) \sim c/n$). In either case, the degrees are highly concentrated around the mean, while the degrees of real life networks tend to obey the "Zipf phenomenon", which means that the tail of the distribution decreases according to a power law (unlike the most familiar distributions like Gaussian, geometric or Poisson, whose tail probability drops exponentially; Figure 1.2).

In 1999 Albert and Barabási [1999, 2002, 2002] created a new random network model. Perhaps the main new feature compared with the Erdős–Rényi graph evolution model is that not only edges, but also nodes are added by natural rules of growing. When a new node is added, it connects itself to a given number d of old nodes, where each neighbor is selected randomly, with probability proportional to its degree. (This random selection is called *preferential attachment*.) The Albert–Barabási graphs reproduce the "heavy tail" behavior of the degree sequences of real-life graphs. Since then a great variety of growing networks were introduced, reproducing this and other empirical properties of real-life networks.

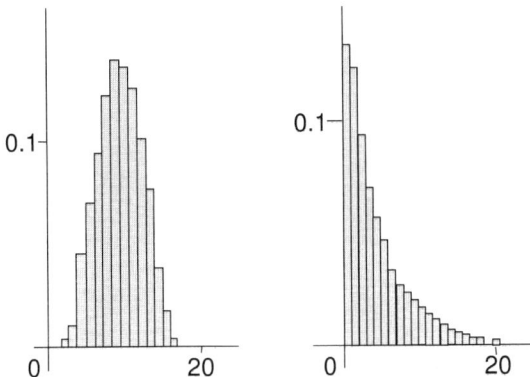

FIGURE 1.2. Degree distributions of an Erdős–Rényi random graph on 100 nodes with edge density .1 (left) and of a real life graph with similar parameters (right). The main feature to observe about the latter is not that the largest frequency is 1, but that it is much more stretched out.

This is perhaps the first point which suggests one of our main tools, namely assigning limits to sequences of graphs. Just as the Law of Large Numbers tells us that adding up more and more independent random variables we get an increasingly deterministically behaving number, these growing graph sequences tend to have a well-defined structure, for almost all of the possible random choices along the way. In the limit, the randomness disappears, and the asymptotic behavior of the sequence can be described by a well-defined limit object. We will return to this in this Introduction in Sections 1.5.3 and 11.3.

EXERCISE 1.2. Prove that the sequence of Paley graphs is quasirandom.

1.5. How to approximate them?

If we want to experiment with a large network (say, try out a new protocol for the internet), then it is good to have a "scaled down" version of it. In other words, we want a compact approximate description of a very large network, from which a network similar to the original, but of suitable size, can be generated. To make this mathematically precise, we need to define what we mean by two graphs to be "similar" or "close", and describe what kind of structures we use for approximation.

1.5.1. The distance of two graphs. There are many ways of defining the distance of two graphs G and G'. Suppose that the two graphs have a common node set $[n]$. Then a natural notion of distance is the *edit distance*, defined as the number of edges to be changed to get from one graph to the other. This could also be viewed as the *Hamming distance* $|E(G)\triangle E(G')|$ of the edge sets ($\triangle$ denotes symmetric difference). Since our graphs are very large, we want to normalize this. If the graphs are dense, then a natural normalization is

$$d_1(G,G') = \frac{|E(G)\triangle E(G')|}{n^2}.$$

While this distance plays an important role in the study of testable graph properties, it does not reflect structural similarity well. To raise one objection, consider two random graphs on $[n]$ with edge density $1/2$. As mentioned in the introduction, these graphs are very similar from almost every aspect, but their normalized edit distance is large (about $1/2$ with high probability). One might try to decrease this by relabeling one of them to get the best overlay minimizing the edit distance; but the improvement would be marginal (tending to 0 if n tends to infinity).

Another trouble with the notion of edit distance is that it is defined only when the two graphs have the same set of nodes. We want to define a notion of distance for two graphs that are so large that we don't even know the number of their nodes, and these numbers might be very different. For example, we want to find that two large random graphs are "close" even if they have a different number of nodes.

One useful way to overcome these difficulties is to base the measurement of distance on sampling. Recall that for a graph G, $\sigma_{G,k}$ is the probability distribution on graphs on $[k] = \{1, 2, \ldots, k\}$ obtained by selecting a random ordered k-subset of nodes and taking the subgraph induced by them. Strictly speaking, this is only defined when $k \leq \mathsf{v}(G)$; but we are interested in taking a small sample from a large graph, not the other way around. To make the definition precise, let us say that the sampling returns the edgeless k-node graph if $k > \mathsf{v}(G)$. (In this case it would be a better solution to sample with repetition, but sampling without repetition is better in other cases, so let us stick to it.)

Now if we have two graphs G and G', we can compare the distributions of k-node samples for any fixed k. We use the *variation distance* between distributions α and β on the same set, defined by

$$d_{\text{var}}(\alpha, \beta) = \sup_X |\alpha(X) - \beta(X)|,$$

where the supremum is taken over all measurable subsets (observable events). If we want to measure the distance of two graphs by a single number, we use a simple trick known from analysis: We define the *sampling distance* of two dense graphs G and G' by

$$(1.2) \qquad \delta_{\text{samp}}(G, G') = \sum_{k=1}^{\infty} \frac{1}{2^k} d_{\text{var}}(\sigma_{G,k}, \sigma_{G',k})$$

(Here the coefficients $1/2^k$ are quite arbitrary, they are there only to make the sum convergent; but the above is a convenient choice.) This distance notion is very suitable for our general goals, since two graphs are close in this distance if and only if random sampling of "small" induced subgraphs does not distinguish them reliably. However, sampling distance has one drawback: it does not directly reflect any structural similarity.

In Chapter 8 we will define a notion of distance, called *cut distance*, between graphs, which will be satisfactory from all these points of view: it will be defined for two graphs with possibly different number of nodes, the distance of two random graphs with the same edge density will be very small, and it will reflect global structural similarity. The definition involves too many technical details to be given here, unfortunately. But it will turn out (and this is one of the main results in this book) that the cut distance is equivalent to the sampling distance in a topological sense.

1.5.2. Approximation by smaller: Regularity Lemmas. Let us return to the question of "scaling down" a huge graph, first in the dense case. The main

1.5. HOW TO APPROXIMATE THEM?

tool for doing so is the "Szemerédi Partition" or "Regularity Lemma". Szemerédi developed the first version of the Regularity Lemma for his celebrated proof of the Erdős–Turán Conjecture on arithmetic progressions in dense sets of integers in 1975. Since then, the Lemma has emerged as a fundamental tool in graph theory, with many applications in extremal graph theory, combinatorial number theory, graph property testing etc., and became a true focus of research in the past years.

Informally, the Regularity Lemma says that every graph can be approximated by a multitype quasirandom graph, where the number of classes depends on the error of the approximation only. This lemma can be viewed as an archetypal example of dichotomy between randomness and structure, where we try to decompose a (large and complicated) object A into a more highly structured object A' with a (quasi)random perturbation (cf. Tao [2006c]). The highly structured part may be easier to handle because of the structure, and the quasirandom part will often be easier to handle due to Laws of Large Numbers.

Pixel pictures. In this introductory part, we want to illustrate the idea of a regularity partition visually. To this end, let us introduce a non-standard way of visualizing graphs. On the left of Figure 1.3 we see a graph (the Petersen graph). In the middle, we see its adjacency matrix. On the right, we see another version of its adjacency matrix, where the 0's are replaced by white squares and the 1's are replaced by black squares. We think of the whole picture as the unit square, so the little squares have side length $1/n$, where n is the number of nodes. The origin is in the upper left corner, following the convention of indexing matrix elements.

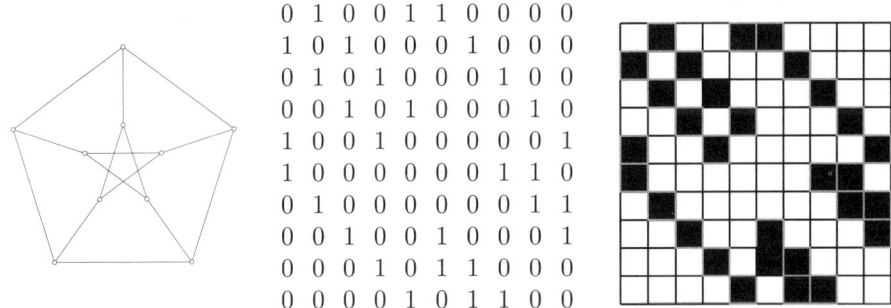

FIGURE 1.3. The Petersen graph, its adjacency matrix, and its pixel picture

It is not clear that this pixel picture reveals more about small graphs than the usual way of drawing them (probably less), but it can be suggestive for large graphs. Figure 1.4 shows the usual drawing and the pixel picture of a *half-graph*, a bipartite graph defined on the set $\{1, \ldots, n, 1', \ldots, n'\}$, where the edges are the pairs (i, j) with $i \leq j'$. For large n, the pixel picture of a half-graph may be more informative, as we will see in the next section.

The left square in Figure 1.5 is the pixel picture of a (reasonably large) random graph. We don't see much structure—and we shouldn't. From a distance, this picture is more-or-less uniformly grey, similar to the second square. The 100×100 chessboard in the third picture is also uniformly grey, or at least it would become so if we increased the number of pixels sufficiently. One might think that it represents

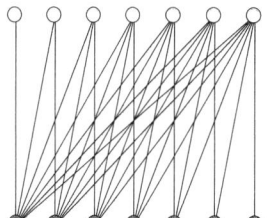 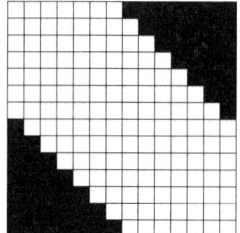

FIGURE 1.4. A half-graph and its pixel picture

a graph that is close to the random graph. But rearranging the rows and columns so that odd indexed columns come first, we get the 2×2 chessboard on the right! So wee see that both the middle and the right side pictures represent a complete bipartite graph. *The pixel picture of a graph depends on the ordering of the nodes.* We can be reassured, however, that a random graph remains random, no matter how we order the nodes, and so the picture on the left remains uniformly grey, no matter how the nodes are ordered.

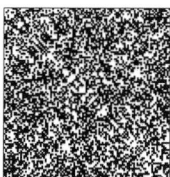   

FIGURE 1.5. A random graph with 100 nodes and edge density $1/2$, a random graph with very many nodes and edge density $1/2$, a chessboard, and the pixel picture obtained by rearranging the rows and columns.

REMARK 1.3. Using pixel pictures to represent graphs, in particular random graphs, goes in a sense in the opposite direction to what was studied in the psychology of vision. Of course, processing images given by pixel pictures has been a fundamental issue in connection with computer graphics and related areas, and we are not going into this issue in this book. But we should mention the work of Julesz, who studied the question of how well the human eye can distinguish random noise (like Figure 1.5(a)) from images that are also uniformly grey but more structured (textured). The chessboard in Figure 1.5(b) would be a trivial example of such an image. Disproving some of his conjectures, Diaconis and Freedman [1981] constructed pixel pictures that are very closely related to our W-random graphs.

The Regularity Lemma. We illustrate the Regularity Lemma by Figure 1.6. The graph on the left side (given by its pixel picture) looks quite random. In the middle we see the same graph, with its nodes ordered differently. In this picture, we see some structure of the graph (even though it is not as clear-cut as in Figure 1.5); what we see is that the upper left corner is denser, and the lower right corner is sparser. If we cut the picture into four equal parts, and average the "blackness" in each, we get the picture on the right. Inside each of the four parts, the arrangement is quite random-like, and further rearrangement would not reveal any additional structure.

Still informally, the Regularity Lemma says the following:

The nodes of every graph can be partitioned into a "small" number of "almost equal" parts in such a way that for "almost all" pairs of partition classes, the bipartite graph between them is "quasirandom".

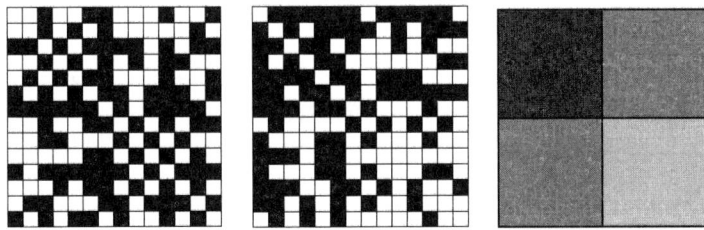

FIGURE 1.6. A random-looking pixel picture, an informative rearrangement, and its regularity partition

Some of the expressions in quotation marks are easy to explain. For the whole theorem, we have an error bound $0 < \varepsilon < 1$ specified in advance. The condition that the parts are "almost equal" means that their sizes differ by at most one: if the graph has n nodes partitioned into k classes, then the size of each class is either $\lfloor n/k \rfloor$ or $\lceil n/k \rceil$. The condition that the number of classes is "small" means that it can be bounded by an explicit function $f(\varepsilon)$ of ε; to exclude trivialities, we also assume that $k \geq 1/\varepsilon$. "Almost all" pairs of classes means that we allow $\varepsilon \binom{k}{2}$ exceptional pairs about which we don't claim anything (we can include the subgraphs induced by the classes among these exceptions). Finally, we need to define what it means to be "random-like": one way to put it is that this bipartite graph is quasirandom with some density p_{ij} (which may be different for different pairs of classes) and with error ε, in the sense introduced (informally) in Section 1.4.2.

Regularity partitions and quasirandomness have a lot to do with each other. Not only is quasirandomness part of the statement of the Regularity Lemma, but the regularity lemma can be used to characterize quasirandomness: Simonovits and Sós [1991] proved that a graph sequence is quasirandom with density p if and only if the graphs have regularity partitions for arbitrarily small $\varepsilon > 0$ such that the densities p_{ij} between the partition classes tend to p.

I have to come back to the "small" number of partition classes. The proof gives a tower $2^{2^{2^{\cdots}}}$ of height $1/\varepsilon^5$, which is a very large number, and which unfortunately cannot be improved too much, since Gowers [1997] constructed graphs for which the smallest number of classes in a Szemerédi partition was at least a tower of height $\log(1/\varepsilon)$. So the tower behavior is a sad fact of life. There are related partitions with a more decent number of classes, as we shall see in Chapter 9, where regularity partitions will be defined formally. We will also discuss situations when the regularity partitions have a very decent size, like $1/\varepsilon^{\text{const}}$ (Sections 13.4 and 16.7). Implicitly or explicitly, regularity partitions will be used throughout this book.

1.5.3. Approximation by infinite: convergence and limits. This idea can be motivated by how we look at a large piece of metal. This is a crystal, that is a really large graph consisting of atoms and bonds between them. But from many points of view (e.g., the use of the metal in building a bridge), it is more useful to consider it as a continuum with a few important parameters (density, elasticity etc.). Its behavior is governed by differential equations relating these parameters. Can we consider a more general very large graph as some kind of a continuum?

Our way to make this intuition precise is to consider a growing sequence (G_n) of graphs whose number of nodes tends to infinity, to define when such a sequence is convergent, and to assign a limit object to convergent graph sequences, which somehow incorporates all the properties we want to be remembered. (We have mentioned this idea in connection with randomly growing graphs, but now we don't assume anything about how the graphs in the sequence are obtained.) This plan is the backbone of this book: we will carry it out both for dense graphs and also for graphs with bounded degree. There will be a good collection of applications of this work.

Our discussion of sampling from a graph suggests a general principle leading to a definition: we consider samples of a fixed size k from G_n, and their distribution. We say that the sequence is *locally convergent* (with respect to the given sampling method) if this distribution tends to a limit as $n \to \infty$ for every fixed k.

For dense graphs, this notion of convergence was defined by Borgs, Chayes, Lovász, Sós, and Vesztergombi [2006, 2008]; some elements of this definition go back to Erdős, Lovász and Spencer [1979]. This notion has many useful properties. Perhaps most important of these is that it can be characterized in terms of the cut distance of graphs. It is not hard to see that the above notion of convergence is equivalent to saying that the graph sequence is a Cauchy sequence in the sampling distance. One of the main results presented in this book is Theorem 11.3, which can be stated informally as follows:

The same graph sequences are convergent (Cauchy sequences) for both the cut distance and the sampling distance.

If we have a notion of convergence, the question arises naturally: what does it converge to? Can we describe a limit object for every convergent graph sequence? The family of limiting sample distributions (one for each k) can be considered as a limit object of the sequence (we call this the "weak limit"). This is not always a helpful representation of the limit object, and a more explicit description is desirable.

A next step is to represent the family of distributions on finite graphs (the samples) by a single probability distribution on countable graphs: we get certain notion of random graphs on the countable set $\mathbb{N}^* = \{1, 2, 3, \dots\}$ (see Theorem 11.52).

More explicit descriptions of these limit objects can also be given, in the form of a two-variable measurable function $W : [0,1]^2 \to [0,1]$, called a *graphon* (Lovász and Szegedy [2006]; see Section 7). These limit objects can be considered as weighted graphs with an underlying set of continuum cardinality. (If you wish, you can also think of these graphons as unweighted graphs on a non-standard model of the unit interval, where $W(x, y)$ is the density of edges between an infinitesimal neighborhood of x and an infinitesimal neighborhood of y; this approach will be explained in Section 11.3.2). Random graphs with edge density $1/2$ converge to the identically 1/2 function (have a look at the two squares on the left of Figure 1.5). Figure 1.7 illustrates that the sequence of half-graphs (discussed in Section 1.5.2) converges to a limit (the function $W(x, y) = \mathbb{1}(y \geq x + 1/2 \text{ or } x \geq y + 1/2)$. It has been observed and used before (see e.g. Sidorenko [1991]) that such functions can be used as generalizations of graphs, and this gives certain arguments a greater analytic flexibility.

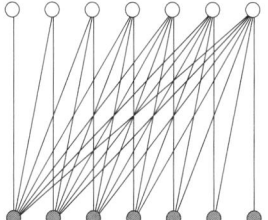 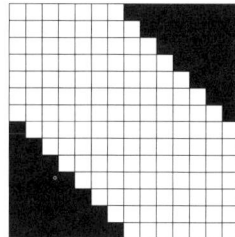 

FIGURE 1.7. A half-graph, its pixel picture, and the limit function

Let us describe another example here (more to follow in Section 11.4.2). The picture on the left side of Figure 1.8 is the adjacency matrix of a graph G with 100 nodes, where the 1's are represented by black squares and the 0's, by white squares. The graph itself is constructed by a simple randomized growing rule: Starting with a single node, we create a new node, and connect every pair of nonadjacent nodes with probability $1/n$, where n is the current number of nodes. (This construction will be discussed in detail in Section 11.4.2.)

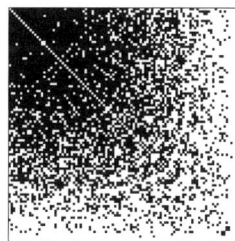 

FIGURE 1.8. A randomly grown uniform attachment graph with 100 nodes, and a (continuous) function approximating it

The picture on the right side is a grayscale image of the function $U(x, y) = 1 - \max(x, y)$. (Recall that the origin is in the upper left corner!) The similarity with the picture on the left is apparent, and suggests that the limit of the graph sequence on the left is this function. This turns out to be the case in a well defined sense. It follows that to approximately compute various parameters of the graph on the left side, we can compute related parameters of the function on the right side. For example, the triangle density of the graph on the left tends (as $n \to \infty$) to the integral

$$(1.3) \qquad \int_{[0,1]^3} U(x,y)U(y,z)U(z,x)\,dx\,dy\,dz$$

(the evaluation of this integral is a boring but easy task). It is easy to see how to generalize this formula to express the limiting density of any fixed graph F.

We hope that the examples above provide motivation for the following fact, which is one of the key results to be discussed in the book (Theorem 11.21):

The limit of any convergent graph sequence can be represented by a graphon, in the sense that the limiting density of any fixed simple graph F is given by an integral of the type (1.3).

Of course, a graphon can be infinitely complicated; but in many cases, limits of growing graph sequences have a limit graphon that is a continuous function described by a simple formula (see some further examples in Section 11.4.2). Such a limit graphon provides a very useful approximation of a large dense graph.

Graphons can be considered as generalizations of graphs, and this way of looking at them is very fruitful. In fact, many results can be stated and proved for graphons in a more natural and cleaner way. In particular, regularity lemmas can be extended to graphons, where we will see that they are statements about approximating general measurable functions by stepfunctions. Approximating graphs by multitype quasirandom graphs is as basic a tool in graph theory as approximating functions by stepfunctions is in analysis.

REMARK 1.4. Much of this book is about finite, countable and uncountable graphs and connections between them. There are two technical limitations of measure theory that we have to work our way around. (a) *One cannot construct more than countably many independent random variables* (in a nontrivial way, neither of them concentrated on a single value). This is the reason while we cannot define a random graph on an uncountable set like $[0,1]$, only on finite and countable subsets of it. (b) *There is no uniform distribution on a countable set* (while there is one on every finite set and then again on sets with continuum cardinality like $[0,1]$). This limitation is connected to the fact that the limit objects for convergent graph sequences will be graphons (which could be considered as graphs defined on a continuum) rather than graphs on a countable set as one would first expect.

I want to emphasize that these difficulties are not just annoying technicalities: they reflect the fact, for example, that the limit object of a convergence graph sequence carries a lot more information than what could be squeezed into a countable graph. Both measure theory and combinatorics force us into the same realm.

1.6. How to run algorithms on them?

1.6.1. Parameter estimation. What can we learn about a huge graph G from sampling? There are several related questions here, depending on what we need as a result. The easiest setup is when we want to compute a numerical parameter of the graph; say, how large is the maximum cut, or what fraction of the triples induce a triangle. We call this problem *parameter estimation*. Most of the time we normalize the parameter to be between 0 and 1. Since, as discussed above, we get information about the graph through random sampling, any answer we can possibly compute will, with some probability, be in error. So we will have to specify an error parameter $\varepsilon > 0$, and will have to accept an answer which, with probability at least $1 - \varepsilon$, will be closer than ε to the true value of the parameter.

An easy example is to estimate the triangle density (number of triangles divided by $\binom{n}{3}$). A trivial algorithm is to pick many random triples of nodes independently, and count how many of them form triangles in the graph. Elementary statistics tells us that if we sample $O(\varepsilon^{-2}|\log \varepsilon|)$ triples, then with probability at least $1 - \varepsilon$, our estimate will be closer than ε to the truth.

A much more interesting and difficult example is that of estimating the density a of the maximum cut (its size divided by $\binom{n}{2}$) in a graph G. One thing we can try is to choose N random nodes (where N depends on the error bound ε), and compute the density X of the maximum cut in the subgraph H they induce. Is X a good estimate for a?

The inequality $X \geq a - \varepsilon$ (for every $\varepsilon > 0$ if N is large enough, with high probability) is relatively easy to prove. The graph G has a cut C with density a, and this cut provides a cut C' in the random induced subgraph H. It is easy to see that the density of C' is the same as the density a of C in expectation, and it takes some routine computation in probability theory to show that it is highly concentrated around this value. The density X of the largest cut in H is at least the density of C', and so with high probability it is at least $a - \varepsilon$ (Figure 1.9).

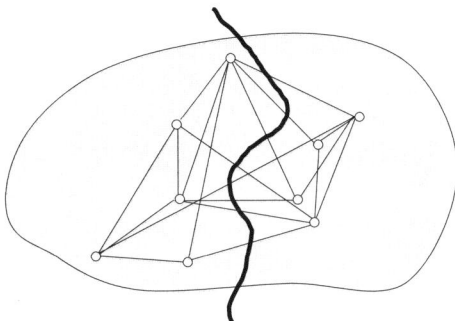

FIGURE 1.9. A dense cut in the large graph gives a dense cut in the sample.

The reverse inequality is much more difficult to prove, at least from scratch, and in fact it is rather surprising. We can phrase the question like this: Suppose that most random induced subgraphs H on N nodes have a cut that is denser than b. Does it follow that G has a cut that is denser than $b - \varepsilon$? It is not clear why this should be so: why should these cuts in these small subgraphs "line up" to give a dense cut in G?

We will see that it does follow that the estimate is correct, once N is large enough (about $\varepsilon^{-4}|\log \varepsilon|$). In fact, one can give general necessary and sufficient conditions under which parameters can be estimated by sampling, as we will see in Section 15.1.

1.6.2. Property testing. A more complicated issue is *property testing*: we want to determine whether the graph has some given property, for example, can it be decomposed into two connected components of equal size, is it planar, or does it contain any triangle. We could consider this as a 0-1 valued parameter, but computing this parameter approximately would not make sense (or rather, it would be requiring too much, since this would be equivalent to exact computation).

A good way of posing this problem was developed by Rubinfeld and Sudan [1996] and Goldreich, Goldwasser and Ron [1998]. In the slightly different context of "additive approximation", closely related problems were studied by Arora, Karger and Karpinski [1995] (see e.g. Fischer [2001] for a survey and the volume edited by Goldreich [2010] for a collection of more recent surveys).

This approach acknowledges that any answer is only approximate. Suppose that we want to test for a property $\mathcal{P}$, and we get information about the graph by taking a bounded size random sample of the nodes, and inspecting the subgraph induced by them. We interpret the answer of the algorithm as follows: If it concludes that the graph has property $\mathcal{P}$, this means that we can change εn^2 edges so that we get a graph with property $\mathcal{P}$; if it concludes that the graph does not have property $\mathcal{P}$, this means that we can change εn^2 edges so that we get a graph without property $\mathcal{P}$.

Again, we have to specify an error parameter $\varepsilon > 0$ in advance, and will have to accept an answer which may be wrong with probability ε, and even if it is "right", it only means that we can change εn^2 edges in the graph so that the answer becomes correct.

Sometimes we can do better and eliminate either false positives or false negatives. As an example, let us try to test whether a given (dense) graph contains a triangle. We take a sample of size $f(\varepsilon)$ (the best function f which is known to work is outrageously large, but let's not worry about this), and check whether they contain a triangle. If they do, then we know that the graph has a triangle. If they don't, then one can prove (see Section 15.3) that with high probability, we can delete εn^2 edges from the graph so that no triangle remains.

REMARK 1.5. We will not be concerned with the sample size as a function of the error bound ε. Sometimes it is polynomial (as in the examples above), but in other cases one uses the Regularity Lemma, which forces tower-size samples, making the algorithms of theoretical interest only. Goldreich [2010], in his survey of property testing, emphasizes the importance of testing with samples of manageable size, and I could not agree more; but this book, being about limit theory, does not address this issue.

Another caveat: Many extensions deal with testing models where we are allowed to sample more than a constant number of nodes of the large graph G. For this, we have to take the number of nodes into account, but usually it is enough to know the order of magnitude of the number of nodes, which in practical situations is easy to do. We do not discuss these important methods in our book.

1.6.3. Computation of a structure. Perhaps the most complex algorithmic task is the *computation of a structure*, where the structure is of size comparable with the graph itself. For example, we want to find a perfect matching in the graph, or a maximum cut (not just its density, but the cut itself), or a regularity partition in a huge dense graph. The conceptual difficulty is that the output of the algorithm is too large to be explicitly produced. What we can do is to carry out some preprocessing whose result can be stored (e.g., label a bounded number of nodes or edges), and give an algorithm which, for given input node or nodes, determines the local part of the structure we are looking for. Usually, this algorithm returns the "status" of a node or edge in the output structure (for example, whether the given edge belongs to matching, or which side of the cut the given node belongs to).

As an example, we will describe in Section 15.4.3 how to compute a maximum cut. We can access the graph by taking a bounded size sample of the nodes, and inspect the subgraph induced by them. For a given $\varepsilon > 0$, we precompute a "representative set" (see next section) together with a bipartition of this set. In addition, we describe a "Placing Algorithm" which has an arbitrary node v as its input, and tells us on which side of the cut v is located. This Placing Algorithm can be called any number of times with different nodes v, and the answers it gives should be consistent with an approximately maximum cut. For example, calling this algorithm many times, we can estimate the density of the maximum cut (but this can be done in an easier way, as we have seen).

The parameter ε is an error bound: the cut computed may be off the true maximum cut by εn^2 edges, the precomputation may be wrong with probability at most ε, and for each query, the answer may be in error with probability at most ε.

1.6.4. Representative set. Szemerédi partitions are closely related to the main ingredient in these algorithms, namely a "representative set". We want to select a (fairly large, but bounded size) subset R of the nodes such that every node is "similar" to one of the nodes in R. To be economical, we don't want to include similar points in R.

We must start with defining what "similar" means; we will do so by defining a "similarity distance" between two nodes of a graph. A first idea would be to use their distance in the graph (the length of the shortest path connecting them). However, this measures something else (the prime minister and the doorman in his office know each other, but their positions in the society are certainly not similar).

We could try considering two nodes similar, if their neighborhoods differ by little. This is certainly a reasonable thing to do, but it is too restrictive for our purposes. For example, if we consider a random graph on n nodes with edge density $1/2$, then the neighborhoods of any two nodes are very different (they have about $n/2$ elements and overlap in about $n/4$), but all nodes of a random graph are alike, so we would like them to be close in the similarity distance.

It turns out (somewhat surprisingly) that it suffices to consider second neighborhoods: we consider two nodes s and t similar, if for most other nodes v, the number of paths of length two from s to v is about the same as the number of paths of length two from t to v. The similarity distance defined this way (for the exact definition, see Section 15.4.1) has many nice properties:

- The similarity distance can be computed by sampling.
- For every $\varepsilon > 0$, every graph has a "representative set" R of nodes, whose size depends on ε only; nodes in this set are at least ε apart, and almost every node is at a distance less than ε from the representative set.
- The representative set can be computed by sampling.
- Borrowing a phrase from geometry, we define the *Voronoi cell of a node* v of the representative set R as the set of all nodes in the whole graph that are closer to v than to any other node of R. The Voronoi cells of the representative set give a Weak Regularity Partition, and vice versa, every Weak Regularity Partition, after deletion of a fraction of ε of the nodes, consists of sets with small diameter in the similarity distance.

The key to many structural computational problems is that first a representative set is computed, and then the status of any node or edge can be computed using the representative set. For example, if we want to compute a Weak Regularity Partition, we compute a representative set, which we consider as a set of representative nodes of the partition classes, which are the Voronoi cells of the nodes. We cannot compute all the Voronoi cells; but if we want to know which class (cell) does a given node belong to, all we need to do is to compute its distance to the nodes in R.

1.7. Bounded degree graphs

Let us discuss briefly how, and to what degree, the above considerations carry over to graphs with bounded degree. (We are doing injustice here to a rich and very active research area; I hope some of this will be rectified in Part 4 of the book. One of the reasons is that the technicalities in the bounded degree case are deeper, and so it is more difficult to state key results, even informally.)

Sampling. In the case of graphs with bounded degree, the subgraph sampling method gives a trivial result: the sampled subgraph will almost certainly be edgeless. Probably the most natural way to fix this is to consider *neighborhood sampling* (Figure 1.1). Let $\mathcal{G}_D$ denote the class of finite graphs with all degrees bounded by D. For $G \in \mathcal{G}_D$, select a random node and explore its neighborhood to a given depth r. This provides a probability distribution $\rho_{G,r}$ on graphs in $\mathcal{G}_D$, with a specified root node, such that all nodes are at distance at most r from the root. We will briefly refer to these rooted graphs as r-balls. Note that the number of possible r-balls is finite if D and r are fixed.

The situation for bounded degree graphs is, however, less satisfactory than for dense graphs, for two reasons. First, a full characterization of what distributions of r-balls the neighborhood sampling procedure can result in is not known (cf. Conjecture 19.8). Second, neighborhood sampling misses some important global properties of the graph, like expansion. In Section 19.2 we will introduce a notion of convergence, called local-global, which is better from this point of view, but it is not based on any implementable sampling procedure.

This suggests looking at further possibilities. Suppose, for example, that instead of exploring the neighborhood of a single random node, we could select two (or more) random nodes and determine simple quantities associated with pairs of nodes, like pairwise distances, maximum flow, electrical resistance, hitting times of random walks (studies of this nature have been performed, for example, on the internet, see e.g. Kallus, Hága, Mátray, Vattay and Laki [2011]). What information can be gained by such tests? Is there a "complete" set of tests that would give enough information to determine the global structure of the graph to a reasonable accuracy? Such questions could lead to different theories of large graphs and their limit objects; at this time, however, they are unexplored.

REMARK 1.6. It is interesting to note that our two sampling methods correspond to the two basic data structures for graph algorithms, adjacency matrix and neighborhood lists. To be more specific, both methods assume that we can choose a uniformly distributed random node, and repeat this a constant number of times. In subgraph sampling, we must be able to determine whether two given nodes are adjacent or not. For a graph that is explicitly given, this is easy if the graph is given by its adjacency matrix. For neighborhood sampling, we have to be able to find all neighbors of a given node. This is easy if the graph is given by neighborhood lists. It would be very time consuming to perform these sampling operations on a graph given by the wrong data structure.

Sampling distance. The construction of the sampling distance can be carried over to graphs with bounded degree, by replacing in (1.2) the sampling distributions $\sigma_{G,k}$ by the neighborhood distributions $\rho_{G,k}$. We must point out, however, that it seems to be difficult to define a notion of distance between two graphs with bounded degree (in analogy with the cut distance) that would reflect global similarity.

Regularity Lemma. This is one of the big unsolved problems for graphs with bounded degree. If we consider regularity lemmas as providing "approximation by the smaller", then there is a simple non-constructive result (Proposition 19.10), which should be proved in a constructive way to be really useful. One can start at many other facets of the Regularity Lemma, but a satisfactory version of bounded degree graphs has turned out most elusive.

Convergence. The notion of a convergent sequence of bounded degree graphs was in fact the first among such convergence notions, introduced by Benjamini and Schramm [2001], motivated in part by earlier work of Aldous [1998]. Our discussion of local convergence of dense graphs above, based on the convergence of the distribution of samples, was modeled on the Benjamini–Schramm definition of convergence of bounded degree graphs.

There are, however, good reasons to try to strengthen this notion. Unlike in the dense case, neighborhood sampling cannot distinguish between bipartite graphs and graphs that are far from being bipartite, cannot estimate the maximum cut etc., which means that locally convergent graph sequences must lose this information in the limit. We will introduce and study a stronger notion of convergence, which we call *local-global*, which passes on these properties and parameters to the limit. However, we don't know if there is any natural and practical algorithmic setup that would correspond to local-global convergence.

Limit objects. For bounded degree graphs, Benjamini and Schramm provide a notion of a limit object (see Section 18). The Benjamini–Schramm limit object can be described as a distribution on rooted countable graphs with a special property called "involution invariance".

Another way of describing a limit object is a "graphing". In a sense, this latter object is what we expect: a bounded degree graph on an infinite (typically uncountable) set, with appropriate measurability and measure preserving conditions. This construction was folklore in an informal way; the first exact statements were published by Aldous and Lyons [2007] and Elek [2007a].

Graphings were invented by group theorists. The idea is to consider a finitely generated group acting on a probability space (for example, rotations of a circle by integer multiples of a given angle). One can construct a graph on the underlying space, by connecting each point to its images under the generators of the group. This construction gives a graph with bounded degree (the set of points is typically of continuum cardinality). It is a beautiful fact that

graphings, representing groups this way, are just right to describe the limit objects of convergent graph sequences with bounded degree.

Depending on personal taste, a graphing may be considered more complicated or less complicated than an involution-invariant random countable rooted graph. But graphings have an important advantage: they can express a richer structure, the limits of graph sequences convergent in the local-global sense.

Algorithms. Here is finally an area where the study of bounded degree graphs can be considered at least as advanced as the study of dense graphs. Let us discuss the task of computing a structure.

Selecting random nodes and exploring their neighborhoods, we see (with high probability) disjoint parts of the graph, and so there is no method to build up a global structure. Still, very nontrivial algorithms can be designed in this model. For example, in Section 22.3.1 we describe an algorithm due to Nguyen and Onak [2008], that constructs an almost maximum matching. The way the output can be described is similar to how the output of a maximum cut algorithm was described in the dense setting: for any node we can tell which other node it is matched to, inspecting a bounded neighborhood only; these assignments will be consistent throughout the graph; and the difference in size from the true maximum matching is only εn, where $\varepsilon > 0$ is an error bound and n is the number of nodes.

There is an equivalent way to describe such algorithms, which may be easier to follow, and this is the model of *distributed computing* (going back to the 1980's). In this case, an agent (or processor) is sitting at each node of the graph, and they cooperate in exploring various properties of it. They can only communicate along the edges. In the case we are interested in (which is in a sense extreme), they are restricted to exchange a bounded number of bits (where the bound may depend on the degree D, on an error bound ε, and of course on the task they are performing, but not on the number of nodes). In some other versions of the model (cellular automata), the amount of communication is not restricted, but the computing power of the agents is. Note that in our model communication between the agents is restricted to a bounded number of bits, and hence they may be assumed to be very stupid, even finite automata.

There is a large literature on distributed computing, both from the practical and theoretical aspect. We will not be able to cover this; we will restrict ourselves to the discussion of the strong connection of this computation model with our approach to large graphs and graph limits.

CHAPTER 2

Large graphs in mathematics and physics

The algorithmic treatment of very large networks is not the only area where the notions of very large graphs and their limits can be applied successfully. Many of the problems and methods in graph limit theory come from extremal graph theory or from statistical physics. Let us give s very brief introduction to these theories.

2.1. Extremal graph theory

Extremal graph theory is one of the oldest areas of graph theory; it has some elegant general results, but also many elementary extremal problems that are still unsolved. Graph limit theory (mostly the related theory of flag algebras by Razborov) has provided powerful tools for the solution of some of these problems. Furthermore, graph limits, along with the algebraic tools that will be introduced soon, will enable us to formulate and (at least partially) answer some very general questions in extremal graph theory (similarly to the general questions for very large graphs posed in the previous chapter).

2.1.1. Edges vs. triangles. Perhaps the first result in extremal graph theory was found by Mantel [1907]. This says that if a graph on n nodes has more than $n^2/4$ edges, then it contains a triangle. Another way of saying this is that if we want to squeeze in the largest number of edges without creating a triangle, then we should split the nodes into two equal classes (if n is odd, then their sizes differ by 1) and insert all edges between the two classes. As another early example, Erdős [1938] proved a bound on the number of edges in a C_4-free bipartite graph (see (2.9) below), as a lemma in a paper about number theory.

Mantel's result is a special case of Turán's Theorem [1941], which is often considered as the work that started the systematic development of extremal graph theory. Turán solved the generalization of Mantel's problem for any complete graph in place of the triangle. We define the *Turán graph* $T(n,r)$ $(1 \leq r \leq n)$ as follows: we partition $[n]$ into r classes as equitably as possible, and connect two nodes if and only if they belong to different classes. Since we are interested in large n and fixed r, the complication that the classes cannot be exactly equal in size (which causes the formula for the number of edges of $T(n,r)$ to be a bit ugly) should not worry us. It will be enough to know that the number of edges in a Turán graph is

$$\mathsf{e}\big(T(n,r)\big) \sim \binom{r}{2}\Big(\frac{n}{r}\Big)^2,$$

and in terms of the homomorphism densities defined in the previous chapter in (1.1), we have $t\big(K_2, T(n,r)\big) \sim 1 - \frac{1}{r}$. For the triangle density we have the similar formula $t\big(K_3, T(n,r)\big) \sim (1 - \frac{1}{r})(1 - \frac{2}{r})$.

THEOREM 2.1 (**Turán's Theorem**). *Among all graphs on n nodes containing no complete k-graph, the Turán graph $T(n, k-1)$ has the maximum number of edges.*

Let us return to triangles, however, and ask for not just their existence, but for their number, when the number of edges is known. All of a sudden, we get to a rather difficult problem with some unexpected complications (which makes the subject fascinating). It is really difficult to think of a simpler question about small subgraphs of a large graph!

Since we are interested in large n, it is natural to normalize, and use homomorphism densities. The Mantel–Turán Theorem says, in this language, that

(2.1) $$t(K_2, G) > 1/2 \Rightarrow t(K_3, G) > 0.$$

Every graph G produces a pair of numbers $\big(t(K_2, G), t(K_3, G)\big)$ this way, which we can consider as a point in the plane. If we plot this point for every graph G, we get a picture as in Figure 2.1(a). To be more precise, we get a countably infinite set of points; the figure shows its closure, which we denote by $D_{2,3}$. (Another motivation for introducing convergent graph sequences and their limit objects: they give a meaning to all points of this figure.)

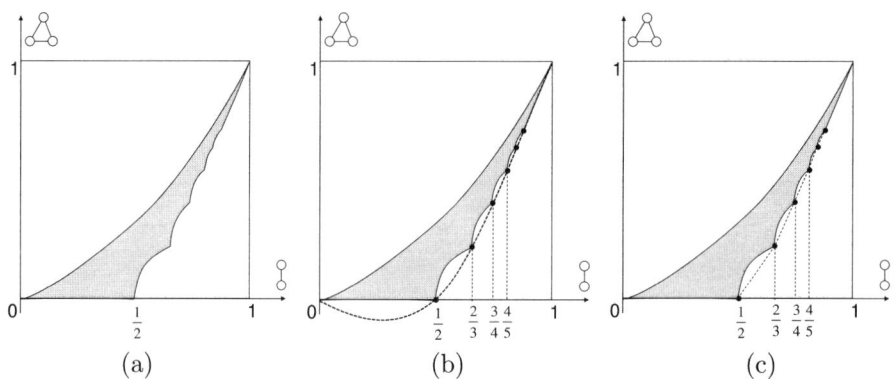

FIGURE 2.1. (a) The closure $D_{2,3}$ of the set of pairs of edge density and triangle density. (b) Goodman's bound. (c) Bollobás's' bound. The picture is a little distorted in order to show its special features better.

Some features of this picture are easy to explain. The lower edge means that there are triangle-free graphs with edge density up to $1/2$, and the Mantel–Turán Theorem says that for larger edge density, the triangle density must be positive. A lower bound for the triangle density was proved by Goodman [1959],

(2.2) $$t(K_3, G) \geq t(K_2, G)(2t(K_2, G) - 1),$$

which corresponds to the parabola shown in 2.1(b).

The upper boundary curve turns out to be given by the equation $y = x^{3/2}$, which is a very special case of the Kruskal–Katona Theorem in extremal hypergraph theory (the full theorem gives the precise value, not just asymptotics, and concerns uniform hypergraphs, not just graphs). In other words, this says that

(2.3) $$t(K_3, G) \leq t(K_2, G)^{3/2}.$$

Both (2.2) and (2.3) are sharp in a sense: Goodman's Theorem is sharp if the edge density is of the form $1/2, 2/3, 3/4, \ldots$ (Turán graphs give equality). In this form of the Kruskal–Katona Theorem equality is not attained except at the points $(0,0)$ and $(1,1)$, but for every point $(x, x^{3/2}$ of the upper boundary curve there are points representing a graph arbitrarily close (just use graphs consisting of a complete graph and isolated nodes).

From our perspective, there is nothing to improve on the upper bound, but can we get arbitrarily close to the lower bound between two special edge density values $1 - 1/k$? Surprisingly, the answer is no. Bollobás [1976] proved in 1976 that the triangle density for a graph with edge density $x \in (1 - \frac{1}{k-1}, 1 - \frac{1}{k})$ is not only above the parabola, but also above the chord of the parabola connecting the special points corresponding to $T(n, k-1)$ and $T(n, k)$.

Lovász and Simonovits [1976, 1983] formulated a conjecture about the exact bounding curve, and proved it in very small neighborhoods of the special edge density values above. One way to state this is that the minimum number of triangles is attained by a complete k-partite graph with unequal color classes. The sizes of the color classes can be determined by solving an optimization problem, which leads to a cubic *concave* curve connecting the two special points. This conjecture turned out quite hard. Lovász and Simonovits proved it in the special case when the edge density x was close to one of the endpoints of the interval. Fisher [1989] proved the conjecture for the first interval $(1/2, 2/3)$. After quite a while, Razborov [2007, 2008] proved the general conjecture. His work was extended by Nikiforov [2011] to bounding the number of complete 4-graphs, and by Reiher [2012] to all complete graphs.

So we know what the lower and upper bounding curves are. Luckily, math plays no further tricks on us: it is easy to see that for every point between the two curves there are points representing graphs arbitrarily close.

I dwelt quite long on this very simple special problem not only to show how complicated it gets (and yet solvable), but also because Razborov's methods for the solution fit quite well in the framework developed in this book, and they will be presented in Chapter 16.

2.1.2. A sampler of classical results. Let us start with some remarks to simplify and to some degree unify the statements of these results. Every algebraic inequality between subgraph densities can be "linearized", using the following multiplicativity of $t(.,G)$:

$$(2.4) \qquad t(F_1 F_2, G) = t(F_1, G) t(F_2, G),$$

where $F_1 F_2$ denotes the disjoint union of F_1 and F_2. (This property will play a very important role in the sequel, but right now it is just a convenient simplification.) For example, we can replace (2.2) by

$$(2.5) \qquad t(K_3, G) \geq 2 t(K_2 K_2, G) - t(K_2, G).$$

We can make the statements (and their proofs, as we will see below) more transparent by two further tricks: first, if a linear inequality between the densities of certain subgraphs $F_1, \ldots, F_k$ holds for all graphs, then we write it as an inequality between $F_1, \ldots, F_k$; and for specific small graphs F_i, we use little pictograms. Goodman's Inequality (2.2) can be expressed as follows:

$$(2.6) \qquad K_3 \geq 2 K_2^2 - K_2$$

or

(2.7) $$\text{[pictogram]} \geq 2\,\text{[pictogram]} - \text{[pictogram]}.$$

The Kruskal–Katona Theorem for triangles is:

(2.8) $$\text{[pictogram]} \leq \text{[pictogram]}.$$

Let us describe some further classical results. Instead of counting complete graphs, we can consider the density of some other graph F in G. Erdős proved the inequality

(2.9) $$t(C_4, G) \geq t(K_2, G)^4,$$

or in pictograms

(2.10) $$\text{[pictogram]} \geq \text{[pictogram]}.$$

Graphs with asymptotic equality here are quasirandom graphs (Section 1.4.2).

Bounding from below the homomorphism density of paths is a more difficult question, but it turns out to be equivalent to theorems of Mulholland and Smith [1959], Blakley and Roy [1965], and London [1966] in matrix theory (applied to the adjacency matrix). If P_k denotes the path with k nodes, then for all $k \geq 2$,

(2.11) $$t(P_k, G) \geq t(K_2, G)^{k-1}.$$

Regular graphs give equality here. The first nontrivial case of inequality (2.11) is

(2.12) $$\text{[pictogram]} \geq \text{[pictogram]}.$$

Translating to homomorphisms, this means that

$$\mathsf{v}(G)\,\mathsf{hom}(P_3, G) \geq \mathsf{hom}(K_2, G)^2.$$

If we count the homomorphisms on the left side by the image of the middle node, we see that it is the sum of the squared degrees of G. Since $\mathsf{hom}(K_2, G) = 2\mathsf{e}(G)$ is the sum of the degrees, this inequality is just the inequality between arithmetic and quadratic means, applied to the sequence of degrees.

Bounding the P_3-density from above in terms of the edge density is more difficult, but it was solved by Ahlswede and Katona [1978]; we formulate this as Exercise 2.4 below.

The next case of inequality (2.11) is

(2.13) $$\text{[pictogram]} \geq \text{[pictogram]},$$

and this is already quite hard, although short proofs with a tricky application of the Cauchy–Schwarz inequality are known.

In Chapter 16 we will return to the question of how far the application of such elementary inequalities takes us in proving inequalities between subgraph densities.

2.1.3. An algebraic "proof" of an extremal theorem. We illustrate the use of the formalism with pictograms for an algebraic proof of Goodman's Inequality 2.2. This will motivate a basic tool to be introduced in Chapter 6, namely graph algebras.

To describe this proof, we extend the pictogram formalism from Section 2.1.2. If we fill a node, this indicates that this node is *labeled*. We should write the label on the node, but to keep the picture simple, let us agree that the labels are $1, 2, \ldots,$

starting from the lower left corner, and going counterclockwise. (It does not really matter.)

The role of the labels is that when taking a "product" of two graphs, we take the disjoint union, but identify nodes with the same label. With this convention, it is easy to check that

$$\left(\text{\raisebox{-2pt}{⋀}} - \text{\raisebox{-2pt}{⋀}} - \text{\raisebox{-2pt}{⋀}} + \text{\raisebox{-2pt}{:.}}\right)^2 = \text{\raisebox{-2pt}{⋀}} - \text{\raisebox{-2pt}{⋀}} - \text{\raisebox{-2pt}{⋀}} + \text{\raisebox{-2pt}{:.}}$$

(this combination is "idempotent") and

$$\left(\text{\raisebox{-2pt}{⋮}} - \text{\raisebox{-2pt}{⋮}}\right)^2 = \text{\raisebox{-2pt}{V}} - 2\,\text{\raisebox{-2pt}{⋮⋮}} + \text{\raisebox{-2pt}{⋮⋮}}$$

Forgetting the labels, adding up, and deleting isolated nodes, we get

$$\left(\text{\raisebox{-2pt}{⋀}} - \text{\raisebox{-2pt}{⋀}} - \text{\raisebox{-2pt}{⋀}} + \text{\raisebox{-2pt}{:.}}\right)^2 + 2\left(\text{\raisebox{-2pt}{⋮}} - \text{\raisebox{-2pt}{⋮}}\right)^2 = \text{\raisebox{-2pt}{⋀}} - 2\,\text{\raisebox{-2pt}{⋮⋮}} + \text{\raisebox{-2pt}{⋮}}.$$

So the right side is a sum of squares, which implies that it is nonnegative:

$$\text{\raisebox{-2pt}{⋀}} - 2\,\text{\raisebox{-2pt}{⋮⋮}} + \text{\raisebox{-2pt}{⋮}} \geq 0,$$

which is just (2.7).

Is this a valid argument? It turns out that it is, and the method can be formalized using the notion of graph algebras. These will be very useful tools in the proofs of characterization theorems of homomorphism functions, and also in some other studies of graph parameters.

2.1.4. General results. Moving from special extremal graph problems to the more general, let us describe some quite general results about extremal graphs, which were obtained quite a long time ago in several papers of Erdős, Stone and Simonovits [1946, 1966, 1968]. We exclude an arbitrary graph L as subgraph of a simple graph G, and want to determine the maximum number of edges of G, given the number of nodes n. Turán's Theorem 2.1 is a special case when L is a complete graph. It turns out that the key quantity that governs the answer is the chromatic number $r = \chi(G)$.

The Turán graph $T(n, r-1)$ is certainly one of the candidates for the extremal graph, since it cannot contain any graph as a subgraph that has chromatic number r. For certain excluded graphs L it is easy to construct examples that have slightly more edges than this Turán graph; however, the gain is negligible: for every graph G on n nodes that does not contain L as a subgraph, we have

$$(2.14) \qquad \mathsf{e}(G) \leq (1+o(1))\mathsf{e}(T(n,r-1)) = \left(1 - \frac{1}{r-1} + o(1)\right)\binom{n}{2}.$$

There is also a "stability" result: For every $\varepsilon > 0$ there is an $\varepsilon' > 0$ (depending on L and ε, but not on G) such that if G is a graph not containing L with at least $\left(1 - 1/(r-1) - \varepsilon'\right)\binom{n}{2}$ edges, then we can change at most $\varepsilon\binom{n}{2}$ edges of G to get a Turán graph $T(n, r-1)$.

We will see that graph limit theory gives very short and elegant proofs for these facts. The idea that extremal graph problems have "continuous versions" (in a sense quite similar to our use of graphons), which are often cleaner and easier to handle, goes back to around 1980, when Katona [1978, 1980, 1985] and Sidorenko [1980, 1982] used this method to generalize graph and hypergraph problems, and also to give applications in probability theory.

REMARK 2.2. If $r = 2$ (which means that L is bipartite), then the main term in (2.14) disappears, and all we get is that the number of edges is $o(n^2)$. Of course, one would like to know the precise order of magnitude of the best upper bound. This is known in several cases (e.g., small complete bipartite graphs and cycles), but in general it seems to be a difficult unsolved problem. The extremal graphs in this case are sparse, and quite complex: for example, C_4-free graphs with maximum edge density are constructed from finite projective planes. Extremal problems for graphs with excluded bipartite graphs do not seem to fit in with the framework developed in this book, but perhaps they can serve as motivation for extending it to sparser graphs.

2.1.5. General questions. We have brought up the idea of introducing graphons (graph limits) in Section 1.5.3 motivated by the goal to approximate very large networks by simpler analytic objects. We have seen that graphons provide cleaner formulations, with no error terms, of some results in graph theory (for example, about quasirandom graphs). We will see in Section 16.7 that extremal graph theory provides another, also quite compelling motivation: Graphons provide a way to state, in an exact way, general questions about the nature of extremal graphs, and also help answering them, at least in some cases. (They have similar uses in the theory of computing; cf. Chapter 15).

Which inequalities between subgraph densities are valid? Given a linear inequality between subgraph densities (like (2.7) above), is it valid for all graphs G? Hatami and Norine [2011] proved recently that this question is algorithmically undecidable. We will describe the proof of this fundamental result in Section 16.6.1. On the other hand, it follows from the results of Lovász and Szegedy [2012a] that if we allow an arbitrarily small "slack", then it becomes decidable (see Section 16.6.2).

Can all linear inequalities between subgraph densities be proved using just Cauchy–Schwarz? We described above a proof of the simple inequality (2.12) using the inequality between arithmetic and quadratic means, or equivalently, the Cauchy–Schwarz Inequality. Many other extremal problems can be proved by using the Cauchy–Schwarz Inequality (often repeatedly and in nontrivial ways). Exercise 2.5 shows that Goodman's Inequality can also be proved by this method. How general a tool is the Cauchy–Schwarz Inequality in this context?

Using the notions of graphons and graph algebras we will be able to give an exact formulation of this question. It will turn out that the answer is negative (Hatami and Norine [2011], Section 16.6.1), but it becomes positive if we allow an arbitrarily small error (Lovász and Szegedy [2012a], Section 16.6.2).

Is there always an extremal graph? Let us consider extremal problems of the form "maximize a linear combination of subgraph densities, subject to fixing other such combinations". For example, "maximize the triangle density subject to a given edge density" (the answer is given by the first nontrivial case of the Kruskal–Katona Theorem (2.8)).

To motivate our approach, consider the following two optimization problems.

CLASSICAL OPTIMIZATION PROBLEM. Find the minimum of $x^3 - 6x$ over all numbers $x \geq 0$.

GRAPH OPTIMIZATION PROBLEM. Find the minimum of $t(C_4, G)$ over all graphs G with $t(K_2, G) \geq 1/2$.

The solution of the classical optimization problem is of course $x = \sqrt{2}$. This means that it has no solution in rationals, but we can find rational numbers that are arbitrarily close to being optimal. If we want an exact solution, we have to go to the completion of the rationals, i.e., to the reals.

The graph optimization problem may take a bit more effort to solve, but (2.9) shows that if the edge density is 1/2, then the 4-cycle density is at least 1/16. With a little effort one can show that equality is never attained here. Furthermore, the 4-cycle-density gets arbitrarily close to 1/16 for appropriate families of graphs: the simplest example is a random graph with edge density 1/2 (cf. also Section 1.4.2).

The analogy with the classical optimization problem above suggests that we should try to enlarge the set of (finite) graphs with new objects so that the appropriate extension of our optimization problem has a solution among the new objects. Furthermore, we want that these new objects should be approximable by graphs, just like real numbers are approximable by rationals. As it turns out, graphons are just the right objects for this.

One can prove that there is always an extremal graphon, which then gives a "template" for asymptotically extremal graphs. This follows from another fact that can be considered one of the basic results treated in this book:

The space of graphons is compact in the cut-distance metric.

(This notion of distance was mentioned in Section 1.5.1, and will be defined in Chapter 8; the compactness of the graphon space will be proved in Section 9.3).

Which graphs are extremal? This is not a good question (every graph is extremal for some sufficiently complicated extremal graph problem), but replacing "graph" by "graphon" makes it mathematically meaningful. Every extremal graphon gives a "template" for asymptotically extremal graphs.

FIGURE 2.2. Templates for optimal solutions to some classical extremal graph results: (a) Turán's Theorem 2.1 and Goodman's Inequality (2.2); (b) the Kruskal–Katona Theorem (2.3); (c) Erdős's inequality (2.9)

In classical extremal graph results, these templates are quite simple (Figure 2.2). A natural guess would be that all templates have the form of a stepfunction, like the rightmost square in Figure 1.6. All of these are indeed templates for appropriate extremal problems, but they is not all the templates: we will see that the limit of half-graphs (the rightmost square in Figure 1.7) is also the template for the extremal graph of a quite simple extremal problem, and there are many other, more complicated, templates. We will prove several results about the structure of these extremal templates (Section 16.7), but no full characterization is known.

EXERCISE 2.3. Prove inequality (2.13)

EXERCISE 2.4. Let G be a simple graph with edge density $d = t(K_2, G)$. Prove that $t(P_3, G) \leq \max(d^{3/2}, 1 - 2d + d^{3/2})$.

EXERCISE 2.5. Translate the "proof" of Goodman's Inequality 2.2 above into a valid proof using the Cauchy–Schwarz inequality twice.

2.2. Statistical physics

One area of research where graph homomorphisms play an important role, and the study of the asymptotic behavior of parameters when tending to infinity with the size of a graph is a main goal, is statistical physics. I am afraid this book will not do justice to this connection; my excuse is that statistical physics is such a large area, with so advanced special methods, that any reasonable treatment would double the size of the book. Nevertheless, I must give a very short introduction to the subject here.

To describe a basic model in statistical physics, suppose that we have a piece of a crystal, where the spin of every atom can point either up or down. We model this by an $n \times n$ grid $G = G_{n \times n}$ (for simplicity, in two dimensions). If we assign to every node of G (every "site") a "state", which can be UP or DOWN, we get a "configuration". The atoms are changing their spins randomly all the time, but not independently of each other: depending on the spins of adjacent atoms, one direction of the spin of an atom may be less likely then the other, or even entirely impossible. We would like to know how a typical configuration looks like: is it random-like as the first picture in Figure 2.3, is it homogeneous as the second (well, maybe with a few exceptions here and there), or is it structured in other ways, as the third?

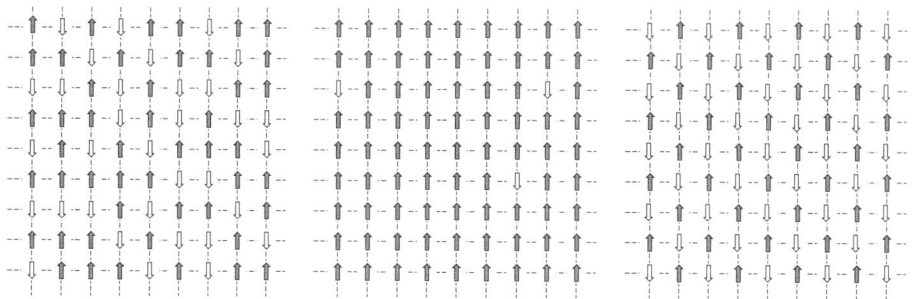

FIGURE 2.3. Three configurations of the Ising model.

Two atoms that are adjacent in the grid have an "interaction energy", which depends on their states. In the simplest version of the basic Ising model, the interaction energy is some number $-J$ if the atoms are in the same state, and J if they are not. The states of an atom can be described by the integers 1 and -1, and so a configuration is a mapping $\sigma : V(G) \to \{1, -1\}$. If σ_u denotes the state of atom u, then the total energy of a given configuration is

$$H(\sigma) = - \sum_{uv \in E(G)} J \sigma_u \sigma_v.$$

Basic physics (going back to Boltzmann) tells us that the system is more likely to be in states with low energy. In formula, the probability of a given configuration is proportional to $e^{-H(\sigma)/T}$, where T is the temperature (from the point of view of the mathematician, just a parameter). Since probabilities must add up to 1, these values must be normalized:

$$\mathsf{P}(\sigma) = \frac{e^{-H(\sigma)/T}}{Z},$$

where the normalizing factor Z is called the *partition function* of the system (it is called a "function" because it depends on the temperature). This is perhaps the most important quantity to know, which contains implicitly many important physical parameters. The partition function is simple to describe:

$$Z = \sum_\sigma e^{-H(\sigma)/T} = \sum_\sigma \exp\Big(\frac{1}{T} \sum_{uv \in E(G)} J\sigma_u \sigma_v\Big),$$

but since the number of terms is enormous, partition functions can be very hard to compute or analyze.

The behavior of the system depends very much on the sign of J. If $J > 0$, then adjacent pairs that are in the same state contribute less to the total energy than those that are in different state, and so the configuration with the lowest energy is attained when all atoms are in the same state. The typical configuration of the system will be close to this, at least as long as the temperature T is small. This is called the *ferromagnetic* Ising model, because it gives an explanation how materials like iron get magnetized. If $J < 0$ (the antiferromagnetic case), then the behavior is different: the chessboard-like pattern minimizes the energy, and no magnetization occurs at any temperature.

One may notice that the temperature T emphasizes the difference between the energy of different configurations when $T \to 0$ (and de-emphasizes it when $T \to \infty$). In the limit when $T \to 0$, all the probability will be concentrated on the states with minimum energy, which are called *ground states*. In the simplest ferromagnetic Ising model, there are two ground states: either all atoms are in state UP, or all of them are in state DOWN. If the temperature increases, disordered states like the left picture in Figure 2.3 become more likely. The transition from the ordered state to the disordered may be gradual (in dimension 1), or it may happen suddenly at a given temperature (in dimensions 2 and higher, for large graphs G); this is called a *phase transition*. This leads us to one of the central problems in statistical physics; alas, we cannot go deeper into the discussion of this issue in our book.

To make the connection to graph homomorphisms, we generalize the Ising model a little. First, we replace the grid by an arbitrary graph G. (From the point of view of physics, other lattices, corresponding to crystals with other structure, are certainly natural. Other materials don't have a simple periodic crystal structure.) Second, we introduce a "magnetic field", which prefers one state over the other: in the simplest case it adds $-\sum_u h\sigma_u$ to the energy function, with some parameter h. Third, we consider not two, but q possible states for every atom, which we label by $1, 2, \ldots, q$ (unlike 1 and -1 before, these should not be considered as numbers: they are just labels). We have to specify an interaction energy J_{ij} for any two states i and j, and a magnetic field energy h_i for every state i. A configuration is now a map $\sigma : V(G) \to [q]$, and the energy of it is

$$H(\sigma) = -\sum_{v \in V(G)} h_{\sigma(v)} - \sum_{uv \in E(G)} J_{\sigma(u),\sigma(v)}.$$

The partition function is
$$Z = \sum_{\sigma: V(G) \to [q]} \exp\left(-\frac{1}{T}\left(\sum_{v \in V(G)} h_{\sigma(v)} + \sum_{uv \in E(G)} J_{\sigma(u),\sigma(v)}\right)\right).$$
We are almost at homomorphisms! For $i, j \in [q]$, let
$$\alpha_i = \exp\left(-\frac{1}{T}h_i\right), \quad \text{and} \quad \beta_{ij} = \exp\left(-\frac{1}{T}J_{ij}\right),$$
then the partition function can be expressed as
$$(2.15) \qquad Z = \sum_{\sigma: V(G) \to [q]} \prod_{v \in V(G)} \alpha_{\sigma(v)} \prod_{uv \in E(G)} \beta_{\sigma(u)\sigma(v)}.$$
Consider the case when $\alpha_i = 1$ for all i, and β_{ij} is 0 or 1 (in the Ising model β_{ij} cannot be zero, but (2.15) allows this substitution). Then every term in (2.15) is either 0 or 1, and a term is 1 if and only if $\beta_{\sigma(u)\sigma(v)} = 1$ for every $uv \in E(G)$. Let us build a graph H with node set $V(H) = [q]$, in which $i, j \in [q]$ are adjacent if and only if $\beta_{ij} = 1$. Then a term in (2.15) is 1 if and only if σ is a homomorphism $G \to H$, and so the sum simply counts these homomorphisms, and gives the value $Z = \text{hom}(G, H)$.

In the case of general values for the α and β, we can define a weighted graph H with nodeweights α_i and edgeweights β_{ij}. Formula (2.15) can then serve as the definition of $\text{hom}(G, H)$, which will be very important for us.

We don't discuss the connections between statistical physics and graph theory (homomorphisms and limits) any further; for an introduction to the connections between statistical physics and graph theory, with more examples, see de la Harpe and Jones [1993].

EXERCISE 2.6. Define a model in statistical physics in which the ground state corresponds to the maximum cut of a graph.

Part 2

The algebra of graph homomorphisms

CHAPTER 3

Notation and terminology

In this book, different areas of mathematics come together (graph theory, probability, algebra, functional analysis), and this makes it difficult to find good notation, and impossible in some cases to stick to standard notation. I tried to find notation that helps readability. For example, when doing computations with small graphs, I often use pictograms instead of introducing dozens of notations for them. When labeling one or more nodes of a graph G, I use $G^\bullet$ or $G^{\bullet\bullet}$, and when adding some loops at the nodes, I use G°. These graphs must be still defined, but perhaps the meaning of the notation is easier to remember keeping this in mind.

3.1. Basic notation

Let $\mathbb{R}, \mathbb{C}, \mathbb{Z}$ denote the sets of real, complex and integer numbers. We denote by $\mathbb{N}$ the set of nonnegative integers, by $\mathbb{N}^*$, the set of positive integers, by $\mathbb{Z}_q$, the set of integers modulo q, and by $\mathbb{R}_+$, the set of nonnegative reals. We use the notation $[n] = \{1, 2, \ldots, n\}$ and $(n)_k = n(n-1)\ldots(n-k+1)$.

If A is a statement, then

$$\mathbb{1}(A) = \begin{cases} 1, & \text{if } A \text{ is true,} \\ 0, & \text{otherwise.} \end{cases}$$

If A is a set, then $\mathbb{1}_A$ is its indicator function: $\mathbb{1}_A(x) = \mathbb{1}(x \in A)$.

If A is a real matrix, then $A \succeq 0$ means that A is positive semidefinite (in particular, symmetric), while $A \geq 0$ means that all its elements are nonnegative. For two matrices $A, B \in \mathbb{R}^{m \times n}$, their *dot product* is defined by

$$A \cdot B = \sum_{i=1}^{m} \sum_{j=1}^{n} A_{ij} B_{ij}.$$

The natural logarithm will be denoted by ln; logarithm of base 2, by log. (There is a recurring dilemma about which logarithm to use. Base 2 is used in information theory, and it is often better suited for combinatorial problems; natural logarithm has simpler analytical formulas. Luckily, the two differ in a constant factor only so the difference is usually irrelevant.) We denote by $\log^* x$ the least n for which the n-times iterated logarithm of x is less than 1. The Lebesgue measure on $\mathbb{R}$ will be denoted by λ.

We will consider partitions of both finite sets and the interval $[0, 1]$. A partition of $[0, 1]$ will be called an *equipartition*, if it has a finite number of measurable classes with the same measure. A partition of a finite set V will be called *equitable*, if $\big||S| - |T|\big| \leq 1$ for any two partition classes.

3.2. Graph theory

We denote by $\mathsf{v}(G) = |V(G)|$ the number of nodes and by $\mathsf{e}(G) = |E(G)|$, the number of edges. The subgraph induced by $S \subseteq V(G)$ is denoted by $G[S]$. For $X, Y \subseteq V(G)$, let $e_G(X, Y)$ denote the number of edges with one endnode in X and another in Y; edges with both endnodes in $X \cap Y$ are counted twice. We denote by $N_G(v)$ the set of neighbors of v in the graph G (which we abbreviate as $N(v)$ if the graph G is understood). We denote by $\nabla(v)$ the set of edges incident with the node v. For every $r \geq 0$ and $v \in V(G)$, we denote by $B_{G,r}(v)$ the subgraph of G induced by those nodes that are at a distance at most r from v. We also call this graph the *r-ball about* v.

For any family C of sets, we denote by $L(C)$ the *intersection graph* of C, i.e. the graph with node set C, where two nodes (sets in C) are connected if and only if they have a nonempty intersection. As a special case, the intersection graph of $E(G)$ (where G is any multigraph) is called the *line-graph* of G, and denoted by $L(G)$.

We have to introduce many types of graphs. A *graph!simple* is a finite graph without loops and multiple edges. A *looped-simple graph* is a finite graph without multiple edges, in which any subset of the nodes can carry a loop; equivalently, this is a symmetric binary relation on a finite set. A *multigraph* is a finite graph (in which loops and multiple edges are allowed). Let $\mathcal{F}_k^{\mathrm{simp}}$ denote the set of simple graphs on node set $[k]$, and $\mathcal{F}_k^{\mathrm{mult}}$, the set of multigraphs on node set $[k]$.

We denote by G^{simp} the simple graph obtained from a multigraph G by deleting loops as well as all but one edge from every parallel class.

Some special graphs need special names: P_n denotes the path with n nodes (note the somewhat unusual indexing; we usually put the number of nodes in the subscript); C_n denotes the cycle with n nodes (this is mostly used for $n \geq 3$, but C_2 and even C_1 (a node with a loop) will be useful occasionally); K_n is the complete graph with n nodes (including the graph K_0 with no nodes and edges); K_n° is the complete graph with n nodes, with a loop added at every node; S_n is the star with n nodes; O_n is the graph on $[n]$ with no edges. The *m-bond* B^m consists of two nodes connected by m edges.

For a simple graph G, we denote by $\mathrm{Conn}(G)$ the set of connected subgraphs, and by $\mathrm{Csp}(G)$, connected spanning subgraphs (note: spanning, not induced!).

Weighted graphs. A *weighted graph* H is a looped-simple graph, with a positive real weight $\alpha_i(H)$ associated with each node i and a real weight $\beta_{i,j}(H)$ associated with each edge ij.

It is often convenient to assume that H is a complete graph with a loop at all nodes; the missing edges can be added with weight 0. Then the weighted graph H is completely described (up to isomorphism) by a nonnegative integer $q = \mathsf{v}(H)$, a positive real vector $a = (\alpha_1, \ldots, \alpha_q) \in \mathbb{R}^q$ of nodeweights and the real symmetric matrix $B = (\beta_{ij}) \in \mathbb{R}^{q \times q}$ of edgeweights. We denote this weighted graph by $H(a, B)$. An *edge-weighted* graph is a weighted graph with nodeweights 1.

A simple graph can be considered as a special edge-weighted graph in which all edge-weights are 0 or 1, and all loops have weight 0. Multigraphs can be considered as edge-weighted graphs in which the nodeweights are 1 and the edgeweights are nonnegative integers (but this is not always the best).

Signed graphs. Suppose that the edges of a graph F are partitioned into two sets E_+ and E_-. The triple $F = (V, E_+, E_-)$ will be called a *signed graph*. (We don't consider this as a weighted graph with edge weights ± 1, because these signs will play a quite different role!)

Partially labeled graphs. This less standard type of graphs will play a crucial role in this book. A *simply k-labeled graph* is a graph in which k of the nodes are labeled by $1, \ldots, k$ (there may be any number of unlabeled nodes). A *k-multilabeled graph* is a graph in which labels $1, \ldots, k$ are attached to some nodes; the same node may carry more than one label (but a label occurs only once). So a k-multilabeled graph is a graph F together with a map $[k] \to V(F)$, and this is k-labeled if this map is injective. We omit "simply" from k-labeled, unless we want to emphasize that it is simply k-labeled. The set of isomorphism types of k-labeled multigraphs will be denoted by $\mathcal{F}_k^\bullet$.

More generally, for every finite set $S \subseteq \mathbb{N}$ of labels we can talk about S-labeled and S-multilabeled graphs. A *partially labeled graph* is an S-labeled graph for some finite set S. A 0-labeled graph (or equivalently an $\emptyset$-labeled graph) is just an unlabeled graph. The set of S-labeled multigraphs will be denoted by $\mathcal{F}_S^\bullet$. A partially labeled graph in which all nodes are labeled will be called *flat* or *fully labeled* or *flat*.

For every partially labeled graph G and $S \subseteq \mathbb{N}$, let $[\![G]\!]_S$ denote the partially labeled graph obtained by removing the labels not in S. For $S = \emptyset$, we denote $[\![G]\!]_\emptyset$ simply by $[\![G]\!]$; this is the unlabeled version of the graph G.

We need some notation for differently labeled versions of some basic graphs (Figure 3.1). We denote by $K_n, K_n^\bullet, K_n^{\bullet\bullet}, \ldots$ the complete graph with $0, 1, 2, \ldots$ nodes labeled. . We denote by $P_n, P_n^\bullet, P_n^{\bullet\bullet}$ the path on n nodes with $0, 1, 2$ endnodes labeled. . The m-bond labeled at both nodes will be denoted by $B^{m\bullet\bullet}$. . We denote by $K_{a,b}, K_{a,b}^\bullet, K_{a,b}^{\bullet\bullet}$ the complete bipartite graph with a nodes in the "first" bipartition class and b nodes in the "second", with no node labeled, the first bipartition class labeled, and all nodes labeled, respectively. In figures, the labeled nodes are denoted by black circles, the labels ordered left-to-right or up-down. The 2-multilabeled graph consisting of a single node will be denoted by $K_1^{\bullet\bullet}$.

The *adjacency matrix* of a multigraph G is the $V(G) \times V(G)$ matrix A_G where $(A_G)_{ij}$ is the number of edges connecting node i and j. In the case of a simple graph, this is a 0-1 matrix. For a weighted graph, we let $(A_G)_{ij}$ denote the weight of the edge ij (the nodeweights can be encoded in a separate vector in $\mathbb{R}^{V(G)}$).

Colored graphs. We will use graphs in which all the edges and all the nodes are colored (so they are colorful objects indeed). To be precise, a colored graph of type (b, c) (where b and c are positive integers) is a multigraph (possibly with loops) $G = (V, E)$, which is node-colored with b colors and edge-colored with c colors.

3.3. Operations on graphs

For the standard notions of edge-deletion, contraction, subdivision, and minor, we refer to any textbook. We will need some less standard operations on graphs.

Twin reduction. Let H be a weighted graph and let $i, j \in V(H)$ be two nodes such that $\beta_{ik} = \beta_{jk}$ for every node k; in particular, this includes that $\beta_{ii} = \beta_{jj} = \beta_{ij}$ (but we allow that $\alpha_i \neq \alpha_j$). Such a pair of nodes will be called *twin nodes*. In the case when H is a simple graph, twin nodes are nonadjacent and have the same neighborhood. Interchanging a twin pair is an automorphism in the unweighted

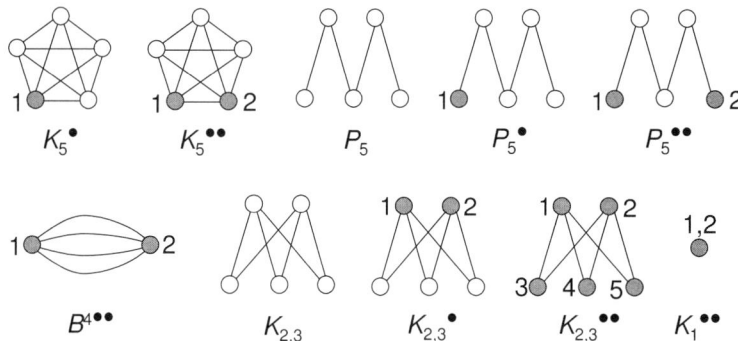

FIGURE 3.1. The most often used partially labeled graphs

case, but not necessarily in the weighted case, since their nodeweights may be different.

Let H' be obtained by identifying two twin nodes i and j, which means that we delete j, and add $\alpha_j(H)$ to $\alpha_i(H)$. We can repeat this operation until we get a weighted graph with no twins. The construction leading to this twin-free weighted graph is called *twin reduction*. It is not hard to see that the twin-free graph obtained from a given graph by twin reduction is uniquely determined.

Quotient. Let $\mathcal{P}$ be a partition of $V(G)$. We denote by $G/\mathcal{P}$ the graph obtained by merging each class of P into a single node. This definition is not precise; in different parts of the book, we need it with edge multiplicities summed, averaged, or maximized. If G is a simple graph (or a looped-simple graph, then one natural interpretation is that $G/\mathcal{P}$ is a looped-simple graph, in which two nodes are adjacent if and only if they have adjacent pre-images. we will call this the *simple quotient.*.

For other versions of the quotient construction, instead of introducing a different notation for each of these versions, we will define how the edges are mapped whenever we use this notation.

Blow-up. We define the *m-blowup* $G(m)$ of a graph G if it is obtained by replacing each node of G by m twin copies ($m \geq 1$). Sometimes we need a blow-up of G with a given number of nodes, and so we need a little more general notion: we say that a graph G' is a *near-blowup* of G if it is obtained by replacing each node of G by m or $m+1$ twin copies for some $m \geq 1$.

Product of graphs. For two looped-simple graphs G_1 and G_2, their categorical (weak) product $G_1 \times G_2$ is defined by $V(G_1 \times G_2) = V(G_1) \times V(G_2)$, and $E(G_1 \times G_2) = \{((u_1, u_2), (v_1, v_2)) : u_1v_1 \in E(G_1), u_2v_2 \in E(G_1)\}$. We denote by $G^{\times k}$ the k-fold categorical product of G with itself.

If G_1 and G_2 are simple, then so is $G_1 \times G_2$. The *strong product* $G_1 \boxtimes G_2$ of two simple graphs can be defined by adding a loop at every node, taking the categorical product, and then removing the loops.

A further operation on graphs is the *Cartesian sum* $G_1 \square G_2$, defined by $V(G_1 \square G_2) = V(G_1) \times V(G_2)$ and $E(G_1 \square G_2) = \{((u_1, u_2), (v_1, v_2)) : u_1v_1 \in E(G_1)$ and $u_2 = v_2$, or $u_1 = v_1$ and $u_2v_2 \in E(G_1)\}$.

CHAPTER 4

Graph parameters and connection matrices

4.1. Graph parameters and graph properties

A *graph parameter* is a function defined on isomorphism types of multigraphs with loops. We will mostly consider real valued graph parameters; we'll say explicitly when complex values are also allowed. A graph parameter f is called *simple* if its value is not changed when loops are removed and edge multiplicities are reduced to 1. Equivalently, we can think of a simple graph parameter as being defined on simple graphs only, but it is often convenient to extend it to multigraphs G by $f(G) = f(G^{\text{simp}})$.

A graph parameter f is *additive*, if $f(G) = f(G_1) + f(G_2)$ whenever G is the disjoint union of G_1 and G_2; it is *multiplicative* if $f(G) = f(G_1)f(G_2)$, and *maxing* if $f(G) = \max\{f(G_1), f(G_2)\}$. We say that a graph parameter is *normalized* if its value on K_1, the graph with one node and no edge, is 1. Note that if a graph parameter is multiplicative and not identically 0, then its value on K_0 (the graph with no nodes and no edges) is 1. We call a graph parameter *isolate-indifferent* if its value does not change when isolated nodes are removed. Every multiplicative and normalized graph parameter is isolate-indifferent.

There are, of course, too many graph parameters to be treated in a unified way. We will need (and say a few words about) those which are everybody's favorites:

• the maximum size of a stable set of nodes $\alpha(G)$: this is additive;

• the maximum number of independent edges (matching number) $\nu(G)$ (additive);

• the number of perfect matchings $\mathsf{pm}(G)$ (multiplicative);

• the size of the maximum clique $\omega(G)$ (maxing);

• the chromatic number $\chi(G)$ (maxing);

Our main objects of study will be graph parameters defined by counting homomorphisms into, and from, given graphs. We discuss those in detail in the next chapter.

A *graph property* is a class of graphs that is invariant under isomorphism. We identify every graph property $\mathcal{P}$ with its indicator function $\mathbb{1}_\mathcal{P}$, so for us, graph properties are just 0-1 valued graph parameters.

Of course, there are almost as many graph properties in the literature as there are graph parameters. For our purposes, some "properties of properties" will be important. In particular we will often consider the following special types of properties.

• *Monotone* property: inherited by subgraphs, i.e., $G \in \mathcal{P}$ implies that $G' \in \mathcal{P}$ for every subgraph G' of G. To be bipartite, triangle-free, or planar are examples of monotone properties.

- *Hereditary* property: inherited by induced subgraphs. All monotone properties are also hereditary. Further (non-monotone) hereditary properties are being perfect, or triangulated, or a line-graph.
- *Minor-closed* property: inherited by minors. Being planar, or series-parallel, or linklessly embedable in 3-space are such properties.

These monotonicity conditions can be extended to real valued graph parameters in a natural way. For example, a graph parameter f is called *minor-monotone*, if $f(G') \leq f(G)$ whenever G' is a minor of G.

An important operation on graph parameters is the Möbius transformation. It is best to introduce this here, because we will need to use more than one kind. Appendix A.1 introduces the Möbius transformation on a general finite lattice. We will need three special cases for graphs:

- The *upper Möbius inverse* of a simple graph parameter f (with respect to the lattice of simple graphs on the given nodes set) is defined by

$$(4.1) \qquad f^{\uparrow}(F) = \sum_{F'} (-1)^{e(F') - e(F)} f(F')$$

(the summation ranges over all simple graphs $F' \supseteq F$ with $V(F') = V(F)$).

- The *lower Möbius inverse* of a multigraph parameter f is defined by

$$(4.2) \qquad f^{\downarrow}(F) = \sum_{F'} (-1)^{e(F) - e(F')} f(F')$$

(the summation ranges over all subgraphs $F' \subseteq F$ with $V(F') = V(F)$).

- The *Möbius inverse* of a graph parameter f, relative to the partition lattice, is defined by

$$(4.3) \qquad f^{\Downarrow}(F) = \sum_{P} \mu_P f(F/P),$$

where P ranges over all partitions of $V(F)$, and μ is the Möbius function of the partition lattice, given by (A.2) (the actual value of μ_P will not be important to us, just that such integers exist). This is in fact the "lower" Möbius inverse on the partition lattice, but thankfully we don't need the upper one in this book. By the general properties of Möbius inversion, we have the relations

$$(4.4)$$
$$f(F) = \sum_{\substack{F' \supseteq F \\ V(F') = V(F)}} f^{\uparrow}(F'), \quad f(F) = \sum_{\substack{F' \subseteq F \\ V(F') = V(F)}} f^{\downarrow}(F'), \quad f(F) = \sum_P f^{\Downarrow}(F').$$

EXERCISE 4.1. Let f be a multiplicative graph parameter. Prove that $f^{\downarrow}$ is multiplicative as well.

EXERCISE 4.2. Let f be an additive graph parameter. Prove that $f^{\downarrow}(G) = 0$ if G is disconnected with at least two non-singleton components.

4.2. Connection matrices

Let F_1 and F_2 be two partially multilabeled graphs. We define their *gluing product* (or often just product, if there is no danger of confusion with any other product notion) $F_1 F_2$ by taking their disjoint union, and then identifying nodes with the same label. Note that this may force further identifications, since labels i and j may occur on the same node u in F_1, but on different nodes v and v' in

F_2, in which case u, v and v' must be identified (Figure 4.1). (If F_1 and F_2 are simply labeled, then this does not happen, and $F_1 F_2$ is also simply labeled. If F_1 and F_2 are k-labeled, then $F_1 F_2$ is also k-labeled.) Another way to describe this construction: form the disjoint union of F_1 and F_2, add edges between nodes with the same label, and contract the new edges. So the new labeled nodes will correspond to the connected components of the graph on the original labeled nodes, formed by the new edges.

Even if F_1 and F_2 are simple graphs, which are k-multilabeled, their product may have loops and parallel edges. If F_1 and F_2 are simply k-labeled and have no loops, then $F_1 F_2$ has no loops, but may have multiple edges. For two 0-labeled (i.e., unlabeled) graphs, $F_1 F_2$ is their disjoint union. Clearly this multiplication is associative and commutative.

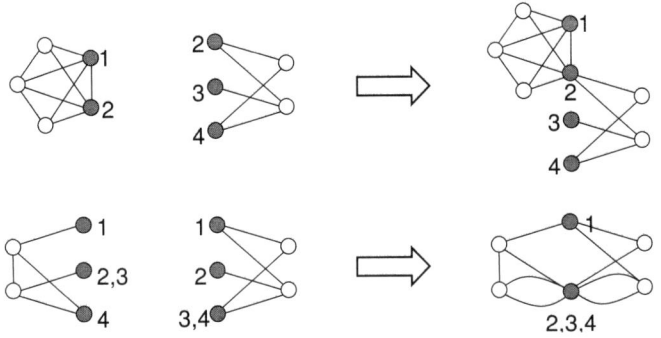

FIGURE 4.1. Top: the product of two simply partially labeled graphs. Bottom: the product of two 4-multilabeled graphs

EXAMPLE 4.3. Consider edgeless fully k-multilabeled graphs. Such a graph is given by a partition of the label set $[k]$. The product of two such graphs is also edgeless, and it corresponds to the join of the partitions in the partition lattice (see Appendix A.1). The (unique) simply labeled graph in this class corresponds to the discrete partition; the graph with one node corresponds to the indiscrete partition.
♦

Our basic tool to study a graph parameter will be the sequence of its *connection matrices*: These are infinite matrices, one for every integer $k \geq 0$, whose linear algebraic properties are closely related to graph-theoretic properties of graph parameters.

Let f be any multigraph parameter and fix an integer $k \geq 0$. We define the k-th *multilabeled connection matrix* of the graph parameter f as the (infinite) symmetric matrix $M^{\text{mult}}(f, k)$, whose rows and columns are indexed by (isomorphism types of) k-multilabeled multigraphs, and the entry in the intersection of the row corresponding to F_1 and the column corresponding to F_2 is $f(\llbracket F_1 F_2 \rrbracket)$. The submatrix corresponding to the simply k-labeled graphs is denoted by $M^{\text{simp}}(f, k)$ or just $M(f, k)$, and will be called simply the k-th connection matrix (Figure 4.2). The submatrix of $M(f, k)$ formed by rows and columns that are fully labeled (flat) will be called the *flat connection matrix* and denoted by $M^{\text{flat}}(f, k)$. If the graph parameter f is a simple graph parameter, then in $M(f,k)$ those rows that correspond to rows indexed by graphs with loops and/or multiple edges are just copies of rows indexed by simple graphs, and similarly for the columns.

Sometimes it is convenient to work with a single connection matrix $M(f,\mathbb{N})$, whose rows and columns are indexed by all partially labeled graphs, and (as before) the entry in row F_1 and column F_2 is $f(\llbracket F_1 F_2 \rrbracket)$. Trivially, $M(f,\mathbb{N})$ contains all the connection matrices $M(f,k)$ as submatrices, but it does not carry substantially more information, at least for parameters that will be most interesting for us: if f is isolate-indifferent, then every finite submatrix of $M(f,\mathbb{N})$ is also a submatrix of $M(f,k)$ for every sufficiently large k.

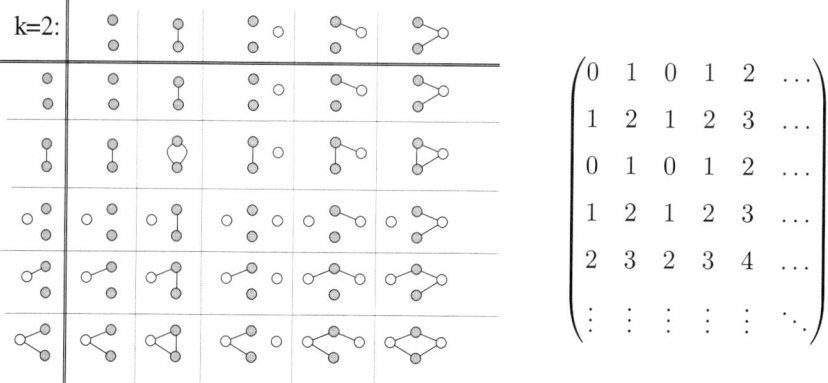

FIGURE 4.2. Some rows and columns of the second connection matrix. The entries are obtained by applying a graph parameter to the graphs shown. If the graph parameter is the number of edges, we get the infinite matrix on the right.

Two possible properties of connection matrices will be particularly important for us. We call the graph parameter f *reflection positive* if all the corresponding connection matrices $M(f,k)$ are positive semidefinite. For isolate-indifferent parameters, this is equivalent to saying that $M(f,\mathbb{N})$ is positive semidefinite. (To be precise, we have to talk about reflection positivity with respect to multilabeled or simply labeled connection matrices. If not said explicitly, we mean simply labeled.) We call the parameter *flatly reflection positive* if its flat connection matrices are positive semidefinite. For simple graphs, these matrices are finite for every fixed k, so flat reflection positivity is a much friendlier notion than general reflection positivity. Nevertheless, they will turn out to be equivalent under mild conditions (Proposition 14.60).

We define the *connection rank function* of a graph parameter as the rank $r(f,k) = \mathrm{rk}(M(f,k))$ as a function of k (again, simply/multilabeled and simple graph/multigraph versions can be defined). This number is infinite in general, but it is finite in a surprisingly large number of cases. Those parameters for which it is finite for all k, which we call *parameters of finite rank*, are of particular interest, and will be discussed next.

EXERCISE 4.4. (a) Show that a multigraph parameter f is multiplicative if and only if $M(f,0)$ is positive semidefinite and has rank 0 or 1. (b) Characterize graph parameters for which $M(f,0)$ has rank 1.

4.3. Finite connection rank

Connection matrices are infinite, and typically they have infinite rank. However, surprisingly many multigraph parameters, including some large classes, have finite connection rank, which makes this finiteness a combinatorially important property.

Finite connection rank will have an important algorithmic consequence: every such parameter can be computed efficiently (in polynomial time) for graphs with bounded treewidth. The exact statement of this fact and the description of the algorithm will be given in Section 6.5.

As a warm-up, we make a few simple observations about operations on multigraph parameters preserving finite connection rank. (These also hold for simple graph parameters, which can be considered as a special case here.)

LEMMA 4.5. *Let f and g be graph parameters and $c \in \mathbb{C}$. If f and g have finite connection rank, then so do cf, $f + g$ and fg.*

Proof. The first two assertions are trivial. The third one follows from the observation that every connection matrix $M(fg, k)$ is a submatrix of the Kronecker product of $M(f, k)$ and $M(g, k)$, and so its rank is at most the product of the ranks of them. □

LEMMA 4.6. *Let f and g be graph parameters with finite rank, and suppose that $f - g$ has finite range. Then $\max(f, g)$ and $\min(f, g)$ have finite connection rank.*

Proof. First, we prove the case when both f and g have finite range. Lemma 4.5 implies that $p(f, g)$ has finite connection rank for every polynomial p in two variables. Since over a finite range, every function of f and g can be expressed as a polynomial of them, the assertion follows.

Second, in the general case we use that $\max(f, g) = f + \max(0, g - f)$. By what we just proved, $\max(0, g - f)$ is a parameter with finite rank, and hence so is $\max(f, g)$. The assertion for the minimum follows similarly. □

4.3.1. Many parameters with finite connection rank.
In this section, we describe connection matrices for a variety of multigraph parameters. Our main concern will be their rank (and sometimes whether they are semidefinite).

EXAMPLE 4.7 (**Nodes and edges**). The number of edges $\mathsf{e}(G)$ in G is an additive parameter: $\mathsf{e}(F_1 F_2) = \mathsf{e}(F_1) + \mathsf{e}(F_2)$ for two unlabeled multigraphs F_1 and F_2. In fact, this holds for two k-labeled graphs as well, and so $M(\mathsf{e}, k)$ is the sum of two matrices of rank 1. Thus $M(e, k)$ has rank 2, so $r(\mathsf{e}, k) = 2$ for all $k \geq 0$. Similarly, the number of nodes has finite connection rank $r(\mathsf{v}, k) = 2$ for all k. ◆

EXAMPLE 4.8 (**Non-parallel edges**). Let $\mathsf{e}'(G) = \mathsf{e}(G^{\text{simp}})$ denote the number of different (i.e., non-parallel) edges in G. For two k-labeled graphs G_1 and G_2, we have $\mathsf{e}'(G_1 G_2) = \mathsf{e}'(G_1) + \mathsf{e}'(G_2) - \frac{1}{2} A_1 \cdot A_1$, where A_i is the adjacency matrix of the subgraph of G_i induced by the labeled nodes. Hence $M(\mathsf{e}', k)$ can be written as the sum of two matrices of rank 1 and one matrix of rank at most $\binom{k}{2}$. Thus $r(\mathsf{e}', k) \leq \binom{k}{2} + 2$, and one can check that this is the exact value. ◆

EXAMPLE 4.9 (**Subgraphs**). Let $\mathsf{sg}(G) = 2^{\mathsf{e}(G)}$ denote the number of spanning subgraphs of G. Then $\mathsf{sg}(G_1 G_2) = \mathsf{sg}(G_1) \mathsf{sg}(G_2)$, and so $M(\mathsf{sg}, k)$ has rank 1. Thus $r(\mathsf{sg}, k) = 1$ for all k. These matrices are trivially positive semidefinite. ◆

EXAMPLE 4.10 (**Simple subgraphs**). Let $\mathsf{sg}'(G) = 2^{\mathsf{e}'(G)}$ denote the number of simple subgraphs of G. Then

$$\mathsf{sg}'(G_1 G_2) = \mathsf{sg}'(G_1)\mathsf{sg}'(G_2)\frac{1}{\mathsf{sg}'(G_1 \cap G_2)}.$$

The first two factors don't change the rank, and the rows of the matrix given by the third factor are determined by the edges induced by the labeled nodes, so the corresponding matrix has at most $2^{\binom{k}{2}}$ different rows. Hence $r(\mathsf{sg}', k) \leq 2^{\binom{k}{2}}$. Again one can check that this is the exact value. ◆

Next we look at some of the less trivial but still common graph parameters, which make more complicated examples.

EXAMPLE 4.11 (**Stability number**). The maximum size $\alpha(G)$ of a stable set of nodes is additive, and has finite connection rank (Godlin, Kotek and Makowski [2008]). This is more difficult to prove. First, we split the rows of the matrix $M(\alpha, k)$ into $2^{\binom{k}{2}}$ classes, according to the subgraph H_i of F_i induced by the labeled nodes. This splits the matrix $M(\alpha, k)$ into $2^{k(k-1)}$ submatrices, and it suffices to show that each of these has finite rank. So let us fix H_1 and H_2. Let $\mathcal{I}$ denote the set of stable sets of nodes in $H_1 \cup H_2$, and let $F_i' = F_i \setminus [k]$ and $F_i^S = F_i' \setminus N_{F_i}(S)$.

For two k-labeled graphs F_1 and F_2 with $F_i[k] = H_i$, we have $\alpha(F_1 F_2) = \max_{S \in \mathcal{I}} \alpha_S(F_1 F_2)$, where

$$\alpha_S(F_1 F_2) = |S| + \alpha(F_1^S) + \alpha(F_2^S)$$

is the maximum size of a stable set in $F_1 F_2$ intersecting $[k]$ in S. The rank of the matrix $(\alpha_S(F_1 F_2))$ is at most 3. Unfortunately, we cannot apply Lemma 4.6 directly, since $\alpha_S(F_1 F_2)$ is not bounded. But we can use that $\alpha(F_1 F_2) \geq \alpha_\emptyset(F_1 F_2) = \alpha(F_1') + \alpha(F_2')$, and hence those sets S for which $\alpha(F_1^S) < \alpha(F_1') - k$ or $\alpha(F_2^S) < \alpha(F_2') - k$ play no role in the maximum. In other words, we can replace α_S by

$$\alpha_S'(F_1 F_2) = |S| + \max\{\alpha(F_1^S), \alpha(F_1') - k\} + \max\{\alpha(F_2^S), \alpha(F_2') - k\},$$

and still have that $\alpha(F_1 F_2) = \max_S \alpha_S'(F_1 F_2)$. The matrices $(\alpha_S'(F_1 F_2))$ have rank at most 3, and for different sets S the corresponding entries differ by at most $3k$, so the same argument as in the proof of Lemma 4.6 implies that α has finite connection rank. ◆

EXAMPLE 4.12 (**Node cover number**). The minimum number of nodes covering all edges, $\tau(G) = \mathsf{v}(G) - \alpha(G)$, has finite connection rank as well, since every connection matrix of τ is the difference of the corresponding connection matrices of the parameters v and α, which both have finite rank. ◆

EXAMPLE 4.13 (**Number of stable sets**). Let $\mathsf{stab}(G)$ denote the number of stable sets in G. This parameter is multiplicative, and has finite connection rank: this can be verified easily by distinguishing stable sets according to their intersection with the set of labeled nodes. ◆

EXAMPLE 4.14 (**Number of perfect matchings**). Let $\mathsf{pm}(G)$ denote the number of perfect matchings in the multigraph G. It is trivial that pm is multiplicative.

Let G be a k-labeled multigraph, let $X \subseteq [k]$, and let $\mathsf{pm}(G, X)$ denote the number of matchings in G that match all the unlabeled nodes, and the nodes with label in X, but not any of the other labeled nodes. Then we have for any two k-labeled multigraphs G_1 and G_2

$$\mathsf{pm}(G_1 G_2) = \sum_{X \subseteq [k]} \mathsf{pm}(G_1, X) \mathsf{pm}(G_2, [k] \setminus X).$$

Hence the matrix $M(\mathsf{pm}, k)$ can be written as the sum of 2^k matrices of rank 1, and its rank is at most 2^k (it is not hard to see that in fact equality holds).

If we consider the matching number as a simple graph parameter (in terms of multigraphs, this means that we don't care which edge in a parallel class matches a given pair of nodes), then the above argument has to be modified, to arrive at a similar conclusion. The details of this are left to the reader as an exercise. ♦

EXAMPLE 4.15 (**Number of Hamilton cycles**). Let $\mathsf{ham}(G)$ be the number of Hamilton cycles in G. For two k-labeled multigraphs G_1 and G_2, every Hamilton cycle H in $G_1 G_2$ defines a cyclic ordering $(i_1, \ldots, i_k)$ of the nodes in $[k]$, and for any two consecutive nodes i_r and i_{r+1}, it defines an index $j_r \in [2]$ which tells us whether the arc of H between i_r and i_{r+1} uses G_1 or G_2. Let us call the cyclic ordering $(i_1, \ldots, i_k)$, together with the indices $(j_1, \ldots, j_k)$, the *trace* of H on the labeled nodes. (If you are living in the set of labeled nodes, and cannot see farther than a small neighborhood of the nodes, then the trace is all that you can see from a Hamilton cycle.)

Given a possible trace $T = (i_1, \ldots, i_k; j_1, \ldots, j_k)$, we denote by $\mathsf{ham}(G_j; T)$ the number of systems of edge-disjoint paths in G_j which connect i_r and i_{r+1} for all r with $j_r = j$, and which cover all unlabeled nodes in G_j. Then

$$\mathsf{ham}(G_1 G_2) = \sum_T \mathsf{ham}(G_1; T) \mathsf{ham}(G_2; T),$$

showing that the rank of $M(\mathsf{ham}, k)$ is bounded by the number of possible traces (which is $2^{k-1}(k-1)!$ by standard combinatorial calculation). ♦

EXAMPLE 4.16 (**Chromatic polynomial**). Every substitution into the chromatic polynomial $\mathsf{chr}(G, x)$ gives a graph parameter (see Appendix A.2). If we substitute a nonnegative integer q for the variable x, we get the number of q-colorings, which is a special case of homomorphism functions (to be discussed in the next Chapter). What about evaluations at other values? The rank of the connection matrices for the general case was determined by Freedman, Lovász and Welsh (see Lovász [2006a]).

Let B_k denote the number of partitions of a k-set (the k-th Bell number), and let $B_{k,q}$ denote the number of its partitions into at most q parts.

PROPOSITION 4.17. *For every fixed x, $\mathsf{chr}(., x)$ is a multiplicative graph parameter. For every $k \geq 0$,*

$$r(\mathsf{chr}(., x), k) = \begin{cases} B_{k,x} & \text{if } x \text{ is a nonnegative integer,} \\ B_k & \text{otherwise.} \end{cases}$$

Furthermore, $M(\mathsf{chr}(., x), k)$ is positive semidefinite if and only if x is a nonnegative integer or $x \geq k - 1$.

Proof. We prove that the right hand side is an upper bound even for the rank of the multi-connection matrix, and a lower bound for the rank of the simple connection matrix. The case $x = 0$ is trivial, so suppose that $x \neq 0$. Using the deletion-contraction relation

(4.5) $$\mathsf{chr}(F, x) = \mathsf{chr}(F - e, x) - \mathsf{chr}(F/e, x),$$

we see that for every k-multilabeled multigraph F with at least one edge, the row of $M(\mathsf{chr}(., x), k)$ corresponding to F is the difference of two earlier rows (where we order the rows so that the number of edges is non-decreasing). So the rank of the whole matrix is the same as the rank of the submatrix formed by k-multilabeled graphs with no edges. Since deleting an unlabeled isolated node just divides the row by x, we may assume that all nodes are labeled. So the rows and columns of the remaining matrix M' correspond to partitions of the label set $[k]$. In the intersection of the row indexed by $\mathcal{P} \in \Pi(k)$ and column indexed by $\mathcal{Q} \in \Pi(k)$, we find the k-multilabeled graph corresponding to $\mathcal{P} \vee \mathcal{Q}$. The chromatic polynomial for this graph is $x^{|\mathcal{P} \vee \mathcal{Q}|}$. Let D denote the diagonal matrix in which the entry in row and column $\mathcal{P}$ is $(x)_{|\mathcal{P}|}$, then by identities (A.3) and (A.1), we have $M' = ZDZ^\mathsf{T}$.

This implies that the rank of M' is the same as the rank of D, and it is positive semidefinite if and only if D is positive semidefinite. If x is not a nonnegative integer, then D has full rank. The same conclusion holds for positive integers $x \geq k$. If $x < k$ is a nonnegative integer, then the number of nonzero diagonal entries in D is $B_{k,x}$. Finally, D is positive semidefinite if and only if $(x)_j \geq 0$ for all $1 \leq j \leq k$, which is clearly equivalent to x being a nonnegative integer or $x \geq k - 1$. □

Note the nontrivial fact that the rank is always finite. If x is a nonnegative integer, then the connection rank is bounded by x^k, but otherwise, as a function of k, it grows faster than c^k for every c. ◆

EXAMPLE 4.18 (**Tutte polynomial**). The cluster expansion version $\mathsf{cep}(G; u, v)$ of the Tutte polynomial generalizes the chromatic polynomial (see again Appendix A.2), and it behaves similarly. It is not hard to show that for $v \neq 0$,

$$r(\mathsf{cep}, k) = \begin{cases} B_{k,u} & \text{if } u \text{ is a nonnegative integer,} \\ B_k & \text{otherwise} \end{cases}$$

(the case $v = 0$ is trivial). Furthermore, $\mathsf{cep}(G; u, v)$ is reflection positive if and only if u is a nonnegative integer. For other versions of the Tutte polynomial (e.g., tut) similar conclusions hold, since they are related to cep by scaling and substitution in the variables (except when the expressions we scale with are 0). ◆

EXAMPLE 4.19 (**Number of spanning trees**). The number of spanning trees $\mathsf{tree}(G)$ of a graph G is obtained by substitution into the Tutte polynomial tut with $x = y = 1$. Since $u = (x-1)(y-1) = 0$, this falls under the exception at the end of the last example. Nevertheless, the arguments can be adjusted appropriately, and we get that $r(\mathsf{tree}, k) = B_k$. ◆

We conclude with a couple of examples of parameters whose connection matrices have infinite rank, but they are still "interesting".

EXAMPLE 4.20 (**Maximum clique**). The size of a maximum clique, $\omega(G)$, is maxing. It does not have finite connection rank. In fact, consider the connection matrix $M(\omega, 0)$, and its submatrix M whose rows and columns are indexed by cliques $K_1, K_2, \ldots$. This looks like

$$M = \begin{pmatrix} 1 & 2 & 3 & 4 & \ldots \\ 2 & 2 & 3 & 4 & \ldots \\ 3 & 3 & 3 & 4 & \ldots \\ 4 & 4 & 4 & 4 & \ldots \\ \vdots & \vdots & \vdots & \vdots & \ddots \end{pmatrix}$$

and has infinite rank. Similar argument shows that *no unbounded maxing graph parameter has finite connection rank* (Exercise 4.34). In particular, the chromatic number has infinite connection rank. ◆

EXAMPLE 4.21 (**Eulerian orientations**). For an undirected multigraph F, let $\overrightarrow{\mathsf{eul}}(F)$ denote the number of eulerian orientations of F (i.e., orientations in which every node has the same outdegree as indegree; for simplicity, let's exclude loops). By Euler's theorem, $\overrightarrow{\mathsf{eul}}(F) = 0$ if and only if F has a node with odd degree. It is clear that the parameter $\overrightarrow{\mathsf{eul}}$ is multiplicative.

Let us define $\overrightarrow{\mathsf{eul}}(G; a_1 \ldots a_k)$ as the number of orientations of a k-labeled graph G such that the unlabeled nodes have equal outdegree and indegree, while for a node $i \in [k]$, the difference between its indegree and outdegree is a_i. Then we have:

$$\begin{aligned}\overrightarrow{\mathsf{eul}}(G_1 G_2) &= \sum_a \overrightarrow{\mathsf{eul}}(G_1; a_1 \ldots a_k) \overrightarrow{\mathsf{eul}}(G_2; -a_1 \cdots -a_k) \\ &= \sum_a \overrightarrow{\mathsf{eul}}(G_1; a_1 \ldots a_k) \overrightarrow{\mathsf{eul}}(G_2; a_1 \ldots a_k).\end{aligned} \tag{4.6}$$

(The sum is finite for every G_1 and G_2, but the number of nonzero terms is not bounded). This implies that the connection matrices $M(\overrightarrow{\mathsf{eul}}, k)$ are positive semidefinite, but it does not follow that they have finite rank; and in fact, they have infinite rank for $k \geq 2$ (see Exercise 4.33). ◆

4.3.2. Minor-closed graph properties. We have seen many examples of graph parameters with finite connection rank. In the next sections, we will describe some very general classes of such parameters. A challenging problem is to determine all such graph parameters.

THEOREM 4.22. *Every minor-closed multigraph property has finite connection rank.*

Proof. We use the very deep theorem of Robertson and Seymour [2004] that such a property can be characterized by a finite number of excluded minors. Let $\mathcal{P}$ be the multigraph property that G does not contain any of $H_1, \ldots, H_m$ as a minor, and let $\mathcal{P}_i$ be the property that G does not contain H_i as a minor. Then $\mathbb{1}_\mathcal{P} = \mathbb{1}_{\mathcal{P}_1} \ldots \mathbb{1}_{\mathcal{P}_m}$, and so it suffices to prove that P_i has finite connection rank. In other words, we may assume that $\mathcal{P}$ is the property of not containing a given graph H as a minor.

Fix k, and consider all graphs H' on at most $\mathsf{v}(H) + k - 1$ nodes that can be contracted to H. In H', select a subset of at most k nodes in all possible ways, and label them by different numbers from $[k]$. Finally, 2-color the edges of H' red and blue so that only labeled nodes can be incident with edges of both colors. Call every partially labeled 2-colored graph obtained this way a *pre-minor*. It is clear that the number of pre-minors is finite.

Let G and G' be two partially labeled edge-colored graphs. We say that G' is a minor of G, if G' can be obtained from G by deleting edges and/or nodes and contracting edges so that no edge with both endnodes labeled is ever contracted. (The remaining edges keep their colors and the remaining labeled nodes keep their labels.)

Consider a product G_1G_2 of two k-labeled graphs, and color the edges of G_1 red and the edges of G_2 blue. It is easy to see that G_1G_2, as an unlabeled uncolored graph, contains H as a minor if and only if it contains, as a labeled edge-colored graph, at least one pre-minor as a minor.

For a pre-minor H', let H'_1 be the subgraph of H' formed by all red edges, their endpoints, and the labeled nodes. We define H'_2 similarly using the blue edges. Then G_1G_2 contains H' as a minor if and only if G_1 contains H'_1 and G_2 contains H'_2 as a minor.

Let G_1 and G'_1 be two k-labeled graphs and suppose that for every pre-minor H', H'_1 is a minor of G_1 if and only if it is a minor of G'_1. Then the rows of $M(\mathbb{1}_\mathcal{P}, k)$ indexed by G_1 and G'_1 are equal. This means that $M(\mathbb{1}_\mathcal{P}, k)$ has only a finite number of different rows, and hence its rank is finite. $\square$

COROLLARY 4.23. *Every nonnegative integer valued bounded minor-monotone multigraph parameter has finite connection rank.*

Proof. Let f be such a parameter, and assume that $f \leq K$. Then
$$f(G) = \mathbb{1}(f(G) \leq 1) + \mathbb{1}(f(G) \leq 2) + \cdots + \mathbb{1}(f(G) \leq K).$$
Since the graph property that $f(.) \leq i$ is minor-closed, each parameter $\mathbb{1}(f(G) \leq i)$ has finite connection rank by Theorem 4.22, and hence so does f. $\square$

4.3.3. Monadic second order formulas. To describe a very rich class of graph properties with finite connection rank (at least for looped-simple graphs), we consider properties defined by certain logical formulas. A *first order formula in graph theory* is composed of primitives "$x = y$" and "$x \sim y$", using logical operations "$\wedge$" (AND), "$\vee$" (OR) and "$\neg$" (NEGATION), and logical quantifiers "$\forall$" and "$\exists$". Every such formula, properly composed, with all variables quantified, defines a property of looped simple graphs, if we interpret the quantified variables as nodes, and the relation $x \sim y$ as x and y being adjacent. For example, the property of being a 2-regular loopless graph can be expressed as

$$(\forall x)(\forall y)(x = y \Rightarrow x \nsim y)$$
$$\wedge\, (\forall x)(\exists y)(\exists z)\big(y \neq z \wedge x \sim y \wedge x \sim z \wedge (\forall u)(x \sim u \Rightarrow (u = y \vee u = z))\big)$$

(to facilitate reading these formulas, we will use some standard conventions like writing $A \Rightarrow B$ instead of $\neg A \vee B$ and $x \neq y$ instead of $\neg(x = y)$).

First order formulas can define rather simple graph properties only, but we get a real jump in generality if we allow quantifying over subsets of nodes and edges. A *monadic second order formula* has three types of variables, which we distinguish using different fonts. Lower case letters denote nodes, upper case letters denote subsets of nodes, and upper case boldface letters denote subsets of the edges. The primitives then also include $x \in X$ and $xy \in \mathbf{Y}$. We call the formula *node-monadic* if quantifying over subsets of edges is not allowed.

This way we get a quite powerful language to express graph formulas, as the following examples show (see also the exercises at the end of the section).

EXAMPLE 4.24. The property of being bipartite (2-colorable) can be expressed as

$$(\exists X)(\exists Y)\big((\forall x)(x \in X \vee x \in Y) \wedge \neg(x \in X \wedge x \in Y)$$
$$\wedge (\forall x)(\forall y)\big(x \sim y \Rightarrow ((x \in X \wedge y \in Y) \vee (x \in Y \wedge y \in X))\big)\big).$$

Colorability by any given number of colors can be expressed similarly. ◆

EXAMPLE 4.25. The existence of a perfect matching can be expressed as

$$(\exists \mathbf{M})\Big((\forall x)(\forall y)\big((xy \in \mathbf{M}) \Rightarrow (x \sim y)\big) \wedge (\forall x)(\exists y)(xy \in \mathbf{M})$$
$$\wedge (\forall x)(\forall y)(\forall z)\big(((xy \in \mathbf{M}) \wedge (xz \in \mathbf{M}) \Rightarrow z = y)\big)\Big).$$

The existence of a Hamilton cycle can also be expressed (see Exercise 4.38). ◆

EXAMPLE 4.26. Planarity of a graph can be expressed by a node-monadic second-order formula. First we construct a formula expressing the property of a graph G that it contains a subdivision of K_5. One way to do so is to look for 5 nodes $v_1, \ldots, v_5$ and 10 subsets of nodes $P_{12}, P_{13}, \ldots, P_{45}$ such that every P_{ij} induces a connected subgraph containing v_i and v_j, and the sets $P_{ij} \setminus \{v_i, v_j\}$ are disjoint. For a given $i < j$, the required properties of P_{ij} can be expressed by

$$\Psi_{ij} : (v_i \in P_{ij}) \wedge (v_j \in P_{ij}) \wedge (\forall S)\Big(((v_i \in S) \wedge (v_j \notin S))$$
$$\Rightarrow (\exists u)(\exists w)\big((u \in P_{ij}) \wedge (w \in P_{ij}) \wedge (u \in S) \wedge (w \notin S) \wedge (u \sim w)\big)\Big).$$

For every pair of pairs $\{\{i,j\},\{k,l\}\}$ with $i < j$, $k < l$, and $\{i,j\} \cap \{k,l\} = \emptyset$ we write

$$\Phi_{i,j,k,l} : (\forall u)\big((u \notin P_{ij}) \vee (u \notin P_{kl})\big).$$

and for every pair of pairs $\{\{i,j\},\{k,l\}\}$ with $i < j$, $k < l$, and $\{i,j\} \cap \{k,l\} = \{m\}$ we write

$$\Phi_{i,j,k,l} : (\forall u)\Big(((u \in P_{ij}) \wedge (u \in P_{kl})) \Rightarrow (u = v_m)\Big).$$

Then the formula

$$\Theta_1 = (\exists v_1)\ldots(\exists v_5)(\exists P_{12})(\exists P_{13})\ldots(\exists P_{45})\Big(\bigwedge_{i<j}\Psi_{ij} \wedge \bigwedge_{\substack{i<j,k<l \\ \{i,j\}\neq\{k,l\}}}\Phi_{i,j,k,l}\Big)$$

expresses that the graph contains a subdivision of K_5. We can construct similarly a formula Θ_2 expressing that the graph contains a subdivision of $K_{3,3}$. Then $\neg\Theta_1 \wedge \neg\Theta_2$ means that the graph is planar.

In a similar way, every minor-monotone property can be expressed by a monadic second order formula, at least for looped-simple graphs (Exercise 4.39). ◆

Godlin, Kotek and Makowski [2008, 2009] proved the following very general sufficient condition for a graph property $\mathcal{P}$ to have finite connection rank.

THEOREM 4.27. *Every looped-simple graph property definable by a monadic second order formula has finite connection rank.*

In fact, they prove a more general result about monadic second order definable graph polynomials, which we do not introduce in this book.

In order to prove Theorem 4.27, we have to look at structures that are more general than simple graphs. A *gaudy graph* of type (a, b, c) ($a, b, c \in \mathbb{Z}$, $a \geq 0$,

$b, c \geq 1$) is a looped-simple graph $G = (V, E)$ that is a-multilabeled, node-colored with b colors, and edge-colored with c colors. In other words, a gaudy graph is a 5-tuple $(V, E, \alpha, \beta, \gamma)$, where (V, E) is an underlying multigraph and $\alpha : [a] \to V$, $\beta : V \to [b]$ and $\gamma : E \to [c]$ are maps. The labels of a gaudy graph will play a role different from the role labels play in most of our discussions, and we will call them *badges*.

Isomorphism of two gaudy graphs of the same type is defined in the natural way, as an isomorphism between the underlying graphs that preserves the badge assignment and coloring maps. A *gaudy graph parameter* of type (a, b, c) is a complex valued function defined on gaudy graphs of type (a, b, c), invariant under isomorphism.

To extend the notion of connection matrices to gaudy graphs is a bit tedious but necessary. Let f be a gaudy graph parameter of type (a, b, c). We will define infinitely many connection matrices (just like in the case of ordinary graph parameters), but these will not be indexed by just a number k, as for ordinary graphs, but we will have a connection matrix $M(f; H, a_1, a_2)$ for every gaudy graph H of type (a_0, b, c), where $a_0 + a_1 + a_2 = a$. Its rows and columns are indexed by gaudy graphs G of type $(a_0 + a_1, b, c)$ and $(a_0 + a_2, b, c)$, respectively, with a fixed embedding of H into G which preserves badges, node colors and edge colors. The product $G_1 G_2$ of two such structures is obtained by taking their disjoint union and then identifying the copies of H in them. The badges and colors are defined in $G_1 G_2$ in a natural way. The entry of $M(f; H, a_1, a_2)$ in row G_1 and column G_2 is $f(G_1 G_2)$.

A gaudy graph parameter f of type (a, b, c) has *finite connection rank* if all connection matrices $M(f; H, a_1, a_2)$ have finite rank.

The two observations in Lemmas 4.5 and 4.6 and their proofs remain valid for gaudy graphs. We now formulate more involved operations, manipulating the badges and colors. Let f be a gaudy graph parameter of type $(a + 1, b, c)$, and define the gaudy graph parameter f^* of type (a, b, c) by

$$(4.7) \qquad f^*(G; \alpha, \beta, \gamma) = \max_{\substack{\alpha' : [a+1] \to V(G) \\ \alpha'|_{[a]} = \alpha}} f(G; \alpha', \beta, \gamma).$$

Let f be a gaudy graph parameter of type $(a, 2b, c)$, and let $\varphi : [2b] \to [b]$. Define the gaudy graph parameter f^{**} of type (a, b, c) by

$$(4.8) \qquad f^{**}(G; \alpha, \beta, \gamma) = \max_{\beta' : \varphi \circ \beta' = \beta} f(G; \alpha, \beta', \gamma).$$

Finally, to formulate an analogous construction for edge-colorings, let f be a gaudy graph parameter of type $(a, b, 2c)$, and let $\psi : [2c] \to [c]$. Define gaudy graph parameter f^{***} of type (a, b, c) by

$$(4.9) \qquad f^{***}(G; \alpha, \beta, \gamma) = \max_{\gamma' : \psi \circ \gamma' = \gamma} f(G; \alpha, \beta, \gamma').$$

The following lemma asserts that these operations preserve finite rank:

LEMMA 4.28. *If f is a gaudy graph parameter with finite connection rank and finite range, then f^*, f^{**} and f^{***} have finite connection rank.*

Proof. We describe the proof for f^*; the proof for f^{**} and f^{***} is similar.

Consider a connection matrix $M = M(f^*; H, a_1, a_2)$, where H is a gaudy graph of type (a_0, b, c), and $a_0 + a_1 + a_2 = a - 1$. Consider a general entry of M, defined

by a row index G_1 and column index G_2, where $G_1G_2 = (G, \alpha, \beta, \gamma)$:
$$M_{G_1,G_2} = f^*(G_1G_2) = \max_{\alpha'|[a]=\alpha} f(G; \alpha', \beta, \gamma).$$

We split this maximum into $\mathsf{v}(H)+2$ parts, according to $\alpha'(a) = v \in V(H)$, $\alpha'(a) \in V(G_1) \setminus V(H)$ and $\alpha'(a) \in V(G_2) \setminus V(H)$. This defines $M = M(f^*; H, a_1, a_2)$ as the maximum of $\mathsf{v}(H) + 2$ matrices A_v ($v \in V(H)$), $\widehat{B}$ and $\widehat{C}$. By the same argument as in the proof of Lemma 4.6, it suffices to prove that these matrices have finite rank. We show that these matrices can be expressed by the connection matrices of f in a way that finite rank is preserved.

First, each of the matrices A_v ($v \in V(H)$) is a connection matrix for f itself (with a new badge added to v), and so it has finite rank.

Second, each entry $\widehat{B}_{G_1G_2}$ is obtained as the maximum of the entries $N_{G'_1 G_2}$ of the matrix $N = M(f; H, a_1 + 1, a_2)$, where G'_1 is obtained from G_1 by attaching badge a to one additional node. Let $L(G_1)$ denote the set of these rows. This is a finite set but there is no common bound on its size. However, we may notice that for a fixed G_1, the columns have a basis with at most $\mathrm{rk}(N)$ elements, and if two rows agree on these basis columns then they agree everywhere. The range of f is finite, say it consists of r elements; hence there are at most $K = r^{\mathrm{rk}(N)}$ different rows in N. Let $L_1(G_1) \subseteq L(G_1)$ be a maximal set of different rows.

Now we create K matrices $\widehat{B}_1, \ldots, \widehat{B}_K$, which are of the same shape as $\widehat{B}$, the row of $\widehat{B}_i$ corresponding to G_1 is the i-th row in $L_1(G_1)$ (we repeat the last row if we run out of rows). These are all submatrices of N, so they have rank at most $\mathrm{rk}(N)$. Furthermore, $\widehat{B}$ is obtained by taking their maximum entry-by-entry, and hence it has finite rank by the argument used in the proof of Lemma 4.6.

It follows by a similar argument that C has finite rank. $\square$

Proof of Theorem 4.27. We prove the theorem more generally for gaudy graph properties definable by a monadic second order formula $\mathfrak{F}$, by induction on the number of quantifiers. If there are no quantifiers in $\mathfrak{F}$, then the assertion is easy. Else, $\mathfrak{F}$ can be written in one of the following forms:

$$(\forall x)\mathfrak{F}_x, \quad (\exists x)\mathfrak{F}_x, \quad (\forall S)\mathfrak{F}_S, \quad (\exists S)\mathfrak{F}_S, \quad (\forall \Psi)\mathfrak{F}_\Psi, \quad (\exists \Psi)\mathfrak{F}_\Psi,$$

Here $\mathfrak{F}_x$ is a monadic second order formula in the language of gaudy graphs with an additional badge x; $\mathfrak{F}_S$ is obtained from $\mathfrak{F}$ using twice as many node colors $\{1, \ldots, 2b\}$, where color i means "colored i and not in S", and color $i + b$ means "colored i and in S". $\mathfrak{F}_\Psi$ is defined analogously, splitting edge colors according to containment in the edge subset Ψ. Replacing the property by its negation if necessary, we can forget about the three versions starting with $\forall$ (see Exercise 4.37).

First, we consider the case when $\mathfrak{F} = (\exists x)\mathfrak{F}_x$. The indicator function $\mathbb{1}_{\mathfrak{F}_x}$ of the gaudy graph property defined by $\mathfrak{F}_x$ has finite connection rank by the induction hypothesis. For the indicator functions, we have $\mathbb{1}_{\mathfrak{F}} = \mathbb{1}^*_{\mathfrak{F}_x}$, so Lemma 4.28 implies that $\mathbb{1}_{\mathfrak{F}}$ has finite connection rank. The two remaining cases follow similarly. $\square$

REMARK 4.29. Further variants of the problem of characterizing graph parameters with finite connection rank ask for the characterization of graph parameters with exponentially bounded connection rank, or polynomially bounded connection rank. Not much is known in this direction. For the analogous "edge-connection" problem, see Schrijver's Theorem 23.6.

EXERCISE 4.30. Prove that for every isolate-indifferent graph parameter f, the connection rank $r(f, k)$ is a monotone non-decreasing function of k.

EXERCISE 4.31. Let $\overline{\chi}(G)$ denote the minimum number of cliques in the graph G covering all nodes (the chromatic number of the complement). Prove that $\overline{\chi}$ has finite connection rank.

EXERCISE 4.32. Prove that the graph parameters $2^{\alpha(G)}$ and $2^{\tau(G)}$ are multiplicative and have finite connection rank.

EXERCISE 4.33. Prove that the multigraph parameter $\overrightarrow{\mathsf{eul}}(G)$ has infinite connection rank for $k \geq 2$. Hint: for $k = 2$, consider the submatrix of $M(\overrightarrow{\mathsf{eul}}, k)$ formed by rows and columns indexed by $(2i-1)$-bonds, $i = 1, 2, \ldots, n$. Verify that this can be written as $2AA^\mathsf{T}$, where $A = \left[\binom{2i-1}{i+j-1}\right]_{i,j=1}^n$ is a lower triangular matrix with 1-s in the diagonal.

EXERCISE 4.34. (a) Prove that no unbounded maxing graph parameter has finite connection rank. (b) Show by an example that a bounded maxing parameter can have finite connection rank.

EXERCISE 4.35. Prove that if $\mathcal{P}$ is a minor-closed property, then the property $\overline{\mathcal{P}}$ defined by $\overline{\mathcal{P}}(G) = \mathcal{P}(\overline{G})$ has finite connection rank.

EXERCISE 4.36. The *Hadwiger number* $\mathsf{had}(G)$ if a graph G is the largest n for which K_n is a minor of G. Prove that the parameter had is minor-monotone, but its connection rank is not finite.

EXERCISE 4.37. Prove that if a graph property has finite connection rank, then so does its negation.

EXERCISE 4.38. Show that the following graph properties can be expressed by monadic second order formulas: (a) G is connected; (b) G is a tree; (c) G is 3-degenerate (i.e., its nodes can be ordered so that every node is connected to no more than 3 earlier nodes); (d) G is Hamiltonian.

EXERCISE 4.39. Show that every minor-monotone graph property can be expressed by a node-monadic second-order formula.

EXERCISE 4.40. The property that the number of nodes is even cannot be expressed by a monadic second order formula.

EXERCISE 4.41. Prove that (a) the property that G is Hamiltonian has finite connection rank; (b) the property that the complement $\overline{G}$ is Hamiltonian does not have finite connection rank. (c) The property that G is Hamiltonian can not be expressed by a node-monadic second order formula.

EXERCISE 4.42. Show that the following graph properties cannot be expressed by monadic second order formulas: (a) G has a nontrivial automorphism; (b) G has a node-transitive automorphism group.

EXERCISE 4.43. Let f be an integer valued graph parameter with finite connection rank, and let m be a positive integer. Prove that $f \mod m$ has finite connection rank.

EXERCISE 4.44. Let f be a bounded graph parameter with finite connection rank. Let $g: \mathbb{R} \to \mathbb{R}$ be an arbitrary function. Prove that $g(f(.))$ has finite connection rank.

EXERCISE 4.45. Let $f(G; x)$ be a graph parameter whose values are analytic functions of a variable x (defined for $|x| \leq 1$). Suppose that for every real x, $f(., x)$ has finite connection rank.

(a) Prove that the k-th connection rank of $f(.; x)$ is uniformly bounded in x.

(b) Prove that $\frac{d}{dx}f(., x)$ has finite connection rank for all x.

EXERCISE 4.46. From a gaudy graph parameter f, define the new parameters f', f'' and f''' by replacing the "max" by "sum" in (4.7), (4.8) and (4.9). Prove that if f has finite connection rank, then so do f', f'' and f''' (no finiteness assumption for the range is needed).

CHAPTER 5

Graph homomorphisms

5.1. Existence of homomorphisms

Let G and H be two simple graphs. An adjacency-preserving map from $V(G)$ to $V(H)$ is called a *homomorphism*. We write $G \to H$ if there is a homomorphism from G into H. The existence of homomorphisms between graphs is an important and highly non-trivial question. For example, $K_n \to G$ means that G contains a clique with n nodes; $G \to K_n$ means that G is n-colorable.

If we fix G, then $G \to H$ can be decided in time that is polynomial in $\mathsf{v}(H)$, by checking all possible maps $V(G) \to V(H)$. What happens if we fix H? If H has no edge, then $G \to H$ if and only if G has no edge. If H is a bipartite graph with at least one edge, then $H \to K_2$ and $K_2 \to H$. Hence $G \to H$ if and only if $G \to K_2$, which is equivalent to saying that G is bipartite.

Unfortunately, these are the only cases when the existence of a homomorphism into a fixed graph H is polynomial time decidable (at least if $P \neq NP$). It was proved by Hell and Nešetřil [1990] that for a (fixed) nonbipartite graph H, the problem whether $G \to H$ is NP-complete. So among the infinitely many problems $G \to H$ (one for each H), some are solvable in polynomial time and the others are NP-hard; there is no case whose complexity would be inbetween these extremes. This is an interesting phenomenon called "dichotomy".

If we consider directed graphs, then the problem becomes more challenging. We don't go into the details in this book; The monograph of Hell and Nešetřil [2004] provides an in-depth treatment of many questions on the existence of homomorphisms; we give a few exercises below to illustrate these questions.

EXERCISE 5.1. Verify that allowing looped-simple graphs would not give any interesting new cases of the homomorphism existence problem.

EXERCISE 5.2. Let $\overrightarrow{C}_n$ denote the directed n-cycle, $\overrightarrow{P}_n$, the directed path on n nodes, and $\overrightarrow{K}_n$, the transitive tournament on n nodes. For any cycle or path in a digraph G, define its *gain* as the difference between the numbers of forward and backward edges (if the cycle or path is traversed in the opposite direction, this number changes sign). Prove that
(a) $G \to \overrightarrow{C}_n$ if and only if the gain of every cycle is a multiple of n;
(b) $G \to \overrightarrow{P}_n$ if and only if the gain of every cycle is 0 and the gain of every path is bounded by $n-1$;
(c) $G \to \overrightarrow{K}_n$ if and only if $\overrightarrow{P}_{n+1} \not\to G$.

EXERCISE 5.3. Prove that $G \to \overrightarrow{C}_n$, $G \to \overrightarrow{P}_n$ and $G \to \overrightarrow{K}_n$ are polynomial time decidable.

5.2. Homomorphism numbers

5.2.1. Versions of homomorphism numbers. Now we come to one of the main notions for this book. Unfortunately, we have to discuss the definition for the different types of graphs introduced in Section 3.2. We could probably give a single very general definition, but this would be too awkward. Rather, we start with homomorphism numbers between simple graphs, and talk about the issues for more complicated types as they arise.

Simple graphs. For two finite simple graphs F and G, let $\mathrm{Hom}(F,G)$ denote the set of homomorphisms of F into G, and let $\hom(F,G)$ be their number. Let $\mathsf{inj}(F,G)$ be the number of injective homomorphisms of F into G, and $\mathsf{ind}(F,G)$, the number of embeddings of F into G as an induced subgraph. In other words, $\mathsf{ind}(F,G)$ counts injective homomorphisms that also preserve non-adjacency. Occasionally, one needs also $\mathsf{surj}(F,G)$, the number of surjective homomorphisms from F to G; here we mean surjective on both the nodes and edges, and refrain from introducing a notation for counting those which are only (say) node-surjective.

Multigraphs. If we want to extend the definition of $\hom(F,G)$ to multigraphs, then there is no agreement any more what it should mean. Most often we will use the notion that a homomorphism must specify not only which node goes on which node, but also which edge goes on which edge. So if $i,j \in V(F)$ are connected by a_{ij} edges, and $u,v \in V(G)$ are connected by b_{uv} edges, then after specifying that $i \mapsto u$ and $j \mapsto v$, we have $b_{uv}^{a_{ij}}$ ways of mapping the i-j edges to the u-v edges. In other words, a *node-and-edge homomorphism* $F \to G$ is a pair of maps $\varphi: V(F) \to V(G)$ and $\psi: E(F) \to E(G)$ such that if $e \in E(F)$ connects i and j, then $\psi(e)$ connects $\varphi(i)$ and $\varphi(j)$. If $b_{uv}, a_{ij} \in \{0,1\}$, then $b_{uv}^{a_{ij}}$ is 1 unless $a_{ij} = 1$ and $b_{uv} = 0$, and so for simple graphs this specializes to the earlier definition. Unless otherwise stated, the number of node-and-edge homomorphisms is what is meant by $\hom(F,G)$.

Another possibility is to define a *node-homomorphism* as a map $\varphi: V(F) \to V(G)$ such that the multiplicity of $\varphi(i)\varphi(j) \in E(G)$ is at least as large as the multiplicity of $ij \in E(F)$. In the case of simple graphs, this too specializes to the old notion. These versions will play an important role in Section 17.1.

We can also consider special node-homomorphisms that preserve edge multiplicities. In this case the role of multiplicities is reduced to distinguishing edges with different multiplicities. These homomorphisms generalize "induced" homomorphisms between simple graphs that preserve both adjacency and non-adjacency, and will be called *induced homomorphisms*

Weighted graphs. The definition of homomorphism numbers can be extended to the case when G is a weighted looped-simple graph with nodeweights $\alpha_v(G)$ and edgeweights $\beta_{uv}(G)$. To every map $\varphi: V(F) \to V(G)$, we assign the weights

$$\alpha_\varphi = \prod_{u \in V(F)} \alpha_{\varphi(u)}(G), \tag{5.1}$$

and

$$\hom_\varphi(F,G) = \prod_{uv \in E(F)} \beta_{\varphi(u)\varphi(v)}(G). \tag{5.2}$$

We define

$$\hom(F,G) = \sum_{\varphi:\, V(F) \to V(G)} \alpha_\varphi \hom_\varphi(F,G), \tag{5.3}$$

and

$$\text{inj}(F, G) = \sum_{\substack{\varphi: \, V(F) \to V(G) \\ \varphi \text{ injective}}} \alpha_\varphi \text{hom}_\varphi(F, G). \tag{5.4}$$

For these definitions to make sense, $\alpha_v(G)$ and $\beta_{uv}(G)$ can be from any commutative ring; we will, however, never need any field other than $\mathbb{R}$ and $\mathbb{C}$, and most of the time $\alpha_v(G)$ will be positive and $\beta_{uv}(G)$ real, and often itself positive.

This definition of $\text{hom}(F, G)$ makes sense if F is a multigraph and G is a weighted graph. If G is an (unweighted) multigraph, then we can consider the weighted simple graph G' in which each edge is weighted by its multiplicity in G. Then $\text{hom}(F, G) = \text{hom}(F, G')$ (in the node-and-edge sense).

One can also define $\text{hom}(F, G)$ when both F and G are weighted, provided these weights satisfy some reasonable conditions. Let us give the formula first. To every map $\varphi: V(F) \to V(G)$, we define the weights

$$\alpha_\varphi = \prod_{u \in V(F)} \alpha_{\varphi(u)}(G)^{\alpha_u(F)}. \tag{5.5}$$

and

$$\text{hom}_\varphi(F, G) = \prod_{uv \in E(F)} \beta_{\varphi(u)\varphi(v)}(G)^{\beta_{uv}(F)} \tag{5.6}$$

We then define

$$\text{hom}(F, G) = \sum_{\varphi: \, V(F) \to V(G)} \alpha_\varphi \text{hom}_\varphi(F, G). \tag{5.7}$$

The exponential $\beta_{\varphi(u)\varphi(v)}(G)^{\beta_{uv}(F)}$ may not be well defined. Mostly (and even this is not very often), we will need this definition when the nodeweights and edgeweights of F are nonnegative integers; then (with the usual convention that $0^0 = 1$) the definition is meaningful. Another case when the homomorphism number is well defined is when all the edgeweights in G are positive.

Note that in the case when F is an unweighted multigraph, we can replace it with a weighted graph on the same set of nodes where the nodeweights are 1 and the edgeweights are equal to the corresponding multiplicities in F. This does not change the homomorphism numbers $\text{hom}(F, G)$.

Signed graphs. There is a convenient way to treat conditions on preservation of edges and also preservation of non-edges together. Let F and G be simple graphs, where $F = (V, E_+, E_-)$ is signed. We define $\text{hom}(F, G)$ as the number of maps $V(F) \to V(G)$ where edges in E_+ must be mapped onto adjacent pairs, and edges in E_- must be mapped onto nonadjacent pairs. The quantity $\text{inj}(F, G)$ is defined analogously. If F is an unsigned simple graph, we construct the signed graph $\widehat{F}$ from the complete graph on $V(F)$ by signing the edges of F positive and the edges not in F negative. Then for every simple graph G,

$$\text{inj}(\widehat{F}, G) = \text{ind}(F, G). \tag{5.8}$$

The definition of homomorphism numbers from signed graphs into simple graphs can be extended to homomorphisms into weighted graphs H. For

$\varphi: V(F) \to V(H)$, we define α_φ by (5.1) as before,

$$\text{(5.9)} \qquad \mathsf{hom}_\varphi(F, H) = \prod_{uv \in E_+} \beta_{\varphi(u)\varphi(v)}(H) \prod_{uv \in E_-} \big(1 - \beta_{\varphi(u)\varphi(v)}(H)\big),$$

and then $\mathsf{hom}(F, H)$ by (5.3) as before.

Partially labeled graphs. Let F and G be a simple graph, $S \subseteq V(F)$, and $\varphi: S \to V(G)$. We denote by $\mathsf{hom}_\varphi(F, G)$ the number of homomorphisms $F \to G$ that extend the mapping φ. Most of the time we use this notation when F is $[k]$-labeled, and $S = [k]$ is the set of labeled nodes. If $\varphi(i) = v_i$ ($i \in [k]$), then we denote hom_φ also by $\mathsf{hom}_{v_1 \ldots v_k}$.

Let H be a weighted graph, and $\varphi: S \to V(H)$ for some $S \subseteq V(F)$. Then $\mathsf{hom}_\varphi(F, H)$ is already defined when $S = \emptyset$ (which is just $\mathsf{hom}(F, H)$), and when $S = V(G)$ (by (5.2)). Extension to the general case is rather natural (but note that nodeweights are used for the unlabeled nodes only):

$$\text{(5.10)} \qquad \mathsf{hom}_\varphi(F, H) = \sum_{\substack{\psi: V(F) \to V(H) \\ \psi \supseteq \varphi}} \prod_{j \in V(F) \setminus S} \alpha_{\psi(j)}(H) \mathsf{hom}_\psi(F, H).$$

Extension of this formula to the case when F is itself weighted or signed is left to the reader.

5.2.2. Homomorphism densities. We often normalize these homomorphism numbers, to get *homomorphism densities*. Setting $n = \mathsf{v}(G)$ and $k = \mathsf{v}(F)$, we define

$$\text{(5.11)} \qquad t(F, G) = \frac{\mathsf{hom}(F, G)}{n^k},$$

which is the probability that a random map of $V(F) \to V(G)$ is a homomorphism. We define similarly

$$\text{(5.12)} \qquad t_{\mathsf{inj}}(F, G) = \frac{\mathsf{inj}(F, G)}{(n)_k}$$

(the probability that a random injection $V(F) \to V(G)$ is a homomorphism), and

$$\text{(5.13)} \qquad t_{\mathsf{ind}}(F, G) = \frac{\mathsf{ind}(F, G)}{(n)_k}$$

(the probability that a random injection $V(F) \to V(G)$ preserves both adjacency and non-adjacency).

For a weighted graph H, we define

$$\alpha_H = \sum_{v \in V(H)} \alpha_\mathsf{v}(H),$$

and

$$t(F, H) = \frac{\mathsf{hom}(F, H)}{\alpha_H^{\mathsf{v}(F)}}.$$

Note that $t(F, H) = \mathsf{hom}(F, H^0)$, where H^0 is obtained from H by dividing all node weights by α_H, so that $\alpha_{H^0} = 1$. The nodeweights in H^0 form a probability distribution, and $t(F, H) = \mathsf{hom}(F, H^0)$ is the expectation of $\mathsf{hom}_\varphi(F, H)$, where φ is the random map $V(F) \to V(H)$ in which the image of each $v \in V(F)$ is chosen independently from the distribution $\alpha(H_0)$.

Bounded degree graphs. The above definition of homomorphism densities works well if the graphs are dense, but they are more-or-less meaningless for sparse graphs. Homomorphism numbers are of course defined in the same way independently of the densities of the graphs, but to introduce meaningful homomorphism densities in the case of bounded degree graphs we must normalize them differently. The best analogue of the dense homomorphism density $t(F, G)$ is the number

$$(5.14) \qquad t^*(F, G) = \frac{\hom(F, G)}{\mathsf{v}(G)},$$

which we consider for connected graphs F. We call this the *homomorphism frequency* of F in G, to distinguish it from the homomorphism densities that are used in the dense case.

We can interpret the homomorphism frequencies as follows. Let us label any node of F by 1, to get a 1-labeled graph F_1. For $v \in V(G)$, the quantity $\hom_v(F_1, G)$ denotes the number of homomorphisms φ of F_1 into G with $\varphi(1) = v$. Now we select a uniform random node v of G. Then $t^*(F, G)$ is the expectation of $\hom_v(F_1, G)$. We can interpret the injective and induced homomorphism frequencies

$$t^*_{\mathsf{inj}}(F, G) = \frac{\mathsf{inj}(F, G)}{\mathsf{v}(G)}, \qquad t^*_{\mathsf{ind}}(F, G) = \frac{\mathsf{ind}(F, G)}{\mathsf{v}(G)}$$

similarly.

For general (not necessarily connected) bounded degree graphs, the order of magnitude of $\hom(F, G)$ (where F is fixed and $\mathsf{v}(G)$ tends to infinity) is $\mathsf{v}(G)^{c(F)}$, where $c(F)$ is the number of connected components of F. But since $\hom(F, G)$ is multiplicative over the connected components of F, we don't lose any information if we restrict the definition of $t^*(F, G)$ to connected graphs F.

REMARK 5.4. Normalizing homomorphism densities as above is not the only reasonable choice. For example, if F is a bipartite graph, then $\hom(F, G)$ will be positive for graphs G with at least one edge, and we may be interested in the order of magnitude of, say, $\hom(C_4, G)$, given $\hom(K_2, G)$. Hence we might look at quotients $\log \hom(F, G) / \log \mathsf{v}(G)$, or more generally, $\log \hom(F_1, G) / \log \hom(F_2, G)$. Such quantities were studied by Kopparty and Rossman [2011] and Nešetřil and Ossona de Mendez [2011]. However, it is fair to say that this interesting area is largely unexplored.

5.2.3. Relations between homomorphism numbers.

Injective and induced homomorphism numbers. These quantities are closely related. For two simple graphs F and G, we have

$$(5.15) \qquad \mathsf{inj}(F, G) = \sum_{F' \supseteq F} \mathsf{ind}(F', G),$$

where F' ranges over all simple graphs obtained from F by adding edges, and

$$(5.16) \qquad \hom(F, G) = \sum_{P} \mathsf{inj}(F/P, G),$$

where P ranges over all partitions of $V(F)$, and F/P is the simple quotient graph. Conversely, ind can be expressed by inj using inclusion-exclusion:

$$\text{ind}(F,G) = \sum_{\substack{F' \supseteq F \\ V(F')=V(F)}} (-1)^{e(F')-e(F)} \text{inj}(F',G). \tag{5.17}$$

We can also express this using the Möbius inverse (recall the definitions (4.1)–(4.3)):

$$\text{ind}(.,G) = \text{inj}^\uparrow(.,G).$$

The inj function, in turn, can be expressed by hom, by considering the values $\text{inj}(F',G)$ in the equations (5.16) as unknowns and solving the system. To give an explicit expression, we use the Möbius inverse of the partition lattice:

$$\text{inj}(F,G) = \sum_P \mu_P \text{hom}(F/P, G), \tag{5.18}$$

where P ranges over all partitions of $V(F)$, and μ_P is defined by (A.2). In other words, $\text{inj}(.,G) = \text{hom}^\Downarrow(.,G)$.

Injective and induced homomorphism densities. We have

$$t_{\text{inj}}(F,G) = \sum_{F' \supseteq F} t_{\text{ind}}(F',G) \tag{5.19}$$

and the inversion formula

$$t_{\text{ind}}(F,G) = \sum_{F' \supseteq F} (-1)^{e(F')-e(F)} t_{\text{inj}}(F',G) = t_{\text{inj}}^\uparrow(F,G). \tag{5.20}$$

For t and t_{inj} the relationship is not so simple, due to the different normalizations in their definitions, but recalling that mostly we are interested in large graphs G, the following inequality is usually enough to relate them:

$$|t_{\text{inj}}(F,G) - t(F,G)| \leq \frac{1}{\mathsf{v}(G)} \binom{\mathsf{v}(F)}{2}. \tag{5.21}$$

(The proof of this inequality is left to the reader as an exercise.) It follows that (for large graphs G, when the error in (5.21) is negligible) subgraph sampling provides the same information as any of the homomorphism densities t, t_{inj} or t_{ind}.

Complementation. Taking induced subgraphs commutes with complementation, which implies

$$\text{ind}(F, \overline{G}) = \text{ind}(\overline{F}, G), \tag{5.22}$$

Counting maps into the complement of a graph can be expressed, via inclusion-exclusion, by numbers of maps into the graph itself. Applying this idea to injective homomorphisms, we get the following identities for every simple graph F:

$$\text{hom}(F, \overline{G}) = \sum_{\substack{F' \subseteq F \\ V(F')=V(F)}} (-1)^{e(F')} \text{hom}(F',G) \tag{5.23}$$

and

$$\text{inj}(F, \overline{G}) = \sum_{\substack{F' \subseteq F \\ V(F')=V(F)}} (-1)^{e(F')} \text{inj}(F',G). \tag{5.24}$$

Averaging. If we know homomorphism numbers from graphs of a given size, then we get homomorphism numbers from smaller graphs; also, we get homomorphism numbers into larger graphs. This idea has many formulations, of which we state a couple. Let $v(F_0) = k \leq t \leq v(G) = n$, then

$$\text{ind}(F_0, G) = \frac{1}{\binom{n-k}{t-k}} \sum_{F:\, v(F)=t} \frac{\text{ind}(F_0, F)\text{ind}(F, G)}{\text{aut}(F)}, \tag{5.25}$$

Let $\mathsf{v}(F) = k \leq t \leq \mathsf{v}(G) = n$, then

$$\text{inj}(F, G) = \frac{1}{\binom{n-k}{t-k}} \sum_{S \in \binom{V(G)}{t}} \text{inj}(F, G[S]). \tag{5.26}$$

We get similar expressions for the induced homomorphism numbers ind. We can also express this in terms of homomorphism densities:

$$t_{\text{inj}}(F, G) = \frac{1}{\binom{n}{t}} \sum_{S \in \binom{V(G)}{t}} t_{\text{inj}}(F, G[S]). \tag{5.27}$$

Graph operations. If F_1 and F_2 are node-disjoint, then

$$\hom(F_1 \cup F_2, G) = \hom(F_1, G)\hom(F_2, G). \tag{5.28}$$

If F is connected and G_1 and G_2 are node-disjoint, then

$$\hom(F, G_1 \cup G_2) = \hom(F, G_1) + \hom(F, G_2). \tag{5.29}$$

About homomorphisms into a product, we have

$$\hom(F, G_1 \times G_2) = \hom(F, G_1)\hom(F, G_2). \tag{5.30}$$

All these identities are straightforward to verify.

5.2.4. Homomorphism numbers and sampling. Our basic way of obtaining information about a graph is sampling (Section 1.3.1): subgraph sampling in the dense case and neighborhood sampling in the bounded degree case. Homomorphism densities and frequencies carry the same information as the appropriate sample distributions. For the dense case, the connection is straightforward:

PROPOSITION 5.5. *For two simple graphs F and G, $t_{\text{ind}}(F, G)$ is the probability that sampling $V(F)$ nodes of G (ordered, without repetition), they induce the graph F (with a fixed labeling of the nodes).* □

In the bounded degree case, homomorphism frequencies contain the same information as the distribution of neighborhood samples, but the proof of this equivalence is a bit trickier. Let us recall that $\rho_{G,r}$ is a probability distribution on rooted r-balls: $\rho_{G,r}(B)$ denotes the probability that selecting a uniform random node of the graph G, its neighborhood of radius r is isomorphic with the ball B.

PROPOSITION 5.6. *Let us fix an upper bound D for the degrees of the graphs we consider.*

(a) Each density $t^*(F, G)$ can be expressed as a linear combination (with coefficients independent of G) of the neighborhood sample densities $\rho_{G,r}$ with $r = \mathsf{v}(F) - 1$.

(b) For every $r \geq 0$ there are a finite number of connected simple graphs $F_1, \ldots, F_m$ such that $\rho_{G,r}$ can be expressed as a linear combination (with coefficients independent of G) of the densities $t^*(F_i, G)$.

Proof. (a) From the interpretation of $t^*(F, G)$ given above, we see that it can be obtained as the expectation of the number of $\mathsf{hom}_{u \to v}(F, \mathbf{B})$, where u is any fixed node of F, and $\mathbf{B}$ is a random ball from the neighborhood sample distribution $\rho_{G,r}$, with center v and radius $r = \mathsf{v}(F) - 1$. This gives the formula

$$t^*(F, G) = \sum_B \rho_{G,r}(B) \mathsf{hom}_{u \to v}(F, B),$$

where the summation extends over all possible r-balls.

(b) To compute the neighborhood sample distributions from the quantities $t^*(F, G)$, we first express the quantities $t^*_{\mathsf{inj}}(F, G)$ via inclusion-exclusion. By a similar argument, we can express the induced densities $t^*_{\mathsf{ind}}(F, G)$. (Since we are normalizing by $\mathsf{v}(G)$ in all cases, we avoid here the difficulty we had in the dense case with expressing t_{inj} by t.)

Next, we count copies of F in G where we also prescribe the degree of each node of F in the whole graph G. To be precise, we consider graphs F together with maps $\delta \colon V(F) \to \{0, \dots, D\}$, and we determine the numbers

$$t^*_{\mathsf{ind}}(F, \delta, G) = \frac{\mathsf{ind}(F, \delta, G)}{\mathsf{v}(G)},$$

where $\mathsf{ind}(F, \delta, G)$ is the number injections $\varphi \colon V(F) \to V(G)$ which embed F in G as an induced subgraph, so that the degree of $\varphi(v)$ is $\delta(v)$. This is again done by an inclusion-exclusion argument.

For a ball B of radius r, we have

$$\rho_{G,r}(B) = \sum_\delta \frac{t^*_{\mathsf{ind}}(B, \delta, G)}{\mathsf{aut}(B)},$$

where the summation extends over all functions δ which assign the degree in B to each node of B at distance less than r from the root, and an arbitrary integer from $[D]$ to those nodes at distance r. This proves that homomorphism densities and neighborhood sampling are equivalent. $\square$

EXERCISE 5.7. Find formulas similar to (5.16) and (5.18), relating hom and surj.

EXERCISE 5.8. Which of the relations (5.15)–(5.30) generalize to weighted graphs?

5.3. What hom functions can express

Homomorphisms of "small" graphs into G are related to sampling, as mentioned earlier. There are many other applications of homomorphism numbers. We start with a number of graph parameters, some unexpected, some not, that can be expressed as homomorphism numbers into and from the given graph.

EXAMPLE 5.9 (**Walks**). A walk in G is a homomorphism of a path into G, so $\mathsf{hom}(P_k, G)$ counts the number of walks with $k - 1$ steps in G. Note that we can express this as the sum of the entries of A^{k-1}, where A is the adjacency matrix of G. ◆

EXAMPLE 5.10 (**Stars and degrees**). Homomorphisms from stars into G give the moments of the degree sequence:

$$\mathsf{hom}(S_k, G) = \sum_{i \in V(G)} \deg(i)^{k-1}.$$

EXAMPLE 5.11 (**Cycles and spectrum**). If C_k denotes the cycle on k nodes, then $\mathsf{hom}(C_k, G)$ is the trace of the k-th power of the adjacency matrix of the graph G. In other words,

$$(5.31) \qquad \mathsf{hom}(C_k, G) = \mathrm{tr}(A^k) = \sum_{i=1}^{n} \lambda_i^k,$$

where $\lambda_1, \ldots, \lambda_n$ are the eigenvalues of the adjacency matrix of G. Knowing this homomorphism number for sufficiently many values of k, the eigenvalues of G recovered; eigenvalues with large absolute value are easier to express. For example, $\mathsf{hom}(C_{2k}, G)^{1/(2k)}$ tends to the largest eigenvalue of G as $k \to \infty$. ◆

Several important graph parameters can be expressed in terms of homomorphisms into fixed "small" graphs.

EXAMPLE 5.12 (**Colorings**). If K_q denotes the complete graph with q nodes (no loops), then $\mathsf{hom}(G, K_q)$ is the number of colorings of the graph G with q colors, satisfying the usual condition that adjacent nodes must get different colors. ◆

EXAMPLE 5.13 (**Stable sets**). Let $H = \begin{smallmatrix}\circ\\|\\\circ\end{smallmatrix}$ be obtained from K_2 by adding a loop at one of the nodes. Then $\mathsf{hom}(G, H)$ is the number $\mathsf{stab}(G)$ of stable sets of nodes in G. ◆

EXAMPLE 5.14 (**Eulerian property**). Recall that a graph is *eulerian*, if all degrees are even. For every loopless graph G, let $\mathsf{Eul}(G) = \mathbb{1}(G \text{ is eulerian})$. This 0-1 valued graph parameter can be represented as a homomorphism function $\mathsf{hom}(., H)$, where $H = (a, B)$ is a weighted graph with two nodes, given by

$$a = \begin{pmatrix} 1/2 \\ 1/2 \end{pmatrix}, \qquad B = \begin{pmatrix} 1 & -1 \\ -1 & 1 \end{pmatrix}$$

This was first noted by de la Harpe and Jones [1993]. (By Theorem 5.54 it will follow that this function is reflection positive, and $r(\mathsf{Eul}, k) \leq 2^k$.) ◆

EXAMPLE 5.15 (**Nowhere-zero flows**). The number of nowhere-zero q-flows is denoted by $\mathsf{flo}(G, q)$ (see Appendix A.2). The choice $q = 2$ gives the special case in example 5.14 (indicator function of eulerian graphs). The parameter flo can be described as a homomorphism function, as will be demonstrated in larger generality in the next example. ◆

EXAMPLE 5.16 (**S-Flows**). Let Γ be a finite abelian group (written additively), and let $S \subseteq \Gamma$ be a subset such that $-S = S$, and let G be a graph. An *S-flow* is an assignment of an element $f(uv) \in S$ to each edge uv with a specified orientation such that $f(uv) = -f(vu)$ for each edge, and $\sum_{u \in N(v)} f(uv) = 0$ for each node v. Let $\mathsf{flo}(G; \Gamma, S)$ denote the number of S-flows. The special case when $\Gamma = \mathbb{Z}_q$ and $S = \Gamma \setminus \{0\}$ gives the number of nowhere zero q-flows in Example 5.15.

For a fixed Γ and S, the graph parameter $\mathsf{flo}(G; \Gamma, S)$ is defined in terms of mappings from the edge set; it is therefore surprising that it can be described as a homomorphism number (which is defined via a function on the node set). Let Γ^* be the character group of Γ, and let H be the complete looped directed graph on Γ^*. Let $\alpha_\chi = 1/|\Gamma|$ for each $\chi \in \Gamma^*$, and let

$$\beta_{\chi, \chi'} = \sum_{s \in S} \chi(-s)\chi'(s),$$

for any two characters $\chi, \chi' \in \Gamma^*$. It follows from our assumption that $-S = S$ that $\beta_{\chi, \chi'}$ is real and $\beta_{\chi', \chi} = \beta_{\chi, \chi'}$. It takes a straightforward computation to show (Freedman, Lovász and Schrijver [2007]) that

$$\text{flo}(G; \Gamma, S) = \text{hom}(G, H). \tag{5.32}$$

◆

EXAMPLE 5.17 (**Tutte polynomial**). The Tutte polynomial is not a homomorphism function in general; Theorem 5.54 will imply that if q is not a positive integer, then it is not. But if q is a positive integer, then it has such a representation: Using the connection between the Tutte polynomial and the Potts model in statistical physics (Welsh and Merino [2000]), one can prove that

$$\text{cep}(G; q, v) = \text{hom}(G, H_{q,v})$$

(here $H_{q,v}$ is a complete graph on q nodes, with a loop added at each node; every node has weight 1, the ordinary edges too have weight 1, but the loops have weight $1 + v$). ◆

EXAMPLE 5.18 (**Maximum cut**). An important graph parameter is the *maximum cut* $\text{Maxcut}(G)$, the maximum number of edges between a set $S \subseteq V(G)$ of nodes and its complement. While finding minimum cuts is perhaps a more natural task at the first sight, the maximum cut problem comes up when we want to approximate general graphs by bipartite graphs, when computing ground states in statistical physics (see Section 2.2), and in many other applications. For our purposes, it will be more convenient to consider the *normalized maximum cut*, defined by

$$\text{maxcut}(G) = \frac{\text{Maxcut}(G)}{|V|^2} = \max_{S \subseteq V} \frac{e_G(S, V \setminus S)}{|V|^2}.$$

The maximum cut cannot be expressed as a homomorphism number, but the following easy fact relates maximum cuts and homomorphism numbers. Let H be the edge-weighted graph on $\{1, 2\}$ with edgeweights 1 except for the non-loop edge, which has weight 2. Then we have the trivial inequalities

$$2^{\text{Maxcut}(G)} \leq \text{hom}(G, H) \leq 2^{\mathsf{v}(G)} 2^{\text{Maxcut}(G)},$$

which upon taking the logarithm and dividing by $\mathsf{v}(G)^2$ become

$$\text{maxcut}(G) \leq \frac{\log \text{hom}(G, H)}{\mathsf{v}(G)^2} \leq \text{maxcut}(G) + \frac{1}{\mathsf{v}(G)}. \tag{5.33}$$

So the homomorphism number into this simple 2-node graph determines $\text{maxcut}(G)$ with an additive error that tends to 0 as $\mathsf{v}(G) \to \infty$. ◆

EXAMPLE 5.19 (**Multicuts**). A natural extension of the maximum cut problem involves partitions into $q \geq 1$ classes instead of 2. Instead of just counting edges between different classes, we specify in advance a symmetric matrix B of coefficients B_{ij} ($i, j \in [q]$). We define the maximum multicut density (with target weights B) as

$$\text{cut}(G, B) = \max \frac{1}{\mathsf{v}(G)^2} \sum_{i,j} B_{ij} e_G(S_i, S_j),$$

where the maximum is taken over all partitions $\{S_1, \ldots, S_q\}$ of $V(G)$. As a special case, the matrix $B = \begin{pmatrix} 0 & 1 \\ 1 & 0 \end{pmatrix}$ defines the maximum cut.

Similarly as above, we construct an edge-weighed graph H on $[q]$ with edgeweights $2^{B_{ij}}$. Then

$$(5.34) \qquad \mathsf{cut}(G,B) \leq \frac{\log \mathsf{hom}(G,H)}{\mathsf{v}(G)^2} \leq \mathsf{cut}(G,B) + \frac{\log q}{\mathsf{v}(G)}.$$

In the terminology of statistical physics, the negative of the value $\mathsf{cut}(G,B)$ would be called the *ground state energy* . ◆

5.3.1. Graph polynomials and homomorphisms. We describe more complex relations between homomorphism functions and some important graph polynomials. These relations will not be used until Chapter 19, but they illustrate the many subtle connections between homomorphism functions and other graph theoretic constructions.

Let G be a graph and let $\mathcal{I}(G)$ denote the set of stable (independent) subsets of $V(G)$. We assign a variable x_i to each node i. For every multiset S of the nodes, let $x_S = \prod_{i \in S} x_i$. We define the *multivariate stable set polynomial* as

$$\mathsf{stab}(G,x) = \sum_{S \in \mathcal{I}(G)} x_S.$$

Note that $\mathsf{stab}(G,1,\ldots,1) = \mathsf{stab}(G) = \mathsf{hom}(G,H)$, where H is the graph on two adjacent nodes, with a loop at one of them (all weights being 1).

We have seen that both the chromatic polynomial and the stable set polynomial can be expressed, at least for special substitutions, as homomorphism numbers. We show that, conversely, homomorphism numbers between graphs can be expressed in terms of the stable set polynomial and also in terms of a the chromatic polynomials of related graphs. Our first lemma expresses the logarithm of the stable set polynomial in terms of the coefficient of the linear term of the chromatic polynomial (see Appendix A.2 for the relevant definitions). We need a natural extension of the notion for an induced subgraph: For a multiset Z of nodes, let $G[Z]$ denote the graph whose nodes are the elements of the multiset Z, and two of them are adjacent iff the corresponding nodes of G are adjacent.

LEMMA 5.20. *Assuming that the series below is absolute convergent, we have*

$$(5.35) \qquad \ln \mathsf{stab}(G,x) = \sum_{m=1}^{\infty} \frac{(-1)^m}{m!} \sum_{v_1,\ldots,v_m \in V(G)} \mathsf{cri}(G[v_1,\ldots,v_m]) x_{v_1} \ldots x_{v_m}.$$

Proof. Let $\mathcal{I}_+$ denote the set of non-empty stable subsets of G. Writing $\mathsf{stab}(G,x) = 1 + \sum_{A \in \mathcal{I}_+} x_A$, we get

(5.36)
$$\mathsf{stab}(G,x)^y = 1 + \sum_{k=1}^{\infty} \binom{y}{k} \Big(\sum_{A \in \mathcal{I}_+} x_A\Big)^k = 1 + \sum_{k=1}^{\infty} \binom{y}{k} \sum_{A_1,\ldots,A_k \in \mathcal{I}_+} x_{A_1} \ldots x_{A_k}.$$

Let $Z = \{v_1, \ldots, v_m\}$ denote the union of the A_i as multiset, so that corresponding term in the last sum is x_Z. Any choice of $\mathcal{S} = \{A_1, \ldots, A_k\}$ that results in the same multiset Z gives rise to a coloring of the graph $G[Z]$ with exactly k colors, and vice versa. We have to take into account that this coloring is not unique: Let r be the number of different nodes v_j, and let $m_1, \ldots m_r$ be the multiplicities, then we can associate $m_1! \ldots m_r!$ colorings of $G[Z]$ with the same family $\mathcal{S} = \{A_1, \ldots, A_k\}$ of stable sets. So the coefficient of x_Z in the last sum in

(5.36) is $\mathsf{chr}_0(G[Z],k)/(m_1!\ldots m_r!)$. We get a nicer formula if instead of multisets, we sum over sequences $(v_1,\ldots,v_m)$ of nodes. Then our multiset Z is counted $m!/(m_1!\ldots m_r!)$ times, so we have to divide by this, to get that the contribution of a sequence $(v_1,\ldots,v_m)$ is

$$\frac{1}{m!}\mathsf{chr}_0(G[v_1,\ldots v_m],k)x_{v_1}\ldots x_{v_m},$$

and summing over k, we get that the contribution of $(v_1,\ldots,v_m)$ is

$$\sum_{k=1}^{\infty}\binom{y}{k}\frac{1}{m!}\mathsf{chr}_0(G[v_1,\ldots v_m],k)x_{v_1}\ldots x_{v_m} = \frac{1}{m!}\mathsf{chr}(G[v_1,\ldots v_m],y)x_{v_1}\ldots x_{v_m}.$$

For $m=0$, we get 1. This implies that

$$(5.37) \qquad \mathsf{stab}(G,x)^y = \sum_{m=0}^{\infty}\sum_{v_1,\ldots,v_m\in V(G)}\frac{1}{m!}\mathsf{chr}(G[v_1,\ldots v_m],y)x_{v_1}\ldots x_{v_m}.$$

Differentiating (5.37) according to y and substituting $y=0$, we get the formula in the lemma. □

Next, we express $\mathsf{hom}(G,H)$ in terms of the intersection graph $\mathcal{G}$ of connected subgraphs of G with at least two nodes. For a weighted graph H, let $\overline{H}$ denote the weighted graph on the same node set and with the same nodeweights as H, and with edgeweights $1-\beta_{ij}(H)$. Recall that $L(C)$ denotes the intersection graph of a family C of sets, and $\mathrm{Conn}(G)$ is the set of connected subgraphs of G with at least one edge.

LEMMA 5.21. *For a simple graph $G=(V,E)$ and weighted graph H, and the vector $t\in\mathbb{R}^{\mathrm{Conn}(G)}$ defined by $t_F = t(F,\overline{H})$, we have $t(G,H) = \mathsf{stab}\bigl(L(\mathrm{Conn}(G)),t\bigr)$.*

Proof. By (5.22), we have

$$t(G,H) = \sum_{E'\subseteq E}(-1)^{|E'|}t(G',\overline{H}),$$

where $G'=(V,E')$. Using that $t(G',\overline{H})$ is multiplicative over the components of G' and that singleton components give a factor of 1, we get

$$(5.38) \qquad t(G,H) = \sum_{E'\subseteq E}\prod_F(-1)^{e(F)}t(F,\overline{H}),$$

where the product extends over all connected components of G' with at least one edge. These components form a stable set in $L(\mathrm{Conn}(G))$, and vice versa, every stable set in $L(\mathrm{Conn}(G))$ corresponds to a subgraph G'. Hence the last sum is just $\mathsf{stab}\bigl(L(\mathrm{Conn}(G)),t\bigr)$. □

Combining this lemma with the previous one, we get a very useful relationship between homomorphism densities and the chromatic invariant.

COROLLARY 5.22. *Assuming the series below is absolute convergent, we have*

$$\ln t(G,H) = \sum_{m=1}^{\infty}\frac{(-1)^m}{m!}\sum_{F_1,\ldots,F_m\in\mathrm{Conn}(G)}\mathsf{cri}\bigl(L(F_1,\ldots,F_m)\bigr)\prod_{j=1}^m t(F_j,\overline{H}).$$

What about all these assumptions about convergence? In fact, can we take the logarithm in Lemma 5.20 at all? A fundamental result about the roots of the stable set polynomial, Dobrushin's Theorem [1996], gives us a sufficient condition for this. Dobrushin's Theorem has many statements in the literature, which are more-or-less equivalent (but not quite), and we choose one that is convenient for our purposes; see e.g. Scott and Sokal [2006] and Borgs [2006].

THEOREM 5.23 (**Dobrushin's Theorem**). *Let $G = (V, E)$ be a simple graph, and let $z \in \mathbb{C}^V$ and $b \in \mathbb{R}_+^V$ satisfy*

$$\sum_{j \in \{i\} \cup N(i)} |z_j| e^{b_j} \leq b_i \tag{5.39}$$

for every node i. Then $\mathsf{stab}(G, z) \neq 0$, *and*

$$|\ln \mathsf{stab}(G, z)| \leq \sum_j |z_j| e^{b_j}.$$

Condition (5.39) defines a multidisc in which $\ln \mathsf{stab}(G, z)$ is analytic, and so by elementary properties of convergence of power series it follows that the series in (5.35) is convergent inside this multidisc. We remark that to get good bounds here, one can combine those terms in (5.38) for which the components of G' give the same partition of $V(G)$; then we get an expression in terms of the intersection graph of connected *induced* subgraphs of G, which has a smaller number of terms; see e.g. Borgs, Chayes, Kahn and Lovász [2012].

Another important graph polynomial is the characteristic polynomial of the adjacency matrix, $A(G, x) = \det(xI - A_G)$. We state a formula from Lyons [2005] for the case when G is a connected D-regular non-bipartite graph.

PROPOSITION 5.24. *Let G be a connected D-regular non-bipartite graph on n nodes. Then for every $x > D$, we have*

$$\ln A(G, x) = n \ln x - \sum_{r=1}^{\infty} \frac{\mathsf{hom}(C_r, G)}{rx^r}, \tag{5.40}$$

and

$$\ln \mathsf{tree}(G) = (n-1) \ln D - \ln n - \sum_{r=1}^{\infty} \frac{\mathsf{hom}(C_r, G) - D^r}{rD^r}. \tag{5.41}$$

The formula for the number of trees extends easily to non-regular graphs, since we can add loops to the nodes to make the graph regular, and adding loops does not change the number of spanning trees (adding a loop to a node increases its degree by 1 in this case). This expression seems to have been first formulated by Lyons [2005].

Proof. We can write this polynomial as

$$A(G, x) = \prod_{k=1}^{n} (x - \lambda_k), \tag{5.42}$$

where $\lambda_1 \geq \lambda_2 \geq \cdots \geq \lambda_n$ are the eigenvalues of the matrix A_G. It is well known that we have $\lambda_1 = D > \lambda_2$ (as G is connected) and $\lambda_n > -D$ (as G is non-bipartite).

To handle this product, we take the logarithm and expand it:

$$\ln \prod_{k=1}^n (x - \lambda_k) = n \ln x + \sum_{k=1}^n \ln(1 - \lambda_k/x) = n \ln x - \sum_{k=1}^n \sum_{r=1}^\infty \frac{1}{r} \frac{\lambda_k^r}{x^r}$$

(5.43)
$$= n \ln x - \sum_{r=1}^\infty \frac{1}{rx^r} \sum_{k=1}^n \lambda_k^r.$$

We can express the last sum using (5.31), to get (5.40).

By the Matrix Tree Theorem, $n\text{tree}(G)$ is the coefficient of the linear term in the determinant $\det(yI + DI - A)$, and hence

$$\ln \text{tree}(G) = \lim_{y \to 0} \bigl(\ln \det(yI + DI - A) - \ln y - \ln n\bigr).$$

Using (5.40),

$$\ln \det(yI + DI - A) = n \ln(y + D) - \sum_{r=1}^\infty \frac{\hom(C_r, G)}{r(y+D)^r}$$

$$= n \ln(y + D) - \sum_{r=1}^\infty \frac{\hom(C_r, G) - D^r}{r(y+D)^r} + \ln \frac{y}{y+D}$$

$$= (n-1) \ln(y + D) - \sum_{r=1}^\infty \frac{\hom(C_r, G) - D^r}{r(y+D)^r} + \ln y.$$

Substituting this in the formula for $\ln \text{tree}(G)$ and letting $y \to 0$, we get (5.41). □

EXERCISE 5.25. Prove identity (5.32).

EXERCISE 5.26. Let $H = H(a, B)$ be a weighted graph, where

$$a = \begin{pmatrix} 1 \\ -1 \end{pmatrix}, \qquad B = \begin{pmatrix} 2 & 1 \\ 1 & 1 \end{pmatrix}$$

(illegal weighting, because there is a negative nodeweight, but the formula defining the hom function makes sense). Prove that $\hom(F, H)$ is the number of those subsets of edges that cover every node.

EXERCISE 5.27. Verify that (5.41) yields the Cayley formula $\text{tree}(K_n) = n^{n-2}$.

EXERCISE 5.28. Prove that $n\text{tree}(G) = \prod_{k=2}^n (D - \lambda_k)$, and show that this implies (5.41).

5.4. Homomorphism and isomorphism

5.4.1. Homomorphism–profiles. We start with a simple but useful observation (Lovász [1967]); various less trivial extensions and generalizations of this fact will play an important role a number of times (cf. Theorems 5.33, 13.9 and 17.5, and Corollaries 5.45 and 10.34).

THEOREM 5.29. *Either one of the simple graph parameters* $\hom(., G)$ *and* $\hom(G, .)$ *determines a simple graph* G.

By the same argument, these parameters defined on looped-simple graphs determine a looped-simple graph G.

Proof. We prove that $\mathsf{hom}(.,G)$ determines G; the argument for $\mathsf{hom}(G,.)$ is similar. The analogous statement for injective homomorphisms is trivial: if G and G' are two simple graphs such that $\mathsf{inj}(F,G) = \mathsf{inj}(F,G')$ for every simple graph F, then in particular $\mathsf{inj}(G',G) = \mathsf{inj}(G',G') > 0$ and $\mathsf{inj}(G,G') = \mathsf{inj}(G,G) > 0$, so G and G' have injective homomorphisms into each other, and hence they are isomorphic.

Now (5.18) expresses injective homomorphism numbers in terms of ordinary homomorphism numbers, which implies that if $\mathsf{hom}(F,G) = \mathsf{hom}(F,G')$ for every simple graph F, then $\mathsf{inj}(F,G) = \mathsf{inj}(F,G')$ for every simple graph F, and hence $G \cong G'$. □

We see from the proof of Theorem 5.29 that in fact G is determined by the values $\mathsf{hom}(F,G)$ where $\mathsf{v}(F) \leq \mathsf{v}(G)$, as well as by the values $\mathsf{hom}(G,F)$ where $\mathsf{v}(F) \leq \mathsf{v}(G)$. It is a long-standing open problem whether, up to trivial exceptions, strictly smaller graphs F are enough:

CONJECTURE 5.30 (**Reconstruction Conjecture**). *If G is a simple graph with $\mathsf{v}(G) \geq 3$, then the numbers $\mathsf{hom}(F,G)$ with $\mathsf{v}(F) < \mathsf{v}(G)$ determine G.*

There is a weaker version, which is also unsolved:

CONJECTURE 5.31 (**Edge Reconstruction Conjecture**). *If G is a simple graph with $\mathsf{e}(G) \geq 4$, then the numbers $\mathsf{hom}(F,G)$ with $\mathsf{e}(F) < \mathsf{e}(G)$ determine G.*

It is known that the Edge Reconstruction Conjecture holds for graphs G with $\mathsf{e}(G) \geq \mathsf{v}(G)\log\mathsf{v}(G)$ (Müller [1977]). We will prove an "approximate" version of the Reconstruction Conjecture (Theorem 10.32): for an arbitrarily large graph G, the numbers $\mathsf{hom}(F,G)$ with $\mathsf{v}(F) \leq k$ determine G up to an error of $O(1/\sqrt{\log k})$ (measured in the cut distance, which was mentioned in the Introduction but will be formally defined in Chapter 8). Unfortunately, this does not seem to bring us closer to the resolution of the Reconstruction Conjecture.

The normalized homomorphism density function $t(.,G)$ does not determine a simple graph G: If $G(p)$ is obtained from G by replacing every node by p twin nodes, then $t(F,G(p)) = t(F,G)$. But this is all that can go wrong:

THEOREM 5.32. *If G_1 and G_2 are simple graphs such that $t(F,G_1) = t(F,G_2)$ for every simple graph F, then there is a third simple graph G and positive integers p_1, p_2 such that $G_1 \cong G(p_1)$ and $G_2 \cong G(p_2)$.*

Proof. Let $n_i = \mathsf{v}(G_i)$, and consider the blowups $G'_1 = G_1(n_2)$ and $G'_2 = G_2(n_1)$. These have the same number of nodes, and hence $t(F,G'_1) = t(F,G_1) = t(F,G_2) = t(F,G'_2)$ implies that $\mathsf{hom}(F,G'_1) = \mathsf{hom}(F,G'_2)$. So by Theorem 5.29, we have $G'_1 \cong G'_2$. It follows that the number of elements in every class of twin nodes of $G'_1 \cong G'_2$ is divisible by both n_1 and n_2, and so it also divisible by $m = \mathsf{lcm}(n_1, n_2)$. So $G'_1 \cong G'_2 \cong G(m)$ for some simple graph G, and hence $p_i = m/n_i$ satisfies the requirements in the theorem. □

For weighted graphs, one must be a little careful. Let H be a weighted graph and let H' be obtained from H by twin reduction. Then $\mathsf{hom}(F,H') = \mathsf{hom}(F,H)$ for every multigraph F, even though H and H' are not isomorphic. Restricting our attention to twin-free graphs, we have an analogue of Theorem 5.29 (Lovász [2006b]):

THEOREM 5.33. *Let H_1 and H_2 be twin-free weighted graphs such that $\hom(F, H_1) = \hom(F, H_2)$ holds for all simple graphs F with at most $2(\mathsf{v}(H_1) + \mathsf{v}(H_2) + 3)^8$ nodes. Then $H_1 \cong H_2$.*

The proof of this theorem is substantially more complicated then that of its unweighted version. A proof without the bound on the size of F will be described in Section 6.4.1, where it will follow easily from the general tools developed there, and the full proof will be postponed until Section 6.4.2

5.4.2. Algebraic properties of graph multiplication. The fact that homomorphism numbers (into it or from it) determine the graph will motivate much in the sequel. Here we describe a few old applications of this fact to some basic algebraic properties of categorical product (Lovász [1967, 1971]).

First, we show that taking k-th root is unique (if it exists at all).

THEOREM 5.34. *If G_1 and G_2 are looped-simple graphs such that $G_1^{\times k} \cong G_2^{\times k}$ for some $k \geq 1$, then $G_1 \cong G_2$.*

Proof. For every looped-simple graph F, we have
$$\hom(F, G_1) = \hom(F, G_1^{\times k})^{1/k} = \hom(F, G_2^{\times k})^{1/k} = \hom(F, G_2),$$
whence by Theorem 5.29, $G_1 \cong G_2$. □

Next we turn to the question of Cancellation Law: does $G_1 \times H \cong G_2 \times H$ (where G_1, G_2 and H are looped-simple graphs) imply that $G_1 \cong G_2$? This is false in general:
$$(5.44) \qquad K_2 \times C_6 \cong K_2 \times (K_3 K_3).$$
But the proof method of Theorem 5.34 almost goes through: for every looped-simple graph F, we have
$$\hom(F, G_1)\hom(F, H) = \hom(F, G_1 \times H) = \hom(F, G_2 \times H)$$
$$= \hom(F, G_2)\hom(F, H);$$
if $\hom(F, H) \neq 0$, then this implies that $\hom(F, G_1) = \hom(F, G_2)$. What to do if $\hom(F, H) = 0$? We can find several simple conditions under which this difficulty can be handled:

PROPOSITION 5.35. *Let G_1, G_2 and H be looped-simple graphs such that $G_1 \times H \cong G_2 \times H$.*

(a) *If H has a loop, then $G_1 \cong G_2$.*

(b) *If both G_1 and G_2 have a homomorphism into H, then $G_1 \cong G_2$.*

(c) *If a looped-simple graph H' has a homomorphism into H, then $G_1 \times H' \cong G_2 \times H'$.*

Since strong product corresponds to having a loop at every node, we get:

COROLLARY 5.36. *Let G_1, G_2 and H be simple graphs such that $G_1 \boxtimes H \cong G_2 \boxtimes H$. Then $G_1 \cong G_2$.*

The proof of Proposition 5.35 is left to the reader as an exercise. With a little more effort, we can characterize cancelable graphs (Lovász [1971]). If H is bipartite, then $\hom(H, K_2) > 0$, and so (5.44) and Proposition 5.35 imply
$$H \times C_6 \cong H \times (K_3 \cup K_3),$$

5.4. HOMOMORPHISM AND ISOMORPHISM

so H is not cancelable. On the other hand, nonbipartite graphs are cancelable:

THEOREM 5.37. *Let G_1, G_2 and H be looped-simple graphs such that $G_1 \times H \cong G_2 \times H$. If H is not bipartite, then $G_1 \cong G_2$.*

The proof depends on the following lemma:

LEMMA 5.38. *Suppose that $G_1 \times H \cong G_2 \times H$. Then there is an isomorphism $\sigma: G_1 \times H \to G_2 \times H$ such that $\sigma(V(G_1) \times \{v\}) = V(G_2) \times \{v\}$ for every $v \in V(H)$.*

Proof. We consider graphs G together with a homomorphism $\pi: G \to H$. We call the pair $\mathbf{G} = (G, \pi)$ an *H-colored graph*. For every graph G, the product $G \times H$ is H-colored in the natural way by the projection onto H. We denote this H-colored graph by G_H.

Two H-colored graphs $\mathbf{F} = (F, \rho)$ and $\mathbf{G} = (G, \pi)$ are *isomorphic*, if there is an isomorphism $\sigma: G_1 \to G_2$ which commutes with the projections to H, i.e., $\pi(\eta(i)) = \rho(i)$ for every $i \in V(F)$. In this language, we want to prove that $G_1 \times H \cong G_2 \times H$ implies that $(G_1)_H \cong (G_2)_H$.

For two H-colored graphs $\mathbf{F} = (F, \rho)$ and $\mathbf{G} = (G, \pi)$, let $\hom(\mathbf{F}, \mathbf{G})$ denote the number of those homomorphisms η from F to G that satisfy $\pi(\eta(i)) = \rho(i)$ for every i. Let $\mathsf{inj}(\mathbf{F}, \mathbf{G})$ denote the number of injective homomorphisms with this property.

We can define the product $\mathbf{F} \times \mathbf{G}$ of two H-colored graphs $\mathbf{F} = (F, \rho)$ and $\mathbf{G} = (G, \pi)$ as the subgraph of $F \times G$ induced by those nodes (i, j) with $\rho(i) = \pi(j)$, together with the homomorphism $\sigma(i, j) = \rho(i)$ into H.

The case when H consists of a single node with a loop is equivalent to just ordinary homomorphism numbers.

Two identities extend quite easily to this more general notion:

$$(5.45) \qquad \hom(\mathbf{F}, \mathbf{G}_1 \times \mathbf{G}_2) = \hom(\mathbf{F}, \mathbf{G}_1) \hom(\mathbf{F}, \mathbf{G}_2),$$

and

$$(5.46) \qquad \mathsf{inj}(\mathbf{F}, \mathbf{G}) = \sum_{\mathbf{F}'} \mu(\mathbf{F}, \mathbf{F}') \hom(\mathbf{F}', G),$$

where we sum over all H-colored graphs $\mathbf{F}'$ on at most $\mathsf{v}(\mathbf{F})$ nodes, with appropriate coefficients $\mu(\mathbf{F}, \mathbf{F}')$. Let us add the easy identity

$$(5.47) \qquad (F \times G)_H \cong F_H \times G_H.$$

From $G_1 \times H \cong G_2 \times H$ it follows that $(G_1 \times H)_H \cong (G_2 \times H)_H$ (as H-colored graphs). By (5.47), this implies that $(G_1)_H \times H_H \cong (G_2)_H \times H_H$, and hence by (5.45),

$$\hom(\mathbf{F}, (G_1)_H) \hom(\mathbf{F}, H_H) = \hom(\mathbf{F}, (G_2)_H) \hom(\mathbf{F}, H_H).$$

But notice that $\hom(\mathbf{F}, H_H) > 0$: if $\mathbf{F} = (F, \sigma)$, then (σ, σ) is a homomorphism $\mathbf{F} \to H_H$. Thus we can divide by $\hom(\mathbf{F}, H_H)$ to get

$$\hom(\mathbf{F}, (G_1)_H) = \hom(\mathbf{F}, (G_2)_H)$$

for every H-colored graph $\mathbf{F}$. From here $(G_1)_H \cong (G_2)_H$ follows just like in the proof of Theorem 5.29. □

Proof of Theorem 5.37. By Proposition 5.35, we may assume that H is an odd cycle with $V(H) = [2r+1]$ and $E(H) = \{ij : j \equiv i+1 \pmod{2r+1}\}$. By Lemma 5.38 there exist bijections $\varphi_1, \ldots, \varphi_{2r+1}: V(G_1) \to V(G_2)$ such that for every $ij \in E(H)$, $\varphi_i(u)\varphi_j(v) \in E(G_2)$ if and only if $uv \in E(G_1)$. (Note that this means a different condition if we interchange i and j.)

We show that φ_1 is an isomorphism between G_1 and G_2. Indeed, we have

$$\varphi_1(u)\varphi_1(v) \in E(G_2) \iff \varphi_2^{-1}\big(\varphi_1(u)\big)v \in E(G_1) \iff \varphi_1(u)\varphi_3(v) \in E(G_2)$$
$$\iff \varphi_4^{-1}\big(\varphi_1(u)\big)v \in E(G_1) \iff \varphi_1(u)\varphi_5(v) \in E(G_2)$$
$$\iff \ldots \iff \varphi_1(u)\varphi_{2r+1}(v) \in E(G_2) \iff uv \in E(G_1)$$

This completes the proof. □

REMARK 5.39. You may have noticed that the proof of Lemma 5.38 followed the lines of the proof of Proposition 5.35, only restricting the notion of homomorphisms to those respecting the H-coloring. This suggests that there is a more general formulation for categories. This is indeed the case, as we will see in Section 23.4.

A further natural question about multiplication is whether prime factorization is unique. This is clearly a stronger property than the Cancellation Law, so let us restrict our attention to the strong product, which satisfies the Cancellation Law. The following example shows that prime factorization is not unique in general. We start with an algebraic identity:

$$(5.48) \qquad (1 + x + x^2)(1 + x^3) = (1 + x)(1 + x^2 + x^4).$$

If we substitute any connected graph G for x, and interpret "+" as disjoint union, we get a counterexample. For example,

$$(5.49) \quad (K_1 \cup K_2 \cup K_4) \boxtimes (K_1 \cup K_8) = (K_1 \cup K_2) \boxtimes (K_1 \cup K_4 \cup K_{16}).$$

But there is a very nice positive result of Dörfler and Imrich [1970] and McKenzie [1971]. (The proof uses different techniques, and we don't reproduce it here.)

THEOREM 5.40. *Prime factorization is unique for the strong product of connected graphs.* □

EXERCISE 5.41. (a) Prove that the strong product of two graphs is connected if and only if both graphs are connected. (b) Show by an example that the categorical product of two connected graphs is not always connected. (c) Characterize all counterexamples in (b).

EXERCISE 5.42. Given two looped-simple digraphs F and G, we define the digraph G^F as follows: $V(G^F) = V(G)^{V(F)}$, $E(G^F) = \{(\varphi,\psi) : \varphi,\psi \in V(G)^{V(F)}, (\varphi(u),\psi(v)) \in E(G) \ (\forall (u,v) \in E(F))\}$. (a) Prove the following identities:

$$(G_1 \times G_2)^F \cong G_1^F \times G_2^F, \qquad G^{F_1 \times F_2} \cong (G^{F_1})^{F_2}, \qquad G^{F_1 F_2} \cong G^{F_1} \times G^{F_2}.$$

(b) Show that $\mathsf{hom}(F,G)$ is the number of loops in G^F. (c) Prove that if adjacency is symmetric both in G and in F, then it is also symmetric in G^F (Lovász [1967]).

5.5. Independence of homomorphism functions

How independent are homomorphism functions $\mathsf{hom}(F,.)$ (in an algebraic sense)? We know that $\mathsf{hom}(F_1 F_2, G) = \mathsf{hom}(F_1, G)\mathsf{hom}(F_2, G)$ for two (unlabeled) graphs F_1 and F_2; is this the only identity relating these functions?

We start with excluding linear relations. For a set of (non-isomorphic) simple graphs $A = \{F_1, \ldots, F_m\}$, we define the matrix

$$M_{\mathsf{hom}}^A = \big(\mathsf{hom}(F_i, F_j)\big)_{i,j=1}^m.$$

The matrices M^A_{inj} and M^A_{surj} are defined analogously. Finally, we also define M^A_{aut} as the matrix with $\mathsf{aut}(F_i) = \mathsf{surj}(F_i, F_i) = \mathsf{inj}(F_i, F_i)$ in the i-th entry of the diagonal and 0 outside the diagonal.

Clearly, M^A_{aut} is a diagonal matrix; if we order the graphs F_i according to increasing number of edges (and arbitrarily for graphs with the same number of edges), then the matrices M^A_{inj} and M^A_{surj} become triangular. All diagonal entries are positive in each case. Hence the matrices M^A_{aut}, M^A_{inj} and M^A_{surj} are nonsingular. With M^A_{hom} the situation is more complicated: it may be singular (Exercise 5.46). However, we have the following simple but useful fact, observed by Borgs, Chayes, Kahn and Lovász [2012]:

PROPOSITION 5.43. *Let A be a family of simple graphs closed under surjective homomorphisms. Then M^A_{hom} is nonsingular.*

In particular, this holds if A consists of all graphs with at most k nodes, or at most k edges, for some $k \geq 0$.

Proof. Under the conditions of the Proposition, the matrices introduced above are related by the following identity:

$$(5.50) \qquad M^A_{\mathsf{hom}} = M^A_{\mathsf{surj}} (M^A_{\mathsf{aut}})^{-1} M^A_{\mathsf{inj}}.$$

Indeed, every homomorphism can be decomposed as a surjective homomorphism followed by an (injective) embedding. By our assumption, the image F of the surjective homomorphism is in A. The decomposition is uniquely determined except for the automorphisms of F. This gives the equation

$$\mathsf{hom}(F_i, F_j) = \sum_{k=1}^m \frac{\mathsf{surj}(F_i, F_k)\mathsf{inj}(F_k, F_j)}{\mathsf{aut}(F_k)},$$

which is just 5.50 written out in coordinates.

It follows that M^A_{hom} is the product of three nonsingular matrices, and hence it is also nonsingular. $\square$

If A is just an arbitrary set of simple graphs, we can still create a nonsingular matrix related to M^A_{hom} (Erdős, Lovász and Spencer [1979]).

PROPOSITION 5.44. *Let $F_1, \ldots, F_k$ be nonisomorphic simple graphs.*

(a) Let H_i be obtained from F_i by weighting its nodes, and suppose that all the weights used are algebraically independent. Then the matrix $\left[\mathsf{hom}(F_i, H_j)\right]_{i,j=1}^k$ is nonsingular.

(b) There are simple graphs $G_1, \ldots, G_k$ such that the matrix $\left[\mathsf{hom}(F_i, G_j)\right]_{i,j=1}^k$ is nonsingular.

(c) If $F_1, \ldots, F_k$ have no isolated nodes, then there are simple graphs $G_1, \ldots, G_k$ such that the matrix $\left[t(F_i, G_j)\right]_{i,j=1}^k$ is nonsingular.

We could use nodeweights in (a) chosen randomly and independently from the uniform distribution on $[0, 1]$ (or form any other atomfree distribution); the matrix will be nonsingular with probability 1.

Proof. (a) Considering the node weights as variables, the determinant of the matrix $\left[\mathsf{hom}(F_i, H_j)\right]_{i,j=1}^{k}$ is a polynomial p with integral coefficients. The multilinear part of p is just the determinant of $\left[\mathsf{inj}(F_i, H_j)\right]_{i,j=1}^{k}$, which is non-zero, since this matrix is upper triangular and the diagonal entries are nonzero polynomials. Hence p is not the zero polynomial, which shows that for an algebraically independent substitution it does not vanish.

(b) Instead of algebraically independent weights, we can also substitute appropriate positive integers in p to get a nonsingular matrix $\left[\mathsf{hom}(F_i, H_j)\right]_{i,j=1}^{k}$, since a nonzero polynomial cannot vanish for all positive integer substitutions. For a graph H_j and a node $v \in V(H_j)$ with weight m_v, we replace v by m_v twin copies of weight 1. Let G_j be the graph obtained this way, then $\mathsf{hom}(F_i, G_j) = \mathsf{hom}(F_i, H_j)$ for all i, and hence $\left[\mathsf{hom}(F_i, G_j)\right]_{i,j=1}^{k} = \left[\mathsf{hom}(F_i, H_j)\right]_{i,j=1}^{k}$ is nonsingular.

(c) Let $n = \max_i \mathsf{v}(F_i)$, and let us add $n - \mathsf{v}(F_i)$ isolated nodes to every F_i. The resulting graphs F_i' are non-isomorphic, and hence there are simple graphs $G_1, \ldots, G_k$ such that the matrix $\left[\mathsf{hom}(F_i', G_j)\right]_{i,j=1}^{k}$ is nonsingular. Since $\mathsf{hom}(F_i', G_j) = \mathsf{v}(G_j)^n t(F_i', G_j)$, we can scale the columns and get that the matrix $\left[t(F_i', G_j)\right]_{i,j=1}^{k}$ is nonsingular. Since clearly $t(F_i, G_j) = t(F_i', G_j)$, this proves the proposition. $\square$

The following corollary of these constructions goes back to Whitney [1932]. We have seen that the homomorphism functions satisfy the multiplicativity relations $\mathsf{hom}(F_1 F_2, G) = \mathsf{hom}(F_1, G)\mathsf{hom}(F_2, G)$ (where $F_1 F_2$ denotes disjoint union). Is there any other algebraic relation between them? Using multiplicativity, we can turn any algebraic relation to a linear relation, so the question is: are the graph parameters $\mathsf{hom}(F, .)$ linearly independent (in the sense that any finite number of them are). Thus (b) above implies:

COROLLARY 5.45. *The simple graph parameters* $\mathsf{hom}(F, .)$ *(where F ranges over simple graphs) are linearly independent. Equivalently, the simple graph parameters* $\mathsf{hom}(F, .)$ *(where F ranges over connected simple graphs) are algebraically independent.*

What about non-algebraic relations? Such relations sound unlikely, and in fact it can be proved (Erdős, Lovász and Spencer [1979]) that they don't exist. To be more precise, for any finite set of distinct connected graphs $A = \{F_1, \ldots, F_k\}$, if we construct the set $T(A)$ of points $(t(F_1, G), \ldots, t(F_k, G)) \in \mathbb{R}^k$, where G ranges over all finite graphs, then the closure $\overline{T(A)}$ has an internal point. We will talk more about these sets $T(A)$ in Chapter 16.

EXERCISE 5.46. Show by an example that M_{hom}^A may be singular.

EXERCISE 5.47. Prove a version of part (a) of Proposition 5.44 in which the edges are weighted (instead of the nodes).

EXERCISE 5.48. Find an upper bound on the number of nodes in the graphs G_i in part (b) and (c) of Proposition 5.44.

EXERCISE 5.49. For every $m \geq 1$, construct a family A of m simple graphs such that the matrix M_{hom}^A is the identity matrix.

EXERCISE 5.50. For every $m \geq 1$ there exist simple graphs $F_1, \ldots, F_m$ such that for every integer vector $a \in \mathbb{N}^m$ there is a simple graph G such that $\mathsf{hom}(F_i, G) = a_i$ for all $i \in [m]$.

EXERCISE 5.51. Let $H_1, \ldots, H_m$ be non-isomorphic simple graphs. Prove that there are no linear relations between the graph parameters $\hom(., H_i)$.

EXERCISE 5.52. (a) Let $H_1, \ldots, H_m$ be non-isomorphic simple connected graphs. Prove that there are no linear relations between the graph parameters $\hom(., H_i)$, even when they are restricted to connected graphs. (b) Show that this is no longer true if we don't assume the connectivity of the H_i.

EXERCISE 5.53. (a) Let $H_1, \ldots, H_m$ be simple nonisomorphic connected nonbipartite graphs. Prove that there is a simple connected graph F such that the homomorphism numbers $\hom(F, G_i)$ are distinct. (b) Show that for every simple graph F, at least two of the numbers $\hom(F, C_6)$, $\hom(F, K_2)$ and $\hom(F, K_3 K_3)$ are equal.

5.6. Characterizing homomorphism numbers

In the previous sections (e.g. in the proof of Theorem 5.34 and related results) our key tool was to associate, with every graph G, the graph parameter $\hom(., G)$. What else can be said about these graph parameters? It turns out that they have an interesting characterization, which will play an important role throughout this book. There are different versions of this characterization, of which we state a sample.

Multigraph parameters of the form $\hom(., H)$, where H is a weighted graph, were characterized by Freedman, Lovász and Schrijver [2007].

THEOREM 5.54. *Let f be a graph parameter defined on multigraphs without loops. Then f is equal to $\hom(., H)$ for some weighted graph H on q nodes if and only if it is reflection positive, $f(K_0) = 1$, and $r(f, k) \leq q^k$ for all $k \geq 0$.*

Let us note that the condition for $k = 0$ says that $r(f, 0) \leq 1$, which implies that f is multiplicative (Exercise 4.4).

In terms of statistical physics, this theorem can be viewed as a characterization of partition functions of vertex coloring models. Theorem 5.54 implies that those graph parameters that can be expressed as homomorphism numbers into fixed weighted graphs are all reflection positive and have exponentially bounded connection rank. It may be instructive to see directly why this is so for the number of nowhere-zero flows.

EXAMPLE 5.55. For two k-labeled graphs G_1 and G_2, the value $\mathsf{flo}(G_1 G_2, q)$ can be computed by a simple formula provided we know, for all $a_1 \ldots, a_k \in \mathbb{Z}_q$, the number $\mathsf{flo}(G_i; a_1, \ldots, a_k)$ of nowhere-zero q-flows in G_i with "surplus" a_i at each node $i \in [k]$; then we have a formula similar to (4.6), except that the summation will range over $a \in \mathbb{Z}_q^k$ instead of $\mathbb{Z}^k$:

$$\mathsf{flo}(G_1 G_2, q) = \sum_a \mathsf{flo}(G_1; a_1, \ldots, a_k) \mathsf{flo}(G_2; -a_1, \ldots, -a_k)$$
(5.51)
$$= \sum_a \mathsf{flo}(G_1; a_1, \ldots, a_k) \mathsf{flo}(G_2; a_1, \ldots, a_k).$$

From this, we see that $M(\mathsf{flo}, k)$ is positive semidefinite and has rank at most q^k.
♦

Schrijver [2009] gave the following characterization of graph parameters representable as homomorphism functions into weighted graphs with node weights 1 and complex edgeweights. Recalling the Möbius inverse on the partition lattice (4.3), we can state the result as follows:

THEOREM 5.56. *Let f be a complex valued graph parameter defined on looped multigraphs. Then $f = \hom(., H)$ for some edge-weighted graph H on q nodes with complex edgeweights if and only if f is multiplicative, $f(K_1) = q$, and $f^{\Downarrow}(G) = 0$ for every graph G with more than q nodes.*

Using this theorem, Schrijver gave a real-valued version, which is more similar to Theorem 5.54.

THEOREM 5.57. *Let f be a real valued graph parameter defined on looped multigraphs. Then $f = \hom(., H)$ for some edge-weighted graph H with real edgeweights if and only if f is multiplicative and, for every integer $k \geq 0$, the multilabeled connection matrix $M^{\mathrm{mult}}(f, k)$ is positive semidefinite.*

Every graph parameter f defined on looped-simple graphs can be extended to looped-multigraphs so that it is invariant under adding parallel edges. Every homomorphism function $\hom(., H)$ where all edge-weights are 0 or 1 defines such a multigraph parameter. Conversely, if $f = \hom(., H)$ (where H is a weighted graph) is invariant under adding parallel edges, then every edge of H must have weight 0 or 1 (Exercise 5.66). In particular, if all nodeweights of H are 1, then H can be viewed as a looped-simple graph itself. Hence Theorem 5.57 implies the following characterization of homomorphism numbers into looped-simple graphs, as noticed by Lovász and Schrijver [2010, 2009]:

COROLLARY 5.58. *Let f be a graph parameter defined on looped-simple graphs. Then $f = \hom(., H)$ for some looped-simple graph H if and only if f is multiplicative and, for every integer $k \geq 0$, the connection matrix $M^{\mathrm{mult}}(f, k)$ is positive semidefinite.*

Note that in Theorem 5.57 and Corollary 5.58 no bound on the connection rank is assumed; in fact (somewhat surprisingly), it follows from the multiplicativity and reflection positivity conditions that f has finite connection rank, and $r(f, k) \leq f(K_1)^k$ for all k. Furthermore, in Corollary 5.58 it also follows from the conditions that the values of f are integers.

Next, we state an analogous (dual) characterization of graph parameters of the form $\hom(F, .)$, defined on looped-simple graphs, where F is also a looped-simple graph (Lovász and Schrijver [2010]). To state the result, we need some definitions. Recall the notion of H-colored graphs and their products from the proof of Lemma 5.38; we need only the rather trivial version where $H = K_q^{\circ}$ is a fully looped complete graph. We define *dual connection matrices* $N(f, q)$ of a graph parameter f: the rows and columns are indexed by K_q°-colored graphs, and the entry in row $\mathbf{G}_1$ and column $\mathbf{G}_2$ is $f(\mathbf{G}_1 \times \mathbf{G}_2)$.

THEOREM 5.59. *Let f be a graph parameter defined on looped-simple graphs. Then $f = \hom(F, .)$ for some looped-simple graph F if and only if f is multiplicative over direct product, and for each $k \geq 1$, the dual connection matrix $N(f, k)$ is positive semidefinite.*

It is interesting to note that "primal" connection matrices of these "dual" homomorphism numbers $\hom(F, .)$ also have finite rank (see Exercise 5.67). However, no characterization of homomorphism numbers in terms of these "primal" connection matrices is known.

5.6.1. Randomly weighted graphs.

The last result to be presented in this line is a characterization of multiplicative and reflection positive multigraph parameters with finite connection rank (Lovász and Szegedy [2012c]). To state the result, we have to generalize the notion of weighted graphs.

A *randomly weighted graph* is a finite graph H (which we may assume to be a looped complete graph) in which the nodes are weighted with positive real numbers (just like in the case of ordinary weighted graphs) and each edge ij is weighted by a random variable $\mathbf{B}_{ij}$ taking values from a finite set of reals. Ordinary weighted graphs can be regarded as randomly weighted graphs in which the edgeweights are random variables concentrated on a single value.

To define homomorphism numbers into a randomly weighted graph takes a little care. A first idea is to define it as the expectation of $\mathsf{hom}(F, \mathbf{H})$ where $\mathbf{H}$ is the weighted graph where the edgeweights are generated randomly and independently from the corresponding distributions. However, this quantity would not be multiplicative. We could start with taking the expectation separately for each edge; this would then give nothing new relative to the homomorphism numbers into weighted graphs. We therefore take a middle ground:

$$(5.52) \qquad \mathsf{hom}(F, H) = \sum_{\varphi: V(F) \to V(H)} \prod_{i \in V(F)} \alpha_{\varphi(i)} \prod_{ij \in E(F^{\mathrm{simp}})} \mathsf{E}(\mathbf{B}_{\varphi(i)\varphi(j)}^{F_{ij}}),$$

where F_{ij} is the multiplicity of the edge ij in F. This quantity is multiplicative, and it specializes to the previously defined homomorphism number when the edge weights are deterministic.

We note two special cases. If F is simple, then we could take the expectation all the way in; in other words, homomorphisms into randomly weighted graphs give no new simple graph parameters. On the other hand, if we consider $\mathsf{inj}(F, H)$ (restricting the summation in (5.52) to injections), then we can take the expectation all the way out, i.e., $\mathsf{inj}(F, H) = \mathsf{E}(\mathsf{inj}(F, \mathbf{H}))$.

EXAMPLE 5.60. Consider the multigraph parameter $f(G) = p^{\mathsf{e}(G^{\mathrm{simp}})}$, where $0 < p < 1$ is fixed. It is not hard to see that this is reflection positive and its connection rank is $2^{\binom{k}{2}}$. We can characterize it as $\mathsf{hom}(G, K_1^\circ[p])$, where $K_1^\circ[p]$ is the randomly weighted graph on a single node with a loop, where the loop is decorated by the probability distribution on $\{0, 1\}$ in which 1 has probability p. We can also characterize it as the expectation of $t_{\mathsf{inj}}(G, \mathbb{G}(n, p))$, where $n \geq \mathsf{v}(G)$. This parameter is multiplicative, reflection positive, and its connection rank if $r(f, k) = 2^{\binom{k}{2}}$, which is finite for every k, but has superexponential growth. ♦

With this generalized notion of homomorphism numbers, we are able to state the theorem announced above:

THEOREM 5.61. *A multigraph parameter f is equal to $\mathsf{hom}(., H)$ for some randomly weighted graph H if and only if it is multiplicative, reflection positive and $r(f, 2)$ is finite.*

It would not be enough to assume that $r(f, 1)$ is finite instead of $r(f, 2)$ (see Exercise 5.65). While in Theorem 5.61 we don't have to assume anything about the higher connection ranks, it does follow that they are all finite. In fact, we have the following "Theorem of Alternatives":

SUPPLEMENT 5.62. *Let f be a multiplicative and reflection positive parameter defined on multigraphs without loops. Then one of three alternative must occur:* (i) $r(f,k)$ *is infinite for all* $k \geq 2$; (ii) $r(f,k)$ *is finite for all* k, *and* $\log r(f,k) = \Theta(k)$; (iii) $r(f,k)$ *is finite for all* k, *and* $\log r(f,k) = \Theta(k^2)$.

It follows that alternative (ii) obtains iff $f = \mathsf{hom}(.,H)$ for some weighted graph H, and alternative (iii) obtains when $f = \mathsf{hom}(.,H)$ for some randomly weighted graph H in which at least one edgeweight has a proper distribution.

It is possible to give a more precise description of the asymptotic behavior of $\log r(\mathsf{hom}(.,H),k)$, but we have to refer to the paper of Lovász and Szegedy [2012c] for details. Let us note that no such conclusion can be drawn without assuming reflection positivity. For example, the chromatic polynomial $\mathsf{chr}(.,x)$ satisfies $\log \mathsf{rk}(\mathsf{chr}(.,x),k) = \Theta(k \log k)$.

REMARK 5.63. Several further improvements, versions and extensions of these results have been obtained, extending them to directed graphs, hypergraphs, semigroups, and indeed, to all categories satisfying reasonable conditions. In this book, related characterizations will be described for homomorphisms into graphons and random graphons (Theorem 11.52 and Proposition 14.60), morphisms in categories (Theorem 23.16), and edge coloring models (Theorem 23.5). One would wish to derive all of these from a single "Master Theorem"; alas, this has not yet been found.

The least appealing feature of these theorems is that the necessary and sufficient condition involves infinite matrices, and in most cases infinitely many of them. While this is clearly unavoidable in a sense (the condition must involve the value of the parameter on all graphs), one can formulate conditions that involve only submatrices with a simpler structure. For example, in Theorem 5.61, it suffices to consider fully and simply labeled graphs, fully labeled edgeless graphs, and fully labeled bonds (cf. also the proof of Theorem 5.57 given is Section 6.6).

5.6.2. About the proofs. The proofs of the theorems above follow at least three different lines. To be more precise, the necessity of the conditions is easy to prove; below we prove the "easy" direction of Theorem 5.54, and the others follow by essentially the same argument. The sufficiency parts will be postponed until some further techniques will be developed:

—The completion of the proof of Theorem 5.54 will be given in Section 6.2.2, after the development of graph algebras. (These algebras will be useful to study other related properties of homomorphism functions, and the technique will also be applied in extremal graph theory.) Corollary 5.58 and its dual, Theorem 5.59, can be proved by a similar technique. This technique extends to a much more general setting, to categories, as we will sketch in Section 23.4.

—The proofs of Theorems 5.56 and 5.57 will be described in Section 6.6, where a general connection to the Nullstellensatz and invariant theory will be developed. This method extends to edge coloring models (see Section 23.2).

—The proof of Theorem 5.61 will use a lot of the analytic machinery to be developed in Part 3 of the book, and will be sketched at the end of that part (Section 17.1.4). For the details of this proof, and for the proof of Supplement 5.62, we refer to Lovász and Szegedy [2012c].

We conclude this section proving the "easy" direction in Theorem 5.54:

PROPOSITION 5.64. *For every weighted graph H, the graph parameter $\hom(.,H)$ is reflection positive and $r(\hom(.,H),k) \leq \mathsf{v}(H)^k$.*

Proof. For any two k-labeled graph F_1 and F_2 and $\varphi: [k] \to V(H)$, we have

(5.53) $$\hom_\varphi(F_1 F_2, H) = \hom_\varphi(F_1, H)\hom_\varphi(F_2, H)$$

(recall the definition of $\hom_\varphi$ from (5.10)). Let $F = [\![F_1 F_2]\!]$, then the decomposition

$$\hom(F,H) = \sum_{\varphi:\ [k]\to V(H)} \alpha_\varphi \hom_\varphi(F,H).$$

writes the matrix $M(\hom(.,H),k)$ as the sum of $\mathsf{v}(H)^k$ matrices, one for each mapping $\varphi: [k] \to V(H)$; (5.53) shows that these matrices are positive semidefinite and have rank 1. This implies Lemma 5.64. □

EXERCISE 5.65. For a multigraph G on $[n]$, let $X_1, \ldots, X_n$ be random points on the unit circle, and let $f(G)$ denote the probability that $X_i \cdot X_j \geq 0$ for every edge ij of G. Prove that f is reflection positive, $r(f,0) = r(f,1) = 1$, but $r(f,2) = \infty$.

EXERCISE 5.66. Let H be a weighted graph for which the looped-multigraph parameter $\hom(.,H)$ is invariant under adding parallel edges. Prove that all edgeweights of H are 0 or 1.

EXERCISE 5.67. Prove that the connection rank $r(\hom(F,.),k)$ is bounded by $(k+2)^{\mathsf{v}(F)}$.

5.7. The structure of the homomorphism set

We have discussed the existence of homomorphisms between two graphs F and G, i.e., the emptiness or non-emptiness of the homomorphism set $\text{Hom}(F,G)$. We considered the size of this set (in fact, much of this book turns around this number). This set has further structure, which is quite interesting and which can be exploited to obtain combinatorial results about graphs. We only give a glimpse of these questions.

5.7.1. The graph of homomorphisms. Let F and G be two simple graphs. The set $\text{Hom}(F,G)$ can be endowed with a graph structure. Brightwell and Winkler [2004] define a graph $Hom(F,G)$ by connecting two nodes, meaning homomorphisms $\varphi, \psi: F \to G$, if they differ only on one node of F.

EXAMPLE 5.68 (**Linegraph**). If $F = K_2$, then we get a version of the linegraph of G: every edge of G will be represented by two nodes in $Hom(K_2,G)$ corresponding to the two orientations of the edge, and two nodes (oriented edges $\overrightarrow{ij}$ and $\overrightarrow{uv}$) will be connected if either $i = u$ or $j = v$. (The ordinary linegraph is obtained by merging the two copies of each edge.) ♦

EXAMPLE 5.69 (**Colorings**). Let the target graph G be a complete q-graph. In this case, nodes are legitimate q-colorings of F, and two of them are adjacent if only one node of F is recolored to get one coloring from the other. This construction is important when analyzing the "heat bath" or "Glauber dynamics" Markov chain in statistical physics, which corresponds to a random walk on this graph. We will need this Markov chain in Section 20.1.2. ♦

Brightwell and Winkler relate properties of the *Hom* graph to a number of important issues in statistical physics, like long-range actions and phase transitions. These would be too difficult to state here, but a corollary of their main result is worth formulating:

THEOREM 5.70. *Suppose that G is a graph such that the graph $Hom(F,G)$ is connected for every connected graph F with maximum degree d. Then the chromatic number of G is at least $d/2 + 1$.* □

(They conjecture that $d/2 + 1$ can be replaced by d.)

5.7.2. The complex of homomorphisms. The set $Hom(F,G)$ can also be equipped with a topological structure. We say that a set of homomorphisms $\varphi_1, \ldots, \varphi_k : F \to G$ is a *cluster* if for every edge $uv \in E(F)$ and any $1 \leq i < j \leq k$, we have $\varphi_i(u)\varphi_j(v) \in E(G)$. It is clear that these clusters form a simplicial complex **Hom**(F,G) (i.e., they are closed under taking subsets). It is quite surprising that topological properties of this complex have graph-theoretic consequences.

What is important about this construction is that it is "functorial", which means that every homomorphism $\psi : G_1 \to G_2$ induces a simplicial (and hence continuous) map $\widehat{\psi} : \mathbf{Hom}(F, G_1) \to \mathbf{Hom}(F, G_2)$ in a canonical way: For every homomorphism $\varphi : F \to G_1$, we define $\widehat{\psi}(\varphi) = \varphi\psi$. It is trivial that this map from $V(\mathbf{Hom}(F, G_1)) = Hom(F, G_1)$ to $V(\mathbf{Hom}(F, G_2)) = Hom(F, G_2)$ maps clusters onto clusters. We also note that the automorphism group of F acts on $\mathbf{Hom}(F,G)$: if α is an automorphism of F, then $\check{\alpha} : \varphi \mapsto \alpha\varphi$ is an automorphism of $\mathbf{Hom}(F,G)$.

We quote two theorems relating properties of these topological spaces to colorability of the graph, and they are important tools in determining the chromatic number of certain graph families. (See Kozlov [2008] and Matoušek [2003] for detailed treatments of this topic.) The first is a re-statement in this language of a result of Lovász [1978].

THEOREM 5.71. *If $\mathbf{Hom}(K_2, G)$ is k-connected as a topological space, then the chromatic number of G is at least $k + 3$.*

The second theorem is due to Babson and Kozlov [2003, 2006, 2007].

THEOREM 5.72. *If $\mathbf{Hom}(C_{2r+1}, G)$ is k-connected as a topological space for some $r \geq 1$, then the chromatic number of G is at least $k + 4$.*

These results suggest the more general assertion that if $\mathbf{Hom}(F,G)$ is k-connected as a topological space, then $\chi(G) \geq k + \chi(F) + 1$. This is, however, false, as shown by Hoory and Linial [2005]. But the relationship between the chromatic numbers of F and G and the topology of the complex $\mathbf{Hom}(F,G)$ is mostly unexplored.

EXERCISE 5.73. Prove that if G is a simple graph with maximum degree d, then for all $q \geq d + 2$ the graph $Hom(G, K_q)$ is connected, i.e., we can transform any q-coloring of G into any other, changing the color of one node at a time, going through legitimate q-colorings.

EXERCISE 5.74. Prove that the graph $Hom(F,G)$ is connected if and only if the simplicial complex $\mathbf{Hom}(F,G)$ is connected.

EXERCISE 5.75. Prove that $\mathbf{Hom}(.,.)$ is a contravariant functor in its first variable: every homomorphism $\xi : F_1 \to F_2$ induces a simplicial map $\check{\varphi} : \mathbf{Hom}(F_2, G) \to \mathbf{Hom}(F_1, G)$. Analogously, $\mathbf{Hom}(.,.)$ is a covariant functor in its second variable.

EXERCISE 5.76. Let α be an automorphism of F which has an orbit on $V(F)$ that is not a stable set. Assume that G has no loops. Then the simplicial map $\check{\alpha}$ is fixed-point-free on the geometric realization of $\mathbf{Hom}(F, G)$.

EXERCISE 5.77. Prove that $\mathbf{Hom}(K_2, K_p)$ is homotopy equivalent to the $(p-2)$-dimensional sphere.

CHAPTER 6

Graph algebras and homomorphism functions

6.1. Algebras of quantum graphs

A *quantum graph* is defined as a formal linear combination of a finite number of multigraphs with real coefficients. To be pedantic, let's add that these coefficients can be zero, but terms with zero coefficient can be deleted without changing the quantum graph. Those graphs that occur with non-zero coefficient are called the *constituents* of x. Quantum graphs form an infinite dimensional linear space, which we denote by $\mathcal{Q}_0$.

Every graph parameter f can be extended to quantum graphs linearly: if $x = \sum_{i=1}^n \lambda_i F_i$, then $f(x) = \sum_{i=1}^n \lambda_i f(F_i)$. In particular, the definition of $\hom(F, G)$ and $t(F, G)$ extends to quantum graphs bilinearly: if $x = \sum_{i=1}^n \alpha_i F_i$ and $y = \sum_{j=1}^m \beta_j G_j$, then we define

$$\hom(x, y) = \sum_{i=1}^n \sum_{j=1}^m \alpha_i \beta_j \hom(F_i, G_j),$$

and similarly for $t(x, y)$. (Most of the time we will use linearity in the first argument only.)

Quantum graphs are useful in expressing various combinatorial situations. For example, for any signed graph F, we consider the quantum graph

(6.1) $$x = \sum_{F'} (-1)^{e(F') - |E_+|} F',$$

where the summation extends over all simple graphs F' such that $V(F') = V(F)$ and $E_+ \subseteq E(F') \subseteq E_+ \cup E_-$. By inclusion-exclusion we see that for any simple graph G, $\hom(F, G) = \hom(x, G)$ is the number of maps $V(F) \to V(G)$ that map positive edges onto edges and negative edges onto non-edges. The equation $\hom(F, G) = \hom(x, G)$ remains valid if G is a weighted graph (one way to see it is to expand the parentheses in definition (5.9)). Due to these nice formulas, we will denote the quantum graph x by F; this will not cause any confusion.

The relationships between homomorphism numbers and injective homomorphism numbers, equations (5.16) and (5.18), can be expressed as follows: For every graph G, let $ZG = \sum_P G/P$ and $MG = \sum_P \mu_P G/P$, where P ranges over all partitions of $V(G)$. Here the quotient graph G/P is defined by merging every class into a single node, and adding up the multiplicities of pre-images of an edge to get its multiplicity in G/P. Then

(6.2) $\qquad \hom(F, G) = \operatorname{inj}(ZF, G),\quad$ and $\quad \operatorname{inj}(F, G) = \hom(MF, G).$

More generally, for any graph parameter f, we have $f(MG) = f^{\Downarrow}(G)$. The operators Z and M extend to linear operators $Z, M : \mathcal{Q}_0 \to \mathcal{Q}_0$. Clearly, they are inverses of each other: $ZM = MZ = \mathrm{id}_{\mathcal{Q}_0}$. (In Appendix A.1 these operators are discussed for general lattices.)

We will see that other important facts, like the contraction/deletion relation of the chromatic polynomial (4.5) can also be conveniently expressed by quantum graphs (cf. Section 6.3).

For any $k \geq 0$, a *k-labeled quantum graph* is a formal linear combination of k-labeled graphs. We say that a k-labeled quantum graph is *simple* [loopless] if all its constituents are simple [loopless].

6.1.1. The gluing algebra. Let $\mathcal{Q}_k$ denote the (infinite dimensional) vector space of k-labeled quantum graphs. We can turn $\mathcal{Q}_k$ into an algebra by using the gluing product $F_1 F_2$ introduced in Section 4.2 as the product of two generators, and then extending this multiplication to the other elements of the algebra by linearity. Clearly $\mathcal{Q}_k$ is associative and commutative. The fully labeled graph O_k on $[k]$ with no edges is the multiplicative unit in $\mathcal{Q}_k$.

Every graph parameter f can be extended linearly to quantum graphs, and defines an inner product on $\mathcal{Q}_k$ by

(6.3) $$\langle x, y \rangle = f(xy).$$

This inner product has nice properties, for example it satisfies the *Frobenius identity*

(6.4) $$\langle x, yz \rangle = \langle xy, z \rangle.$$

Let $\mathcal{N}_k(f)$ denote the kernel (annihilator) of this inner product, i.e.,

$$\mathcal{N}_k(f) = \{x \in \mathcal{Q}_k : f(xy) = 0 \,\, \forall y \in \mathcal{Q}_k\}.$$

Note that it would be equivalent to require this condition for (ordinary) k-labeled graphs only in place of y. Sometimes we write this condition as $x \equiv 0 \pmod{f}$, and then use $x \equiv y \pmod{f}$ if $x - y \equiv 0 \pmod{f}$. We define the factor algebra

$$\mathcal{Q}_k/f = \mathcal{Q}_k/\mathcal{N}_k(f).$$

Formula (6.3) still defines an inner product on $\mathcal{Q}_k/f$, and identity (6.4) remains valid. While the algebra $\mathcal{Q}_k$ is infinite dimensional, the factor algebra $\mathcal{Q}_k/f$ is finite dimensional for many interesting graph parameters f.

PROPOSITION 6.1. *The dimension of $\mathcal{Q}_k/f$ is equal to the rank of the connection matrix $M(f, k)$. The inner product (6.3) is positive semidefinite on $\mathcal{Q}_k$ if and only if $M(f, k)$ is positive semidefinite.*

So if the parameter f is reflection positive, then the inner product is positive semidefinite on every $\mathcal{Q}_k$; equivalently, it is positive definite on $\mathcal{Q}_k/f$. It follows that the examples in section 4.3 provide several graph parameters for which the algebras $\mathcal{Q}_k/f$ have finite dimension. This means that in these cases our graph algebra $\mathcal{Q}_k/f$ is a *Frobenius algebra* (see Kock [2003]). For a reflection positive parameter, the inner product is positive definite on $\mathcal{Q}_k/f$, so it turns $\mathcal{Q}_k/f$ into an inner product space.

EXAMPLE 6.2 (**Number of perfect matchings**). Consider the number $\mathsf{pm}(G)$ of perfect matchings in the graph G. It is a basic property of this value that subdividing an edge by two nodes does not change it. This can be expressed as

$$\begin{matrix} \bullet\!\!-\!\!\circ\!\!-\!\!\circ\!\!-\!\!\bullet \end{matrix} \equiv \begin{matrix} \bullet\!\!-\!\!\bullet \end{matrix} \quad (\mathrm{mod}\ \mathsf{pm}),$$

where the black nodes are labeled. ◆

Sometimes it will be convenient to put all k-labeled graphs into a single structure as follows. Recall the notion of partially labeled graphs from Section 4.2, and also the notion of their gluing product. Let $\mathcal{Q}_\mathbb{N}$ denote the (infinite dimensional) vector space of formal linear combinations (with real coefficients) of partially labeled graphs. We can turn $\mathcal{Q}_\mathbb{N}$ into an algebra by using the product $G_1 G_2$ introduced above (gluing along the labeled nodes) as the product of two generators, and then extending this multiplication to the other elements linearly. Clearly $\mathcal{Q}_\mathbb{N}$ is associative and commutative, and the empty graph is a unit element.

A graph parameter f defines an inner product on the whole space $\mathcal{Q}_\mathbb{N}$ by (6.3), and we can consider the kernel $\mathcal{N}(f) = \{x \in \mathcal{Q}_\mathbb{N} : \langle x, y \rangle = 0 \ \forall y \in \mathcal{Q}_\mathbb{N}\}$ of this inner product. It is not hard to see that $\mathcal{N}_k(f) = \mathcal{Q}_k \cap \mathcal{N}(f)$.

For every finite set $S \subseteq \mathbb{N}$, the set of all formal linear combinations of S-labeled graphs form a subalgebra $\mathcal{Q}_S$ of $\mathcal{Q}_\mathbb{N}$. We set $\mathcal{Q}_S/f = \{x/f : x \in \mathcal{Q}_S\}$. Clearly $\mathcal{Q}_S/f$ is a subalgebra of $\mathcal{Q}_\mathbb{N}/f$, and it is not hard to see that $\mathcal{Q}_S/f \cong \mathcal{Q}_{|S|}/f$. The graph with $|S|$ nodes labeled by the elements of S and no edges, which we denote by O_S, is a unit in the algebra $\mathcal{Q}_S$.

6.1.2. The concatenation algebra. There is another algebra on these vector spaces. For two 2-multilabeled multigraphs F and G, we define their *concatenation* by identifying node 2 of F with node 1 of G, and unlabeling this merged node. We denote the resulting 2-labeled graph by $F \circ G$. It is easy to check that this operation is associative (but not commutative). We extend this operation linearly over $\mathcal{Q}_2$.

This algebra has a $*$ (conjugate) operation: for a 2-labeled graph F, we define F^* by interchanging the two labels. Clearly $(F \circ G)^* = G^* \circ F^*$. We can also extend this linearly over $\mathcal{Q}_2$.

Let f be a graph parameter. It is easy to see that if $x \equiv 0 \pmod{f}$ then $x^* \equiv 0 \pmod{f}$, so the $*$ operator is well defined on elements of $\mathcal{Q}_2/f$. A further important property of concatenation is that for any three 2-labeled graph F, G and H,
$$(F \circ G) H \cong F(H \circ G^*),$$
and hence
$$(6.5) \qquad f\big((x \circ y) z\big) = f\big(x(z \circ y^*)\big),$$
for any three elements $x, y, z \in \mathcal{Q}_2$. It follows that if $x \equiv 0 \pmod{f}$ then $x \circ y \equiv 0 \pmod{f}$ for every $y \in \mathcal{Q}_2$ and thus concatenation is well defined on the elements of $\mathcal{Q}_2/f$. It is easy to see that $(\mathcal{Q}_2/f, \circ)$ is an associative (but not necessarily commutative) algebra.

We can think of a 2-labeled graph as a graph having one labeled node on its left side and one on its right side. Then concatenation means that we identify the right labeled node of one graph with the left labeled node of another. This suggests a generalization: Instead of a single node, we consider graphs that have k labeled nodes on each side. Let's say the labels are $1, \ldots, k$ on both sides, so each label occurs twice, once on the left and once on the right. It is convenient to allow that one and the same node gets a left label and a right label. Such a graph will be called (k, k)-labeled, and we denote their set by $\mathcal{F}_{k,k}$.

We can define a multiplication on $\mathcal{F}_{k,k}$, denoted by $\circ$, in which we identify each right labeled node of the first graph with the left labeled node of the second graph with the same label. We can take the space $\mathcal{Q}_{k,k}$ of "quantum bi-labeled graphs", i.e., formal linear combinations of graphs in $\mathcal{F}_{k,k}$. The graph O_k on $[k]$ with no edges, with its nodes labeled $1,\ldots,k$ from both sides, is a unit in the algebra.

This algebra is associative, but not commutative. It has a "conjugate" operation, which we denote by $*$, of interchanging "left" and "right". This is related to multiplication through the identity $(A \circ B)^* = B^* \circ A^*$.

Given a graph parameter f defined on looped-multigraphs, we can define the inner product of two (k,k)-labeled graphs as before: we consider them as multilabeled graphs (where left label i is different from right label i), form their gluing product, and evaluate the parameter on the resulting multigraphs. (We have to work with multilabeled graphs, since a node is allowed to have two labels. As a consequence, the gluing product can have loops.)

A further natural generalization involves graphs with possibly different numbers of labeled nodes on the left and on the right. Let $\mathcal{F}_{k,m}$ denote the set of multigraphs with k labeled nodes on the left and m labeled nodes on the right. We cannot form the product of any two graphs, but we can multiply a graph $F \in \mathcal{F}_{k,m}$ with a graph $G \in \mathcal{G}_{m,n}$ to get a graph $F \circ G \in \mathcal{F}_{k,n}$. So bi-labeled graphs form the morphisms of a category, in which the objects are the natural numbers. The star operation (interchanging left and right) maps $\mathcal{F}_{k,m}$ onto $\mathcal{F}_{m,k}$. Any graph parameter f defines a scalar product on every $\mathcal{F}_{k,m}$ by $\langle F, G \rangle = f(FG)$, where FG is defined by identifying nodes with the same left-label as well as nodes with the same right-label in the disjoint union of F and G.

Just as above, the operations $\circ$, $*$, and $\langle .,. \rangle$ extend linearly to the linear spaces $\mathcal{Q}_{k,m}$ of formal linear combinations of graphs in $\mathcal{F}_{k,m}$. This leads us to semisimple categories and topological quantum field theory (see Witten [1988]), which topics are beyond the limits of this book.

6.1.3. Unlabeling.
Having defined the graph algebras we need, we are going to describe the relationship between algebras of labeled graphs using different label sets. There is nothing terribly deep or surprising here; but it might serve as a warm-up, illustrating how combinatorial and algebraic constructions correspond to each other.

The *unlabeling operator* $G \mapsto [\![G]\!]_S$ extends to $\mathcal{Q}$ by linearity. We note that for any two partially labeled graphs G and H, $[\![[\![G]\!]_S H]\!] \cong [\![[\![G]\!]_S [\![H]\!]_S]\!] \cong [\![G[\![H]\!]_S]\!]$, and hence we get the identity

$$(6.6) \qquad \langle [\![x]\!]_S, y \rangle = \langle [\![x]\!]_S, [\![y]\!]_S \rangle = \langle x, [\![y]\!]_S \rangle \qquad (x, y \in \mathcal{Q}).$$

By a similar argument we get that if $S, T \subset \mathbb{N}$ are finite sets, then

$$(6.7) \qquad \langle x, y \rangle = \langle [\![x]\!]_{S \cap T}, [\![y]\!]_{S \cap T} \rangle \qquad (x \in \mathcal{Q}_S,\ y \in \mathcal{Q}_T)$$

One consequence of identity (6.6) is that if some $x \in \mathcal{Q}$ is congruent modulo f to some S-labeled quantum graph $y \in \mathcal{Q}_S$, then such a y can be obtained by simply removing the labels outside S:

$$(6.8) \qquad x - y \in \mathcal{N}(f) \Longrightarrow x - [\![x]\!]_S \in \mathcal{N}(f).$$

Indeed, for any $z \in \mathcal{Q}$, we have

$$\begin{aligned}\langle x - [\![x]\!]_S, z \rangle &= \langle x, z \rangle - \langle [\![x]\!]_S, z \rangle = \langle y, z \rangle - \langle [\![x]\!]_S, z \rangle \\ &= \langle y, [\![z]\!]_S \rangle - \langle x, [\![z]\!]_S \rangle = \langle y - x, [\![z]\!]_S \rangle = 0.\end{aligned}$$

As a special case, we get that $[\![x]\!]_S \in \mathcal{N}(f)$ for all $x \in \mathcal{N}(f)$. This implies that the operator $x \mapsto [\![x]\!]_S$ is defined on the factor algebra $\mathcal{Q}/f$, and in fact it gives the orthogonal projection of $\mathcal{Q}/f$ to the subalgebra $\mathcal{Q}_S/f$. Indeed, by (6.6)

$$\langle [\![x]\!]_S, x - [\![x]\!]_S \rangle = \langle x, [\![x - [\![x]\!]_S]\!]_S \rangle = \langle x, [\![x]\!]_S - [\![x]\!]_S \rangle = 0.$$

Another consequence of (6.8) is that for every $x \in \mathcal{Q}$ there is a unique smallest set $S \subset \mathbb{N}$ such that $x \equiv [\![x]\!]_S \pmod{f}$.

For the rest of this section, we assume that f is multiplicative and normalized so that $f(K_1) = 1$. (This latter condition is usually easily achieved by replacing $f(G)$ by $f(G)/f(K_1)^{v(G)}$.)

One important consequence of this assumption is that deleting isolated nodes (labeled or unlabeled) from a graph G does not change $f(G)$. This implies that it does not change G/f either. Indeed, let F denote the graph obtained from G by deleting some isolated nodes, then for every partially labeled graph H, the products FH and GH differ only in isolated nodes, and hence $f(FH) = f(GH)$, showing that $F/f = G/f$. In particular, every graph with no edges has the same image in $\mathcal{Q}/f$, which is the unit element of $\mathcal{Q}/f$.

LEMMA 6.3. *Let f be a multiplicative and normalized graph parameter, and let $S \subseteq T$ be finite subsets of $\mathbb{N}$.*

(a) *If $S \subseteq T$, then $\mathcal{Q}_S/f$ has a natural embedding into $\mathcal{Q}_T/f$.*

(b) *For any two $S, T \subseteq \mathbb{N}$, we have $\mathcal{Q}_S/f \cap \mathcal{Q}_T/f = \mathcal{Q}_{S \cap T}/f$.*

(c) *If $S \cap T = \emptyset$, then $\mathcal{Q}_S \mathcal{Q}_T \cong \mathcal{Q}_S \otimes \mathcal{Q}_T$ and $(\mathcal{Q}_S/f)(\mathcal{Q}_T/f) \cong \mathcal{Q}_S/f \otimes \mathcal{Q}_S/f$.*

Proof. (a) Every S-labeled graph G can be turned into a T-labeled graph G' by adding $|T \setminus S|$ new isolated nodes, and label them by the elements of $T \setminus S$. (This is equivalent to multiplying it by U_T). As remarked above, $G - G' \in \mathcal{N}(f)$, and so $G/f = G'/f$.

(b) The containment $\supseteq$ follows from (a). To prove the other direction, we consider any $z \in \mathcal{Q}_S/f \cap \mathcal{Q}_T/f$. Then we have an $x \in \mathcal{Q}_S$ with $x/f = z$, and a $y \in \mathcal{Q}_T$ with $y/f = z$. So $x - y = x - [\![y]\!]_T \in \mathcal{N}(f)$, and so by (6.8), $x - [\![x]\!]_T \in \mathcal{N}(f)$. But we can write this as $[\![x]\!]_T - [\![x]\!]_S \in \mathcal{N}(f)$, and then by the same reasoning $[\![x]\!]_T - [\![[\![x]\!]_T]\!]_S = [\![x]\!]_T - [\![x]\!]_{T \cap S} \in \mathcal{N}(f)$, showing that $x - [\![x]\!]_{T \cap S} \in \mathcal{N}(f)$, and so $z = x/f \in \mathcal{Q}_{S \cap T}/f$.

(c) The first relation is trivial, since the partially labeled graphs FG, $F \in \mathcal{F}_S^\bullet$, $G \in \mathcal{F}_T^\bullet$ are different generators of $\mathcal{Q}_{S \cup T}$. To prove the second, let $a_1, a_2, \dots$ be any basis of $\mathcal{Q}_S/f$ and $b_1, b_2, \dots$, any basis of $\mathcal{Q}_T/f$. Consider the map $a_i \otimes b_j \mapsto a_i b_j$ (which is defined on a basis of $\mathcal{Q}_S/f \otimes \mathcal{Q}_T/f$), and extend it linearly to a map $\Phi: \mathcal{Q}_S/f \otimes \mathcal{Q}_T/f \to (\mathcal{Q}_S/f)(\mathcal{Q}_T/f)$. We show that Φ is an isomorphism between $\mathcal{Q}_S/f \otimes \mathcal{Q}_T/f$ and $(\mathcal{Q}_S/f)(\mathcal{Q}_T/f)$.

It is straightforward to check that Φ preserves product in the algebra and also the unit element. It is also clear that $(\mathcal{Q}_S/f)(\mathcal{Q}_T/f)$ is generated by the elements $a_i b_j$, so Φ is surjective. To prove that Φ is injective, suppose that there are real numbers c_{ij} of which a finite but positive number is nonzero such that $\sum_{i,j} c_{ij} a_i b_j = 0$. Then for every $x \in \mathcal{Q}_S/f$ and $y \in \mathcal{Q}_T/f$, we have by multiplicativity

$$\sum_{i,j} c_{ij} f(xa_i) f(yb_j) = \sum_{i,j} c_{ij} f(xya_i b_j) = \sum_{i,j} c_{ij} \langle xy, a_i b_j \rangle = \left\langle xy, \sum_{i,j} c_{ij} a_i b_j \right\rangle = 0.$$

Writing this equation as $\langle y, \sum_{i,j} c_{ij} f(xa_i) b_j \rangle = 0$, we see that $\sum_{i,j} c_{ij} f(xa_i) b_j = 0$. Since the b_i are linearly independent, this means that for every $1 \leq j \leq m$,

$$\left\langle x, \sum_i c_{ij} a_i \right\rangle = \sum_i c_{ij} f(xa_i) = 0.$$

This implies that $\sum_i c_{ij} a_i = 0$, and since the a_i are linearly independent, it follows that $c_{ij} = 0$ for all i and j. $\square$

COROLLARY 6.4. *For every multiplicative graph parameter f with finite rank, $r(f, k)$ is a supermultiplicative function of k in the sense that*

$$r(f, k+l) \geq r(f, k) r(f, l).$$

Proof. It follows from Lemma 6.3(c) that for any two disjoint finite sets S and T there is an embedding

(6.9) $$\mathcal{Q}_S/f \otimes \mathcal{Q}_T/f \hookrightarrow \mathcal{Q}_{S \cup T}/f.$$

Considering the dimensions, the assertion follows. $\square$

EXERCISE 6.5. Prove that if all nodes of a simple graph F are labeled, then both F and the quantum graph $\widehat{F}$ introduced above are idempotent in the algebra of simple partially labeled graphs: $F^2 = F$ and $\widehat{F}^2 = \widehat{F}$.

EXERCISE 6.6. Let f be a graph parameter for which $r(f, 2) = r$ is finite.
(a) Prove that every path labeled at its endpoints can be expressed, modulo f, as a linear combination of paths of length at most r.
(b) Prove that a 2-labeled m-bond $B^{m\bullet\bullet}$ can be expressed, modulo f, as a linear combination of 2-labeled k-bonds with $k \leq r - 1$.
(c) A *series-parallel graph* is a 2-labeled graph obtained from $K_2^{\bullet\bullet}$ by repeated application of the gluing and concatenation operations. Prove that every series-parallel graph can be expressed, modulo f, as a linear combination of series-parallel graphs with at most 2^{r-1} edges.

6.2. Reflection positivity

In this section we assume that f is reflection positive, multiplicative, normalized and has finite connection rank. We are going to prove that the gluing algebras (modulo f) have a very tight structure.

6.2.1. The idempotent basis. Let S be a finite subset of $\mathbb{N}$. If f is reflection positive and the dimension of $\mathcal{Q}_S/f$ is a finite number, then the factor algebra $\mathcal{Q}_S/f$ is a finite dimensional commutative Frobenius algebra: it has a commutative and associative product as well as a positive definite inner product, related by the Frobenius identity (6.4).

This implies that $\mathcal{Q}_S/f$ has a very simple structure. Every element $x \in \mathcal{Q}_S/f$ defines a linear transformation $A_x : \mathcal{Q}_S/f \to \mathcal{Q}_S/f$ by $A_x y = xy$. Clearly $x \mapsto A_x$ is an algebra homomorphism, and the fact that the inner product is definite on $\mathcal{Q}_S/f$ implies that $x \mapsto A_x$ is injective. Commutativity and the Frobenius identity imply that A_x is symmetric, and that any two transformations A_x commute. This implies that there is an orthonormal basis in which all the A_x are simultaneously diagonal. Counting dimensions shows that every diagonal matrix is of the form A_x, and so $\mathcal{Q}_S/f$ is isomorphic with the algebra of diagonal matrices. Another way of saying this is that $\mathcal{Q}_S/f$ is isomorphic to $\mathbb{R}^m$ endowed with the coordinate-wise product and the usual inner product (where $m = \dim(\mathcal{Q}_S/f) = r(f, k)$).

The algebra elements corresponding to the standard basis vectors form a (uniquely determined) basis $B_S = \{p_1^S, \ldots, p_r^S\}$ such that $(p_i^S)^2 = p_i^S$ (the basis elements are idempotent in the algebra) and $p_i^S p_j^S = 0$ for $i \neq j$. We call this the *idempotent basis* of $\mathcal{Q}_S/f$.

If $p \in \mathcal{Q}/f$ is any nonzero idempotent, then
$$f(p) = f(pp) = \langle p, p \rangle > 0.$$
In particular, $f(p_i^S) > 0$. Note, however, that $f(p_i^S) \neq 1$ in general, so the algebra isomorphism between $\mathcal{Q}_S/f$ and $\mathbb{R}^m$ does not preserve the inner product.

This purely algebraic construction carries a lot of combinatorial information, as we shall see. But first, let us work out an example.

EXAMPLE 6.7 (**Eulerian property**). Let us compute the algebras associated with $\mathsf{Eul}(G)$, the indicator function of G being eulerian (example 4.21). We know that this is a homomorphism function into a 2-node weighted graph again, hence it is reflection positive and $r(\mathsf{Eul}, k) \leq 2^k$, but this last inequality will not hold with equality for $k \geq 1$. The space $\mathcal{Q}_0/\mathsf{Eul}$ is 1-dimensional. To determine $\mathcal{Q}_1/\mathsf{Eul}$, we note that for two 1-labeled graphs G_1 and G_2, the product $G_1 G_2$ is eulerian if and only if both G_1 and G_2 are eulerian (G_1 and G_2 cannot have a single node with odd degree!), and so $\mathsf{Eul}(G_1 G_2) = \mathsf{Eul}(G_1)\mathsf{Eul}(G_2)$ holds for 1-labeled graphs as well. This shows that $\mathcal{Q}_1/\mathsf{Eul}$ is also 1-dimensional.

Next, for general k, let $\mathrm{Odd}(G)$ denote the set of nodes of G with odd degree. Clearly $G \equiv 0 \pmod{\mathsf{Eul}}$ for any graph with $\mathrm{Odd}(G) \not\subseteq [k]$. Since $|\mathrm{Odd}(G)|$ is even, the set $\mathrm{Odd}(G)$ is uniquely determined by the intersection $\mathrm{Odd}'(G) = [k-1] \cap \mathrm{Odd}(G)$. For two k-labeled graphs G_1 and G_2, the product $G_1 G_2$ is eulerian if and only if $\mathrm{Odd}'(G_1) = \mathrm{Odd}'(G_2)$ (which implies that the unlabeled nodes have even degree). Furthermore, $\mathrm{Odd}'(G_1 G_2) = \mathrm{Odd}'(G_1) \triangle \mathrm{Odd}'(G_2)$, and hence $G \mapsto \mathrm{Odd}'(G)$ induces an algebra isomorphism between $\mathcal{Q}_k/\mathsf{Eul}$ and the group algebra of Z_2^{k-1}. Hence $r(\mathsf{Eul}, k) = 2^{k-1}$ for $k \geq 1$.

The idempotents of a finite abelian group are determined by its characters, which in this simple case means that they are indexed by subsets $S \subseteq [k-1]$, and the idempotent is the group algebra element
$$p_S = \frac{1}{2^{k-1}} \sum_{X \subseteq [k-1]} (-1)^{|S \cap X|} X.$$

The set $\mathrm{Odd}(G)$ can be expressed in this basis by discrete Fourier inversion:
$$\mathrm{Odd}(G) = \frac{1}{2^{k-1}} \sum_{X \subseteq [k-1]} (-1)^{|S \cap X|} p_X.$$

It follows that
$$G \mapsto \left((-1)^{|S \cap \mathrm{Odd}(G)|} : S \subseteq [k-1]\right)$$
defines an algebra isomorphism between $\mathcal{Q}_k/\mathsf{Eul}$ and $\mathbb{R}^{2^{k-1}}$. ◆

For two idempotents p and q in $\mathcal{Q}/f$, we say that q *resolves* p, if $pq = q$. It is clear that this relation is transitive.

LEMMA 6.8. *Let r be any idempotent element of $\mathcal{Q}_S/f$. Then r is the sum of those idempotents in B_S that resolve it.*

Proof. Indeed, we can write $r = \sum_{p \in B_S} \mu_p p$ with some scalars μ_p. Using that r is idempotent, we get that

$$r = r^2 = \sum_{p,p' \in B_S} \mu_p \mu_{p'} pp' = \sum_{p \in B_S} \mu_p^2 p,$$

which shows that $\mu_p^2 = \mu_p$ for every p, and so $\mu_p \in \{0, 1\}$. So r is the sum of some subset $X \subseteq B_S$. It is clear that $rp = p$ for $p \in X$ and $rp = 0$ for $p \in B_S \setminus X$, so X consists of exactly those elements of B_S that resolve q. $\square$

As a special case, we see that

(6.10) $$u = \sum_{p \in B_S} p$$

is the unit element of $\mathcal{Q}_S$ (this is the image of the edgeless graph U_S), and also the unit element of the whole algebra $\mathcal{Q}$.

LEMMA 6.9. *Let $S \subset T$ be two finite sets. Then every $q \in B_T$ resolves exactly one element of B_S.*

Proof. We have by (6.10) that

$$u = \sum_{p \in B_S} p = \sum_{p \in B_S} \sum_{\substack{q \in B_T \\ q \text{ resolves } p}} q,$$

and also

$$u = \sum_{q \in B_T} q,$$

so by the uniqueness of the representation we get that every q must resolve exactly one p. $\square$

LEMMA 6.10. *If $p \in B_S$ and q resolves p, then $[\![q]\!]_S = \frac{f(q)}{f(p)} p$.*

Proof. Clearly $\frac{f(q)}{f(p)} p \in \mathcal{Q}_S/f$. Furthermore,

$$\left\langle q - \frac{f(q)}{f(p)} p, p \right\rangle = f(qp) - \frac{f(q)}{f(p)} f(p^2) = f(q) - \frac{f(q)}{f(p)} f(p) = 0,$$

and for every other basic idempotent $p' \in B_S$, we have

$$\left\langle q - \frac{f(q)}{f(p)} p, p' \right\rangle = f(qp') - \frac{f(q)}{f(p)} f(pp') = f(qp') - \frac{f(q)}{f(p)} f(pp') = 0.$$

This shows that $\frac{f(q)}{f(p)} p$ is the orthogonal projection of q to $\mathcal{Q}_S/f$. Since $[\![q]\!]_S$ has the same characterization, the lemma follows. $\square$

LEMMA 6.11. *Let $S, T \subset \mathbb{N}$ be finite sets, let $p \in B_{S \cap T}$, and let $q \in B_S$ resolve p. Then for any $x \in \mathcal{Q}_T/f$ we have $f(p)f(qx) = f(q)f(px)$.*

Indeed, by Lemma 6.10 and (6.7),

$$f(qx) = f([\![q]\!]_{S \cap T} x) = \frac{f(q)}{f(p)} f(px).$$

LEMMA 6.12. *If both idempotents $q \in B_S$ and $r \in B_T$ resolve the same idempotent $p \in B_{S \cap T}$, then $qr \neq 0$.*

6.2. REFLECTION POSITIVITY

Indeed, by Lemma 6.11,
$$f(qr) = \frac{f(q)}{f(p)}f(pr) = \frac{f(q)}{f(p)}f(r) > 0.$$

Let $S \subset \mathbb{N}$ and $p \in B_S$. For $u \in \mathbb{N} \setminus S$, let $q_1^u, \ldots, q_D^u$ denote the elements of $B_{S \cup \{u\}}$ resolving p. Note that for $u, v \in \mathbb{N} \setminus S$, there is a natural isomorphism between $\mathcal{Q}_{S \cup \{u\}}/f$ and $\mathcal{Q}_{S \cup \{v\}}/f$ (induced by the map that fixes S and maps u onto v), and we may choose the labeling so that q_i^u corresponds to q_i^v under this isomorphism.

Let $T \supset S$ and $V = T \setminus S$. For every map $\varphi : V \to [D]$, let

(6.11) $$q_\varphi = \prod_{v \in V} q_{\varphi(v)}^v.$$

LEMMA 6.13. *The algebra elements q_φ are nonzero idempotents in $\mathcal{Q}_T/f$ such that $q_\varphi q_\psi = 0$ if $\varphi \neq \psi$.*

It is clear that the q_φ are idempotents. By Lemma 6.11,

(6.12) $$f(q_\varphi) = f(\prod_{v \in V} q_{\varphi(v)}^v) = \Big(\prod_{v \in V} \frac{f(q_{\varphi(v)}^v)}{f(p)}\Big) f(p) \neq 0,$$

and so $q_\varphi \neq 0$. Finally, if $\varphi \neq \psi$, then there is a $v \in V$ such that $\varphi(v) \neq \psi(v)$, and then $q_{\varphi(v)}^v q_{\psi(v)}^v = 0$, which implies that $q_\varphi q_\psi = 0$.

If $p \in B_S$ and $S \subset T$, $|T| = |S| + 1$, then the number of elements in B_T that resolve p will be called the *degree* of p, and denoted by $\deg(p)$. Obviously this value is independent of which $(|S|+1)$-element superset T of S we are considering.

LEMMA 6.14. *If $S \subset T$, and $q \in B_T$ resolves $p \in B_S$, then $\deg(q) \geq \deg(p)$.*

It suffices to show this in the case when $|T| = |S| + 1$. Let $T = S \cup \{u\}$ and $v = \mathbb{N} \setminus T$. Let $q_1^u, \ldots, q_D^u$ denote the elements of B_T resolving p, where (say) $q = q_1^u$, and let $q_1^v, \ldots, q_D^v$ be the basic idempotents in $B_{S \cup \{v\}}$ resolving p. Then by Lemma 6.13, the elements $q_1^u q_i^v$, $i \in [D]$ are nonzero idempotents in $\mathcal{Q}_{S \cup \{u,v\}}/f$ resolving q, such that the product of any two of them is 0. Writing every such q_{1i} as a sum of basic idempotents we see that $\deg(q) \geq D$.

6.2.2. Homomorphisms into weighted graphs. After all these preparations, we are able to complete the proof of Theorem 5.54. Our goal is to construct a weighted graph H for which $f(G) = \hom(G, H)$ for every loopless multigraph G.

The non-degenerate case. We start with sketching the proof in the non-degenerate case, when $\dim(\mathcal{Q}_k/f) = q^k$ for all k. (This is in fact the generic case, in the sense that it occurs with probability 1 if $f = \hom(., H)$ for a weighted graph H with randomly chosen edge and nodeweights; see Section 6.4.1.) This implies that the embedding $\mathcal{Q}_S/f \otimes \mathcal{Q}_T/f \hookrightarrow \mathcal{Q}_{S \cup T}/f$ in (6.9) is an isomorphism. In this case, the idempotent bases of the algebras $\mathcal{Q}_k/f$ are easy to construct explicitly: if $p_1, \ldots, p_q$ is the idempotent basis of $\mathcal{Q}_1/f$, then the elements $p_{i_1} \otimes \cdots \otimes p_{i_k}$ ($i_j \in [q]$) form the idempotent basis of $\mathcal{Q}_k/f$.

We can define a weighted complete graph H on $[q]$ as follows: let $\alpha_i = f(p_i) = f(p_i^2) > 0$ and define β_{ij} by expressing the graph k_2 (a single edge with both nodes labeled) in the idempotent basis:

$$k_2 = \sum_{i,j \in [q]} \beta_{ij} (p_i \otimes p_j) \tag{6.13}$$

To show that the weighted graph H obtained this way satisfies $f(G) = \hom(G, H)$ for any multigraph G, we may assume that $V(G) = [k]$ and all nodes of G are labeled. Then we can write

$$G = \prod_{uv \in E(G)} K_{uv}, \tag{6.14}$$

where K_{uv} is the graph on k labeled nodes, with a single edge connecting u and v. Defining $p_\varphi = p_{\varphi(1)} \otimes \cdots \otimes p_{\varphi(k)}$ for $\varphi \colon [k] \to [q]$, the k-labeled quantum graphs p_φ form a basis of $\mathcal{Q}_k / f$ consisting of idempotents, and hence

$$O_k = \sum_{\varphi \colon [k] \to [q]} p_\varphi. \tag{6.15}$$

By the definition of the β_{ij}, we have

$$K_{uv} p_\varphi = \beta_{\varphi(u)\varphi(v)} p_\varphi. \tag{6.16}$$

We want to evaluate $f(G) = \langle G, O_k \rangle$. Substituting from (6.14) and (6.15), and using (6.16) and the Frobenius identity (6.4) repeatedly, we get that

$$f(G) = \sum_{\varphi \colon [k] \to [q]} \prod_{uv \in E(G)} \beta_{\varphi(u)\varphi(v)} \prod_{u \in V(G)} \alpha_{\varphi(u)} = \hom(G, H).$$

The general (and degenerate) case takes more work, but we are ready to give it now.

Proof of Theorem 5.54. The idea is that we find a basic idempotent $p \in B_S$ for a sufficiently large finite set $S \subseteq \mathbb{N}$, with the property that the subalgebra $p\mathcal{Q}/f$ behaves like the whole algebra behaved in generic case. So the idempotent bases in it, and from these the weighted graph H, can be constructed explicitly.

Bounding the expansion. If a basic idempotent $p \in B_S$ has degree D, then by Lemma 6.14, there are D basic idempotents in B_T with $|T| = |S| + 1$ with degree $\geq D$ that resolve p. Hence if $|T| \geq |S|$, then the dimension of $\mathcal{Q}_T$ is at least $D^{|T \setminus S|}$. It follows that the degrees of basic idempotents are bounded by q. Let us choose S and $p \in B_S$ so that $D = \deg(p)$ is maximum degree. Then it follows by Lemma 6.14 that all basic idempotents resolving p have degree exactly D.

Describing the idempotents. Let us fix a set S and a basic idempotent $p \in B_S$ with maximum degree D. For $u \in \mathbb{N} \setminus S$, let $q_1^u, \ldots, q_D^u$ denote the elements of $B_{S \cup \{u\}}$ resolving p.

We can describe, for a finite set $T \supset S$, all basic idempotents in B_T that resolve p. Let $V = T \setminus S$, and for every map $\varphi \colon V \to \{1, \ldots, D\}$, let

$$q_\varphi = \prod_{v \in V} q_{\varphi(v)}^v. \tag{6.17}$$

Note that by Lemma 6.11,

$$f(q_\varphi) = f\Big(\prod_{v \in V} q_{\varphi(v)}^v\Big) = \Big(\prod_{v \in V} \frac{f(q_{\varphi(v)}^v)}{f(p)}\Big) f(p) \neq 0, \tag{6.18}$$

and so $q_\varphi \neq 0$.

6.2. REFLECTION POSITIVITY

CLAIM 6.15. *The basic idempotents in $\mathcal{Q}_T/f$ resolving p are exactly the algebra elements of the form q_φ, $\varphi \in \{1,\ldots,D\}^V$.*

We prove this by induction on $|T \setminus S|$. For $|T \setminus S| = 1$ the assertion is trivial. Suppose that $|T \setminus S| > 1$. Let $u \in T \setminus S$, $U = S \cup \{u\}$ and $W = T \setminus \{u\}$; thus $U \cap W = S$. By the induction hypothesis, the basic idempotents in B_W resolving p are elements of the form q_ψ ($\psi \in \{1,\ldots,D\}^{V \setminus \{u\}}$).

Let r be one of these. By Lemma 6.12, $rq_i^u \neq 0$ for any $1 \leq i \leq D$, and clearly resolves r. We can write rq_i^u as the sum of basic idempotents in B_T, and it is easy to see that these also resolve r. Furthermore, the basic idempotents occurring in the expression of rq_i^u and rq_j^u ($i \neq j$) are different. But r has degree D, so each rq_i^u must be a basic idempotent in B_T itself.

Since the sum of the basic idempotents rq_i^u ($r \in B_{W,p}$, $1 \leq i \leq D$) is p, it follows that these are all the elements of $B_{T,p}$. This proves the Claim.

It is immediate from the definition that an idempotent q_φ resolves q_i^v if and only if $\varphi(v) = i$. Hence it also follows that

$$q_i^v = \sum_{\varphi:\ \varphi(v)=i} q_\varphi. \tag{6.19}$$

Constructing the target graph. Now we can define H as follows. Let H be the looped complete graph on $V(H) = \{1,\ldots,D\}$. We have to define the node weights and edge weights.

Fix any $u \in \mathbb{N} \setminus S$. For every $i \in V(H)$, let $\alpha_i = f(q_i^u)/f(p)$ be the weight of the node j. Clearly $\alpha_i > 0$.

Let $u, v \in \mathbb{N} \setminus S$, $v \neq u$, and let $W = S \cup \{u,v\}$. Let K_{uv} denote the graph on W which has only one edge connecting u and v, and let k_{uv} denote the corresponding element of $\mathcal{Q}_W$. We can express pk_{uv} as a linear combination of elements of $B_{W,p}$ (since for any $r \in B_W \setminus B_{W,p}$ one has $rp = 0$ and hence $rpk_{u,v} = 0$):

$$pk_{uv} = \sum_{i,j} \beta_{ij} q_i^u q_j^v.$$

This defines the weight β_{ij} of the edge ij. Note that $\beta_{ij} = \beta_{ji}$, since $pk_{uv} = pk_{vu}$.

Verifying the target graph. We prove that this weighted graph H gives the right homomorphism function: $f(G) = \hom(G, H)$ for every multigraph G. By (6.19), we have for each pair u, v of distinct elements of $V(G)$

$$pk_{uv} = \sum_{i,j \in V(H)} \beta_{i,j} q_i^u q_j^v = \sum_{i,j \in V(H)} \beta_{i,j} \sum_{\substack{\varphi:\ \varphi(u)=i \\ \varphi(v)=j}} q_\varphi = \sum_{\varphi \in V(H)^V} \beta_{\varphi(u),\varphi(v)} q_\varphi.$$

Consider any V-labeled graph G with $V(G) = V \subseteq \mathbb{N} \setminus S$, and let g be the corresponding element of $\mathcal{Q}/f$. Then

$$pg = \prod_{uv \in E(G)} pk_{uv} = \prod_{uv \in E(G)} \Big(\sum_{\varphi \in V(H)^V} \beta_{\varphi(u),\varphi(v)} q_\varphi\Big)$$
$$= \sum_{\varphi: V \to V(H)} \Big(\prod_{uv \in E(G)} \beta_{\varphi(u),\varphi(v)}\Big) q_\varphi.$$

Since $p \in \mathcal{Q}_S/f$ and $g \in \mathcal{Q}_V/f$ where $S \cap V = \emptyset$, we have $f(p)f(g) = f(pg)$, and so by (6.12),

$$f(p)f(g) = f(pg) = \sum_{\varphi \in V(H)^V} \Big(\prod_{uv \in E(G)} \beta_{\varphi(u),\varphi(v)} \Big) f(q_\varphi)$$

$$= \sum_{\varphi: V \to V(H)} \Big(\prod_{uv \in E(G)} \beta_{\varphi(u),\varphi(v)} \Big) \Big(\prod_{v \in V(G)} \alpha_{\varphi(v)} \Big) f(p) = \mathsf{hom}(G,H) f(p).$$

The factor $f(p) > 0$ can be cancelled from both sides, completing the proof of the theorem. □

6.3. Contractors and connectors

We study the existence and properties of two special elements in graph algebras. These will serve as important building blocks for other constructions, like a more explicit description of the idempotent basis.

6.3.1. Contractors and connectors for general graph parameters. In the algebra $\mathcal{Q}_2$ of 2-multilabeled graphs, multiplication by the single node with two labels (denoted by $K_1^{\bullet\bullet}$) results in identifying nodes labeled 1 and 2. In the factor algebra modulo a graph parameter, it may be the case that multiplication by some other graph has essentially the same result. For example, for the chromatic polynomial $\mathsf{chr}(.,x)$, the contraction-deletion identity (A.10) can be written like this:

(6.20) $$K_1^{\bullet\bullet} \equiv O_2^{\bullet\bullet} - K_2^{\bullet\bullet} \quad (\mathrm{mod}\ \mathsf{chr}(.,x)).$$

or in pictograms

(6.21) $$\overset{12}{\bullet} \equiv \overset{\bullet}{\bullet} - \overset{\bullet}{\bullet} \quad (\mathrm{mod}\ \mathsf{chr}(.,x)).$$

We say that the 2-labeled quantum graph on the right side is a *contractor* for the graph parameter $\mathsf{chr}(.,x)$. Starting with any multilabeled quantum graph, we can apply this identity repeatedly to construct a simply labeled quantum graph which represents the same element in $\mathcal{Q}/f$.

A consequence of the contraction-deletion relation is the identity

(6.22) $$\overset{\bullet}{\bullet} \equiv \circ\overset{\bullet}{\bullet} - \circ\overset{\bullet}{\bullet} \quad (\mathrm{mod}\ \mathsf{chr}(.,x)),$$

which does not involve multiple labels. One way to read this is that the edge (with both endnodes labeled) can be replaced by the difference of two simple graphs in which the two labeled nodes are nonadjacent. We say that the 2-labeled quantum graph on the right is a *connector* for the graph parameter $\mathsf{chr}(.,x)$. One important consequence of this identity is that applying it repeatedly, we can eliminate all edge multiplicities.

To state the definition formally, let us say that a quantum graph x is a *proper expansion* of a partially labeled graph F modulo a graph parameter f, if $x \equiv F$ $(\mathrm{mod}\ f)$, and no constituent of x is of the form FG for some partially labeled graph G. Then a *contractor* is a proper expansion of $K_1^{\bullet\bullet}$, and a *connector* is a proper expansion of $K_2^{\bullet\bullet}$. Also, the remarks after (6.20) and (6.22) can be formalized like this:

PROPOSITION 6.16. *A graph parameter f has a contractor if and only if every multilabeled quantum graph is congruent modulo f to a simply labeled quantum graph. A graph parameter f has a simple connector if and only if every simply labeled quantum graph is congruent modulo f to a simply labeled simple quantum graph with no edge connecting the labeled nodes.*

We can put the notion of a contractor in a different context. Let $\mathcal{F}_k^{\text{stab}}$ denote the set of k-labeled multigraphs in which the labeled nodes form an independent (stable) set, and let $\mathcal{Q}_k^{\text{stab}}$ denote the subalgebra of $\mathcal{Q}_k$ generated by them. For a 2-labeled graph $F \in \mathcal{F}_2^{\text{stab}}$, let F' denote the graph obtained by identifying the two labeled nodes and labeling it by 1. (So F' is obtained from the product $FK_1^{\bullet\bullet}$ by removing the label 2.) The map $F \mapsto F'$ maps 2-labeled graphs to 1-labeled graphs. We can extend it linearly to get an algebra homomorphism $x \mapsto x'$ from $\mathcal{Q}_2^{\text{stab}}$ into $\mathcal{Q}_1$.

The map $x \mapsto x'$ does not in general preserve the inner product or even its kernel; we say that the graph parameter f is *contractible*, if for every $x \in \mathcal{Q}_2^{\text{stab}}$, $x \equiv 0 \pmod{f}$ implies $x' \equiv 0 \pmod{f}$; in other words, $x \mapsto x'$ factors to a linear map $\mathcal{Q}_2^{\text{stab}}/f \to \mathcal{Q}_1/f$. With this notation, $z \in \mathcal{Q}_2$ is a contractor for f if and only if for every $x \in \mathcal{Q}_2^{\text{stab}}$, we have $f(xz) = f(x')$.

Contractors also relate to the algebra of concatenations:

PROPOSITION 6.17. *A contractor for f is the multiplicative identity for the operation $\circ$ modulo f.*

Proof. We have to verify that if z is a contractor, then

$$z \circ x \equiv x \pmod{f} \tag{6.23}$$

for all $x \in \mathcal{Q}_2$. This is equivalent to $f((z \circ x)y) = f(xy)$ for all $x, y \in \mathcal{Q}_2$. Using (6.5) and that in $x \circ y$ the labeled nodes are nonadjacent, we obtain:

$$f((z \circ x)y) = f(z(y \circ x^*)) = f((y \circ x^*)') = f(xy),$$

which proves (6.23). $\square$

Note that in every constituent of $x \circ y$ the labeled nodes are nonadjacent for all $x, y \in \mathcal{Q}_2$. It follows that if the algebra $(\mathcal{Q}_2/f, \circ)$ has a multiplicative identity (in particular, if it has a contractor), then every $y \in \mathcal{Q}_2/f$ can be represented by a 2-labeled quantum graph with nonadjacent labeled nodes.

While the existence of a contractor, the existence of a connector, and contractibility are three different properties of a graph parameter, there is some connection, as expressed in the following propositions.

PROPOSITION 6.18. *If a graph parameter has a contractor, then it is contractible.*

Proof. Let w be a contractor for f. Suppose that $x \in \mathcal{Q}_2$ satisfies $x \equiv 0 \pmod{f}$, and let $y \in \mathcal{Q}_1$. Choose a $z \in \mathcal{Q}_2$ such that $z' = y$. Then

$$f(x'y) = f(x'z') = f((xz)') = f((xz)w) = f(x(zw)) = 0,$$

showing that $x' \equiv 0 \pmod{f}$. $\square$

PROPOSITION 6.19. *If a graph parameter has a contractor, then it has a connector; if it has a simple contractor, then it has a simple connector.*

Proof. Let z be a contractor. We claim that $z \circ P_2^{\bullet\bullet}$ is a connector. Indeed, for any 2-labeled graph G, we have $(z \circ P_2^{\bullet\bullet})G \cong z(G \circ P_2^{\bullet\bullet})$, and by the definition of a contractor, $f\bigl(z(G \circ P_2^{\bullet\bullet})\bigr) = f\bigl(K_1^{\bullet\bullet}(G \circ P_2)\bigr) = f(P_2^{\bullet\bullet}G)$. So $z \circ P_2^{\bullet\bullet} \equiv P_2^{\bullet\bullet}$ (mod f), and thus $z \circ P_2^{\bullet\bullet}$ is a connector. The second assertion is trivial by the same construction. □

PROPOSITION 6.20. *If f is contractible, has a connector, and $r(f, 2)$ is finite, then f has a contractor.*

Proof. Since $\langle x, y \rangle = f(xy)$ is a symmetric (possibly indefinite) bilinear form that is not singular on $\mathcal{Q}_2/f$, there is a basis $p_1, \ldots, p_N$ in $\mathcal{Q}_2/f$ such that $f(p_i p_j) = 0$ if $i \neq j$ and $f(p_i p_i) \neq 0$. By the assumption that f has a connector, we may represent this basis by quantum graphs with nonadjacent labeled nodes; then the contracted quantum graphs p'_i have no loops. Let

$$z = \sum_{i=1}^{N} \frac{f(p'_i)}{f(p_i^2)} p_i.$$

We claim that z is a contractor. Indeed, let $x \in \mathcal{Q}_2$ be a quantum graph with nonadjacent labeled nodes, and write $x \equiv \sum_{i=1}^{N} a_i p_i$ (mod f). Then we have

$$f(xz) = \sum_{i=1}^{N} a_i \frac{f(p'_i)}{f(p_i^2)} f(p_i^2) = \sum_{i=1}^{N} a_i f(p'_i).$$

On the other hand, contractibility implies that $x' \equiv \sum_{i=1}^{N} a_i p'_i$ (mod f), and so

$$f(x') = \sum_{i=1}^{N} a_i f(p'_i) = f(xz). \qquad \square$$

PROPOSITION 6.21. *If $M(f, 2)$ is positive semidefinite and has finite rank r, and f is contractible, then f has a connector whose constituents are paths of length at most $r + 1$.*

Proof. Since $\mathcal{Q}_2/f$ is finite dimensional, there is a linear dependence between $P_2^{\bullet\bullet}, P_3^{\bullet\bullet}, \ldots, P_{r+2}^{\bullet\bullet}$ in $\mathcal{Q}_2/f$. Hence there is a (smallest) $k \geq 2$ such that $P_k^{\bullet\bullet}$ can be expressed as

(6.24) $$P_k^{\bullet\bullet} \equiv \sum_{i=1}^{r} a_i P_{k+i}^{\bullet\bullet} \pmod{f}$$

with some real numbers $a_1, \ldots, a_r$. The assertion is equivalent to saying that $k = 2$.

Let $x = P_2^{\bullet\bullet} - \sum_{i=1}^{r} a_i P_{2+i}^{\bullet\bullet}$. Then (6.24) can be written as $x \circ P_{k-1}^{\bullet\bullet} \equiv 0$ (mod f). If $k = 3$, then this implies that $x \circ P_{2+i}^{\bullet\bullet} \equiv 0$ (mod f) for all $i \geq 0$, and hence $x \circ x \equiv 0$ (mod f). Using contractibility we obtain that $0 = f\bigl((x \circ x)'\bigr) = f(x^2)$. Now semidefiniteness of $M(f, 2)$ shows that $x \equiv 0$ (mod f). So (6.24) holds with $k = 2$ as well, a contradiction.

Suppose that $k > 3$, then

$$(x \circ P_{k-2}^{\bullet\bullet})^2 = (x \circ P_{k-1}^{\bullet\bullet})(x \circ P_{k-3}^{\bullet\bullet}) \equiv 0 \pmod{f},$$

and so by the assumption that $M(f, 2)$ is positive semidefinite, we get that $x \circ P_{k-2}^{\bullet\bullet} \equiv 0$ (mod f), which contradicts the minimality of k again. □

The following statement is a corollary of Propositions 6.20 and 6.21.

COROLLARY 6.22. *If $M(f,2)$ is positive semidefinite and has finite rank, and f is contractible, then f has a contractor.*

We conclude this section with a number of examples of connectors and contractors.

EXAMPLE 6.23 (**Perfect matchings**). Recall that $\mathsf{pm}(G)$ denotes the number of perfect matchings in the graph G. We have seen that $\mathrm{rk}(\mathsf{pm}, k) = 2^k$ is exponentially bounded, but pm is not reflection-positive, and thus $\mathsf{pm}(G)$ cannot be represented as a homomorphism function.

On the other hand: pm has a contractor: a path of length 2, and also a connector: a path of length 3. ♦

EXAMPLE 6.24 (**Number of triangles**). The graph parameter $\mathsf{hom}(K_3, .)$ has no connector. Indeed, suppose that $x \in \mathcal{Q}_2$ is a connector, then we must have
$$\mathsf{hom}(K_3, xP_3^{\bullet\bullet}) = \mathsf{hom}(K_3, K_2^{\bullet\bullet} P_3^{\bullet\bullet}) = \mathsf{hom}(K_3, K_3) = 6,$$
and also
$$\mathsf{hom}(K_3, xP_4^{\bullet\bullet}) = \mathsf{hom}(K_3, K_2^{\bullet\bullet} P_4^{\bullet\bullet}) = \mathsf{hom}(K_3, C_4) = 0.$$
On the other hand,
$$\mathsf{hom}(K_3, xP_3^{\bullet\bullet}) = \mathsf{hom}(K_3, xP_4^{\bullet\bullet}),$$
since x has with nonadjacent labeled nodes, so no homomorphism from K_3 touches the edges of the $P_3^{\bullet\bullet}$ factor. This contradiction shows that $\mathsf{hom}(K_3, .)$ has no connector.

A similar argument shows that $\mathsf{hom}(K_3, .)$ is not contractible (and so it has no contractor). ♦

EXAMPLE 6.25 (**S-Flows**). The number $\mathsf{flo}(G)$ of flows on G with values from a given subset of a finite abelian group can be described as a homomorphism function (Example 5.16). It has a trivial connector, a path of length 2 (which is an algebraic way of saying that if we subdivide an edge, then the flows don't change essentially). In the case of nowhere-0 flows, $K_2^{\bullet\bullet} + U_2^{\bullet\bullet}$ is a contractor (which amounts to the contraction-deletion identity for the flow polynomial). In general, it is more difficult to describe the contractors, but it is possible (Garijo, Goodall and Nešetřil [2011]). ♦

EXAMPLE 6.26 (**Density in a random graph**). Recall the multigraph parameter from Example 5.60: $f(G) = p^{\mathsf{e}(G^{\mathrm{simp}})}$ $(0 < p < 1)$, which is reflection positive, multiplicative, and has finite connection rank. This parameter has neither a contractor nor a connector; it is not even contractible. We have

$$\vcenter{\hbox{⚬⚬}} \equiv \vcenter{\hbox{⚬⚬}} \qquad (\mathrm{mod}\ f),$$

but identifying the labeled nodes produces a pair of parallel edges in the first graph but not in the second, so they don't remain congruent. ♦

EXAMPLE 6.27 (**Eulerian orientations**). Recall that $\overrightarrow{\mathsf{eul}}(G)$ denotes the number of eulerian orientations of the graph G. We have seen that the graph parameter $\overrightarrow{\mathsf{eul}}$ is reflection positive, but has infinite connection rank (so it is not a homomorphism function). Similarly as in Example 6.25, a path of length 2 is a connector. Furthermore, this graph parameter is contractible, but has no contractor (see Exercise 6.33). ♦

6.3.2. Contractors and connectors for homomorphism functions.
Many graph parameters have contractors and/or connectors, as we have seen.
Our next theorem asserts this for homomorphism functions $f = \mathsf{hom}(.,H)$ for any
weighted graph H. It is easy to check from the definitions that a 2-labeled quantum
graph z is a contractor for $\mathsf{hom}(.,H)$ if and only if

(6.25) $$\mathsf{hom}_{ij}(z,H) = (1/\alpha_i)\mathbb{1}(i=j)$$

for every $i,j \in V(H)$. It is a connector for $\mathsf{hom}(.,H)$ if and only if z is simple, it
has nonadjacent labeled nodes, and

(6.26) $$\mathsf{hom}_{ij}(z,H) = \beta_{ij}$$

for every $i,j \in V(H)$.

THEOREM 6.28. *Let $f = \mathsf{hom}(.,H)$ for some weighted graph H. Then f has a simple contractor and a simple connector.*

We can in fact construct connectors and contractors of a rather simple form.
To state the result, we define classes of 2-labeled multigraphs as follows. We start
with $\mathcal{K} = \{K_2^{\bullet\bullet}\}$. The (gluing) products of at most a of members of $\mathcal{K}$ form the
class $\mathcal{K}(a)$ (so these are 2-labeled bonds). Concatenations of at most b bonds form
the class $\mathcal{K}(a,b)$. The gluing products of at most c members of $\mathcal{K}(a,b)$ form the
class $\mathcal{K}(a,b,c)$. Concatenations of at most d members of $\mathcal{K}(a,b,c)$ form the class
$\mathcal{K}(a,b,c,d)$ etc (Fig. 6.1).

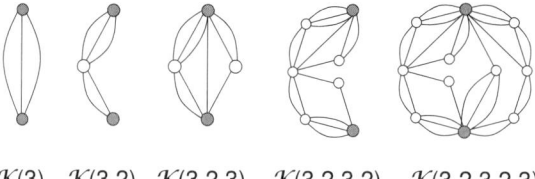

$\mathcal{K}(3)$ $\mathcal{K}(3,2)$ $\mathcal{K}(3,2,3)$ $\mathcal{K}(3,2,3,2)$ $\mathcal{K}(3,2,3,2,3)$

FIGURE 6.1. Graphs in classes $\mathcal{K}$

SUPPLEMENT 6.29. *Let $q = \mathsf{v}(H)$.*

(a) *The graph parameter $f = \mathsf{hom}(.,H)$ has a connector whose constituents are paths $P_k^{\bullet\bullet}$ with $3 \le k \le q+3$.*

(b) *The graph parameter $f = \mathsf{hom}(.,H)$ has a contractor whose constituents are in the class $\mathcal{K}(q^2-1, 2, q^2-1, 2, q)$. In particular, the number of nodes in this contractor is at most $2q^3$.*

All graphs in the classes $\mathcal{K}(a,b,\dots)$ are series-parallel. So in particular, the
graph parameter $f = \mathsf{hom}(.,H)$ has a contractor whose constituents are series-
parallel (Lovász and Szegedy [2009]). The Supplement above is a refinement of this
statement. For us, the bound on the number of nodes will be more important than
the structure. If we get rid of the parallel edges in the contractor by replacing each
edge by the simple connector constructed above, every constituent will still be a
series-parallel graph, and the number of nodes in any constituent will be bounded
by $2q^6$.

6.3. CONTRACTORS AND CONNECTORS

Proof of Theorem 6.28 and Supplement 6.29. We may assume that H is twin-free, since identifying twins does not change the graph parameter $\mathsf{hom}(.,H)$.

(a) To construct a simple connector, let $B = (\beta_{ij})$ be the (weighted) adjacency matrix of H, and let $D = \mathrm{diag}(\sqrt{\alpha_1},\ldots,\sqrt{\alpha_m})$. Let $\lambda_1,\ldots,\lambda_t$ be the nonzero eigenvalues of the matrix DBD (which are real as DBD is symmetric), and consider the polynomial $\rho(z) = z\prod_{i=1}^{t}(1-z/\lambda_i)$. Then $\rho(DBD) = 0$ (since the eigenvalues of DBD are roots of ρ). Since the constant term in $\rho(z)$ is 0 and the linear term is z, this expresses DBD as a linear combination of higher powers of DBD: $DBD = \sum_{s=2}^{t} a_s(DBD)^s$, or

$$(6.27) \qquad B = \sum_{s=2}^{t} a_s(BD^2)^{s-1}B.$$

For all $i,j \in V(H)$, we have

$$(6.28) \qquad \mathsf{hom}_{ij}(P_s^{\bullet\bullet},H) = \left((BD^2)^{s-2}B\right)_{ij}.$$

Let $y = \sum_{s=2}^{t} a_s P_{s+1}^{\bullet\bullet}$, then (6.27) and (6.28) imply that y is a connector, with the structure described in the Supplement.

(b) The existence of a contractor, a 2-labeled quantum graph z satisfying (6.25), follows by Theorem 6.38. We can replace each edge of the contractor by a simple connector, to make it simple.

The construction of a contractor of the special type described in the Supplement takes more work. We consider the values $\beta_{ij} = \mathsf{hom}_{ij}(K_2^{\bullet\bullet},H)$. Replacing K_2 by the m-bond, we get $\mathsf{hom}_{ij}(B_m^{\bullet\bullet},H) = \beta_{ij}^m$. (Note that this relation also holds for $m=0$.) Hence for every real polynomial $p \in \mathbb{R}[x]$ we get a 2-labeled quantum graph y_p that is a quantum bond (a linear combination of bonds), such that $\mathsf{hom}_{ij}(y_p,H) = p(\beta_{ij})$. Since every real function equals to a polynomial of degree less than q^2 on the q^2 values β_{ij}, we get that for every real function g there is quantum graph y_g with constituents from $\mathcal{K}(q^2-1)$, such that $\mathsf{hom}_{ij}(y_g,H) = g(\beta_{ij})$.

Let $\mathbb{K}$ denote the field obtained from $\mathbb{Q}$ by adjoining the node weights of H. We choose the function g above so that the values $\gamma_{ij} = g(\beta_{ij})$ are algebraically independent over $\mathbb{K}$ for different values of β_{ij}.

Next, we form the concatenation $u_g = y_g \circ y_g^*$ (which is a linear combination of graphs in $\mathcal{K}(q^2-1,2)$), and consider the values $\mathsf{hom}_{ij}(u_g,H)$. We claim that a "diagonal" value $\mathsf{hom}_{ii}(u_g,H)$ can never be equal to an "off-diagonal" value $\mathsf{hom}_{jk}(u_g,H)$, $i \neq j$. Indeed, if they are equal, then we have

$$\sum_{r=1}^{q} \alpha_r \gamma_{ir}^2 = \sum_{r=1}^{q} \alpha_r \gamma_{jr}\gamma_{kr}.$$

Since different γ's are algebraically independent over $\mathbb{K}$ (which contains the coefficients α_r), the two sides must be equal formally. In particular, every product $\gamma_{jr}\gamma_{kr}$ must occur on the left side, which implies that $\gamma_{jr} = \gamma_{kr}$. By the definition of γ, this means that $\beta_{jr} = \beta_{kr}$ for every r, and so nodes j and k are twins, which was excluded.

It follows that we can find a polynomial h of degree at most $q^2 - 1$ such that $h(\gamma_{ii}) = 1$ but $h(\gamma_{ij}) = 0$ if $i \neq j$. Hence we get a quantum graph w such that

$$(6.29) \qquad \mathsf{hom}_{ij}(w,H) = \mathbb{1}(i=j).$$

Every constituent of w is the (gluing) product of at most q^2-1 graphs in $\mathcal{K}(q^2-1,2)$, so it is in class $\mathcal{K}(q^2-1,2,q^2-1)$.

Consider the quantum graph $w' = w \circ w$. The constituents of w' are graphs in class $\mathcal{K}(q^2-1,2,q^2-1,2)$. Using (6.29), we get

$$\text{hom}_{ij}(w',H) = \sum_k \alpha_k \text{hom}_{ik}(w,H) \text{hom}_{kj}(w,H) = \alpha_i \mathbb{1}(i=j). \tag{6.30}$$

Expressing the function $i(t) = \frac{1}{t}\mathbb{1}(t \neq 0)$ on the values of $\text{hom}_{ij}(w',H)$ by a polynomial of degree at most q as above, we can construct a quantum graph z in class $\mathcal{K}(q^2-1,2,q^2-1,2,q)$ satisfying (6.25). □

Using the notion of a contractor, we can give the following characterization of homomorphism functions, which does not involve the finiteness of the connection rank.

THEOREM 6.30. *A graph parameter f can be represented in the form $f = \text{hom}(.,H)$ for some weighted graph H if and only if it is multiplicative, reflection positive and has a contractor.*

Proof. The necessity of the conditions follows by Theorem 6.28.

To prove the sufficiency of the conditions, it suffices to prove that there exists a $q > 0$ such that $r(f,k) \leq q^k$ for all $k \geq 0$, and then invoke Theorem 5.54. Note that reflection positivity is used twice: the existence of a contractor does not in itself imply an exponential bound on the connection rank (cf. our introductory example of the chromatic polynomial).

Let g_0 be a contractor for f; we show that $q = f(g_0^2)$ provides the appropriate upper bound on the connection rank. Since f is multiplicative, we already know this for $k = 0$.

We may normalize f so that $f(K_1) = 1$. If the conclusion is false (with $q = f(g_0^2)$) then for some integer $k > 0$ we have (possibly infinite) $r(f,k) > q^k$ and hence for $N = \lfloor q^k + 1 \rfloor$ there are mutually orthogonal unit vectors $q_1, \ldots, q_N$ in the algebra $\mathcal{Q}_k/f$. Let $q_i \otimes q_i$ denote the $(2k)$-labeled quantum graph obtained from $2k$ labeled nodes by attaching a copy of q_i at $\{1,\ldots,k\}$ and another copy of q_i at $\{k+1,\ldots,2k\}$. Let h denote the $(2k)$-labeled quantum graph obtained from $2k$ labeled nodes by attaching a copy of g_0 at $\{i,k+i\}$ for each $i = 1,\ldots,k$. Consider the quantum graph

$$x = \sum_{i=1}^N q_i \otimes q_i - h.$$

By reflection positivity, we have $f(x^2) \geq 0$. But

$$f(x^2) = \sum_{i=1}^N \sum_{j=1}^N \langle q_i \otimes q_i, q_j \otimes q_j \rangle - 2 \sum_{i=1}^N \langle q_i \otimes q_i, h \rangle + \langle h, h \rangle.$$

Here $\langle q_i \otimes q_i, q_j \otimes q_j \rangle = \langle q_i, q_j \rangle^2 = \mathbb{1}(i=j)$, and so

$$\sum_{i=1}^N \sum_{j=1}^N \langle q_i \otimes q_i, q_j \otimes q_j \rangle = N.$$

Furthermore, by the definition of g_0 and h, we have
$$\sum_{i=1}^{N}\langle q_i \otimes q_i, h\rangle = \sum_{i=1}^{N}\langle q_i, q_i\rangle = N.$$
Finally, by the definition of h and the multiplicativity of f, we have $\langle h, h\rangle = f(g_0^2)^k$. Thus $f(x^2) \geq 0$ implies that $N \leq f(g_0^2)^k = q^k$, a contradiction. □

EXERCISE 6.31. Let z be a contractor for a graph parameter f, and let F be a k-labeled graph. Let us delete an edge $1i$ from F, and add the edge $2i$, to obtain another k-labeled graph F'. Prove that $zF \equiv zF' \pmod{f}$.

EXERCISE 6.32. Let z be a contractor for a graph parameter f, and let $\overline{z} = O_2 - z$. Construct a 3-labeled quantum graph x by gluing a copy of z on nodes 1 and 2, a copy of z on nodes 2 and 3, and a copy of $\overline{z}$ on nodes 1 and 3. Prove that $x \equiv 0 \pmod{f}$.

EXERCISE 6.33. Prove that the number of eulerian orientations $\overrightarrow{\text{eul}}$ is contractible, but has no contractor.

EXERCISE 6.34. For every multigraph G, let $\check{G}$ be obtained by repeatedly contracting edges with multiplicity larger than 1 until a simple graph is obtained, and define $f(G) = \text{hom}(K_3, \check{G})$. (a) Prove that $\check{G}$ is uniquely determined. (b) Show that f has a contractor. (c) Prove that f has no connector.

6.4. Algebras for homomorphism functions

We have proved that reflection positive graph parameters with exponentially bounded connection rank are homomorphism functions into weighted graphs. In this section we continue the study of such parameters, now using their representation as homomorphism functions. A key step in the proof was to construct an idempotent basis in the appropriate graph algebra. What is the exact size of this basis? How large graphs we need to represent the basis elements as quantum graphs? Our goal is to prove an exact answer to the first question, and to give good bounds for the second.

6.4.1. The connection rank of homomorphism functions. One can give an exact formula for the connection rank $r(f, k)$ of homomorphism functions $f = \text{hom}(., H)$ (Lovász [2006b]); to state it, we need a definition. Two nodes i and j in the weighted graph H are *twins*, if $\beta_H(i, k) = \beta_H(j, k)$ for every node $k \in V(H)$. (Note that this applies also to $k = i$ and $k = j$. On the other hand, twin nodes may have different nodeweights.) Twin nodes can be merged (adding up their weights) without changing the homomorphism functions $t(., H)$ and $\text{hom}(., H)$.

If H is a weighted graph, then we denote the factor algebra $\mathcal{Q}_k/\text{hom}(., H)$ simply by $\mathcal{Q}_k/H$. The proof of the following characterization of the kernels of the algebras is left to the reader as an exercise:

PROPOSITION 6.35. *If H is a weighted graph and x is a k-labeled quantum graph, then $x \equiv 0 \pmod{\text{hom}(., H)}$ if and only if $\text{hom}_\varphi(x, H) = 0$ for every $\varphi : [k] \to V(H)$.* □

This fact allows us to consider the functions hom_φ as functions on the factor algebra $\mathcal{Q}_k/H$.

THEOREM 6.36. *If H is a twin-free weighted graph, then $r(\text{hom}(., H), k)$ is the number of orbits of the automorphism group of H on the ordered k-tuples of nodes.*

COROLLARY 6.37. *Let H be a weighted graph that has no twins and no proper automorphisms. Then $r(\hom(.,H),k) = \mathsf{v}(H)^k$ for every k.*

Theorem 6.36 has a number of essentially equivalent formulations, which are interesting on their own right. One of these characterizes homomorphism functions of the form $\hom_\varphi(F, H)$.

THEOREM 6.38. *Let H be a twin-free weighted graph and $h : V(H)^k \to \mathbb{R}$. Then there exists a k-labeled quantum graph z such that $\hom_\varphi(z, H) = h(\varphi)$ for every $\varphi \in V(H)^k$ if and only if h is invariant under the automorphisms of H: for every $\varphi \in V(H)^k$ and every automorphism σ of H, $h(\sigma \circ \varphi) = h(\varphi)$.*

Another variant of these theorems gives a combinatorial description of the basic idempotents $p_1, \ldots, p_n$ in the algebra $\mathcal{Q}_k/H$, which played an important role in the proof of the characterization theorem. For every $\varphi \in V(H)^k$, we have
$$\hom_\varphi(p_i, H) = \hom_\varphi(p_i^2, H) = \hom_\varphi(p_i, H)^2,$$
and hence $\hom_\varphi(p_i, H) \in \{0, 1\}$. Furthermore, for $i \neq j$, we have
$$\hom_\varphi(p_i, H)\hom_\varphi(p_j, H) = \hom_\varphi(p_i p_j, H) = 0,$$
and hence the sets $\Phi_i = \{\varphi \in V(H)^k : \hom_\varphi(p_i, H) = 1\}$, which we call *idempotent supports*, are disjoint. Since
$$\sum_\varphi \alpha_\varphi(H)\hom_\varphi(p_i, H) = \hom(p_i, H) = \hom(p_i^2, H) > 0,$$
the idempotent supports are nonempty. We have $\sum_i p_i = O_k$, and hence
$$\sum_{i=1}^n \hom_\varphi(p_i, H) = \hom_\varphi(O_k, H) = 1,$$
and so the idempotent supports form a partition of $V(H)^k$. Since $p_1, \ldots, p_n$ form a basis of $\mathcal{Q}_k/H$, it follows that functions $\varphi \mapsto \hom_\varphi(x, H)$ ($x \in \mathcal{Q}_k$) are exactly those that are constant on every idempotent support.

The following characterization of the partition B_k into idempotent supports is a further equivalent version of Theorem 6.36.

THEOREM 6.39. *For a twin-free weighted graph H, the idempotent supports are exactly the orbits of the automorphism group of H on $V(H)^k$.*

We call two maps $\varphi, \psi \in V(H)^k$ *equivalent* (in notation $\varphi \sim \psi$) if $\hom_\varphi(F, H) = \hom_\psi(F, H)$ for every k-labeled graph F. It follows from the discussion above that this means that they belong to the same idempotent support. With this notion finally we come to the version of these equivalent theorems that we are going to prove first.

THEOREM 6.40. *Two maps $\varphi, \psi \in V(H)^k$ are equivalent if and only if there exists an automorphism σ of H such that $\psi = \sigma \circ \varphi$.*

Proof. The "if" part is trivial, so let do the "only if" part. For any map $\varphi : [k] \to [q]$, let φ' denote its restriction to $[k-1]$. We start with some easy facts about equivalence of maps.

CLAIM 6.41. *If two maps φ, ψ are equivalent, then so are φ' and ψ'.*

Indeed, for any $(k-1)$-labeled graph F, and the graph F_1 obtained from F by adding a new isolated node labeled k, we have $\hom_{\varphi'}(F, H) = \hom_\varphi(F_1, H) = \hom_\psi(F_1, H) = \hom_{\psi'}(F, H)$.

CLAIM 6.42. *Suppose that $\varphi, \psi \in [q]^k$ are equivalent. Then for every $\mu \in [q]^{k+1}$ such that $\varphi = \mu'$ there exists a $\nu \in [q]^{k+1}$ such that $\psi = \nu'$ and μ and ν are equivalent.*

Indeed, let μ belong to the support of the basic idempotent $p \in \mathcal{Q}_{k+1}/H$, then for every $\nu \in V(H)^{k+1}$ we have $\hom_\nu(p, H) = \mathbb{1}(\nu \sim \mu)$. Let p' be obtained by unlabeling $k+1$ in p. Then

$$(6.31) \qquad \hom_\varphi(p', H) = \sum_{\eta: \eta' = \varphi} \alpha_{\eta(k+1)}(H) \hom_\eta(p, H) = \sum_{\substack{\eta: \eta' = \varphi \\ \eta \sim \mu}} \alpha_{\eta(k+1)}(H),$$

and similarly

$$(6.32) \qquad \hom_\psi(p', H) = \sum_{\substack{\eta: \eta' = \psi \\ \eta \sim \mu}} \alpha_{\eta(k+1)}(H).$$

These two numbers are equal since $\varphi \sim \psi$. Since the right side of (6.31) is positive, this implies that the sum in (6.32) is nonempty, and hence there is a map ν such that $\nu' = \psi$ and $\nu \sim \mu$.

The next observation makes use of the twin-free assumption.

CLAIM 6.43. *Every map $\sigma : [q] \to [q]$ such that $\beta_{\sigma(i)\sigma(j)} = \beta_{ij}$ for every $i, j \in [q]$ is bijective.*

To prove this, note that the mapping σ has some power $\gamma = \sigma^s$ that is idempotent. Then for all $i, j \in [q]$, we have $\beta_{ij} = \beta_{\gamma(i)\gamma(j)} = \beta_{\gamma^2(i)\gamma(j)} = \beta_{\gamma(i)j}$, which shows that i and $\gamma(i)$ are twins for all $i \in [q]$. Since H is twin-free, this implies that γ is the identity, and so σ must be bijective.

After this preparation, we prove the theorem for larger and larger classes of mappings.

Case 1: φ is bijective. Then $k = q$. We may assume that the nodes of H are labeled so that φ is the identity, and then we want to prove that ψ (viewed as a map of $V(H)$ into itself) is an automorphism of H. First, we show that

$$(6.33) \qquad \beta_{ij} = \beta_{\psi(i)\psi(j)}$$

for every $i, j \in [k]$. Indeed, let k_{ij} be the k-labeled graph consisting of k nodes and a single edge connecting nodes i and j. Then $\beta_{ij} = \hom_\varphi(k_{ij}, H) = \hom_\psi(k_{ij}, H) = \beta_{\psi(i)\psi(j)}$. It follows by Claim 6.43 that ψ is also bijective.

Second, we show that for every $j \in [k]$,

$$(6.34) \qquad \alpha_j = \alpha_{\psi(j)}.$$

It suffices to prove this for the case $j = k$. For the graph O_{k-1} consisting of $k-1$ isolated labeled nodes,

$$\hom_{\varphi'}(O_{k-1}, H) = \prod_{j=1}^{k-1} \alpha_j,$$

and since ψ is bijective,
$$\mathsf{hom}_{\psi'}(O_{k-1}, H) = \prod_{j=1}^{k-1} \alpha_{\psi(j)} = \frac{1}{\alpha_{\psi(k)}} \prod_{j=1}^{k} \alpha_j.$$
Since $\psi' \sim \varphi'$ by Claim 6.41, equation (6.34) follows.

Case 2: φ is surjective. By permuting the labels $1, \ldots, k$ if necessary, we may assume that $\varphi(1) = 1, \ldots, \varphi(q) = q$. Claim 6.41 implies that the restriction of ψ to $[q]$ is equivalent to the restriction of φ to $[q]$, and so by Case 1, there is an automorphism σ of H such that $\psi(i) = \sigma(i)$ for $i = 1, \ldots, q$.

Consider any $q + 1 \leq j \leq k$, and let $\varphi(j) = r$. We claim that $\psi(j) = \psi(r)$. Indeed, the restriction of φ to $\{1, \ldots, r-1, r+1, \ldots, q, j\}$ is bijective, and equivalent to the restriction of ψ to this set; hence the restriction of ψ to this set must be bijective, which implies that $\psi(j) = \psi(r)$. This implies that for every $1 \leq i \leq k$, $\psi(i) = \sigma(\varphi(i))$ as claimed.

Case 3. φ is arbitrary. We can extend φ to a mapping $\mu \colon [\ell] \to [q]$ ($\ell \geq k$) which is surjective. By Claim 6.42, there is a mapping $\nu \colon [\ell] \to [q]$ extending ψ such that μ and ν are equivalent. Then by Case 2, there is an automorphism σ of G such that $\nu = \sigma \circ \mu$. Restricting this map to $[k]$, the assertion follows. $\square$

The other theorems stated above are easy to derive now.

Proof of Theorems 6.36, 6.38, and 6.39. Theorem 6.39 is trivially equivalent to Theorem 6.40 by the description of idempotent supports. The "only if" part of Theorem 6.38 is also trivial. To prove the "if" part, notice that every function $h \colon V(H)^k \to \mathbb{R}$ invariant under automorphisms can be written as a linear combination of indicator functions of the orbits of the automorphism group. By Theorem 6.39, this means that it is a linear combination of the functions $\mathsf{hom}_\varphi(p_i, H)$, and hence it is of the form $\mathsf{hom}_\varphi(z, H)$ with some $z \in \mathcal{Q}_k$.

Finally, it follows that the number of orbits of the automorphism group of H on $V(H)^k$ is the number of the idempotents p_i, which is $r(f, k)$, which proves Theorem 6.36. $\square$

These results describe an interesting isomorphism between the graph algebras defined by a homomorphism function $\mathsf{hom}(., H)$ and algebras of functions on a twin-free weighted graph H: $\mathcal{Q}_k/H$ is isomorphic to the algebra of those functions $V(H)^k \to \mathbb{R}$ that are invariant under the automorphisms of H. The isomorphism is defined by the map $F \mapsto \mathsf{hom}_{v_1 \ldots v_k}(F, H)$, where $\mathsf{hom}_{v_1 \ldots v_k}(F, H)$ is viewed as a function of $v_1, \ldots, v_k$. This correspondence between quantum graphs and functions on $V(H)$ is useful in constructing quantum graphs with special properties.

As an application of the tools developed in this section, we are now able to prove a weaker version of Theorem 5.33, without the bounds on the size of the graph.

COROLLARY 6.44. *If H_1 and H_2 are twin-free weighted graphs such that $\mathsf{hom}(F, H_1) = \mathsf{hom}(F, H_2)$ holds for all simple graphs F, then $H_1 \cong H_2$.*

Proof. Let H be the graph obtained by taking the disjoint union of H_1 and H_2, creating two new nodes v_1 and v_2, and connecting v_i to all nodes of H_i. Also add loops at both v_i. The new nodes and new edges have weight 1, except for the loops at v_1 and v_2, which get some weight β different from all other edgeweights. This last trick is needed to make sure that the graph H is twin-free.

We claim that for every 1-labeled graph F

(6.35) $$\text{hom}_{v_1}(F, H) = \text{hom}_{v_2}(F, H).$$

Indeed, if F is not connected, then those components not containing the labeled node contribute the same factors to both sides. So it suffices to prove (6.35) when F is connected. Then we have

$$\text{hom}_{v_1}(F, H) = \sum_{v_1 \in S \subseteq V(F)} \beta^{e_F(S)} \text{hom}(F \setminus S, H_1).$$

Indeed, if we fix the set $S = \varphi^{-1}(v_1)$, then the restriction φ' of φ to $V(F) \setminus S$ is a map into $V(H_1)$ (else, the contribution of the map to $\text{hom}(F, H)$ is 0), and the contribution of φ to $\text{hom}_{v_1}(F, H)$ is the product of contributions from the edges induced by S and the contribution of φ' to $\text{hom}(F \setminus S, H)$.

Since $\text{hom}_{v_2}(F, H)$ can be expressed by a similar formula, and the sums on the right hand sides are equal by hypothesis, this proves (6.35).

Now (6.35) can be phrased as the maps $1 \mapsto v_1$ and $1 \mapsto v_2$ are equivalent, and so Theorem 6.40 implies that there is an automorphism of H mapping v_1 to v_2. This automorphism gives an isomorphism between H_1 and H_2. □

6.4.2. The size of basis graphs. Every element of the factor algebra $\mathcal{Q}_k/H$ has many representations as a quantum graph in $\mathcal{Q}_k$. The following theorem asserts that it has a representation whose constituents are (in a sense) small.

THEOREM 6.45. *Let H be a weighted graph with $V(H) = [q]$. The algebra $\mathcal{Q}_k/H$ is generated by simple k-labeled graphs with at most $2(k+q^2)q^6$ nodes, in which the labeled nodes form a stable set.*

Proof. Let $F = (V, E)$ be any k-labeled graph; we construct a simple k-labeled quantum graph x, where each constituent has no more than $2(k+q^2)q^6$ nodes, and $F \equiv x \pmod{H}$.

Let z be a 2-labeled quantum graph such that $\text{hom}_\varphi(z, H) = \mathbb{1}\big(\varphi(1) = \varphi(2)\big)$ for all $\varphi : \{1,2\} \to [q]$. (So z is very similar to a contractor. We have $z^2 = z$, but $z \circ z \neq z$.) We may assume that every constituent of z has at most $2q^6$ nodes (Supplement 6.29 and the Remark after it). Let $\overline{z} = O_2 - z$. Let w be a simple connector; we can assume that w is a linear combination of paths of length at least 3 and at most $q+3$ by Exercise 6.46.

Let us glue a copy of U_2 on every pair of distinct nodes of V; this does not change F. But we can expand every O_2 as $O_2 = z + \overline{z}$, and obtain a representation of F as a sum of quantum graphs x_ℓ ($\ell = 1, \ldots, 2^{\binom{|V|}{2}}$), each of which is obtained from F by gluing either z or $\overline{z}$ on every pair of nodes in V.

Many of these terms will be 0. For any term x_ℓ, let G_ℓ denote the graph on S in which two nodes are connected if and only if they have a copy of z glued on. If (i,j) and (j,k) have z glued on, but (i,k) has $\overline{z}$, then the union of these three is 0 as a quantum graph in $\mathcal{Q}_3$ (this is easy to check; cf. Exercise 6.32). Hence if x_i is a nonzero term, then adjacency must be transitive in G_ℓ, and so G_ℓ consists of disjoint complete graphs. If G_ℓ has more than q components, then any map $V \to V(H)$ will collapse two nodes of V on which a $\overline{z}$ is glued, and hence $x_\ell = 0$. So we are left with only those terms in which G_ℓ consists of at most q disjoint complete graphs. Let $V = V_1 \cup \cdots \cup V_r$ be the partition onto the node sets of these components ($r \leq q$).

Let us select a representative node v_i from every V_i. It is easy to see that deleting the copies of z except those which are attached to a v_i, and also the copies of $\bar{z}$ except those connecting two nodes v_i, does not change x_ℓ.

If $uv \in E$ with $u \in V_i$ and $v \in V_j$ ($i \neq j$), then we can "slide" this edge to $v_i v_j$ without changing x_ℓ (cf. Exercise 6.31). If $u, v \in V_i$, then we replace the edge uv by a simple connector w in which the labeled nodes are at a distance at least 3 (cf. Exercise 6.46), and then slide both attachment nodes to v_i, to get a copy of w' hanging from v_i.

Each constituent of the resulting quantum graph consists of a "core", the set of the nodes v_i and the set of labeled nodes, at most $k+q$ nodes altogether. What is not bounded is the sets of edges connecting a v_i and a v_j, the sets of copies of w' hanging from a v_i, and the copies of z connecting v_i to other nodes in V_i.

However, we can get rid of these unbounded multiplicities. First, a set of q^2 or more parallel edges can be replaced by a linear combination of sets of parallel edges with multiplicity at most $q^2 - 1$, by Exercise 6.6(b). By a similar argument, a set of q or more copies of w' hanging from the same node v_i can be expressed as a linear combination of sets of at most $q-1$ copies. Finally, again by the same argument, a set of q or more copies of z connecting v_i to unlabeled nodes can be expressed as a linear combination of sets of at most $q-1$ copies. So we are left with at most $\binom{q}{2}(q^2-1)$ edges that may be parallel to others, at most $q(q-1)$ hanging copies of w', and at most $k + q(q-1)$ copies of z.

We get rid of the edge multiplicities by replacing each edge between core nodes by a simple connector w.

After that, each constituent will be a simple graph. By the choice of z and q, the number of nodes in each constituent will be bounded by $k + q + (q+2)\binom{q}{2}(q^2 - 1) + (q+2)q(q-1) + (k + q(q-1))(2q^6) < 2(k+q^2)q^6$. $\square$

As an application of the previous theorem, we prove Theorem 5.33 in its full strength, including the bounds on the sizes of the graphs needed.

Proof of Theorem 5.33. Following the proof of Corollary 6.44, we have to show that (6.35) holds. We do know that it holds for every F with at most $2(\mathsf{v}(H_1) + \mathsf{v}(H_2) + 3)^8$ nodes. Since this includes all basis graphs of $\mathcal{Q}_1/H$ by Theorem 6.45, it follows that (6.35) holds for all simple 1-labeled graphs F. From here, the proof is unchanged. $\square$

EXERCISE 6.46. Prove that for every weighted graph H with q nodes and every $t \geq 2$, $\mathsf{hom}(., H)$ has a connector whose constituents are $P_t^{\bullet\bullet}, P_{t+1}^{\bullet\bullet}, \ldots, P_{t+q}^{\bullet\bullet}$.

EXERCISE 6.47. Prove that for every weighted graph H, $\mathsf{hom}(., H)$ has a contractor whose constituents are series-parallel graphs.

6.5. Computing parameters with finite connection rank

As an application of graph algebras, we prove the theorem announced before, namely that graph parameters with finite connection rank can be computed in polynomial time for certain classes of graphs. Of course, the algorithm could be described without reference to algebras, but I feel that the essence is better shown through this tool.

Suppose you want to evaluate a graph parameter on a graph G. There is a cut of size k in a graph, and while you know everything about one side of the cut, you have to pay for information about the other side. How much information do you need about the other side? To avoid the trivial solution "just tell me the value of the parameter, if my side looks like this", let us assume that the information about the other side must be independent from what is on our side, and it is encoded in the form of an m-tuple of real numbers. Furthermore, the answer must be obtained by taking an appropriate linear combination of these m numbers, with coefficients that depend only on the graph on our side.

As an example, let $k = 1$ (so we have a cutset $\{v\}$ with one node), and suppose that we want to compute the number of independent sets in the whole graph. Then we need to know two data about the other side: the number a_0 of independent sets not containing v, and the number a_1 of independent sets containing v. We determine the analogous numbers b_0, b_1 for our side, and then the number of independent sets in the whole graph is $a_0 b_0 + a_1 b_1$.

One reason to be interested in the finiteness of connection rank is the fact that such a graph parameter can be evaluated in polynomial time for graphs with bounded treewidth, based on the idea explained above. The treewidth of a graph G is defined as follows. A *tree-decomposition* of a graph G is determined by a tree T and a family $(G_i)_{i \in V(T)}$ of induced subgraphs of G such that $G = \cup_i G_i$ and whenever i is on the path from j to k in T $(i, j, k \in V(T))$, then $V(H_i) \supset V(H_j) \cap V(H_k)$. The *tree-width* of a graph G is the smallest integer k such that G has a tree-decomposition into subgraphs of size at most $k + 1$.

THEOREM 6.48. *Let f be a graph parameter and $k \geq 0$. If $r(f, k)$ is finite, then f can be computed in polynomial time for graphs with treewidth at most k.*

Proof. We describe a dynamic programing algorithm to compute the parameter. If the connection matrix $M(f, k)$ has finite rank m, then the graph algebra $\mathcal{Q}_m / f$ is finite dimensional for all $m \leq k$ (Exercise 4.30). We need to do some (large, but finite) precomputation.

First, we compute a basis B_m, consisting of (ordinary) m-labeled graphs, for each of the algebras $\mathcal{Q}_m / f$. We also express the product of any two basis elements in this basis (the "Schur constants").

Second, let H be an l-labeled graph with at most $k + 1$ nodes ($l \leq k$), and for every ordered subset $S \subseteq V(H)$ with $|S| \leq k$, let a basis graph $F_S \in B_{|S|}$ be assigned. Let us glue the labeled nodes of F_S onto the set S (erasing the labels in F_S at the same time), to get an l-labeled graph H'. We compute the representation of H' in the basis B_l. We do this precomputation for every H and every assignment of basis graphs F_S.

Third, we compute the values $f(G)$ for every $G \in B_0$.

Let G be a graph with treewidth at most k, and let $(G_i)_{i \in V(T)}$ be a tree-decomposition of the graph with $\mathsf{v}(G_i) \leq k + 1$. Designate any leaf r of T as its root, and for $i \in V(T) \setminus \{r\}$, let i' denote its parent.

For every node $i \in V(T) \setminus \{r\}$, the set $S_i = V(G_i) \cap V(G_{i'})$ is a cutset in G with $k_i \leq k$ nodes. Let F_i denote the union of all graphs G_j where j is a descendant of i (including i), in which the k_i nodes of S_i are labeled. The algorithm will consist of expressing every F_i in the basis B_{k_i}, starting from the leaves and working our way up to the root.

Suppose that such an expression has been computed for every proper descendant of i. The k_i-labeled graph F_i is obtained from G_i by attaching different branches F_j at the sets S_j. We already know how to express each F_j in the basis B_{k_j}; let us substitute this expression for F_j, to get a representation of F_i as a linear combination of graphs, each of which consists of G_i with some number of basis graphs attached at various subsets $S \subseteq V(G_i)$ with $|S| \leq k$. If two or more basis graphs are attached at the same set S, we can replace them by one, since we have precomputed products of basis graphs. But then we have a linear combination of k_i-labeled graphs of the type we have already expressed in the basis B_{k_i}.

When we get to the root, we consider it 0-labeled, and get an expression for G in the basis B_0, which yields the value $f(G)$. □

6.6. The polynomial method

In this section we describe a method of proving representation theorems for graph parameters, which depends on commutative algebra (properties of multivariate polynomials). This method was developed by B. Szegedy [2007] for the proof of the characterization of graph parameters that are partition functions of edge coloring models (we will describe this result without proof in Chapter 23.2). Here we use an adaptation of this method to prove Theorem 5.57, due to Schrijver [2009].

The basic idea is to treat the edge weights of the target graph H as variables. Then homomorphism numbers into H will be polynomials in these variables. One treats this as a polynomial-valued graph parameter, works out the corresponding graph algebras, and then proves that one can find a substitution for the variables that reproduces the given graph parameter.

Let H be a weighted graph on $[q]$, in which the nodeweights are 1, and the edgeweights are different variables x_{ij} (where $x_{ij} = x_{ji}$). It will be convenient to arrange these variables into a symmetric $q \times q$ matrix X. Every substitution of complex numbers for the x_{ij} gives a complex valued homomorphism function into an edge-weighted graph on $[q]$, and vice versa. The homomorphism number

$$\text{hom}(G, H) = \sum_{\varphi: V(G) \to [q]} \prod_{ij \in E(G)} x_{\varphi(i)\varphi(j)}$$

is a polynomial in the $\binom{q}{2} + q$ variables x_{ij}, which we denote by $\text{hom}(G, X)$. We define the polynomial $\text{inj}(G, X)$ analogously. Clearly $\text{hom}(K_0, X) = 1$, $\text{hom}(K_1, X) = q$, and $\text{hom}(G_1 G_2, X) = \text{hom}(G_1, X)\text{hom}(G_2, X)$. In particular, $\text{hom}(GK_1, X) = q\text{hom}(G, X)$ for every graph G. We extend the definition linearly to quantum graphs, to get polynomials $\text{hom}(g, X)$, $\text{inj}(g, X) \in \mathbb{C}[X]$ associated with every quantum graph g.

We start with describing the range and kernel of the map $g \mapsto \text{hom}(g, X)$ (where q is fixed). Clearly, $\text{hom}(g, X)$ is invariant under the permutations of $[q]$. To be more precise, if $\sigma \in S_q$ and we define $X^\sigma = (x_{\sigma(i),\sigma(j)} : i, j \in [q])$, then trivially $\text{hom}(., X) = \text{hom}(., X^\sigma)$. Let $\mathbb{C}[X]^{S_q}$ denote the space of polynomials in $\mathbb{C}[X]$ that are invariant under S_q in this sense.

LEMMA 6.49. *The polynomials $\text{hom}(g, X)$, where $g \in \mathcal{Q}_0$, form the space $\mathbb{C}[X]^{S_q}$.*

Proof. Let $X^a = \prod_{i \leq j} x_{ij}^{a_{ij}}$ be any monomial, and let G denote the multigraph on $[q]$ in which nodes i and j are connected by a_{ij} edges. Then $\text{inj}(G, X) =$

$\sum_{\sigma \in S_q}(X^\sigma)^a$. Since every polynomial in $\mathbb{C}[X]^{S_q}$ can be written as a linear combination (with constant coefficients) of such special polynomials, it follows that every polynomial in $\mathbb{C}[X]^{S_q}$ can be written as $\mathsf{inj}(g, X)$ for some quantum graph g. By identity (5.18) (which remains valid if the graph G is replaced by the matrix X), this implies the Lemma. □

Next, we describe quantum graphs g with $\mathsf{hom}(g, X) = 0$ (identically 0 as a polynomial in the entries of X). Note that if we remove an isolated node to a constituent of any quantum graph g, and multiply its coefficient by q, then we get a quantum graph g' such that $\mathsf{hom}(g, X) = \mathsf{hom}(g', X)$. Let us call the repeated application of this operation an *isolate removal*.

LEMMA 6.50. *A quantum graph g satisfies $\mathsf{hom}(g, X) = 0$ if and only if there is a quantum graph h in which all constituents have more than q nodes such that removing isolates from g we obtain Mh.*

Proof. If $g = Mh$, where all constituents of h have more than q nodes, then $\mathsf{hom}(g, X) = \mathsf{inj}(h, X) = 0$. Isolate removal does not change the value of $\mathsf{hom}(g, X)$.

Conversely, suppose that $\mathsf{hom}(g, X) = 0$. We may assume that the constituents of g have no isolated nodes. We have $\mathsf{inj}(Zg, X) = \mathsf{hom}(g, X) = 0$. If Zg has a constituent with at most q nodes, then this produces in $\mathsf{inj}(Zg, X)$ a term which does not cancel (here we use that the constituent has no isolated nodes). So all constituents of Zg have more than q nodes, and we can take $h = Zg$. □

Now we are ready to prove Theorems 5.56 and 5.57.

Proof of Theorem 5.56. Multiplicativity implies that $f(K_0) = 1$ (since f is not identically 0), and $f(GK_1) = qf(G)$.

We want to prove that $f = \mathsf{hom}(., A)$ for an appropriate symmetric complex matrix A; in other words, we want to show that the polynomial equations

(6.36) $\quad \mathsf{hom}(G, X) - f(G) = 0 \quad$ (for all looped multigraphs G)

are solvable for the variables x_{ij} ($1 \leq i, j \leq q$) over the complex numbers. We are going to use Hilbert's Nullstellensatz for this, but we need some preparation. We begin with relating the kernel of the map $\mathsf{hom}(., X)$ to the kernel of f.

CLAIM 6.51. *If $\mathsf{hom}(g, X) = c$ (a constant polynomial) for some quantum graph g, then $f(g) = c$.*

First we consider the case when $c = 0$. We may assume that g has no isolated nodes, since isolate removal does not change the values $\mathsf{hom}(g, X)$ and $f(g)$. By Lemma 6.50, $g = Mh$ for some quantum graph h in which all constituents have more than q nodes. But then $f(Mh) = 0$ by the hypothesis of the Theorem.

The case of general constant c follows easily: we have $\mathsf{hom}(g - cK_0, X) = \mathsf{hom}(g, X) - c = 0$, and hence $f(g) = f(g - cK_0) + c = c$.

CLAIM 6.52. *The ideal generated by the polynomials $\mathsf{hom}(g, X)$ with $f(g) = 0$ does not contain the constant polynomial 1.*

Suppose that we have a representation

$$1 = \sum_{i=1}^{N} p_i(X) \mathsf{hom}(g_i, X),$$

where $f(g_i) = 0$, and the p_i are arbitrary polynomials in $\mathbb{C}[X]$. Let us apply a permutation $\sigma \in S_q$ to the variables, and sum over all σ. We get:

$$q! = \sum_{i=1}^{N} \sum_{\sigma \in S_q} p_i(X^\sigma) \hom(g_i, X^\sigma) = \sum_{i=1}^{N} \Big(\sum_{\sigma \in S_q} p_i(X^\sigma)\Big) \hom(g_i, X).$$

The expression in the large parenthesis is a polynomial in $\mathbb{C}[X]^{S_q}$, and hence by Lemma 6.49, it is a polynomial of the form $\hom(h_i, X)$. Hence we get

$$q! = \sum_{i=1}^{N} \hom(h_i, X) \hom(g_i, X) = \hom\Big(\sum_{i=1}^{N} h_i g_i, X\Big).$$

By Claim 6.51, this implies that

$$f\Big(\sum_{i=1}^{N} h_i g_i\Big) = q!.$$

But we have, using the multiplicativity of f,

$$f\Big(\sum_{i=1}^{N} h_i g_i\Big) = \sum_{i=1}^{N} f(h_i g_i) = \sum_{i=1}^{N} f(h_i) f(g_i) = 0,$$

a contradiction. So Claim 6.52 is proved.

Now it is easy to finish the proof of the theorem. Claim 6.52 and the Nullstellensatz imply that there are complex numbers a_{ij} such that $a_{ij} = a_{ji}$ ($1 \leq i, j \leq q$), and $\hom(g, A) = 0$ for every quantum graph g for which $f(g) = 0$ (where A is the matrix with entries a_{ij}). Applying this to the quantum graph $G - f(G)K_0$ (where G is an arbitrary multigraph), we get that $\hom(G - f(G)K_0, A) = 0$, and hence $f(G) = \hom(G, A)$. □

Proof of Theorem 5.57. We will apply Theorem 5.56, but first we need to show a couple of properties of f following from reflection positivity.

CLAIM 6.53. *$f(K_1)$ is a nonnegative integer.*

Let $q = f(K_1)$. For $k \geq 1$, and for a partition $P = (S_1, \ldots, S_m)$ of $[k]$, let U_P denote the k-multilabeled graph on $[m]$ with no edges, where node i is labeled by the elements of S_i. Consider the submatrix M of $M^{\text{mult}}(f, k)$ formed by those rows and columns indexed by the graphs U_P.

Let $h_k = \sum_P \mu_P U_P$. Then using identity (A.5) for the Möbius function, we get

$$(6.37) \quad \langle h_k, h_k \rangle = \sum_{P,Q} \mu_P \mu_Q f(U_{P \vee Q}) = \sum_{P,Q} \mu_P \mu_Q q^{|P \vee Q|} = q(q-1) \cdots (q-k+1).$$

Since f is reflection positive, this value must be nonnegative for every k, which implies that q is a nonnegative integer.

CLAIM 6.54. *If G is a multigraph with $k = \mathsf{v}(G) > q$, then $f(MG) = 0$.*

Let us label the nodes of G by $[k]$, to get a k-labeled graph. Then $MG = [\![h_k G]\!]$, and so $f(MG) = \langle h_k, G \rangle$. Equation (6.37) implies that $\langle h_k, h_k \rangle = 0$, which (using reflection positivity again) implies that $\langle h_k, G \rangle = 0$, which proves the Claim.

So Theorem 5.56 applies, and we get that there exists a symmetric matrix $A \in \mathbb{C}^{q \times q}$ for which $f = \hom(., A)$. To complete the proof, we have to show:

CLAIM 6.55. *The matrix A is real.*

This does not follow just from the assumption that f is real (see Exercise 6.56); we have to use reflection positivity again. Suppose that A has an entry a_{uv} which is not real. There is a polynomial $p = \sum_{j=0}^{N} c_j z^j \in \mathbb{C}[z]$ such that

$$p(z) = \begin{cases} i & \text{if } z = a_{uv}, \\ -i & \text{if } z = \overline{a}_{uv}, \\ 0 & \text{if } z \in \{a_{st}, \overline{a}_{st}\} \text{ for some entry } a_{st} \neq a_{uv}, \overline{a}_{uv}. \end{cases}$$

(so p takes pure imaginary values on the entries of A and on their conjugates). This polynomial may have complex coefficients, but it is easy to see that its complex conjugate $\overline{p}(x)$ satisfies the same conditions, and hence, replacing p by $(p + \overline{p})/2$ if necessary, we may assume that p has real coefficients.

Consider the 2-labeled quantum graph $g = \sum_j c_j B_j^{\bullet\bullet}$. We have

$$\langle g, g \rangle = \sum_{k,j} c_k c_j \, \mathsf{hom}(B_{k+j}, A) = \sum_{k,j} c_k c_j \sum_{u,v} a_{uv}^{k+j} = \sum_{u,v} p(a_{uv})^2 < 0,$$

which contradicts the assumption that f is reflection positive. □

EXERCISE 6.56. Show by an example that $\mathsf{hom}(G, Z)$ can be real for every multigraph G for a non-real matrix Z.

Part 3

Limits of dense graph sequences

CHAPTER 7

Kernels and graphons

The aim of this Chapter is to introduce certain analytic objects, which will serve as limit objects for graph sequences in the dense case. In the Introduction (Section 1.5.3) we already gave an informal description of how these graphons enter the picture as limit objects; however, for the next few chapters we will not talk about graph sequences, but we treat graphons as generalizations of graphs, to which many graph-theoretic definitions and results can be extended. Quite often, the formulation and even the proof of these more general facts are easier in this analytic setting. We will define the cut norm and cut-distance of these objects, state and prove regularity lemmas for them, and prove basic properties of sampling from them. These results will enable us to show that these are just the right objects to represent the limits of convergent dense graph sequences.

7.1. Kernels, graphons and stepfunctions

Let $\mathcal{W}$ denote the space of all bounded symmetric measurable functions $W : [0,1]^2 \to \mathbb{R}$. The elements of $\mathcal{W}$ will be called *kernels* (the name refers to the fact that they give rise to kernel operators on function spaces on $[0,1]$; we will return to this connection in Section 7.5). Let $\mathcal{W}_0$ denote the set of all kernels $W \in \mathcal{W}$ such that $0 \leq W \leq 1$. The elements of $\mathcal{W}_0$ will be called *graphons* (the name comes from the contraction of graph-function). Sometimes we will also need to consider the set of all functions $W \in \mathcal{W}$ such that $-1 \leq W \leq 1$; this will be denoted by $\mathcal{W}_1$.

As usual, we will not distinguish functions that are almost everywhere equal (... most of the time). Then the space $\mathcal{W}$ is just the space of symmetric functions in $L_\infty([0,1]^2)$, which we could identify with the space $L_\infty(T)$, where T is the triangle $\{(x,y) \in [0,1]^2 : x \leq y\}$. We introduced a separate notation for it because we want to consider a number of different norms on $\mathcal{W}$, of which the L_∞ norm will play a relatively minor role.

A graphon whose values are 0 and 1 can be considered as a graph on node set $[0,1]$. In this case, we can talk about its subgraphs, induced subgraphs, complement, and so on. Such 0-1 valued graphons will come up in our discussions repeatedly; however, they would not be sufficient for our main goal, namely, describing limit objects for convergent graph sequences.

Kernels generalize weighted graphs in the following sense. A function $W \in \mathcal{W}$ is called a *stepfunction*, if there is a partition $S_1 \cup \cdots \cup S_k$ of $[0,1]$ into measurable sets such that W is constant on every product set $S_i \times S_j$. The sets S_i are the *steps* of W. For every weighted graph H (on node set $V(H) = [n]$), we define a stepfunction $W_H \in \mathcal{W}$ as follows: Split $[0,1]$ into n intervals $J_1, \ldots, J_n$ of length $\lambda(J_i) = \alpha_i/\alpha_H$, and for $x \in J_i$ and $y \in J_j$, let $W_H(x,y) = \beta_{ij}(H)$. Note that the function W_H depends on how the nodes of H are labeled.

Conversely, every stepfunction U corresponds to a weighted graph: if $S_1,\ldots,S_k$ are its steps, then the graph is defined on $[k]$, and the edge ij has weight $U(x,y)$, where $x \in S_i$ and $y \in S_j$.

If H is a weighted graph with nodeweights 1 and weighted adjacency matrix A, then we write $W_A = W_H$.

If the edgeweights of H are from the interval $[0,1]$, then W_H is a graphon. In particular, for every simple (unweighted) graph G, W_G is a 0-1 valued graphon (recall Figure 1.3). In this sense, simple graphs can be considered as special 0-1-valued graphons.

This correspondence with simple graphs suggests how to extend some basic quantities associated with graphs to kernels (or at least to graphons). Most important of these is the (normalized) *degree function*

$$(7.1) \qquad d_W(x) = \int_0^1 W(x,y)\,dy.$$

(If the graphon is associated with a simple graph G, this corresponds to the scaled degree $d_G(x)/\mathsf{v}(G)$.) We will see more such quantities in the next sections.

Instead of the interval $[0,1]$, we can consider any probability space $(\Omega, \mathcal{A}, \pi)$ with a symmetric measurable function $W: \Omega \times \Omega \to [0,1]$. This would not provide substantially greater generality, but it is sometimes useful to represent graphons by probability spaces other than $[0,1]$. We'll discuss this in detail in Chapter 13, but will use this different way of representing a graphon throughout.

Graphons will come up in several quite different forms in our discussions. In Theorem 11.52 we will collect the many disguises in which they occur.

7.2. Generalizing homomorphisms

Homomorphism densities in graphs extend to homomorphism densities in graphons and, more generally, in kernels. For every $W \in \mathcal{W}$ and multigraph $F = (V, E)$ (without loops), define

$$t(F,W) = \int_{[0,1]^V} \prod_{ij \in E} W(x_i, x_j) \prod_{i \in V} dx_i$$

We can think of the interval $[0,1]$ as the set of nodes, and of the value $W(x,y)$ as the weight of the edge xy. Then the formula above is an infinite analogue of weighted homomorphism numbers. We get weighted graph homomorphisms as a special case when W is a stepfunction: For every unweighted multigraph F and weighted graph G,

$$(7.2) \qquad t(F,G) = t(F,W_G).$$

Of the two modified versions of homomorphism densities (5.12) and (5.13), the notion of the injective density t_{inj} has no significance in this context, since a random assignment $i \mapsto x_i$ ($i \in V(F)$, $x_i \in [0,1]$) is injective with probability 1. In other words, $t_{\text{inj}}(F,W) = t(F,W)$ for any kernel W and any graph F. But the induced subgraph density is worth defining, and in fact it can be expressed by a rather simple integral:

$$(7.3) \quad t_{\text{ind}}(F, W) = \int_{[0,1]^V} \prod_{ij \in E} W(x_i, x_j) \prod_{ij \in \binom{V}{2} \setminus E} (1 - W(x_i, x_j)) \prod_{i \in V} dx_i.$$

We have an analogue of the inclusion-exclusion formula (5.20), which follows by expanding the parentheses in the integrand (7.3):

$$(7.4) \quad t_{\text{ind}}(F, W) = \sum_{\substack{F' \supseteq F \\ V(F') = V(F)}} (-1)^{e(F') - e(F)} t(F', W) = t^{\uparrow}(F, W).$$

We should point out that $t_{\text{inj}}(F, W_H) \neq t_{\text{inj}}(F, H)$ and $t_{\text{ind}}(F, W_H) \neq t_{\text{ind}}(F, H)$ in general. We have seen that $t_{\text{inj}}(F, W_H) = t(F, W_H) = t(F, H)$. For the induced density, $t_{\text{ind}}(F, W_H)$ has a combinatorial meaning if H is a looped-simple graph: it is the probability that a random map $V(F) \to V(H)$ (not necessarily injective) preserves both adjacency and nonadjacency.

Many other basic properties of homomorphism numbers extend to graphons, often to kernels, in a straightforward way, like (5.19) generalizes to

$$(7.5) \quad t(F, W) = \sum_{F' \supseteq F} t_{\text{ind}}(F', W),$$

and (5.28) generalizes to the identity

$$(7.6) \quad t(F_1 F_2, W) = t(F_1, W) t(F_2, W).$$

We can also generalize homomorphism numbers from partially labeled graphs. Let $F = (V, E)$ be a k-labeled multigraph. Let $V_0 = V \setminus [k]$ be the set of unlabeled nodes. For $W \in \mathcal{W}$ and $x_1, \ldots, x_k \in [0, 1]$, we define

$$t_{x_1, \ldots, x_k}(F, W) = \int_{x \in [0,1]^{V_0}} \prod_{ij \in E} W(x_i, x_j) \prod_{i \in V_0} dx_i$$

(this is a function of $x_1, \ldots, x_k$). In particular, we have $t_x(K_2^\bullet, W) = d_W(x)$. It is often convenient to use the notation $t_\mathbf{x}$, where $\mathbf{x} = (x_1, \ldots, x_k)$. The product of two k-labeled graphs F_1 and F_2 satisfies

$$(7.7) \quad t_\mathbf{x}(F_1 F_2, W) = t_\mathbf{x}(F_1, W) t_\mathbf{x}(F_2, W)$$

If F' arises from F by unlabeling node k (say), then

$$(7.8) \quad t_{x_1, \ldots, x_{k-1}}(F', W) = \int_{[0,1]} t_{x_1, \ldots, x_k}(F, W) \, dx_k.$$

By repeated application of this equation, we get that if F is a k-labeled multigraph, then

$$(7.9) \quad t([\![F]\!], W) = \int_{[0,1]^k} t_\mathbf{x}(F, W) \, d\mathbf{x}.$$

Further versions of homomorphism densities treated before can be extended to homomorphism densities in kernels in a straightforward way. Homomorphism densities of quantum graphs in kernels are defined simply by linearity. Densities of signed graphs can be defined generalizing expression 7.3 for the induced subgraph densities. Explicitly, let $F = (V, E^+, E^-)$ be a signed graph and $W \in \mathcal{W}$, then we define

$$(7.10) \qquad t(F, W) = \int_{[0,1]^V} \prod_{ij \in E_+} W(x_i, x_j) \prod_{ij \in E_-} \bigl(1 - W(x_i, x_j)\bigr) \prod_{i \in V} dx_i.$$

From this definition, it follows that if W is a graphon, then $0 \leq t(F, W) \leq 1$ for every signed graph F. We can also express $t(F, W)$ as

$$(7.11) \qquad t(F, W) = \sum_{Y \subseteq E_-} (-1)^{|Y|} t\bigl((V, E_+ \cup Y), W\bigr).$$

This shows that we can still identify a signed graph $F = (V, E^+, E^-)$ with the quantum graph $\sum_{Y \subseteq E_-} (-1)^{|Y|}(V, E_+ \cup Y)$.

If all edges are signed "+", then $t(F, W)$ is the same as for unsigned graphs. If $\widehat{F}$ is the signed complete graph, obtained from an unsigned simple graph F on the same node set, in which the edges of F are signed positive and the edges of the complement are signed negative, then we get the following identity, equivalent to (7.4):

$$(7.12) \qquad t(\widehat{F}, W) = t_{\mathsf{ind}}(F, W).$$

We define the induced density $t_{\mathsf{ind},\mathbf{x}}(F, W)$ of a k-labeled graph F, and the density of a k-labeled signed graph or quantum graph in the obvious way.

The following proposition states some main properties of subgraph densities in kernels.

PROPOSITION 7.1. *The graph parameter $t(., W)$ is multiplicative and reflection positive for every kernel $W \in \mathcal{W}$. The corresponding simple graph parameter is also multiplicative, and it is reflection positive if $W \in \mathcal{W}_0$.*

Proof. The second assertion is more difficult to prove, and we describe the proof in this case only. Multiplicativity is trivial. To prove that $t(., W)$ is reflection positive, consider any finite set $F_1, \ldots, F_m$ of k-labeled graphs, and real numbers $y_1, \ldots, y_m$. We want to prove that

$$\sum_{p,q=1}^{m} t(\llbracket F_p F_q \rrbracket, W) y_p y_q \geq 0.$$

For every k-labeled graph F with node set $[n]$, let F' denote the subgraph of F induced by the labeled nodes, and F'' denote the graph obtained from F by deleting the edges spanned by the labeled nodes. Then we have

$$(7.13) \qquad \sum_{p,q=1}^{m} y_p y_q t(\llbracket F_p F_q \rrbracket, W)$$

$$= \int_{[0,1]^k} \sum_{p,q=1}^{m} y_p y_q t_{\mathbf{x}}(F_p'', W) t_{\mathbf{x}}(F_p'', W) t_{\mathbf{x}}(F_p' \cup F_q', W) \, d\mathbf{x}.$$

We substitute $t_{\mathbf{x}}(F_p' \cup F_q', W) = \sum_H t_{\mathsf{ind},\mathbf{x}}(H, W)$, where the summation extends over all graphs on $[k]$ containing $F_p' \cup F_q'$ as a subgraph. Interchanging summation, we get

(7.14) $$\sum_{p,q=1}^{m} y_p y_q t(\llbracket F_p F_q \rrbracket, W)$$
$$= \int_{[0,1]^k} \sum_{H} \sum_{F_p, F_q \subseteq H} y_p y_q t_{\mathbf{x}}(F_p'', W) t_{\mathbf{x}}(F_p'', W) t_{\mathsf{ind}, \mathbf{x}}(H, W) \, d\mathbf{x}$$

For a fixed H, the integrand can be written as

$$\sum_{F_p, F_q \subseteq H} y_p y_q t_{\mathbf{x}}(F_p'', W) t_{\mathbf{x}}(F_p'', W) t_{\mathsf{ind}, \mathbf{x}}(H, W) = \Big(\sum_{F_p \subseteq H} y_p t_{\mathbf{x}}(F_p'', W) \Big)^2 t_{\mathsf{ind}, \mathbf{x}}(H, W),$$

which is nonnegative by the assumption that $0 \leq W \leq 1$. $\square$

We will see (Theorem 11.52) that multiplicativity and reflection positivity, together with the trivial condition that $t(K_1) = 1$, characterize simple graph parameters of the form $t(F, W)$.

For a general graphon W, the graph parameter $t(., W)$ cannot be represented as a homomorphism number into a (finite) weighted graph: it is multiplicative and reflection positive, but it may have infinite connection rank. We will see (Corollary 13.48) that $t(., W)$ has finite connection rank if and only if W is equal to a stepfunction almost everywhere.

The multigraph parameter $t(., W)$ is contractible, but has no contractor. This will follow from Theorem 6.30 together with the uniqueness of representation of a parameter in the form $t(., W)$ (Theorem 13.10).

EXAMPLE 7.2 (**Eulerian orientations revisited**). We have seen that the number of eulerian orientations $\mathsf{eul}(G)$ is not a homomorphism function. However, it can be expressed as a homomorphism density in a kernel:

(7.15) $$\overrightarrow{\mathsf{eul}}(F) = t\big(F, 2\cos(2\pi(x-y))\big).$$

Indeed, we can write

$$2\cos(2\pi(x-y)) = e^{2\pi i(x-y)} + e^{2\pi i(y-x)},$$

so if we expand the product

$$\prod_{uv \in E(F)} (e^{2\pi i(x_u - x_v)} + e^{2\pi i(x_v - x_u)}),$$

then every term corresponds to an orientation $\overrightarrow{F}$ of F, where selecting $e^{2\pi i(x_v - x_u)}$ corresponds to orienting the edge uv from u to v. Thus

$$\prod_{uv \in E(F)} 2\cos(2\pi(x_u - x_v)) = \sum_{\overrightarrow{F}} \prod_{uv \in E(\overrightarrow{F})} e^{2\pi i(x_v - x_u)}$$
$$= \sum_{\overrightarrow{F}} \prod_{u \in V(F)} e^{2\pi i(d^+_{\overrightarrow{F}}(u) - d^-_{\overrightarrow{F}}(u))x_u}.$$

If we integrate over all the x_u, every term cancels in which the orientation is not eulerian, i.e., where any of the nodes u has $d^+_{\overrightarrow{F}}(u) - d^-_{\overrightarrow{F}}(u) \neq 0$. Those terms corresponding to eulerian orientations contribute 1. So the sum counts eulerian orientations. ♦

REMARK 7.3. There is probably no good way to define homomorphism numbers from graphons into graphs or into other graphons. The parameters related to such homomorphisms that extend naturally to graphons are defined by maximization, like the normalized maximum cut, and more generally, restricted maximum multicuts. We will discuss these in Chapter 12.

We can generalize the functional $t(F, W)$ further (believe me, not for the sake of generality). Let $\mathcal{A}$ be a set of kernels. An $\mathcal{A}$-*decorated graph* is a finite simple graph $F = (V, E)$ in which every edge $e \in E$ is labeled by a function $W_e \in \mathcal{A}$. We write $w = (W_e : e \in E)$. For every $\mathcal{W}$-decorated graph (F, w) we define

$$(7.16) \qquad t(F, w) = \int_{[0,1]^V} \prod_{ij \in E} W_{ij}(x_i, x_j) \prod_{i \in V} dx_i.$$

For a fixed graph F, the functional $t(F, w)$ is linear in every edge decoration W_e. So it may be considered as linear functional on the tensor product $\mathcal{W} \otimes \cdots \otimes \mathcal{W}$ (one factor for every edge of F), or equivalently, as a tensor on $\mathcal{W}$ with $\mathsf{e}(F)$ slots.

This definition contains some of the previous variations on homomorphism numbers, and it can be used to express homomorphism densities in sums of kernels.

EXAMPLE 7.4. Let $F = (V, E^+, E^-)$ be a signed graph and $W \in \mathcal{W}_0$. Let us decorate each edge in E^+ by W, and each edge in E^- by $1 - W$. Let F_0 be the unsigned version of F. Then for the $\mathcal{W}$-decorated graph (F, w) obtained this way, we have $t(F_0, w) = t(F, W)$. ♦

EXAMPLE 7.5. For $W_1, \ldots, W_k \in \mathcal{W}$, we have

$$t(F, W_1 + \cdots + W_k) = \sum_w t(F, w),$$

where w ranges over all $\{W_1, \ldots, W_k\}$-decorations of F. ♦

EXERCISE 7.6. Let F and G be two simple graphs, and let W be a graphon such that $t(F, G) > 0$ and $t(G, W) > 0$. Prove that $t(F, W) > 0$. [Hint: Use the Lebesgue Density Theorem.]

EXERCISE 7.7. Prove that for any two simple graphs F and G with $\mathsf{v}(F) \leq \mathsf{v}(G)$ we have

$$\left| t_{\mathsf{ind}}(F, G) - t_{\mathsf{ind}}(F, W_G) \right| \leq \frac{\binom{\mathsf{v}(F)}{2}}{\mathsf{v}(G)}.$$

EXERCISE 7.8. Let us generalize the construction of graph integrals by adding "nodeweights": for every graph F and bounded measurable functions $\alpha : [0, 1] \to \mathbb{R}$ and $W : [0, 1]^2 \to \mathbb{R}$ (where W is symmetric), we define

$$t(F, \alpha, W) = \int_{[0,1]^{V(F)}} \prod_{i \in V(F)} \alpha(x_i) \prod_{ij \in E(F)} W(x_i, x_j) \, dx.$$

Show that if we require that $\alpha \geq 0$, then $t(F, \alpha, W)$ can be expressed as $c^{\mathsf{v}(F)} t(F, U)$ with some $c \geq 0$ and $U : [0, 1]^2 \to \mathbb{R}$, where c and U depend on α and W, but not on F.

EXERCISE 7.9. Prove that the number of perfect matchings in a graph $G = (V, E)$ can be expressed as $t(G, e^{-2\pi i x}, 1 + e^{2\pi i (x+y)})$.

7.3. Weak isomorphism I

One complication caused by moving to infinite objects is that isomorphism does not have an obvious (and unique) definition any more. We can of course talk about two kernels U, W to be equal as functions, but this is not very useful. More in the spirit of functional analysis, we will talk about the two kernels to be equal *almost everywhere*, i.e., $W(x,y) = U(x,y)$ for almost all $(x,y) \in [0,1]^2$ (with respect to the Lebesgue measure).

This notion, however, is not what we mean by two kernels being "essentially the same": it corresponds to the equality of labeled graphs, not to isomorphism of unlabeled graphs, which involves finding the right bijection between the node sets. In terms of graphons (or kernels), we can define this as follows: two kernels $U, W \in \mathcal{W}$ are *isomorphic up to a null set* if there is an invertible measure preserving map $\varphi : [0,1] \to [0,1]$ such that $U(\varphi(x), \varphi(y)) = W(x,y)$ almost everywhere. (See Appendix A.3 and the book of Sinai [1976] for the basics of measure preserving maps.) Since the inverse of an invertible measure preserving map $\varphi : [0,1] \to [0,1]$ is also measure preserving, isomorphism up to a null set is an equivalence relation.

However, there is a weaker notion of isomorphism, which will be more important for us. The motivation for this notion is the fact that a measure preserving map need not be invertible. Let $W \in \mathcal{W}$ and let $\varphi : [0,1] \to [0,1]$ be a measure preserving map. We define a kernel W^φ by

$$W^\varphi(x,y) = W(\varphi(x), \varphi(y)).$$

From the point of view of using these functions as continuous analogues of graphs, the functions W and W^φ are not essentially different. For example, we have the following important fact:

PROPOSITION 7.10. *Let $W \in \mathcal{W}$ and let $\varphi : [0,1] \to [0,1]$ be a measure preserving map. Then for every multigraph $F = (V, E)$, we have $t(F, W^\varphi) = t(F, W)$.*

Proof. This follows from the fact that $(x_1, \ldots, x_n) \mapsto (\varphi(x_1), \ldots, \varphi(x_n))$ is a measure preserving map $[0,1]^n \to [0,1]^n$, and hence for every integrable function $f : [0,1]^n \to \mathbb{R}$ we have

$$\int_{[0,1]^n} f(\varphi(x_1), \ldots, \varphi(x_n)) \, dx_1 \ldots dx_n = \int_{[0,1]^n} f(x_1, \ldots, x_n) \, dx_1 \ldots dx_n$$

by (A.16) in the Appendix. Applying this equation to the function $f(x_1, \ldots, x_n) = \prod_{ij \in E} W(x_i, x_j)$, we get the assertion. $\square$

We want to say that W and W^φ are "weakly isomorphic". One has to be a little careful though, because measure preserving maps are not necessarily invertible, and so the relationship between W and W^φ in Proposition 7.10 is not symmetric (see Example 7.11). For the time being, we take the easy way out, and call two kernels U and W *weakly isomorphic* if $t(F, U) = t(F, W)$ for every simple graph F. We will come back to a characterization of weakly isomorphic kernels in terms of measure preserving maps (in other words, proving a certain converse of Proposition 7.10) in Sections 10.7 and 13.2. It will also follow that in this case the equation $t(F, U) = t(F, W)$ holds for all multigraphs F (see Exercise 7.18 for a direct proof).

Weak isomorphism of kernels is clearly an equivalence relation, and we can identify kernels that are weakly isomorphic. This identification will play an important role in our discussions.

EXAMPLE 7.11. The map $\varphi_2 : x \mapsto 2x \pmod 1$ is measure preserving. For every kernel W, the kernel W^{φ_2} consists of four "copies" of W (see Figure 7.1). Similarly, $\varphi_3 : x \mapsto 3x \pmod 1$ is measure preserving, and W^{φ_3} consists of nine "copies" of W. The kernels W, W^{φ_2} and W^{φ_3} are weakly isomorphic, but there is no measure preserving map transforming W^{φ_2} to W^{φ_3} (Exercise 7.13). ◆

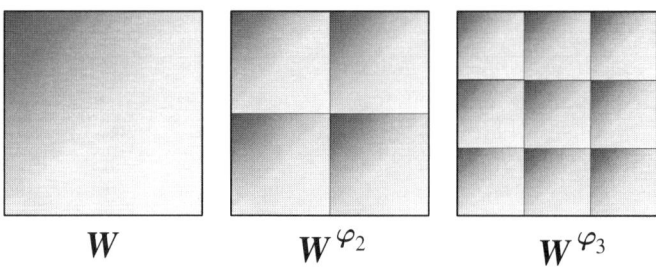

FIGURE 7.1. Gray-scale images of three graphons that are weakly isomorphic, but not isomorphic up to a null set. Recall that the origin is in the upper left corner.

This example illustrates that weak isomorphism is not a very easy notion. We will return to it and develop more and more information about it when we introduce distances between graphons, sampling, twin reduction, and other tools in the theory of graphons.

EXERCISE 7.12. Suppose that two kernels U and W are weakly isomorphic. Prove that so are the kernels $aU + b$ and $aW + b$ ($a, b \in \mathbb{R}$).

EXERCISE 7.13. Prove that the kernels W, W^{φ_2} and W^{φ_3} in Example 7.11 are weakly isomorphic, but not isomorphic up to a null set.

7.4. Sums and products

Perhaps the first tool we use in graph theory is the decomposition into connected components. For kernels, a similar decomposition exists, but one must be a bit careful with 0-sets. This was worked out by Janson [2008].

Let $W_1, W_2, \ldots$ be a finite or countably infinite family of kernels, and let $a_1, a_2 \ldots$ be positive real numbers with $\sum_i a_i = 1$. We define the *direct sum* of the W_i with weights a_i, in notation $W = a_1 W_1 \oplus a_2 W_2 \oplus \ldots$, as follows. We split the interval $[0,1]$ into intervals $J_1, J_2, \ldots$ of length $a_1, a_2, \ldots$, consider the monotone affine maps φ_i mapping J_i onto $[0,1]$, and let

$$W(x,y) = \begin{cases} W_i(\varphi_i(x), \varphi_i(y)), & \text{if } x,y \in J_i,\ i=1,2,\ldots, \\ 0, & \text{otherwise.} \end{cases}$$

A kernel will be called *connected*, if it is not isomorphic up to a null set to the direct sum of two kernels. This is equivalent to saying that for every subset $S \subseteq [0,1]$ with $0 < \lambda(S) < 1$, we have

$$\int_{S \times ([0,1] \setminus S)} |W(x,y)|\, dx\, dy > 0.$$

Every kernel can be written as the direct sum of connected kernels and perhaps the 0 kernel. (We have to allow the 0 kernel, which cannot be written as the sum of connected kernels.) This decomposition is unique (up to zero sets); see Bollobás, Janson and Riordan [2007] and Janson [2008] for more.

Somewhat confusingly, we can introduce three "product" operations on kernels, and we will need all three of them. Let $U, W \in \mathcal{W}$. We denote by UW their (pointwise) product as functions, i.e.,

$$(UW)(x,y) = U(x,y)W(x,y).$$

We denote by $U \circ W$ their *operator product* (the name refers to the fact that this is the product of U and W as kernel operators, see Section 7.5)

$$(U \circ W)(x,y) = \int_0^1 U(x,z)W(z,y)\,dz.$$

We note that $U \circ W$ is not symmetric in general, but it will be in the cases we use this operation (for example, when $U = W$).

Finally, we denote by $U \otimes W$ their *tensor product*; this is defined as a function $[0,1]^2 \times [0,1]^2 \to [0,1]$ by

$$(U \otimes W)(x_1, x_2, y_1, y_2) = U(x_1, y_1)W(x_2, y_2).$$

This function is not defined on $[0,1]^2$ and hence it is not in $\mathcal{W}$; however, we can consider any measure preserving map $\varphi \colon [0,1] \to [0,1]^2$, and define the kernel

$$(U \otimes W)^\varphi(x,y) = (U \otimes W)\bigl(\varphi(x), \varphi(y)\bigr).$$

It does not really matter which particular measure preserving map we use here: these kernels obtained from different maps φ are weakly isomorphic by the same computation as used in the proof of Proposition 7.10, and so we can call any of them the tensor product of U and W.

We note that the tensor product has the nice property that

(7.17) $$t(F, U \otimes W) = t(F,U)t(F,W)$$

for every multigraph F.

We denote the n-th power of a kernel according to these three multiplications by W^n (pointwise power), $W^{\circ n}$ (operator power), and $W^{\otimes n}$ (tensor power).

There are many other properties and constructions for graphs that can be generalized to graphons in a natural way. For example, we call a graphon W *bipartite*, if there is a partition $V(G) = V_1 \cup V_2$ such that $W(x_1, x_2) = 0$ for almost all $(x_1, x_2) \in V_1 \times V_2$. We can define k-colorable kernels similarly. We call a graphon *triangle-free*, if $t(K_3, W) = 0$. Simple facts like "every bipartite graphon is triangle-free" can be proved easily. Often one faces minor complications because of exceptional nullsets; a rather general remedy for this problem, called pure graphons, will be introduced in Section 13.3.

EXERCISE 7.14. Show that for every simple graph F, $t(F, W^{\circ n}) = t(F', W)$, where F' is obtained from F by subdividing each edge by $n-1$ new nodes.

EXERCISE 7.15. Prove that connectivity of a graphon is invariant under weak isomorphism.

EXERCISE 7.16. Prove that a graphon W is bipartite if and only if $t(C_{2k+1}, W) = 0$ for all $k \geq 1$.

7.5. Kernel operators

Every function $W \in \mathcal{W}$ defines an operator $T_W : L_1[0,1] \to L_\infty[0,1]$, by

$$(T_W f)(x) = \int_0^1 W(x,y) f(y) \, dy. \tag{7.18}$$

Sometimes it will be useful to consider T_W as an operator $L_\infty[0,1] \to L_1[0,1]$ or $L_2[0,1] \to L_2[0,1]$; the formula is meaningful in each of these cases.

If we consider $T_W : L_2[0,1] \to L_2[0,1]$, then it is a Hilbert-Schmidt operator, and the rich theory of such operators can be applied. It is a compact operator, which has a discrete spectrum, i.e., a countable multiset $\mathrm{Spec}(W)$ of nonzero (real) eigenvalues $\{\lambda_1, \lambda_2 \ldots\}$ such that $\lambda_n \to 0$. In particular, every nonzero eigenvalue has finite multiplicity. Furthermore, it has a spectral decomposition

$$W(x,y) \sim \sum_k \lambda_k f_k(x) f_k(y), \tag{7.19}$$

where f_k is the eigenfunction belonging to the eigenvalue λ_k with $\|f_k\|_2 = 1$. The series on the right may not be almost everywhere convergent (only in L_2), but one has

$$\sum_{k=1}^\infty \lambda_k^2 = \int_{[0,1]^2} W(x,y)^2 \, dx \, dy = \|W\|_2^2 \leq \|W\|_\infty^2.$$

A useful consequence of this bound is that if we order the λ_i by decreasing absolute value: $|\lambda_1| \geq |\lambda_2| \geq \ldots$, then

$$|\lambda_k| \leq \frac{\|W\|_2}{\sqrt{k}}. \tag{7.20}$$

It also follows that for every other kernel U on the same probability space, the inner product can be computed from the spectral decomposition:

$$\langle U, W \rangle = \int_{[0,1]^2} U(x,y) W(x,y) \, dx \, dy = \sum_k \lambda_k \int_{[0,1]^2} U(x,y) f_k(x) f_k(y) \, dx \, dy \tag{7.21}$$

$$= \sum_k \lambda_k \langle f_k, U f_k \rangle$$

(where the series on the right is absolute convergent). The spectral decomposition is particularly useful if we need to express operator powers: The spectral decomposition of the n-th operator power is

$$W^{\circ n}(x,y) = \sum_k \lambda_k^n f_k(x) f_k(y),$$

and the series on the right hand side converges to the left hand side almost everywhere if $n \geq 2$.

PROPOSITION 7.17. *The eigenfunctions f_k belonging to a nonzero eigenvalue λ_k of any function $W \in \mathcal{W}$ are bounded.*

Proof. Indeed,

$$|f_k(x)| = \frac{1}{|\lambda_k|} \left| \int_0^1 W(x,y) f_k(y) \, dy \right| \leq \frac{1}{|\lambda_k|} \|W\|_\infty \|f_k\|_1. \qquad \square$$

Some subgraph densities have nice expressions in terms of this spectrum. Generalizing (5.31), we have
$$t(C_n, W) = \int_{[0,1]^n} W(x_1, x_2) \cdots W(x_{n-1}, x_n) W(x_n, x_1) \, dx_1 \ldots dx_n = \sum_k \lambda_k^n. \tag{7.22}$$

This expression is also valid for $n = 2$:
$$t(C_2, W) = \int_{[0,1]^2} W(x, y)^2 \, dx \, dy = \|W\|_2^2 = \sum_k \lambda_k^2. \tag{7.23}$$

Furthermore, for every $n \geq 3$,
$$t_{xy}(P_n^{\bullet\bullet}, W) = \sum_k \lambda_k^{n-1} f_k(x) f_k(y) \tag{7.24}$$

almost everywhere. For $n = 2$ the left side is just W, and so we don't always get pointwise equality, only convergence in L_2.

For a general multigraph $F = (V, E)$, we can express its density in a kernel W by a rather hairy spectral formula (Lovász and Szegedy [2011]), which is nevertheless useful. Substituting (7.19) in the definition of $t(F, W)$ and expanding, we get
$$t(F, W) = \sum_{\chi: E \to \mathbb{N}^*} \prod_{e \in E} \lambda_{\chi(e)} \prod_{v \in V} M_\chi(v), \tag{7.25}$$

where
$$M_\chi(v) = \int_0^1 \prod_{u: \, uv \in E} f_{\chi(uv)}(x) \, dx. \tag{7.26}$$

(One has to be careful, since (7.19) only converges in L_2, not necessarily almost everywhere. But using (7.21) we can substitute for the values $W(x_i, x_j)$ one by one.) This representation expresses $t(F, W)$ in an infinite "edge-coloring model", which is analogous to homomorphism numbers with the role of nodes and edges interchanged (see Section 23.2 for a discussion of finite edge-coloring models): we sum over all colorings of the edges with $\mathbb{N}$; for every coloring, we take the product of nodeweights and the product of edgeweights; the edgeweights are just the eigenvalues, and the weight of a node is computed from the colors of the edges incident with it.

One consequence of (7.22) is that the cycle densities in W determine the spectrum of T_W and vice versa. In fact, we don't have to know all cycle densities: any "tail" $(t(C_k, W) : k \geq k_0)$ is enough. This follows from Proposition A.21 in the Appendix. In particular, we see that $t(C_2, W) = \|W\|_2^2$ is determined by the cycle densities $t(C_k, W)$, $k \geq 3$.

EXERCISE 7.18. (a) Let $F = (V, E)$ be a multigraph without loops, and let us subdivide each edge $e \in E$ by $m(e) \geq 0$ new nodes, to get a multigraph F'. Show that using (7.24) the density of F' in W can be expressed by a formula similar to (7.25). (b) Show that the densities of simple graphs in a kernel determine the densities of multigraphs.

EXERCISE 7.19. Let W be a graphon. Prove that (a) all eigenvalues of T_W are contained in the interval $[-1, 1]$; (b) the largest eigenvalue is also largest in absolute value; (c) at least one of the eigenvectors belonging to the largest eigenvalue is nonnegative almost everywhere.

CHAPTER 8

The cut distance

We have announced in the Introduction that we are going to define the distance of two arbitrary graphs, so that this distance will reflect structural similarity. The definition is quite involved, and we will approach the problem in several steps: starting with two graphs on the same node set, then moving to graphs with the same number of nodes (but on unrelated sets of nodes), then moving to the general case. Finally, we extend the definition to kernels, where it will turn out simpler (at least in words) than in the finite case.

In this section we consider dense graphs. The definitions are of course valid for all graphs, but they give a distance of $o(1)$ between two graphs with edge density $o(1)$, so they are not useful in that setting.

8.1. The cut distance of graphs

8.1.1. Norms of a matrix. Let A be an $n \times n$ matrix. There are a number of norms that come up in various studies. We will need the ℓ_1-norm

$$\|A\|_1 = \frac{1}{n^2} \sum_{i,j=1}^{n} |A_{ij}|, \tag{8.1}$$

the ℓ_2 or Frobenius norm

$$\|A\|_2 = \Big(\frac{1}{n^2} \sum_{i,j=1}^{n} A_{ij}^2\Big)^{1/2}, \tag{8.2}$$

and the ℓ_∞-norm

$$\|A\|_\infty = \max_{i,j} |A_{ij}|. \tag{8.3}$$

(Note the normalization for the ℓ_1 and ℓ_2 norms: when A an adjacency matrix, all these norms are between 0 and 1.)

Our main tool will be a less standard norm, called the *cut norm*, which was introduced by Frieze and Kannan [1999]. This is defined by

$$\|A\|_\square = \frac{1}{n^2} \max_{S,T \subseteq [n]} \Big| \sum_{i \in S, j \in T} A_{ij} \Big|. \tag{8.4}$$

It is clear that

$$\|A\|_\square \leq \|A\|_1 \leq \|A\|_2 \leq \|A\|_\infty. \tag{8.5}$$

EXAMPLE 8.1. Let A be an $n \times n$ matrix, whose entries are independent random ± 1's (with expectation 0). Then $\|A\|_1 = \|A\|_2 = \|A\|_\infty = 1$. On the other hand, the expectation of $\sum_{i \in S, j \in T} A_{ij}$ is 0, and the variance is $\Theta(n^2)$, and so the expectation of $\big|\sum_{i \in S, j \in T} A_{ij}\big|$ is $\Theta(n)$. The expectation of the maximum in (8.4) is more difficult to compute, but using the Chernoff–Hoeffding inequality, one gets that $\|A\|_\square < 4n^{-1/2}$ with high probability. ♦

Alon and Naor [2006] relate the cut norm of a symmetric matrix to its *Grothendieck norm* (well known in functional analysis). It follows by the results of Grothendieck that the cut norm is between two absolute constant multiples of the Grothendieck norm. The Grothendieck norm can be viewed as a semidefinite relaxation of the cut norm, and it is polynomial time computable to an arbitrary precision. So we can compute, in polynomial time, an approximation of the cut norm with a multiplicative error less than 2. We don't go into the details of these results here; in our setting it will be more important to approximate the cut norm by a randomized sampling algorithm, to be described in Section 10.3.

We'll say more about approximation of the cut norm in the more general setting of graphons in Section 14.1.

8.1.2. Two graphs on the same set of nodes. Let G and G' be two graphs with a common node set $[n]$. From any of the matrix norms introduced above, the norm of the difference of their adjacency matrices defines a distance between two graphs. Two of these distances have special significance.

The ℓ_1 distance
$$d_1(G, G') = \frac{|E(G) \triangle E(G')|}{n^2} = \|A_G - A_{G'}\|_1$$
is also called the *edit distance* (usually without the normalization). It can be thought of as the fraction of pairs of nodes whose adjacency we have to toggle to get from one graph to the other.

The *cut metric* derived from the cut norm can be described combinatorially as follows. For an unweighted graph $G = (V, E)$ and sets $S, T \subseteq V$, let $e_G(S, T)$ denote the number of edges in G with one endnode in S and the other in T (the endnodes may also belong to $S \cap T$; so $e_G(S, S) = 2e_G(S)$ is twice the number of edges spanned by S). For two graphs G and G' on the same node set $[n]$, we define their *cut distance* (as labeled graphs) by
$$d_\square(G, G') = \max_{S,T \subseteq V(G)} \frac{|e_G(S,T) - e_{G'}(S,T)|}{n^2} = \|A_G - A_{G'}\|_\square.$$
In this setting dividing by $|S| \times |T|$ instead of n^2 might look more natural. However, dividing by $|S| \times |T|$ would emphasize small sets too much, and the maximum would be attained when $|S| = |T| = 1$. With our definition, the contribution of a pair S, T is at most $|T||S|/n^2$ (for simple graphs).

It is easy to see that $d_\square(G, G') \leq d_1(G, G')$, and in general the two distances are quite different. For example, if $\mathbf{G}$ and $\mathbf{G}'$ are two independent random graphs on $[n]$ with edge probability $1/2$, then with high probability $d_1(\mathbf{G}, \mathbf{G}') \approx 1/2$ but $d_\square(\mathbf{G}, \mathbf{G}') = O(1/\sqrt{n})$.

We will have to define the distance of two weighted graphs G and G' on the same node set V, but with possibly different nodeweights. In this case, we have to add a term accounting for the difference in their node weighting. To simplify notation, let $\alpha_i = \alpha_i(G)/\alpha_G$, $\alpha'_i = \alpha_i(G')/\alpha_{G'}$, $\beta_{ij} = \beta_{ij}(G)$ and $\beta'_{ij} = \beta_{ij}(G')$. Then we define

(8.6) $$d_1(G, G') = \sum_{i \in V} |\alpha_i - \alpha'_i| + \sum_{i,j \in V} |\alpha_i \alpha_j \beta_{ij} - \alpha'_i \alpha'_j \beta'_{ij}|$$

and
$$(8.7) \quad d_\square(G,G') = \sum_{i\in V} |\alpha_i - \alpha_i'| + \max_{S,T\subseteq V}\Big|\sum_{i\in S, j\in T}(\alpha_i\alpha_j\beta_{ij} - \alpha_i'\alpha_j'\beta_{ij}')\Big|.$$

It is easy to check that these formulas define metrics, and they specialize to the "old" definitions when the nodeweights are 1 and the edgeweights are 0 or 1. Another special case worth mentioning is when the nodeweights of the two graphs are the same: in this case, the first term in both definitions disappears, and inside the second term, we get the slightly simpler expression $\alpha_i\alpha_j(\beta_{ij} - \beta_{ij}')$. We note, furthermore, that since G and G' can be represented as points in the same finite dimensional space, all usual distance functions on the set of weighted graphs on the same set of nodes would give the same topology.

EXAMPLE 8.2. Let H_n denote the complete graph on $[n]$, where all nodes have weight 1 and all edges have weight $1/2$. Then for a random graph $\mathbb{G} = \mathbb{G}(n, 1/2)$ on the same node set, we have $d_\square(\mathbb{G}, H_n) = o(1)$ with high probability. ◆

8.1.3. Two graphs with the same number of nodes. If G and G' are unlabeled unweighted graphs on possibly different node sets but of the same cardinality n, then we define their distance by

$$(8.8) \quad \widehat{\delta}_\square(G, G') = \min_{\widehat{G},\widehat{G}'} d_\square(\widehat{G}, \widehat{G}'),$$

where $\widehat{G}$ and $\widehat{G}'$ range over all labelings of G and G' by $1, \ldots, n$, respectively. (Of course, it would be enough to fix a labeling for one of the graphs and minimize over all labelings of the other.)

The hat above the δ indicates that the "ultimate" definition will be somewhat different. Indeed, handling of this quantity $\widehat{\delta}_\square(G, G')$ is quite difficult, due to the min-max in the definition.

8.1.4. Two arbitrary graphs. Let $G = (V, E)$ and $G' = (V', E')$ be two graphs with (say) $V = [n]$ and $V' = [n']$. To define their distance, recall that for every graph G and positive integer m, the graph $G(m)$ is obtained from G by replacing each node of G by m nodes, where two new nodes are connected if and only if their predecessors were. Using this operation, we can change the graphs so that they have the same number of nodes, by replacing them with $G(n')$ and $G'(n)$, or more generally, by $G(kn')$ and $G'(kn)$ for any $k \in \mathbb{N}$. Now we can use the distance $\widehat{\delta}_\square$ to define the distance

$$\delta_\square(G, G') = \lim_{k\to\infty} \widehat{\delta}_\square\big(G(kn'), G'(kn)\big).$$

A more complicated but "finite" definition of the same quantity can be given as follows (cf. Exercise 8.5). A *fractional overlay* of G and G' is a nonnegative $n \times n'$ matrix $X = (X_{iu})$ such that $\sum_{u=1}^{n'} X_{iu} = \frac{1}{n}$ and $\sum_{i=1}^{n} X_{iu} = \frac{1}{n'}$. If $n = n'$ and $\sigma: V \to V'$ is a bijection, then $X_{iu} = \frac{1}{n}\mathbb{1}(\sigma(i) = u)$ is a fractional overlay (which in this case is an honest-to-good overlay). We denote by $\mathcal{X}(G, G')$ the set of all fractional overlays.

Fixing a fractional overlay X, we can define a generalization of the labeled cut distance:

$$(8.9) \quad d_\square(G, G', X) = \max_{Q,R\subseteq V\times V'}\Big|\sum_{iu\in Q,\ jv\in R} X_{iu}X_{jv}\big(\mathbb{1}(ij\in E) - \mathbb{1}(uv\in E')\big)\Big|.$$

The distance of the two graphs can be described by optimizing over fractional overlays:

$$\delta_\square(G, G') = \min_{X \in \mathcal{X}(G,G')} d_\square(G, G', X) \tag{8.10}$$

One can generalize this to weighted graphs. Let $G = (V, E)$ and $G' = (V', E')$ be two weighted graphs with normalized nodeweights $\alpha_i = \alpha_i(G)$ and $\alpha'_u = \alpha_u(G')$ (so that $\alpha_G = \alpha_{G'} = 1$), and edgeweights $\beta_{ij} = \beta_{ij}(G)$ and $\beta'_{ij} = \beta_{ij}(G')$. A *fractional overlay* of G and G' is defined as a nonnegative $n \times n'$ matrix X such that $\sum_{u=1}^{n'} X_{iu} = \alpha_i(G)$ and $\sum_{i=1}^{n} X_{iu} = \alpha_u(G')$. We define

$$d_\square(G, G', X) = \max_{Q, R \subseteq V \times V'} \left| \sum_{iu \in Q,\, jv \in R} X_{iu} X_{jv} (\beta_{ij} - \beta'_{uv}) \right| \tag{8.11}$$

and then $\delta_\square(G, G')$ can be defined by the same formula (8.10). This formula can be rephrased as follows, using two more $V \times V'$ matrices Y and Z:

$$\delta_\square(G, G') = \min_{X \in \mathcal{X}(G,G')} \max_{0 \leq Y, Z \leq X} \left| \sum_{i,j \in V, u, v \in V'} Y_{iu} Z_{jv} (\beta_{ij} - \beta'_{uv}) \right|. \tag{8.12}$$

Indeed, the absolute value on the right is a convex function of the entries of Y and Z, and so it is maximized when every entry is equal to either 0 or to the corresponding entry of X.

To illuminate definition (8.10) a little, we can think of a fractional overlay as a probability distribution χ on $V \times V'$ whose marginals are uniform. In other words, it is a coupling of the uniform distribution on V with the uniform distribution on V'. Select two pairs (i, u) and (j, v) from the distribution χ. Then (8.9) expresses some form of correlation between ij being an edge and uv being an edge.

One word of warning: $\delta_\square$ is only a pseudometric, not a true metric, because $\delta_\square(G, G')$ may be zero for different graphs G and G'. This is the case e.g. if $G' = G(k)$ for some k (cf. Exercise 8.6).

We have to discuss a technical problem, for which only partial results are available (but these will be enough for our purposes). If G and G' have the same number of nodes, then the definition of $\delta_\square$ may give a value different from their $\widehat{\delta}_\square$ distance. It is trivial that

$$\delta_\square(G, G') \leq \widehat{\delta}_\square(G, G'),$$

but how much larger can the right side be? It may be larger (see Exercise 8.8. Perhaps the increase is never larger than a factor of 2, but this is open. To prove anything nontrivial requires tools to be developed later; in Section 9.4 we are going to prove, among others, the (rather weak) inequality

$$\widehat{\delta}_\square(G, G') \leq \frac{45}{\sqrt{-\log \delta_\square(G, G')}}.$$

(One important consequence of this weak inequality will be that any Cauchy sequence of graphs in the $\delta_\square$ distance is also a Cauchy sequence in the $\widehat{\delta}_\square$ distance.)

EXAMPLE 8.3. Let K denote the graph with a single node of weight 1, endowed with a loop with weight $1/2$. Then for a random graph $\mathbb{G} = \mathbb{G}(n, 1/2)$, we have $\delta_\square(\mathbb{G}, K) = o(1)$ with high probability. ♦

EXERCISE 8.4. Let A be a symmetric matrix. Show that restricting the pairs (S, T) in the definition (8.4) of the cut norm in any of the following ways will decrease it by a small factor only: (a) $T = S$, by at most 2; (b) $T \cap S = \emptyset$, by at most 4; (c) $T = [n] \setminus S$, by at most 6; (d) $|S|, |T| \geq n/2$, by at most 4.

EXERCISE 8.5. Prove that the definitions of $\delta_\square(G, G')$ through blow-ups and through fractional overlays lead to the same value.

EXERCISE 8.6. Let G_1 and G_2 be two simple graphs with $\delta_\square(G_1, G_2) = 0$. Prove that there is a simple graph G and $n_1, n_2 \geq 1$ such that $G_i \cong G(n_i)$.

EXERCISE 8.7. Let A be a symmetric $n \times n$ matrix with all entries in $[-1, 1]$. Let A' be obtained from A by deleting a row and the corresponding column. Prove that
$$\left| \|A\|_\square - \|A'\|_\square \right| \leq \frac{2}{n}.$$

EXERCISE 8.8. (a) Let H denote the graph on two nonadjacent nodes, with a loop at each of them. Prove that $\widehat{\delta}_\square(H, K_2) = 1/4$ but $\delta_\square(H, K_2) = 1/8$. (b) Prove that if n is odd, then $\widehat{\delta}(K_{n,n}, \overline{K}_{n,n}) > \delta(K_{n,n}, \overline{K}_{n,n})$.

8.2. Cut norm and cut distance of kernels

After the rather heavy going with the cut distance for graphs, it sounds frightening that we want to extend all this to kernels. But in fact, the definitions become simpler and more transparent. (This is not the last time when graphons will provide a more user-friendly environment.)

8.2.1. Cut norm. We define the *cut norm* on the linear space $\mathcal{W}$ of kernels by

$$(8.13) \qquad \|W\|_\square = \sup_{S,T \subseteq [0,1]} \left| \int_{S \times T} W(x,y)\, dx\, dy \right|$$

where the supremum is taken over all measurable subsets S and T. It is sometimes convenient to use the corresponding metric $d_\square(U, W) = \|U - W\|_\square$.

The cut norm is a norm; this is easy to prove using standard analysis. Similarly as in the case of matrices, we have the trivial inequalities between the most important norms of a kernel in $\mathcal{W}_1$:

$$(8.14) \qquad \|W\|_\square \leq \|W\|_1 \leq \|W\|_2 \leq \|W\|_\infty \leq 1.$$

In the opposite direction, we have trivially $\|W\|_2 \leq \|W\|_1^{1/2}$ (showing that $\|.\|_1$ and $\|.\|_2$ define the same topology on $\mathcal{W}_1$), but the other two norms in the formula above define different topologies. However, for a stepfunction U with k steps we have the trivial inequality

$$(8.15) \qquad \|U\|_1 \leq k^2 \|U\|_\square.$$

It can be shown, in fact, that the coefficient k^2 can be replaced by $\sqrt{2k}$ (see Janson [2010], Remark 9.8, and also our Exercise 8.18); but the inequality above will be enough for us.

There is some natural notation that goes with this norm. For every set $\mathcal{R} \subseteq \mathcal{W}_0$, we define its ε-neighborhood in the cut-norm

$$B_\square(\mathcal{R}, \varepsilon) = \{W \in \mathcal{W}_0 :\ d_\square(W, \mathcal{R}) < \varepsilon\} = \{W \in \mathcal{W}_0 :\ (\exists U \in \mathcal{R})\ d_\square(W, U) < \varepsilon\}.$$

We define the ε-neighborhood $B_1(\mathcal{R}, \varepsilon)$ in the L_1-norm analogously. (We defined all this in the graphon space $\mathcal{W}_0$, where we need this notation. One could of course take other sets of kernels as the universe.)

8.2.2. Cut distance of unlabeled kernels. Kernels, defined on the fixed set $[0,1]$, correspond to labeled graphs. Just as for graphs, we introduce an "unlabeled" version of the cut norm, by finding the best overlay of the underlying sets. Let $\overline{S}_{[0,1]}$ denote the set of measure preserving maps $[0,1] \to [0,1]$, and let $S_{[0,1]}$ denote the set of all invertible measure preserving maps $[0,1] \to [0,1]$ (the inverse of such a map is known to be measure preserving as well, so $S_{[0,1]}$ is a group; see Appendix A.3.2). We define the *cut distance* of two kernels by

$$(8.16) \qquad \delta_\square(U, W) = \inf_{\varphi \in S_{[0,1]}} d_\square(U, W^\varphi),$$

(where $W^\varphi(x,y) = W(\varphi(x), \varphi(y))$). It is easy to see that either one of the following expressions could be used to define the cut distance:

$$(8.17) \qquad \delta_\square(U, W) = \inf_{\varphi \in S_{[0,1]}} d_\square(U^\varphi, W) = \inf_{\varphi \in \overline{S}_{[0,1]}} d_\square(U, W^\varphi)$$
$$= \inf_{\varphi, \psi \in \overline{S}_{[0,1]}} d_\square(U^\psi, W^\varphi).$$

We will prove the much less trivial fact that in the last expression the infimum is attained: Theorem 8.13 below establishes this in larger generality, for all norms satisfying some natural conditions.

The distance $\delta_\square$ of kernels is only a pseudometric, since different kernels can have distance zero. (Such pairs of kernels will turn out exactly the weakly isomorphic pairs, but this will take more work to prove.) We can identify two kernels whose cut distance is 0, to get the set $\widetilde{\mathcal{W}}$ of *unlabeled kernels*. We define the sets $\widetilde{\mathcal{W}}_0$ and $\widetilde{\mathcal{W}}_1$ analogously.

Going into all the complications with using the cut norm and then minimizing over measure preserving transformations is justified by the important fact that the metric $\delta_\square$ defines a compact metric space on graphons. We will state and prove this fact in Section 9.3.

One main advantage in using graphons instead of graphs is that many formulas and proofs become much simpler and more transparent. (Just compare the definition (8.16) of the distance of two graphons with the definition (8.12) of the analogous quantity for two weighted graphs!) When going from graphs to graphons via the correspondence $G \mapsto W_G$, we may pay a prize by having to estimate how much error we make by this. This will indeed require extra work in some cases, but in other cases we will be lucky, and no error will be made. For example, equation (7.2) shows that homomorphism numbers "from the left" don't change when we replace G by W_G. The next lemma shows that the situation is similar with the $\delta_\square$ distance. (We will not always be so lucky; Section 12.4.4 will be devoted to estimating this kind of error for multicuts.)

LEMMA 8.9. *For any two weighted graphs H and H'*

$$\delta_\square(H, H') = \delta_\square(W_H, W_{H'}).$$

Proof. Let $\varphi: [0,1] \to [0,1]$ be a measure preserving map. Let $(S_i : i \in V(H))$ and $(T_u : u \in V(H'))$ be the partitions of $[0,1]$ into the steps of W_H and

$W_{H'}$. Define $X_{iu} = \lambda\big(S_i \cap \varphi(T_u)\big)$, then the matrix (X_{iu}) is a fractional overlay of H and H'. Conversely, every fractional overlay can be obtained from a measure preserving map this way.

We claim that for this measure preserving map and the corresponding fractional overlay we have

$$(8.18) \quad \max_{Q,R\subseteq V\times V'} \Big| \sum_{iu\in Q,\, jv\in R} X_{iu} X_{jv} (\beta_{ij} - \beta'_{uv}) \Big| = \sup_{Y,Z\subseteq [0,1]} \Big| \int_{Y\times Z} (W_H - W^{\varphi}_{H'}) \Big|.$$

For every $Q \subseteq V \times V'$, let $S_Q = \cup_{(i,u)\in Q} S_i \cap \varphi(T_u)$. Then for a fixed $Q, R \subseteq V \times V'$, it is easy to check that

$$\sum_{iu\in Q,\, jv\in R} X_{iu} X_{jv} (\beta_{ij} - \beta'_{uv}) = \int_{S_Q \times S_R} (W_H - W^{\varphi}_{H'}).$$

On the other hand, if $Z_{iu} = \lambda(Z \cap S_i \cap \varphi(T_u))$ and $Y_{iu} = \lambda(Y \cap S_i \cap \varphi(T_u))$, then $0 \le Y_{iu}, Z_{iu} \le X_{iu}$, and

$$\int_{Y\times Z} (W_H - W^{\varphi}_{H'}) = \sum_{i,j\in V, u,v\in V'} Y_{iu} Z_{jv} (\beta_{ij} - \beta'_{uv}).$$

So the definition (8.10) of $\delta_\square(H, H')$ implies the direction $\le$ in (8.18), while formula (8.12) implies reverse direction. This proves (8.18), from which the Lemma follows. □

8.2.3. Maxima versus suprema: cut norm. One price we have to pay for working with infinite objects like graphons is that when maximizing a function over an infinite set of objects (e.g. subsets), we don't necessarily have a maximum, only a supremum; hence we have to work with approximate optima. With two important definitions, the cut norm and the cut distance, we don't have this difficulty. (The Compactness Theorem 9.23 will provide another powerful tool to avoid such problems in many cases.) Next we prove this for the cut norm, and at the end of this chapter, for the cut distance. This would not be absolutely necessary: in most cases, we could just carry along an arbitrarily small error term. Nevertheless, it makes sense to include these facts in this book: if you want to work with these notions, you might as well work with them as conveniently as possible. The next lemma also provides a useful expression for the cut norm.

LEMMA 8.10. *For any kernel $W \in \mathcal{W}$, the optima*

$$(8.19) \quad \sup_{S,T\subseteq [0,1]} \Big| \int_{S\times T} W(x,y)\, dx\, dy \Big|$$

and

$$(8.20) \quad \sup_{f,g:\,[0,1]\to[0,1]} \Big| \int_{[0,1]^2} f(x)g(y) W(x,y)\, dx\, dy \Big|$$

are attained, and they are both equal to $\|W\|_\square$.

The sets S, T and the functions f, g are tacitly assumed to be measurable. We can write the expression to be maximized in (8.19) as $\langle \mathbb{1}_S, T_W \mathbb{1}_T \rangle$, and in (8.20), as $\langle f, T_W g \rangle$ (where T_W is the operator defined by (7.18)). The assertion of the lemma is equivalent to saying that the optimum in (8.20) is attained, and it is attained by 0-1 valued functions f and g. I am grateful to Svante Janson for suggesting a simplification of the proof that follows.

Proof. Let $D = \sup_{f,g} \langle f, T_W g \rangle$. We start with proving that this supremum is attained by appropriate functions f and g. Let $f_n, g_n : [0,1] \to [0,1]$ ($n = 1, 2, \ldots$) be functions such that $\langle f_n, T_W g_n \rangle \to D$. The set of functions $[0,1] \to [0,1]$ are weak*-compact, which means that by selecting a subsequence, we may assume that it tends to a limit $f : [0,1] \to [0,1]$ in the sense that $\langle f_n, h \rangle \to \langle f, h \rangle$ for every $h \in L_1[0,1]$. Similarly, we can go to a further subsequence to assume that g_n converges to a function g in the same sense. It is easy to see that f and g are bounded (perhaps after changing them on a null set). Now we claim that

$$\int_{[0,1]^2} f_n(x) g_n(y) W(x,y) \, dx \, dy \longrightarrow \int_{[0,1]^2} f(x) g(y) W(x,y) \, dx \, dy.$$

This convergence is trivial when $W = \mathbb{1}_{S \times T}$ for two measurable sets $S, T \subseteq [0,1]$. Hence it follows when W is stepfunction, since stepfunctions are linear combinations of a finite number of functions of the type $\mathbb{1}_{S \times T}$. Hence it follows for every kernel, since every kernel can be approximated by stepfunctions in $L_1([0,1]^2)$, and the factors f_n, g_n, f, g are bounded. This implies that $\langle f, T_W g \rangle = D$.

Next we show that the maximizing functions f and g can be chosen to be 0-1 valued. Let $S = \{x : 0 < f(x) < 1\}$, and suppose that $\lambda(S) > 0$. Define

$$f_s(x) = f(x) + s \min(f(x), 1 - f(x)).$$

Then for $-1 \leq s \leq 1$, the function f_s satisfies $0 \leq f_s \leq 1$, and hence, by the maximality property of f, we have $\langle f_s, T_W g \rangle \leq \langle f, T_W g \rangle$. Since $\langle f_s, T_W g \rangle$ is a linear function of s and equality holds for $s = 0$, we must have equality for all values of s, in particular for $s = 1$, and so we can replace f by $f_1(x) = \min(1, 2f(x))$. Repeating this construction, we get a sequence of optimizing functions that monotone converges to the 0-1 valued function $\bar{f} = \mathbb{1}(f(x) > 0)$. So we can replace f by $\bar{f}$, and similarly we can replace g by a 0-1 valued function $\bar{g}$. □

8.2.4. Operator norms and cut norm. While the cut norm is best suited for combinatorial purposes, it is equivalent to more traditional norms, such as the operator norm of T_W as an operator $L_\infty \to L_1$, as the following simple lemma shows:

LEMMA 8.11. *For every kernel W, we have*

$$\|W\|_\square \leq \|T_W\|_{\infty \to 1} \leq 4\|W\|_\square.$$

Proof. By definition,

$$\|T_W\|_{\infty \to 1} = \sup_{-1 \leq g \leq 1} \|T_W g\|_1 = \sup_{-1 \leq f, g \leq 1} \langle f, T_W g \rangle = \sup_{-1 \leq f, g \leq 1} |\langle f, T_W g \rangle|.$$

Comparing this expression with (8.20), we get the first inequality. For the second, we write

$$\|T_W\|_{\infty \to 1} = \sup_{0 \leq f, f', g, g' \leq 1} \langle f - f', T_W(g - g') \rangle.$$

Here

$$\langle f - f', T_W(g - g') \rangle = \langle f, T_W g \rangle - \langle f', T_W g \rangle - \langle f, T_W g' \rangle + \langle f', T_W g' \rangle \leq 4\|T\|_\square. \quad \square$$

There are many other variations on the definition which give norms that are some constant factor away from the cut norm; these are useful since in some proofs they come up more directly than the cut norm. Some of these are stated as exercises at the end of this section.

There are other well-studied operator norms that are topologically equivalent to the cut norm (even though they are not equivalent up to a constant factor). The *Schatten p-norm* $S_p(T_W)$ of a kernel operator T_W is defined as the ℓ_p-norm of the sequence of its eigenvalues. For an even integer p, these can be expressed in terms of homomorphism densities:

$$S_p(T_W) = t(C_p, W)^{1/p}.$$

(It is not trivial that $t(C_{2r}, U)^{1/(2r)}$ is a norm, i.e., it is subadditive (the other defining properties of a norm are easy). In Proposition 14.2 we'll describe a method to prove that Schatten norms are indeed norms, along with certain more general norms defined by graphs.)

These norms define the same topology on $\mathcal{W}_1$ as the cut norm. We prove the explicit relationship for the case $p = 4$, which we need.

LEMMA 8.12. *For every graphon* $U \in \mathcal{W}_1$, $\|U\|_\square^4 \leq t(C_4, U) \leq 4\|U\|_\square$.

Proof. The second inequality is a special case of Lemma 10.23. To prove the first inequality, we use

$$\|U\|_\square = \sup_{0 \leq f, g \leq 1} \langle f, T_U g \rangle,$$

where

$$\langle f, T_U g \rangle \leq \|f\|_2 \|T_U g\|_2 \leq \|T_U g\|_2 = \langle T_U g, T_U g \rangle^{1/2} = \langle g, T_U^2 g \rangle^{1/2}$$
$$= \langle g, T_{U \circ U} g \rangle^{1/2} \leq \|g\|_2 \|T_{U \circ U}\|_{2 \to 2}^{1/2} \leq \|T_{U \circ U}\|_{2 \to 2}^{1/2} \leq \|U \circ U\|_2^{1/2}$$
$$= t(C_4, U)^{1/4}. \qquad \square$$

8.2.5. Minima versus infima: cut distance. The last result in this section is of a similar nature as Lemma 8.10: we prove that the "inf" in the last quantity in formula 8.17 above is in fact a "min". This was proved by Bollobás and Riordan [2009]. An analogous result for the L_1-norm was proved by Pikhurko [2010]. With later applications in mind, we prove it in greater generality.

The construction that gives the cut distance $\delta_\square$ from the cut norm can be applied to any other norm on $\mathcal{W}$ that is invariant under maps $W \mapsto W^\varphi$ for all $\varphi \in S_{[0,1]}$. We will call such a norm *invariant*. For an invariant norm N on the linear space $\mathcal{W}$, we define

$$\delta_N(U, W) = \inf_{\varphi \in S_{[0,1]}} N(U - W^\varphi).$$

We call this function the *distance derived from* N. The distances δ_N will be interesting for us mainly in the cases when $N = \|.\|_\square$, $N = \|.\|_1$ and $N = \|.\|_2$. The corresponding unlabeled distances are $\delta_\square$, δ_1 and δ_2.

Since the norm is invariant under measure preserving bijections, we have $N(U - W^\varphi) = N(U^{\varphi^{-1}} - W)$, implying that $\delta_N(U, W) = \delta_N(W, U)$. It is trivial that the triangle inequality holds for δ_N, so it is a semimetric (and clearly it is not a true metric, since $\delta_N(U, U^\varphi) = 0$ for every measure preserving map $\varphi \in S_{[0,1]}$).

We call a norm N *smooth*, if it is continuous in the topology of pointwise convergence in $\mathcal{W}$. In other words, for every sequence of kernels (W_n) such that

$W_n \in \mathcal{W}_1$ and $W_n \to 0$ almost everywhere, we have $N(W_n) \to 0$. This implies that if $W_n \to W$ almost everywhere, then $N(W_n) \to N(W)$. The L_1, L_2 and cut norms are smooth, but L_∞ is not.

We have defined invariance of a norm using measure preserving bijections, but (at least for smooth norms) this implies invariance under all measure preserving maps $\varphi : [0,1] \to [0,1]$. This is easy to see for stepfunctions W, since for any measure preserving map W^φ is a stepfunction with the same number of steps, same size of steps, and same function value on these steps as W, and hence there is a *bijective* measure preserving map ψ such that $W^\varphi = W^\psi$. For a general kernel $W \in \mathcal{W}$, we have a sequence of stepfunctions W_n such that $W_n \to W$ almost everywhere, and then also $W_n^\varphi \to W^\varphi$ almost everywhere. By the smoothness of N this implies that $N(W_n) \to N(W)$ and $N(W_n^\varphi) \to N(W^\varphi)$. Since we know that $N(W_n^\varphi) = N(W_n)$, it follows that $N(W^\varphi) = N(W)$.

Let us note that an invariant norm N on $\mathcal{W}$ defines a norm on bounded symmetric measurable functions on any standard probability space. Indeed, if $(\Omega, \mathcal{A}, \pi)$ is such a space, then there is a measure preserving map $\psi : [0,1] \to \Omega$, and then we can define $N(W) = N(W^\psi)$ for every bounded symmetric measurable function $W : \Omega \times \Omega \to \mathbb{R}$. This value will not depend on the choice of ψ, which follows easily from the invariance of N.

One can also give a more probabilistic description of the distance δ_N, using coupling measures (see Appendix A.3). For every coupling measure μ between two copies of $[0,1]$, the two projection maps $\pi, \rho : [0,1]^2 \to [0,1]$ (where $[0,1]^2$ is equipped with the measure μ and $[0,1]$, with the Lebesgue measure) are measure preserving. So for every kernel U, the function U^π is a kernel on the probability space $([0,1]^2, \mathcal{B}, \mu)$, and similarly for the projection ρ. As remarked above, N defines a norm on kernels on $([0,1]^2, \mathcal{B}, \mu)$; we denote this norm by N_μ. It is easy to see that for every kernel U on $[0,1]$, we have

(8.21) $$N_\mu(U^\pi) = N(U).$$

After this explanation, we can state the theorem:

THEOREM 8.13. *Let N be a smooth invariant norm on $\mathcal{W}$. Then we have the following alternate expressions for the unlabeled distance derived from N:*

(8.22) $$\delta_N(U, W) = \inf_{\varphi \in S_{[0,1]}} N(U - W^\varphi) = \inf_{\varphi \in \overline{S}_{[0,1]}} N(U - W^\varphi)$$
$$= \inf_{\psi \in S_{[0,1]}} N(U^\psi - W) = \inf_{\psi \in \overline{S}_{[0,1]}} N(U^\psi - W)$$
$$= \inf_{\varphi, \psi \in S_{[0,1]}} N(U^\psi - W^\varphi) = \min_{\varphi, \psi \in \overline{S}_{[0,1]}} N(U^\psi - W^\varphi),$$

and

(8.23) $$\delta_N(U, W) = \min_\mu N_\mu(U^\pi - W^\rho),$$

where μ ranges over all coupling measures on $[0,1]^2$.

Proof. The equality of the first expressions in each line of (8.22) follows from the fact that invertible measure preserving maps form a group.

First, let U and W be stepfunctions. As used before, the kernel W^φ for any measure preserving map φ can be realized by an invertible measure preserving map, which implies that in each line of 8.22, the two expressions are equal. Equation (8.23) follows similarly easily in this case.

Second, we consider arbitrary functions $U, W \in \mathcal{W}$, and prove the formulas with the two occurrences of "min" replaced by "inf". Let (U_n) and (W_n) be sequences of stepfunctions converging almost everywhere to U and W, respectively. Then $N(U_n - U) \to 0$ by the smoothness of N, and similarly for W. Since $N(U_n^\varphi - U^\varphi) = N(U_n - U)$ for every measure preserving map φ, this implies that

$$\inf_{\varphi \in \overline{S}_{[0,1]}} N(U_n - W_n^\varphi) = \inf_{\varphi \in S_{[0,1]}} N(U_n - W_n^\varphi) \to \inf_{\varphi \in S_{[0,1]}} N(U - W^\varphi) = \delta_N(U, W),$$

and also that

$$\inf_{\varphi \in \overline{S}_{[0,1]}} N(U_n - W_n^\varphi) \to \inf_{\varphi \in \overline{S}_{[0,1]}} N(U - W^\varphi),$$

which proves the equality in the first line of (8.22). The other equations follow similarly.

However, this argument only gives an "inf" in the last two expressions for δ_N. To prove that it is in fact a minimum, we begin with (8.23). The space of coupling measures is compact in the weak topology, so it suffices to show that $N_\mu(U^\pi - W^\rho)$, as a function of μ, is lower semicontinuous. This means that if $\mu_n \to \mu$ weakly (where μ and μ_n are coupling measures), then for every two kernels U and W, we have

(8.24) $$\liminf_n N_{\mu_n}(U^\pi - W^\rho) \geq N_\mu(U^\pi - W^\rho).$$

As a first step, we prove that $N_{\mu_n}(V) \to N_\mu(V)$ for every continuous function V. Let f_n and f be the functions representing the measures μ_n and μ as in Proposition A.6(iv). Then $N_{\mu_n}(V) = N(V^{f_n})$, and $N_\mu(V) = N(V^f)$. Since V is continuous, we have $V^{f_n}(x,y) = V(f_n(x), f_n(y)) \to V(f(x), f(y)) = V^f(x,y)$ for almost all $x, y \in [0,1]^2$. By our assumption on the norm N, this implies that $N_{\mu_n}(V) \to N_\mu(V)$. As a special case, we get (8.24) for continuous kernels U and W.

Let $U, W : [0,1] \times [0,1] \to \mathbb{R}$ be arbitrary kernels, and fix any $\varepsilon > 0$. There are continuous kernels U_k and W_k ($k = 1, 2, \dots$) such that $U_k \to U$ and $W_k \to W$ almost everywhere. By the smoothness of N, we can fix k large enough so that $N(U_k - U) \leq \varepsilon$ and $N(W^k - W) \leq \varepsilon$.

By the special case proved above, we know that

$$N_{\mu_n}(U_k^\pi - W_k^\rho) \to N_\mu(U_k^\pi - W_k^\rho) \qquad (n \to \infty),$$

and we can fix n so that $|N_{\mu_n}(U_k^\pi - W_k^\rho) - N_\mu(U_k^\pi - W_k^\rho)| \leq \varepsilon$. Then, using (8.21),

$$N_\mu(U^\pi - W^\rho) \leq N_\mu(U_k^\pi - W_k^\rho) + N_\mu(U_k^\pi - U^\pi) + N_\mu(W_k^\rho - W^\rho)$$
$$= N_\mu(U_k^\pi - W_k^\rho) + N(U_k - U) + N(W_k - W)$$
$$\leq N_\mu(U_k^\pi - W_k^\rho) + 2\varepsilon.$$

Here, by the choice of n,

$$N_\mu(U_k^\pi - W_k^\rho) \leq N_{\mu_n}(U_k^\pi - W_k^\rho) + \varepsilon$$
$$\leq N_{\mu_n}(U^\pi - W^\rho) + N_{\mu_n}(U_k^\pi - U^\pi) + N_{\mu_n}(W_k^\rho - W^\rho) + \varepsilon$$
$$= N_{\mu_n}(U^\pi - W^\rho) + N(U_k - U) + N(W_k - W) + \varepsilon$$
$$\leq N_{\mu_n}(U^\pi - W^\rho) + 3\varepsilon.$$

Combining these inequalities, we get that $N_\mu(U^\pi - W^\rho) \leq N_{\mu_n}(U^\pi - W^\rho) + 5\varepsilon$ if n is large enough. This proves (8.24) and thereby the existence of the minimum in (8.23).

The existence of the minimum in (8.22) follows easily now. Let μ be a coupling measure such that $\delta_N(U, W) = N_\mu(U^\pi - W^\rho)$. Let σ be a measure preserving bijection from $[0, 1]$ with the Lebesgue measure into $[0, 1]^2$ with the measure μ, and let π and ρ be the projections of $[0, 1]^2$ to the two coordinates. The fact that μ is a coupling measure implies that the compositions $\varphi = \sigma\pi$ and $\psi = \sigma\rho$ are measure preserving, and
$$N(U^\varphi - W^\psi) = N_\mu(U^\pi, W^\rho) = \delta_N(U, W). \qquad \square$$

This theorem has an important corollary:

COROLLARY 8.14. *For any smooth and invariant norm N on $\mathcal{W}$, we have $\delta_N(U, W) = 0$ if and only if there exist maps $\varphi, \psi \in \overline{S}_{[0,1]}$ such that $U^\psi = W^\varphi$ almost everywhere.*

This corollary allows us to consider the distances δ_1 and δ_2 as defined on $\widetilde{\mathcal{W}}$ (just as $\delta_\square$). In other words, the condition $\delta_N(U, W) = 0$ is independent of N, and identifying such pairs of kernels gives the same space for every smooth and invariant norm N.

EXERCISE 8.15. Let H and H' be two weighted graphs on the same set with $\alpha_H = \alpha_{H'} = 1$, with the same edgeweights, but different nodeweights. Prove that
$$\delta_1(H, H') \leq \|\alpha(H) - \alpha(H')\|_1.$$

EXERCISE 8.16. Prove that for every kernel $W \in \mathcal{W}_1$ and $k \geq 2$, we have
$$t(C_4, W) \leq t(C_{2k}, W)^{1/(2k)} \leq t(C_4, W)^{1/4}.$$

EXERCISE 8.17. Let $\sigma : [0, 1] \to [0, 1]$ range over maps that can be obtained as follows: we split $(0, 1]$ into the intervals $I_k = (\frac{k-1}{n}, \frac{k}{n}]$ ($k = 1, \ldots, n$) and permute these intervals arbitrarily. Prove that for every smooth and invariant norm N we have
$$\delta_N(U, W) = \inf_\sigma N(U - W^\sigma).$$

EXERCISE 8.18. Improve the coefficient in (8.15) to (a) $2k$, (b) $2\sqrt{k}$ (this is not easy!).

EXERCISE 8.19. Show that if we use the formula $\sup_S \int_{S \times S} W$ to define a norm (which is only a constant factor off the cut norm), then the supremum is not always attained (Laczkovich [1995]).

EXERCISE 8.20. Show by examples that one could not replace any of the "inf"-s by "min" in Theorem 8.13.

EXERCISE 8.21. Show that if N is the L_∞ norm on $\mathcal{W}$, then even the "easy" part of Theorem 8.13 fails: there are two kernels U and W such that $\inf_{\varphi \in \overline{S}_{[0,1]}} \|U - W^\varphi\|_\infty = 0$ but $\inf_{\varphi \in S_{[0,1]}} \|U - W^\varphi\|_\infty > 0$.

8.3. Weak and L_1-topologies

We end this discussion of graphon distances with a further somewhat technical issue. The topology on $\mathcal{W}$ defined by the cut norm is certainly different from the topology defined by the L_1-norm; there are, however, some nontrivial relationships between them. We will discuss these in larger generality and detail in Section 14.2, but a few simple facts can be proved here easily, and we will need some of them soon.

The key to relating the cut norm to other topologies is the following lemma.

LEMMA 8.22. *Suppose that* $\|W_n\|_\square \to 0$ *as* $n \to \infty$ ($W_n \in \mathcal{W}_1$). *Then for every function* $Z \in L_1([0,1]^2)$, $\|ZW_n\|_\square \to 0$. *In particular,* $\langle Z, W_n \rangle \to 0$ *and* $\int_S W_n \to 0$ *for every measurable set* $S \subseteq [0,1]^2$.

Proof. If Z is the indicator function of a rectangle, these conclusions follow from the definition of the $\|.\|_\square$ norm. Hence the conclusion follows for stepfunctions, since they are linear combinations of a finite number of indicator functions of rectangles. Then it follows for all integrable functions, since they are approximable in $L_1([0,1]^2)$ by stepfunctions. $\square$

A uniformly bounded sequence of kernels $W_n \in \mathcal{W}$ is called *weak* convergent* to a kernel W if $\langle W_n, U \rangle \to \langle W, U \rangle$ for every integrable function $U : [0,1]^2 \to \mathbb{R}$. This is equivalent to requiring that $\int_{S \times T} W_n \to \int_{S \times T} W$ for all measurable sets S and T. This sound almost like convergence in the cut norm, but it is not the same! Lemma 8.22 implies that convergence in the $\|.\|_\square$ norm implies weak* convergence. However, weak* convergence does not imply convergence in the cut norm (Exercise 8.26; an interesting counterexample follows from Example 11.41).

Since $\|.\|_\square \leq \|.\|_1$, the cut norm is continuous with respect to the L_1-norm. The converse is not true (recall the example of random graphs from the Introduction, Figure 1.5), but the following fact, proved and used by Lovász and Szegedy [2010a], shows that it is at least lower semicontinuous:

PROPOSITION 8.23. *Let* $W_n \to W$ *in the cut norm* ($W_n, W \in \mathcal{W}_1$). *Then*
$$\liminf_{n \to \infty} \|W_n\|_1 \geq \|W\|_1.$$

Proof. Let $Y = \mathrm{sgn}(W)$. Then by Lemma 8.22,
$$\|W_n\|_1 \geq \langle W_n, Y \rangle \to \langle W, Y \rangle = \|W\|_1. \quad \square$$

As noted above, we cannot claim in Proposition 8.23 that $\|W_n\|_1 \to \|W\|_1$. However, as further applications of Lemma 8.22, we can state two weaker facts in this direction (Lovász and Szegedy [2010a]); the first of these was also proved (in a slightly different form) by Pikhurko [2010].

PROPOSITION 8.24. *Let W be a 0-1 valued graphon and let (W_n) be a sequence of graphons such that* $\|W_n - W\|_\square \to 0$. *Then* $\|W_n - W\|_1 \to 0$.

Proof. By Lemma 8.22, we have
$$\|W_n - W\|_1 = \int_{\{W=0\}} W_n + \int_{\{W=1\}} (1 - W_n) \to \int_{\{W=0\}} W + \int_{\{W=1\}} (1-W) = 0.$$
$\square$

PROPOSITION 8.25. *Suppose that* $U_n \to U$ *in the cut norm as* $n \to \infty$ ($U, U_n \in \mathcal{W}_0$). *Then for every* $W \in \mathcal{W}_0$ *there is a sequence of graphons* $W_n \in \mathcal{W}_0$ *such that* $W_n \to W$ *in the cut norm, and* $\|U_n - W_n\|_1 \to \|U - W\|_1$.

It is important that we want $W_n \in \mathcal{W}_0$; if we only wanted kernels, we could take simply $W_n = W + U_n - U$.

Proof. First we consider the case when $U \geq W$. Let
$$Z(x,y) = \begin{cases} W(x,y)/U(x,y) & \text{if } U(x,y) > 0, \\ 0 & \text{otherwise}. \end{cases}$$
and define $W_n = ZU_n$. Trivially $W_n \in \mathcal{W}_0$, $W = ZU$, and
$$\|W - W_n\|_\square = \|Z(U - U_n)\|_\square \to 0$$
by Lemma 8.22. Furthermore, using that $U_n \geq W_n$ and $U \geq W$, we get
$$\big|\|U_n - W_n\|_1 - \|U - W\|_1\big| = \big|\|U_n - W_n\|_\square - \|U - W\|_\square\big| \leq \|U - U_n\|_\square + \|W - W_n\|_\square.$$
This implies that $\|U_n - W_n\|_1 \to \|U - W\|_1$ as $n \to \infty$.

The case when $U \leq W$ follows by a similar argument, replacing U, W, U_n by $1 - U, 1 - W, 1 - U_n$.

Finally, in the general case, consider the graphon $V = \max(U, W)$. Then clearly $\|U - V\|_1 + \|V - W\|_1 = \|U - W\|_1$. Since $U \leq V$, there exists a sequence (V_n) of graphons such that $\|V_n - V\|_\square \to 0$ and $\|V_n - U_n\|_1 \to \|V - U\|_1$. Since $V \geq W$, there is a sequence (W_n) of graphons such that $\|W_n - W\|_\square \to 0$ and $\|W_n - V_n\|_1 \to \|W - V\|_1$. Hence
$$\limsup_{n\to\infty} \|U_n - W_n\|_1 \leq \limsup_{n\to\infty} \|U_n - V_n\|_1 + \limsup_{n\to\infty} \|V_n - W_n\|_1$$
$$= \|U - V\|_1 + \|V - W\|_1 = \|U - W\|_1.$$

Using Proposition 8.23, the lemma follows. □

EXERCISE 8.26. Show that weak* convergence of a sequence of graphons does not imply convergence in the cut norm.

EXERCISE 8.27. Show that $\|W_n\|_\square \to 0$ ($W_n \in \mathcal{W}_1$) does not imply that $\|W_n\|_1 \to 0$.

CHAPTER 9

Szemerédi partitions

One of the most important tools in understanding large dense graphs is the Regularity Lemma of Szemerédi [1975, 1978] and its extensions. This lemma has many interesting connections to other areas of mathematics, including analysis and information theory (see Lovász and Szegedy [2007], Bollobás and Nikiforov [2008], Tao [2006a]). It also has weaker (but more effective) and stronger versions. Here we survey as much as we need from this rich theory, extend it to graphons (as it happens quite often, this leads to simpler, more elegant formulations), and prove a very general version of it using the space of graphons.

9.1. Regularity Lemma for graphs

9.1.1. Homogeneous bipartite graphs and the original lemma. For a graph $G = (V, E)$ and for $X, Y \subseteq V$, let $e_G(X, Y)$ denote the number of edges with one endnode in X and another in Y; edges with both endnodes in $X \cap Y$ are counted twice. We denote by

$$d_G(X,Y) = \frac{e_G(X,Y)}{|X||Y|}$$

the density of edges between X and Y. If X and Y are disjoint, we denote by $G[X, Y]$ the bipartite graph on $X \cup Y$ obtained by keeping just those edges of G that connect X and Y.

Let $\mathcal{P} = \{V_1, \ldots, V_k\}$ be a partition of V. We define the weighted graph $G_\mathcal{P}$ on V by taking the complete graph and weighting its edge uv by $d_G(V_i, V_j)$ if $u \in V_i$ and $v \in V_j$. A related, but different construction is that of the *template graph* of the partition $\mathcal{P}$. This weighted quotient graph $G/\mathcal{P}$ is defined on $[k]$: node i gets nodeweight $|S_i|/|V|$, and the edge ij gets edgeweight $e_G(S_i, S_j)/(|S_i||S_j|)$ (we allow loops here).

The Regularity Lemma says, roughly speaking, that the node set of every graph has an equitable partition $\mathcal{P}$ into a "small" number of classes such that $G_\mathcal{P}$ is "close" to G. Various (non-equivalent) forms of this lemma can be proved, depending on what we mean by "close".

Let G be a bipartite graph G with bipartition $\{U, W\}$. On the average, we expect that for $X \subseteq U$ and $Y \subseteq W$,

$$e_G(X, Y) \approx d_G(U, V)|X||Y|.$$

For two arbitrary subsets of the nodes, $e_G(X, Y)$ may be very far from this "expected value", but if G is a random graph, or at least "random-like", then it will be close; random graphs are very "homogeneous" in this respect. We say that G is ε-*homogeneous*, if

(9.1) $$\big|e_G(X,Y) - d_G(U,V)|X||Y|\big| \leq \varepsilon |U||W|$$

holds for all subsets $X \subseteq U$ and $Y \subseteq W$.

REMARK 9.1. Here we diverge from the usual statement of the Regularity Lemma: one usually considers ε-*regular* bipartite graphs, where the stronger condition

$$(9.2) \qquad \left| e_G(X,Y) - d_G(U,V)|X||Y| \right| \leq \varepsilon |X||Y|$$

is required for all $Y \subseteq U$ and $Y \subseteq W$ such that $|X| > \varepsilon|U|$ and $|Y| > \varepsilon|W|$. (Clearly we could not require condition (9.2) to hold for small X and Y: for example, if both have one element, then the quotient $e_G(X,Y)/(|X||Y|)$ is either 0 or 1.) The properties of ε-homogeneity and ε-regularity are essentially equivalent (see Exercise 9.6). In either version, this property can be viewed as a quantitative version of quasirandomness discussed in the Introduction.

With these definitions, the Regularity Lemma can be stated as follows:

LEMMA 9.2 (**Regularity Lemma, almost original form**). *For every $\varepsilon > 0$ there is an $S(\varepsilon) \in \mathbb{N}$ such that every graph $G = (V,E)$ has an equitable partition $\{V_1, \ldots, V_k\}$ ($1/\varepsilon \leq k \leq S(\varepsilon)$) such that for all but εk^2 pairs of indices $1 \leq i < j \leq k$, the bipartite graph $G[V_i, V_j]$ is ε-homogeneous.*

The important point is that the bound $S(\varepsilon)$ on the number of classes is independent of the graph G. Note that the Regularity Lemma does not say anything about the internal structure of the classes V_i. The lower bound $k \geq 1/\varepsilon$ guarantees that the number of edges inside the classes is bounded by $k(n/k)^2 \leq \varepsilon n^2$. The exceptional pairs of classes contain at most $\varepsilon k^2(n/k)^2 = \varepsilon n^2$ edges, so all these edges can be considered as "error terms". If we need information about the internal structure of the classes, we have to appeal to the Strong Regularity Lemma to be discussed below.

One feature of the Regularity Lemma, which unfortunately forbids practical applications, is that the upper bound $S(\varepsilon)$ it provides on the number of classes is very large: standard proofs give a tower $2^{2^{2^{\cdots}}}$ of height about $1/\varepsilon^2$, and unfortunately this is not far from the truth, as was shown by Gowers [2006] (for a simpler recent construction, see Conlon and Fox [2011]).

9.1.2. Weak Regularity Lemma. A version of the Regularity Lemma with a weaker conclusion but with a more reasonable error bound was proved by Frieze and Kannan [1999]. This is the form that we use most of the time in this book.

LEMMA 9.3 (**Weak Regularity Lemma**). *For every $k \geq 1$ and every graph $G = (V,E)$, V has a partition $\mathcal{P}$ into k classes such that*

$$d_\square(G, G_\mathcal{P}) \leq \frac{2}{\sqrt{\log k}}.$$

Note that we do not require here that $\mathcal{P}$ be an equitable partition; it is not hard to see that this version implies that there is also an equitable partition with similar property, just we have to increase the error bound to $4/\sqrt{\log k}$ (see Exercise 9.7).

To see the connection with the original lemma, we note that if G_0 is an ε-homogeneous bipartite graph, and H is the weighted complete bipartite graph with the same bipartition $\{U, W\}$ and with edge weights $d = \mathbf{e}(G_0)/(|U||W|)$, then (9.1) says that $d_\square(G_0, H) \leq \varepsilon$. Hence if $\mathcal{P}$ is a Szemerédi partition in the sense of Lemma

9.2, then the distance between the bipartite subgraph of G induced by V_i and V_j, and the corresponding weighted bipartite subgraph of $G_\mathcal{P}$, is at most ε for all but εk^2 pairs (i,j), and at most 1 for the remaining εk^2 pairs. This implies that the cut distance between G and $G_\mathcal{P}$ is at most 2ε. So the partition in Lemma 9.3 has indeed weaker properties than the partition in Lemma 9.2. This is compensated for by the relatively decent number of partition classes.

The Weak Regularity Lemma implies that there is a partition $\mathcal{P}$ such that the template graph satisfies

$$(9.3) \qquad \delta_\square(G, G/\mathcal{P}) \leq d_\square(G, G_\mathcal{P}) \leq \frac{2}{\sqrt{\log k}}.$$

9.1.3. Strong Regularity Lemma. Other versions of the Regularity Lemma strengthen, rather than weaken, the conclusion (of course, at the cost of replacing the tower function by an even more formidable value). Such a "super-strong" Regularity Lemma was proved by Alon, Fischer, Krivelevich and Szegedy [2000]. To state this lemma, we need a further definition. Let $\mathcal{P}$ be an equitable partition of $V(G)$, and let $\mathcal{Q}$ be an equitable refinement of it. Following Conlon and Fox [2011], we say that $\mathcal{Q}$ is ε-*close* to $\mathcal{P}$, if for almost every pair $S \neq T \in \mathcal{P}$ (with at most $\varepsilon|\mathcal{P}|^2$ exceptions), for almost every pair $X, Y \in \mathcal{Q}$ (with at most $(|\mathcal{Q}|/|\mathcal{P}|)^2$ exceptions), we have

$$\left| \frac{e_G(X,Y)}{|X||Y|} - \frac{e_G(S,T)}{|S||T|} \right| \leq \varepsilon.$$

LEMMA 9.4 (**Very Strong Regularity Lemma**). *For every sequence $\boldsymbol{\varepsilon} = (\varepsilon_0, \varepsilon_1, ...)$ of positive numbers there is a positive integer $S(\boldsymbol{\varepsilon})$ such that for every graph $G = (V, E)$, the node set V has an equitable partition $\mathcal{P}$ and an equitable refinement $\mathcal{Q}$ of $\mathcal{P}$ such that $|\mathcal{Q}| \leq S(\boldsymbol{\varepsilon})$, $\mathcal{P}$ is ε_0-regular, $\mathcal{Q}$ is $\varepsilon_{|\mathcal{P}|}$-regular, and $\mathcal{Q}$ is ε_0-close to $\mathcal{P}$.*

While this Very Strong Regularity Lemma has many important applications, it is not easy to explain its significance at this point. One important feature is that through the second partition $\mathcal{Q}$, it carries information about the inside of the partition classes of $\mathcal{P}$.

A somewhat weaker (but essentially equivalent) version, which is simpler to state but more difficult to apply, was proved by Tao [2006b] and by Lovász and Szegedy [2007].

LEMMA 9.5 (**Strong Regularity Lemma**). *For every sequence $\boldsymbol{\varepsilon} = (\varepsilon_0, \varepsilon_1, ...)$ of positive numbers there is a positive integer $S(\boldsymbol{\varepsilon})$ such that for every graph $G = (V, E)$, there is a graph G' on V, and V has a partition $\mathcal{P}$ into $k \leq S(\boldsymbol{\varepsilon})$ classes such that*

$$(9.4) \qquad d_1(G, G') \leq \varepsilon_0 \quad \text{and} \quad d_\square(G', (G')_\mathcal{P}) \leq \varepsilon_k.$$

Note that the first inequality involves the normalized edit distance, and so it is stronger than a similar condition with the cut distance would be. The second error bound ε_k in (9.4) can be thought of as being very small. If we choose $\varepsilon_k = \varepsilon/2$ for all k, we get the Weak Regularity Lemma 9.3 (without an explicit bound on the number of classes). Choosing $\varepsilon_k = \varepsilon_0^2/k^2$, the partition obtained satisfies the requirements of the Original Regularity Lemma 9.2.

We can replace ε_k by the much smaller number $\varepsilon_k/(k^2 S(\varepsilon_k)^2)$, where S is the bound in the Original Regularity Lemma. Then we can apply the Original

Regularity Lemma to each of the partition classes obtained in Lemma 9.5, to get the Very Strong Regularity Lemma 9.4. (The details of this derivation are left to the reader as an exercise.)

We will formulate the Strong Regularity Lemma for kernels, and prove it in that version, in Section 9.3.

EXERCISE 9.6. Show that (a) if a bipartite graph is ε-regular, then it is ε-homogeneous; (b) if a bipartite graph is ε^3-homogeneous, then it is ε-regular.

EXERCISE 9.7. Prove that for every $k \geq 1$ and every graph $G = (V, E)$, V has an equitable partition $\mathcal{P}$ into k classes such that $d_\square(G, G_\mathcal{P}) \leq 4/\sqrt{\log k}$.

9.2. Regularity Lemma for kernels

The Weak Regularity Lemma extends to kernels, and this is the form we are going to prove first. While stating and proving the Lemma directly would be quite easy, we make a detour by introducing the "stepping operator" formally and stating some basic properties. These will be useful later on.

9.2.1. The stepping operator. Let $W \in \mathcal{W}$ and let $\mathcal{P} = (S_1, \ldots, S_q)$ be a partition of $[0,1]$ into a finite number of measurable sets. (When we speak of a partition of $[0,1]$, we always mean such a partition.) We define the function $W_\mathcal{P}$ by

$$W_\mathcal{P}(x,y) = \frac{1}{\lambda(S_i)\lambda(S_j)} \int_{S_i \times S_j} W(x,y)\, dx\, dy \qquad (x \in S_i, y \in S_j).$$

So $W_\mathcal{P}$ is obtained by averaging W over the "steps" $S_i \times S_j$; it is a stepfunction with steps in $\mathcal{P}$. If $\lambda(S_i) = 0$ or $\lambda(S_j) = 0$, then we define $W_\mathcal{P}(x,y) = 0$ (this is just to have a complete definition; sets of measure zero in the partition can usually be ignored). We call this construction a *stepping* of W; it will be used throughout this book.

Analytically, the stepping operator is the orthogonal projection of the Hilbert space $L_2([0,1]^2)$ onto the subspace of stepfunctions with $\mathcal{P}$-steps. In probability language, it is the conditional expectation relative to the (finite) sigma-algebra generated by the sets in $\mathcal{P}$. These remarks may make some of the basic properties below easier to understand, but we will use the more elementary direct formulation.

Most of the information about the stepfunction $W_\mathcal{P}$ is contained in a finite weighted graph, the *quotient graph* $W/\mathcal{P}$. This is a weighted graph on $[q]$, with nodeweights $\alpha_i(W/\mathcal{P}) = \lambda(S_i)$ and edgeweights $\beta_{ij}(W/\mathcal{P}) = W_\mathcal{P}(x,y)$ for any $x \in S_i, y \in S_j$.

On the space $\mathcal{W}$, the stepping operator $W \mapsto W_\mathcal{P}$ is a linear operator which is idempotent and symmetric:

(9.5) $$\langle U_\mathcal{P}, W_\mathcal{P} \rangle = \langle U_\mathcal{P}, W \rangle = \langle U, W_\mathcal{P} \rangle.$$

Stepfunctions with steps in a fixed partition $\mathcal{P}$ form a finite dimensional linear space, and the stepping operator is the orthogonal projection onto this space, which is shown by the simple identity

(9.6) $$\langle W_\mathcal{P}, W - W_\mathcal{P} \rangle = 0.$$

This implies that stepping operator is contractive with respect to the L_2 norm:

(9.7) $$\|W_\mathcal{P}\|_2^2 = \|W\|_2^2 - \|W - W_\mathcal{P}\|_2^2 \leq \|W\|_2^2.$$

It is not hard to see that the stepping operator is also contractive with respect to the cut norm (Exercise 9.17). In fact, we will see in Section 14.2.1 that stepping is contractive with respect to any other reasonable norm on $\mathcal{W}$.

9.2.2. Weak Regularity Lemma. It is a basic fact from analysis that every kernel W can be approximated arbitrarily well by stepfunctions in the L_1 norm. The approximating stepfunctions can be obtained by averaging over "steps":

PROPOSITION 9.8. *Let $(\mathcal{P}_n)$ be a sequence of measurable partitions of $[0,1]$ such that every pair of points is separated by all but a finite number of partitions $\mathcal{P}_n$. Then $W_{\mathcal{P}_n} \to W$ almost everywhere for every $W \in \mathcal{W}$.* □

The Weak Regularity Lemma for kernels, proved by Frieze and Kannan [1999] (and in particular its Corollary 9.13 below), is a related statement about approximation by stepfunctions in the cut norm (instead of in the sense of almost everywhere convergence).

LEMMA 9.9 (**Weak Regularity Lemma for Kernels**). *For every function $W \in \mathcal{W}$ and $k \geq 1$ there is stepfunction U with k steps such that*

$$\|W - U\|_\square < \frac{2}{\sqrt{\log k}}\|W\|_2.$$

Roughly speaking, this Lemma says that every kernel can be approximated well in the cut norm by stepfunctions (in fact, by its steppings). Proposition 9.8 asserts something similar about approximating in the L_1-norm. Since $\|W\|_\square \leq \|W\|_1$, approximating in the L_1 norm seems to be a stronger result. However, the error in the L_1-norm approximation depends not only on the number of steps, but on W as well. The crucial fact about Lemma 9.9 is that the error tends to 0 as $k \to \infty$, uniformly in W.

The error bound in Lemma 9.9 is only attractive when compared with the error bound in the stronger versions; for a prescribed error ε, the number of partition classes we need is still exponential in $1/\varepsilon^2$. Frieze and Kannan give a stronger form of this result that provides a polynomial size description of the approximating stepfunction.

LEMMA 9.10. *For every kernel $U \in \mathcal{W}_1$ and $k \geq 1$ there are k pairs of subsets $S_i, T_i \subseteq [0,1]$ and k real numbers a_i such that*

$$\|U - \sum_{i=1}^{k} a_i \mathbb{1}_{S_i \times T_i}\|_\square < \frac{1}{\sqrt{k}}.$$

It is clear that the function $\sum_i a_i \mathbb{1}_{S_i \times T_i}$ is a stepfunction; we can make it symmetric by taking the average with $\sum_i a_i \mathbb{1}_{T_i \times S_i}$, getting $2k$ terms. This symmetric stepfunction has at most 2^{2k} steps, so Lemma 9.9 follows from Lemma 9.10 (replacing k by 2^{2k}).

We have mentioned the significance of the interplay between the cut norm and other kernel norms. The proof of the Regularity Lemma is the first point where this is apparent. For later reference, we state the key observation in the proof of the Weak Regularity Lemma separately, in two versions.

LEMMA 9.11. (a) *For every $U \in \mathcal{W}$ there are two sets $S, T \subseteq [0,1]$ and a real number $0 \leq a \leq \|U\|_\infty$ such that*

$$\|U - a\mathbb{1}_{S \times T}\|_2^2 \leq \|U\|_2^2 - \|U\|_\square^2.$$

(b) *Let $U \in \mathcal{W}$ and let $\mathcal{P}$ be a measurable k-partition of $[0,1]$. Then there is a partition $\mathcal{Q}$ refining $\mathcal{P}$ with at most $4k$ classes such that*

$$\|U - U_\mathcal{P}\|_\square = \|U_\mathcal{Q} - U_\mathcal{P}\|_\square.$$

Proof. Let S and T be measurable subsets of $[0,1]$ such that

$$\|U\|_\square = \left| \int_{S \times T} U \right| = |\langle U, \mathbb{1}_{S \times T} \rangle|,$$

where we may assume that $\langle U, \mathbb{1}_{S \times T} \rangle \geq 0$. Let $a = \frac{1}{\lambda(S)\lambda(T)} \|U\|_\square$. Then

(9.8) $$\|U - a\mathbb{1}_{S \times T}\|_2^2 = \|U\|_2^2 - \frac{1}{\lambda(S)\lambda(T)} \|U\|_\square^2 \leq \|U\|_2^2 - \|U\|_\square^2.$$

This proves (a).

The proof of (b) is similar. The inequality $\|U - U_\mathcal{P}\|_\square \geq \|U_\mathcal{Q} - U_\mathcal{P}\|_\square$ follows by the contractivity of the stepping operator (Exercise 9.17). To prove the other direction, let S and T be measurable subsets of $[0,1]$ such that $|\langle U - U_\mathcal{P}, \mathbb{1}_{S \times T} \rangle| = \|U - U_\mathcal{P}\|_\square$, and let $\mathcal{Q}$ denote the partition generated by $\mathcal{P}$, S and T. Clearly $\mathcal{Q}$ has at most $4k$ classes. Using (9.5), we get $\langle U, \mathbb{1}_{S \times T} \rangle = \langle U_\mathcal{Q}, \mathbb{1}_{S \times T} \rangle$, and hence

$$\|U - U_\mathcal{P}\|_\square = |\langle U - U_\mathcal{P}, \mathbb{1}_{S \times T} \rangle| = |\langle U_\mathcal{Q} - U_\mathcal{P}, \mathbb{1}_{S \times T} \rangle| \leq \|U_\mathcal{Q} - U_\mathcal{P}\|_\square.$$

This completes the proof. $\square$

Proof of Lemma 9.10. We apply Lemma 9.11(a) repeatedly, to get pairs of sets S_i, T_i and real numbers a_i such that the "remainders" $W_j = U - \sum_{i=1}^{j} a_i \mathbb{1}_{S_i \times T_i}$ satisfy

$$\|W_j\|_2^2 \leq \|U\|_2^2 - \sum_{i=0}^{j-1} \|W_i\|_\square^2.$$

Since the right hand side remains nonnegative, it follows that for every k there is a $0 \leq i < k$ with $\|W_i\|_\square^2 \leq 1/k$. Changing $a_{i+1}, \ldots, a_k$ to 0, we get the lemma. $\square$

The stepfunction approximating a given graphon W in Lemma 9.9 is usually not a stepping of W. Is the optimally approximating stepfunction necessarily a stepping of W? While this looks plausible, the answer is negative (see Exercise 9.18). But, as noted by Frieze and Kannan [1999], such steppings are almost optimal:

LEMMA 9.12. *Let $W \in \mathcal{W}_1$, let U be a stepfunction, and let $\mathcal{P}$ denote the partition of $[0,1]$ into the steps of U. Then*

$$\|W - W_\mathcal{P}\|_\square \leq 2\|W - U\|_\square.$$

Proof. Using that $U = U_\mathcal{P}$ and the contractivity of the stepping operator with respect to the cut norm, we get

$$\|W - W_\mathcal{P}\|_\square \leq \|W - U\|_\square + \|U - W_\mathcal{P}\|_\square = \|W - U\|_\square + \|U_\mathcal{P} - W_\mathcal{P}\|_\square$$
$$= \|W - U\|_\square + \|(U - W)_\mathcal{P}\|_\square \leq 2\|W - U\|_\square.$$

$\square$

Lemmas 9.9 and 9.12 imply:

COROLLARY 9.13. *For every function $W \in \mathcal{W}_1$ and $k \geq 1$ there is a partition $\mathcal{P}$ of $[0,1]$ into at most k sets with positive measure for which*
$$\|W - W_\mathcal{P}\|_\square \leq \frac{2}{\sqrt{\log k}}. \quad \square$$

A partition $\mathcal{P}$ of $[0,1]$ such that $\|W - W_\mathcal{P}\|_\square \leq \varepsilon$ will be called a *weak regularity partition of w with error ε*.

Lemma 9.11(b) provides an alternative way of getting this corollary. It is easy to check that $\langle U_\mathcal{Q} - U_\mathcal{P}, U_\mathcal{P} \rangle = \langle U - U_\mathcal{P}, U_\mathcal{P} \rangle = 0$ if $\mathcal{Q}$ is a refinement of $\mathcal{P}$. Hence
$$\|U - U_\mathcal{P}\|_\square^2 = \|U_\mathcal{Q} - U_\mathcal{P}\|_\square^2 \leq \|U_\mathcal{Q} - U_\mathcal{P}\|_2^2 = \|U_\mathcal{Q}\|_2^2 - \|U_\mathcal{P}\|_2^2$$
$$= \|U - U_\mathcal{P}\|_2^2 - \|U - U_\mathcal{Q}\|_2^2.$$

So $U_\mathcal{Q}$ is a better approximation of U in L_2 than $U_\mathcal{P}$, and the gain is at least as large as $\|U - U_\mathcal{P}\|_\square^2$. From here we can conclude just as in the proof above.

Coming back to the approximation described in Lemma 9.10, it is often useful to have a bound on the numbers a_i. With the notation of its proof, looking at the proof carefully, we see that
$$a_i = \frac{1}{\lambda(S_i)\lambda(T_i)}\|W_i\|_\square,$$
and
$$\frac{1}{\lambda(S_i)\lambda(T_i)}\|W_i\|_\square^2 = \|W_i\|_2^2 - \|W_{i+1}\|_2^2,$$
whence
$$\sum_{i=1}^k \lambda(S_i)\lambda(T_i)a_i^2 = \sum_{i=1}^k (\|W_i\|_2^2 - \|W_{i+1}\|_2^2) = \|U\|_2^2 - \|W_k\|_2^2 \leq \|U\|_2^2.$$

This bound allows a_i to be large when $\lambda(S_i)\lambda(T_i)$ is small, but it is easy to fix the argument to get a more useful bound. Instead of choosing the optimal sets S_i and T_i, we choose a pair S_i, T_i such that $\lambda(S_i), \lambda(T_i) \geq 1/2$, and
$$\langle W_i, \mathbb{1}_{S_i \times T_i} \rangle \geq \frac{1}{4}\|W_i\|_\square.$$

It is easy to see that such a pair exists (cf. Exercise 8.4). Then we get the following:

LEMMA 9.14. *For every kernel $W \in \mathcal{W}_1$ and $k \geq 1$ there are k pairs of subsets $S_i, T_i \subseteq [0,1]$ and k real numbers a_i such that $\sum_i a_i^2 \leq 4$ and*
$$\|W - \sum_{i=1}^k a_i \mathbb{1}_{S_i \times T_i}\|_\square < \frac{4}{\sqrt{k}}.$$

We can easily add other requirements in Lemma 9.9.

LEMMA 9.15. *Let $W \in \mathcal{W}_1$ and $1 \leq m < k$.*

(a) *For every m-partition $\mathcal{Q}$ of $[0,1]$ there is k-partition $\mathcal{P}$ refining $\mathcal{Q}$ such that*
$$\|W - W_\mathcal{P}\|_\square \leq \frac{2}{\sqrt{\log k/m}}.$$

(b) *For every m-partition $\mathcal{Q}$ of $[0,1]$ there is an equipartition $\mathcal{P}$ with k classes such that*
$$\|W - W_\mathcal{P}\|_\square \leq 2\|W - W_\mathcal{Q}\|_\square + \frac{2m}{k}.$$

Proof. Statement (a) follows by the same argument as Lemma 9.9, just starting with $\mathcal{Q}$ instead of the indiscrete partition. To prove (b), we partition each class of $\mathcal{Q}$ into classes of measure $1/k$, with at most one exceptional class of size less then $1/k$. Keeping all classes of size $1/k$, let us take the union of exceptional classes, and repartition it into classes of size $1/k$, to get a partition $\mathcal{P}$.

To analyze this construction, let us also consider the common refinement $\mathcal{R} = \mathcal{P} \wedge \mathcal{Q}$. Then $W_\mathcal{R}$ and $W_\mathcal{P}$ differ on a set of measure less than $2(m/k)$, and so

$$\|W - W_\mathcal{P}\|_\square \leq \|W - W_\mathcal{R}\|_\square + \frac{2m}{k}.$$

Lemma 9.12 implies that $\|W - W_\mathcal{R}\|_\square \leq 2\|W - W_\mathcal{Q}\|_\square$, which completes the proof. $\square$

9.2.3. Strong Regularity Lemma.
The Strong Regularity Lemma too has a "continuous" version:

LEMMA 9.16 (**Strong Regularity Lemma for Kernels**). *For every sequence $\boldsymbol{\varepsilon} = (\varepsilon_0, \varepsilon_1, ...)$ of positive numbers there is a positive integer $S(\boldsymbol{\varepsilon})$ such that for every graphon W, there is another graphon W', and a stepfunction $U \in \mathcal{W}_0$ with $k \leq S(\boldsymbol{\varepsilon})$ steps such that*

$$\|W - W'\|_1 \leq \varepsilon_0 \quad \text{and} \quad \|W' - U\|_\square \leq \varepsilon_k. \tag{9.9}$$

We will give a proof of this Lemma, deriving it from an even more general theorem, in the next section. Here we sketch how to derive the graph version 9.5 from the kernel version. Let $(\varepsilon_0, \varepsilon_1, ...)$ be a sequence of positive numbers, which we may assume is monotone decreasing. Let G be a simple graph on $[n]$. We apply Lemma 9.16 with $\varepsilon_k/2$ to W_G, to get a threshold S' (depending only on $(\varepsilon_0, \varepsilon_1, ...)$), a kernel W' and a partition $\mathcal{P}$ of $[0,1]$ such that $|\mathcal{P}| \leq S'$, $\|W_G - W'\|_1 \leq \varepsilon_0/2$ and $\|W' - W'_\mathcal{P}\|_\square \leq \varepsilon_k/2$.

First, we have to turn W' into a graph G'. This can be done by randomization. Let $I_i = ((i-1)/n, i/n]$ and $R_{ij} = I_i \times I_j$. We connect i and j with probability $n^2 \int_{R_{ij}} W'$. The probability that this edge will be in the symmetric difference of $E(G)$ and $E(G')$ is at most $n^2 \int_{R_{ij}} |W_G - W'|$, and hence the expected (normalized) edit distance between G and G' is at most $\|W_G - W'\|_1 \leq \varepsilon_0/2$. Markov's inequality gives that with probability at least $1/2$, the distance $d_1(G, G') \leq \varepsilon_0$.

Next, we have to turn the partition $\mathcal{P}$ of $[0,1]$ into a partition $\mathcal{Q}$ of $[n]$. We do this randomly again, by selecting a uniform random point $X_i \in I_i$ ($i = 1, ..., n$), and putting i into the m-th class of $\mathcal{Q}$ if X_i is in the m-th class of $\mathcal{P}$. A bit trickier computation with second moments (which is similar to the proof of Proposition 12.19, but simpler, and is not given here) shows that with high probability, $d_\square(G', (G')_\mathcal{Q}) \leq \varepsilon_k/2 + 10/\sqrt{n}$.

Now we choose $k_0 = \max(k'_0, 400/\varepsilon^2_{k'_0})$. If $n \leq k_0$, then we can take $G = G'$ and partition $[n]$ into singletons. If $n > k_0$, then with positive probability the partition $\mathcal{Q}$ constructed above satisfies $|\mathcal{Q}| = k \leq k'_0 \leq k_0$, $d_1(G, G') \leq \varepsilon_0$ and $d_\square(G', (G')_\mathcal{Q}) \leq \varepsilon_k/2 + 10/\sqrt{n} \leq \varepsilon_k$.

EXERCISE 9.17. Prove that the stepping operator is contractive with respect to the L_1 norm and the cut norm.

EXERCISE 9.18. Show by an example that the best approximation in the cut norm of a function $W \in \mathcal{W}_1$ by a stepfunction with a given number of steps is not necessarily a stepping of W. Is stepping the best approximation in the L_2 or in the L_1 norm?

EXERCISE 9.19. Are analogues of Lemma 9.12 valid for the L_1 L_2 and L_∞ norms?

EXERCISE 9.20. Formulate and prove the original Regularity Lemma for kernels.

EXERCISE 9.21. Give a proof of the Strong Regularity Lemma 9.16 along the lines of the proof of Lemma 9.9.

EXERCISE 9.22. Let $\mathcal{K}_1, \mathcal{K}_2, \ldots$ be arbitrary nonempty subsets of a Hilbert space $\mathcal{H}$. Prove for every $\varepsilon > 0$ and $f \in \mathcal{H}$ there is an integer $1 \leq m \leq \lceil 1/\varepsilon^2 \rceil$ and a vector $f_0 = \alpha_1 f_1 + \cdots + \alpha_m f_m$ ($\alpha_i \in \mathbb{R}, f_i \in \mathcal{K}_i$ such that for every $g \in \mathcal{K}_{m+1}$ we have $|\langle g, f - f_0 \rangle| \leq \varepsilon \|g\| \|f\|$. Derive the weak, original, and strong lemmas by choosing the sets $\mathcal{K}_i$ appropriately.

9.3. Compactness of the graphon space

In this section we prove a theorem of Lovász and Szegedy [2007] that is equivalent (at least in a non-effective sense) to all versions of the Regularity Lemma. To be more precise, we will derive the theorem from the Weak Regularity Lemma, and then we will show that the Strong Regularity Lemma can be derived from it quite easily. In a sense, this theorem can be considered as the strongest form of regularity.

THEOREM 9.23. *The space $(\widetilde{\mathcal{W}_0}, \delta_\square)$ is compact.*

Proof. In a metric space, it suffices to prove that every sequence $W_1, W_2, \ldots$ of graphons has a convergent subsequence.

For every $n \geq 1$, we can construct the partitions $\mathcal{P}_{n,k}$ of $[0,1]$ ($k = 1, 2, \ldots$), using Lemma 9.15, such that these partitions and the corresponding stepfunctions $W_{n,k} = (W_n)_{\mathcal{P}_{n,k}} \in \mathcal{W}_0$ satisfy the following conditions:

(i) $\|W_n - W_{n,k}\|_\square \leq 1/k$,

(ii) The partition $\mathcal{P}_{n,k+1}$ refines $\mathcal{P}_{n,k}$,

(iii) $|\mathcal{P}_{n,k}| = m_k$ depends only on k.

Once we have such partitions, we can rearrange the points of $[0,1]$ for every fixed n by a measure preserving bijection so that every partition class in every $\mathcal{P}_{n,k}$ is an interval.

CLAIM 9.24. *We can replace the sequence (W_n) by a subsequence so that for every k, the sequence $W_{n,k}$ converges almost everywhere to a stepfunction U_k with m_k steps as $n \to \infty$.*

Indeed, we can select a subsequence of the W_n for which the length of the i-th interval of $W_{n,1}$ converges for every i, and also the value of $W_{n,1}$ on the product of the i-th and j-th intervals converges for every i and j (as $n \to \infty$). It follows then that the sequence $W_{n,1}$ converges to a limit U_1 almost everywhere, which itself is a stepfunction with m_1 steps that are intervals.

We repeat this for $k = 2, 3, \ldots$, to get subsequences for which $W_{k,n} \to U_k$ almost everywhere, where U_k is a stepfunction with m_k steps that are intervals. As usual, we always keep the k-th function after the k-th step. This yields the subsequence with the properties in the Claim.

Let $\mathcal{P}_k$ denote the partition of $[0,1]$ into the steps of U_k. For every $k < l$, the partition $\mathcal{P}_{n,l}$ is a refinement of the partition $\mathcal{P}_{n,k}$, and hence $W_{n,k} = (W_{n,l})_{\mathcal{P}_{n,k}}$. It is easy to see that this kind of relation is inherited by the limiting stepfunctions:

$$(9.10) \qquad U_k = (U_l)_{\mathcal{P}_k}.$$

Let (X,Y) be a random point in $[0,1]^2$ chosen uniformly, then (9.10) implies that the sequence $(U_1(X,Y), U_2(X,Y), \dots)$ is a martingale. Since the random variables $U_i(X,Y)$ remain bounded, the Martingale Convergence Theorem A.12 implies that this sequence is convergent with probability 1. In other words, the sequence of functions $(U_1, U_2, \dots)$ is convergent almost everywhere. Let U be its limit; we show that $\|U - W_n\|_\square \to 0$.

Fix any $\varepsilon > 0$. Then there is a $k > 3/\varepsilon$ such that $\|U - U_k\|_1 < \varepsilon/3$. Fixing this k, there is an n_0 such that $\|U_k - W_{n,k}\|_1 < \varepsilon/3$ for all $n \geq n_0$. Then

$$\delta_\square(U, W_n) \leq \delta_\square(U, U_k) + \delta_\square(U_k, W_{n,k}) + \delta_\square(W_{n,k}, W_n)$$
$$\leq \|U - U_k\|_1 + \|U_k - W_{n,k}\|_1 + \delta_\square(W_{n,k}, W_n) \leq \frac{\varepsilon}{3} + \frac{\varepsilon}{3} + \frac{\varepsilon}{3} = \varepsilon.$$

This completes the proof of Theorem 9.23. $\square$

The theorem remains valid if we replace $\widetilde{\mathcal{W}}_0$ by any uniformly bounded subset of $\widetilde{\mathcal{W}}$, closed in the $\delta_\square$ distance. For example, the space $(\widetilde{\mathcal{W}}_1, \delta_\square)$ is also compact. This can be proved by the same argument, or by noticing that $W \mapsto 2W - 1$ is a mapping $(\widetilde{\mathcal{W}}_0, \delta_\square) \to (\widetilde{\mathcal{W}}_1, \delta_\square)$ that is continuous and surjective, and so it preserves compactness.

An easy consequence of Theorem 9.23 is the following:

COROLLARY 9.25. *For every $\varepsilon > 0$ there is an integer $k(\varepsilon) \geq 1$ such that simple graphs with k_ε nodes form an ε-net in $(\mathcal{W}_0, \delta_\square)$.*

We conclude this section with showing that Theorem 9.23 implies the Strong Regularity Lemma quite easily.

Proof of Lemma 9.16. Every graphon W is the limit of stepfunctions in the $\|.\|_1$ norm, hence there is a stepfunction $U \in \mathcal{W}_0$ with $\|W - U\|_1 \leq \varepsilon_0$. This means that the sets $B_1(U, \varepsilon_0)$, where U is a stepfunction, cover the whole space $\mathcal{W}_0$. Unfortunately, these sets are not open in the $d_\square$ metric. Therefore, we take a little larger sets. Let $k(U)$ denote the number of steps of a stepfunction U, and define

$$A_U = \{W \in \mathcal{W}_0 : (\exists V \in \mathcal{W}_0) \, \|U - V\|_\square < \varepsilon_{k(U)}, \, \|V - W\|_1 < \varepsilon_0\}.$$

CLAIM 9.26. *The set A_U is open in the cut norm.*

Indeed, let $W \in A_U$ and $W_n \in \mathcal{W}_0$ such that $\|W - W_n\|_\square \to 0$. By the definition of A_U, there is a graphon V such that $\|U - V\|_\square < \varepsilon_{k(U)}$ and $\|V - W\|_1 < \varepsilon_0$. By Proposition 8.25, there are graphons V_n such that $\|V - V_n\|_\square \to 0$ and $\|W_n - V_n\|_1 \to \|V - W\|_1 < \varepsilon_0$. So if n is large enough, we have $\|U - V_n\|_\square < \varepsilon_{k(U)}$ and $\|V_n - W_n\|_1 < \varepsilon_0$, showing that $W_n \in A_U$.

Now we have to go to the factor space $\widetilde{\mathcal{W}}_0$. For every stepfunction U, we consider the sets

$$\widetilde{A}_U = \{W \in \widetilde{\mathcal{W}}_0 : (\exists V \in \widetilde{\mathcal{W}}_0) \, \delta_\square(U, V) < \varepsilon_{k(U)}, \, \delta_1(V, W) < \varepsilon_0\}.$$

It is easy to see, using Claim 9.26, that $\widetilde{A}_U$ is open in $(\widetilde{\mathcal{W}}_0, \delta_\square)$. The sets $\widetilde{A}_U$ cover the whole space, so by the compactness of the space (Theorem 9.23), we obtain a finite set of stepfunctions $U_1, \ldots, U_t$ such that $\cup_{i=1}^t \widetilde{A}_{U_i} = \widetilde{\mathcal{W}}_0$.

We claim that we can choose $S(\varepsilon) = \max_{i \leq t} k(U_i)$ to satisfy the requirements of the lemma. Indeed, for every graphon W there is a stepfunction U_i ($1 \leq i \leq t$) such that $W \in \widetilde{A}_U$, which means that there is a graphon V such that $\delta_\square(U_i, V) < \varepsilon_{k(U_i)}$ and $\delta_1(V, W) < \varepsilon_0$. We can apply a measure preserving bijections φ, ψ to get $d_1(V^\varphi, W) < \varepsilon_0$ and $\delta_\square(U_i^{\psi \circ \varphi}, V^\varphi) < \varepsilon_{k(U_i)}$. Since $U_i^{\psi \circ \varphi}$ is a stepfunction with $k(U_i)$ steps, we can take $U = U_i^{\psi \circ \varphi}$ and $W' = V^\varphi$ to complete the proof. □

EXERCISE 9.27. Prove that for every $\varepsilon > 0$ there is a positive integer $S(\varepsilon)$ such that for every graphon W there is another graphon W' and a stepfunction $U \in \mathcal{W}_0$ with $k \leq S(\varepsilon)$ steps such that $\|W - W'\|_\square \leq \varepsilon/k!$ and $\|W' - U\|_1 \leq \varepsilon$.

9.4. Fractional and integral overlays

Using the Regularity Lemma, we are now ready to discuss the problem of comparing the two distances $\delta_\square$ and $\widehat{\delta}_\square$, as raised in Section 8.1.4. If two graphs G_1 and G_2 have the same number of nodes, then the inequality

$$\delta_\square(G_1, G_2) \leq \widehat{\delta}_\square(G_1, G_2),$$

is easy, but what can we say in the other direction? (This is admittedly a very technical issue, but it comes up in all sorts of arguments.) Perhaps the following very close connection is true:

CONJECTURE 9.28. For any two simple graphs G and G' on n nodes, $\widehat{\delta}_\square(G, G') \leq 2\delta_\square(G, G')$.

Unfortunately, I can only offer some weaker bounds (but these will be sufficient for the applications later). We will need these results for weighted graphs too. The following theorem is a combination of results of Borgs, Chayes, Lovász, Sós and Vesztergombi [2008] and Alon [unpublished].

THEOREM 9.29. *For any two edge-weighted graphs H_1 and H_2 with the same number n of nodes, with edgeweights in $[0, 1]$, we have the following inequalities:*

$$(9.11) \qquad \widehat{\delta}_\square(H_1, H_2) \leq n^6 \delta_\square(H_1, H_2),$$

$$(9.12) \qquad \widehat{\delta}_\square(H_1, H_2) \leq \delta_\square(H_1, H_2) + \frac{17}{\sqrt{\log n}},$$

$$(9.13) \qquad \widehat{\delta}_\square(H_1, H_2) \leq \frac{45}{\sqrt{-\log \delta_\square(H_1, H_2)}}.$$

Proof. The first inequality is quite easy. Let (X_{ui}) be an optimal fractional overlay of H_1 and H_2. We claim that there is a bijection $\pi : V_1 \to V_2$ such that $X_{u,\pi(u)} \geq 1/n^3$ for all $u \in V(H_1)$. This follows from the Marriage Theorem: if there is no such bijection, then there are two sets $S \subseteq V_1$ and $T \subseteq V_2$ such that $|S| + |T| > n$ and $X_{st} < 1/n^3$ for all $s \in S$ and $t \in T$. Then $X(S, T) \leq |S||T|/n^3 < 1/n$. On the other hand,

$$X(S, T) = X(S, V_2) - X(S, V_2 \setminus T) \geq \frac{|S|}{n} - \frac{|V_2 \setminus T|}{n} = \frac{|S| + |T| - n}{n} \geq \frac{1}{n},$$

a contradiction.

Let Y be the fractional overlay corresponding to the bijection π. Then $Y \leq n^3 X$, and hence
$$\widehat{\delta}_\square(H_1, H_2) \leq d_\square(H_1, H_2, Y) \leq n^6 d_\square(H_1, H_2, X) = n^6 \delta_\square(H_1, H_2).$$
This proves (9.11).

To prove the second inequality (9.12), let $k = \lfloor n^{1/3} \rfloor$. By the Weak Regularity Lemma 9.3, there are partitions $\mathcal{P} = \{V_1, \ldots, V_k\}$ of $V(H_1)$ and $\mathcal{Q} = \{U_1, \ldots, U_k\}$ of $V(H_2)$ into k almost equal classes so that
$$d_\square(H_1, (H_1)_\mathcal{P}),\ d_\square(H_2, (H_2)_\mathcal{Q}) \leq \frac{4}{\sqrt{\log k}}.$$
For the weighted k-node "template" graphs $H_1/\mathcal{P}$ and $H_2/\mathcal{Q}$ we have
$$\delta_\square(H_1/\mathcal{P}, H_2/\mathcal{Q}) = \delta_\square((H_1)_\mathcal{P}, (H_2)_\mathcal{Q})$$
$$\leq \delta_\square(H_1, H_2) + \delta_\square(H_1, (H_1)_\mathcal{P}) + \delta_\square(H_2, (H_2)_\mathcal{Q})$$
$$\leq \delta_\square(H_1, H_2) + \frac{8}{\sqrt{\log k}}.$$

Let $(X_{ij})_{i,j=1}^k$ be an optimal fractional overlay of $H_1/\mathcal{P}$ and $H_2/\mathcal{Q}$. We define a bijection $\varphi : V(H_1) \to V(H_2)$ by mapping $\lfloor X_{ij} n \rfloor$ nodes of V_i to U_j arbitrarily. This is possible, since
$$\sum_{j=1}^k \lfloor X_{ij} n \rfloor \leq \sum_{j=1}^k X_{ij} n = |V_i| \quad \text{and} \quad \sum_{i=1}^k \lfloor X_{ij} n \rfloor \leq \sum_{i=1}^k X_{ij} n = |U_j|.$$
The nodes left in the two graphs are matched with each other arbitrarily.

The bijection φ between $V(H_1) = V((H_1)_\mathcal{P})$ and $V(H_2) = V((H_2)_\mathcal{Q})$ defines a fractional overlay Y between $H_1/\mathcal{P}$ and $H_2/\mathcal{Q}$, such that
$$d_\square(\varphi((H_1)_\mathcal{P}), (H_2)_\mathcal{Q}) = d_\square(H_1/\mathcal{P}, H_2/\mathcal{Q}, Y).$$

The fractional overlays X and Y are very close: $|X_{ij} - Y_{ij}| \leq 1/n$ for every $1 \leq i, j \leq k$. Hence it follows that
$$d_\square(H_1/\mathcal{P}, H_2/\mathcal{Q}, Y) \leq d_\square(H_1/\mathcal{P}, H_2/\mathcal{Q}, X) + \frac{k^2}{n} = \delta_\square(H_1/\mathcal{P}, H_2/\mathcal{Q}) + \frac{k^2}{n}.$$
Combining, we get
$$\widehat{\delta}(H_1, H_2) \leq d_\square(\varphi(H_1), H_2) \leq d_\square(\varphi((H_1)_\mathcal{P}), (H_2)_\mathcal{Q}) + \frac{8}{\sqrt{\log k}}$$
$$= d_\square(H_1/\mathcal{P}, H_2/\mathcal{Q}, Y) + \frac{8}{\sqrt{\log k}} \leq \delta_\square(H_1/\mathcal{P}, H_2/\mathcal{Q}) + \frac{k^2}{n} + \frac{8}{\sqrt{\log k}}$$
$$\leq \delta_\square(H_1, H_2) + \frac{k^2}{n} + \frac{16}{\sqrt{\log k}}.$$

Recalling the choice of k, we get (9.12).

Finally, the third inequality (9.13) follows easily from the first two. If $n < \delta_\square(H_1, H_2)^{-1/7}$, then (9.11) implies that
$$\widehat{\delta}(H_1, H_2) \leq \delta_\square(H_1, H_2)^{1/7} < \frac{55}{\sqrt{-\log \delta_\square(H_1, H_2)}},$$

while for $n \geq \delta_\square(H_1, H_2)^{-1/7}$, (9.12) gives that

$$\widehat{\delta}_\square(H_1, H_2) \leq \delta_\square(H_1, H_2) + \frac{17}{\sqrt{\log(\delta_\square(H_1, H_2)^{-1/7})}} < \frac{45}{\sqrt{-\log \delta_\square(H_1, H_2)}}. \quad \square$$

The first and third inequalities in the theorem are very weak (at least, in comparison with the conjectured bound of $\widehat{\delta}_\square \leq 2\delta_\square$). Nevertheless, (9.13) will be important for us, since it implies that $\delta_\square$ and $\widehat{\delta}_\square$ define the same Cauchy sequences of graphs. Borgs, Chayes, Lovász, Sós and Vesztergombi [2008] prove a stronger inequality of this nature:

(9.14) $$\widehat{\delta}_\square(H_1, H_2) \leq 32\delta_\square(H_1, H_2)^{1/67}.$$

Since this is still far from Conjecture 9.28, we don't reproduce the proof here.

We can ask the same question about any of the unlabeled distances: if two graphs have the same number of nodes, is the distance between them defined through optimal overlay essentially the same as the distance defined by going to the associated graphons and considering their distance? For the edit distance, an affirmative answer was proved by Pikhurko [2010]; this bound is much stronger than 9.14: it is optimal except for the constant 3. We prove it in a bit more general form, for weighted graphs, since we will need it.

THEOREM 9.30. *For any two edge-weighted graphs H_1 and H_2 on $[n]$ we have the following inequalities:*

$$\delta_1(H_1, H_2) \leq \widehat{\delta}_1(H_1, H_2) \leq 3\delta_1(H_1, H_2).$$

Proof. The first inequality is trivial. Let A and B be the adjacency matrices of H_1 and H_2, respectively. Then we have

$$\widehat{\delta}_1(H_1, H_2) = \min_P \|A - PBP\|_1,$$

where P ranges over all permutation matrices. To express the distance δ_1 is a bit more complicated:

$$\delta_1(H_1, H_2) = \min_X d_1(H_1, H_2, X),$$

where

$$d_1(H_1, H_2, X) = \sum_{i,j,u,v \in [n]} X_{iu} X_{jv} |A_{ij} - B_{uv}|,$$

and X ranges over all $n \times n$ fractional overlays. We want to prove that

(9.15) $$\widehat{\delta}_1(H_1, H_2) \leq 3 d_1(H_1, H_2, X)$$

for every fractional overlay X. It suffices to prove this for overlays X of the form

$$X = \frac{1}{mn} \sum_{k=1}^m P_k,$$

where the P_k are $n \times n$ permutation matrices, because matrices of this form are dense among all fractional overlays by the Birkhoff–von Neumann Theorem. Then

$$d_1(H_1, H_2, X) = \frac{1}{m^2 n^2} \sum_{k,l=1}^m \sum_{i,j=1}^n |A_{ij} - B_{P_k(u), P_l(v)}| = \frac{1}{m^2} \sum_{k,l=1}^m \|A - P_k B P_l\|_1.$$

We need the following simple inequality for three permutation matrices P, Q and R:

(9.16) $\qquad \|A - PBP\|_1 \leq \|A - PBQ\|_1 + \|A - QBR\|_1 + \|A - RBP\|_1.$

Indeed, by the triangle inequality and the invariance of the $\|.\|_1$-norm under permutations of rows and columns, we have

$$\|A - PBP\|_1 \leq \|A - PBQ\|_1 + \|PR^{-1}A - PBQ\|_1 + \|PR^{-1}A - PBP\|_1$$
$$= \|A - PBQ\|_1 + \|A - RBQ\|_1 + \|A - RBP\|_1.$$

Transposing the middle term, we get (9.16).

Averaging this inequality with P, Q, R ranging over $\{P_1, \ldots, P_m\}$, we get

$$\frac{1}{m}\sum_{k=1}^{m} \|A - P_k B P_k\|_1 \leq \frac{3}{m^2} \sum_{k,l=1}^{m} \|A - P_k B P_l\|_1 = 3 d_1(H_1, H_2, X).$$

Since $(1/m)\sum_k \|A - P_k B P_k\|_1 \geq \widehat{\delta}(H_1, H_2)$, this completes the proof. $\square$

EXERCISE 9.31. Show by an example that $\delta_1 \neq \widehat{\delta}_1$ in general.

9.5. Uniqueness of regularity partitions

Is the regularity partition of a graph (weak, original or strong) uniquely determined? Perhaps uniquely determined up to some error?

The answer to this question is negative. The image of any kind of regularity partition under any automorphism of a graph G is another regularity partition with the same properties. Typically, we get this way many different partitions. For example, consider the graph G on $[n]$ in which every node i is connected to $i \pm 1, i \pm 2, \ldots i \pm \lceil n/4 \rceil$ (modulo m). A regularity partition (in the original sense, and also in the strong sense) is obtained by splitting $[n]$ into k blocks of consecutive numbers of approximately equal size, and rotations give many different partitions of this kind.

Giving up the uniqueness of the partition itself, we can ask about the approximate uniqueness of the template graph $G/\mathcal{P}$. For partitions in the weak or in the original sense, even this does not hold. Let (F_n) be a quasirandom graph sequence with $\mathsf{v}(F_n) = n$. Consider the direct product $G = K_n \times F_n$, and the two n-partitions $\mathcal{P}_1$ and $\mathcal{P}_2$ induced by the projections onto the two factors. The bipartite graph between any two classes of $\mathcal{P}_1$ is $K_2 \times F_n$, which is ε-homogeneous for an arbitrarily small ε if n is large enough. So $\mathcal{P}_1$ is a regularity partition with template graph K_n, where every edge has weight $1/2$. On the other hand, the subgraph of G between two classes of $\mathcal{P}_2$ is either edgeless, or it is a complete bipartite graph with a perfect matching removed. Both of these are ε-homogeneous if n is large enough, and the template graph is F_n.

But for strong regularity partitions, Alon, Shapira and Stav [2009] did prove a uniqueness result. We state it for graphons, and for our version of the Strong Regularity lemma; for other versions, we refer to the paper.

THEOREM 9.32. *Let $\varepsilon > 0$, let W, W_1, W_2 be graphons and let $\mathcal{P}_1, \mathcal{P}_2$ be equipartitions of $[0,1]$ into k parts so that $\|W - W_i\|_1 \leq \varepsilon$ and $\|W_i - (W_i)_{\mathcal{P}_i}\|_\square \leq \varepsilon/k^4$. Then $\delta_1(W_1/\mathcal{P}_1, W_2/\mathcal{P}_2) \leq 8\varepsilon$.*

9.5. UNIQUENESS OF REGULARITY PARTITIONS

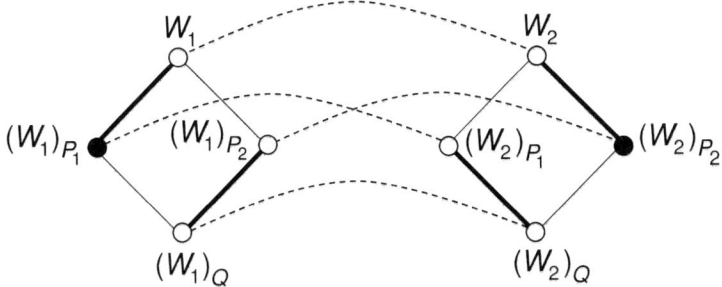

FIGURE 9.1. Proving uniqueness of strong regularity partitions. Heavy lines mean small cut distance, broken lines mean small L_1-distance.

Proof. Let $\mathcal{Q} = \mathcal{P}_1 \vee \mathcal{P}_2$ denote the common refinement of $\mathcal{P}_1$ and $\mathcal{P}_2$; clearly $\mathcal{Q}$ has at most k^2 classes, and $\|W_1 - W_2\| \le 2\varepsilon$. By the contractivity of the stepping operator (Exercise 9.17 of Proposition 14.13), we have

$$\|(W_1)_{\mathcal{P}_1} - (W_2)_{\mathcal{P}_1}\|_1 \le 2\varepsilon, \quad \|(W_1)_{\mathcal{P}_2} - (W_2)_{\mathcal{P}_2}\|_1 \le 2\varepsilon, \quad \|(W_1)_{\mathcal{Q}} - (W_2)_{\mathcal{Q}}\|_1 \le 2\varepsilon,$$

and

$$\|(W_1)_{\mathcal{P}_2} - (W_1)_{\mathcal{Q}}\|_\square \le \|W_1 - (W_1)_{\mathcal{P}_1}\|_\square \le \varepsilon/k^4,$$
$$\|(W_2)_{\mathcal{P}_1} - (W_1)_{\mathcal{Q}}\|_\square \le \|W_2 - (W_2)_{\mathcal{P}_2}\|_\square \le \varepsilon/k^4$$

(see Figure 9.1 for the chain of small distances followed by the proof). Hence

$$\delta_1(W_1/\mathcal{P}_1, W_2/\mathcal{P}_2) = \delta_1((W_1)_{\mathcal{P}_1}, (W_2)_{\mathcal{P}_2}) \le \|(W_1)_{\mathcal{P}_1} - (W_2)_{\mathcal{P}_2}\|_1$$
$$\le \|(W_1)_{\mathcal{P}_1} - (W_2)_{\mathcal{P}_1}\|_1 + \|(W_2)_{\mathcal{P}_1} - (W_1)_{\mathcal{Q}}\|_1$$
$$+ \|(W_1)_{\mathcal{Q}} - (W_2)_{\mathcal{Q}}\|_1 + \|(W_2)_{\mathcal{P}_1} - (W_1)_{\mathcal{Q}}\|_1$$
$$+ \|(W_1)_{\mathcal{P}_2} - (W_2)_{\mathcal{P}_2}\|_1.$$

By the trivial inequality (8.15),

$$\|(W_2)_{\mathcal{P}_1} - (W_1)_{\mathcal{Q}}\|_1 \le k^4 \|(W_2)_{\mathcal{P}_1} - (W_1)_{\mathcal{Q}}\|_\square \le \varepsilon,$$

and similarly

$$\|(W_1)_{\mathcal{P}_2} - (W_2)_{\mathcal{Q}}\|_1 \le \varepsilon.$$

Substituting these bounds, the theorem follows. □

The proof above bounds the fractional edit distance of the two template graphs $(W_1)/\mathcal{P}_1$ and $(W_2)/\mathcal{P}_2$. Using Pikhurko's Theorem 9.30, we could replace it by the integral version of the edit distance $\widehat{\delta}_1$, at the cost of another factor of 3. Using Exercise 8.18, we could replace the bound $\|W_i - (W_i)_{\mathcal{P}_i}\|_\square \le \varepsilon/k^4$ by $\|W_i - (W_i)_{\mathcal{P}_i}\|_\square \le \varepsilon/(2k)$ Also, we could derive similar bounds for the weighted graphs $W/\mathcal{P}_1$ and $W/\mathcal{P}_2$.

CHAPTER 10

Sampling

We turn to the analysis of sampling from a graph, our basic method of gathering information about very large dense graphs. In fact, most of the time we prove our results in the framework of sampling from a graphon. We start with describing what it means to sample from a graphon.

10.1. W-random graphs

A graphon W gives rise to a way of generating random graphs that are more general than the Erdős–Rényi graphs. This construction was introduced independently by Diaconis and Freedman [1981], Boguñá and Pastor-Satorras [2003], Lovász and Szegedy [2006], and Bollobás, Janson and Riordan [2007], and quite probably implicitly by others.

Given a graphon W and an ordered set $S = (x_1, \ldots, x_n)$, where $x_i \in [0,1]$, we define a weighted graph $\mathbb{H}(S, W)$ on node set $[n]$ by assigning weight $W(x_i, x_j)$ to edge ij $(i, j \in [n], i \neq j)$. We give weight 0 to the loops.

Every weighted graph H with edgeweights $\beta_{ij}(H) \in [0,1]$ gives rise to a random simple graph $\mathbb{G}(H)$ on $V(H)$: we connect nodes i and j with probability $\beta_{ij}(H)$, making an independent decision for distinct pairs (i, j) $(i, j \in [n], i \neq j)$. In particular, we can construct a random simple graph $\mathbb{G}(S, W) = \mathbb{G}\big(\mathbb{H}(S, W)\big)$.

For an integer $n > 0$, we define the random weighted graph $\mathbb{H}(n, W) = \mathbb{H}(S, W)$, and the random simple graph $\mathbb{G}(n, W) = \mathbb{G}(S, W)$, where S is an ordered n-tuple of independent uniform random points from $[0, 1]$.

To mention some special cases, if W is the identically p function, we get "ordinary" random graphs $\mathbb{G}(k, p)$. If $W = W_G$ for some simple graph G, then $\mathbb{G}(k, W_G) = \mathbb{H}(k, W_G)$ is "almost" the same as the random induced subgraph $\mathbb{G}(k, G)$ of G. To be more precise, if we condition on $x_1, \ldots, x_k$ belonging to different steps of W_G, then $\mathbb{G}(k, W_G)$ is a random k-node induced subgraph. The set this condition excludes, namely sequences $x_1, \ldots, x_k$ containing repetitions, has a measure at most $\binom{k}{2}/\mathsf{v}(G)$. Hence

$$(10.1) \qquad d_{\mathrm{var}}\big(\mathbb{G}(k, G), \mathbb{G}(k, W_G)\big) \leq \binom{k}{2}\frac{1}{\mathsf{v}(G)}.$$

It is straightforward to extend this construction to generating a countable random graph $\mathbb{G}(W)$ on $\mathbb{N}$: We generate an infinite sequence $X_1, X_2, \ldots$ of independent uniformly distributed random points from $[0, 1]$, and (as before) connect nodes i and j with probability $W(X_i, X_j)$.

REMARK 10.1. There are two ways of thinking about a graphon as a generalized graph. First, we can consider it as a weighted graph with node set $[0, 1]$. Second, we may think of each element $x \in [0, 1]$ as an infinite set S_x of nodes with infinitesimally small measure, where there is a random bipartite graph $G_{x,y}$ between S_x and S_y with density $W(x,y)$. These random bipartite graphs must be independent as random variables, which makes this impossible to construct in standard measure theory (one can construct such an object in non-standard analysis, cf. Section 11.3.2). But often this is a useful informal way of thinking of a graphon. The two random samples $\mathbb{H}(n,W)$ and $\mathbb{G}(n,W)$ correspond to these two ways of looking at graphons.

The definition of the sampling distance can also be extended from simple graphs to graphons (recall (1.2) for graphs):

$$(10.2) \qquad \delta_{\text{samp}}(U,W) = \sum_{k=1}^{\infty} \frac{1}{2^k} d_{\text{var}}\big(\mathbb{G}(k,U), \mathbb{G}(k,W)\big).$$

Using the fact that for any graphon U and simple graph F on node set $[k]$, the probability that $\mathbb{G}(k,U) = F$ is just $t_{\text{ind}}(F,U)$, we have for all $U,W \in \mathcal{W}_0$

$$(10.3) \qquad d_{\text{var}}\big(\mathbb{G}(k,U), \mathbb{G}(k,W)\big) = \frac{1}{2} \sum_{F \in \mathcal{F}_k^{\text{simp}}} |t_{\text{ind}}(F,U) - t_{\text{ind}}(F,W)|.$$

Hence

$$(10.4) \qquad \delta_{\text{samp}}(U,W) = \sum_F 2^{-\mathsf{v}(F)-1} |t_{\text{ind}}(F,U) - t_{\text{ind}}(F,W)|,$$

where F ranges through all finite graphs with $V(F) = \{1, \ldots, \mathsf{v}(F)\}$. By (10.1) the distributions of $\mathbb{G}(k,G)$ and $\mathbb{G}(k,W_G)$ are almost the same if $\mathsf{v}(G)$ is large, and hence

$$(10.5) \qquad \big|\delta_{\text{samp}}(F,G) - \delta_{\text{samp}}(W_F, W_G)\big| \leq \frac{4}{\mathsf{v}(G)}.$$

While the sampling procedure described above is the most natural and most often used, we sometimes need to sample in other ways. In Lemma 10.18 we will describe a sampling method where the random selection of the nodes is more restricted, but which is still good enough to get the same information about W (however, we need much larger samples).

There are other uses of graphons and kernels in generating random graphs. Bollobás, Borgs, Chayes and Riordan [2010] and Bollobás, Janson and Riordan [2007] study sparse random graphs generated from a nonnegative kernel W by constructing a (W/n)-random graph on n nodes. Bollobás, Janson and Riordan [2010] and Bollobás and Riordan [2009] study random trees generated from a graphon. Palla, Lovász and Vicsek [2010] construct sparse random graphs as $(W^{\otimes n})$-random graphs with n' nodes, where n and n' are chosen so as to keep the average degree constant. We will not go into the details of these constructions.

10.2. Sample concentration

If we take a bounded size sample from a graph, we can see very different graphs. For a sufficiently large random graph, for example, we can see anything. The natural way to use the sample $G[S]$ is to compute some graph parameter $f(G[S])$. But this parameter can vary wildly with the choice of the sample, so what information do we get?

The following theorem asserts that every reasonably smooth parameter of a sample is highly concentrated. (Note: we don't say anything here about the connection between the value of the parameter on the whole graph and on the sample! We return to this question in Chapter 15.) Let us define a *reasonably smooth graph parameter* as a parameter f satisfying $|f(G) - f(G')| \leq 1$ for any two graphs G and G' on the same node set whose edge sets differ only in edges incident with a single node. More generally, we define parameter of edge-weighted graphs as *reasonably smooth* if $|f(H) - f(H')| \leq 1$ for any two edge-weighted graphs on the same node set that differ only in the weights of edges incident with a single node.

THEOREM 10.2 (**Sample Concentration for Graphs**). *Let f be a reasonably smooth graph parameter, let G be a graph, and let $1 \leq k \leq \mathsf{v}(G)$. Let $f_0 = \mathsf{E}(f(\mathbb{G}(k, G)))$, then for every $t \geq 0$,*

$$\mathsf{P}\Big(f(\mathbb{G}(k,G)) \geq f_0 + \sqrt{2tk}\Big) \leq e^{-t}.$$

The result extends to graphons. We formulate two versions, corresponding to the two sampling methods defined above.

THEOREM 10.3 (**Sample Concentration for Graphons**). (a) *Let f be a reasonably smooth simple graph parameter, let $W \in \mathcal{W}_0$, and let $k \geq 1$. Let $f_0 = \mathsf{E}(f(\mathbb{G}(k, W)))$, then for every $t \geq 0$,*

$$\mathsf{P}\Big(f(\mathbb{G}(k,W)) \geq f_0 + \sqrt{2tk}\Big) \leq e^{-t}.$$

(b) *Let f be a reasonably smooth parameter of edge-weighted graphs. Let $W \in \mathcal{W}$, let $k \geq 1$, and let $f_0 = \mathsf{E}(f(\mathbb{H}(k, W)))$, then for every $t > 0$,*

$$\mathsf{P}\Big(f(\mathbb{H}(k,W)) \geq f_0 + \sqrt{2tk}\Big) \leq e^{-t}.$$

In both theorems, we can apply the same inequality to the function $-f$, to obtain a bound on the probability of a large deviation from the mean in the other direction.

Proof. The function $f(\mathbb{G}(\{x_1, \ldots, x_k\}, W))$ (as a function of $x_1, \ldots, x_k \in [0,1]$) satisfies the conditions of Corollary A.15 of Azuma's Inequality, and hence applying the inequality with $n = k$ and $\varepsilon = (2t/k)^{1/2}$, the inequality in (a) follows. The proof of (b) is essentially the same. $\square$

Applying this theorem with $f(G) = (\mathsf{v}(G)/\mathsf{v}(F))t_{\mathrm{inj}}(F, G)$ (which is reasonably smooth), and combining it with (5.21), we get the following concentration inequalities for subgraph densities:

COROLLARY 10.4. *Let $W \in \mathcal{W}_0$, $n \geq 1$, $0 < \varepsilon < 1$, and let F be a simple graph, then the W-random graph $\mathbb{G} = \mathbb{G}(n, W)$ satisfies*

$$\mathsf{P}\big(|t_{\mathrm{inj}}(F, \mathbb{G}) - t(F, W)| > \varepsilon\big) \leq 2 \exp\left(-\frac{\varepsilon^2 n}{2\mathsf{v}(F)^2}\right)$$

and

$$\mathsf{P}\big(|t(F, \mathbb{G}) - t(F, W)| > \varepsilon\big) \leq 2 \exp\left(-\frac{\varepsilon^2 n}{8\mathsf{v}(F)^2}\right). \quad \square$$

We will see in Section 10.4 that not only numerical parameters of subgraph samples are concentrated, but the samples themselves are concentrated in the cut distance.

10.3. Estimating the distance by sampling

10.3.1. The main sampling lemma. Among the main technical tools used in this book are a couple of probabilistic theorems, which relate sampling to cut distance. The first of these theorems is due to Alon, Fernandez de la Vega, Kannan and Karpinski [2003], with an improvement by Borgs, Chayes, Lovász, Sós and Vesztergombi [2008]. Its proof will be quite involved. Its main implication is that the $d_\square$-distance of two graphs on the same set of nodes can be estimated by sampling.

LEMMA 10.5 (**First Sampling Lemma for Graphs**). *Let G and H be weighted graphs with $V(G) = V(H)$, with the same node weights, and with edge weights in $[0,1]$. Let $k \leq \mathsf{v}(G)$ be a positive integer, and let S be chosen uniformly from all subsets of $V(G)$ of size k. Then with probability at least $1 - 4e^{-\sqrt{k}/10}$,*

$$\left| d_\square(G[S], H[S]) - d_\square(G, H) \right| \leq \frac{8}{k^{1/4}}.$$

This Lemma extends to kernels, and this is the form which we prove. For $U \in \mathcal{W}$ and $X = (X_1, \ldots, X_k) \subseteq [0,1]$, let $U[X]$ denote the symmetric $k \times k$ matrix defined by $(U[X])_{ij} = U(X_i, X_j)$.

LEMMA 10.6 (**First Sampling Lemma for Kernels**). *Let $U \in \mathcal{W}_1$ and let X be a random ordered of k-subset of $[0,1]$. Then with probability at least $1 - 4e^{-\sqrt{k}/10}$,*

$$-\frac{3}{k} \leq \|U[X]\|_\square - \|U\|_\square \leq \frac{8}{k^{1/4}}.$$

Not only are the lower and upper bounds in this lemma different, they are also quite different in difficulty. To prove the lower bound is rather straightforward, but the proof of the upper bound will need a couple of lemmas about a tricky sampling procedure estimating the sum of entries of a matrix.

It will be more convenient to work with the following one-sided version of the cut norm:

$$\|A\|_\square^+ = \frac{1}{n^2} \max_{S,T \subseteq [n]} \sum_{i \in S, j \in T} A_{ij}$$

for an $n \times n$ matrix A, and

$$\|W\|_\square^+ = \sup_{S,T \subseteq [0,1]} \int_{S \times T} W(x,y) \, dx \, dy$$

for a kernel W. We note that $\|A\|_\square = \max\{\|A\|_\square^+, \|-A\|_\square^+\}$, and similarly for the cut norm of kernels. In terms of this norm, we are going to prove the following similar bounds:

LEMMA 10.7. *Let $U \in \mathcal{W}_1$ and let X be a random ordered of k-subset of $[0,1]$. Then with probability at least $1 - 2e^{-\sqrt{k}/10}$,*

$$-\frac{3}{k} \leq \|U[X]\|_\square - \|U\|_\square \leq \frac{8}{k^{1/4}}.$$

Let $B = U[X]$. For any set Q_1 of rows and any set Q_2 of columns, we set $B(Q_1, Q_2) = \sum_{i \in Q_1, j \in Q_2} B_{ij}$. We denote by Q_1^+ the set of columns $j \in [k]$ for which $B(Q_1, \{j\}) > 0$. We define the set of columns Q_1^- and the sets of rows Q_2^+, Q_2^- analogously. Note that $B(Q_1, Q_1^+), B(Q_2^+, Q_2) \geq 0$ by this definition.

We start with proving an inequality for the case when only a random subset Q of columns is selected.

LEMMA 10.8. *Let $S_1, S_2 \subseteq [k]$, and let Q be random q-subset of $[k]$ ($1 \leq q \leq k$). Then*

$$B(S_1, S_2) \leq \mathsf{E}_Q\big(B((Q \cap S_2)^+, S_2)\big) + \frac{k^2}{\sqrt{q}}.$$

Proof. The inequality is clearly equivalent to the following:

(10.6) $$\mathsf{E}_Q\big(B((Q \cap S_2)^-, S_2)\big) \leq \frac{k^2}{\sqrt{q}}.$$

Note that there is no absolute value on the left side: the expectation of $B(Q \cap S_2)^-, S_2)$ can be very negative, but not very positive. The lemma says that the set $Q \cap S_2)^-$ tends to pick out those rows whose sum is small.

Consider row i of B. Let $m = |S_2|$, $b_i = \sum_{j \in S_2} B_{ij}$, $c_i = \sum_{j \in S_2} B_{ij}^2$ and $A_i = \sum_{j \in Q \cap S_2} B_{ij}$. The contribution of row i to the left side is b_i if $A_i \leq 0$ (i.e., $i \in (Q \cap S_2)^-$), and 0 otherwise. So the expected contribution of row i is $\mathsf{P}(A_i \leq 0)b_i$.

If $b_i \leq 0$, then this contribution is nonpositive. Else, we use Chebyshev's inequality to estimate the probability of $A_i \leq 0$. We have $\mathsf{E}(A_i) = qb_i/k$ and $\mathsf{Var}(A_i) < qc_i/k$. Hence

$$\mathsf{P}(A_i \leq 0) \leq \mathsf{P}\left(\left|A_i - \frac{qb_i}{k}\right| \geq \frac{qb_i}{k}\right) \leq \frac{k^2 \mathsf{Var}(A_i)}{q^2 b_i^2} < \frac{kc_i}{qb_i^2}.$$

The probability on the left is at most 1, and so we can bound it from above by its square root:

$$\mathsf{P}(A_i \leq 0) \leq \sqrt{\mathsf{P}(A_i \leq 0)} \leq \frac{\sqrt{kc_i}}{\sqrt{q}b_i}.$$

So the contribution of row i to $\mathsf{E}_Q\big(B((Q \cap S_2)^-, S_2)\big)$ is $\mathsf{P}(A_i \leq 0)b_i \leq \sqrt{kc_i/q} \leq k/\sqrt{q}$. Summing over all $i \in S_1$, inequality (10.6) follows. □

The following lemma gives an upper bound on the one-sided cut norm, using the sampling procedure from the previous lemma.

LEMMA 10.9. *Let $S_1, S_2 \subseteq [k]$, and let Q_1 and Q_2 be random q-subsets of $[k]$, ($1 \leq q \leq k$). Then*

$$\|B\|_\square^+ \leq \frac{1}{k^2} \mathsf{E}_{Q_1, Q_2}\Big(\max_{R_i \subseteq Q_i} B(R_2^+, R_1^+)\Big) + \frac{2}{\sqrt{q}}.$$

The Lemma estimates the (one-sided) cut norm by maximizing only over certain rectangles (at the cost of averaging these estimates). the main point for our purposes will be that (for a fixed Q_1 and Q_2), the number of rectangles to consider is only 4^q, as opposed to 4^k in the definition of the cut norm.

Proof. Fix any two sets $S_1, S_2 \subseteq [k]$. By Lemma 10.8,

(10.7) $$B(S_1, S_2) \leq \mathsf{E}_{Q_2}\big(B((Q_2 \cap S_2)^+, S_2)\big) + \frac{k^2}{\sqrt{q}}.$$

We apply Lemma 10.8 again, interchanging the roles of rows and columns:

$$B((Q_2 \cap S_2)^+, S_2) \leq \mathsf{E}_{Q_1}\left(B((Q_2 \cap S_2)^+, (Q_1 \cap (Q_2 \cap S_2)^+)^+)\right) + \frac{k^2}{\sqrt{q}}$$

$$\leq \mathsf{E}_{Q_1}\left(\max_{R_i \subseteq Q_i} B(R_1^+, R_2^+)\right) + \frac{k^2}{\sqrt{q}}.$$

Substituting in (10.7), the Lemma follows. $\square$

Now we can turn to the main part of the proof.

Proof of Lemma 10.7. To bound the difference $\|B\|_\square^+ - \|U\|_\square^+$, we first bound its expectation. For any two measurable subsets $S_1, S_2 \subset [0,1]$, we have

$$\|B\|_\square^+ \geq \frac{1}{k^2} U(S_1 \cap X, S_2 \cap X),$$

(where $U(Z_1, Z_2) = \sum_{x \in Z_1, y \in Z_2} U(x,y)$ for finite subsets $Z_1, Z_2 \subset [0,1]$). Choosing the set X randomly, we get

$$\mathsf{E}_X\left(\|B\|_\square^+\right) \geq \frac{1}{k^2} \mathsf{E}_X\left(U(S_1 \cap X, S_2 \cap X)\right)$$

$$= \frac{k-1}{k} \int_{S_1 \times S_2} U(x,y)\, dx\, dy + \frac{1}{k} \int_{S_1 \cap S_2} U(x,x)\, dx$$

$$\geq \int_{S_1 \times S_2} U(x,y)\, dx\, dy - \frac{2}{k}.$$

Taking the supremum of the right side over all measurable sets S_1, S_2 we get

$$\mathsf{E}_X\left(\|B\|_\square^+\right) \geq \|U\|_\square^+ - \frac{2}{k}.$$

From here, the bound follows by sample concentration (Theorem 10.3).

To prove an upper bound on the difference $\|B\|_\square^+ - \|U\|_\square^+$, let Q_1 and Q_2 be random q-subsets of $[k]$, where $q = \lfloor \sqrt{k}/4 \rfloor$. Lemma 10.9 say that for every X,

$$\|B\|_\square^+ \leq \frac{1}{k^2} \mathsf{E}_{Q_1, Q_2}\left(\max_{R_i \subseteq Q_i} B(R_2^+, R_1^+)\right) + \frac{2}{\sqrt{q}}.$$

Next we take expectation over the choice of X. More precisely, we fix the sets $R_i \subseteq Q_i \subseteq [k]$, and also those points $X_i \in [0,1]$ for which $i \in Q = Q_1 \cup Q_2$. Define $Y_1 = \{y \in [0,1] : \sum_{i \in R_1} U(X_i, y) > 0\}$, and define Y_2 analogously. Let $X' = (X_i : i \in [k] \setminus Q)$, then for every $i \in S_1 \setminus Q$ and $j \in S_2 \setminus Q$, the contribution of the term $U(X_i, X_j)$ to $\mathsf{E}_{X'} B(R_2^+, R_1^+)$ is $\int_{Y_1 \times Y_2} U \leq \|U\|_\square^+$. The contribution of the remaining terms $U(X_i, X_j)$ with either $i \in Q$ or $j \in Q$ is at most $2k|Q| \leq 4kq$ in absolute value. Hence

(10.8) $$\mathsf{E}_{X'} B(R_2^+, R_1^+) \leq k^2 \|U\|_\square^+ + 4kq.$$

Next we show that the value of $B(R_2^+, R_1^+)$ is highly concentrated around its expectation. This is a function of the independent random variables X_i, $i \in [k] \setminus Q$, and if we change the value of one of these X_i, the sum $B(R_2^+, R_1^+)$ changes by at most $4k$ (there are fewer than $2k$ entries that may change, and each of them by at most 2). We can apply Corollary A.15 of Azuma's Inequality, and conclude that with probability at least $1 - e^{-1.9q}$, we have

$$B(R_2^+, R_1^+) \leq \mathsf{E}_{X'} B(R_2^+, R_1^+) + 7.9k\sqrt{kq} \leq k^2 \|U\|_\square^+ + 4kq + 7.9k\sqrt{kq}.$$

The number of possible pairs of sets R_1 and R_2 is 4^q, and hence with probability at least $1 - 4^q e^{-1.9q} > 1 - e^{-q/2}$, this holds for all $R_1 \subseteq Q_1$ and $R_2 \subseteq Q_2$, and so it holds for the maximum. Taking expectation over Q_1 and Q_2 does not change this, so we get that with probability (over X) at least $1 - e^{-q/2}$, we have

$$\|B\|_\square^+ \leq \|U\|_\square^+ + \frac{2}{\sqrt{q}} + \frac{4q}{k} + \frac{7.9\sqrt{q}}{\sqrt{k}}.$$

This implies the upper bound in the lemma by simple computation (if k large enough). $\square$

Proof of Lemma 10.6. Applying Lemma 10.7 to both kernels U and $-U$, with probability at least $1 - 4e^{-\sqrt{k}/10}$ all four inequalities will hold, and in this case so do the inequalities in the Lemma. $\square$

10.3.2. First applications. We can apply the First Sampling lemma when $U = W_1 - W_2$ is a difference of two graphons. Considering $W_i[X]$ as the edge-weighted graph $\mathbb{H}(X, W_i)$, Lemma 10.6 implies the following:

COROLLARY 10.10. *Let $W_1, W_2 \in \mathcal{W}_0$ and let X be a sequence of $k \geq 1$ random points of $[0, 1]$ chosen independently from the uniform distribution. Then with probability at least $1 - 4e^{-\sqrt{k}/10}$,*

$$\left| d_\square(\mathbb{H}(X, W_1), \mathbb{H}(X, W_2)) - \|W_1 - W_2\|_\square \right| \leq \frac{8}{k^{1/4}}.$$

In terms of the random weighted graphs $\mathbb{H}(k, W_1)$ and $\mathbb{H}(k, W_2)$ this means that they can be coupled so that $d_\square\big(\mathbb{H}(k, W_1), \mathbb{H}(k, W_2)\big) \approx \delta_\square(W_1, W_2)$ with high probability. We will see that more is true: $\mathbb{H}(k, W)$ will be close to W in the cut distance with high probability. (However, quantitatively "closeness" will be much weaker.)

We have seen that the cut distance of two samples $\mathbb{H}(k, W_1)$ and $\mathbb{H}(k, W_2)$ is close to the distance of W_1 and W_2 (if coupled appropriately). How about the simple graphs $\mathbb{G}(k, W_1)$ and $\mathbb{G}(k, W_2)$? The following simple lemma shows that if k is large enough, then $\mathbb{G}(k, W)$ is close to $\mathbb{H}(k, W)$, so similar conclusions hold.

LEMMA 10.11. *For every edge-weighted graph H with edgeweights in $[0, 1]$, and for every $\varepsilon \geq 10/\sqrt{q}$,*

$$\mathsf{P}(d_\square(\mathbb{G}(H), H) > \varepsilon) \leq e^{-\varepsilon^2 q^2/100}.$$

Applying this inequality with $\varepsilon = 10/\sqrt{q}$ and bounding the distance by 1 in the exceptional cases, we get the inequality

(10.9) $$\mathsf{E}(d_\square\big(\mathbb{G}(H), H\big)) \leq \frac{11}{\sqrt{q}}.$$

Note that no similar assertion would hold for the distances d_1 or d_2. For example, if all edgeweights of H are $1/2$, then $d_1(\mathbb{G}(H), H) = d_2(\mathbb{G}(H), H) = 1/2$ for any instance of $\mathbb{G}(H)$.

Proof. For $i, j \in [q]$, define the random variable $X_{ij} = \mathbb{1}\big(ij \in E(\mathbb{G}(H))\big)$. Let S and T be two disjoint subsets of $[q]$. Then the X_{ij} ($i \in S, j \in T$) are independent, and $\mathsf{E}(X_{ij}) = \beta_{ij}(H)$, which gives that

$$e_{\mathbb{G}(H)}(S,T) - e_H(S,T) = \sum_{i \in S,\, j \in T} \big(X_{ij} - \mathsf{E}(X_{ij})\big).$$

Let us call the pair (S,T) *bad*, if $|e_{\mathbb{G}(H)}(S,T) - e_H(S,T)| > \varepsilon q^2/4$. The probability of this can be estimated by the Chernoff–Hoeffding Inequality:

$$\mathsf{P}\Big(\Big|\sum_{i \in S,\, j \in T} \big(X_{ij} - \mathsf{E}(X_{ij})\big)\Big| > \frac{1}{4}\varepsilon q^2\Big) \leq 2\exp\Big(-\frac{\varepsilon^2 q^4}{32|S||T|}\Big) \leq 2\exp\Big(\frac{-\varepsilon^2 q^2}{32}\Big).$$

The number of disjoint pairs (S,T) is 3^q, and so the probability that there is a bad pair is bounded by $2 \cdot 3^q e^{-\varepsilon^2 q^2/32} < e^{-\varepsilon^2 q^2/100}$. If there is no bad pair, then it is easy to see that $d_\square(\mathbb{G}(H), H) \leq \varepsilon$ (cf. Exercise 8.4). This completes the proof. $\square$

This lemma implies that the weighted sample in the First Sampling Lemma can be replaced by a simple graph at little cost. We state one corollary:

COROLLARY 10.12. *Let $W_1, W_2 \in \mathcal{W}_0$ and $k \geq 1$. Then the random graphs $\mathbb{G}(k, W_1)$ and $\mathbb{G}(k, W_2)$ can be coupled so that with probability at least $1 - 5e^{-\sqrt{k}/10}$,*

$$\Big|d_\square\big(\mathbb{G}(k, W_1), \mathbb{G}(k, W_2)\big) - \|W_1 - W_2\|_\square\Big| \leq \frac{10}{k^{1/4}}. \quad \square$$

EXERCISE 10.13. Derive the First Sampling Lemma for graphs (Lemma 10.5) from the graphon version (Lemma 10.6). Attention: sampling from a graph G and sampling from W_G does not quite give the same distribution!

EXERCISE 10.14. Prove the (much easier) analogue of the First Sampling Lemma for the edit distance: Let G and H be simple graphs $V(G) = V(H)$. Let $k \leq \mathsf{v}(G)$ be a positive integer, and let S be chosen uniformly from all ordered subsets of $V(G)$ of size k. Then

$$\mathsf{E}\big(d_1(G[S], H[S])\big) = \frac{(k-1)n}{k(n-1)} d_1(G, H),$$

and for every $\varepsilon > 0$, with probability at least $1 - 2e^{-k\varepsilon^2/2}$,

$$\Big|d_1(G[S], H[S]) - d_1(G, H)\Big| \leq \varepsilon.$$

10.4. The distance of a sample from the original

A second lemma about sampling that will be used very often, due to Borgs, Chayes, Lovász, Sós and Vesztergombi [2008], shows that a sample is close to the original graph (or graphon) with high probability. Note that here we have to use the $\delta_\square$ distance, rather than the $d_\square$ distance, since the graphs have different number of nodes, and no overlaying is given a priori. Also note that the bound on the distance is much weaker than in the previous lemma (but it does tend to 0 with the sample size).

LEMMA 10.15 (**Second Sampling Lemma for Graphs**). *Let $k \geq 1$, and let G be a simple graph on at least k nodes. Then with probability at least $1 - \exp\big(-k/(2\log k)\big)$,*

$$\delta_\square\big(G, \mathbb{G}(k, G)\big) \leq \frac{20}{\sqrt{\log k}}.$$

10.4. THE DISTANCE OF A SAMPLE FROM THE ORIGINAL

The Second Sampling Lemma also extends to graphons, which can be stated in terms of the W-random graphs $\mathbb{H}(k,W)$ and $\mathbb{G}(k,W)$.

LEMMA 10.16 (**Second Sampling Lemma for Graphons**). *Let $k \geq 1$, and let $W \in \mathcal{W}_0$ be a graphon. Then with probability at least $1 - \exp(-k/(2\log k))$,*

$$\delta_\square(\mathbb{H}(k,W), W) \leq \frac{20}{\sqrt{\log k}},$$

and

$$\delta_\square(\mathbb{G}(k,W), W) \leq \frac{22}{\sqrt{\log k}}.$$

Proof. First we prove that these inequalities hold in expectation. Let $m = \lceil k^{1/4} \rceil$. By Lemma 9.15, there is an equipartition $\mathcal{P} = \{V_1, \ldots, V_m\}$ of $[0,1]$ into m classes such that

$$d_\square(W, W_\mathcal{P}) \leq \frac{8}{\sqrt{\log k}}.$$

Let S be a random k-subset of $[0,1]$, then by the First Sampling Lemma 10.6, we have

$$\big|d_\square(W[S], W_\mathcal{P}[S]) - d_\square(W, W_\mathcal{P})\big| \leq \frac{8}{k^{1/4}}$$

with high probability. This implies that

$$\mathsf{E}\big(\big|d_\square(W[S], W_\mathcal{P}[S]) - d_\square(W, W_\mathcal{P})\big|\big) \leq \frac{10}{k^{1/4}}$$

(k is large enough for this, else the bound in the lemma is trivial), and so

$$\mathsf{E}\big(d_\square(W[S], W_\mathcal{P}[S])\big) \leq \mathsf{E}\big(\big|d_\square(W[S], W_\mathcal{P}[S]) - d_\square(W, W_\mathcal{P})\big|\big) + d_\square(W, W_\mathcal{P})$$
$$\leq \frac{9}{\sqrt{\log k}}.$$

So it suffices to prove that $\delta_\square(W_\mathcal{P}, W_\mathcal{P}[S])$ is small on the average.

Let $H = W_\mathcal{P}[S]$. The graphons $W_\mathcal{P}$ and W_H are almost the same: both are stepfunctions with m steps, with the same function values on corresponding steps. The only difference is that the measure of the i-th step V_i in $W_\mathcal{P}$ is $1/m$, while the measure of the i-th step in W_H is $|V_i \cap S|/k$, which is expected to be close to $1/m$ if k is large enough.

Write $|V_i \cap S|/k = 1/m + r_i$, then it is easy to see that $\delta_\square(W_\mathcal{P}, W_H) \leq \sum_i |r_i|$. Hence it is easy to estimate the expectation of this distance, using elementary probability theory:

$$\mathsf{E}\big(\delta_\square(W_\mathcal{P}, W_H)\big) \leq \sum_i \mathsf{E}(|r_i|) = m\mathsf{E}(|r_1|) \leq m\sqrt{\mathsf{E}(r_1^2)} = \sqrt{\frac{m-1}{k}} < \frac{1}{k^{3/8}}.$$

Hence

$$\mathsf{E}(\delta_\square(W, W[S])) \leq \delta_\square(W, W_\mathcal{P}) + \mathsf{E}\big(\delta_\square(W_\mathcal{P}, W_\mathcal{P}[S])\big) + \mathsf{E}\big(\delta_\square(W_\mathcal{P}[S], W[S])\big)$$
$$\leq \frac{8}{\sqrt{\log k}} + \frac{1}{k^{3/8}} + \frac{9}{\sqrt{\log k}} \leq \frac{18}{\sqrt{\log k}}.$$

A similar estimate for $\delta_\square\big(W, \mathbb{G}(k,W)\big)$ follows if we invoke inequality (10.9):

$$\mathsf{E}\big(\delta_\square(W, \mathbb{G}(k,W))\big) \leq \mathsf{E}\big(\delta_\square(W, \mathbb{H}(k,W))\big) + \mathsf{E}\big(\delta_\square(\mathbb{H}(k,W), \mathbb{G}(k,W))\big)$$
$$\leq \frac{18}{\sqrt{\log k}} + \frac{11}{\sqrt{k}} < \frac{20}{\sqrt{\log k}}.$$

Now the Lemma follows by the Sample Concentration Theorem 10.3 applied to the graph parameter $f(G) = v(G)\delta_\square(G,W)$. □

We have seen that the (weak) Regularity Lemma implies that we can approximate every simple graph G by a weighted graph on k nodes with error $O(1/\sqrt{\log k})$. As an application of the Second Sampling Lemma, we get that we can also approximate every simple graph G by an (unweighted) simple graph H on k nodes, at the cost of a constant factor in the error:

COROLLARY 10.17. *For every $k \geq 1$ and simple graph G, there is a simple graph H with k nodes such that*

$$\delta_\square(G, H) \leq \frac{10}{\sqrt{\log k}}.$$

We need a version of the Second Sampling Lemma for the modified sampling procedure mentioned above in Remark 10.1. Let W be a graphon and $n \geq 1$. Let $S = (s_1, \ldots, s_n)$, where s_i is a random uniform point from the interval $[(i-1)/n, i/n]$. We denote the random graph $\mathbb{G}(S, W)$ by $\mathbb{G}'(n, W)$. The following bound was proved by Lovász and Szegedy [2010a].

LEMMA 10.18. *For every graphon W and positive integer k, we have with probability at least $1 - 5/\sqrt{k}$,*

$$\delta_\square(\mathbb{G}'(k, W), W) < \frac{176}{\sqrt{\log k}}.$$

Proof. The trick is to do a second sampling: we choose a random r-tuple T of nodes of $G' = \mathbb{G}'(k, W)$, where $r = \lceil k^{1/4} \rceil$. We may assume that $k > 25$, else there is nothing to prove. The Second Sampling Lemma implies that with probability at least $1 - 2\exp(-r/(2\log r))$, we have

$$\delta_\square(G'[T], G') \leq \frac{22}{\sqrt{\log r}}.$$

Now we can generate $G'[T] = \mathbb{G}(r, G')$ in the following way: we choose a random sequence X of r independent uniform points in $[0, 1]$; if they belong to different intervals $J_i = [(i-1)/k, i/k]$, then we return $\mathbb{G}(X, W)$; else, we try again. This gives us a coupling between $\mathbb{G}(r, W)$ and $\mathbb{G}(r, G')$ such that

$$\mathsf{P}\big(\mathbb{G}(r, W) \neq \mathbb{G}(r, G')\big) \leq \mathsf{P}(\exists i : |X \cap J_i| \geq 2) \leq \frac{r(r-1)}{k}.$$

Invoking the Second Sampling Lemma again, with probability at least $1 - 2\exp(-r/(2\log r))$ we have

$$\delta_\square(\mathbb{G}(r, W), W) \leq \frac{22}{\sqrt{\log r}},$$

and hence with probability at least

$$1 - 4\exp\left(-\frac{r}{2\log r}\right) - \frac{r(r-1)}{k} \geq 1 - \frac{5}{\sqrt{k}}$$

we have

$$\delta_\square(G', W) \leq \delta_\square(G', G'[T]) + \delta_\square(\mathbb{G}(r, W), W) \leq \frac{44}{\sqrt{\log r}} \leq \frac{176}{\sqrt{\log k}}. \quad \square$$

EXERCISE 10.19. Consider the template graph H of a weak regularity partition, with k almost equal classes, of a large graph G, and turn it to a simple graph by the method of Lemma 10.11. Prove that the (random) simple graph $\mathbb{G}(H)$ obtained this way satisfies, with high probability, $\delta_\square(G, \mathbb{G}(H)) \leq 10/\sqrt{\log k}$.

EXERCISE 10.20. Let $k \geq 1$, let W be a graphon, and let S_1 and S_2 be two independent random k-subsets of $[0,1]$. Then with probability at least $1 - 2^{1-k}$,
$$\widehat{\delta}_\square(\mathbb{G}(S_1, W), \mathbb{G}(S_2, W)) \leq \frac{22}{\sqrt{\log k}}$$
(note the "hat" over the δ).

EXERCISE 10.21. Let f be a graph parameter and assume that $|f(G) - f(G')| \leq d_\square(G, G')$ for any two graphs on the same node set. Then for every graph G and $1 \leq k \leq \mathsf{v}(G)$ there is a value f_0 such that if $S \subseteq V(G)$ is a random k-subset, then
$$|f(G[S]) - f_0| < \frac{22}{\sqrt{\log k}}$$
with probability at least $1 - o(1)$.

10.5. Counting Lemma

It is time to relate the two main quantities we introduced to study large dense graphs and graphons: homomorphism densities (which are equivalent to sample distributions) and the cut distance. The following simple but fundamental relation between them, due to Lovász and Szegedy [2006], is a generalization of the "Counting Lemma" in the theory of Szemerédi partitions. (Lemma 10.32, which will be more difficult to prove, will state a certain converse of this fact.)

We start with a combinatorial formulation.

LEMMA 10.22 (**Counting Lemma for Graphs**). *For any three simple graphs F, G and G'*
$$|t(F, G) - t(F, G')| \leq \mathsf{e}(F)\, \delta_\square(G, G').$$

The Lemma extends to graphons:

LEMMA 10.23 (**Counting Lemma for Graphons**). *Let F be a simple graph and let $W, W' \in \mathcal{W}_0$. Then*
$$|t(F, W) - t(F, W')| \leq \mathsf{e}(F)\delta_\square(W, W').$$

This lemma shows that for any simple graph F, the function $W \mapsto t(F, W)$ is Lipschitz-continuous on $\mathcal{W}_0$ in the metric $\delta_\square$. At the end of this section, we state several further versions of the Counting Lemma as exercises. The proof of the lemma will be given in the more general setting of $\mathcal{W}_0$-decorated graphs (which actually makes the proof simpler!). Recall that a $\mathcal{W}_0$-decorated graph is a simple graph in which a graphon W_e is assigned to each edge e. Also recall the definition (7.16) of homomorphism density of such a decorated graph.

LEMMA 10.24 (**Counting Lemma for decorated graphs**). *Let (F, w) and (F, w') be two $\mathcal{W}_0$-decorated graphs with the same underlying simple graph, where $w = (W_e : e \in E)$ and $w' = (W'_e : e \in E)$. Then*
$$|t(F, w) - t(F, w')| \leq \sum_{e \in E(F)} \|W_e - W'_e\|_\square.$$

Proof. It suffices to prove this bound for the case when $W_e = W'_e$ for all edges but one. Let $F = (V, E)$, and let uv be the edge with $W_{uv} \neq W'_{uv}$. Then

$$t(F, w) - t(F, w') = \int_{[0,1]^V} \prod_{ij \in E(F) \setminus \{uv\}} W_{ij}(x_i, x_j) \bigl(W_{uv}(x_u, x_v) - W'_{uv}(x_u, x_v)\bigr)\,dx$$

$$= \int_{[0,1]^V} f(x) g(x) \bigl(W_{uv}(x_u, x_v) - W'_{uv}(x_u, x_v)\bigr)\,dx,$$

where

$$f(x) = \prod_{ij \in \nabla(u) \setminus uv} W_{ij}(x_i, x_j)$$

does not depend on x_v, and satisfies $0 \leq f \leq 1$. Similarly,

$$g(x) = \prod_{ij \in E \setminus \nabla(u)} W_{ij}(x_i, x_j)$$

does not depend on x_u, and satisfies $0 \leq g \leq 1$. Fixing all variables except x_u and x_v, we get the following estimate by Lemma 8.10:

$$\left| \int_{[0,1]^2} f(x) g(x) \bigl(W_{uv}(x_u, x_v) - W'_{uv}(x_u, x_v)\bigr)\,dx_u\,dx_v \right| \leq \|W_{uv} - W'_{uv}\|_\square.$$

Integrating over the remaining variables, we get that

$$|t(F, w) - t(F, w')| \leq \|W_{uv} - W'_{uv}\|_\square. \qquad \square$$

From this lemma, along with (10.3) and (7.4), it is easy to derive a relationship between the variation distance of the distributions of the random graphs $\mathbb{G}(k, U)$ and $\mathbb{G}(k, W)$, and the cut distance of U and W.

COROLLARY 10.25. *Let U and W be two graphons, then for every $k \geq 2$, we have*

$$d_{\mathrm{var}}\bigl(\mathbb{G}(k, U), \mathbb{G}(k, W)\bigr) \leq 2^{k^2} \delta_\square(U, W).$$

EXERCISE 10.26. Show that the Counting Lemma does not hold for multigraphs, not even for $F = C_2$.

EXERCISE 10.27. Let F be a simple graph with m edges and let $W, W' \in \mathcal{W}_1$. Then

$$|t(F, W) - t(F, W')| \leq 4m \delta_\square(W, W').$$

EXERCISE 10.28. Let F be a simple graph with m edges and let $W \in \mathcal{W}_1$. Then

$$|t(F, W)| \leq 4m \|W\|_\square.$$

EXERCISE 10.29. For every $\mathcal{W}_1$-decorated graph (F, w),

$$t(F, w) \leq 4 \min_{e \in E(F)} \|W_e\|_\square.$$

EXERCISE 10.30. Prove the following "induced" version of the Counting Lemma: If F is a simple graph and $U, W \in \mathcal{W}_0$, then

$$|t_{\mathrm{ind}}(F, U) - t_{\mathrm{ind}}(F, W)| \leq 4 \binom{k}{2} \|U - W\|_\square.$$

Use this to improve the coefficient 2^{k^2} in Corollary 10.25.

10.6. Inverse Counting Lemma

Our goal is to establish a converse to the Counting Lemma: if two "large" graphs are locally close (in the sense of sampling or homomorphism densities) then they are globally close (in the sense of cut distance). This treatment is based on Borgs, Chayes, Lovász, Sós and Vesztergombi [2008]. We prove two versions, both of which will play an important role later on.

LEMMA 10.31. *Let U and W be two graphons and suppose that for some $k \geq 2$, we have*
$$d_{\mathrm{var}}\big(\mathbb{G}(k,U), \mathbb{G}(k,W)\big) < 1 - 2\exp\Big(-\frac{k}{2\log k}\Big).$$
Then
$$\delta_\square(U,W) \leq \frac{50}{\sqrt{\log k}}.$$

Note that the bound on the variation distance of the distributions of the random subgraphs $\mathbb{G}(k,U)$ and $\mathbb{G}(k,W)$ is very weak: a tiny overlap between them already implies that the graphons U and W are close. Applying the lemma to W_{G_1} and W_{G_2} gives a similar result for two large graphs.

Proof. The assumption implies that we can couple $\mathbb{G}(k,U)$ and $\mathbb{G}(k,W)$ so that $\mathbb{G}(k,U) = \mathbb{G}(k,W)$ with probability larger than $2\exp\big(-k/(2\log k)\big)$. The Second Sampling Lemma 10.16 implies that with probability at least $1 - \exp\big(-k/(2\log k)\big)$, we have
$$\delta_\square\big(U, \mathbb{G}(k,U)\big) \leq \frac{22}{\sqrt{\log k}},$$
and similar assertion holds for W. It follows that with positive probability all three happen, and then we get
$$\delta_\square(U,W) \leq \delta_\square\big(U, \mathbb{G}(k,U)\big) + \delta_\square\big(W, \mathbb{G}(k,W)\big) \leq \frac{50}{\sqrt{\log k}}. \qquad \square$$

LEMMA 10.32 (**Inverse Counting Lemma**). *Let k be a positive integer, let $U, W \in \mathcal{W}_0$, and assume that for every simple graph F on k nodes, we have*
$$|t(F,U) - t(F,W)| \leq 2^{-k^2}.$$
Then
$$\delta_\square(U,W) \leq \frac{50}{\sqrt{\log k}}.$$

Proof. Assume that $U, W \in \mathcal{W}_0$ satisfy
$$|t(F,U) - t(F,W)| \leq 2^{-k^2}$$
for every graph F with k nodes. This implies (by inclusion-exclusion) that
$$|t_{\mathsf{ind}}(F,U) - t_{\mathsf{ind}}(F,W)| \leq 2^{\binom{k}{2}} 2^{-k^2} = 2^{-\binom{k+1}{2}}.$$
In terms of the W-random graphs $\mathbb{G}(k,U)$ and $\mathbb{G}(k,W)$,
$$\big|\mathsf{P}\big(\mathbb{G}(k,U) = F\big) - \mathsf{P}\big(\mathbb{G}(k,W) = F\big)\big| \leq 2^{-\binom{k+1}{2}}.$$

Hence

$$d_{\mathrm{var}}\big(\mathbb{G}(k,U),\mathbb{G}(k,W)\big) = \sum_{F}\big|\mathsf{P}\big(\mathbb{G}(k,U)=F\big) - \mathsf{P}\big(\mathbb{G}(k,W)=F\big)\big| \leq 2^{\binom{k}{2}}2^{-\binom{k+1}{2}}$$

$$= 2^{-k} < 1 - 2\exp\Big(-\frac{k}{2\log k}\Big).$$

An application of Lemma 10.31 completes the proof. $\square$

EXERCISE 10.33. *Prove that for any two graphons U and W,*

$$\sqrt{\log\frac{1}{\delta_{\square}(U,W)}} \leq \log\frac{1}{\delta_{\mathrm{samp}}(U,W)} \leq \exp\Big(\frac{400}{\delta_{\square}(U,W)^2}\Big).$$

10.7. Weak isomorphism II

An important consequence of (10.4) is that two weakly isomorphic graphons have sampling distance 0, i.e., they are indistinguishable by sampling. The converse of this assertion follows by the same kind of argument.

The significance of this easy remark is that it allows us to relate weak isomorphism to the cut distance via the sampling distance. We start with a more general fact, showing the topological equivalence of the sampling distance and the cut distance on the space of graphons.

We have noted that two graphons U and W are weakly isomorphic (i.e., $t(F,U) = t(F,W)$ for every simple graph) if and only if their sampling distance is 0. The Counting Lemma and the Inverse Counting Lemma imply that two graphons are weakly isomorphic if and only if their cut distance is 0. It is easy to see that this implies the same conclusion for general kernels:

COROLLARY 10.34. *Two kernels U and W are weakly isomorphic if and only if $\delta_{\square}(U,W) = 0$.*

Since $\delta_{\square}(U,W) = 0$ expresses the existence of a correspondence between the points of the two graphons, this theorem can be considered as a generalization of Theorem 5.29 (which can be derived from it with some effort).

The proof of Corollary 10.34, if we include the proofs of the Counting Lemma and the Inverse Counting Lemma, is quite long, and in particular the proof of the Inverse Counting Lemma, which builds on the First Sampling Lemma, is quite involved. One can get a more direct proof using only rather standard analysis; see Exercise 11.27.

Theorem 8.13 and its Corollary 8.14, for the special case of the cut norm and cut-distance 0, imply the following further characterizations of weak isomorphism:

COROLLARY 10.35. *(a) Two kernels U and W are weakly isomorphic if and only if there exist measure preserving maps $\varphi, \psi : [0,1] \to [0,1]$ such that $U^{\varphi} = W^{\psi}$ almost everywhere.*

(b) Two kernels U and W are weakly isomorphic if and only if there exists a coupling measure μ on $[0,1]^2$ such that for two random samples (x_1, y_1) and (x_2, y_2) from μ, we have $U(x_1, x_2) = W(y_1, y_2)$ with probability 1.

As a further corollary we get the following fact (stated before as Exercise 7.18):

COROLLARY 10.36. *If two kernels $U, W \in \mathcal{W}$ are weakly isomorphic, then $t(F,U) = t(F,W)$ holds for all multigraphs F.*

EXERCISE 10.37. Construct the coupling measures in Theorem 8.13 for the cut distance of the three weakly isomorphic graphons in Example 7.11.

EXERCISE 10.38. Show by an example that the sampling distance and the cut distance do not define the same topology on the set of finite graphs.

CHAPTER 11

Convergence of dense graph sequences

Finally we have come to the central topic of this book: convergent graph sequences and their limits. The two key elements, namely sampling and graphons, have been introduced in the Introduction. Here we take our time to look at them from various aspects.

11.1. Sampling, homomorphism densities and cut distance

Recall from the introduction that we can define a notion of convergence if we fix a *sampling method*. For dense graphs, we use subgraph sampling: We select uniformly a random k-element subset of $V(G)$, and return the subgraph induced by it. The probability that we see a given graph F is the quantity $t_{\mathsf{ind}}(F, G)$ introduced in (5.13). A sequence of graphs (G_n) with $\mathsf{v}(G_n) \to \infty$ is *convergent* if the induced subgraph densities $t_{\mathsf{ind}}(F, G_n)$ converge for every finite graph F.

It is often more convenient to define convergence using the homomorphism densities $t(F, G_n)$ or the subgraph densities $t_{\mathsf{inj}}(F, G_n)$. This does not change the notion of convergence as introduced above in terms of sampling. Indeed, subgraph densities can be expressed as linear combinations of induced subgraph densities and vice versa (we have discussed such relations in Section 5.2.3), and hence $t_{\mathsf{inj}}(F, G_n)$ tends to a limit as $n \to \infty$ if and only if $t_{\mathsf{ind}}(F, G_n)$ does. For the homomorphism densities the argument is a bit more involved: we know that $t(F, G) - t_{\mathsf{inj}}(F, G) = O\big(1/\mathsf{v}(G)\big)$, and so this difference tends to 0 if $\mathsf{v}(G) \to \infty$. Hence $t(F, G_n)$ tends to a limit as $n \to \infty$ if and only if $t_{\mathsf{inj}}(F, G_n)$ does.

This notion of convergence of dense graphs is often called *left-convergence*, since it is based on homomorphisms "from the left". In the case of dense graphs, this notion is rather robust (it seems to be the only reasonable way to define convergence), and hence we call it simply "convergence". The parallel notion of right-convergence (which will turn out to be equivalent, at least if defined properly) will be discussed in Chapter 12.

Many examples of convergent graph sequences will be shown in Section 11.4.2, but let us describe a couple of very simple ones here.

EXAMPLE 11.1. The sequence of complete graphs is convergent, since a random induced k-node subgraph is always a complete graph itself. ♦

EXAMPLE 11.2. Fix any $0 \leq p \leq 0$, and generate a random graph $G_n = \mathbb{G}(n, p)$ for every n. This sequence will be convergent with probability 1. Indeed, for large n, a random induced k-subgraph $G_n[S]$ of G_n will be very close in distribution to $\mathbb{G}(k, p)$, for most choices of G_n. (Note that what we mean here is that G_n is fixed, the randomness comes from the choice of the k-subset S). This is not hard to verify directly, using elementary probability theory; it also follows from the much stronger results in Chapter 10. ♦

The definition of convergence can be reformulated using the notion of sampling distance (1.2): a sequence (G_n) of simple graphs with $\mathsf{v}(G_n) \to \infty$ is convergent if for every graph F, $(t_{\mathsf{ind}}(F, G_n) : n = 1, 2, \dots)$ is a Cauchy sequence (equivalently, $(t(F, G_n) : n = 1, 2, \dots)$ is a Cauchy sequence). This is equivalent to saying that the graph sequence is Cauchy in the δ_{samp} metric (1.2). The following theorem of Borgs, Chayes, Lovász, Sós and Vesztergombi [2006, 2008], which is one of the main results in this theory, justifies the use of the cut metric $\delta_\square$.

THEOREM 11.3. *A sequence (G_n) of simple graphs with $\mathsf{v}(G_n) \to \infty$ is convergent if and only if it is a Cauchy sequence in the metric $\delta_\square$.*

Proof. The Counting Lemma 10.22 implies that every Cauchy sequence in the metric $\delta_\square$ is convergent. The Inverse Counting Lemma 10.32 (applied to the graphons W_{G_n}) implies the converse. $\square$

REMARK 11.4. This proof builds on a fairly long chain of previous results, some of which, like the First Sampling Lemma 10.6, were quite involved. The advantage of this proof is that it gives a quantitative form of the equivalence of two convergence notions. As pointed out by Schrijver, a weaker qualitative form is easier to prove, inasmuch we can replace the use of the Inverse Counting Lemma by the characterization of weak isomorphism (for which a simple direct proof is sketched in Exercise 11.27). Indeed, consider the two metric spaces $(\widetilde{\mathcal{W}}_0, \delta_\square)$ (the graphon space) and $[0,1]^{\mathcal{F}}$ (the space of graph parameters with values in $[0,1]$). Both of these are compact (one by the Compactness Theorem 9.23, the other by Tychonoff's Theorem). The map $W \mapsto t(.,W)$ is continuous by the Counting Lemma, and injective by Corollary 10.34, and hence its inverse is also continuous. For a convergent sequence of graphs, this means precisely that the graphons W_{G_n} form a convergent sequence in $(\widetilde{\mathcal{W}}_0, \delta_\square)$.

Theorem 11.3 can be generalized to characterize convergence in the space $\mathcal{W}$. The proof is the same, except that the graphon versions of the Counting Lemmas must be used.

THEOREM 11.5. *Let (W_n) be a sequence of graphons in $\mathcal{W}_0$ and let $W \in \mathcal{W}_0$. Then $t(F, W_n)$ converges for all finite simple graphs F if and only if W_n is a Cauchy sequence in the $\delta_\square$ distance. Furthermore, $t(F, W_n) \to t(F, W)$ for all finite simple graphs F if and only if $\delta_\square(W_n, W) \to 0$.* $\square$

11.2. Random graphs as limit objects

Once we have defined convergence of a graph sequence, we would like to answer the question: what does it converge to? I have told you (and justified by some pictures) that the answer is "graphons", but let us dwell on this question in a more abstract setting for a while. In an abstract sense, we can assign a "limit object" to every convergent sequence: we say that two convergent sequences are "equivalent" if interlacing them we get a convergent sequence, and the limit objects can be defined as equivalence classes of convergent sequences. This abstract definition is not of much help; we are going to describe much more explicit representations for the limit objects. Our favorite one is the graphon, but we'll see other, equivalent representations in the form of random graph models, reflection positive graph parameters, and more.

11.2.1. Finite random graph models. The first construction of a limit object, which we call the *weak limit*, is actually quite general: it can be constructed for convergence of any reasonable sequence of structures for which we have any reasonable sampling process. (We will see later, for example, how it works for graphs with bounded degree.)

Given a simple graph G and $k \in [\mathsf{v}(G)]$, the random sample $\mathbf{G}(k,G)$ is a random graph on k labeled nodes; we denote its distribution by $\sigma_{G,k}$. Clearly $\sigma_{G,k}(F) = t_{\mathsf{ind}}(F,G)$. If $(G_1, G_2, \dots)$ is a convergent graph sequence, then the distributions $\sigma_{G_n,k}$ tend to some distribution σ_k on k-node labeled graphs. Conversely, if the distributions $\sigma_{G_n,k}$ tend to a limit for every k, then the graph sequence is convergent. (The distribution $\sigma_{G_n,k}$ may be undefined for a finite number of indices n for every fixed k.) So the sequence of limit distributions $(\sigma_1, \sigma_2, \dots)$ encodes the "limit" of the convergent graph sequence. Which sequences of distributions arise this way?

A *random graph model* is a probability distribution σ_k on simple graphs on node set $[k]$, for every $k \geq 1$, which is invariant under the reordering of the nodes. In other words, it is a sequence of random variables $\mathbf{G}_k$, whose values are simple graphs on $[k]$, and isomorphic graphs have the same probability. We say that a random graph model is *consistent* if deleting node k from $\mathbf{G}_k$, the distribution of the resulting graph is the same as the distribution of $\mathbf{G}_{k-1}$. In formulas, this means that for every graph H on $k-1$ nodes,

$$(11.1) \qquad \sigma_{k-1}(H) = \sum_{F:\ F'=H} \sigma_k(F),$$

where F' denotes the graph obtained by deleting node k from F. We say that the model is *local*, if for two disjoint subsets $S, T \subseteq [k]$, the subgraphs of $\mathbf{G}_k$ induced by S and T are independent as random variables.

We note that consistency, together with the invariance under reordering the nodes, implies that for every simple graph F on k nodes, the expectation $\mathsf{E}(t_{\mathsf{ind}}(F, \mathbf{G}_n)) = \sigma_k(F)$ is independent of n once $n \geq k$.

EXAMPLE 11.6. For every graphon W, the random graph model $\mathbf{G}_k = \mathbb{G}(k, W)$ is both consistent and local, which is trivial to check. (It will turn out that this example represents all such models.) ♦

THEOREM 11.7. *If a graph sequence $(G_1, G_2, \dots)$ is convergent, then the distributions $\sigma_k = \lim_{n \to \infty} \sigma_{k, G_n}$ form a consistent and local random graph model. Conversely, every consistent and local random graph model arises this way.*

Before proving this theorem, we need some preparation. Let G be a graph and $k \leq \mathsf{v}(G)$. The sequence of distributions $(\sigma_{G,1}, \sigma_{G,2}, \dots)$ is not quite consistent, because it breaks down for $k > \mathsf{v}(G)$; but it is consistent for the values of k for which it is defined. There is a more serious problem with locality: selecting i distinct random nodes of G will bias the selection of the remaining $k-i$, if we insist on selecting distinct points. So locality will be only approximately true.

We can fix both problems if we consider the slightly modified distributions $\sigma'_{G,k}(F) = t_{\mathsf{ind}}(F, W_G)$. The sequence $(\sigma'_{G,1}, \sigma'_{G,2}, \dots)$ is consistent. The random graphs corresponding to this sequence of distributions are $\mathbb{G}(k, W_G)$. (We could also generate this from G by selecting the k random nodes with replacement.) The difference between $\sigma_{G,k}$ and $\sigma'_{G,k}$ is very small if G is large: if we sample $\mathbb{G}(k, W_G)$

and keep it iff the sampled points correspond to different nodes of G, and otherwise resample, then we get a sample from the distribution $\mathbb{G}(k,G)$. This shows that

$$(11.2) \qquad d_{\text{var}}(\sigma'_{G,k}, \sigma_{G,k}) \leq 1 - \frac{n(n-1)\ldots(n-k+1)}{n^k} < \frac{1}{n}\binom{k}{2}.$$

As discussed in the introduction, random graphs satisfy quite strong laws of large numbers in the sense that two large random graphs are very much alike; this translates to the fact that a sequence of independently generated random graphs $\mathbb{G}(n,p)$ is convergent with probability 1. The next lemma shows that all local and consistent random graph models have a similar property.

LEMMA 11.8. *Let $(\sigma_1, \sigma_2, \ldots)$ be a local consistent random graph model, and generate a graph $\mathbf{G}_n$ from every σ_n, independently for different values of n. Then the sequence $(\mathbf{G}_1, \mathbf{G}_2, \ldots)$ is convergent with probability 1.*

Proof. First we note that for every simple graph F on $[k]$ and $n \geq k$, we have

$$(11.3) \qquad \mathsf{E}\big(t_{\text{ind}}(F, \mathbf{G}_n)\big) = \sigma_k(F).$$

Indeed, consider any injective map $\varphi : V(F) \to V(\mathbf{G}_n)$. It follows from the isomorphism invariance of σ_n that the probability that φ is an induced embedding is the same for every map φ, so it suffices to compute this probability when φ is the identity map on $[k]$. By the consistency of the model, this probability is $\mathsf{P}(\mathbf{G}_k = F) = \sigma_k(F)$.

Next we show that $t_{\text{ind}}(F, \mathbf{G}_n)$ is concentrated around its expectation $\sigma_k(F)$. We could compute second moments, but this would not give a sufficiently good bound. So (sigh!) we compute the fourth moment.

Let S_1, S_2, S_3, S_4 be independent random ordered k-subsets of $[n]$ (we assume that $n > k^2$). Define $X_i = \mathbb{1}(\mathbf{G}_n[S_i] = F) - \sigma_k(F)$. Note that $\mathsf{E}(X_i) = 0$ by (11.3), even if we condition on the choice of the S_i, since the distribution of $\mathbf{G}_n[S]$ is the same for every ordered k-set $S \subseteq [n]$. Furthermore,

$$(11.4) \qquad \mathsf{E}(X_1 X_2 X_3 X_4) = \mathsf{E}\big((t_{\text{ind}}(F, \mathbf{G}_n) - \sigma_k(F))^4\big),$$

since for a fixed $\mathbf{G}_n$ the variables X_i are independent, and $\mathsf{E}(X_i \,|\, \mathbf{G}_n) = t_{\text{ind}}(F, \mathbf{G}_n) - \sigma_k(F)$.

Let A denote the event that every S_i meets at least one other S_j. The key observation is that

$$\mathsf{E}(X_1 X_2 X_3 X_4 \,|\, \overline{A}) = 0.$$

This follows since if the S_i are fixed so that (say) S_4 does not meet the others, then X_4 is independent of $\{X_1, X_2, X_3\}$, and its expectation is 0. (This is where we use the assumption that our random graph model is local!) Thus

$$\mathsf{E}(X_1 X_2 X_3 X_4) \leq \mathsf{E}(X_1 X_2 X_3 X_4 \,|\, \overline{A})\mathsf{P}(\overline{A}) + \mathsf{E}(X_1 X_2 X_3 X_4 \,|\, A)\mathsf{P}(A)$$
$$\leq \mathsf{P}(A) \leq \frac{7k^4}{n^2}.$$

(The last inequality follows by elementary combinatorics.) Thus we get that

$$\mathsf{E}\big((t_{\text{ind}}(F, \mathbf{G}_n) - \sigma_k(F))^4\big) \leq \frac{7k^4}{n^2},$$

and hence by Markov's Inequality

(11.5) $\quad \mathsf{P}(|t_{\mathsf{ind}}(F, \mathbf{G}_n) - \sigma_k(F)| > \varepsilon) = \mathsf{P}\big((t_{\mathsf{ind}}(F, \mathbf{G}_n) - \sigma_k(F))^4 > \varepsilon^4\big)$

$$\leq \frac{1}{\varepsilon^4} \mathsf{E}\big((t_{\mathsf{ind}}(F, \mathbf{G}_n) - \sigma_k(F))^4\big) \leq \frac{7k^4}{\varepsilon^4 n^2}.$$

If we sum (11.5) for $n \geq 1$ with a fixed $\varepsilon > 0$, then the sum of the right hand sides is convergent, so it follows by the Borel–Cantelli Lemma that with probability 1, $|t_{\mathsf{ind}}(F, \mathbf{G}_n) - \sigma_k(F)| > \varepsilon$ holds for a finite number of values of n only, and so $t_{\mathsf{ind}}(F, \mathbf{G}_n) \to \sigma_k(F)$ with probability 1. Hence the graph sequence $\mathbf{G}_n$ converges with probability 1. $\square$

With this lemma at hand, our theorem is easy to prove.

Proof of Theorem 11.7. First, consider a convergent graph sequence $(G_1, G_2, \ldots)$ and the probability distributions σ_k defined by it. Consistency and locality follow by the consistency and locality of the distributions σ'_{k, G_n}.

Second, consider consistent and local random graph model $(\sigma_1, \sigma_2, \ldots)$, and a sequence of random graphs $\mathbf{G}_n$ ($n = 1, 2, \ldots$), which are independently generated from distribution σ_n for different indices n. It follows from Lemma 11.8 that with probability 1, this graph sequence is convergent. Equation 11.3 implies that it reproduces the right random graph model. $\square$

11.2.2. Countable random graph models. We can arrange all labeled simple graphs in a locally finite rooted tree, where the empty graph is the root, and F' is the parent of F. If $(\sigma_1, \sigma_2, \ldots)$ is a consistent sequence of distributions, then σ_k is a probability distribution on the k-th level of the tree, and the probability of each node is the sum of probabilities of its children.

From this setup, we can combine all the distributions σ_k into a single probability distribution on all infinite paths starting at the root. To be more precise, let Ω denote the set of such paths, and let Ω_F denote the set of paths passing through the node F. Then the sets Ω_F generate a sigma-algebra $\mathcal{A}$ on Ω. The Kolmogorov Extension Theorem implies that there is a (unique) probability measure σ on $(\Omega, \mathcal{A})$ such that $\sigma(\Omega_F) = \sigma_k(F)$ for every F.

This is so far an abstract construction. We can, however, make explicit sense of the elements of Ω. A path in the tree starting at the root is a sequence $(F_0, F_1, \ldots)$ of graphs such that $F_k = F'_{k+1}$. Hence the path gives rise to the countable graph $F = \cup_n F_n$ on the set of positive integers $\mathbb{N}^*$. Conversely, every graph on $\mathbb{N}^*$ corresponds to a path in the tree starting at the origin.

Thus the points of Ω can be identified with the graphs on $\mathbb{N}^*$. The sets Ω_F are obtained by fixing adjacency between a finite number of nodes. Thus σ can be thought of as a probability distribution on graphs on $\mathbb{N}^*$.

A *countable random graph model* is a probability distribution σ on $(\Omega, \mathcal{A})$, invariant under permutations of $\mathbb{N}^*$. Such a random graph can also be considered as a symmetric exchangeable array of 0-1 valued random variables (we will come back to this way of looking at them in Section 11.3.3). The countable random graph model is *local* if for any two finite disjoint subsets $S_1, S_2 \subseteq \mathbb{N}^*$, the subgraphs induced by S_1 and S_2 are independent (as random variables). The discussion above shows that every consistent random graph model defines a countable random graph model.

There is a way to go back from a countable random graph model σ to finite consistent random graph models, which is even simpler: to get a random graph on $[n]$, generate a random graph $\mathbf{G}$ from σ, and take the induced subgraph $\mathbf{G}[n]$. It is easy to see that this random graph model is consistent. Furthermore, a consistent random graph model is local if and only if the corresponding countable random graph model is local. To sum up,

PROPOSITION 11.9. *There is a bijection between consistent random graph models and countable random graph models. This bijection preserves locality.*

It follows that local countable random graph models can serve as representations of the limit objects for convergent graph sequences.

EXAMPLE 11.10 (**Cliques and stable sets**). Let $\mathbf{G}$ be either the complete graph or the edgeless graph on $\mathbb{N}^*$, each with probability $1/2$. Clearly $\mathbf{G}$ does not depend on the ordering of $\mathbb{N}^*$, so it is a countable random graph model. The subgraph $\mathbf{G}[n]$ is also complete with probability $1/2$ and edgeless with probability $1/2$ (at least for $n > 1$). This defines a consistent random graph model. However, this model is not local: the subgraph induced by $\{0,1\}$ is not independent (in the probabilistic sense) from the subgraph induced by $\{2,3\}$; to the contrary, they are always the same. ♦

EXAMPLE 11.11 (**The Rado graph**). We construct a random graph on $\mathbb{N}^*$ by connecting a pair of distinct integers $i, j \in \mathbb{N}^*$ with probability $1/2$, independently for different pairs. The resulting random graph $\mathbb{G}(\mathbb{N}^*, 1/2)$, called the *Rado graph*, has many interesting properties (some of these are stated in the exercises at the end of the section), but right now, what is important for us is that it is local by construction. The corresponding local and consistent random graph model is the ordinary random graph $\mathbb{G}(n, 1/2)$. ♦

EXAMPLE 11.12 (**Triangle-free random graph model**). It is easy to construct a countable triangle-free graph that contains every finite triangle-free graph as an induced subgraph (Exercise 11.18). In a theory developed independently from ours, Petrov and Vershik [2010] prove that there exist a local countable random graph model that is triangle-free with probability 1. They also construct an appropriate graphon, which turns out to be 0-1 valued. ♦

EXAMPLE 11.13 (**Infinite W-random graph**). We can extend the definition of W-random graphs (Section 10.1) to get a countable random graph $\mathbb{G}(\mathbb{N}^*, W)$. Given a graphon W, we select a sequence of independent random points $(X_1, X_2, \dots)$ from $[0,1]$, and connect i and j ($i, j \in \mathbb{N}^*$) with probability $W(X_i, X_j)$. This construction generalizes the Rado graph. It is immediate that the distribution of the infinite W-random graph is invariant under permutations of $\mathbb{N}^*$, and it is also local: for two disjoint sets $S, T \subseteq \mathbb{N}^*$, the graph we construct on S is independent (in the probability sense) from the graph we construct on T. We will see that all local countable random graph models can be constructed this way. ♦

We conclude this section with a construction of a convergent graph sequence from a countable random graph model (without the locality assumption) given independently by Lovász and Szegedy [2012a] and Diaconis and Janson ([2008], Theorem 5.3).

11.2. RANDOM GRAPHS AS LIMIT OBJECTS

PROPOSITION 11.14. *Let $\mathbf{G}$ be a random graph on $\mathbb{N}^*$ drawn from a countable random graph model. Let $\mathbf{G}[n]$ denote the subgraph induced by $[n]$. Then the sequence $(\mathbf{G}[1], \mathbf{G}[2], \dots)$ is convergent with probability 1.*

Proof. For every fixed simple graph F, the sequence $(t_{\mathsf{inj}}(F, \mathbf{G}[n]) : n = 1, 2, \dots)$ is a reverse martingale for $n \geq \mathsf{v}(F)$ in the sense that $\mathsf{E}\big(t_{\mathsf{inj}}(F, \mathbf{G}[n-1]) \mid \mathbf{G}[n]\big) = t_{\mathsf{inj}}(F, \mathbf{G}[n])$ (this follows by the simple averaging principle (5.27)). By the Reverse Martingale Convergence Theorem A.17, it follows that this sequence is convergent with probability 1. Hence with probability 1, $(t_{\mathsf{inj}}(F, \mathbf{G}[n]) : n = 1, 2, \dots)$ is convergent for every F. $\square$

This last proposition may sound similar to the construction in Lemma 11.8, but there is a significant difference: in this construction, locality is not needed. Unlike in Lemma 11.8, $\mathbf{G}[n]$ and $\mathbf{G}[m]$ are not independently generated. If we apply the construction in Lemma 11.8 twice, and then pick the even-indexed graphs from one sequence and interlace them with the odd-indexed graphs from the other, we get a sequence that is constructed in the same way, and so it is convergent with probability 1. This means that almost all sequences generated by Lemma 11.8 (for a fixed consistent and local random graph model) have the same limit. In contrast to this, running the construction in Proposition 11.14 twice we could not necessarily interlace the resulting sequences into a single convergent sequence: in Example 11.10, we get a sequence of growing cliques with probability $1/2$ and a sequence of growing edgeless graphs with probability $1/2$. Both of these sequences are convergent, but they don't have the same limit. Sequences constructed from one and the same countable random graph model are almost always convergent, but they may converge to different limits.

In view of Examples 11.6 and 11.13, we also get:

COROLLARY 11.15. *For every graphon W, generating a W-random graph $\mathbb{G}(n, W)$ for $n = 1, 2, \dots$ we get a convergent sequence with probability 1, whose limiting countable random graph model is $\mathbb{G}(\mathbb{N}^*, W)$.*

EXERCISE 11.16. (a) Prove that the Rado graph almost surely has the *extension property*: for any two disjoint finite subsets $S, T \subseteq \mathbb{N}^*$ there is a node connected to all nodes in S but to no node in T.
(b) Prove that every countable graph with the extension property is isomorphic to the Rado graph.
(c) Prove that if you generate two Rado graphs independently, they will be isomorphic with probability 1.

EXERCISE 11.17. (a) We can define a random graph $\mathbb{G}(\mathbb{N}^*, p)$ for all $0 < p < 1$. Prove that with probability 1, this random graph will be isomorphic to the Rado graph for any p.
(b) More generally, if W is a graphon with $0 < W(x, y) < 1$ for all $x, y \in [0, 1]$, then $\mathbb{G}(n, W)$ is almost always isomorphic to the Rado graph.
(c) Construct a graphon W such that two independent countable W-random graphs are almost surely *non*-isomorphic [Gábor Kun].

EXERCISE 11.18. Construct a universal triangle-free graph: a countable graph containing every finite triangle-free graph as an induced subgraph.

EXERCISE 11.19. Show that without the assumption of locality, Lemma 11.8 does not remain valid.

EXERCISE 11.20. Prove that if we generate two sequences as in Proposition 11.14 from a *local* countable random graph model, then interlacing the two sequences, we get a convergent sequence.

11.3. The limit graphon

In this section we give a more explicit description of limit objects for convergent graph sequences: we show that graphons (up to weak isomorphism) are precisely the structures that are needed.

11.3.1. Existence. Let (G_n) be a convergent graph sequence, so that the densities $t(F, G_n)$ tend to a limit $t(F)$ for every finite simple graph F. We know that a limit object can be described as a consistent and local random graph model (finite or countable). The main motivation behind introducing graphons is that they provide a much more explicit representation for this limit object, as the following theorem shows (Lovász and Szegedy [2006]).

THEOREM 11.21. *For any convergent sequence (G_n) of simple graphs there exists a graphon W such that $t(F, G_n) \to t(F, W)$ for every simple graph F.*

We say that this graphon W is the *limit* of the graph sequence, and write $G_n \to W$.

The reader might wonder if one really needs complicated objects like integrable functions to describe limits of graph sequences; would perhaps piecewise linear, or monotone, or continuous functions suffice? It turns out that (up to weak isomorphism) all measurable functions are needed: Every graphon W can be obtained as the limit of a convergent sequence of simple graphs; this follows by Corollary 11.15.

There are three quite different ways to prove Theorem 11.21. The original one by Lovász and Szegedy [2006] uses Szemerédi partitions and the Martingale Convergence Theorem. This is very closely related to the proof of the compactness of the graphon space (Theorem 9.23). Below we will use compactness to prove Theorem 11.21, but we could as well go the other way, since Theorem 11.21 easily implies the compactness of the graphon space (Exercise 11.28).

A more recent proof by Elek and Szegedy [2012] constructs a different limit object first, in the form of a graph on a very large sigma-algebra, by taking an ultraproduct; then obtains the graphon as an appropriate projection of this. This proof technique is quite general, it extends to hypergraphs and many other structures.

As a third route to prove Theorem 11.21, it was shown by Diaconis and Janson [2008] that it can be derived from results of Aldous [1981] and Hoover [1979] on exchangeable random variables, pointing out a basic connection to probability theory. We will sketch these alternative proofs in the next two sections.

Proof of Theorem 11.21. By Theorem 9.23, the metric space $(\widetilde{\mathcal{W}}_0, \delta_\square)$ is compact, and hence the sequence $(W_n = W_{G_n} : n = 1, 2, \dots)$ has a convergent subsequence $(W_{n_j} : j = 1, 2, \dots)$ with limit $W \in \widetilde{\mathcal{W}}_0$. By the Counting Lemma 10.23, we have for every simple graph F

$$|t(F, W_{n_j}) - t(F, W)| \leq \mathsf{e}(F)\, \delta_\square(W_{n_j}, W) \longrightarrow 0 \qquad (j \to \infty),$$

and so $t(F, W_{n_j}) = t(F, G_{n_j}) \to t(F, W)$. Since $(t(F, G_n) : n = 1, 2, \dots)$ is a Cauchy sequence, this implies that $t(F, G_n) \to t(F, W)$ for every simple graph F. □

There may be several graphons W representing the limit of a convergent graph sequence. Of course, we may change the value of W on a set of measure 0; but more generally, we can replace W by any other kernel weakly isomorphic with W. Conversely, of W and W' both represent the limit of a convergent graph sequence, then they are weakly isomorphic by the definition of weak isomorphism. (Weak isomorphism has been discussed in Sections 7.3 and 10.7, and we will say more about it in Section 13.2.)

Convergence to the limit object can also be characterized by the distance function:

THEOREM 11.22. *For a sequence (G_n) of graphs with $\mathsf{v}(G_n) \to \infty$ and graphon W, we have $G_n \to W$ if and only if $\delta_\square(W_{G_n}, W) \to 0$.*

Proof. If $\delta_\square(W_{G_n}, W) \to 0$, then $G_n \to W$ follows by the Counting Lemma just like in the proof of Theorem 11.21 above. Conversely, suppose that $G_n \to W$, so $t(F, G_n) \to t(F, W)$ for every simple graph F, and hence for every fixed k, the Inverse Counting Lemma 10.32 implies that

$$\delta_\square(W_{G_n}, W) \leq \frac{20}{\sqrt{\log k}}$$

if n is large enough. Hence $\delta_\square(W_{G_n}, W) \to 0$ as claimed.

(This proof gives an explicit connection between the rates of convergence in $G_n \to W$ and $\delta_\square(W_{G_n}, W) \to 0$. If we don't care about this, the theorem follows by abstract arguments: the space $(\widetilde{\mathcal{W}}_0, \delta_\square)$ is compact, the map $W \mapsto (t(F, W) : F \in \mathcal{F})$ is continuous and injective, hence its inverse is also continuous.) $\square$

11.3.2. Ultralimit and limit. The theory of ultraproducts and ultralimits provides a general way to construct limit objects. This proof technique has some drawbacks: it is non-constructive, and requires advanced special techniques from model theory (outlined in Appendix A.5). On the other hand, its advantage is that it is very flexible: one can define the limit of virtually any kind of sequence of structures this way, and only have to wonder later whether this limit object can be "brought down to earth" to have combinatorial (or algebraic, or arithmetic) significance.

Let ω be an ultrafilter on $\mathbb{N}^*$. Recall that we call the sets in ω "Large", the other subsets of $\mathbb{N}^*$, "Small". As a first try, let us define the limit of a graph sequence $(G_1, G_2, \dots)$ as the ultraproduct $\prod_\omega G_n$.

This assigns a limit object to every graph sequence, not just to those that are convergent. Unfortunately, this ultraproduct will depend on the ultrafilter we use, even for convergent graph sequences (so the situation is not as simple as for numerical sequences, see Exercise 11.31). For example, let H_n be the edgeless graph on n nodes if n is even, and let it be a graph consisting of a clique of size $\lfloor\sqrt{n}\rfloor$ together with $n - \lfloor\sqrt{n}\rfloor$ isolated nodes if n is odd. This graph sequence is clearly convergent. On the other hand, $\prod_\omega G_n$ has no edge if the set of odd integers is Small, and it consists of a clique of continuum size and continuum many isolated nodes if the set of odd integers is Large (and either can happen).

While it is of the same cardinality as the whole ultraproduct, this clique should occupy a negligible part of this huge product graph. To make this precise, we have to introduce a measure on the product. Let $(G_1, G_2, \dots)$ be a sequence of finite graphs. For every graph $G_n = (V_n, E_n)$, we can consider the (finite) sigma-algebra

$\mathcal{A}_n = 2^{V_n}$ of all subsets of V_n, and the uniform probability measure π_n on V_n. Then the ultraproduct $G = (V, E) = \prod_\omega G_n$ is a graph whose node set V is also equipped with a sigma-algebra $\mathcal{A} = \prod_\omega \mathcal{A}_n$ and a probability measure $\pi = \prod_\omega \pi_n$ on it.

This answers our concern with the example above: the ultraproduct $\prod_\omega H_n$ becomes edgeless if we delete an appropriate set of nodes of measure zero.

However, the construction is still not satisfactory. First, the sigma-algebra $\mathcal{A}$ is an ugly one: for example, it is not separable in general. Second, the set E of edges of the product, as a subset of $V \times V$, may not be measurable with respect to the product sigma-algebra $\mathcal{A} \times \mathcal{A}$. This is really serious: we would like to be able to assert, for example, that if the graphs G_n have edge density $1/2$, then the limit has edge density $1/2$, which means that $(\pi \times \pi)(E) = 1/2$. However, this assertion does not even make sense!

The way out of this trouble is to do some fiddling with the ultraproduct. Let us start with taking the ultraproduct $\prod_\omega (V_n \times V_n)$. This ultraproduct can be identified with $V \times V$ in a natural way: if $a_n, b_n \in V_n$, then

$$(11.6) \qquad [(a_1, b_1), (a_2, b_2), \ldots] \leftrightarrow ([a_1, a_2, \ldots], [b_1, b_2, \ldots]).$$

It is easy to see that if the bracket on the left side is represented by another equivalent sequence of pairs, then the brackets on the right don't change, and vice versa.

As before, sets of the form $\prod_\omega S_n$ ($S_n \subseteq V_n \times V_n$) form a set algebra on $V \times V$, which generates a sigma-algebra $\mathcal{A}'$. In particular, we can take the ultraproduct of the edge sets of the graphs G_n, to get $E = \prod_\omega E_n \in \mathcal{A}'$. Furthermore, the ultraproduct η of the uniform measures on $V_n \times V_n$ is a probability measure on $(V \times V, \mathcal{A}')$.

It is easy to see that the sigma-algebra $\mathcal{A} \times \mathcal{A}$ is generated by sets of the form $\prod_\omega (S_n \times T_n)$, where $S_n, T_n \subseteq V_n$. Hence $\mathcal{A} \times \mathcal{A} \subseteq \mathcal{A}'$. In general, we don't have equality (this is why E is not measurable in $\mathcal{A} \times \mathcal{A}$). But the following lemma shows that in an important way, $\mathcal{A}'$ is not too much larger than $\mathcal{A} \times \mathcal{A}$:

LEMMA 11.23. *For every set $B \in \mathcal{A}'$ and every $x \in V$, the neighborhood $B(x) = \{y \in V : (x, y) \in B\}$ belongs to $\mathcal{A}$.*

Proof. First, we consider the case when $B = \prod_\omega B_n$, and let $x = [x_1, x_2, \ldots]$. We claim that $B(x) = \prod_\omega B_n(x_n)$ (which of course implies that $B(x) \in \mathcal{A}$). Indeed: $y = [y_1, y_2, \ldots] \in \prod_\omega B_n(x_n)$ if and only if $y_n \in B_n(x_n)$ for a large set of indices n, if and only if $(x_n, y_n) \in B_n$ for a large set of indices n, if and only if $[x, y] \in B$.

Now taking complements and countable intersections of sets $B \in \mathcal{A}'$ corresponds to carrying out the same operations on the sets $B(x)$ (where $x \in V$ is fixed), so these sets stay in $\mathcal{A}$. $\square$

The measure μ, when restricted to $\mathcal{A} \times \mathcal{A}$, gives the probability measure $\pi \times \pi$. Hence we can take the conditional expectation $W = \mathsf{E}(\mathbb{1}_E \,|\, \mathcal{A} \times \mathcal{A})$, which is a function on $V \times V$, measurable with respect to $\mathcal{A} \times \mathcal{A}$, and has the property that

$$(11.7) \qquad \int_X W(x, y) \, d\pi(x) \, d\pi(y) = \mu(X \cap E)$$

for every set $X \in \mathcal{A} \times \mathcal{A}$. This results in a probability space $(V, \mathcal{A}, \pi)$ with a symmetric measurable function $W : V \times V \to [0, 1]$.

We show that W can serve as the limit graphon (at least as long as we ignore the ugliness of the underlying sigma-algebra):

PROPOSITION 11.24. *For every simple graph F and any sequence $(G_n : n = 1, 2, \dots)$ of graphs,*
$$\lim_\omega t(F, G_n) = t(F, W).$$

If the graph sequence $(G_1, G_2, \dots)$ is convergent, then the ultralimit on the left side is equal to $\lim_{n\to\infty} t(F, G_n)$, independently from ω.

Proof. Let $V(F) = [k]$. As a first step, we express the left hand side in the ultraproduct space. We have to introduce more sigma-algebras for this. For every set $U \subseteq [k]$, we take the set V_n^U of all maps $U \to V_n$. The set $\prod_\omega V_n^U$ can be identified with V^U, just like (11.6) identifies $\prod_\omega V_n \times V_n$ with $V \times V$. The ultraproduct of the Boolean algebra of all subsets of V_n^U gives a sigma-algebra $\mathcal{A}_U$ on V^U. The ultraproduct of the uniform measures on the sets V_n^U gives a measure τ_U on the product sigma-algebra on V^U. We abbreviate $\tau_{[k]}$ by τ_k.

The set $\mathrm{Hom}(F, G_n)$ of homomorphisms from F into G_n is a subset of V_n^k, and so $\prod_\omega \mathrm{Hom}(F, G_n)$ is a subset of $\prod_\omega V_n^k = V^k$. The set $\prod_\omega \mathrm{Hom}(F, G_n)$ can be identified with the set $\mathrm{Hom}(F, G) \subseteq V^k$ of homomorphisms of F into G. Furthermore,
$$\lim_\omega t(F, G_n) = \tau_k\Big(\prod_\omega \mathrm{Hom}(F, G_n)\Big) = \tau_k\big(\mathrm{Hom}(F, G)\big)$$
by the definition of the ultraproduct of measures. So it suffices to prove that
$$(11.8) \qquad \tau_k\big(\mathrm{Hom}(F, G)\big) = t(F, W).$$

Let $(X_1, \dots, X_k) \in V^k$ be a random node chosen from the distribution τ_k. Then we can rephrase the equality to be proved as
$$(11.9) \qquad \mathsf{E}\Big(\prod_{ij \in E(F)} \mathbb{1}_E(X_i, X_j)\Big) = \mathsf{E}\Big(\prod_{ij \in E(F)} W(X_i, X_j)\Big).$$

(It is easy to check that the functions $\mathbb{1}_E(X_i, X_j)$ are measurable with respect to the sigma-algebra $\mathcal{A}_{[k]}$.)

If the random variables $\mathbb{1}_E(X_i, X_j)$ were independent, we could take the expectation factor-by-factor, and we would be done. But of course they are not. The trick in the proof is to replace the factors $\mathbb{1}_E(X_i, X_j)$ by $W(X_i, X_j)$ one by one. Consider any edge uv of F; we show that
$$(11.10) \qquad \mathsf{E}\Big(\prod_{ij \in E(F)} \mathbb{1}_E(X_i, X_j)\Big) = \mathsf{E}\Big(\prod_{ij \neq uv} \mathbb{1}_E(X_i, X_j) W(X_u, X_v)\Big).$$

This will show that we can replace $\mathbb{1}_E(X_u, X_v)$ by $W(X_u, X_v)$ without changing the expectation, and repeating a similar argument for all edges of F, we get (11.9).

For notational convenience, assume that $u = 1$ and $v = 2$. The main difficulty in the rest of the argument is to be careful about measurability, because we have several sigma-algebras floating around. Using Lemma 11.23, it is not hard to argue that fixing X_1 and X_2, the functions $\mathbb{1}_E(X_i, X_j)$ are measurable with respect to $\mathcal{A}_{[k]}$, and so the expectation
$$f(X_1, X_2) = \mathsf{E}_{X_3, \dots, X_k} \prod_{ij \in E(F) \{i,j\} \neq \{1,2\}} \mathbb{1}_E(X_i, X_j)$$

is well defined. Furthermore, again by Lemma 11.23, if we fix $X_3,\ldots,X_k$, then every function $\mathbb{1}_E(X_i,X_j)$ ($\{i,j\}\neq\{1,2\}$) becomes either constant, or $\mathcal{A}_{\{1\}}$-measurable (if the edge ij is incident with 1) or $\mathcal{A}_{\{2\}}$-measurable (if ij is incident with 2). Hence it follows that $f(x_1,x_2)$ is $\mathcal{A}_{\{1\}}\times\mathcal{A}_{\{2\}}$-measurable.

By the definition of W, this implies that

$$\int_{V\times V} f(x_1,x_2)\mathbb{1}_E(x_1,x_2)\,d\tau_{\{1,2\}}(x_1,x_2) = \int_{V\times V} f(x_1,x_2)W(x_1,x_2)\,d\pi(x_1)\,d\pi(x_2),$$

which proves (11.10). □

To finish the construction of the limit graphon, one has to map this big sigma-algebra $(V,\mathcal{A})$ onto $[0,1]$, define the appropriate image of W, and show that it represents the same subgraph densities. We refer to the paper of Elek and Szegedy [2012] for these details.

REMARK 11.25. If you compare this construction of the limit object with the construction given before, some parallelism between their elements is apparent. For example, (11.7), applied to a generator set $X=S\times T$, asserts that the density of edges of G between S and T is the same as the integral of W on $S\times T$, so W and $\mathbb{1}_E$ have distance 0 in the cut norm. Perhaps further exploration of this parallelism could shed some light on the nature of the use of non-constructive infinite methods like ultraproducts in the theory of (very large, but) finite graphs.

In the last proof, one needs to handle various sigma-algebras; indeed, a little "calculus of sigma-algebras" was used. Elek and Szegedy push this much further, and develop more from this combinatorial theory of sigma-algebras, which enables them to extend this construction to hypergraphs; cf. Section 23.3.

11.3.3. Exchangeable random variables. Let $(G_1,G_2,\ldots)$ be a convergent graph sequence. We start with the weak limit of the sequence in the form of a local countable random graph model σ. Let $\mathbf{G}$ be a graph from this distribution, and let $(X_{ij})_{i,j=1}^\infty$ be its adjacency matrix; in other words, $X_{ij}=\mathbb{1}(ij\in E(\mathbf{G}))$.

It follows from the invariance of σ under permutations that the random variables X_{ij} have the same distribution: $\mathsf{P}(X_{ij}=1)=\lim_{n\to\infty}t(K_2,G_n)$. They are not independent, but have the following property, which is called *symmetrically exchangeable*: if α is a permutation of $\mathbb{N}^*$, then for every $k\geq 1$, the joint distribution of $(X_{ij}:1\leq i,j\leq n)$ is the same as the joint distribution of $(X_{\alpha(i)\alpha(j)}:1\leq i,j\leq n)$. This is of course just a reformulation of the condition that σ is invariant.

Now symmetrically exchangeable random variables have a representation theorem due to Aldous [1981] and Hoover [1979]: every such system can be represented as a mixture of random variables of the form $X_{ij}=W(Y_i,Y_j)$, where Y_i ($i\in\mathbb{N}^*$) are independent random variables with values in $[0,1]$ and $W:[0,1]^2\to[0,1]$ is a symmetric measurable function in two variables. Furthermore, as Diaconis and Janson [2008] show, if the random graph $\mathbf{G}$ is local, then you don't get a real mixture: local exchangeable distributions are precisely the extreme points of the space of symmetrically exchangeable random variables, in other words, random variables of the form $X_{ij}=W(Y_i,Y_j)$. So the representation theorem of Aldous and Hoover provides the limit graphon W directly!

For the details of this theory and its application here, we refer to the monograph of Kallenberg [2005] and the paper [2008]. See also Austin [2008] for a survey of the many other applications of this approach.

EXERCISE 11.26. Prove that if a graph sequence (G_n) satisfies $\mathsf{v}(G_n) \to \infty$, then it is Cauchy in the cut distance if and only if it is Cauchy in the sampling distance.

EXERCISE 11.27. Prove the following facts: (a) For every stepfunction W, $\delta_1(W, \mathbb{H}(n,W)) \to 0$ as $n \to \infty$ with probability 1. (b) For every graphon W, $\delta_1(W, \mathbb{H}(n,W)) \to 0$ as $n \to \infty$ with probability 1. (c) For every graphon W, $\delta_\square(\mathbb{G}(n,W), \mathbb{H}(n,W)) \to 0$ as $n \to \infty$ with probability 1. (d) If U and W are weakly isomorphic graphons, then $\mathbb{G}(n,U)$ and $\mathbb{G}(n,W)$ have the same distribution. (e) If U and W are weakly isomorphic graphons, then $\delta_\square(U,W) = 0$ (A. Schrijver).

EXERCISE 11.28. Show that Theorems 11.21, 11.22 and Corollary 11.15 imply that the space $(\widetilde{\mathcal{W}_0}, \delta_\square)$ is compact.

EXERCISE 11.29. Prove that the following properties of graphs are inherited to their ultraproduct: (a) 3-regular; (b) all degrees bounded by 10; (c) triangle-free; (d) containing a triangle; (e) bipartite; (f) disconnected.

EXERCISE 11.30. Prove that the following properties of graphs are not inherited to their ultraproduct: (a) connected; (b) all degrees are even; (c) non-bipartite.

EXERCISE 11.31. (a) Prove that every bounded sequence of real numbers has a unique ultralimit. (b) Prove that the $\lim_\omega (a_i + b_i) = \lim_\omega a_i + \lim_\omega b_i$. (c) Prove that the ultralimit $\lim_\omega a_i$ is independent of the choice of the ultrafilter ω if and only if the sequence is convergent in the classical sense.

11.4. Proving convergence

It is not always easy to show that a certain graph sequence is convergent. We have several characterizations (in terms of subgraph densities, sample distribution, cut distance) and sometimes one, sometimes the other condition is easier to apply. First we develop some useful sufficient conditions for convergence, and then apply these to give a number of examples of interesting convergent graph sequences.

11.4.1. Convergence of sampling methods. We start with a supplement to Corollary 11.15:

PROPOSITION 11.32. *For every graphon W, generating a W-random graph $\mathbb{G}(n,W)$ for $n = 1, 2, \ldots$ we get a graph sequence such that $\mathbb{G}(n,W) \to W$ with probability 1.*

Proof. It is straightforward to check that for every simple graph F and $n \geq \mathsf{v}(F)$,
$$\mathsf{E}(t_{\mathsf{inj}}(F, \mathbb{G}(n,W))) = t(f, W).$$
Since $t_{\mathsf{inj}}(F, \mathbb{G}(n,W))$ is highly concentrated around its expectation (by Theorem 10.2 or by the proof of Lemma 11.8), it follows that $t_{\mathsf{inj}}(F, \mathbb{G}(n,W)) \to t(f,W)$ with probability 1. $\square$

Sometimes we need that other sequences constructed by similar, but different sampling from a graphon are convergent. We describe one lemma of this type by Borgs, Chayes, Lovász, Sós and Vesztergombi [2008] [2011].

We start with a definition. For every $n \geq 1$, let $S_n \subseteq [0,1]$ be a finite set such that $|S_n| \to \infty$. We say that the sequence (S_n) is *well distributed*, if $|S_n \cap J|/|S_n| \to \lambda(J)$ for every interval J as $n \to \infty$. Equivalently, the uniform measure on S_n converges weakly to the uniform measure on $[0,1]$ (see Billingsley [1999] for more on related notions like uniform distribution of sequences).

LEMMA 11.33. *Let $W \in \mathcal{W}_0$ be almost everywhere continuous, and let S_n be a well distributed sequence of sets. Then $\mathbb{G}(S_n, W) \to W$ with probability 1.*

It is clear that such a conclusion cannot hold without some assumption on W, since a general measurable function could be changed on the sets $S_n \times S_n$ arbitrarily without changing its subgraph densities. For an extension to graphons on general metric spaces, see Exercise 13.8. Note that there need not be any randomness in the sequence S_n.

Proof. Consider a partition $\{Z_1, \ldots, Z_m\}$ of $[0, 1]$ into m intervals of equal length. Since (S_n) is well distributed, we have $1/(m+1) \leq |S_n \cap Z_j|/|S_n| \leq 1/(m-1)$ for every j if $n \geq n_0(m)$. Given any n that is large enough, we choose the largest m for which $n \geq n_0(m)$, and partition each set Z_j into $|S_n \cap Z_j|$ sets of equal measure, each containing exactly one point of $S_n \cap Z_j$, to get the partition $\mathcal{Q}_n$. This partition has the properties that $|\mathcal{Q}_n| = |S_n|$, every partition class contains exactly one point of S_n, the maximum diameter of partition classes tends to 0, and $\max\{|\lambda(Q)|S_n| - 1| : Q \in \mathcal{Q}_n\} \to 0$.

For $s \in S_n$, let Q_s^n be the partition class of $\mathcal{Q}_n$ containing s. Define the function W_n as follows: for $s, s' \in S_n$ and $(x, y) \in Q_s^n \times Q_{s'}^n$, let $W_n(x, y) = W(s, s')$. Then $W_n(x, y) \to W(x, y)$ in every point (x, y) where W is continuous, in particular $W_n \to W$ almost everywhere. This implies that

$$(11.11) \qquad \|W_n - W\|_1 \longrightarrow 0 \qquad (n \to \infty).$$

We can view W_n as the graphon W_{H_n} associated with a weighted graph H_n with $V(H_n) = S_n$, where the weight of node $s \in S_n$ is $\lambda(Q_s^n)$, and the weight of edge ss' ($s, s' \in S$) is $W(s, s')$. Note that H_n is almost the same weighted graph as $\mathbb{H}_n = \mathbb{H}(S_n, W)$: they are defined on the same set of nodes, the edges have the same weights, and the nodeweight $\lambda(Q_s^n)$ is asymptotically $1/|S_n|$ by the construction of $\mathcal{Q}_n$. Given $\varepsilon > 0$, we have $|\lambda(Q_s^n) - 1/|S_n|| < \varepsilon/|S_n|$ if n is large enough. Hence there is a measure preserving bijection $\varphi : [0, 1] \to [0, 1]$ and a set $R \subseteq [0, 1]$ of measure ε such that

$$W_{H_n}(x, y) = W_{\mathbb{H}_n}^{\varphi}(x, y) \qquad (x, y \notin R).$$

This implies that

$$(11.12) \qquad \delta_1(H_n, \mathbb{H}_n) \longrightarrow 0 \qquad (n \to \infty).$$

Formulas (11.11) and (11.12) imply that $\mathbb{H}(S_n, W) \to W$, which in turn implies that $\mathbb{G}(S_n, W) \to W$ with probability 1 (cf. Lemma 10.11). $\square$

COROLLARY 11.34. *Let W be a graphon that is almost everywhere continuous. Then $\mathbb{G}(S_n, W) \to W$ with probability 1, where*

(a) $S_n \subseteq [0, 1]$ *is obtained by selecting a uniform random point from every interval $[j/n, (j+1)/n]$, $n = 0, \ldots, n-1$;*

(b) $S_n = \{1/n, 2/n, \ldots, n/n\}$.

11.4.2. Examples: Convergent graph sequences. We discuss a variety of examples of convergent graph sequences. It turns out that it is not always easy to prove that these are convergent and determine the limit graphon. In some examples, the limit can be guessed and the proof of convergence in the cut distance is easy. In other examples, it is quite tricky to guess the limit graphon. In other cases, the convergence of subgraph densities can be proved. We will also see a randomly

11.4. PROVING CONVERGENCE

growing graph sequence which is convergent with probability 1, but if we run it again, it converges to a different limit!

We start with two easy examples.

EXAMPLE 11.35 (**Complete bipartite graphs**). It is natural to guess, and easy to prove, that complete bipartite graphs $K_{n,n}$ converge to the graphon $W(x,y) = \mathbb{1}(0 \leq x \leq 1/2 \leq y \leq 1) + \mathbb{1}(0 \leq y \leq 1/2 \leq x \leq 1)$. ♦

EXAMPLE 11.36 (**Simple threshold graphs**). These graphs are defined on the set $\{1,\ldots,n\}$ by connecting i and j if and only if $i + j \leq n$. These graphs converge to the graphon defined by $\mathbb{1}(x+y \leq 1)$, which we call the *simple threshold graphon*. ♦

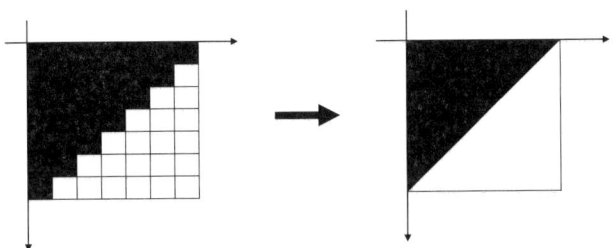

FIGURE 11.1. Simple threshold graphs and their limits

A more interesting example is the following.

EXAMPLE 11.37 (**Quasirandom graphs**). A sequence of graphs tending to the identically-p function is exactly what we called a "quasirandom sequence" with density p (by the second property in Section 1.4.2). In particular, the Paley graphs (Example 1.1) converge to the graphon $W \equiv 1/2$. ♦

EXAMPLE 11.38 (**Multitype quasirandom graphs**). Generalizing the previous example, we consider a *multitype quasirandom graph sequence* (G_n) with a template graph H. This means that (assuming that $V(H) = [q]$ and $V(G_n) = [n]$) $V(G_n)$ has a partition $(V_1,\ldots,V_q)$ such that $|V_i| = \alpha_i(H)n + o(n)$ and for every fixed $i,j \in [q]$, the bipartite graphs $G_n[V_i,V_j]$ form a quasirandom bipartite graph sequence with edge density $\beta_{ij}(H)$ (in the case when $i = j$, the induced subgraphs $G_n[V_i]$ form a quasirandom graph sequence).

A multitype quasirandom graph sequence with template graph H tends to the graphon W_H, and vice versa. The first statement follows easily, since $\delta_\square(W_{G_n}, W_H) \to 0$ by the definition of a multitype quasirandom sequence. The converse is less trivial; if $\delta_\square(W_{G_n}, W_H) \to 0$, then there is a way to label the nodes of G_n so that $\|W_{G_n} - W_H\|_\square \to 0$ (this follows from Theorem 11.59 below), and then we can partition the nodes of G_n by putting node u in class i ($u \in [n], i \in [q]$) iff $\alpha_1(H) + \cdots + \alpha_{i-1}(H) \leq u/n < \alpha_1(H) + \cdots + \alpha_i(H)$. It is not hard to verify that this partition has the right properties to guarantee that the sequence (G_n) is multitype quasirandom with template H. ♦

We continue with several examples, given by Borgs, Chayes, Lovász, Sós and Vesztergombi [2011], of convergent graph sequences obtained by random growing processes.

EXAMPLE 11.39 (**Growing uniform attachment graphs**). We generate a randomly growing graph sequence G_n^{ua} as follows. We start with a single node. At the n-th iteration, a new node is born, and then every pair of nonadjacent nodes is connected with probability $1/n$. We call this graph sequence a *uniform attachment graph sequence*; see Figure 1.8.

Let us do some simple calculations. After n steps, let $\{0, 1, \ldots, n-1\}$ be the nodes (born in this order). The probability that nodes $i < j$ are not connected is $\frac{j}{j+1} \cdot \frac{j+1}{j+2} \cdots \frac{n-1}{n} = \frac{j}{n}$. These events are independent for all pairs (i,j). The expected degree of j is

$$\sum_{i=0}^{j-1} \frac{n-j}{n} + \sum_{i=j+1}^{n-1} \frac{n-i}{n} = \frac{n-1}{2} - \frac{j(j-1)}{2n}.$$

The expected number of edges is

$$\frac{1}{2} \sum_{j=0}^{n-1} \left(\frac{n-1}{2} - \frac{j(j-1)}{2n} \right) = \frac{n^2-1}{6}.$$

To figure out the limit graphon, note that the probability that nodes i and j are connected is $1 - \max(i,j)/n$. If $i = xn$ and $j = yn$, then this is $1 - \max(x,y)$. This motivates the following:

PROPOSITION 11.40. *The sequence G_n^{ua} tends to the limit function $1 - \max(x,y)$ with probability 1.*

Proof. For a fixed n, the events that nodes i and j are connected are independent for different i, j, and so by the computation above, G_n^{ua} has the same distribution as $\mathbb{G}(S_n, 1 - \max(x,y))$, where $S_n = \{0, 1/n, \ldots, (n-1)/n\}$. It is easy to see that this sequence is well distributed in $[0,1]$, and so the Proposition follows by Lemma 11.33. □

One can get a good explicit bound on the convergence rate by estimating the cut-distance of $W_{G_n^{\mathrm{ua}}}$ and $1 - \max(x,y)$, using the Chernoff-Hoeffding bound. ◆

EXAMPLE 11.41 (**Prefix attachment graphs**). In this construction, it will be more convenient to label the nodes starting with 1. At the n-th iteration, a new node n is born, a node z is selected at random, and node n is connected to nodes $1, \ldots, z-1$. We denote the n-th graph in the sequence by G_n^{pfx}, and call this graph sequence a *prefix attachment graph sequence* (Figure 11.2).

Again we start with some simple calculations. The probability that nodes $i < j$ are connected is $\frac{j-i}{j}$ (but these events are not independent in this case!). The expected degree of j is therefore

$$\sum_{i=1}^{j-1} \frac{j-i}{j} + \sum_{i=j+1}^{n} \frac{i-j}{i} = n - \frac{j}{2} + j \ln \frac{n}{j} + o(n).$$

The expected number of edges is $n(n-1)/4$.

Looking at the picture, it seems that it tends to some function, which we can try to figure out similarly as in the case of uniform attachment graphs. The probability that i and j are connected can be written in a symmetric form as $|j-i|/\max(i,j)$. If $i = xn$ and $j = yn$, then this is $|x-y|/\max(x,y)$.

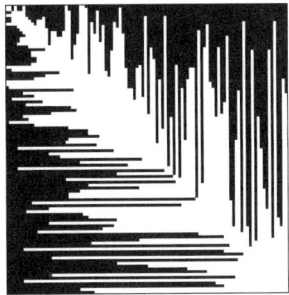 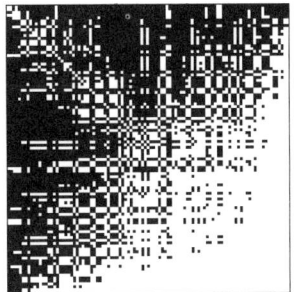

FIGURE 11.2. A randomly grown prefix attachment graph with 100 nodes, and the same graph with nodes ordered by their degrees.

Does this mean that the graphon $U(x,y) = |x-y|/\max(x,y)$ is the limit? Somewhat surprisingly, the answer is negative, which we can see by computing triangle densities. The probability that three nodes $i < j < k$ form a triangle is $\left(1 - \frac{j}{k}\right)\left(1 - \frac{i}{j}\right)$ (since if k is connected to j, then it is also connected to i). Hence the expected number of triangles is

$$\sum_{i<j<k} \left(1 - \frac{j}{k}\right)\left(1 - \frac{i}{j}\right) = \frac{1}{6}\binom{n}{3}.$$

Hence

$$t(K_3, G_n) = \frac{1}{n^3}\binom{n}{3} \longrightarrow \frac{1}{6}.$$

On the other hand,

$$t(K_3, U) = \int_{[0,1]^3} \frac{|x-y|}{\max(x,y)} \cdot \frac{|x-z|}{\max(x,z)} \cdot \frac{|y-z|}{\max(y,z)} \, dx \, dy \, dz.$$

Since the integrand is independent of the order of the variables, we can compute this easily:

$$t(K_3, U) = 6 \int_{0 \le x < y < z \le 1} \left(1 - \frac{x}{y}\right)\left(1 - \frac{x}{z}\right)\left(1 - \frac{y}{z}\right) dx \, dy \, dz = \frac{5}{36}.$$

So U is not the limit of the sequence G_n^{pfx}.

Perhaps ordering the nodes by degrees helps? The second pixel picture in Figure 11.2 suggests that after this reordering, the functions $W_{G_n^{\text{pfx}}}$ converge to some other continuous function. But again this convergence is only in the weak* topology, not in the $\delta_\square$ distance. We'll see that no continuous function can represent the "right" limit: the limit graphon is 0-1 valued, and it is uniquely determined up to measure preserving transformations by Theorem 13.10, which do not change this property.

Is this graph sequence convergent at all? Our computation of the triangle densities above can be extended to computing the density of any subgraph, and it follows that the sequence of densities $t(F, G_n^{\text{pfx}})$ is convergent for every n. How to figure out the limit?

Let us label a node born in step k, connected to $\{1, \ldots, m\}$, by $(k/n, m/k) \in [0,1] \times [0,1]$. Then we can observe that *nodes with label* (x_1, y_1) *and* (x_2, y_2) *are connected if and only if either* $x_1 < x_2 y_2$ *or* $x_2 < x_1 y_1$.

This suggests a description of the limit in the following form: Consider the function $W^{\mathrm{pfx}} : [0,1]^2 \times [0,1]^2 \to [0,1]$, given by

$$W^{\mathrm{pfx}}\big((x_1, y_1), (x_2, y_2)\big) = \mathbb{1}(x_1 < x_2 y_2 \text{ or } x_2 < x_1 y_1).$$

(As remarked before, we can consider 2-variable functions on other probability spaces, not just $[0, 1]$; in this case, $[0, 1]^2$ is a more convenient representation. In the proof below, we use an analogue of Lemma 11.33, adopted to this case. For a general statement containing both, see Exercise 13.8.)

PROPOSITION 11.42. *The prefix attachment graphs* G_n^{pfx} *tend to* W^{pfx} *almost surely.*

Proof sketch. Let S_n be the (random) set of points in $[0,1]^2$ of the form $(i/n, z_i/i)$ where $i = 1, \ldots, n$ and z_i is a uniformly chosen random integer in $[i]$. Then $G_n^{\mathrm{pfx}} = \mathbb{G}(S_n, W^{\mathrm{pfx}}) = \mathbb{H}(S_n, W^{\mathrm{pfx}})$.

Furthermore, with probability 1, the sets S_n are well distributed in $[0,1]^2$ in the sense that $|S_n \cap A|/|S_n| \to \lambda(A)$ for every open set A. It suffices to verify this for the case when $A = J_1 \times J_2$, where J_1, J_2 are open intervals, and it will be also convenient to assume that J_1 does not start at 0. The assertion is then easily verified, based on the fact that the first coordinates (i/n) are well distributed in $[0,1]$, and the second coordinates are uniformly distributed random points in $\{1/i, \ldots, i/i\}$. Thus the generalized version of Lemma 11.33 applies and proves the Proposition. □

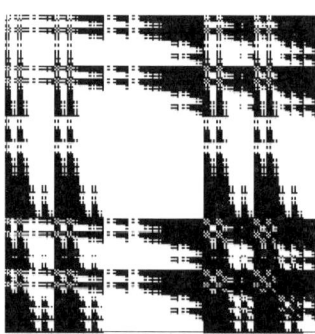

FIGURE 11.3. The limit of randomly grown prefix attachment graphs as a function on $[0,1]^2$.

Proposition 11.42 gives a nice and simple representation of the limit object with the underlying probability space $[0,1]^2$ (with the uniform measure). If we want a representation by a graphon on $[0,1]$, we can map $[0,1]$ into $[0,1]^2$ by a measure preserving map φ; then $W^{\varphi}_{\mathrm{pfx}}(x,y) = W^{\mathrm{pfx}}\big(\varphi(x), \varphi(y)\big)$ gives a representation of the same graphon as a 2-variable function. For example, using the map φ that separates even and odd bits of x, we get the fractal-like picture in Figure 11.3.

It is interesting to note that the graphs $\mathbb{G}(n, W)$ form another (different) sequence of random graphs tending to the same limit W with probability 1.

A final remark on this graph sequence. It is not hard to verify that for the graphon $U(x,y) = |x-y|/\max(x,y)$, we have

$$\int_{S \times T} (W_{G_n^{\text{pfx}}} - U) \longrightarrow 0 \tag{11.13}$$

for every $S, T \subseteq [0,1]$. (Indeed, it is enough to prove this for sets S, T from a generating set of the sigma-algebra of Borel sets, e.g. rational intervals. Since there is only a countable number of these intervals, it suffices to prove that (11.13) holds with probability 1 for any two rational intervals S and T. This is a rather straightforward computation in probability.)

So $W_{G_n^{\text{pfx}}} \to U$ in the weak* topology of $L_\infty([0,1]^2)$, but not in our sense. We will see (Lemma 8.22) that our convergence implies weak* convergence, but not the other way around. This example also shows that had we defined convergence of a graph sequence by weak* convergence (after appropriate relabeling), the limit would not be unique. The uniqueness of the limit graphon is a nontrivial fact! ◆

So far, our randomly grown sequences tended to well-defined limit graphons with probability 1. Now we turn to examples of randomly grown sequences that are convergent with probability 1, but if we run the process again, they may converge to a different limit. There is in fact a very simple sequence with this property.

EXAMPLE 11.43 (**Cloning**). Given a simple graph G_0, we select a uniform random node v, and create a twin of v (a new node v' connected to the same nodes as v; v and v' are not connected). Repeating this we get a sequence of graphs $G_0, G_1, G_2, \ldots$.

We claim that this sequence is convergent with probability 1. Let $\mathsf{v}(G_0) = k$. Note that each G_n is determined by the sequence $(n_i : i \in V(G_0))$, where n_i is the number of clones of i we created, including i itself. Clearly $\sum_i n_i = n + k$, and the probability that node i will be cloned in the next step is $n_i/(n+k)$. So the development of the sequence $(n_i : i \in V(G_0))$ follows a Pólya urn model (see e.g. Grimmett and Stirzaker [1982]), which implies that with probability 1, every ratio $n_i/(n+k)$ tends to some real number x_i. Clearly $x_i \geq 0$ and $\sum_i x_i = 1$. So $G_n \to W_H$, where H is obtained from G_0 by weighting node i with weight x_i (the edges remain unweighted).

What are these values x_i? They can be anything (as long as they are nonnegative and sum to 1). In fact, it follows from the theory of the Pólya urn that the vector (x_i) is uniformly distributed over the simplex $\{x \in \mathbb{R}^{V(G_0)} : x_i \geq 0, \sum_i x_i = 1\}$. ◆

Let us close this section with a more interesting example with similar property.

EXAMPLE 11.44 (**Growing preferential attachment graphs**). This randomly growing graph sequence G_n^{pa} is generated as follows. We start with a single node. At the n-th step (when we already have a graph with n nodes), a new node labeled $n+1$ is created. This new node is connected to each old node i with probability $(d_n(i) + 1)/(n+1)$, independently for different nodes i, where $d_n(i)$ is the current degree of node i. (Adding 1 in the numerator and denominator is needed in order to generate anything other than empty graphs.)

The behavior of the graph sequence G_n^{pa} is somewhat unexpected: it is convergent with probability 1, but the limit is not determined. More precisely:

PROPOSITION 11.45. *With probability 1, the sequence G_n^{pa} is quasirandom, i.e., it converges to a constant function.*

It is perhaps surprising that if we build a new preferential attachment graph sequence by this method, it may tend to a different constant function. The distribution of limit density is not known.

Proof. Set $X_n = \mathsf{e}(G_n^{\mathrm{pa}})$. Then

$$\mathsf{E}(X_n \mid G_{n-1}^{\mathrm{pa}}) = X_{n-1} + \sum_{i=1}^{n-1} \frac{d_{n-1}(i)+1}{n} = X_{n-1} + \frac{2}{n}X_{n-1} + \frac{n-1}{n}.$$

Hence

$$\frac{1}{(n+2)(n+1)}\mathsf{E}(2X_n + 2n + 1 \mid X_{n-1}) = \frac{1}{(n+1)n}(2X_{n-1} + 2n - 1),$$

which shows that the values $Y_n = (2X_n + 2n + 1)/((n+2)(n+1))$ form a martingale. Since they are obviously bounded, the Martingale Convergence Theorem implies that with probability 1 there is a value a such that $Y_n \to a$. Clearly, $Y_n \sim t(K_2, G_n^{\mathrm{pa}})$, and so $t(K_2, G_n^{\mathrm{pa}}) \to a$.

Given G_{n-1}^{pa}, the degree of node n when it is born is $\sum_{i=1}^{n-1} X_i$, where the X_i are independent 0-1 random variables with $\mathsf{E}(X_i) = (d_{n-1}(i)+1)/(n+1)$. Hence

$$\mathsf{E}(d_n(n) \mid G_{n-1}^{\mathrm{pa}}) = \sum_{i=1}^{n-1} \frac{d_{n-1}(i)+1}{n} = \frac{2}{n}\mathsf{e}(G_{n-1}^{\mathrm{pa}}) + \frac{n-1}{n},$$

and hence $(d_n(n)+1)/(n+1)$ will be heavily concentrated around a. In particular, $(d_n(n)+1)/(n+1) \to a$ as $n \to \infty$.

Next, observe that the development of $d_n(i)$, for a fixed i, follows a Pólya Urn model with $d_i(i) + 1$ red and $i - d_i(i)$ green balls, whence $(d_n(i)+1)/(n+1)$ is a martingale converging to the beta distribution with parameters $d_i(i) + 1$ and $i - d_i(i)$. So for large i, $(d_n(i)+1)/(n+1)$ will be heavily concentrated around its expectation $(d_i(i)+1)/(i+1)$, which in turn is heavily concentrated around a. So for large n, most nodes will have degree around an.

It follows that the process is almost the same as $\mathbb{G}(n, a)$, where we can also think of the nodes created one-by-one and joined to each previous node with probability a. We can couple the two processes to show that with probability 1, they converge to the same limit, which is clearly the identically-a function.

Note that by the Martingale Stopping Theorem A.11,

$$\mathsf{E}(a \mid G_n^{\mathrm{pa}}) = Y_n = \frac{2\mathsf{e}(G_n^{\mathrm{pa}}) + 2n + 1}{(n+2)(n+1)}.$$

Since G_n^{pa} can be any simple graph on n nodes with positive probability, it follows that a is not determined, and with a more careful computation one can see that a falls into any interval with positive probability. □

♦

REMARK 11.46. In several examples above (11.35, 11.36, 11.41, 11.43), the limit graphing is 0-1 valued. Consider, for example, the case of prefix attachment graphs. It follows by Proposition 8.24 that $W_{G_n^{\mathrm{pfx}}} \to W^{\mathrm{pfx}}$ with probability 1 in the edit distance, not just in the cut distance. This means that while the graphs G_n^{pfx} are random, they are very highly concentrated: two instances of G_n^{pfx} differ in

$o(n^2)$ edges only, if overlayed properly (not in the original ordering of the nodes!). Informally, they have a relatively small amount of randomness in them, which disappears as $n \to \infty$. Indeed, G_n^{pfx} is generated using only $O(n \log n)$ bits, as opposed to, say, $\mathbb{G}(n, 1/2)$, which is generated using $\binom{n}{2}$ bits. We'll further explore this phenomenon in Section 14.3.2.

EXERCISE 11.47. We define a randomly growing graph sequence $\mathbf{G}_n$ as follows. We start with a single node. At the n-th iteration, a new node is born, it is connected to node i with probability $1 - i/n$, and every pair of nonadjacent nodes is connected with probability $2/n$. Prove that the sequence $\mathbf{G}_n$ tends to the graphon $1 - xy$ with probability 1.

EXERCISE 11.48. Let us modify the cloning sequence as follows: at each step, we select one of the original nodes (uniformly) and clone it. What is the limit of the resulting graph sequence?

EXERCISE 11.49. Find $\lim_{n \to \infty} t(F, G_n^{\text{pfx}})$ for a general simple graph F.

EXERCISE 11.50. A *random threshold graph* is obtained by the following simple procedure: starting with a single node, at each step we create a new node, flip a coin, and connect the new node either to all previous nodes, or to none of them, depending on the outcome of the coin flip. Prove that with probability 1, a random threshold graph sequence converges to the simple threshold graphon (Diaconis, Holmes and Janson [2008]).

EXERCISE 11.51. Verify that the uniform attachment graphon $1 - \max(x, y)$ in Example 11.39 is the operator square of the simple threshold graphon $\mathbb{1}(x+y \le 1)$ in Example 11.36. Find a combinatorial interpretation of this fact.

11.5. Many disguises of graph limits

We have seen that graphons, homomorphism numbers into graphons, certain finite random graph models, as well as certain infinite random graph models can all be considered as limit objects for convergent graph sequences. All of these carry the same information, so it is a matter of taste which of these is called "the" limit object. The following theorem summarizes a number of structures that are equivalent to graphons. All these are familiar from our discussions.

THEOREM 11.52. *The following are equivalent (cryptomorphic):*

(a) *a graphon, up to weak isomorphism;*

(b) *a multiplicative, normalized simple graph parameter that is nonnegative on signed graphs;*

(c) *a consistent and local random graph model;*

(d) *a local random countable graph model;*

(e) *a point in the completion of the space of finite graphs with the cut distance.*

What we mean by equivalent (or cryptomorphic) is that given one of these objects, one can construct an object from each of the other types (a)–(e), so that these constructions are consistent (going around a cycle we get back to the original object). Below we describe these constructions. This is a summary, most of the work has been done in earlier parts of this book.

The conditions on these objects have useful alternative forms; for example, the condition in (b) that the graph parameter is nonnegative on signed graphs could be replaced by reflection positivity. We will discuss these equivalences in the more general context of random graphons in Section 14.5.2.

Proof. (a)→(b): Every graphon W gives rise to the simple graph parameter $t(.,W)$, which is, as we have seen, multiplicative, normalized, and nonnegative on signed graphs.

(b)→(c): Let f be a multiplicative, normalized simple graph parameter that is nonnegative on signed graphs. The conditions imply that the value of f does not change if isolated nodes are added or deleted. We consider the signed graph $\widehat{F}$ obtained from a simple graph F by signing its edges with $+$, and the edges of its complement with $-$. It is easy to check that $\sum_{F \in \mathcal{F}_k^{\mathrm{simp}}} \widehat{F} = O_k$ (the graph with no edges), and hence $\sum_{F \in \mathcal{F}_k^{\mathrm{simp}}} f(\widehat{F}) = f(O_k) = f(O_1)^k = 1$. So the values $f(\widehat{F})$ form a probability distribution σ_k on $\mathcal{F}_k^{\mathrm{simp}}$. It is clear that this distribution is invariant under isomorphism, so we get a random graph model (σ_k).

It is also easy to check that $\sum_{H:\ H'=F} \widehat{H} = FK_1$, and so
$$f(\widehat{F}) = f(\widehat{FK_1}) = \sum_{H:\ H'=F} f(\widehat{H}).$$
This means that generating a random graph from σ_{k+1}, and deleting its last node, we get a random graph from σ_k. So the model is consistent.

To show that the model is local, let S and T be disjoint subsets of $[k]$, and let F_S and F_T be two simple graphs on S and T, respectively. Let $\mathbb{G}$ be a random graph from σ_k. Then
$$\mathsf{P}\big(\mathbb{G}[S] = F_S, \mathbb{G}[T] = F_T\big) = \sum_{\substack{V(H)=S\cup T \\ H[S]=F_S, H[T]=F_T}} \mathsf{P}\big(\mathbb{G}[S\cup T] = H\big)$$
$$= \sum_{\substack{V(H)=S\cup T \\ H[S]=F_S, H[T]=F_T}} f(\widehat{H}) = f(\widehat{F_S F_T}) = f(\widehat{F_S})f(\widehat{F_T})$$
$$= \mathsf{P}\big(\mathbb{G}[S]=F_S\big)\mathsf{P}\big(\mathbb{G}[T]=F_T\big).$$
Thus the model is local.

(c)↔(d): This follows by Proposition 11.9 and the discussion before it.

(c)→(a): Generate a random graph $\mathbb{G}_n$ from the consistent local random graph model. By Lemma 11.8, we get a convergent graph sequence with probability 1.

(d)↔(e): We have seen that a graph sequence is convergent if and only if it is Cauchy in the cut distance. Every point in the completion is defined by a Cauchy sequence, which tends to a graphon W. Two Cauchy sequences define the same point of the completion if and only if merging them we get a Cauchy sequence, which implies that they have the same limit graphon (up to weak isomorphism). Conversely, every graphon is the limit of a Cauchy sequence (for example, the sequence of W-random graphs), and so it corresponds to a point in the completion. □

11.6. Convergence of spectra

We have seen in Section 7.5 that graphons can be considered as kernel operators and they have a discrete spectrum such that there are only a finite number of eigenvalues outside any neighborhood of 0; in particular, every nonzero eigenvalue has finite multiplicity. All these eigenvalues are contained in the interval $[-1,1]$. Magnitude does not order these eigenvalues into a single sequence: there may be an infinite number of positive ones and an infinite number of negative ones, forming two sequences $\lambda_1(W) \geq \lambda_2(W) \geq \cdots \geq 0$ and $\lambda'_1(W) \leq \lambda'_2(W) \leq \cdots \leq 0$ converging to 0. One or the other of these sequences may be finite; in this case, we still define these infinite sequences, padding them with 0's at the end.

It was proved by Borgs, Chayes, Lovász, Sós and Vesztergombi [2012] that if a sequence of graphs converges to a graphon W, then their spectra converge to the spectrum of W in an appropriate sense. (The more general Theorem 11.54 below also implies a similar result for weighted graphs.)

Let G be a simple graph with n nodes, and let $\mu_1 \geq \mu_2 \geq \cdots \geq \mu_n$ be the eigenvalues of its adjacency matrix. We normalize these eigenvalues to get $\lambda_i = \lambda_i(G) = \mu_i/n$, $(i = 1, \ldots, n)$. It will be useful to consider the sequence ordered increasingly too, and denote it by $\lambda'_1 \leq \lambda'_2 \leq \ldots$.

THEOREM 11.53. *Let $(G_1, G_2, \ldots)$ be a sequence of simple graphs converging to a graphon W. Then for every fixed $i \geq 1$ and $n \to \infty$,*
$$\lambda_i(G_n) \to \lambda_i(W) \quad \text{and} \quad \lambda'_i(G_n) \to \lambda'_i(W).$$

One may be puzzled by the fact that $\lambda_i(W)$ is nonnegative, while for a finite graph G_n, we have no control over the sign of $\lambda_i(G_n)$. For example, if $G_n = K_n$, then every eigenvalue except the largest is negative. The theorem implies that $\lambda_i(G_n)$ must have a nonnegative limit for every fixed i. This is of course not hard to see directly (see Exercise 11.58).

This observation allows us to re-formulate this theorem in terms of the graphons $W_n = W_{G_n}$. It is easy to see that the spectrum of W_n consists of the normalized spectrum of G_n, together with infinitely many 0's. So $\lambda_1(W_n) \geq \lambda_2(W_n) \geq \cdots$ starts with the positive (or nonnegative) part of the spectrum of G_n, followed by 0-s, and similar description can be given of the negative eigenvalues. Hence $\lambda_i(G_n) \leq \lambda_i(W_n)$, but by the remark above, $\lambda_i(W_n) - \lambda_i(G_n) \to 0$ for every fixed i. So the following theorem implies Theorem 11.53.

THEOREM 11.54. *Let $(W_1, W_2, \ldots)$ be a sequence of graphons, converging to a graphon W in the $\delta_\square$ distance. Then for every fixed $i \geq 1$ and $n \to \infty$,*
$$\lambda_i(W_n) \to \lambda_i(W) \quad \text{and} \quad \lambda'_i(W_n) \to \lambda'_i(W).$$

Proof. By choosing a subsequence, we may assume that the limits
$$\alpha_i = \lim_{n \to \infty} \lambda_i(W_n) \quad \text{and} \quad \alpha'_i = \lim_{n \to \infty} \lambda'_i(W_n)$$
exist for every $j \geq 1$. We claim that for every $k \geq 3$,

(11.14)
$$\sum_{i=1}^\infty \lambda_i(W_n)^k \longrightarrow \sum_{i=1}^\infty \alpha_i^k.$$

Indeed, every term on the left converges to the corresponding term on the right, and the series are uniformly majorized by the convergent series $\sum_i 1/i^{k/2}$ by (7.20). Similar arguments can be applied to the negative eigenvalues, and we get that
$$\sum_{\lambda \in \mathrm{Spec}(W_n)} \lambda^k \longrightarrow \sum_{i=1}^\infty \alpha_i^k + \sum_{i=1}^\infty (\alpha'_i)^k \qquad (n \to \infty).$$

For the sum of powers of the eigenvalues we have the graph-theoretic expression
$$\sum_{\lambda \in \mathrm{Spec}(W_n)} \lambda^k = t(C_k, W_n) \quad \text{and} \quad \sum_{\lambda \in \mathrm{Spec}(W)} \lambda^k = t(C_k, W).$$

Now we know that $W_n \to W$ and hence $t(C_k, W_n) \to t(C_k, W)$. Thus we get

$$\sum_{\lambda \in \mathrm{Spec}(W)} \lambda^k = \sum_{i=1}^{\infty} \alpha_i^k + \sum_{i=1}^{\infty} (\alpha_i')^k$$

for every $k \geq 3$. By Proposition A.21 in the Appendix, this implies that $\{\alpha_1, \alpha_2, \dots\}$ and $\{\lambda_1(W), \lambda_2(W), \dots\}$ are equal as multisets, and similar conclusion holds for $\{\alpha_1', \alpha_2', \dots\}$ and $\{\lambda_1'(W), \lambda_2'(W), \dots\}$. □

The functionals $\lambda_i(W)$ and $\lambda_i'(W)$ are invariant under measure preserving transformations, and so they can be considered as a functional on the space $(\widetilde{\mathcal{W}}, \delta_\square)$. The theorem shows that these functionals are continuous. Of course, similar conclusion holds for the eigenvalues of kernels in $\widetilde{\mathcal{W}}_1$. By compactness, these maps are uniformly continuous, which can be stated as follows:

COROLLARY 11.55. *For $\varepsilon > 0$ and every $i \geq 1$, there is a $\delta_i > 0$ such that if $U, W \in \mathcal{W}_1$ and $\delta_\square(U, W) \leq \delta_i$, then*

$$|\lambda_i(U) - \lambda_i(W)| \leq \varepsilon \quad \text{and} \quad |\lambda_i'(U) - \lambda_i'(W)| \leq \varepsilon.$$

EXAMPLE 11.56. If (G_n) is a quasirandom sequence with density p, then the largest normalized eigenvalue of G_n tends to p, while the others tend to 0. The limiting graphon, the identically-p function, has one nonzero eigenvalue (namely p). ♦

The last example suggests that perhaps convergent graph sequences can be characterized through the convergence of their spectra, since if (G_n) is a sequence of graphs such that the edge density on G_n tends to p, the largest normalized eigenvalue of G_n tends to p, and all the other eigenvalues tend to 0, then (G_n) is quasirandom. There is no real hope for this, however:

EXAMPLE 11.57. Consider two non-isomorphic graphs G_1 and G_2 with the same spectrum (for example, the incidence graphs of two non-isomorphic finite projective planes of the same order). Consider the blow ups $G_1(n)$ and $G_2(n)$, $n = 1, 2, \dots$, and merge them into a single sequence. This sequence is not convergent, but all graphs in it have the same spectra except for the multiplicity of 0. ♦

EXERCISE 11.58. Prove that for any finite simple graph G, $\lambda_i(G) \geq -(i-1)/(\mathsf{v}(G) - i + 1)$.

11.7. Convergence in norm

We discuss a somewhat technical issue that is important and also often convenient when working with convergence of graph and graphon sequences.

Let $G_n \to W$, so $\delta_\square(W_{G_n}, W) \to 0$. Of course, this does not mean that $\|W_{G_n} - W\|_\square \to 0$: the function W_{G_n} depends on the labeling of the nodes of G_n (the distance $\delta_\square(W_{G_n}, W)$ does not, since relabeling G_n results in weak isomorphism of W_{G_n}). By the definition of the $\delta_\square$ distance, we have measure preserving bijections $\varphi_n : [0,1] \to [0,1]$ such that $\|W_{G_n}^{\varphi_n} - W\|_\square \to 0$. The functions $W_{G_n}^{\varphi_n}$ can be quite complicated for general measure preserving bijections φ_n, so it is nice to know that the maps φ_n can be chosen to be just permutations of the steps of W_{G_n}. In other words, choosing the labeling of the graphs G_n appropriately, we can achieve convergence in norm.

THEOREM 11.59. *Let (G_n) be a sequence of graphs such that $G_n \to W$. Then the graphs G_n can be labeled so that $\|W_{G_n} - W\|_\square \to 0$.*

Proof. Let $\mathcal{P}_n$ be a partition of $[0,1]$ into consecutive intervals of length $1/\mathsf{v}(G_n)$. By Proposition 9.8, we have that $\|W - W_{\mathcal{P}_n}\|_\square \to 0$, so combined with the assumption that $\delta_\square(W, W_{G_n}) \to 0$ we see that $\delta_\square(W_{\mathcal{P}_n}, W_{G_n}) \to 0$. Here $\delta_\square(W_{\mathcal{P}_n}, W_{G_n}) = \delta_\square(W/\mathcal{P}_n, G_n)$ can be thought of as the distance of two weighted graphs on the same number of nodes, so by Theorem 9.29, we get that $\widehat{\delta}_\square(W/\mathcal{P}_n, G_n) \to 0$. This means that the graphs in the sequence (G_n) can be relabeled to get a graph sequence (G'_n) such that
$$\|W_{\mathcal{P}_n} - W_{G'_n}\|_\square = d_\square(W/\mathcal{P}_n, G'_n) \to 0.$$
Since $\|W - W_{\mathcal{P}_n}\|_\square \to 0$, this proves the Theorem. $\square$

EXERCISE 11.60. Let us extend the definition of the distance $\widehat{\delta}$ to the case when one of the arguments is a graphon U: $\widehat{\delta}_\square(U, G) = \min_{G'} \|U - W_{G'}\|_\square$, where G' ranges over all relabeled versions of G. (a) Prove that if $G_n \to U$, then $\widehat{\delta}_\square(U, G_n) \to 0$. (b) Show by an appropriate construction that the following stronger statement is not true: there exists a function $f : [0,1] \times \mathbb{N}^* \to [0,1]$ such that $f(x, n) \to 0$ if $x \to 0$ and $n \to \infty$, and $\widehat{\delta}_\square(U, G) \leq f(\delta_\square(U, G), \mathsf{v}(G))$.

EXERCISE 11.61. Prove that a sequence of graphs is convergent if and only if they have weak regularity partitions with convergent templates. More precisely, (G_n) is convergent if and only if for every $k \in \mathbb{N}^*$ $V(G_n)$ has a k-partition $\mathcal{P}_{k,n}$ such that (a) for every n, we have $d_\square(G_n, (G_n)_{\mathcal{P}_{n,k}}) \leq 10/\sqrt{\log k}$, and (b) for every k, the template graphs $G_n/\mathcal{P}_{n,k}$ converge to some weighted graph H_k on k nodes as $n \to \infty$.

11.8. First applications

As a first illustration how our graph limits help proving theorems about finite graphs, we describe two proofs working with graph limits of two important results: a characterization of quasirandom graphs (see Section 1.4.2) and the Removal Lemma (see Section 11.8.2). This should illustrate not only that graph limits are useful, but also the method of obtaining a limit graphon from a sequence of counterexamples.

11.8.1. Quasirandom graphs. We start with quasirandom graphs. Let (G_n) be a quasirandom sequence with density p (see Examples 11.37 and 11.56). This means that $t(F, G_n) \to p^{e(F)}$ for every simple graph F, or even simpler, $G_n \to p$ (the identically-p graphon). We mentioned in the introduction the surprising fact, due to Chung, Graham and Wilson [1989], that it is enough to require this relation for $F = K_2$ and $F = C_4$:

THEOREM 11.62. *If (G_n) is a sequence of simple graphs such that $\mathsf{v}(G_n) \to \infty$, $t(K_2, G_n) \to p$, and $t(C_4, G_n) \to p^4$, then (G_n) is quasirandom with density p.*

Proof. Suppose that (G_n) is not quasirandom, i.e., there is a simple graph F such that $t(F, G_n) \not\to p^{e(F)}$. We can select a subsequence for which $t(F, G_n) \to c \neq p^{e(F)}$, and then we can select a convergent subsequence. Let W be its limit graphon; then $t(K_2, W) = p$, $t(C_4, W) = p^4$, and $W \neq pJ$ as $t(F, W) = c \neq p^{e(F)}$.

To get a contradiction, it suffices to prove:

CLAIM 11.63. *If W is a kernel such that $t(K_2, W) = p$ and $t(C_4, W) = p^4$ for some real number p, then $W = pJ$.*

We start with two applications of the Cauchy–Schwarz inequality:

$$t(\text{\tiny □□}, W) = \langle t_{xy}(\text{\tiny △}, W), t_{xy}(\text{\tiny △}, W) \rangle \geq \langle t_{xy}(\text{\tiny △}, W), J \rangle^2 = t(\text{\tiny △}, W)^2$$
$$= \langle t_x(\text{\tiny ↑}, W), t_x(\text{\tiny ↑}, W) \rangle^2 \geq \langle t_x(\text{\tiny ↑}, W), J \rangle^4 = t(\text{\tiny ○}, W)^4.$$

Since we have equality, the function $t_{xy}(\text{\tiny △}, W)$ must be constant, and by integration we see that its value is p^2. This means that $W \circ W = t_{xy}(\text{\tiny △}, W) = p^2 J$. This means that the operator $T_{W \circ W} = T_W^2$ has a single nonzero eigenvalue p^2 with eigenfunction $\equiv 1$. But then trivially T_W has a single eigenvalue $\pm p$ with the same eigenfunction, i.e., $W \equiv p$ or $W \equiv -p$. The condition $t(K_2, W) = p$ rules out the second alternative. □

11.8.2. Removal Lemma. One of the first, most famous, and in a sense infamous consequences of the Regularity Lemma was proved by Ruzsa and Szemerédi [1976].

LEMMA 11.64 (**Removal Lemma**). *For every $\varepsilon > 0$ there is an $\varepsilon' > 0$ such that if a simple graph G with n nodes has at most $\varepsilon' n^3$ triangles, then we can delete εn^2 edges from G so that the remaining graph has no triangles.*

This lemma sounds innocent, almost like a trivial average computation. This is far from the truth! No simple proof is known, and (worse) all the known proofs give a terrible dependence of ε' on ε. The best bound, due to Fox [2011] gives an ε' such that $1/\varepsilon'$ is a tower $2^{2^{2^{\cdots}}}$ of height about $\log(1/\varepsilon)$. The original proof gives a tower of height about $1/\varepsilon^2$. Perhaps this looks friendlier (?) if we write it as $\varepsilon \approx 1/\sqrt{\log^*(1/\varepsilon')}$. The proof given below does not give any explicit bound, but it illustrates the way graph limit theory can be used.

Proof. Suppose that the lemma is false. This means that there is an $\varepsilon > 0$ and a sequence of graphs (G_n) such that $t(K_3, G_n) \to 0$ but deleting any set of εn^2 edges, the remaining graph will contain a triangle. By selecting a subsequence, we may assume that $t(F, G_n)$ is convergent for every simple graph F, and then there is a graphon W such that $G_n \to W$. We have then $t(K_3, W) = \lim_{n \to \infty} t(K_3, G_n) = 0$.

The condition on the deletion of edges is harder to deal with, because it does not translate directly to any property of the limit graphon W. What we can do is to "pull back" information from W to the graphs G_n. By Theorem 11.59, we may assume that $\|W_{G_n} - W\|_\square \to 0$. (This step is not absolutely necessary, but convenient.)

Let $S = \{(x, y) \in [0, 1]^2 : W(x, y) > 0\}$. By Lemma 8.22, we have

$$\int_{[0,1]^2} (1 - \mathbb{1}_S) W_{G_n} \to \int_{[0,1]^2} (1 - \mathbb{1}_S) W = 0,$$

so we can choose n large enough so that $\int (1 - \mathbb{1}_S) W_{G_n} < \varepsilon/4$. Let $V(G_n) = [N]$, $J_i = [(i-1)/N, i/N]$, and $R_{ij} = J_i \times J_j$.

We modify G_n by deleting the edge ij if $\lambda(S \cap R_{ij}) < 3/(4N^2)$.

CLAIM 11.65. *The remaining graph G'_n is triangle-free.*

Indeed, suppose that i, j, k are three nodes such that $\lambda(S \cap R_{ij}) \geq 3/(4N^2)$, $\lambda(S \cap R_{jk}) \geq 3/(4N^2)$ and $\lambda(S \cap R_{ik}) \geq 3/(4N^2)$. Observe that $t(K_3, W) = 0$

implies that $t(K_3, \mathbb{1}_S) = 0$. But we have

$$t(K_3, \mathbb{1}_S) \geq \int\limits_{J_i \times J_j \times J_k} \mathbb{1}_S(x,y) \mathbb{1}_S(y,z) \mathbb{1}_S(x,z) \, dx \, dy \, dz$$

$$\geq \frac{1}{N^3} - \frac{1}{N} \lambda(R_{ij} \setminus S) - \frac{1}{N} \lambda(R_{jk} \setminus S) - \frac{1}{N} \lambda(R_{ik} \setminus S)$$

$$\geq \frac{1}{N^3} - 3\frac{1}{4N^3} = \frac{1}{4N^3} > 0,$$

which is a contradiction.

What is left is to bound the number m of edges deleted. This is easy: if edge ij is deleted, then $W_{G_n} = 1$ on R_{ij}, and so

$$\int\limits_{R_{ij}} (1 - \mathbb{1}_S) W_{G_n} = \int\limits_{R_{ij}} (1 - \mathbb{1}_S) = \lambda(R_{ij} \setminus S) \geq \frac{1}{4N^2},$$

We know that the number of deleted edges must be at least εN^2, and so

$$\int\limits_{[0,1]^2} (1 - \mathbb{1}_S) W_{G_n} \geq \frac{\varepsilon N^2}{4N^2} = \frac{\varepsilon}{4},$$

which contradicts the choice of n. $\square$

If you know the usual derivation of the Removal Lemma from the Regularity Lemma, or have worked it out yourself by solving Exercise 11.66, you may feel that the two proofs are analogous; and you are quite right. One may even say that it is the same proof told in a different language, or at least, that it points out a nontrivial connection between the Regularity Lemma and measure theory. Indeed, other much deeper versions and applications of the Regularity Lemma (like the Regularity Lemma for hypergraphs or Szemerédi's Theorem on arithmetic progressions) can be proved using mainly measure theoretic arguments; see Elek and Szegedy [2012] [2012] for details.

EXERCISE 11.66. (a) Let G be a graph on n nodes, and consider a Szemerédi partition of G with k classes and error bound ε as in the Original Regularity Lemma. Let us delete all edges within the classes, between exceptional pairs of classes, and between classes where the edge-density is less than $100\varepsilon^{1/3}$. Prove that if the remaining graph contains any triangle, then it contains at least $\varepsilon(n/k)^3$ triangles. (b) Prove the Removal Lemma, based on (a).

EXERCISE 11.67. Extend the proof of the Removal Lemma above to the following more general theorem: For every simple graph F and every $\varepsilon > 0$ there is an $\varepsilon' > 0$ such that if a simple graph G with n nodes has $t(F, G) \leq \varepsilon'$, then we can delete εn^2 edges from G so that F has no homomorphism into the remaining graph.

CHAPTER 12

Convergence from the right

Recall from the Introduction the formula
$$F \longrightarrow G \longrightarrow H,$$
which referred to the framework in which we study large graphs from "both sides": by mapping small graphs into them, and mapping them into small graphs. So far, we have used homomorphisms into the large graph to define convergence. We have seen many examples, however, of homomorphism numbers from the large graph into fixed target graphs H which were very interesting: the number of k-colorings of a graph; Ising and Potts models in statistical physics; approximating maximum cuts. We have seen that homomorphism functions $\hom(G,.)$ have a characterization (Theorem 5.59) perfectly analogous to the characterization of homomorphism functions $\hom(., G)$ (Corollary 5.58). Does this duality extend to convergence?

In this chapter we set out to characterize convergence of a graph sequence $(G_1, G_2, \ldots)$ in terms of mappings "to the right", using homomorphisms from the graphs G_n into some fixed graph H, based on the results of Borgs, Chayes, Lovász, Sós and Vesztergombi [2012].

The first and most natural approach is not going to work. It is clear that considering simple graphs H would not give sufficient information: if the chromatic number of the graphs in the sequence tends to infinity, then $\hom(G_n, H)$ is eventually 0 for every fixed simple graph H. We will see that even counting homomorphisms into weighted graphs H would not suffice to characterize convergence. On the other hand, we show that a modification of the notion of homomorphism numbers, or replacing counting by maximization (in other words, considering restricted multicuts), does lead to a characterizations of convergence. It will turn out that the convergence conditions hold for general weighted target graphs H, but it is enough to require them for simple graphs (in the maximization version) of for weighted graphs with just two positive edgeweights (in the counting case).

We will talk informally about left-convergence and right-convergence. Left-convergence of a sequence (G_n) means our notion of convergence as defined and studied in the previous chapter. Right-convergence means a number of possible convergence notions defined in terms of homomorphisms from the graph G_n into fixed smaller graphs.

12.1. Homomorphisms to the right and multicuts

12.1.1. Naive right-convergence. Consider homomorphisms $G \to H$, where we think of G as a very large simple graph and H is a small weighted graph. It will be convenient to scale the nodeweights of H so that $\alpha_H = 1$; this only scales the values $\hom(G, H)$ by an easily computable factor. We will assume that $V(G) = [n]$ and $V(H) = [q]$. (We refer to Borgs, Chayes, Lovász, Sós and

Vesztergombi [2012] for a treatment of the case when G is also weighted, and also for consequences of these results in statistical physics.)

Recall that for a fixed weighted graph H, the value $\hom(G,H)$ grows exponentially with n^2, and so a reasonable normalization is to consider the (dense) *homomorphism entropy*

$$\mathsf{ent}(G,H) = \frac{\log \hom(G,H)}{\mathsf{v}(G)^2}.$$

The first, "naive" notion of right-convergence would be to postulate that these homomorphism entropies converge for all weighted graphs H with (say) positive edgeweights. This is at least a necessary condition for convergence:

PROPOSITION 12.1. *Let (G_n) be a convergent graph sequence. Then for every weighted graph H with positive edgeweights, the sequence $\mathsf{ent}(G_n, H)$ is convergent.*

To prove this proposition, let us recall from Example 5.19 that the homomorphism entropy can be approximated by the maximum weighted multicut density: Let G be a simple graph on $[n]$, and H, a weighted graph on $[q]$ with positive edgeweights and $\alpha_H = 1$. Define $B_{ij} = \log \beta_{ij}(H)$, then

$$(12.1) \qquad \mathsf{cut}(G,B) \leq \frac{\log \hom(G,H)}{n^2} \leq \mathsf{cut}(G,B) + \frac{\log q}{n},$$

where $\mathsf{cut}(G,B)$ is the maximum weighted multicut density

$$\mathsf{cut}(G,B) = \max_{(S_1,\ldots,S_q) \in \Pi_n} \frac{1}{n^2} \sum_{i,j \in [q]} B_{ij} e_G(S_i, S_j).$$

We need a couple of facts about weighted multicut densities. First, they are invariant under blow-ups:

LEMMA 12.2. *For a simple graph G, symmetric matrix $B \in \mathbb{R}^{q \times q}$ and integer $k \geq 1$, we have*

$$\mathsf{cut}(G(k), B) = \mathsf{cut}(G, B).$$

Proof. The inequality $\mathsf{cut}(G(k), B) \geq \mathsf{cut}(G, B)$ is clear, since every q-partition of $V(G)$ can be lifted to a q-partition of $V(G(k))$, contributing the same value to the maximization in the definition of $\mathsf{cut}(G(k), B)$. To prove the reverse inequality, let $(S_1, \ldots, S_q)$ be the q-partition of $V(G(k))$ attaining the maximum in the definition of $\mathsf{cut}(G(k), B)$. For every node $v \in V(G)$, we pick a random element $v' \in V(G(k))$ uniformly from the set of twins of v created when blowing it up, and let $T = \{v' : v \in V(G)\}$. Let $G' = G(k)[T]$. Then $G' \cong G$, and

$$\mathsf{E}\left(\frac{1}{n^2} \sum_{i,j \in [q]} B_{ij} e_{G'}(S_i \cap T, S_j \cap T)\right) = \frac{1}{(nk)^2} \sum_{i,j \in [q]} B_{ij} e_{G(k)}(S_i, S_j)$$
$$= \mathsf{cut}(G(k), B).$$

It follows that for at least one choice of the nodes v', we have

$$\mathsf{cut}(G', B) \geq \frac{1}{n^2} \sum_{i,j \in [q]} B_{ij} e_{G'}(S_i \cap T, S_j \cap T) \geq \mathsf{cut}(G(k), B). \qquad \square$$

The following lemma is superficially similar to the Counting Lemma 10.22; it is in fact quite a bit simpler.

LEMMA 12.3. *For two simple graphs G and G' and symmetric matrix $B \in \mathbb{R}^{q \times q}$, we have*
$$|\mathsf{cut}(G, B) - \mathsf{cut}(G', B)| \leq q^2 \delta_\square(G, G').$$

Proof. We start with proving the weaker inequality
$$|\mathsf{cut}(G, B) - \mathsf{cut}(G', B)| \leq q^2 \widehat{\delta}_\square(G, G') \tag{12.2}$$
in the case when $\mathsf{v}(G) = \mathsf{v}(G') = n$. We may assume that G and G' are optimally overlayed, so that $V(G) = V(G') = [n]$ and $\widehat{\delta}_\square(G, G') = d_\square(G, G')$. Then for every partition $(S_1, \ldots, S_q) \in \Pi_n$, we have
$$\left| \frac{1}{n^2} \sum_{i,j \in [q]} B_{ij} e_G(S_i, S_j) - \frac{1}{n^2} \sum_{i,j \in [q]} B_{ij} e_{G'}(S_i, S_j) \right|$$
$$\leq \frac{1}{n^2} \sum_{i,j \in [q]} B_{ij} |e_G(S_i, S_j) - e_{G'}(S_i, S_j)| \leq q^2 d_\square(G, G').$$

This proves (12.2). To get the more general inequality in the lemma, we apply (12.2) to the graphs $G(n'k)$ and $G'(nk)$, where $n = \mathsf{v}(G)$, $n' = \mathsf{v}(G')$, and k is a positive integer. The left side equals $|\mathsf{cut}(G, B) - \mathsf{cut}(G', B)|$ for any k by Lemma 12.2, while the right side tends to $q^2 \delta_\square(G, G')$ if $k \to \infty$ by the definition of $\delta_\square$. □

Proof of Proposition 12.1. By the Theorem 11.3, we have $\delta_\square(G_n, G_m) \to 0$ as $n, m \to \infty$; by Lemma 12.3, this implies that the sequence of numbers $\mathsf{cut}(G_n, B)$ is a Cauchy sequence; by (12.1), it follows that the values $\mathsf{ent}(G_n, H)$ form a Cauchy sequence. □

It would be a natural idea here to define convergence of a graph sequence in terms of the convergence of the homomorphism entropies $\mathsf{ent}(G_n, H)$. However, this notion of convergence would not be equivalent to left-convergence, and it would allow sequences that we would not like to consider "convergent", as Example 12.4 below shows. (Some suspicion could have been raised by (12.1) already: the nodeweights of H disappeared, which indicated loss of information.)

EXAMPLE 12.4. Let (F_n) be a quasirandom graph sequence with edge density p, and let (G_n) be a quasirandom graph sequence of density $2p$, where (to keep the notation simple), we assume that $\mathsf{v}(F_n) = \mathsf{v}(G_n) = n$. Then we have, for every weighted graph H with positive edgeweights,
$$\mathsf{ent}(F_n, H) = \frac{1}{n^2} \max_{(S_1,\ldots,S_q) \in \mathcal{P}_n} \sum_{i,j \in [q]} B_{ij} e_{F_n}(S_i, S_j) + O\left(\frac{1}{n}\right)$$
$$= \max_{(S_1,\ldots,S_q) \in \mathcal{P}_n} \sum_{i,j \in [q]} B_{ij} \left(p \frac{|S_i|}{n} \frac{|S_j|}{n} + o(1)\right) + O\left(\frac{1}{n}\right)$$
$$= p \max\{x^\mathsf{T} B x : x \in \mathbb{R}^q_+, x^\mathsf{T} \mathbb{1} = 1\} + o(1).$$
Applying the same computation to G_n, we get that for the graphs G_n^2 (disjoint union of two copies of G_n), we have
$$\mathsf{ent}(G_n^2, H) = \frac{\log \mathsf{hom}(G_n^2, H)}{(2n)^2} = \frac{\log(\mathsf{hom}(G_n, H)^2)}{4n^2} = \frac{\log \mathsf{hom}(G_n, H)}{2n^2}$$
$$= \frac{1}{2} \mathsf{ent}(G_n, H) = p \max\{x^\mathsf{T} B x : x \in \mathbb{R}^q_+, x^\mathsf{T} \mathbb{1} = 1\} + o(1).$$

So merging the sequences (F_n) and (G_n^2) we get a graph sequence for which the quantities $\mathsf{ent}(G_n, H)$ converge for every H, but which is clearly not convergent (check the triangle density!). ♦

12.1.2. Typical homomorphisms. Let us try to take the nodeweights of H into account. The values $\alpha_\varphi = \prod_{v \in V(G)} \alpha_{\varphi(v)}$ form a probability distribution on the maps $\varphi \colon [n] \to [q]$, where (by the Law of Large Numbers) we have $|\varphi^{-1}(i)| \approx \alpha_i n$ with high probability, if n is large. However, this information becomes irrelevant as $n \to \infty$, and only the largest term will count rather than the "typical". It turns out that it is often advantageous to restrict ourselves to maps that are "typical", by forcing φ to divide the nodes in the given proportions. Let $\Pi(n, \alpha)$ denote the set of partitions $(V_1, \ldots, V_q)$ of $[n]$ into q parts with $\lfloor \alpha_i n \rfloor \leq |V_i| \leq \lceil \alpha_i n \rceil$, , and consider the set of maps

$$\Phi(n, \alpha) = \{ \varphi \in [q]^n \,:\, \lfloor \alpha_i n \rfloor \leq |\varphi^{-1}(i)| \leq \lceil \alpha_i n \rceil \text{ for all } i \in [q] \}.$$

(We could be less restrictive and allow, say $\varphi^{-1}(i) \in [\alpha_i n - \sqrt{n}, \alpha_i n + \sqrt{n}]$. This would not change the considerations below in any significant way.)

We define a modified homomorphism number, by summing only over the "typical" homomorphisms:

$$\mathsf{hom}^*(G, H) = \sum_{\varphi \in \Phi(n, \alpha)} \alpha_\varphi \prod_{uv \in E(G)} \beta_{\varphi(u)\varphi(v)}.$$

From this, we get the *typical homomorphism entropy*

$$\mathsf{ent}^*(G, H) = \frac{\log \mathsf{hom}^*(G, H)}{n^2}.$$

12.1.3. Restricted multicuts. In maximum multicut problems, it is quite natural to fix the proportion into which a multicut separates the node set. For example, the "maximum bisection problem" asks for the maximum size of a cut that separates the nodes into two almost equal parts. We can formulate the *restricted multicut problem* as follows. We specify (in addition to the coefficients B_{ij}), q further numbers $\alpha_1, \ldots, \alpha_q > 0$ with $\alpha_1 + \cdots + \alpha_q = 1$.

It will convenient to consider the parameters α_i and B_{ij} as the nodeweights and edge weights of a weighted graph H' with $V(H') = [q]$. Then we are interested in the value

$$(12.3) \qquad \mathsf{cut}(G, H') = \max \frac{1}{n^2} \sum_{i,j} B_{ij} e_G(S_i, S_j),$$

where $\{S_1, \ldots, S_q\}$ ranges over all partitions of $V(G)$ such that

$$(12.4) \qquad \lfloor \alpha_i n \rfloor \leq |S_i| \leq \lceil \alpha_i n \rceil \quad (i = 1, \ldots, q).$$

This quantity is called a *maximum restricted multicut*, or in terms of statistical physics, a *microcanonical ground state energy*. This quantity can be defined for all graphs H with positive nodeweights, by scaling the nodeweights so that they sum to 1.

The same simple computation as in Example 5.19 gives the following formula:

$$(12.5) \qquad \mathsf{ent}^*(G, H) = \mathsf{cut}(G, H') + O(\frac{1}{n}).$$

Here H' is obtained from H by replacing all edgeweights by their binary logarithms (while keeping the same nodeweights).

In the study of subgraph densities, the identity $t(F,G) = t(F,W_G)$ guaranteed an easy transition between graphs and graphons. A finite consequence of this is the identity $t(F,G) = t(F,G(m))$ (where $G(m)$ is the m-blowup of the graph G). Lemma 12.2 above shows that a similar identity holds for multicuts. However, for restricted multicuts or homomorphisms to the right such a simple identity does not hold any more. We will have to estimate the error we are making when we replace a graph G by the associated graphon W_G (luckily, this error will be small if G is large enough).

EXERCISE 12.5. Show that if $\alpha_i = 1/q$ for all i and $n < q$, then $\mathsf{hom}^*(G,H) = t_{\mathsf{inj}}(G,H)$.

EXERCISE 12.6. Let (G_n) be a quasirandom sequence with edge density p, and let F be a simple graph such that $2e(F[S]) \leq q|S|^2$ for every subset $S \subseteq V(F)$. Prove that $\mathsf{cut}(G_n, F) \leq pq + o(1)$ $(n \to \infty)$.

12.2. The overlay functional

The main advantage of the maximum-cut type functions introduced in the previous section is that it is easy to extend their definitions to the case when the graph G is replaced by a graphon. (Let me repeat that there does not seem to be any reasonable extension of the hom function to graphons.) For a probability distribution α on $[q]$, let $\Pi(\alpha)$ denote the set of partitions $(S_1, \ldots, S_q)$ of $[0,1]$ into q measurable sets with $\lambda(S_i) = \alpha_i$. For every graphon U and weighted graph H on node set $[q]$, we define

$$\mathcal{C}(U, H) = \sup_{(S_1,\ldots,S_q) \in \Pi(\alpha)} \sum_{i,j \in [q]} B_{ij} \int_{S_i \times S_j} U(x,y) \, dx \, dy$$

This notion does not quite extend the maximum restricted weighted multicut of graphs G; the reason is that in a graph, we cannot partition the set of nodes in exactly the desired proportions. But the difference is small; we will come back to this question in Section 12.4.1.

We can generalize even further and define, for two kernels U and W,

$$\mathcal{C}(U,W) = \sup_{\varphi \in S_{[0,1]}} \langle U, W^\varphi \rangle = \sup_{\varphi \in S_{[0,1]}} \int_{[0,1]^2} U(x,y) W\bigl(\varphi(x), \varphi(y)\bigr) \, dx \, dy.$$

It is easy to see that this extends the definition of maximum restricted weighted multicuts in the sense that if U is any graphon and H is a weighted graph, then

(12.6) $$\mathcal{C}(U, H) = \mathcal{C}(U, W_H).$$

The functional $\mathcal{C}(U, W)$, which we call the *overlay functional*, has many good properties. It follows just like the similar statement for norms in Theorem 8.13 that

(12.7) $$\mathcal{C}(U,W) = \sup_{\varphi \in S_{[0,1]}} \langle U, W^\varphi \rangle = \sup_{\varphi \in S_{[0,1]}} \langle U^\varphi, W \rangle = \sup_{\varphi, \psi \in \overline{S}_{[0,1]}} \langle U^\varphi, W^\psi \rangle$$
$$= \sup\bigl\{\langle U_0, W_0 \rangle : (\exists \varphi, \psi \in \overline{S}_{[0,1]}) \, U = U_0^\varphi, W = W_0^\psi \bigr\}.$$

Hence it follows that the overlay functional is invariant under measure preserving transformations of the kernels, i.e., it is a functional on the space $\widetilde{\mathcal{W}}_0 \times \widetilde{\mathcal{W}}_0$. It is also immediate from the definition that this quantity has the (somewhat unexpected) symmetry property $\mathcal{C}(U,W) = \mathcal{C}(W,U)$, and satisfies the inequalities

(12.8) $\qquad \langle U, W \rangle \leq \mathcal{C}(U,W) \leq \|U\|_2 \|W\|_2, \quad \mathcal{C}(U,W) \leq \|U\|_\infty \|W\|_1.$

This suggests that $\mathcal{C}(.,.)$ behaves like some kind of inner product. This analogy is further supported by the following identity, reminiscent of the cosine theorem, relating it to the distance δ_2 derived from the L_2-norm:

$$\text{(12.9)} \qquad \mathcal{C}(U,W) = \frac{1}{2}\Big(\|U\|_2^2 + \|W\|_2^2 - \delta_2(U,W)^2\Big)$$
$$= \frac{1}{2}\Big(\delta_2(U,0)^2 + \delta_2(W,0)^2 - \delta_2(U,W)^2\Big).$$

Indeed,

$$\delta_2(U,W)^2 = \inf_{\varphi \in S_{[0,1]}} \|U - W^\varphi\|_2^2 = \|U\|_2^2 + \|W\|_2^2 - 2 \sup_{\varphi \in S_{[0,1]}} \langle U, W^\varphi \rangle$$
$$= \|U\|_2^2 + \|W\|_2^2 - 2\mathcal{C}(U,W).$$

We have to be a bit careful: the functional $\mathcal{C}(U,W)$ is not bilinear, only subadditive in each variable:

(12.10) $\qquad \mathcal{C}(U+V, W) \leq \mathcal{C}(U,W) + \mathcal{C}(V,W).$

It is homogeneous for positive scalars: if $\lambda > 0$, then

(12.11) $\qquad \mathcal{C}(\lambda U, W) = \mathcal{C}(U, \lambda W) = \lambda \mathcal{C}(U,W).$

We have $\mathcal{C}(U,W) = \mathcal{C}(-U,-W)$, but $\mathcal{C}(U,W)$ and $\mathcal{C}(-U,W)$ are not related in general.

A less trivial property of the overlay functional is that it is continuous in each variable (with respect to the $\delta_\square$ distance). This does not follow from (12.9), since the distance $\delta_2(U,W)$ is not continuous with respect to $\delta_\square$ (only lower semicontinuous; see Section 14.2.1).

LEMMA 12.7. *If $\delta_\square(U_n, U) \to 0$ as $n \to \infty$ $(U, U_n \in \mathcal{W}_1)$, then for every $W \in \mathcal{W}_1$ we have $\mathcal{C}(U_n, W) \to \mathcal{C}(U,W)$.*

Proof. By subadditivity (12.10), we have

$$-\mathcal{C}(U - U_n, W) \leq \mathcal{C}(U_n, W) - \mathcal{C}(U,W) \leq \mathcal{C}(U_n - U, W),$$

and hence it is enough to prove that $\mathcal{C}(U_n - U, W), \mathcal{C}(U - U_n, W) \to 0$. In other words, it suffices to prove the lemma in the case when $U = 0$.

By definition, we have $\mathcal{C}(U_n, W) \geq \langle U_n, W \rangle$, and the right side tends to 0 by Lemma 8.22. Hence $\liminf_n \mathcal{C}(U_n, W) \geq 0$.

To prove the opposite inequality, we start with the case when W is a stepfunction. Write $W = \sum_{i=1}^m a_i \mathbb{1}_{S_i \times T_i}$, then using (12.10) and (12.11), we get

$$\mathcal{C}(U_n, W) \leq \sum_{i=1}^m \mathcal{C}(U_n, a_i \mathbb{1}_{S \times T}) = \sum_{i=1}^m \mathcal{C}(a_i U_n, \mathbb{1}_{S \times T})$$
$$\leq \sum_{i=1}^m \|a_i U_n\|_\square = \sum_{i=1}^m |a_i| \|U_n\|_\square \to 0.$$

Now if W is an arbitrary kernel, then for every $\varepsilon > 0$ we can find a stepfunction W' such that $\|W - W'\|_1 \leq \varepsilon/2$. We know that $\mathcal{C}(U_n, W') \to 0$, and hence $\mathcal{C}(U_n, W') \leq \varepsilon/2$ if n is large enough. But then

$$\mathcal{C}(U_n, W) \leq \mathcal{C}(U_n, W - W') + \mathcal{C}(U_n, W') \leq \|U_n\|_\infty \|W - W'\|_1 + \varepsilon/2 \leq \varepsilon.$$

This shows that $\limsup_n \mathcal{C}(U_n, W) \leq 0$, and completes the proof. $\square$

While the functional $\mathcal{C}(U, W)$ is continuous in each variable, it is not continuous as a functional on the product space $(\widetilde{\mathcal{W}}_1, \delta_\square) \times (\widetilde{\mathcal{W}}_1, \delta_\square)$. Let (G_n) be any quasirandom graph sequence and let $W_n = U_n = 2W_{G_n} - 1$. Then $U_n, W_n \to 0$ in the cut norm (and so also in $\delta_\square$), but $\mathcal{C}(U_n, W_n) = 1$ for all n.

EXERCISE 12.8. Define the following functional on $\mathcal{W}_0 \times \mathcal{W}_0$:
$$\mathcal{C}^*(U, W) = \sup_{\varphi:\ [0,1] \to [0,1]} \langle U, W^\varphi \rangle$$
where φ is measurable, but not necessarily measure preserving. Prove the formulas
$$\mathsf{maxcut}(G) = \mathcal{C}^*(W_G, W_{K_2}),$$
and
$$\|U\|_\square \approx \left| \mathcal{C}^*\left(U, \mathbb{1}(x, y \leq 1/2)\right) \right|,$$
where the $\approx$ sign means equality up to a factor of 2.

12.3. Right-convergent graphon sequences

As we have experienced before, many results are easier and cleaner when formulated for graphons. This is particularly true for results about right-convergence. the goal of this section is to formulate various characterizations of convergent graphon sequences in terms of quantities defined by maps *from* the underlying set of a graphon. Then in the next section we give a characterization of convergent graph sequences, which will sound almost identical, but whose prof will be much more tedious.

12.3.1. Quotient sets of graphons. If W is a kernel and $\mathcal{P} = (S_1, \ldots, S_q)$ is a measurable q-partition of $[0, 1]$, we have defined the template (quotient) graph $W/\mathcal{P}$: it is a weighted graph on $[q]$, with node weights $\alpha_i(W/\mathcal{P}) = \lambda(S_i)$ and edge weights
$$\beta_{ij}(W/\mathcal{P}) = \frac{1}{\lambda(S_i)\lambda(S_j)} \int_{S_i \times S_j} W.$$

The Regularity Lemma (Lemma 9.13 and its versions) said that there is always a template that is close to the original graphon (or graph in the finite case) in the cut norm. In order to get right-convergence criteria, we will study *all* templates of a given graphon. For a kernel W and probability distribution $\mathbf{a}$ on $[q]$, we denote by $\mathcal{Q}_\mathbf{a}(W)$ the set of templates $L = W/\mathcal{P}$ with $\alpha(L) = \mathbf{a}$. We denote by $\mathcal{Q}_q(W)$ the set of all q-partitions.

We will consider these quotient sets as subsets of the $\binom{q}{2} + q$ dimensional real space. The quotient set $\mathcal{Q}_q(W)$ is not always closed, but it is closed if W is a stepfunction (see Exercise 12.13). The closure can be described in terms of "fractional partitions", also discussed in exercises.

Quotient sets can be used to express the overlay functional, at least if one of the kernels involved is a weighted graph. For every weighted graph H with $\alpha(H) = \mathbf{a}$

and every kernel W, we have

$$(12.12) \qquad \mathcal{C}(W,H) = \max_{L\in\mathcal{Q}_{\mathbf{a}}(W)} \sum_{i,j\in[q]} \alpha_i(L)\alpha_j(L)\beta_{ij}(H)\beta_{ij}(L).$$

In this case, we have $\alpha_i(L) = \alpha_i(H)$ for all i.

We need a classical definition: if (X,d) is a metric space, then the *Hausdorff metric* is defined on the set of subsets of X by the formula

$$(12.13) \qquad d^{\text{Haus}}(A,B) = \inf\{c : d(a,B) \leq c \ \forall a \in A \text{ and } d(b,A) \leq c \ \forall b \in B\}$$

(here $d(a,B)$ denotes the distance of point a from set B). The special case we need is when A,B are sets of weighted graphs on $[q]$. we can use the edit distance or the cut distance, and denote the corresponding Hausdorff distance of two sets A and B by $d_1^{\text{Haus}}(A,B)$ or $d_\square^{\text{Haus}}(A,B)$. Most of the time it does not make much difference which one we use, because of the trivial inequalities

$$(12.14) \qquad d_\square^{\text{Haus}}(A,B) \leq d_1^{\text{Haus}}(A,B) \leq q^2 d_\square^{\text{Haus}}(A,B).$$

We start with a lemma relating quotient sets with different node weights.

LEMMA 12.9. *For any $W \in \mathcal{W}_1$ and any two probability distributions $\mathbf{a}, \mathbf{a}'$ on $[q]$, we have $d_1^{\text{Haus}}(\mathcal{Q}_\mathbf{a}(W), \mathcal{Q}_{\mathbf{a}'}(W)) \leq 3\|\mathbf{a}-\mathbf{a}'\|_1$.*

Proof. Let $L = W/\mathcal{P} \in \mathcal{Q}_\mathbf{a}(W)$, where $\mathcal{P} = \{S_1, \ldots, S_q\} \in \Pi(\mathbf{a})$. It is easy to construct a partition $\mathcal{P}' = \{S_1', \ldots, S_q'\} \in \Pi(\mathbf{a}')$ such that either $S_i \subseteq S_i'$ or $S_i' \subseteq S_i$ for every i. Let $L' = W/\mathcal{P}' \in \mathcal{Q}_{\mathbf{a}'}(W)$. By definition,

$$d_1(L,L') = \|\mathbf{a}-\mathbf{a}'\|_1 + \sum_{i,j\in[q]} \left| \int_{S_i\times S_j} W - \int_{S_i'\times S_j'} W \right|.$$

Here

$$\left| \int_{S_i\times S_j} W - \int_{S_i'\times S_j'} W \right| \leq \lambda\big((S_i\times S_j)\triangle(S_i'\times S_j')\big)$$

$$\leq |a_i - a_i'|\max(a_j, a_j') + |a_j - a_j'|\min(a_i, a_i').$$

Summing over all i and j, we get

$$d_1(L,L') \leq \|\mathbf{a}-\mathbf{a}'\|_1 \left(1 + \sum_j \max(a_j, a_j') + \sum_i \min(a_i, a_i')\right)$$

$$= \|\mathbf{a}-\mathbf{a}'\|_1 \left(1 + \sum_j (a_j + a_j')\right) = 3\|\mathbf{a}-\mathbf{a}'\|_1.$$

This proves the Lemma. $\square$

We need a couple of lemmas about the Hausdorff distance of quotient sets of different graphons.

LEMMA 12.10. *For any two graphons U and W and any integer $q \geq 1$, we have*

$$d_\square^{\text{Haus}}(\mathcal{Q}_q(U), \mathcal{Q}_q(W)) \leq \delta_\square(U,W).$$

Proof. The quotient sets are invariant under weak isomorphisms, and hence we may assume that U and W are optimally overlayed, so that $\delta_\square(U,W) = \|U-W\|_\square$. The contractivity of the stepping operator (Exercise 9.17) asserts that $d_\square(U/\mathcal{P}, W/\mathcal{P}) \leq \|U - W\|_\square$ for any q-partition $\mathcal{P}$ of $[0,1]$. By the definition of Hausdorff distance, this implies that

$$d_\square^{\text{Haus}}(\mathcal{Q}_q(U), \mathcal{Q}_q(W)) \leq \|U - W\|_\square = \delta_\square(U, W). \qquad \square$$

LEMMA 12.11. *For any two graphons U and W and any integer $q \geq 1$, we have*

$$d_\square^{\text{Haus}}(\mathcal{Q}_q(U), \mathcal{Q}_q(W)) \leq \sup_{\mathbf{a}} d_\square^{\text{Haus}}(\mathcal{Q}_\mathbf{a}(U), \mathcal{Q}_\mathbf{a}(W)) \leq 4 d_\square^{\text{Haus}}(\mathcal{Q}_q(U), \mathcal{Q}_q(W))$$

(where $\mathbf{a}$ ranges over all probability distributions on $[q]$).

Proof. The first inequality is easy: let $H \in \mathcal{Q}_q(U)$, then $H \in \mathcal{Q}_\mathbf{b}(U)$ for the distribution $\mathbf{b} = \alpha(H)$. Hence

$$d_\square(H, \mathcal{Q}_q(W)) \leq d_\square(H, \mathcal{Q}_\mathbf{b}(W)) \leq d_\square^{\text{Haus}}(\mathcal{Q}_\mathbf{b}(U), \mathcal{Q}_\mathbf{b}(W))$$
$$\leq \sup_{\mathbf{a}} d_\square^{\text{Haus}}(\mathcal{Q}_\mathbf{a}(U), \mathcal{Q}_\mathbf{a}(W)).$$

Since this holds for every $H \in \mathcal{Q}_q(U)$, and analogously for every graph in $\mathcal{Q}_q(W)$, the inequality follows by the definition of the Hausdorff distance.

To prove the second inequality, let $\mathbf{a}$ be any probability distribution on $[q]$ and $H \in \mathcal{Q}_\mathbf{a}(U)$. For every $\varepsilon > 0$, there is a quotient $L \in \mathcal{Q}_q(W)$ such that $d_\square(H, L) \leq d_\square^{\text{Haus}}(\mathcal{Q}_q(U), \mathcal{Q}_q(W)) + \varepsilon$. Let $L \in \mathcal{Q}_\mathbf{b}(W)$, then $\|\mathbf{a} - \mathbf{b}\|_1 \leq d_\square(H, L)$ by the definition of $d_\square(H, L)$. By Lemma 12.9, there is a quotient $L' \in \mathcal{Q}_\mathbf{a}(W)$ such that $d_\square(L, L') \leq d_1(L, L') \leq 3|\mathbf{a} - \mathbf{b}| + \varepsilon \leq 3 d_\square(H, L) + \varepsilon$. Thus

$$d_\square(H, L') \leq d_1(H, L) + d_\square(L, L') \leq 4 d_\square(H, L) + \varepsilon \leq 4 d_\square^{\text{Haus}}(\mathcal{Q}_q(U), \mathcal{Q}_q(W)) + 5\varepsilon.$$

Since ε was arbitrary, this proves the lemma. $\square$

12.3.2. Graphon convergence from the right. After this preparation, we are ready to characterize convergence of a graphon sequence in terms of homomorphisms into fixed weighted graphs.

THEOREM 12.12. *For any sequence (W_n) of graphons, the following are equivalent:*

(i) *the sequence (W_n) is convergent in the cut distance $\delta_\square$;*

(ii) *the overlay functional values $\mathcal{C}(W_n, U)$ are convergent for every kernel U;*

(iii) *the restricted multicut densities $\mathcal{C}(W_n, H)$ are convergent for every simple graph H;*

(iv) *the quotient sets $\mathcal{Q}_q(W_n)$ form a Cauchy sequence in the $d_\square^{\text{Haus}}$ Hausdorff metric for every $q \geq 1$.*

It follows from conditions (ii) and (iii) that it would be equivalent to assume the convergence of the sequence $\mathcal{C}(W_n, H)$ for every weighted graph H. Lemma 12.11 implies that we could require in (iv) the convergence of $\mathcal{Q}_\mathbf{a}(W_n)$ for every $q \geq 1$ and probability distribution $\mathbf{a}$ on $[q]$. In fact, it would be enough to require this for the uniform distribution (see Exercise 12.24). In (iv), we could use the d_1^{Haus} Hausdorff metric as well.

Proof. (i)$\Rightarrow$(ii) by Lemma 12.7. (ii)$\Rightarrow$(iii) is trivial. (i)$\Rightarrow$(iv) by Lemma 12.10.

(iii)$\Rightarrow$(i): Let (W_n) be a sequence of graphons that is not convergent in the cut distance. By the compactness of the graphon space, it has two subsequences

(W_{n_i}) and (W_{m_i}) converging to different unlabeled graphons W and W'. There is a graphon U such that $\mathcal{C}(W, U) \neq \mathcal{C}(W', U)$; in fact, (12.9) implies

$$\big(\mathcal{C}(W', W') - \mathcal{C}(W', W)\big) + \big(\mathcal{C}(W, W) - \mathcal{C}(W', W)\big) = \delta_2(W', W)^2 > 0,$$

and so either $\mathcal{C}(W', W') \neq \mathcal{C}(W, W')$ or $\mathcal{C}(W, W') \neq \mathcal{C}(W, W)$, and we can take either $U = W$ or $U = W'$. Furthermore, we can choose U of the form $U = W_H$, where H is a simple graph. This follows using the fact that simple graphs are dense in the graphon space, and the continuity of the overlay functional (Lemma 12.7). Since $\mathcal{C}(W_{n_i}, H) \to \mathcal{C}(W, H)$ and $\mathcal{C}(W_{m_i}, H) \to \mathcal{C}(W', H)$ by Lemma 12.7, it follows that the values $\mathcal{C}(W_n, H)$ cannot form a convergent sequence, contradicting (ii).

(iv)$\Rightarrow$(iii): Fix any simple graph H on $[q]$, and let $\mathbf{a}$ be the uniform distribution on $[q]$. Let $n, m \geq 1$, then we have

$$\mathcal{C}(W_n, H) = \sup_{L \in \mathcal{Q}_\mathbf{a}(W_n)} \frac{1}{q^2} \sum_{\substack{i,j \\ ij \in E(H)}} \beta_{ij}(L) = \max_{L \in \overline{\mathcal{Q}}_\mathbf{a}(W_n)} \frac{1}{q^2} \sum_{\substack{i,j \\ ij \in E(H)}} \beta_{ij}(L).$$

Let $L_n \in \overline{\mathcal{Q}}_\mathbf{a}(W_n)$ attain the maximum. By the definition of Hausdorff distance, there is an $L' \in \overline{\mathcal{Q}}_\mathbf{a}(W_m)$ such that $d_\square(L_n, L') \leq d_\square^{\text{Haus}}\big(\mathcal{Q}_\mathbf{a}(W_n), \mathcal{Q}_\mathbf{a}(W_n)\big)$. The definition of the $\mathcal{C}$ functional implies that $\mathcal{C}(W_m, H) \geq (1/q^2) \sum_{ij \in E(H)} \beta_{ij}(L')$. Hence

$$\mathcal{C}(W_n, H) - \mathcal{C}(W_m, H) \leq \frac{1}{q^2} \sum_{ij \in E(H)} \beta_{ij}(L_n) - \frac{1}{q^2} \sum_{\substack{i,j \\ ij \in E(H)}} \beta_{ij}(L')$$

$$\leq \frac{1}{q^2} \sum_{i,j} |\beta_{ij}(L_n) - \beta_{ij}(L')| = d_1(L_n, L') \leq q^2 d_\square(L_n, L')$$

$$\leq q^2 d_\square^{\text{Haus}}\big(\mathcal{Q}_\mathbf{a}(W_n), \mathcal{Q}_\mathbf{a}(W_n)\big).$$

By Lemma 12.11, we have

$$d_\square^{\text{Haus}}\big(\mathcal{Q}_\mathbf{a}(W_n), \mathcal{Q}_\mathbf{a}(W_n)\big) \leq 4 d_\square^{\text{Haus}}\big(\mathcal{Q}_q(W_n), \mathcal{Q}_q(W_n)\big),$$

which tends to 0 as $n, m \to \infty$ by hypothesis. This implies that

$$\limsup_n \big(\mathcal{C}(W_n, H) - \mathcal{C}(W_m, H)\big) \leq 0.$$

Since a similar conclusion holds with n and m interchanged, we get that $\big(\mathcal{C}(W_n, H) : n = 1, 2, \dots\big)$ is a Cauchy sequence. $\square$

Some of the arguments in the proof of Theorem 12.12, most notably the proof of (iii)$\Rightarrow$(i), were not effective. One can in fact prove explicit inequalities between the different distance measures that occur. We refer to Borgs, Chayes, Lovász, Sós and Vesztergombi [2012] for the details.

EXERCISE 12.13. Show by an example that the set $\mathcal{Q}_q(W)$ is not closed in general, but it is closed if W is a stepfunction.

EXERCISE 12.14. Show by an example that $\mathcal{Q}_\mathbf{a}(W)$ is not convex in general, even if W is a stepfunction.

EXERCISE 12.15. A *fractional partition* of $[0,1]$ into q parts is an ordered q-tuple of measurable functions $\rho_1, \ldots, \rho_q : S \to [0,1]$ such that for all $x \in [0,1]$, we have $\rho_1(x) + \cdots + \rho_q(x) = 1$. For a fractional partition ρ of $[0,1]$ and a kernel $W \in \mathcal{W}$, we define the *fractional quotient graph* W/ρ as a weighted graph on $[q]$ with $\alpha_i(W/\rho) = \|\rho_i\|_1$ and

$$\beta_{ij}(W/\rho) = \frac{1}{\|\rho_i\|_1 \|\rho_j\|_1} \int_{[0,1]^2} \rho_i(x)\rho_j(y) W(x,y) \, dx \, dy.$$

Prove that the operation $W \mapsto W/\rho$ is contractive for the L_1 and L_2.

EXERCISE 12.16. Let ρ be a fractional q-partition of $[0,1]$. Prove that $W/\rho \in \overline{\mathcal{Q}}_q(W)$. Also proved that every weighted graph in $\overline{\mathcal{Q}}_q(W)$ can be represented this way.

12.4. Right-convergent graph sequences

Our basic plan is to apply theorem 12.12 to graphons W_{G_n} to get characterizations of convergent graph sequences in terms of homomorphisms to the right. There are two difficulties in the way. First, in the restricted multicut density condition (iii), thee is no way in general to partition the nodes of a graph in given proportions; we have to allow some rounding of the prescribed sizes for the partition classes. Second, a partition of $[0,1]$ obtained when we overlay W_{G_n} and H optimally may not correspond to any partition of $V(G_n)$. (It corresponds to a "fractional partition", which we will have to define because of this.) This problem is more serious, and we have to work harder to obtain true partitions of $V(G_n)$.

12.4.1. Restricted quotients. Let G be a simple graph with nodeset $[n]$ and let $\mathcal{P} = (S_1, \ldots, S_q)$ be a partition of $[n]$. We consider the quotient graph $G/\mathcal{P}$ as a weighted graph on $[q]$, with node weights $\alpha_i(G/\mathcal{P}) = |S_i|/n$ ($i \in [q]$), and edge weights $\beta_{ij}(G/\mathcal{P}) = e_G(S_i, S_j)/|S_i||S_j|$ ($i,j \in [q]$).

The set of all weighted graphs $G/\mathcal{P}$, where $\mathcal{P}$ ranges over all q-partitions of $[n]$, will be called the *quotient set* of G (of size q), and will be denoted by $\mathcal{Q}_q(G)$. For a graph G and a probability distribution $\mathbf{a}$ on $[q]$, to define the *restricted quotient set* $\mathcal{Q}_\mathbf{a}(G)$, we have to allow the relative sizes of the partition classes to deviate a little from the prescribed values $\mathbf{a}$: we consider the set of quotients $G/\mathcal{P}$, where $\mathcal{P} \in \Pi(n, \mathbf{a})$.

Quotient sets can be used to express multicut functions. For every weighted graph H,

$$\mathsf{cut}(G, H) = \max_{L \in \mathcal{Q}_\mathbf{a}(G)} \sum_{i,j \in [q]} \alpha_i(L)\alpha_j(L)\beta_{ij}(L)\beta_{ij}(H). \tag{12.15}$$

Note that the nodeweights of H and L are not the same in general, but almost: $|\alpha_i(L) - \alpha_i(H)| \leq \frac{1}{n}$.

REMARK 12.17. The quotient sets are in a sense dual to the (multi)sets of induced subgraphs of a given size, which was one of the equivalent ways of describing what we could see by sampling. Instead of gaining information about a large graph by taking a small subgraph, we take a small quotient.

However, there are substantial differences. On the set of induced subgraphs of a given size, we had a probability distribution, which carried the relevant information. We can also introduce a probability distribution on quotients of a given size of a graph G, by taking a random partition. This would be quite relevant to statistical physics, but we would run into difficulties when tending to infinity with the size of G. The probability distributions would concentrate more and more on boring average quotients, while the real information would be contained in the outliers. To be more specific, a random induced subgraph (of a fixed, but sufficiently large size) approximates the original graph well, but a random quotient does not carry this information. In other words, it is the *set* of quotients that characterizes the convergence of a graph sequence, and not the *distribution* on it.

12.4.2. Fractional partitions. A *fractional partition* of a set S is an ordered q-tuple of functions $\rho_1, \ldots, \rho_q : S \to [0,1]$ such that for all $x \in [0,1]$, we have $\rho_1(x) + \cdots + \rho_q(x) = 1$. An ordinary partition corresponds to the special case when every ρ_i is 0-1 valued. For every vector $\alpha \in [0,1]^q$, we denote by $\Pi^*(n,\alpha)$ the set of fractional partitions ρ of $[n]$ with $\|\rho_i\|_1 = \alpha_i n$.

We extend the notion of quotients to fractional partitions. For every fractional partition ρ of $[n]$, we consider the fractional quotient graph G/ρ, which is a weighted graph on $[q]$, with node weights

$$\alpha_i(G/\rho) = \frac{1}{n}|\rho_i| = \frac{1}{n}\sum_{u \in [n]} \rho_i(u) \qquad (i \in [q])$$

and edge weights

$$\beta_{ij}(G/\rho) = \sum_{u,v \in [n]} \rho_i(u)\rho_j(v)/|\rho_i||\rho_j| \qquad (i,j \in [q])$$

In the special case when every value $\rho_i(u)$ is 0 or 1, then the supports of the functions ρ_i form a partition $\mathcal{P}$, and $G/\mathcal{P} = G/\rho$.

We also introduce *fractional quotient sets*, replacing partitions by fractional partitions. The set of all fractional q-quotients of a graph G is denoted by $\mathcal{Q}_q^*(G)$, and the set of all fractional q-quotients G/ρ for which $\alpha(G/\rho)$ is a fixed distribution $\mathbf{a}$ on $[q]$, by $\mathcal{Q}_\mathbf{a}^*(G)$.

12.4.3. Relations between quotient sets. Our goal is to use quotient sets to characterize convergence of a graph sequence. But before doing so, we have to formulate and prove a number of rather technical relationships between different quotient sets.

For any simple graph G and positive integer q, we have two quotient sets: the set $\mathcal{Q}_q(G)$ of quotients $G/\mathcal{P}$, and the set $\mathcal{Q}_q^*(G)$ of fractional quotients G/ρ. In addition, we have the restricted versions $\mathcal{Q}_\mathbf{a}$ of both of these. The quotient sets $\mathcal{Q}_q(W_G)$ and $\mathcal{Q}_\mathbf{a}(W_G)$ will also come up; but it is easy to see that these are just the same as $\mathcal{Q}_q^*(G)$ and $\mathcal{Q}_\mathbf{a}^*(G)$.

Turning to the quotient set $\mathcal{Q}_q(G)$ (which is of course the most relevant from the combinatorial point of view), it follows immediately from the definition that $\mathcal{Q}_q(G) \subseteq \mathcal{Q}_q^*(G)$. Note, however, that $\mathcal{Q}_\mathbf{a}(G)$ and $\mathcal{Q}_\mathbf{a}^*(G)$ are in general not comparable. The first set is finite, the second is typically infinite. On the other hand, $\mathcal{Q}_\mathbf{a}(G)$ contains graphs whose nodeweight vector is only approximately equal to $\mathbf{a}$, and so it is not contained in $\mathcal{Q}_\mathbf{a}^*(G)$.

In the rest of this section we are going to prove that the "true" quotient sets and their fractional versions are not too different, at least if the graph is large. We will need the following version of Lemma 12.9, which can be proved along the same lines.

LEMMA 12.18. *For any simple graph G and any two probability distributions $\mathbf{a}, \mathbf{a}'$ on $[q]$, we have $d_1^{\text{Haus}}(\mathcal{Q}_{\mathbf{a}}(G), \mathcal{Q}_{\mathbf{a}'}(G)) \leq 3\|\mathbf{a} - \mathbf{a}'\|_1$.*

The two kinds of quotient sets of the same graph are related by the following proposition.

PROPOSITION 12.19. *For every simple graph G on $[n]$, integer $q \geq 1$, and probability distribution $\mathbf{a}$ on $[q]$,*

$$d_1^{\text{Haus}}(\mathcal{Q}_q^*(G), \mathcal{Q}_q(G)) \leq \frac{4q}{\sqrt{n}} \quad \text{and} \quad d_1^{\text{Haus}}(\mathcal{Q}_{\mathbf{a}}^*(G), \mathcal{Q}_{\mathbf{a}}(G)) \leq \frac{16q}{\sqrt{n}}.$$

Proof. We start with the first inequality, whose proof gets somewhat technical. Since $\mathcal{Q}_q(G) \subseteq \mathcal{Q}_q^*(G)$, it suffices to prove that if H is a fractional q-quotient of G, then there exists a q-quotient L of G such that $d_1(H, L) \leq 4q/\sqrt{n}$. We may assume that $q \geq 2$ and $n > 9q^2 \geq 36$ (else, the assertion is trivial).

Let $\rho \in \Pi^*(n, \alpha)$ be a fractional partition such that $G/\rho = H$. We want to "round" the values $\rho_i(u)$ to $r_i(u) \in \{0, 1\}$ so that we get an integer partition in $\Pi(n, \alpha)$ with "almost" the same quotient. Let A denote the adjacency matrix of G, and define $F_{ij}(r) = \sum_{u,v \in [n]} A_{uv} r_i(u) r_j(v)$, then we want

(12.16) $$\sum_i r_i(u) = 1, \quad \lfloor \alpha_i n \rfloor \leq \sum_u r_i(u) \leq \lceil \alpha_i n \rceil, \quad F_{ij}(r) \approx F_{ij}(\rho)$$

for all $i, j \in [q]$.

We do random rounding: for each $u \in [n]$, let Z_u be chosen randomly from the distribution $(\rho_1(u), \ldots, \rho_q(u))$, and let $R_i(u) = \mathbb{1}(Z_u = i)$. So $\mathsf{P}(R_i(u) = 1) = \rho_i(u)$, and for different nodes u of G the random variables $R_i(u)$ are independent.

We have $\sum_i R_i(u) = 1$ for every $u \in [n]$, and so the R_i define a partition $\mathcal{P}$. Let $L = G/\mathcal{P}$. For the other two conditions we get the right value at least in expectation: $\mathsf{E}\big(\sum_u R_i(u)\big) = \alpha_i n$ and $\mathsf{E}(F_{ij}(R)) = F_{ij}(\rho)$ (in the last equation we use that $R_i(u)$ and $R_j(v)$ are independent if $u \neq v$, and $A_{uv} = 0$ if $u = v$). We consider the errors $X_i = \sum_u R_i(u) - \alpha_i n$ and $Y_{ij} = F_{ij}(R) - F_{ij}(\rho)$, and use a second moment argument to show that these are small. We have

$$\mathsf{Var}(X_i) = \sum_u \mathsf{Var}(R_i(u)) = \sum_u (\rho_i(u) - \rho_i(u)^2) < \alpha_i n,$$

and hence

$$\mathsf{E}(\sum_i X_i^2) = \sum_i \mathsf{Var}(X_i) < n.$$

Furthermore,

(12.17)
$$\mathsf{Var}(Y_{ij}) = \mathsf{Var}(F_{ij}(R)) = \sum_{u,v,u',v' \in [n]} A_{uv} A_{u'v'} \mathsf{cov}(R_i(u) R_j(v), R_i(u') R_j(v')).$$

Each covariance in this sum depends on which of u, v, u', v' and also which of i and j are equal, but each case is easy to treat. The covariance term is 0 if the edges uv and $u'v'$ are disjoint. If $i \neq j$, we get:

$$\begin{cases} \rho_i(u)\rho_j(v) - \rho_i(u)^2 \rho_j(v)^2 < \rho_i(u)\rho_j(v), & \text{if } u = u' \text{ and } v = v', \\ -\rho_i(u)\rho_i(v)\rho_j(u)\rho_j(v) < 0, & \text{if } u = v', v = u', \\ \rho_j(v)\rho_j(v')(\rho_i(u) - \rho_i(u)^2) < \rho_i(u)\rho_j(v)\rho_j(v'), & \text{if } u = u' \text{ and } v \neq v', \\ -\rho_i(u)\rho_j(v)\rho_i(v)\rho_j(v') < 0, & \text{if } u = v', u' \neq v. \end{cases}$$

(The other possibilities are covered by symmetry.) Summing over u, v, u', v', we get that the sum in (12.17) is at most $(\alpha_i n)(\alpha_j n) + 0 + (\alpha_i n)(\alpha_j n)^2 + (\alpha_i n)^2(\alpha_j n) + 0$. The case when $i = j$ can be treated similarly, and we get that the sum in (12.17) is at most $2(\alpha_i n)^2 + 4(\alpha_i n)^3$. Hence, summing over all i and j, we get

$$\sum_{i,j} \mathsf{E}(Y_{ij}^2) = \sum_{i,j} \mathsf{Var}(Y_{ij}) \leq n^2 + (n^2 + 2n^3)\sum_i \alpha_i^2 + 4n^3 \sum_i \alpha_i^3 \leq 6n^3 + 2n^2.$$

By Cauchy–Schwarz,

$$d_1(H, L)^2 = \Big(\sum_i \frac{|X_i|}{n} + \sum_{i,j} \frac{|Y_{ij}|}{n^2}\Big)^2 \leq (q^2 + q)\Big(\frac{1}{n^2}\sum_i X_i^2 + \frac{1}{n^4}\sum_{i,j} Y_{ij}^2\Big),$$

and so

$$\mathsf{E}(d_1(H, L)^2) = (q^2 + q)\Big(\frac{1}{n} + \frac{6}{n} + \frac{2}{n^2}\Big) < \frac{16q^2}{n}.$$

Hence with positive probability, $d_1(H, L) \leq 4q/\sqrt{n}$.

The second inequality in the proposition is quite easy to prove now, except that we cannot use containment in either direction, and so we have to prove two "almost containments". Let $H \in \mathcal{Q}_{\mathbf{a}}(G)$ and let $\mathbf{b} = \alpha(H)$, $\|\mathbf{a} - \mathbf{b}\|_1 \leq q/n$, and $H \in \mathcal{Q}_{\mathbf{b}}^*(G)$. By Lemma 12.18, there is an $L \in \mathcal{Q}_{\mathbf{a}}^*(G)$ such that $d_1(H, L) \leq 3\|\mathbf{a} - \mathbf{b}\|_1 \leq 3q/n < 16q/\sqrt{n}$.

Conversely, let $H \in \mathcal{Q}_{\mathbf{a}}^*(G)$, then by part (a), there exists a q-quotient $H' \in \mathcal{Q}_q(G)$ such that $d_1(H, H') \leq 4q/\sqrt{n}$. Lemma 12.18 implies that there exists an $L \in \mathcal{Q}_{\mathbf{a}}^*(G)$ such that

$$d_1(L, H') \leq 3|\mathbf{a} - \alpha(H')| = 3|\alpha(H) - \alpha(H')| \leq 3d_1(H, H'),$$

and so $d_1(L, H) \leq d_1(L, H') + d_1(H', H) \leq 4d_1(H, H') \leq 16q/\sqrt{n}$. □

12.4.4. Right-convergent graph sequences. In a sense, right-convergence of a graph sequence is a special case of right-convergence of a graphon sequence. However, quantities like multicuts associated with a graphon of the form W_G are only approximations of the analogous combinatorial quantities associated with the corresponding graph G. (This is in contrast with the homomorphism densities from the left, recall e.g. (7.2).) In this section we prove that these approximations are good enough for the equivalent characterizations of convergence of graphon sequences to carry over to graph sequences.

We prove the following characterization of convergence of a dense graph sequence, analogous to the characterization of convergence of a graphon sequence given in Theorem 12.12.

THEOREM 12.20. *Let (G_n) be a sequence of simple graphs such that $\mathsf{v}(G_n) \to \infty$ as $n \to \infty$. Then the following are equivalent:*

(i) *the sequence (G_n) is convergent;*

(ii) *the overlay functional values $\mathcal{C}(W_{G_n}, U)$ are convergent for every kernel U;*

(iii) *the restricted multicut densities $\mathsf{cut}(G_n, H)$ are convergent for every simple graph H;*

(iv) *the quotient sets $\mathcal{Q}_q(G_n)$ are Cauchy in the Hausdorff metric for every $q \geq 1$.*

Clearly, conditions (ii) and (iii) are also equivalent to the convergence of $\mathsf{cut}(G_n, H)$ for every weighted graph H. By (12.5), this is equivalent to the convergence of typical homomorphism entropies $\mathsf{ent}^*(G, J)$ for every weighted graph J with positive edgeweights. By our discussion in Section 2.2, we could talk about microcanonical ground state energies instead of restricted multicuts. By the results of the previous section, we could replace $\mathcal{Q}_q(G_n)$ by $\mathcal{Q}_q^*(G_n)$, or we could require the convergence of $\mathcal{Q}_{\mathbf{a}}(G_n)$ for every $q \geq 1$ and probability distribution $\mathbf{a}$ on $[q]$. On the other hand, it would be enough to require this for the uniform distribution (see Exercise 12.24).

Proof. If we replace G_n by W_{G_n}, then (i) and (ii) do not change. In (iii), we have
$$\mathcal{C}(G_n, H) = \max_{L \in \mathcal{Q}_{\mathbf{a}}(G)} \sum_{i,j \in [q]} \alpha_i(L) \alpha_j(L) \beta_{ij}(L) \beta_{ij}(H),$$
and
$$\mathcal{C}(W_{G_n}, H) = \max_{L \in \mathcal{Q}_{\mathbf{a}}^*(G)} \sum_{i,j \in [q]} \alpha_i(L) \alpha_j(L) \beta_{ij}(L) \beta_{ij}(H).$$
Since $d_1^{\mathrm{Haus}}\big(\mathcal{Q}_{\mathbf{a}}^*(G), \mathcal{Q}_{\mathbf{a}}(G)\big) \leq 9q/\sqrt{n}$ by Proposition 12.19, it follows that $\mathcal{C}(G_n, H)$ is convergent as $n \to \infty$ if and only if $\mathcal{C}(W_{G_n}, H)$ is. By a similar argument, the validity of (iv) does not change if we replace G_n by W_{G_n}. So the theorem follows by Theorem 12.12. □

The last theorem in this chapter describes the limiting values in Theorem 12.20. The proof is contained in our previous considerations.

THEOREM 12.21. *Let G_n be a convergent sequence of simple graphs such that $G_n \to W$ ($W \in \mathcal{W}_0$). Let H be a weighted graph with positive edgeweights, and let J be obtained from H by replacing every edgeweight by its binary logarithm. Then $\mathsf{cut}(G_n, J) \to \mathcal{C}(W, J)$ and $\mathsf{ent}^*(G_n, H) \to \mathcal{C}(W, J)$.*

REMARK 12.22. How far can we push convergence results for right-homomorphism parameters of a convergent graph sequence? Suppose that (G_n) is a convergent graph sequence. Does it follow that $\log t(G_n, W)/\mathsf{v}(G_n)^2$ tends to a limit for every graphon W? Or at least, for every graphon $W > 0$? or at least for every graphon W with $1/2 \leq W \leq 1$? Theorem 12.20(ii) suggests an affirmative answer, but this is false: the example in Exercise 12.25 (which is hard!) shows that right-convergence in this more general sense does not follow from left-convergence.

EXERCISE 12.23. Prove that $\max_H |\mathcal{C}(U, H) - \mathcal{C}(W, H)|$, where the maximum is taken over all weighted graphs H on $[q]$ with nodeweight vector $\mathbf{a}$ and edgeweights in $[-1, 1]$, is equal to the Hausdorff distance $d_1^{\mathrm{Haus}}\big(\mathsf{conv}(\mathcal{Q}_{\mathbf{a}}(U)), \mathsf{conv}(\mathcal{Q}_{\mathbf{a}}(U))\big)$.

EXERCISE 12.24. Prove that a sequence (W_n) of graphons is convergent if and only if the quotient sets $\mathcal{Q}_{\mathbf{u}}(W_n)$ are convergent in the Hausdorff metric for every $q \geq 1$, where $\mathbf{u}$ is the uniform distribution on $[q]$.

EXERCISE 12.25. Let (G_n) be a quasirandom graph sequence with edge density $1/2$, such that $\mathsf{v}(G_n) = k_n$ is a sufficiently fast increasing sequence of integers. Define $W(x, y) = 1 + \sum_{n=1}^{\infty} W_{G_{2n}}(k_{2n}x, k_{2n}y)$ (where $W_{G_n}(x, y) = 0$ if $x \notin [0, 1]$ or $y \notin [0, 1]$). Prove that the kernel W is 1-2 valued, and
$$\frac{\log t(G_{2n}, W)}{k_{2n}^2} = \frac{1}{2} + o(1), \qquad \frac{\log t(G_{2n+1}, W)}{k_{2n+1}^2} = \frac{1}{4} + o(1).$$

CHAPTER 13

On the structure of graphons

13.1. The general form of a graphon

A probability space $J = (\Omega, \mathcal{A}, \pi)$ together with a symmetric function $W : J \times J \to \mathbb{R}$, measurable with respect to the completion of the sigma-algebra $\mathcal{A} \times \mathcal{A}$, will be called a *kernel*, and if the range of W is contained in $[0,1]$, a *graphon*.

So far, we have assumed that J is the unit interval $[0,1]$ with the Lebesgue measure, and this was good enough to get a limit object for every convergent graph sequence. But recall Example 11.41, where it took an artificial step of applying a measure preserving map to carry the graphon structure over to the unit interval. A similar step was needed in the definition of the tensor product of two kernels (Section 7.4). In both cases, allowing more general underlying sets leads to a simpler and cleaner situation. If we have to make this distinction, a graphon on $[0,1]$ with the Lebesgue measure will be called a *graphon on $[0,1]$*.

Subgraph densities can be defined in any graphon by the same formula as in the special case of $[0,1]$: For a multigraph $F = (V, E)$, we define

$$t(F, W) = \int_{\Omega^V} \prod_{ij \in E} W(x_i, x_j) \prod_{i \in V} d\pi(x_i).$$

Hence we can define weak isomorphism of kernels as before.

We can sample $\mathbb{H}(n, W)$ and $\mathbb{G}(n, W)$ from any graphon. If we fix the underlying space, we can talk about kernel norms as before. If $(\Omega, \mathcal{A}, \pi)$ and $(\Omega', \mathcal{A}', \pi')$ are two probability spaces, $(\Omega, \mathcal{A}, \pi, W)$ is a kernel, and $\varphi : \Omega' \to \Omega$ is a measure preserving map, then the pullback W^φ can be defined as before:

$$W^\varphi(x, y) = W(\varphi(x), \varphi(y)).$$

It is clear that $(\Omega', \mathcal{A}', \pi', W^\varphi)$ will be a kernel, which we call the *pullback* of (J', W'). Simple computation shows that for every graph F,

(13.1) $$t(F, W) = t(F, W^\varphi),$$

so $(\Omega', \mathcal{A}', \pi', W^\varphi)$ is weakly isomorphic to $(\Omega, \mathcal{A}, \pi, W)$.

To define the $\delta_\square$ distance of two kernels, we have to be a little careful, since at this point we allow the underlying spaces to have atoms, and then the difficulties in the definition of the cut-distance for graphs (Sections 8.1.2–8.1.4) resurface. Let $A_1 = (J_1, W_1)$ and $A_2 = (J_2, W_2)$ be two kernels. To avoid the difficulty with atoms, we consider all kernels (J, W) and all pairs of measure preserving maps $\varphi_i : J \to J_i$. We define

$$\delta_\square(A_1, A_2) = \inf \|W_1^{\varphi_1} - W_2^{\varphi_2}\|_\square,$$

where the infimum ranges over all choices of J, W, φ_1 and φ_2.

At this point, there are some rather technical questions to address. Do we want to assume that J is a standard probability space (see Appendix A.3)? Do we want to consider $\mathcal{A}$ as a complete sigma-algebra with respect to the probability measure (like Lebesgue measurable sets in $[0,1]$), or to the contrary, do we want to assume that it is countably generated (like Borel sets in $[0,1]$)?

It was shown by Borgs, Chayes and Lovász [2010] that a kernel on an arbitrary probability space can be transformed, by very simple steps, into a kernel on a standard probability space, which is equivalent for all practical purposes (in particular, weakly isomorphic). The steps of such a transformation are described in Exercises 13.11–13.13 below. This implies that we can work with standard probability spaces whenever necessary (or just convenient). We call a kernel *standard*, if the underlying probability space is standard.

From the point of view of subgraph densities, multicuts, etc. the underlying space does not matter much, as we shall see; but choosing the underlying probability space appropriately may lead to a simpler form for the function W and to simpler computations. If the space J is finite, we get just a weighted graph with normalized nodeweights. Let us see a number of further examples where allowing this more general form is very useful (and so Example 11.41 was not an isolated occurrence).

EXAMPLE 13.1 (**Limits of interval graphs**). An *interval graph* is a graph obtained from a finite set of intervals on a line, where the intervals are the nodes, and two intervals are connected by an edge if and only if they have a point in common. Diaconis, Holmes and Janson [2011] give the following description of limits of interval graphs.

Let $J = \{(x,y) \in [0,1]^2 : x \leq y\}$, and define $W\big((x_1,y_1),(x_2,y_2)\big) = \mathbb{1}\big([x_1,y_1] \cap [x_2,y_2] \neq \emptyset\big)$. So (J,W) can be considered as an infinite interval graph, whose nodes are all sub-intervals of $[0,1]$. We can take any probability measure on the Borel sets in J: we always obtain a graphon that is the limit of interval graphs, and all interval graph limits arise this way. ◆

EXAMPLE 13.2. Fix some $d \geq 2$, and let V_n be a set of n unit vectors in $\mathbb{R}^d$, chosen independently from the uniform distribution on the unit sphere. Connect two elements $x,y \in V_n$ by an edge if and only if $x^\mathsf{T} y \geq 0$, to get a graph $G_n = (V_n, E_n)$. The sequence $(G_n : n = 1, 2, \dots)$ is convergent, and its limit is the graphon whose underlying set is S^{d-1}, with the uniform distribution, and $W(x,y) = \mathbb{1}(x^\mathsf{T} y \geq 0)$. ◆

13.1.1. Atomfree and twin-free kernels. There are still ways to further simplify a kernel (J,W) on a standard probability space $(\Omega, \mathcal{A}, \pi)$. One possibility is to get rid of the atoms by a procedure generalizing the construction of the kernel W_H from a weighted graph H, by assigning to each atom a an interval I_a of length $\pi(a)$, and an interval I to the atom-free part of Ω, so that these intervals partition $[0,1]$. This defines a measure preserving map $\psi : [0,1] \to \Omega$, and the pullback W^φ will define a kernel on $[0,1]$ that is weakly isomorphic to (J,W).

This procedure takes us to a very familiar domain (two-variable real functions), but the kernel on $[0,1]$ is still not uniquely determined by its weak isomorphism class, as we have seen in Example 7.11. To really standardize a kernel, we go the opposite way, by creating and merging atoms as much as we can. To be more precise, we need some definitions.

13.1. THE GENERAL FORM OF A GRAPHON

Let (J, W) be a kernel. Two points $x, x' \in J$ are called *twins* if $W(x, y) = W(x', y)$ for almost all $y \in J$. This defines an equivalence relation on J. We call the kernel H *twin-free* if no two distinct points in J are twins.

PROPOSITION 13.3. *For every kernel (J, W) there is a twin-free kernel (J_1, W_1) and a measure preserving map $\varphi \colon J \to J_1$ such that $W = W_1^\varphi$ almost everywhere. If J is standard, then we can require that J_1 be standard as well.*

Proof. The twin-free kernel and the measure preserving map φ will be easy to define, but we have to work on verifying their properties. First, we only modify the sigma-algebra of the probability space $J = (\Omega, \mathcal{A}, \pi)$. Let us define a new sigma-algebra $\mathcal{A}'$ consisting of those sets in $\mathcal{A}$ that do not separate any twin points. Let $J' = (\Omega, \mathcal{A}', \pi)$ (this is a little abuse of notation, since we should restrict π to $\mathcal{A}'$). Define $W' = \mathsf{E}(W \mid \mathcal{A}' \times \mathcal{A}')$.

CLAIM 13.4. $W = W'$ *almost everywhere.*

It suffices to show that

$$\tag{13.2} \int_{A \times B} W \, d\pi \times d\pi = \int_{A \times B} W' \, d\pi \times d\pi$$

for all $A, B \in \mathcal{A}$. (Note that this holds for $A, B \in \mathcal{A}'$ by the definition of conditional probability.) Consider the functions

$$U_A = \int_A W(., y) \, d\pi(y), \qquad g_A = \mathsf{E}(\mathbb{1}_A \mid \mathcal{A}'), \qquad V_A = \int_\Omega W(., y) g_A(y) \, d\pi(y).$$

These functions are $\mathcal{A}'$-measurable by the definition of twins. Furthermore, g_A is the orthogonal projection of $\mathbb{1}_A$ into the space of $\mathcal{A}'$-measurable functions, $g_{AB}(x, y) = g_A(x) g_B(y)$ is the orthogonal projection of $\mathbb{1}_{A \times B}$ into the space of $\mathcal{A}' \times \mathcal{A}'$-measurable functions, and by definition, W' is the orthogonal projection of W into this space. Using these observations, we have

$$\int_{A \times B} W \, d\pi \times d\pi = \langle \mathbb{1}_B, U_A \rangle = \langle g_B, U_A \rangle = \langle V_B, \mathbb{1}_A \rangle = \langle V_B, g_A \rangle$$

$$= \langle W, g_{AB} \rangle = \langle W', g_{AB} \rangle = \langle W', \mathbb{1}_{A \times B} \rangle = \int_{A \times B} W' \, d\pi \times d\pi.$$

This proves (13.2) and the Claim.

It seems that instead of eliminating twin points, we made them even more "twin-like": they are now not separated by any set from $\mathcal{A}'$. But this is good, because then identifying them does not change anything: Let Ω_1 denote the set of equivalence classes of "being twins" on Ω, let $\varphi(x)$ be the equivalence class containing $x \in \Omega$, let $\mathcal{A}_1 = \{\varphi(X) : X \in \mathcal{A}'\}$, and define $\pi_1(X) = \pi(\varphi^{-1}(X))$ for $X \in \mathcal{A}_1$. Then $J_1 = (\Omega_1, \mathcal{A}_1, \pi_1)$ is a probability space. Furthermore, W' is constant on $S \times T$ for two equivalence classes, and hence $W_1(S, T) = W'(x, y)$ ($x \in S, y \in T$) is well defined. Trivially, $W_1^\varphi = W'$, so by Claim 13.4, we have $W_1^\varphi = W$ almost everywhere.

For the proof of the second statement of the Proposition, we need the following.

CLAIM 13.5. *If $\mathcal{A}$ is countably generated, then $\mathcal{A}_1$ is countably separated.*

Let $\mathcal{R}$ be a countable generating set in $\mathcal{A}$. For every $R \in \mathcal{R}$ and rational number r, consider the set

$$S_{R,r} = \Big\{ x \in \Omega : \int_R W(x,y) \, d\pi(y) \geq r \Big\}.$$

Clearly $S_{R,r} \in \mathcal{A}'$. Furthermore, if x and x' are not twins, then $W(x,.)$ and $W(x',.)$ differ on a set of positive measure, and so there is a set $R \in \mathcal{R}$ such that $\int_R W(x,y) \, d\pi(y) \neq \int_R W(x',y) \, d\pi(y)$. Assume that (say) $\int_R W(x,.) > \int_R W(x',.)$, then for any rational number r with $\int_R W(x,.) > r > \int_R W(x',.)$ we have $x \in S_{R,r}$ but $x' \notin S_{R,r}$. So the countable family of sets $S_{R,r}$ separates any two points of Ω that are not twins. It follows that the sets $\varphi(S_{R,r}) \in \mathcal{A}_1$ separate any two points of Ω_1.

Claim 13.5 and Proposition A.4 in the Appendix complete the proof. □

EXERCISE 13.6. Show that for the set J and function W in Example 13.1, several different measures on J can yield the same—isomorphic—graphons.

EXERCISE 13.7. Consider two graphons $U = (\Omega, \mathcal{A}, \pi, W)$ and $U' = (\Omega, \mathcal{A}, \pi', W)$ which only differ in their probability measures. Prove that $\delta_1(U, U') \leq 2 d_{\text{var}}(\pi, \pi')$. [Hint: use Exercise 8.15.]

EXERCISE 13.8. Suppose that a graphon $(\Omega, \mathcal{A}, \pi, W)$ is defined on a metric space (Ω, d), where $\mathcal{A}$ is the set of Borel sets, π is atom-free, and W is almost everywhere continuous. Suppose that the sequence $S_n \subseteq J$ is well distributed in the sense that $|S_n \cap U|/|S_n| \to \pi(U)$ for every open set U. Then $t(F, \mathbb{G}(S_n, W)) \to t(F, W)$ for every simple graph F with probability 1.

13.2. Weak isomorphism III

We give a characterization of weakly isomorphic kernels first in the twin-free case, and then use this to complete our arsenal of characterizations in general.

THEOREM 13.9. *If two standard twin-free kernels are weakly isomorphic, then they are isomorphic up to a nullset.*

Proof. Let (J_1, W_1) and (J_2, W_2) be two weakly isomorphic twin-free kernels, where $J_i = (\Omega_i, \mathcal{A}_i, \pi_i)$ $(i = 1, 2)$ are standard probability spaces. By Corollary 10.35, there is a third kernel (J, W) $(J = (\Omega, \mathcal{A}, \varphi))$ and measure preserving maps $\varphi_i : J_i \to J$ such that $W_i = W^{\varphi_i}$ almost everywhere.

Let Ω'_i be the set of all elements $u \in \Omega_i$ for which $W_i(u, v) = W^{\varphi_i}(u, v)$ for almost all $v \in \Omega_i$. By the definition of W and φ_i, we must have $\pi_i(\Omega_i \setminus \Omega'_i) = 0$. So we can delete the elements of $\Omega_i \setminus \Omega'_i$ from Ω_i for $i = 1, 2$. In other words, we may assume that for every $u \in \Omega_i$, $W_i(u, v) = W^{\varphi_i}(u, v)$ for almost all $v \in \Omega_i$.

This implies that φ_i is injective. Indeed, the set $\varphi_i^{-1}(x)$ consists of twins in the kernel (J_i, W^{φ_i}), which remain twins in (J_i, W_i). Since (J_i, W_i) is twin-free, it follows that $\varphi_i^{-1}(x)$ has only one element.

An injective measure preserving map from a standard probability space into another one is almost bijective: the set of points with no inverse image has measure 0 (Proposition A.4 in the Appendix). Hence $\Omega^* = \varphi_1(\Omega_1) \cap \varphi_2(\Omega_2)$, $\Omega_1^* = \varphi_1^{-1}(\Omega^*)$ and $\Omega_2^* = \varphi_2^{-1}(\Omega^*)$ have measure 1 in the corresponding graphons, and we can restrict these kernels to them. But then φ_1 and φ_2 are isomorphisms between these three kernels. □

This theorem adds a further characterization of weak isomorphism of standard kernels, formulated before. The following theorem summarizes these characterizations: Corollaries 10.34, 10.35 and 10.36 established the equivalence of (a) with (c), (e), (f) and (b); Proposition 13.3 and Theorem 13.9 imply the nontrivial part of (d).

THEOREM 13.10. *For two standard kernels (J_1, W_1) and (J_2, W_2) the following are equivalent:*

(a) $t(F, W_1) = t(F, W_2)$ *for every simple graph F (i.e., (J_1, W_1) and (J_2, W_2) are weakly isomorphic);*

(b) $t(F, W_1) = t(F, W_2)$ *for every loopless multigraph F;*

(c) $\delta_N(W_1, W_2) = 0$ *for every (or just for one) smooth invariant norm;*

(d) *there exist a standard kernel (J_0, W_0) and measure preserving maps $\varphi_i : J_0 \to J_i$ such that $W_i = W_0^{\varphi_i}$ almost everywhere;*

(e) *there exist a standard kernel (J_0, W_0) and measure preserving maps $\varphi_i : J_i \to J_0$ such that $W_i^{\varphi_i} = W_0$ almost everywhere;*

(f) *there exists a coupling measure μ between J_1 and J_2 such that $W_1(X_1, Y_1) = W_2(X_2, Y_2)$ for almost all pairs (X_1, X_2) and (Y_1, Y_2) selected independently from the distribution μ.* □

There are (at least) four quite different ways to prove Theorem 13.10 (not counting trivial differences like the order in which various conditions are proved).

• The main steps in the route followed here was to establish the equivalence of (a) and (c) (which was proved by Borgs, Chayes, Lovász, Sós and Vesztergombi [2008]) and then use Theorem 8.13 (due to Bollobás and Riordan [2009]).

• The original proof by Borgs, Chayes and Lovász [2010] is more direct but a lot longer; it is built on the natural idea to bring every kernel to a "canonical form", so that weakly isomorphic kernels would have identical canonical forms. In the case of functions in a single variable, a canonical form that works in many situations is "monotonization" (see the Monotone Reordering Theorem A.19 in the Appendix). For kernels there does not seem to exist such a canonical form, but one can construct, for every kernel W, a "canonical ensemble": a probability distribution π_W on the set of kernels such that two kernels U and W are weakly isomorphic if and only if the distributions π_U and π_W are identical.

• Diaconis and Janson [2008] showed that theorem 13.10 also follows from results of Kallenberg [2005] in the theory of exchangeable random variables [2005].

• The proof of Janson [2010] is based on the idea of pure kernels (see Section 13.3) and Theorem 13.9.

EXERCISE 13.11. Let $J = (\Omega, \mathcal{A}, \pi)$ be a probability space and $W : \Omega \times \Omega \to \mathbb{R}$, a symmetric function measurable with respect to the completion of $\mathcal{A} \times \mathcal{A}$. Show that W can be changed on a set of measure 0 so that it becomes measurable with respect to $\mathcal{A} \times \mathcal{A}$.

EXERCISE 13.12. Let (J, W) be a kernel on a (non-standard) probability space $J = (\Omega, \mathcal{A}, \pi)$. Show that there is a countably generated σ-algebra $\mathcal{A}_0 \subseteq \mathcal{A}$ on Ω such that W is measurable with respect to the completion of $\mathcal{A}_0 \times \mathcal{A}_0$.

EXERCISE 13.13. Let (J, W) be a kernel on a (non-standard) countably generated separating probability space $J = (\Omega, \mathcal{A}, \pi)$. Show that there is a standard probability space $J_0 = (\Omega_0, \mathcal{A}_0, \pi_0)$, a kernel (J_0, W_0), and an injective measure preserving map $\varphi : J \to J_0$ such that $W = W_0^{\varphi}$ almost everywhere.

EXERCISE 13.14. Consider two probability spaces $(\Omega, \mathcal{A}, \pi)$ and $(\Omega', \mathcal{A}', \pi')$, a kernel $(\Omega, \mathcal{A}, \pi, W)$, and a measure preserving map $\varphi \colon \Omega \to \Omega'$ (note: it goes in the opposite direction than in the definition of the pullback!). (a) Show that one can construct a "push-forward" kernel $(\Omega', \mathcal{A}', \pi', W_\varphi$ such that
$$(W_\varphi)^\varphi = \mathsf{E}\big(W \mid \varphi^{-1}(\mathcal{A}') \times \varphi^{-1}(\mathcal{A}')\big).$$
(b) show by an example that $t(F, W_\varphi) = t(F, W)$ does not hold in general.

13.3. Pure kernels

In the spirit of classical analysis, there is no room to further standardize a kernel: We have seen that every kernel is equivalent to a twin-free kernel, and two weakly isomorphic twin-free kernels are isomorphic up to a nullset, and usually we don't care about nullsets. But it turns out that cleaning up these nullsets is worth the trouble.

13.3.1. Purifying kernels. We introduce a distance notion on the points of a kernel. Let (J, W) be a kernel. We can endow the space J with the distance function
$$r_W(x, y) = \|W(x, .) - W(y, .)\|_1 = \int_J |W(x, z) - W(y, z)| \, dz.$$

This function is defined for almost all pairs x, y; we can delete those points from J where $W(x, .) \notin L_1(J)$ (a set of measure 0), to have r_W defined on all pairs. It is clear that r_W is a pseudometric (it is symmetric and satisfies the triangle inequality). We call r_W the *neighborhood distance* on W.

EXAMPLE 13.15 (**Stepfunctions**). For stepfunctions, the underlying metric space is finite. ♦

EXAMPLE 13.16 (**Spherical distance**). Let S^d denote the unit sphere in $\mathbb{R}^{d+1}$, consider the uniform probability measure on it, and let $W(x, y) = 1$ if $x \cdot y \geq 0$ and $W(x, y) = 0$ otherwise. Then (S^d, W) is a graphon, in which the neighborhood distance of two points $a, b \in S^d$ is just their spherical distance (normalized by dividing by π). ♦

EXAMPLE 13.17. Let (M, d) be a metric space, and let π be a Borel probability measure on M. Then d can be viewed as a kernel on (M, d). For $x, y \in M$, we have
$$r_d(x, y) = \int_M |d(x, z) - d(y, z)| \, d\pi(z) \leq \int_M d(x, y) \, d\pi(z) = d(x, y),$$

so the identity map $(M, d) \to (M, r_d)$ is contractive. This implies that if (M, d) is compact, and/or finite dimensional (in many senses of dimension), then so is (M, r_d). For most "everyday" metric spaces (like segments, spheres, or balls) $r_d(x, y)$ can be bounded from below by $\Omega(d(x, y))$, in which case (M, d) and (M, r_d) are homeomorphic.

More generally, if $F \colon [0, 1] \to \mathbb{R}$ is a continuous function, then $W(x, y) = F(d(x, y))$ defines a kernel, and the identity map $(M, d) \to (M, r_W)$ is continuous. ♦

A kernel (J, W) is *pure* if (J, r_W) is a complete separable metric space and the probability measure has full support (i.e., every open set has positive measure).

This definition includes that $r_W(x, y)$ is defined for all $x, y \in J$ and $r_W(x, y) > 0$ if $x \neq y$, i.e., the kernel has no twins.

THEOREM 13.18. *Every twin-free kernel is isomorphic, up to a null set, to a pure kernel.*

Proof. Let (J, W) be a twin-free kernel. Let T be the set of functions $f \in L_1(J)$ such that for every L_1-neighborhood U of f, the set $\{x \in J : W(x,.) \in U\}$ has positive measure. Clearly T is a closed subset of $L_1(J)$, and it is complete and separable in the L_1-metric. Let J' be the set of points in J for which $W(x,.) \in T$, and let $T' = \{W(x,.) : x \in J'\}$. The map $\varphi : J' \to T'$ defined by $x \mapsto W(x,.)$ is bijective, since (J, W) is twin-free. The set T inherits a probability measure $\pi' = \pi \circ \varphi^{-1}$ from J. It is easy to see from the construction that J' and T' are measurable.

We claim that

$$\pi(J \setminus J') = 0. \tag{13.3}$$

It is clear that for almost all $x \in J$, $W(x,.) \in L_1(J)$. Every function $g \in L_1(J) \setminus T$ has an open neighborhood U_g in $L_1(J)$ such that $\pi\{x \in J : W(x,.) \in U_g\} = 0$. Let $U = \bigcup_{g \notin T} U_g$. Since $L_1(J)$ is separable, U equals the union of some countable subfamily $\{U_{g_i} : i \in \mathbb{N}\}$ and thus $\pi\{x \in J : W(x,.) \in U\} = 0$. Since $J \setminus J' \subseteq U$, this proves (13.3).

The functions $W(x,.)$ ($x \in J'$) are everywhere dense in T and have measure 1. So T is a complete separable metric space with a probability measure on its Borel sets. It also follows from the definition of T that every open set has positive measure, and (13.3) implies that $\pi'(T \setminus T') = 0$.

We define a kernel $W' : T \times T \to [0, 1]$ as follows. Let $f, g \in T$. If $f \in T'$, then $f = W(x,.)$ for some $x \in J$, and we define $W'(f, g) = g(x)$. Similarly, if $g \in T'$, then $g = W(.,y)$ and we define $W'(f, g) = f(y)$. Note that if both $f, g \in T'$, then this definition is consistent: $W'(f, g) = f(y) = g(x) = W(x, y)$. If $f, g \notin T'$, then we define $W'(f, g) = 0$. We note that f and g are determined up to a zero set only; we can choose any function representing them, and since we are changing W on a set of measure 0 only, it remains measurable.

The kernel (T, W') is pure; indeed, we just have to check that $r_{W'}$ coincides with the L_1 metric on T; then T will have all the right properties. For $f, g \in T$, we have

$$r_{W'}(f, g) = \int_T |W'(f, y) - W'(g, y)| \, d\pi'(y) = \int_{T'} |W'(f, y) - W'(g, y)| \, d\pi'(y)$$
$$= \int_{J'} |f(y) - g(y)| \, d\pi(y) = \int_J |f(y) - g(y)| \, d\pi(y) = \|f - g\|_1.$$

The kernels (J, W) and (T, W') are isomorphic up to a 0-set; indeed, we can get the kernel $(J', W|_{J'})$ from (J, W) and the kernel $(T', W'|_{T'})$ from the kernel (T, W') by deleting appropriate 0-sets, and φ is an isomorphism between $(J', W|_{J'})$ and $(T', W'|_{T'})$. This proves the theorem. □

There is still some freedom left: given a pure kernel (J, W), we can change the value of W on a symmetric subset of $J \times J$ that intersects every fiber $J \times \{v\}$ in a set of measure 0. We can take the integral of W (which is a measure ω on $J \times J$), and then the derivative of ω with respect to $\pi \times \pi$ wherever this exists. This way we get back W almost everywhere, and a well defined value for some further points.

After all these changes, where W is left undefined is the set of "essential discontinuities" of W (of measure 0). It would be interesting to relate this set to combinatorial properties of W.

13.3.2. Density functions on pure kernels. Now we come to the utilization of our work with purifying kernels. The following technical lemma will be very useful in the study of homomorphism densities on pure graphons.

LEMMA 13.19. *Let (J, W) be a pure graphon and let $F = (V, E)$ be a k-labeled multigraph with nonadjacent labeled nodes. Then*

$$|t_x(F, W) - t_x(F, W)| \leq \mathsf{e}(F) \max_{u \in [k]} r_W(x_u, x'_u)$$

for all $x, x' \in J^k$.

It follows that the functions t are Lipschitz (and hence continuous).

Proof. Let $E = \{u_1 v_1, \ldots u_m v_m\}$, where we may assume that v_i is unlabeled. For each $v \in V \setminus [k]$, let $y_v = x_v = x'_v$ be a variable. Then, using a telescoping sum,

$$t_x(F, W) - t_{x'}(F, W) = \int_{J^{V \setminus [k]}} \prod_{i=1}^{m} W(x_{u_i}, x_{v_i})\, dy - \int_{J^{V \setminus [k]}} \prod_{i=1}^{m} W(x'_{u_i}, x'_{v_i})\, dy$$

$$= \sum_{j=1}^{m} \int_{J^{V \setminus [k]}} \prod_{i<j} W(x_{u_i}, x_{v_i}) \big(W(x_{u_j}, x_{v_j}) - W(x'_{u_j}, x'_{v_j})\big) \prod_{j>i} W(x'_{u_i}, x'_{v_i})\, dy$$

and hence

$$|t_x(F, W) - t_{x'}(F, W)| \leq \sum_{j=1}^{m} \int_{J^{V \setminus [k]}} |W(x_{u_j}, x_{v_j}) - W(x'_{u_j}, x'_{v_j})|\, dy.$$

By the assumption that v_i is unlabeled, we have $x_{v_j} = x'_{v_j}$ for every j, and so

$$|t_x(F, W) - t_{x'}(F, W)| \leq \sum_{j=1}^{m} \int_{J^{V \setminus [k]}} |W(x_{u_j}, x_{v_j}) - W(x'_{u_j}, x_{v_j})|\, dy$$

$$\leq \sum_{j=1}^{m} r_W(x_{u_j}, x'_{u_j}) \leq \mathsf{e}(F) \max_{u \in [k]} r_W(x_u, x'_u),$$

which proves the assertion. $\square$

COROLLARY 13.20. *Let (J, W) be a pure kernel, and let $F = (V, E)$ be a k-labeled graph with nonadjacent labeled nodes. Then $t_x(F, W)$ is a continuous function of $x \in J^S$ with respect to the metric r_W.*

In the case when F is a path of length 2, we get a corollary that will be important in the next section.

COROLLARY 13.21. *For every pure kernel (J, W), $W \circ W$ is a continuous function (in two variables) on the metric space (J, r_W).*

Most applications of Corollary 13.20 use the following consequence:

COROLLARY 13.22. *Let (J, W) be a pure kernel, and let $F_1, \ldots, F_m$ be $(k+n)$-labeled multigraphs with nonadjacent labeled nodes. Let $a_1, \ldots, a_m$ be real numbers and $x \in J^k$, such that the equation*

$$\sum_{i=1}^{m} a_i t_{x,y}(F_i, W) = 0 \tag{13.4}$$

holds for almost all $y \in J^k$. Then it holds for all $y \in J^k$.

Proof. By Corollary 13.20, the left side of (13.4) is a continuous function of (x, y), and so it remains a continuous function of y if we fix x. Hence the set where it is not 0 is an open subset of J^k. Since the graphon is pure, it follows that this set is either empty or has positive measure. □

Going to pure graphons is a good proof method, which can lead to nontrivial results. We illustrate this by discussing properties of $t(., W)$ ($W \in \mathcal{W}$) from the point of view of Section 6.3.2. This parameter is multiplicative and reflection positive. If W is not a stepfunction, then $t(., W)$ has no contractor (else, 6.30 would imply that it is a homomorphism function). On the other hand, it is contractible. We prove this in a more general form.

Let F be a k-labeled multigraph, and let $\mathcal{P} = \{S_1, \ldots, S_m\}$ be a partition of $[k]$. We say that $\mathcal{P}$ is *legitimate* for F, if each set S_i is stable in F. If this is the case, then the m-labeled multigraph $F/\mathcal{P}$ (obtained by identifying the nodes in each S_i, and labeling the obtained node with i) has no loops. For a k-labeled quantum graph g, we say that the partition $\mathcal{P}$ of $[k]$ is *legitimate for g* if it is legitimate for every constituent. Then we can define $g/\mathcal{P}$ by linear extension.

PROPOSITION 13.23. *Let g be a k-labeled quantum graph and $\mathcal{P}$, a legitimate partition for g. Let $W \in \mathcal{W}$, and suppose that $t_x(g, W) = 0$ almost everywhere on $[0, 1]^k$. Then $t_y(g/\mathcal{P}, W) = 0$ for almost all $y \in [0, 1]^{|\mathcal{P}|}$.*

Proof. We may assume that W is pure. If $t_x(g, W) = 0$ almost everywhere, then it holds everywhere by Corollary 13.20. In particular, it holds for every substitution where the variables corresponding to the same class of $\mathcal{P}$ are identified, which means that $t_y(g/\mathcal{P}, W)$ is identically 0 on $[0, 1]^{|\mathcal{P}|}$. □

COROLLARY 13.24. *The multigraph parameter $t(., W)$ is contractible for every kernel W.*

EXERCISE 13.25. Show that using properties of pure kernels, one gets a very short proof of the statement of Exercise 7.6.

13.4. The topology of a graphon

We have seen that every graphon is weakly isomorphic to a pure graphon (which is unique up to changing the function on special nullsets), which has an underlying complete metric space J. So we could ask: what topological properties does J have for special graphons, and are they related to the combinatorial properties of graph sequences converging to W? It turns out that these questions are more interesting if we ask them for a different, but related topology on the underlying set, and this is what we are going to introduce now.

13.4.1. The similarity distance.

It was noted by Lovász and Szegedy [2007, 2010b] that for a pure graphon (J, W), the distance function $\bar{r}_W = r_{W \circ W}$ defined by the operator square of W is also closely related to combinatorial properties of a graphon. We call this the *similarity distance*. In the special case of finite graphs, this notion was defined in the Introduction, where the motivation for its name was also explained. We will use the graph version to design algorithms in Section 15.4.1. In explicit terms, we have

$$\bar{r}_W(a,b) = r_{W \circ W}(a,b) = \int_J \left| \int_J W(a,y) W(y,x) \, dy - \int_J W(b,y) W(y,x) \, dy \right| dx$$

$$(13.5) \qquad = \int_J \left| \int_J W(x,y)\big(W(a,y) - W(b,y)\big) \, dy \right| dx.$$

(We write here and in the sequel dx instead of $d\pi(x)$, where π is the probability measure of the graphon.)

LEMMA 13.26. *If (J, W) is a pure graphon, then the similarity distance $\bar{r}_W$ is a metric.*

Proof. The only nontrivial part of this lemma is that $\bar{r}_W(a,b) = 0$ implies that $a = b$. The condition $\bar{r}_W(a,b) = 0$ implies that for almost all $x \in J$ we have

$$\int_J W(x,y)\big(W(a,y) - W(b,y)\big) \, dy = 0.$$

Using that (J, W) is pure, Corollary 13.22 implies that this holds for every $x \in J$. In particular, it holds for $x = a$ and $x = b$. Substituting these values and taking the difference, we get that

$$\int_J \big(W(a,y) - W(b,y)\big)^2 \, dy = 0,$$

and hence $W(a,y) = W(b,y)$ for almost all z. Using again that (J, W) is pure, we conclude that $a = b$. $\square$

So $(J, \bar{r}_W)$ is a metric space, and hence Hausdorff. We have to be careful though, since the metric space $(J, \bar{r}_W)$ is not necessarily complete. We will work with its completion $(\overline{J}, \bar{r}_W)$, but first we define it differently.

Let us say that a sequence of points $x_n \in J$ is *weakly convergent* if

$$\int_A W(x_n, y) \, dy \to \int_A W(x, y) \, dy$$

for every measurable set $A \subseteq J$. We call this the *weak topology* on J. (We need this name only temporarily, since we are going to show that $\bar{r}_W$ gives a metrization of the weak topology.) It is well known that this topology is metrizable.

Let $\overline{J}$ denote the completion of J in the weak topology. The map $x \mapsto W(x, .)$ embeds J into $L_1(J)$, and weak convergence corresponds to weak* convergence of functions in $L_1(J)$. Hence $\overline{J}$ corresponds to the weak* closure of J. It follows in particular that $\overline{J}$ is a compact separable metric space (compactness follows by Aleoglu's Theorem, since $\overline{J}$ is a closed subset of the unit ball of $L_1(J)$).

We can extend the definition of W to $\overline{J}\times\overline{J}$: for $x\in J$ and $y\in\overline{J}\setminus J$, we define $\overline{W}(x,y) = x(y)$ (recall that the elements of $\overline{J}$ can be identified with functions on J), and we define $\overline{W}(x,y) = 0$ if $x,y\in\overline{J}\setminus J$. We also extend the measure π by $\overline{\pi}(\overline{J}\setminus J) = 0$.

THEOREM 13.27. *For any pure graphon, the metric $\overline{r}_W$ defines exactly the weak topology on $\overline{J}$.*

Proof. First we show that the weak topology is finer than the topology of $(\overline{J}, \overline{r}_W)$. Suppose that $x_n \to x$ in the weak topology, this means that the functions $W(x_n, .)$ converge weakly to $W(x, .)$. Consider
$$\overline{r}_W(x_n, x) = \int \left| \int_J (W(x_n, y) - W(x,y))W(y,z)\,dy \right| dz.$$
Here the inner integral tends to 0 for every z, by the weak convergence $x_n \to x$. Since it also remains bounded, it follows that the outer integral tends to 0. This implies that $x_n \to x$ in $(J, \overline{r}_W)$. (Let us note that since $\overline{\pi}(\overline{J}\setminus J) = 0$, it does not matter whether we integrate over J or over $\overline{J}$.)

From here, the equality of the two topologies follows by general arguments: the weak topology on $\overline{J}$ is compact, and the coarser topology of $\overline{r}_W$ is Hausdorff, which implies that they are the same. $\square$

COROLLARY 13.28. *For every pure graphon (J, W), the space $(\overline{J}, \overline{r}_W)$ is compact.*

Another useful corollary of these considerations concerns continuity in the similarity metric. We have seen that $W \circ W$ is continuous in the metric r_W; one might hope that the function W is continuous as a function on $(\overline{J}, \overline{r}_W)$, but this would be too much to ask for (the half-graphon is an easy example). However, integrating out one of the variables we get a continuous function. To be more precise (and more general):

COROLLARY 13.29. *For every pure graphon (J, W), and every function $g \in L_1(J)$, the function*
$$(T_W g)(.) = \int_J W(., y)g(y)\,dy$$
is continuous on $(\overline{J}, \overline{r}_W)$.

In particular, it follows that every eigenfunction of T_W is continuous on $(\overline{J}, \overline{r}_W)$.

Proof. Let $x_n \to x$ in the $\overline{r}_W$ metric. Then $W(x_n, .) \to W(x, .)$ in the weak topology by Theorem 13.27, and hence
$$\int_J W(X_n, y)g(y)\,dy \to \int_J W(x, y)g(y)\,dy. \qquad \square$$

We conclude this section with an example in which the topologies defined by r_W and $\overline{r}_W$ are different. Note that for any two points $x, y \in J$, we have

(13.6) $$\overline{r}_W(x, y) \leq r_W(x, y),$$

which implies that the topology of (J, r_W) is finer than the topology of $(J, \overline{r}_W)$. The two topologies may be different. Graphons for which the finer space (J, r_W) is also compact seem to have special importance in combinatorics.

EXAMPLE 13.30. For $y \in [0,1)$, let $y = 0.y_1 y_2 \ldots$ be the binary expansion of y. Let us decompose $[0,1)$ into the intervals $I_k = [1 - 2^{-k}, 1 - 2^{-k-1})$. Define $U(x,y) = y_k$ for $0 \leq y \leq 1$ and $x \in I_k$. Define $W(1,y) = 1/2$ for all y. This function is not symmetric, so we put it together with a reflected copy to get a graphon:

$$W(x,y) = \begin{cases} U(2x, 2y-1), & \text{if } x \leq 1/2 \text{ and } y \geq 1/2, \\ U(2y, 2x-1), & \text{if } x \geq 1/2 \text{ and } y \leq 1/2, \\ 0, & \text{otherwise.} \end{cases}$$

(This is rather difficult to parse, but it is an important example. Perhaps Figure 13.1 helps.) Selecting one point from each interval $[1 - 2^{-k}, 1 - 2^{-k-1})$, we get an infinite number of points in $[0,1)$ mutually at r_W-distance $1/4$; so this sequence is not convergent even in the completion of this graphon. (In particular, (J, r_W) is not compact.) On the other hand, this same sequence converges in $(J, \overline{r}_W)$. So the two topologies are different. ♦

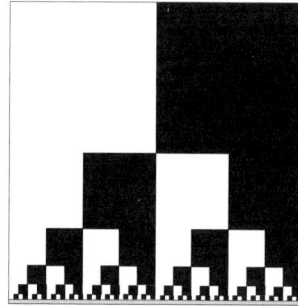 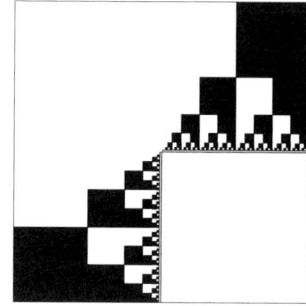

FIGURE 13.1. A graphon defined on a space whose completion in r_W is not compact. The picture on the left shows just one half.

13.4.2. Similarity distance and regularity partitions. Now we come to the results that first motivated the introduction of a second distance notion defined by a graphon.

Let (J, d) be a metric space and let π be a probability measure on its Borel sets. We say that a set $S \subseteq J$ is an *average ε-net*, if $\int_J d(x, S) \, d\pi(x) \leq \varepsilon$.

Let $S \subseteq J$ be a finite set and $s \in S$. The *Voronoi cell* of S with center s is the set of all points $x \in J$ for which $d(x, s) \leq d(x, y)$ for all $y \in S$. Clearly, the Voronoi cells of S cover J. (We can break ties arbitrarily to get a partition.)

THEOREM 13.31. *Let (J, W) be a pure graphon, and $\varepsilon > 0$.*

(a) *Let S be an average ε-net in the metric space $(S, \overline{r}_W)$. Then the Voronoi cells of S form a weak regularity partition $\mathcal{P}$ with error at most $8\sqrt{\varepsilon}$.*

(b) *Let $\mathcal{P} = \{J_1, \ldots, J_k\}$ be a weak regularity partition with error ε. Then there are points $v_i \in J_i$ such that the set $S = \{v_1, \ldots, v_k\}$ is an average (4ε)-net in the metric space $(S, \overline{r}_W)$.*

Proof. (a) Let $\mathcal{P}$ be the partition into the Voronoi cells of S. Let us write $R = W - W_\mathcal{P}$. We want to show that $\|R\|_\square \leq 8\sqrt{\varepsilon}$. It suffices to show that for any

0-1 valued function f,

(13.7) $$\langle f, Rf\rangle \leq 2\sqrt{\varepsilon}.$$

Let us write $g = f - f_{\mathcal{P}}$, where $f_{\mathcal{P}}(x)$ is obtained by replacing $f(x)$ by the average of f over the class of $\mathcal{P}$ containing x. Clearly $\langle f_{\mathcal{P}}, Rf_{\mathcal{P}}\rangle = 0$, and so

(13.8) $\langle f, Rf\rangle = \langle g, Rf\rangle + \langle f_{\mathcal{P}}, Rf\rangle = \langle f, Rg\rangle + \langle f_{\mathcal{P}}, Rg\rangle \leq 2\|Rg\|_1 \leq 2\|Rg\|_2.$

For each $x \in J$, let $\varphi(x) \in S$ be the center of the Voronoi cell containing x, and define $W'(x,y) = W\big(x, \varphi(y)\big)$ and similarly $R'(x,y) = R\big(x, \varphi(y)\big)$. Then using that $(W-R)g = W_{\mathcal{P}}g = 0$, $W - W' = R - R'$ and $R'g = 0$, we get

$$\|Rg\|_2^2 = \langle Rg, Rg\rangle = \langle Wg, (R-R')g\rangle = \langle Wg, (W-W')g\rangle = \langle g, W(W-W')g\rangle$$
$$\leq \|W(W-W')\|_1 = \int_{J^2}\bigg|\int_J W(x,y)\big(W(y,z) - W(y,\varphi(z))\big)\,dy\bigg|\,dx\,dz$$
$$= \int_J \overline{r}_W\big(z, \varphi(z)\big) = \mathsf{E}_{\mathbf{x}}(\overline{r}_W(\mathbf{x}, S)) \leq \varepsilon.$$

This proves (13.7).

(b) Suppose that $\mathcal{P}$ is a weak regularity partition with error ε. Let $R = W - W_{\mathcal{P}}$, then we know that $\|R\|_\square \leq \varepsilon$. For every $x \in [0,1]$, define

$$F(x) = \int_J\bigg|\int_J R(x,y)W(y,z)\,dy\bigg|\,dz = \int_{J^2} s(x,z)R(x,y)W(y,z)\,dy\,dz,$$

where $s(x,z)$ is the sign of $\int R(x,y)W(y,z)\,dy$. Lemma 8.10 implies that for every $z \in J$,

$$\int_{J^2} s(x,z)R(x,y)W(y,z)\,dx\,dy \leq 2\|R\|_\square \leq 2\varepsilon,$$

and hence

(13.9) $$\int_J F(x)\,dx \leq 2\varepsilon.$$

Let $x, y \in J$ be two points in the same partition class of $\mathcal{P}$. Then $W_{\mathcal{P}}(x,s) = W_{\mathcal{P}}(y,s)$ for every $s \in J$, and hence

(13.10) $$\overline{r}_W(x,y) = \int_J\bigg|\int_J (W(x,s) - W(y,s))W(s,z)\,ds\bigg|\,dz$$
$$= \int_J\bigg|\int_J (R(x,s) - R(y,s))W(s,z)\,ds\bigg|\,dz$$
$$\leq \int_J\bigg|\int_J R(x,s)W(s,z)\,ds\bigg|\,dz + \int_J\bigg|\int_J R(y,s)W(s,z)\,ds\bigg|\,dz$$
$$= F(x) + F(y).$$

For every set $T \in \mathcal{P}$, let $v_T \in T$ be a point "below average" in the sense that

$$F(v_T) \leq \frac{1}{\pi(T)}\int_T F(x)\,dx,$$

and let $S = \{v_T : T \in \mathcal{P}\}$. Then using (13.9),

$$\mathsf{E}_\mathbf{x} \bar{r}_W(\mathbf{x}, S) \leq \sum_{T \in \mathcal{P}} \int_T \bar{r}_W(x, v_T) \, dx \leq \sum_{T \in \mathcal{P}} \int_T \big(F(x) + F(v_T)\big) \, dx$$

$$\leq \int_J F(x) \, dx + \sum_{T \in \mathcal{P}} \lambda(T) F(v_T) \leq 2 \int_J F(x) \, dx \leq 4\varepsilon. \qquad \square$$

By the Weak Regularity Lemma, we get that every graphon has an average ε-net of size $2^{O(1/\varepsilon^2)}$. How about a "true" ε-net? By a standard trick, if we take a maximum set R of points such that any two are at a distance at least ε, then every point is at a distance of at most ε from R. But is such a set necessarily finite? And if so, how can we bound its size?

It turns out that one can give a bound that is similar to the bound on the size of an average ε-net derived from Theorem 13.31. The following result is due to Alon [unpublished].

PROPOSITION 13.32. *Let (J, W) be a graphon and let $R \subseteq J$ be a set such that $\bar{r}_W(s,t) \geq \varepsilon$ for all $s, t \in R$ ($s \neq t$). Then $|R| \leq (16/\varepsilon^2)^{257/\varepsilon^2}$.*

The bound on the size of R is somewhat worse than for the average ε-net, but the main point is that it depends on ε only. There are examples showing that an exponential dependence on $1/\varepsilon$ is unavoidable (Exercise 13.41).

Proof. Consider the Frieze–Kannan decomposition of W provided by Lemma 9.14:

$$(13.11) \qquad W = \sum_{i=1}^k a_i \mathbb{1}_{S_i \times T_i} + U,$$

where $k = \lceil 256/\varepsilon^2 \rceil$, the sets $S_i, T_i \subseteq J$ are measurable, $\sum_i a_i^2 \leq 4$, and $\|U\|_\square \leq 4/\sqrt{k} \leq \varepsilon/4$. For $s, t \in R$, we have

$$\bar{r}_W(s,t) = \int_J \Big| \int_J \big(W(s,y) - W(t,y)\big) W(y,z) \, dy \Big| \, dz$$

$$= \int_{J \times J} \sigma_{st}(z) \big(W(s,y) - W(t,y)\big) W(y,z) \, dy \, dz,$$

where $\sigma_{st}(z)$ is the sign of $\int \big(W(s,y) - W(t,y)\big) W(y,z) \, dy$. Substituting from (13.11) for the last occurrence of W, we get

$$\bar{r}_W(s,t) = \sum_{i=1}^k a_i \int_{J \times J} \sigma_{st}(z) \big(W(s,y) - W(t,y)\big) \mathbb{1}_{S_i \times T_i}(y,z) \, dy \, dz$$

$$(13.12) \qquad + \int_{J \times J} \sigma_{st}(z) \big(W(s,y) - W(t,y)\big) U(y,z) \, dy \, dz.$$

Here the last term is small by Lemma 8.10 and the choice of U:

$$\Big| \int_{J \times J} \sigma_{st}(z) \big(W(s,y) - W(t,y)\big) U(y,z) \, dy \, dz \Big| \leq 2\|U\|_\square \leq \frac{\varepsilon}{2}.$$

To bound a term from the first sum, we do a little computation:

$$\int\int_{T_i\ S_i} \sigma_{st}(z)(W(s,y)-W(t,y))\,dy\,dz = \Big(\int_{T_i}\sigma_{st}(z)\,dz\Big)\Big(\int_{S_i} W(s,y)-W(t,y)\,dy\Big),$$

and hence, with $f_i(s) = \int_{S_i} W(s,y)\,dy$, we have

$$\Big|\int\int_{T_i\ S_i}\sigma_{st}(z)(W(s,y)-W(t,y))\,dy\,dz\Big|$$
$$\leq \Big|\int_{S_i} W(s,y)-W(t,y)\,dy\Big| = |f_i(s)-f_i(t)|.$$

Now if $|R| > (16/\varepsilon^2)^{257/\varepsilon^2} \geq (4\sqrt{k}/\varepsilon)^k$, then there is a pair of points $s,t \in R$ such that $|f_i(s)-f_i(t)| \leq \varepsilon/(4\sqrt{k})$ for all i, and for this choice of s and t we have

$$\sum_{i=1}^{k} a_i \int_{J\times J} \sigma_{st}(z)(W(s,y)-W(t,y))\mathbb{1}_{S_i\times T_i}(y,z)\,dy\,dz < \sum_{i=1}^{k} |a_i|\frac{\varepsilon}{4\sqrt{k}} \leq \frac{\varepsilon}{2}.$$

(In the last step we used that $\sum_i a_i^2 \leq 4$ and the inequality between arithmetic and quadratic means.) By (13.12) this implies that $\overline{r}_W(s,t) < \varepsilon$, a contradiction. □

13.4.3. Finite dimensional graphons. The main reason to be interested in this topology is the following consequence of Theorem 13.31. We define the *(upper) Minkowski dimension* of a metric space (M,d) as

$$\limsup_{\varepsilon\to 0}\frac{\log N(\varepsilon)}{\log(1/\varepsilon)},$$

where $N(\varepsilon)$ is the maximum number of points in M mutually at distance at least ε. This dimension is finite if and only if there is a $d \geq 0$ such that every set of points mutually at distance at least ε has at most ε^{-d} elements.

COROLLARY 13.33. *If (J,W) is a graphon for which the space $(\overline{J},\overline{r}_W)$ has finite Minkowski dimension d, then for every $\varepsilon > 0$ the graphon has a weak regularity partition with $O\big((1/\varepsilon)^d\big)$ classes.* □

Which graphons are finite dimensional in this sense? Almost all of those we have met so far are 1 or at most 2-dimensional (see Exercise 13.44). We will formulate some conjectures later, in Section 16.7.1; right now we describe an interesting combinatorially defined class with this property.

We say that a graphon W *misses* a signed graph F if $t(F,W) = 0$. Trivially, if W misses F, then the complementary graphon $1-W$ misses the signed graph F^- obtained from F by negating the signs of the edges. We will only consider bipartite graphs F; extension of these results to non-bipartite graphs is open.

Graphons missing a signed bipartite graph can be characterized in terms of the Vapnik–Chervonenkis dimension (see Appendix A.6 for basic information about the VC-dimension). This characterization is not much more than a reformulation of the definition, but useful nonetheless.

PROPOSITION 13.34. *A pure graphon (J,W) misses some signed bipartite graph with k nodes in the smaller bipartition class if and only if W is 0-1 valued almost everywhere and the VC-dimension of the family of neighborhoods $\mathcal{R}_W = \{\operatorname{supp}(W(x,.)) : x \in J\}$ is less than k.*

Proof. First, suppose that (J,W) misses a signed bipartite graph F with bipartition $V_1 \cup V_2$, where $V_1 = [k]$ and $V_2 = \{1', \ldots, m'\}$. We start with showing that W is 0-1 valued almost everywhere. Let $F^\bullet$ be obtained by labeling all nodes of V_1. Then for almost all $x \in J^k$, we have $t_x(F^\bullet, W) = 0$. By Corollary 13.22, it follows that $t_x(F^\bullet, W) = 0$ for *every* $x \in J^k$. In particular, $t_{z\ldots z}(F^\bullet, W) = 0$ for all $z \in J$. But for this substitution,

$$t_{z\ldots z}(F^\bullet, W) = \int_{J^m} \prod_{j=1}^m W(z,y_j)^{d^+(j)} \bigl(1 - W(z,y_j)\bigr)^{d^-(j)} dy_1 \ldots dy_m$$

(where $d^+(j)$ and $d^-(j)$ are the numbers of positive and negative edges of F incident with j, respectively). If there is a $z \in J$ such that $0 < W(y,z) < 1$ for all $y \in Y$, where Y has positive measure, then the part of the integral over Y^m is already positive, so $t_{z\ldots z}(F^\bullet, W) > 0$, a contradiction.

Next, we show that the VC-dimension of $\mathcal{R}_W$ is less than k. Suppose not, then there is a set $S = \{x_1, \ldots, x_k\} \subseteq J$ with $|S| = k$ such that the family $\mathcal{H} = \{\operatorname{supp}(W(x,.)) : x \in S\}$ is qualitatively independent (this means that for every $\mathcal{H}' \subseteq \mathcal{H}$ there is a point contained in all sets of $\mathcal{H}'$ but in no set of $\mathcal{H} \setminus \mathcal{H}'$). This implies that $t_{x_1 \ldots x_k}(F^\bullet, W) > 0$. By the purity of (J, W) and Corollary 13.22, the set of points $(y_1, \ldots, y_k) \in [0,1]^k$ for which $t_{y_1 \ldots y_k}(F^\bullet, W) > 0$ has positive measure. Hence $t(F, W) > 0$.

Conversely, suppose that W is 0-1 valued (we may assume everywhere), and $\dim_{VC}(\mathcal{R}_W) < k$. Let F denote the signed complete bipartite graph with k nodes in one class U and 2^k nodes in the other class U', in which each node in U' is connected to a different set of nodes in U by positive edges. Let $F^\bullet$ be obtained by labeling the nodes in U. Then any choice of $x_1, \ldots, x_k$ for which $t_{x_1 \ldots x_k}(F^\bullet, W) > 0$ gives k points with qualitatively independent neighborhoods, which is impossible. So we must have $t(F, W) = 0$. □

Our main goal is to connect the VC-dimension of neighborhoods to the dimension of J. The following theorem was proved (in a slightly more general form) by Lovász and Szegedy [2010b].

THEOREM 13.35. *If a pure graphon (J, W) misses some signed bipartite graph F, then*

(a) *W is 0-1 valued almost everywhere,*

(b) *(J, r_W) is compact, and*

(c) *it has Minkowski dimension at most $10\mathsf{v}(F)$.*

Proof. (a) is just repeated from Proposition 13.34. To prove (b), we start with studying weakly convergent sequences of functions $W(x,.)$. Let $(x_1, x_2, \ldots)$ be a sequence of points in J and suppose that there is a function $f \in L_1(J)$ such that

$$\int_S W(x_n, y) \, dy \longrightarrow \int_S f(y) \, dy$$

for every measurable set $S \subseteq J$.

CLAIM 13.36. *The weak limit function f is almost everywhere 0-1 valued.*

Suppose not, then there is an $\varepsilon > 0$ and a set $Y \subseteq J_2$ with positive measure such that $\varepsilon \leq f(x) \leq 1-\varepsilon$ for $x \in Y$. Let $S_n = \mathrm{supp}\big(W(x_n,.)\big) \cap Y$. We select, for every $k \geq 1$, k indices $n_1, \ldots n_k$ so that the Boolean algebra generated by $S_{n_1}, \ldots S_{n_k}$ (as subsets of Y) has 2^k atoms of positive measure. If we have this for some k, then for every atom A of the Boolean algebra

$$\lambda(A \cap S_n) = \int_A W(x, y_n)\, dx \longrightarrow \int_A f(x)\, dx \qquad (n \to \infty),$$

and so if n is large enough, then

$$\frac{\varepsilon}{2}\lambda(A) \leq \lambda(A \cap S_n) \leq \left(1 - \frac{\varepsilon}{2}\right)\lambda(A).$$

If n is large enough, then this holds for all atoms A, and so S_n cuts every previous atom into two sets with positive measure, and we can choose $n_{k+1} = n$.

But this means that the VC-dimension of the supports of the $W(x,.)$ is infinite, contradicting Proposition 13.34. This proves Claim 13.36.

CLAIM 13.37. *The convergence $W(x_n,.) \to f$ also holds in L_1.*

Indeed, we know that $f(x) \in \{0, 1\}$ for almost all x, and hence

$$\|f - W(x_n,.)\|_1 = \int_{\{f=1\}} \big(1 - W(x_n, y)\big)\, dy + \int_{\{f=0\}} W(x_n, y)\, dy \longrightarrow 0.$$

Now it is easy to prove that (J, r_W) is compact. Consider any infinite sequence $(x_1, x_2, \ldots)$ of points of J. By Alaoglu's Theorem, this has a subsequence for which the functions $W(x_n,.)$ converge weakly to a function $f \in L_1(J)$. By Claim 13.37, they converge to f in L_1. This implies that they form a Cauchy sequence in L_1, and so $(x_1, x_2, \ldots)$ is a Cauchy sequence in (J, r_W). Since (J, r_W) is a complete metric space, this sequence has a limit in J.

To prove (c), let F be a signed bipartite graph such that $t(F, W) = 0$, and let (V_1, V_2) be a bipartition of F with $|V_1| = k$, where we may assume that $k \leq v(F)/2$. We may assume that F is complete bipartite, since adding edges (with any signs) does not change the condition that $t(F, W) = 0$. Let $F^\bullet$ be obtained from F by labeling the nodes in V_1.

We want to show that the Minkowski dimension of (J, r_W) is at most $20k$. It suffices to show that every finite set $Z \subseteq J$ such that the r_W-distance of any two elements is at least ε, is bounded by $|Z| \leq c(k)\varepsilon^{-20k}$. Let $\mathcal{H} = \{\mathrm{supp}(W(x,.)) : x \in Z\}$. Since W is 0-1 valued, the condition on Z means that

(13.13) $$\pi(X \triangle Y) \geq \varepsilon$$

for any two distinct sets $X, Y \in \mathcal{H}$.

We do a little clean-up: Let A be the union of all atoms of the set algebra generated by $\mathcal{H}$ that have measure 0. Clearly A itself has measure 0, and hence the family $\mathcal{H}' = \{X \setminus A : X \in \mathcal{H}\}$ still has property (13.13).

We claim that $\mathcal{H}'$ has VC-dimension less than k. Indeed, suppose that $J \setminus A$ contains a shattered k-set S. To each $j \in V_1$, we assign a point $q_j \in S$ bijectively. To each $i \in V_2$, we assign a point $p_i \in Z$ such that $q_j \in \mathrm{supp}(W(p_i,.))$ if and only if $ij \in E^+$. (This is possible since S is shattered.) Now fixing the p_i, for each j there is a subset of J of positive measure whose points are contained in exactly the same members of $\mathcal{H}'$ as q_j, since $q_j \notin A$. This means that the function $t_{x_1\ldots x_k}(F^\bullet, W)$ is positive for $x_i = p_i$. Corollary 13.22 implies that $t_{x_1\ldots x_k}(F^\bullet, W) > 0$ for a positive fraction of the choices of $x_1, \ldots x_k \in J$, and hence $t(F, W) > 0$, a contradiction.

Applying Proposition A.30 we conclude that $|Z| = |\mathcal{H}| \leq (80k)^{10k}\varepsilon^{-20k}$. This proves that the Minkowski dimension of (J, r_W) is bounded by $20k$. □

The results in this section do not remain true if the signed graph we exclude is nonbipartite. For example, if we exclude any non-bipartite graph, then any bipartite graph satisfies the condition, but some bipartite graphs are known to need an exponential (in $1/\varepsilon$) number of classes in their weak regularity partitions.

EXERCISE 13.38. Two metrics d_1 and d_2 on the same set are called *uniformly equivalent*, if there is a function $f : R_+ \to R_+$ such that $f(x) \searrow 0$ if $x \searrow 0$, $d_1(x, y) \leq f(d_2(x, y))$ and $d_2(x, y) \leq f(d_1(x, y))$. Prove that for a pure kernel (J, W), the space (J, r_W) is compact if and only if the metrics r_W are $\bar{r}_W$ are uniformly equivalent.

EXERCISE 13.39. Figure out the completions of the spaces $([0, 1], r_W)$ and $([0, 1], \bar{r}_W)$ for the graphon W in Example 13.30.

EXERCISE 13.40. For the graphon (S^d, W) defined in Example 13.16, show that the similarity distance of two points $a, b \in S^d$ is $\Omega(\angle(a, b)/\sqrt{d})$.

EXERCISE 13.41. Show that the graphon in the previous exercise, with an appropriate choice of d, contains $2^{\Omega(1/\varepsilon^2)}$ points mutually at least ε apart in the similarity distance.

EXERCISE 13.42. Let W be a graphon such that (J, r_W) can be covered by m balls of radius ε. Prove that there exists a stepfunction U with $m(1/\varepsilon)^m$ steps such that $\|W - U\|_1 \leq 2\varepsilon$.

EXERCISE 13.43. Let $M(\varepsilon)$ denote the minimum number of sets of diameter at most ε covering a metric space (S, d), and define the *covering dimension* of (S, d) by $\limsup_{\varepsilon \to 0} (\log M(\varepsilon))/(\log(1/\varepsilon))$. Prove that this is the same as the Minkowski dimension.

EXERCISE 13.44. (a) Check that all graphons constructed in Section 11.4.2 are at most 2-dimensional. (b) Prove that a graphon W on $[0, 1]$ that is a continuous function is at most 1-dimensional. (c) Find the dimension of the graphon in Example 13.16. (d) Construct an infinite dimensional graphon.

EXERCISE 13.45. Let W be a graphon such that $t(F, W) = 0$ for a signed bipartite graph $F = (V, E)$. Prove that for every $0 < \varepsilon < 1$, there exists a 0-1 valued stepfunction U with $O(\varepsilon^{-10\mathsf{v}(F)^2})$ steps such that $\|W - U\|_1 \leq \varepsilon$.

13.5. Symmetries of graphons

An *automorphism* of a graphon W on $[0, 1]$ is an invertible measure preserving map $\sigma : [0, 1] \to [0, 1]$ such that $W^\sigma = W$ almost everywhere. Clearly, the automorphisms of W form a group $\text{Aut}(W)$. An example with many automorphisms is a setfunction W: here $\text{Aut}(W)$ contains the group of all invertible measure preserving transformations that leave the steps invariant, and it contains all the automorphisms of the corresponding weighted graph. Note, however, that if we purify a stepfunction, then we get a finite weighted graph, so the large and ugly subgroups consisting of measure preserving transformations of the steps disappear.

We can endow $\text{Aut}(W)$ with the topology of pointwise convergence in the $\bar{r}_W$ metric. Szegedy observed that if W is pure, then $\text{Aut}(W)$ is compact in this topology. This follows from the facts that every automorphism of a graphon is an

13.5. SYMMETRIES OF GRAPHONS

isometry of the compact metric space $(J, \overline{r}_W)$ (this is trivial), and those isometries that correspond to automorphisms form a closed subgroup (this takes some work to prove; see Lovász [Notes]).

We will not go into the detailed study of $\mathrm{Aut}(W)$ in this book, even though it has interesting and nontrivial properties. We restrict our treatment to generalizing the easy direction of Theorem 6.36, and to an application of the results of this chapter to characterizing when $t(., W)$ has finite connection rank.

The group $\mathrm{Aut}(W)$ acts on J^k for any k. The number of orbits of this action can be estimated from below as follows.

PROPOSITION 13.46. *The number of orbits of the automorphism group of W on $[0,1]^k$ is at least $r(t(., W), k)$.*

Proof. Suppose that $\mathrm{Aut}(W)$ has a finite set of orbits $O_1, \ldots O_m$ on $[0,1]^k$. Let F and F' be two k-labeled graphs. Then

$$t(\llbracket FF' \rrbracket, W) = \int_{[0,1]^k} t_{x_1 \ldots x_k}(F, W) t_{x_1 \ldots x_k}(F', W) \, dx_1 \ldots dx_n.$$

The functions $t_{x_1 \ldots x_k}(F, W)$ and $t_{x_1 \ldots x_k}(F', W)$ are constant on every orbit, and hence

$$t(\llbracket FF' \rrbracket, W) = \sum_{j=1}^m \lambda(O_j) t_{x_{j,1} \ldots x_{j,k}}(F, W) t_{x_{j,1} \ldots x_{j,k}}(F', W),$$

where $(x_{j,1} \ldots x_{j,k})$ is any representative point of O_j. This shows that $M(t(., W), k)$ is the sum of m matrices of rank 1, and so it has rank at most m. $\square$

To be able to say something about the finiteness of the number of orbits of the automorphism group on k-tuples of points of a graphing, we need the following theorem.

THEOREM 13.47. *Let W be a graphon such that $r(t(., W), 2)$ is finite. Then W is a stepfunction.*

Proof. First we show that T_W has finite rank. It is clear that a kernel W has at most m different nonzero eigenvalues if and only if there are real numbers $a_0, \ldots, a_m$, not all 0, such that

$$(13.14) \qquad \sum_{k=0}^m a_k W^{\circ(k+2)} = 0$$

almost everywhere (so that all eigenvalues of W will be roots of the polynomial $\sum_k a_k x^{k+2}$). We claim that this is equivalent to requiring that

$$(13.15) \qquad \sum_{k=0}^m a_k \langle W^{\circ(k+2)}, W^{\circ(l+2)} \rangle = 0 \qquad (l = 0, \ldots, m).$$

Indeed, (13.14) clearly implies (13.15) for every l; on the other hand, (13.15) implies that

$$(13.16) \qquad \Big\langle \sum_{k=0}^m a_k W^{\circ(k+2)}, \sum_{k=0}^m a_k W^{\circ(k+2)} \Big\rangle = 0,$$

which implies (13.14). Using (7.22), equation (13.15) can be rewritten as

$$\sum_{k=0}^{m} a_k t(C_{k+l+4}, W) = 0 \qquad (l = 0, \ldots, m). \tag{13.17}$$

This is a system of $m+1$ homogeneous linear equations in $m+1$ variables a_k, which is solvable if and only if its determinant vanishes:

$$\begin{vmatrix} t(C_4, W) & t(C_5, W) & \ldots & t(C_{m+4}, W) \\ t(C_5, W) & t(C_6, W) & \ldots & t(C_{m+5}, W) \\ \vdots & & & \vdots \\ t(C_{m+4}, W) & t(C_{m+5}, W) & \ldots & t(C_{2m+4}, W) \end{vmatrix} = 0. \tag{13.18}$$

Since this matrix is a submatrix of $M(f, 2)$, this determinant will certainly vanish if $m \geq r(f, 2)$. It follows that the number of distinct nonzero eigenvalues of the operator T_W is at most $r(f, 2)$. Since every eigenvalue has finite multiplicity, it follows that T_W has finite rank.

Next, we show that the range of $W \circ W$ is finite (up to a set of measure 0). Consider its moments *as a single variable function* on the probability space $[0,1]^2$: $M_k(W \circ W) = \int_{[0,1]^2} (W \circ W)^k$, and the corresponding moment matrix $M(W \circ W) = \bigl(M_{k+l}(W \circ W)\bigr)_{k,l=0}^{\infty}$. Note that $M_k(W \circ W) = t(K_{2,k}, W)$, and so $M_{k+l}(W \circ W) = t(K_{2,k}^{\bullet\bullet} K_{2,l}^{\bullet\bullet}, W)$. It follows that $M(W \circ W)$ is a submatrix of $M(f, 2)$, and hence its rank is finite. By Theorem A.22, the range of $W \circ W$ is finite (up to a set of measure 0).

The fact that T_W has finite rank implies that $T_{W \circ W} = T_W^2$ has finite rank. This, together with the fact that the range of W is finite, implies that W is a stepfunction. Indeed, the row space of W is finite dimensional, so we can select a finite set of points $x_1, \ldots, x_r$ so that every row $W(x, .)$ is a linear combination of the functions $W(x_i, .)$. Since W has finite range, the functions $W(x_i, .)$ are stepfunctions. There is a finite partition $[0, 1] = S_1 \cup \cdots \cup S_p$ such that every function $W(x_i, .)$ is constant on every S_i, and hence every row is constant on every S_i. By symmetry, this implies that W is constant on every rectangle $S_i \times S_j$, i.e., it is a stepfunction. □

It is easy to derive from this the following analogue of Theorem 5.54.

COROLLARY 13.48. *Let f be a reflection positive, multiplicative, and normalized simple graph parameter. Then either $r(f, k)$ is infinite for all $k \geq 2$, or there is a twinfree weighted graph H such that $f = t(., H)$, $r(f, k)$ is finite for all $k \geq 0$, and $r(f, k)^{1/k} \to \mathsf{v}(H)$.*

This result is not a strengthening of Theorem 5.54, because it concerns simple graph parameters. The extension to multigraph parameters is more complicated, and we will return to it later, in Chapter 17.

Proof. By Theorem 11.52 and Proposition 14.61, there is a graphon W such that $f = t(., W)$. If $r(f, 2) = \infty$, then trivially $r(f, k) = \infty$ for all $k \geq 2$. Suppose that $r(f, 2) < \infty$, then by Theorem 13.47, W is a stepfunction, and so there is a weighted graph H such that $f = t(., H)$. The conclusion follows by theorem 6.36. □

For the graph parameter $f = t(., W)$, we have $r(f, 0) = 1$ by multiplicativity, so this is always finite. The rank $r(f, 1)$ may be finite; this happens if W has an automorphism group that has a finite number of orbits with positive measure. Proposition 13.46 and Theorem 13.47 imply that if $\mathrm{Aut}(W)$ has a finite number of orbits on pairs, then W is a stepfunction. It follows in particular that $\mathrm{Aut}(W)$ has a finite number of orbits on k-tuples of points for every k. This is somewhat surprising in view of the fact that there are arbitrarily large finite graphs whose automorphism group has only three orbits on pairs, but an unbounded number of orbits on k-tuples for $k \geq 3$ (for example, the Paley graphs in Example 1.1).

CHAPTER 14

The space of graphons

The space of graphons is the stage where many acts of interaction between graph theory and analysis take place. This chapter collects a number of questions about the structure of this space that arise naturally and that have at least partial answers.

14.1. Norms defined by graphs

We mentioned, and through the results in the last chapters also illustrated, that the cut norm and the cut-distance are best suited for measuring structural similarity of two graphons. However, other norms are also important; the connection between norms on the graphon space and our theory is twofold: first, homomorphism densities give rise to interesting norms, and second, norms with some natural properties are closely related to the cut norm. We start with a discussion of norms defined by homomorphism densities, based on the work of Hatami [2010].

We have seen that many Schatten norms of a kernel operator can be expressed by the homomorphism densities of even cycles. Here we consider a more general question: for which graphs F is $|t(F,W)|^{1/e(F)}$ a norm? We call such a graph F *norming*. This condition can be relaxed in two directions: (1) We can ask whether the functional $W \mapsto |t(F,W)|^{1/e(F)}$ is a seminorm on $\mathcal{W}$ (i.e., whether it is subadditive, but could be 0 even if W is not identically 0; homogeneity is trivial). We call F *seminorming* if this holds. (2) We can ask whether $W \mapsto t(F,|W|)^{1/e(F)}$ is a norm (we moved the absolute value signs in); we call F *weakly norming* if this holds. This is equivalent to asking: which graphs F have the property that the subadditivity inequality $t(F, W_1 + W_2)^{1/e(F)} \leq t(F, W_1)^{1/e(F)} + t(F, W_2)^{1/e(F)}$ holds for all $W_1, W_2 \in \mathcal{W}_0$? (There is no fourth version: if the functional $W \mapsto t(F, |W|)^{1/e(F)}$ is a seminorm, then it is a norm; see Exercise 14.9).

A related property is that the graph F satisfies $t(F, W) \geq 0$ for every kernel W. Such graphs are called *positive*. If $F = [\![F_1^2]\!]$ for some k-labeled graph F_1 with nonadjacent labeled nodes, then F is positive, but the converse is not known. Exercises 14.3 and 14.4 state some of the known properties of positive graphs.

Returning to graphs that are norming in one sense or the other, we collect some (easy) facts about them, due to Hatami [2010] and Kunszenti-Kovács [unpublished]. Every seminorming or weakly norming graph is bipartite (Exercise 14.7). Being seminorming is almost equivalent to being norming: every seminorming graph that is not norming is a star with an even number of edges. Every seminorming graph is positive (and hence we don't need the absolute value in the definition; Exercise 14.8), but not every positive graph is seminorming. Graphs with an odd number of edges cannot be seminorming, but they can be weakly norming, as the example of K_2 shows.

There are several classes of graphs F with norming properties. Besides cycles, complete bipartite graphs with an even number of nodes in each bipartition class are norming, and all complete bipartite graphs are weakly norming. For more properties and examples of graphs with norming properties, see Exercises 14.5–14.8.

Norming properties are closely related to Hölder-type inequalities for homomorphism densities, which can be stated using the notion of $\mathcal{W}$-decorated graphs introduced in Section 7.2. A simple graph $F = (V, E)$ has the *Hölder property*, if for every $\mathcal{W}$-decoration $w = (w_e : e \in E(F))$ of F,

$$(14.1) \qquad t(F, w)^{\mathsf{e}(F)} \leq \prod_{e \in E} t(F, w_e).$$

It has the *weak Hölder property*, if this inequality holds for every $\mathcal{W}_0$-decoration of F (equivalently, for every $\mathcal{W}$-decoration with nonnegative functions).

Hatami [2010] gives the following characterizations of seminorming and weakly norming graphs in terms of Hölder properties.

THEOREM 14.1. *A simple graph is seminorming if and only if it has the Hölder property. It is weakly norming if and only if it has the weak Hölder property.*

Proof. We prove the second assertion; the proof of the first is similar. In the "if" direction, suppose that a simple graph $F = (V, E)$ with m edges has the weak Hölder property. Let $W_1, W_2 \in \mathcal{W}_0$. We have

$$t(F, W_1 + W_2) = \sum_w t(F, w),$$

where the summation extends to all $\{W_1, W_2\}$-decorations w of F. So by the weak Hölder property,

$$t(F, W_1 + W_2) \leq \sum_w \prod_{e \in E} t(F, w_e)^{1/m} = \sum_{k=0}^m \binom{m}{k} t(F, W_1)^{k/m} t(F, W_2)^{(m-k)/m}$$
$$= \left(t(F, W_1)^{1/m} + t(F, W_2)^{1/m} \right)^m,$$

which shows that the functional $t(F, W)^{1/m}$ is subadditive on $\mathcal{W}_0$.

The proof of the "only if" direction is trickier. Suppose that F is weakly norming, and let (F, w) be a $\mathcal{W}_0$-decoration of F. Inequality (14.1) is homogeneous of degree m in each graphon w_e, so we may scale those and assume that $t(F, w_e) = 1$ for every edge. We want to prove that $t(F, w) \leq 1$, but first we prove the weaker inequality

$$(14.2) \qquad t(F, w) \leq m^m.$$

Indeed, if $W = \sum_e w_e$, then using subadditivity, we get

$$t(F, w) \leq t(F, W) \leq \left(\sum_e t(F, w_e)^{1/m} \right)^m = m^m.$$

To conclude, we use a method called *tensoring*. Let $n \geq 1$, and let us decorate every edge e by the tensor product $w_e^{\otimes n}$. Then one has

$$t(F, w^{\otimes n}) = t(F, w)^n, \qquad t(F, w_e^{\otimes n}) = t(F, w_e)^n = 1,$$

and hence 14.2 implies that $t(F, w) \leq m^{m/n}$. Since this holds for every n, it follows that $t(F, w) \leq 1$. □

Using this Theorem, one can prove about some graphs that they are norming (Hatami [2010]). A characterization of such graphs is open.

PROPOSITION 14.2. (a) *Even cycles are norming.*

(b) *Hypercubes are weakly norming.*

(c) *Deleting a perfect matching from a complete bipartite graph $K_{n,n}$, we get a weakly norming graph.*

Proof. We describe the proof of (b); the proofs of (a) and (c) are similar (in fact, simpler). Consider the d-dimensional hypercube graph Q^d. We consider its node set as $V = \{0,1\}^d$, and its edge set as $E = \{xy : x, y \in V, x_i = y_i \text{ for all but one } i\}$.

By Theorem 14.1, it is enough to prove that (14.1) holds for $F = Q^d$ and any decoration with graphons. Let $\mathcal{A}$ be the set of graphons that occur. We may assume that $\mathcal{A}$ does not contain the graphon that is almost everywhere 0 (else, the inequality is trivial). We say that an $\mathcal{A}$-decoration W of Q^d is *pessimal*, if $t(Q^d, W)^{\mathsf{e}(Q^d)} \big/ \prod_{e \in E} t(Q^d, W_e)$ is maximal among all $\mathcal{A}$-decorations. Since there are only a finite number of such decorations, at least one of them is pessimal. In these terms, inequality (14.1) means that there is a pessimal $\mathcal{A}$-decoration with all decorating graphons equal.

Let S_1 denote the set of nodes x of Q^d with $x_1 = 1$, $x_2 = 0$; let S_2 be the set of nodes x with $x_1 = 0$, $x_2 = 1$; and let $T = V \setminus S_1 \setminus S_2$. Note that T separates S_1 and S_2. Let E_i be the set of edges incident with any node in S_i, and let E_0 be the set of edges spanned by T.

We can write
$$t(F,W) = \int_{[0,1]^V} \prod_{ij \in E} W_{ij}(x_i, x_j)\, dx = \int_{[0,1]^V} \prod_{ij \in E_0} \prod_{ij \in E_1} \prod_{ij \in E_2}.$$

Considering the first factor as a weight function (here we use that $W \geq 0$), we can apply the Cauchy–Schwarz Inequality to get
$$t(Q^d, W) \leq \left(\int_{[0,1]^V} \prod_{ij \in E_0} \Big(\prod_{ij \in E_1} \Big)^2 \right)^{1/2} \left(\int_{[0,1]^V} \prod_{ij \in E_0} \Big(\prod_{ij \in E_2} \Big)^2 \right)^{1/2}.$$

Interchanging x_1 and x_2 in every $x \in V$ (in other words, reflecting in the hyperplane $x_1 = x_2$) is an automorphism σ of Q^d which maps E_1 onto E_2, and therefore
$$\int_{[0,1]^V} \prod_{ij \in E_0} \Big(\prod_{ij \in E_1} \Big)^2 = t(Q^d, W'),$$
where $W'_e = W_e$ if $e \in E_1 \cup E_0$, and $W'_e = W_{\sigma(e)}$ if $e \in E_2$. Similarly,
$$\int_{[0,1]^V} \prod_{ij \in E_0} \Big(\prod_{ij \in E_2} \Big)^2 = t(Q^d, W''),$$
where $W''_e = W_e$ if $e \in E_2 \cup E_0$, and $W'_e = W_{\sigma(e)}$ if $e \in E_1$. Thus we get
$$t(Q^d, W) \leq t(Q^d, W')^{1/2} t(Q^d, W'')^{1/2}.$$

By the definition of pessimal decoration, we must have equality here, and the decorations W' and W'' must also be pessimal. Here (say) W' is a pessimal decoration which is invariant under interchanging the first two entries in every $x \in V$. We call this *symmetrization* with respect to the hyperplane $x_1 = x_2$. We can symmetrize similarly with respect to the hyperplane $x_1 + x_2 = 1$.

Now consider a pessimal decoration W and a face Z of the cube such that all edges of Z are decorated by the same graphon W. Suppose that Z is not the whole cube. We may assume that Z is the face defined by $x_1 = x_2 = \cdots = x_k = 0$, where $0 < k < d$. Let Z' be the face obtained by reflecting Z in the hyperplane $x_k = x_{k+1}$. The intersection of Z and Z' is the face defined by $x_1 = x_2 = \cdots = x_k = x_{k+1} = 0$. The smallest face Z'' containing both Z and Z' is defined by $x_1 = x_2 = \cdots = x_{k-1} = 0$.

Let us symmetrize with respect to the hyperplane $x_k = x_{k+1}$. The decoration of the edges of Z does not change, but the decoration of the edges of Z' also becomes U. Symmetrizing with respect to $x_k + x_{k+1} = 1$, we get a pessimal decoration in which all edges of Z'' have the same decoration. Repeating this procedure, we get a pessimal decoration with all edges decorated by the same graphon and we are done. $\square$

EXERCISE 14.3. A graph F is positive if and only if every connected component of F that is not positive occurs with even multiplicity.

EXERCISE 14.4. Let F be a positive simple graph. (a) $F \times G$ is positive for every simple graph G. (b) There is a homomorphism $F \to F$ such that every edge has an even number of pre-images (c) [Harder] If F is positive, then there is a homomorphism $F \to G$ into a simple graph G with $\mathsf{v}(G) \geq \mathsf{v}(F)/2$ such that every edge has an even number of pre-images (Camarena, Csóka, Hubai, Lippner and Lovász [2012]).

EXERCISE 14.5. Prove that (a) complete bipartite graphs with an even number of nodes in both color classes are norming, (b) complete bipartite graphs are weakly norming, (c) stars with an even number of edges are seminorming, (c) $K_{2,3}$ is not seminorming (but weakly norming).

EXERCISE 14.6. Let T be a tree that is seminorming. (a) Prove that if $U, W \in \mathcal{W}$ and $\int_0^1 U(x,y)\,dx = \int_0^1 W(x,y)\,dx$ for every y, then $t(T, U) = t(T, W)$. (b) Prove that T is a star.

EXERCISE 14.7. (a) Every seminorming or weakly norming graph is bipartite. (b) Every seminorming graph is either a star, or eulerian. (c) Every seminorming graph that is not norming is a star.

EXERCISE 14.8. Let F be a seminorming graph. Prove that (a) kernels with $t(F, W) = 0$ form a linear space; (b) $\mathsf{e}(F)$ is even; (c) F is positive.

EXERCISE 14.9. Prove that if the functional $W \mapsto t(F, |W|)^{1/\mathsf{e}(F)}$ is a seminorm, then it is a norm.

14.2. Other norms on the kernel space

The topologies on $\mathcal{W}_1$ defined by the cut norm, L_1-norm, weak convergence etc. are different, but there are some subtle, nonobvious relationships between them. This turns out to be quite important for graph-theoretic applications: the interplay between the cut norm and L_2-norm is crucial in the proof of the Regularity Lemma (Section 9.1.2), and the relationship between the cut norm and L_1-norm is the key to the analytic theory of property testing (Section 15.3) and to the stability theory of extremal graphs (Section 16.4). We are not going to explore all the connections between these norms, just those that have graph theoretical significance.

14.2. OTHER NORMS ON THE KERNEL SPACE

Almost all norms on $\mathcal{W}$ (in short, norms for this section) that we need have some natural properties. Recall from Section 8.2 that a norm N is called *invariant*, if $N(W^\varphi) = N(W)$ for every measure preserving transformation $\varphi \in S_{[0,1]}$, and *smooth*, if for every sequence $W_n \in \mathcal{W}_1$ of kernels such that $W_n \to 0$ almost everywhere, we have $N(W_n) \to 0$. The norms L_1, L_2, the cut norm, and the graph norms from the previous section share these properties. (But the L_∞-norm is not smooth!)

Recall the obvious inequalities

(14.3) $$\|W\|_\square \leq \|W\|_1 \leq \|W\|_2.$$

For $W \in \mathcal{W}_1$, we have $\|W\|_2 \leq \|W\|_1^{1/2}$, and hence these two norms define the same topology on $\mathcal{W}_1$. Trivially, the cut norm is continuous in this topology. How about the other way around? There are easy examples showing that $\|W_n\|_\square \to 0$ does not imply that $\|W_n\|_1 \to 0$ or $\|W_n\|_2 \to 0$: let (G_n) be a quasirandom graph sequence with edge density $1/2$, and $W_n = 2W_{G_n} - 1$. Then $\|W_n\|_\square \to 0$, but $\|W_n\|_1 = \|W_n\|_2 = 1$.

The main goal in this section is to establish the following picture about smooth invariant norms.

THEOREM 14.10. (a) *Every smooth invariant norm (as a function on $\widetilde{\mathcal{W}_1}$) is continuous with respect to the L_1 norm, and the cut norm is continuous with respect to any smooth invariant norm.*

(b) *Any smooth invariant norm is lower semicontinuous with respect to any other smooth invariant norm.*

We also prove an analogous (but not equivalent!) theorem about the distances δ_N on $\widetilde{\mathcal{W}}$ defined by smooth invariant norms N. Let us call these, for brevity, *delta-metrics*.

THEOREM 14.11. (a) *Every delta-metric is continuous (as a function on $\widetilde{\mathcal{W}_1} \times \widetilde{\mathcal{W}_1}$) with respect to δ_1, and $\delta_\square$ is continuous with respect to any delta-metric.*

(b) *Any delta-metric is lower semicontinuous with respect to any other delta-metric.*

The fact that we prove continuity (or lower semicontinuity) as a function in two variables, and not just separately in each variable, is significant. As an example of a different nature, recall that the overlay functional $\mathcal{C}(U, W)$ is continuous in each variable, but not as a 2-variable function (Section 12.2).

Some of the above statements are trivial, and some follow easily from each other. Along the lines, we are going to prove a couple of facts that will be useful in other contexts too.

14.2.1. Smooth and invariant norms. As a technical preparation, we have to prove some simple facts about smooth and invariant norms.

LEMMA 14.12. *Every smooth norm N is uniformly continuous with respect to the L_1 norm on $\mathcal{W}_1$.*

Proof. Suppose not, then there exists an $\varepsilon > 0$ and a sequence of kernels $W_n \in \mathcal{W}_1$ such that $\|W_n\|_1 \to 0$ but $N(W_n) > 0$. By selecting a subsequence, we may assume that $W_n \to 0$ almost everywhere, contradicting the assumption that N is smooth. $\square$

Next we prove a basic property of the stepping operator.

PROPOSITION 14.13. *The stepping operator is contractive with respect to any smooth invariant norm.*

Proof. Let $\mathcal{P}$ be any finite measurable partition of $[0,1]$, and let N be an invariant norm. We want to prove that $N(W_\mathcal{P}) \leq N(W)$ for every $W \in \mathcal{W}$. By the invariance of N, we may assume that the partition classes of $\mathcal{P}$ are intervals. We may also assume that $W \in \mathcal{W}_1$.

For every interval $I = [a,b] \in \mathcal{P}$, let $\varphi_I : I \to I$ denote a measure preserving map $x \mapsto a + 2(x-a) \pmod{b-a}$. Then every map $(x,y) \mapsto (\varphi_I(x), \varphi_J(y))$ is ergodic on $I \times J$. Let $\varphi : [0,1] \to [0,1]$ denote the map that acts on $I \in \mathcal{P}$ as φ_I.

For $n \geq 1$, define
$$U_n(x,y) = \frac{1}{n} \sum_{k=0}^{n-1} W^{\varphi^k}(x,y) = \frac{1}{n} \sum_{k=0}^{n-1} W(\varphi^k(x), \varphi^k(y)).$$

Using the subadditivity and invariance of N, we get
$$N(U_n) \leq \frac{1}{n} \sum_{k=0}^{n-1} N(W^{\varphi^k}) = N(W).$$

On the other hand, the Ergodic Theorem implies that $U_n \to W_\mathcal{P}$ almost everywhere as $n \to \infty$. Since, trivially, $U_n \in \mathcal{W}_1$ and N is smooth, this implies that $N(W_\mathcal{P}) = \lim_{n\to\infty} N(U_n) \leq N(W)$. □

Next we give a useful representation of smooth invariant norms. By the Hahn–Banach Theorem, we can represent any norm on $\mathcal{W}$ that is continuous in the L_∞ norm as

(14.4) $$N(W) = \sup_{\ell \in \mathcal{L}} \ell(W),$$

where $\mathcal{L}$ is an appropriate set of linear functionals on $\mathcal{W}$, continuous in the L_∞ norm. We show that for our norms, the linear functionals in $\mathcal{L}$ can be represented as inner products with functions in $\mathcal{W}$.

PROPOSITION 14.14. *For every smooth and invariant norm N there is a set $K \subseteq \mathcal{W}$ such that*
$$N(W) = \sup_{U \in K} \langle U, W \rangle$$
for every $W \in \mathcal{W}$.

Proof. Define $K = \{Y \in \mathcal{W} : \langle Y, U \rangle \leq N(U) \; \forall U \in \mathcal{W}\}$. Let $W \in \mathcal{W}$, then we want to prove that $N(W) = \sup_{Y \in K} \langle Y, W \rangle$. Suppose not, then we may assume that $N(W) > 1 > \sup_{Y \in K} \langle Y, W \rangle$.

First, we assume that W is a stepfunction. Let $\mathcal{P}$ be the partition of $[0,1]$ into the steps of W. The linear space $\mathcal{W}_\mathcal{P}$ of kernels with steps in $\mathcal{P}$ is finite dimensional, and $B = \{U \in \mathcal{W}_\mathcal{P} : N(U) \leq 1\}$ is a convex set in it. Since $W \notin B$, and we are working in a finite dimensional space, there is a hyperplane of the form $\langle Y, . \rangle = 1$ ($Y \in \mathcal{W}_\mathcal{P}$) through the point W such that $\langle Y, X \rangle \leq 1$ for all $X \in B$. Then for any $U \in \mathcal{W}$, using Proposition 14.13, we get

$$\langle Y, U \rangle = \langle Y, U_\mathcal{P} \rangle = N(U_\mathcal{P}) \Big\langle Y, \frac{1}{N(U_\mathcal{P})} U_\mathcal{P} \Big\rangle \leq N(U_\mathcal{P}) \leq N(U),$$

which shows that $Y \in K$. Since $\langle Y, W \rangle = 1$, this is a contradiction.

Second, let W be an arbitrary kernel. Proposition 9.8 implies that there is a stepfunction W' such that $N(W-W') < N(W)-1$. Then $N(W-W') < N(W)-1$, and hence $N(W') > 1$. We know already that there is a stepfunction $Y \in K$ with the same steps as W' such that $\langle Y, W' \rangle = 1$. Since $\langle Y, W \rangle = \langle Y, W' \rangle$, this contradicts the choice of W, and completes the proof. $\square$

The following lemma is a special case of the theorem, but it is best to formulate and prove it separately.

LEMMA 14.15. *Let N be any smooth invariant norm, and let $W_n \to W$ in the cut norm ($W_n, W \in \mathcal{W}_1$). Then*
$$\liminf_{n \to \infty} N(W_n) \geq N(W).$$

Proof. By Proposition 14.14, the norm N can be represented as $N(X) = \sup_{Y \in K} \langle X, Y \rangle$ for some $K \subseteq \mathcal{W}$. Let $\varepsilon > 0$ and choose a function $Y \in K$ such that $\langle Y, W \rangle \geq N(W) - \varepsilon$. Then by Lemma 8.22, $N(W_n) \geq \langle Y, W_n \rangle \to \langle Y, W \rangle \geq N(W) - \varepsilon$. Since $\varepsilon > 0$ was arbitrary, this proves the Proposition. $\square$

We can now give the proof of the first main theorem in this section.

Proof of Theorem 14.10. (a) Lemma 14.12 proves the first statement. To prove the second, let N be a smooth invariant norm. Suppose that the cut norm is not continuous with respect to N, then there is a sequence of kernels $W_n \in \mathcal{W}_1$ such that $N(W_n) \to 0$ but $\|W_n\|_\square \geq c > 0$ for all n. By the compactness of the graphon space, we may also assume that $\delta_\square(W_n, U) \to 0$ for some nonzero graphon U. This means that there are invertible measure preserving transformations φ_n such that $\|W_n^{\varphi_n} - U\|_\square \to 0$. Lemma 14.15 implies that $\liminf_n N(W_n^{\varphi_n}) = \liminf_n N(W_n) \geq N(U) > 0$. Since N is invariant, this contradicts the assumption $N(W_n) \to 0$.

(b) Let N_1 and N_2 be two smooth invariant norms on $\mathcal{W}$, and let $W_1, W_2, \ldots, W \in \mathcal{W}_1$ such that $N_1(W_n - W) \to 0$. Then $\|W_n - W\|_\square \to 0$ by (a). Hence $\liminf_n N_2(W_n) \geq N(W)$ by Lemma 14.15. $\square$

14.2.2. Delta-distances. We start with a fact similar to Lemma 14.15, but more difficult to prove, with the distances δ_N and $\delta_\square$ replacing the norm N and the cut norm. For the case when N is the L_1-norm, this was proved by Lovász and Szegedy [2010a].

LEMMA 14.16. *Let N be a smooth invariant norm on $\mathcal{W}$. Let $\delta_\square(U_n, U) \to 0$ and $\delta_\square(W_n, W) \to 0$ as $n \to \infty$ ($U, W, U_n, W_n \in \mathcal{W}_1$). Then*
$$\liminf_{n \to \infty} \delta_N(W_n, U_n) \geq \delta_N(W, U).$$

In other words, the distance δ_N is lower semicontinuous on the compact metric space $(\widetilde{\mathcal{W}_1}, \delta_\square)$.

Proof. Applying appropriate measure preserving transformations to the kernels U_n and W_n, we may assume that $\|U_n - U\|_\square \to 0$ and $\|W_n - W\|_\square \to 0$ when $n \to \infty$. Fix an $\varepsilon > 0$. Let $\mathcal{P}$ and $\mathcal{Q}$ denote finite partitions of $[0,1]$ such that $N(W - W_\mathcal{P}) \leq \varepsilon$ and $N(U - U_\mathcal{Q}) \leq \varepsilon$. For any positive integer n, there are measure preserving transformations $\varphi_n, \psi_n : [0,1] \mapsto [0,1]$ such that $\delta_N(W_n, U_n) = N(W_n^{\varphi_n} - U_n^{\psi_n})$. The difficulty (why we cannot apply Lemma 14.15) is that these transformations φ_n and ψ_n may depend on n.

But as a next step, we fix n as follows. Let $\mathcal{R} = \{R_1, \ldots, R_m\}$ denote the common refinement of the partitions $\mathcal{P}$ and $\mathcal{Q}$. We claim that if n is large enough then $N\big((W_n)_\mathcal{R} - W_\mathcal{R}\big) \leq \varepsilon$. By Lemma 14.12, there is an $\varepsilon' > 0$ such that it is enough to guarantee that $\|(W_n)_\mathcal{R} - W_\mathcal{R}\|_1 \leq \varepsilon'$. By (8.15), this follows if we have $\|(W_n)_\mathcal{R} - W_\mathcal{R}\|_\square \leq \varepsilon'/m^2$. By Proposition 14.13, this will hold if $\|W_n - W\|_\square \leq \varepsilon'/m^2$, which holds for every n that is large enough by the assumption that $\|W_n - W\|_\square \to 0$ when $n \to \infty$. Similarly, $N\big((U_n)_\mathcal{R} - U_\mathcal{R}\big) \leq \varepsilon$ holds if we choose n large enough.

We consider n fixed from now on, and so we can replace W_n, W and $\mathcal{P}$ by $W_n^{\varphi_n}$, W^{φ_n} and $\varphi_n(\mathcal{P})$, and replace U_n, U and $\mathcal{Q}$ by $U_n^{\psi_n}$, U^{ψ_n} and $\psi(\mathcal{Q})$. With this new notation, we have $\delta_N(W_n, U_n) = N(W_n - U_n)$. Then

$$\delta_N(W, U) \leq N(W - U) \leq N(W - W_\mathcal{R}) + N(W_\mathcal{R} - U_\mathcal{R}) + N(U_\mathcal{R} - U).$$

By the choice of $\mathcal{P}$ and by Proposition 14.13

$$N(W - W_\mathcal{R}) \leq N(W - W_\mathcal{P}) + N(W_\mathcal{P} - W_\mathcal{R}) = N(W - W_\mathcal{P}) + N\big((W_\mathcal{P} - W)_\mathcal{R}\big) \leq 2\varepsilon,$$

and using the analogous estimate for U, we get that

$$\delta_N(W, U) \leq N(W_\mathcal{R} - U_\mathcal{R}) + 4\varepsilon.$$

Using Proposition 14.13 again and the fact that the kernels $W_\mathcal{P}$ and $U_\mathcal{Q}$ are both constant on the rectangles $R_i \times R_j$, we get

$$\begin{aligned}\delta_N(W_n, U_n) = N(W_n - U_n) &\geq N\big((W_n)_\mathcal{R} - (U_n)_\mathcal{R}\big) \\ &\geq N(W_\mathcal{R} - U_\mathcal{R}) - N\big((W_n)_\mathcal{R} - W_\mathcal{R}\big) - N\big(U_\mathcal{R} - (U_n)_\mathcal{R}\big) \\ &\geq N(W_\mathcal{R} - U_\mathcal{R}) - 2\varepsilon \geq \delta_N(W, U) - 6\varepsilon.\end{aligned}$$

Since this holds for every $\varepsilon > 0$ if n is large enough, the assertion follows. $\square$

Proof of Theorem 14.11. The theorem follows from Lemma 14.16 similarly as Theorem 14.10 followed from Lemma 14.15. The details are not repeated. $\square$

A consequence of Lemma 14.16 (or Theorem 14.11) is worth formulating.

COROLLARY 14.17. *Let N be a smooth invariant norm. Let $\mathcal{R} \subseteq \widetilde{\mathcal{W}}_1$ be compact with respect to the $\delta_\square$ distance. Then the functional $\delta_N(., \mathcal{R})$ is lower semicontinuous on $(\widetilde{\mathcal{W}}_1, \delta_\square)$.*

Proof. Suppose that $U_n \to U$ in the $\delta_\square$ distance; we want to prove that

$$\liminf_{n \to \infty} \delta_N(U_n, \mathcal{R}) \geq \delta_N(U, \mathcal{R}).$$

By selecting an appropriate subsequence, we may assume that the limes inferior is actually a limit. For every n, let $W_n \in \mathcal{R}$ be such that $\delta_N(U_n, W_n) \leq \delta_N(U_n, \mathcal{R}) + 1/n$. Again by going to a subsequence, using the compactness of $\mathcal{R}$, we may assume that $W_n \to W$ for some $W \in \mathcal{R}$. By Lemma 14.16,

$$\liminf_{n \to \infty} \delta_N(U_n, \mathcal{R}) \geq \liminf_{n \to \infty} \left(\delta_N(U_n, W_n) - \frac{1}{n}\right) \geq \delta_N(U, W) \geq \delta_N(U, \mathcal{R}). \quad \square$$

EXERCISE 14.18. Prove that the statement of Lemma 14.12 remains valid if the assumption of smoothness is replaced by the assumption that N has a representation as in Proposition 14.14.

EXERCISE 14.19. Construct a norm N such that the stepping operator is not contractive with respect to N.

EXERCISE 14.20. Let W_1 and W_2 be graphons that are monotone increasing in both variables. Prove that (a) $\delta_1(W_1, W_2) = \|W_1 - W_2\|_1$; (b) $\delta_\square(W_1, W_2) = \|W_1 - W_2\|_\square$; (c) $\|W_1 - W_2\|_1 \leq 10\|W_1 - W_2\|_\square^{2/3}$ (Bollobás, Janson and Riordan [2012]).

14.3. Closures of graph properties

A class of graphons closed under weak isomorphism is called a *graphon property*. Since every graphon is weakly isomorphic to a graphon on $[0, 1]$, we can usually restrict our attention to graphons on $[0, 1]$ when studying graphon properties. In this section, we study graphon properties obtained as closures of graph properties, and graphon properties defined by equations.

A graph property $\mathcal{P}$ is a class of finite graphs closed under isomorphism. Let $\overline{\mathcal{P}}$ is the set of graphons (J, W) that arise as limits of graph sequences in $\mathcal{P}$. Geometric and topological properties of the closure reveal important information about the graph property, as we shall see.

14.3.1. Hereditary properties.
Recall that a graph property is called *hereditary*, if whenever $G \in \mathcal{P}$, then every induced subgraph is also in $\mathcal{P}$. For every graphon W, let $\mathcal{I}(W)$ denote the set of its "induced subgraphs", i.e., the set of those graphs F for which $t_{\mathsf{ind}}(F, W) > 0$. Clearly, $\mathcal{I}(W)$ is a hereditary graph property.

Let $\mathcal{P}$ be a hereditary property of graphs. Then

(14.5) $$\cup_{W \in \overline{\mathcal{P}}} \mathcal{I}(W) \subseteq \mathcal{P}.$$

Indeed, if $F \notin \mathcal{P}$, then $t_{\mathsf{ind}}(F, G) = 0$ for every $G \in \mathcal{P}$, since $\mathcal{P}$ is hereditary. This implies that $t_{\mathsf{ind}}(F, W) = 0$ for all $W \in \overline{\mathcal{P}}$, and so $F \notin \mathcal{I}(W)$.

Equality does not always hold in (14.5). For example, we can always add a graph G and all its induced subgraphs to $\mathcal{P}$ without changing $\overline{\mathcal{P}}$. As a less trivial example, consider all graphs with degrees bounded by 10. This property is hereditary, and $\overline{\mathcal{P}}$ consists of a single graphon (the identically 0 function), so the left hand side of (14.5) consists of edgeless graphs only. Equality in (14.5) can be characterized by assuming that $\mathcal{P}$ is closed not only under induced subgraphs, but also under a certain version of multiplying points (Exercise 14.28).

The closure of the set of triangle-free graphs is the set of *triangle-free graphons*, which can be characterized by the property $t(K_3, W) = 0$. More generally:

PROPOSITION 14.21. *Let $\mathcal{P}$ be a hereditary graph property. Then its closure is characterized by the (infinitely many) equations $t_{\mathsf{ind}}(F, W) = 0$ for all $F \notin \mathcal{P}$.*

Proof. By the definition of hereditary properties, the equations $t_{\mathsf{ind}}(F, G) = 0$ hold for all $G \in \mathcal{P}$ and $F \notin \mathcal{P}$, which implies that $t_{\mathsf{ind}}(F, W) = 0$ for every $W \in \overline{\mathcal{P}}$.

Conversely, suppose that W has the property that $t_{\mathsf{ind}}(F, W) = 0$ for all $F \notin \mathcal{P}$. This means that $\mathsf{P}(\mathbb{G}(k, W) \in \mathcal{P}) = 1$ for every k. Since $\mathbb{G}(k, W) \to W$ with probability 1, this implies that $W \in \overline{\mathcal{P}}$. $\square$

14.3.2. Random-free properties.
A graph property $\mathcal{P}$ is *random-free*, if every $W \in \overline{\mathcal{P}}$ is 0-1 valued almost everywhere. (For an explanation of the name, see Exercise 14.29.) By Proposition 8.24, if (G_n) is a convergent random-free sequence of graphs with limit graphon W, then $W_{G_n} \to W$ in the L_1 norm. Example 11.41 illustrated how random-freeness was related to a small amount of randomness in a randomly generated random-free graph sequence. We can also point out that if W

is 0-1 valued, then $\mathbb{G}(n,W) = \mathbb{H}(n,W)$, and so to generate $\mathbb{G}(n,W)$, we don't need randomness to get the edges (of course, we still need randomness to generate the nodes).

Among hereditary properties, it is quite easy to characterize random-free properties.

LEMMA 14.22. *A hereditary graph property $\mathcal{P}$ is random-free if and only if there is a signed bipartite graph F such that $t(F,W) = 0$ for all $W \in \overline{\mathcal{P}}$.*

The proof will show that it would be enough to assume that for every graphon $W \in \overline{\mathcal{P}}$ there is a signed bipartite graph with $t(F,W) = 0$.

Proof. Suppose that for every signed bipartite graph F there is a graphon $W \in \overline{\mathcal{P}}$ such that $t(F,W) > 0$. Let (F_n) be a quasirandom sequence of bipartite graphs with bipartition $V(F_n) = V_n' \cup V_n''$, with edge density $1/2$, and with $|V_n'| = |V_n''|$. Consider the signed bipartite graphs $\widehat{F}_n$, obtained from $K_{n,n}$ by signing the edges of F_n with $+$, the other edges with $-$. Let $W_n \in \overline{\mathcal{P}}$ be a graphon such that $t(\widehat{F}_n, W_n) > 0$. It is easy to see that this means that there is a simple graph G_n obtained from F_n by adding edges within the color classes such that $t_{\mathsf{ind}}(G_n, W_n) > 0$. By Proposition 14.21, this implies that $G_n \in \mathcal{P}$.

By selecting a subsequence we may assume that the graph sequences $G_n' = (G_n[V_n'])$ and $G_n'' = (G_n[V_n''])$ are convergent. By Theorem 11.59, we can order the nodes in V_n' and in V_n'' so that $W_{G_n'}$ converges to a graphon W' on $[0,1]$ in the cut norm, and similarly $W_{G_n''}$ converges to a graphon W'' on $[0,1]$. If we order the nodes of G_n so that the nodes in V_n' precede the nodes in V_n'', and keep the above ordering inside V_n' and V_n'', then W_{G_n} converges to the graphon

$$U(x,y) = \begin{cases} W'(2x, 2y) & \text{if } x,y < 1/2, \\ W''(2x-1, 2y-1) & \text{if } x,y > 1/2, \\ 1/2 & \text{otherwise.} \end{cases}$$

So $U \in \overline{\mathcal{P}}$ is not 0-1 valued, and hence $\mathcal{P}$ is not random-free.

Conversely, suppose that $\mathcal{P}$ is not random-free, and let $W \in \overline{\mathcal{P}}$ be a graphon that is not 0-1 valued almost everywhere. Then by Theorem 13.35(a), $t(F,W) > 0$ for every signed bipartite graph F. □

COROLLARY 14.23. *If a hereditary property of bipartite graphs does not contain all bipartite graphs, then it is random-free.*

Using Theorem 13.35(c), we can associate a finite dimension with every nontrivial hereditary property of bipartite graphs. It would be interesting to find further combinatorial properties of this dimension.

The natural analogue of this corollary for properties of nonbipartite graphs fails to hold.

EXAMPLE 14.24. Let $\mathcal{P}$ be the property of a graph that it is triangle-free. Then every bipartite graphon is in its closure, but such graphons need not be 0-1 valued. ♦

For more characterizations of hereditary and random-free properties, and for more on their connection, see Janson [2011b].

14.3.3. Flexible properties. A graphon U is called a *flexing* of a graphon W, if $U(x,y) = W(x,y)$ for all x,y with $W(x,y) \in \{0,1\}$ (so we may change the values of W that are strictly between 0 and 1; we may change them to 0 or to 1, so the relation is not symmetric). We say that a graphon property is *flexible* if it is preserved under flexing.

Every random-free graphon property is trivially flexible. It is also clear that the intersection and union of any set of flexible graphon properties is flexible. If $\mathcal{R}$ is a flexible graphon property, then so are its "complement" $\{1 - W : W \in \mathcal{R}\}$, its "downward closure" $\{U \in \mathcal{W}_0 : (\exists W \in \mathcal{R})\, U \leq W\}$, and its "upward closure" defined analogously.

For every signed graph F, the graphon property $\{W \in \mathcal{W}_0 : t_{\mathsf{ind}}(F, W) = 0\}$ is flexible, since the condition means that for almost all vectors $(x_i : i \in V(F))$, at least one of the factors in $\prod_{ij \in E(F)} W(x_1, x_j) \prod_{ij \in E(\overline{F})} (1 - W(x_1, x_j))$ is 0, which is preserved if values strictly between 0 and 1 are changed. By Proposition 14.21, the closure of any hereditary property can be defined by conditions of the form $t_{\mathsf{ind}}(F, W) = 0$. This implies:

PROPOSITION 14.25. *The closure of a hereditary property is flexible.* □

There are other, non-hereditary graph properties whose closure is flexible. Some of these are described in Exercise 14.31

Every flexible property has the following interesting geometric feature:

PROPOSITION 14.26. *If $\mathcal{R} \subseteq \mathcal{W}_0$ is flexible, then $\mathcal{W}_0 \setminus \mathcal{R}$ is convex.*

Proof. Indeed, let $W_1, W_2 \in \mathcal{W}_0 \setminus \mathcal{R}$, and suppose that a convex combination $W = \alpha_1 W_1 + \alpha_2 W_2 \in \mathcal{R}$. Then for every $x, y \in [0,1]$ with $W(x,y) \in \{0,1\}$ we have $W_1(x,y) = W_2(x,y) = W(x,y)$, and so by the definition of flexibility, we must have $W_1, W_2 \in \mathcal{R}$, a contradiction. □

COROLLARY 14.27. *If $\mathcal{P}$ is a hereditary graph property, then $\mathcal{W}_0 \setminus \overline{\mathcal{P}}$ is convex.*

We will discuss an application of this fact in extremal graph theory in Section 16.5.1.

EXERCISE 14.28. Prove that for a hereditary property $\mathcal{P}$ of graphs equality holds in (14.5) if and only if for every graph $G \in \mathcal{P}$ and $v \in V(G)$, if we add a new node v' and connect it to all neighbors of v, then at least one of the two graphs obtained by joining or not joining v and v' has property $\mathcal{P}$.

EXERCISE 14.29. Prove that a graph property is not random-free if it contains large quasirandom bipartite graphs in the following sense: for every $\varepsilon > 0$ there is $\delta > 0$, a sequence of graphs $G_1, G_2, \cdots \in \mathcal{P}$, and disjoint sets $S_n, T_n \subseteq V(G_n)$ with $|S_n| = |T_n| \geq \delta \mathsf{v}(G_n)$ such that the bipartite graphs $G_n[S_n, T_n]$ form a quasirandom sequence with error ε.

EXERCISE 14.30. Prove that a graph property $\mathcal{P}$ is random-free if and only if for every $\varepsilon > 0$ there is an $n \in \mathbb{N}$ such that for every graph $G \in \mathcal{P}$ with $\mathsf{v}(G) \geq n$ and every ε-regular k-partition of G, all but εk^2 pairs of partition classes span bipartite graphs whose edge density is at most ε or at least $1 - \varepsilon$.

EXERCISE 14.31. Prove that the closure of the following graph properties is flexible: (a) G is clique of size $\lceil |V(G)|/2 \rceil$ together with $\lfloor |V(G)|/2 \rfloor$ isolated nodes; (b) $\omega(G) \geq |V(G)|/2$; (c) $\alpha(G) \geq |V(G)|/2$; (c) there is a labeling of the nodes by $\{1, \ldots, n\}$ such that all (i,j) with $i + j \leq n$ are connected by an edge.

14.4. Graphon varieties

The topic of this section is reminiscent of the setup of algebraic geometry: we study subsets of $\mathcal{W}$, called varieties, defined by equations specifying linear (equivalently, algebraic) dependence between subgraph densities. These equations are invariant under weak isomorphism, so we may work in $\widetilde{\mathcal{W}}$ if we wish. Conditions like this play a role in extremal graph theory, and a better understanding of these varieties seems to be an important direction in the study of graphons.

A set of kernels satisfying a condition of the form $t(g, W) = 0$, where g is a quantum graph, will be called a *kernel variety*. We get, of course, different versions by putting restrictions on g and on W. A *simple variety* is defined by a simple quantum graph g. A *graphon variety* is the intersection of a kernel variety with $\mathcal{W}_0$. It is often convenient to restrict our attention to pure graphons; since every graphon is weakly isomorphic to a pure graphon, this is not an essential restriction.

It is clear that every kernel/graphon variety is closed under weak isomorphism. Every simple graphon variety is closed in the cut distance, and hence it can be considered as a closed (and hence compact) subset of the graphon space $(\widetilde{\mathcal{W}}_0, \delta_\square)$. However, this does not hold for non-simple varieties (see Example 14.36). The union and intersection of two [simple] graphon varieties are [simple] graphon varieties (Exercise 14.50).

We could try to be more general and consider the common solutions (in W) of a system of constraints $t(f_1, W) = 0, \ldots, t(f_m, W) = 0$. However, this could always be replaced by the single condition $t(f_1^2 + \cdots + f_m^2, W) = 0$.

While a general theory of graphon varieties is not at hand, there are some interesting examples, which will be needed later on. We will see some less trivial varieties in Section 14.4.2, but this will need some preparation in Section 14.4.1.

EXAMPLE 14.32 (**Constants**). As an immediate application of Claim 11.63, we get that every constant function $W = J_p$ forms a simple kernel variety. Indeed, this kernel can be defined by the equations $t(K_2, W) = p$ and $t(C_4, W) = p^4$. ♦

EXAMPLE 14.33 (**Complete graphs**). We have mentioned in the introduction that the densities of triangles and edges in a graph G satisfy the inequality $t(K_3, G) \geq 2t(K_2, G)^2 - t(K_2, G)$, and equality holds if and only if G is a blow-up of the complete graph. This is equivalent to saying that (up to weak isomorphism) the graphon variety defined by the equation $t(K_3 - 2K_2K_2 + K_2, W) = 0$ consists of the countably many graphons W_{K_n} ($n = 1, 2, \ldots$) and the identically-1 graphon. ♦

EXAMPLE 14.34 (**Regularity**). We call a kernel d-regular, if $\int_0^1 W(x,y)\,dy = d$ for almost all $0 \leq x \leq 1$. This kernel variety can be defined by two subgraph density constraints: $t(K_2, W) = d$ and $t(P_3, W) = d^2$. (This can be shown by a simpler version of the argument in the proof of Claim 11.63.) Regular kernels (without specifying the degree d) can be defined by the constraint $t(P_3 - K_2K_2, W) = 0$. ♦

EXAMPLE 14.35 (**Hadamard kernels**). Our next examples show that graphon varieties can encode quite substantial combinatorial complications. A symmetric $n \times n$ Hadamard matrix B gives rise to a kernel W_B, which we alter a little to get a graphon $U_B = (W_B + 1)/2$. We call U_B an *Hadamard graphon*.

Hadamard graphons, together with the graphon $J_{1/2}$, form a simple graphon variety. Indeed, the condition

(14.6) $\qquad t\bigl(K_3 - 2K_2K_2 + K_2, 1 - (2U - 1) \circ (2U - 1)\bigr) = 0$

implies that $1 - (2U - 1) \circ (2U - 1)$ is a kernel that is either identically 1 or it corresponds to a complete graph (Example 14.33). Let $W = 2U - 1$, then either $W \circ W = 0$ or $W \circ W = W_I$ where I is an identity matrix of some size n. In the first case, $W = 0$ and so $U = J_{1/2}$. In the second case, we note that every eigenvector of $T_{W \circ W} = T_W^2$ is a stepfunction with steps $[0, 1/n), \ldots, [(n-1)/n, 1)$, and hence so are the eigenvectors of T_W. It follows that W is a stepfunction with these steps, and so $W = W_B$ for some $n \times n$ matrix B. Furthermore, $W \circ W = W_I$ implies that $B^2 = nI$. Since U is a graphon, we have $-1 \leq W \leq 1$, and so every entry of B is in $[-1, 1]$. The condition $B^2 = nI$ implies that $\sum_i B_{ij}^2 = n$ for every j, which implies that every entry of B is either 1 or -1, and so B is an Hadamard matrix.

We must add that (14.6) can be expanded into a subgraph density condition on U, using the fact that $t(F, U \circ U) = t(F', U)$ (where F' is the subdivision of F). ♦

EXAMPLE 14.36 (**Zero-one valued graphons**). It is not hard to see that $W \in \mathcal{W}$ is 0-1 valued almost everywhere if and only if $t(B_4 - 2B_3 + B_2, W) = 0$ (one approach is to note that W is 0-1 valued iff $t_{xy}(B_2^{\bullet\bullet}, W) = t_{xy}(B_1^{\bullet\bullet}, W)$, and use Lemma 14.37 below). Hence 0-1 valued graphons form a kernel variety. However, the variety of 0-1 valued graphons is not simple, because it is not closed in the cut distance: for a quasirandom graph sequence (G_n) the associated graphons W_{G_n} are 0-1 valued, but $W_{G_n} \to J_{1/2}$ in the $\|.\|_\square$ norm. ♦

14.4.1. Unlabeling. Before describing more complicated graphon varieties, we introduce a tool that is very useful in constructing varieties. Instead of prescribing subgraph densities, we can try to define graphon or kernel varieties by a (seemingly) more general condition on the density function of a k-labeled graph or quantum graph: such conditions can be written as $t_x(g, W) = 0$ (for all $x \in [0, 1]^k$) for some k-labeled quantum graph g. However, there is a way to translate labeled constraints to unlabeled constraints. This fact will be convenient in constructions, since it is often easier to describe a property by the density of a labeled quantum graph.

LEMMA 14.37. *For every k-labeled quantum graph f there is an unlabeled quantum graph g such that for any $W \in \mathcal{W}$, $t(g, W) = 0$ if and only if $t_{x_1 \ldots x_k}(f, W) = 0$ almost everywhere. If f is simple, then we can require that g is simple, and the labeled nodes form a stable set in every constituent of g.*

Proof. The first assertion is trivial: $t_{x_1 \ldots x_k}(f, W) = 0$ almost everywhere if and only if $t(\llbracket f^2 \rrbracket, W) = 0$. This construction works for the second statement as well, provided the labeled nodes form a stable set in every constituent of f.

To prove the second statement for every simple k-labeled quantum graph g, define $\mathsf{Lb}(g)$ as the disjoint union of the subgraphs of the constituents induced by the labeled nodes (note that these subgraphs all have the same node set $[k]$). We use induction on the chromatic number $\chi(\mathsf{Lb}(f))$. If $\chi(\mathsf{Lb}(f)) = 1$, then the labeled nodes are nonadjacent in every constituent, and the trivial construction above works.

Suppose that $\chi(\mathsf{Lb}(f)) = r > 1$, let $[k] = S_1 \cup \cdots \cup S_r$ be an r-coloring of $\mathsf{Lb}(f)$, and let $q = |S_r|$. We may suppose that $S_r = \{k-q+1, \ldots, k\}$. We glue together two copies of f along S_r. Formally, let f_1 be obtained from f by increasing the labels in S_r by $k-q$ (the labels not in S_r are not changed). Let f_2 be obtained from f by increasing all labels by $k-q$. So the product $f_1 f_2$ is a $(2k-q)$-labeled quantum graph, in which the nodes of S_r are labeled $2k-2q+1, \ldots, 2k-q$. Let h be obtained from $f_1 f_2$ by unlabeling the nodes in S_r.

CLAIM 14.38. *For every $W \in \mathcal{W}$, $t_{x_1 \ldots x_k}(f, W) = 0$ almost everywhere if and only if $t_{x_1 \ldots x_{2k-2q}}(h, W) = 0$ almost everywhere.*

The "only if" part is obvious, since

$$t_{x_1 \ldots x_k}(f, W) = 0 \;\Rightarrow\; t_{x_1 \ldots x_{2k-q}}(f_1, W) = t_{x_1 \ldots x_{2k-q}}(f_2, W) = 0$$
$$\Rightarrow\; t_{x_1 \ldots x_{2k-q}}(f_1 f_2, W) = 0 \;\Rightarrow\; t_{x_1 \ldots x_{2k-2q}}(h, W) = 0.$$

To prove the "if" part, note that two labeled nodes whose labels correspond to the same label in f are never adjacent, so we can identify these labels in h to get f^2 (with the labels in S_r removed). So $t_{x_1 \ldots x_{2k-2q}}(h, W) = 0$ almost everywhere implies by Proposition 13.23 that $t([\![f^2]\!], W) = 0$, and hence we get that $t_{x_1 \ldots x_k}(f, W) = 0$ almost everywhere. This proves the Claim.

Thus it suffices to express the constraint $t_{x_1 \ldots x_{2k-2q}}(h, W) = 0$ by an appropriate unlabeled constraint. This can be done by induction, since $\chi(\mathsf{Lb}(h)) \leq r-1$. □

In some cases, the following simple observation suffices to go between labeled and unlabeled conditions.

LEMMA 14.39. *Let F be a k-labeled signed graph. Then in $\mathcal{W}_0$, the constraints $t_{x_1 \ldots x_k}(F, W) = 0$ and $t([\![F]\!], W) = 0$ define the same graphon variety.*

Proof. Clearly $t_{x_1 \ldots x_k}(F, W) = 0$ implies that $t([\![F]\!], W) = 0$. Conversely, in the constraint

$$t([\![F]\!], W) = \int_{[0,1]^{V(F)}} \prod_{ij \in E_+} W(x_i, x_j) \prod_{ij \in E_-} \bigl(1 - W(x_i, x_j)\bigr) dx = 0$$

the integrand is nonnegative, so it must be 0 almost everywhere. Thus integrating only over the unlabeled nodes, we also get 0 almost everywhere. □

EXAMPLE 14.40. The identically-p graphon J_p is defined by the constraint $t_{xy}(K_2^{\bullet\bullet}, U) = p$. Following the above construction, we get that it can be defined by the constraint $t(C_4(p), U) = 0$, where $C_4(p)$ is obtained from C_4 by replacing each edge by the quantum graph $K_2^{\bullet\bullet} - pO_2$. It is a good exercise to verify that this is equivalent to the conditions $t(K_2, U) = p$ and $t(C_4, U) = p^4$. ◆

We have seen that 0-1 valued graphons do not form a simple variety. On the other hand, there are many interesting varieties whose elements are 0-1 valued. As a first application of Lemma 14.37, we describe a rather general sufficient condition, generalizing Theorem 13.35(a) from graphons to kernels.

LEMMA 14.41. *Let F be a signed bipartite graph on n nodes, all labeled. Suppose that for some $W \in \mathcal{W}$ we have*

(14.7) $$t_{x_1 x_2 \ldots x_n}(F, W) = 0$$

almost everywhere. Then $W(x, y) \in \{0, 1\}$ almost everywhere.

Proof. By Proposition 13.23, (14.7) implies that for the 2-labeled signed bond $B^{\bullet\bullet}$ obtained by identifying each color class of F, we have $t_{xy}(B^{\bullet\bullet}, W) = 0$. This clearly implies that W is 0-1 valued almost everywhere. □

It follows by Lemma 14.39 that if $W \in \mathcal{W}_0$, then it is enough to assume that $t(\llbracket F \rrbracket, W) = 0$, and we get another proof of Theorem 13.35(a).

EXAMPLE 14.42 (**Threshold graphons**). Let $\alpha : [0,1] \to [0,1]$ be a "weight function", and consider the graphon $U_\alpha(x,y) = \mathbb{1}(\alpha(x) + \alpha(y) < 1)$. We call these graphons *threshold graphons*. They have been studied by Diaconis, Holmes and Janson [2008] as limits of threshold graphs.

Every such weight function α has a "monotone reordering" in the form of a measure preserving function $\varphi : [0,1] \to [0,1]$ such that $\alpha \circ \varphi$ is monotone increasing (see Proposition A.19). Then $U_{\alpha \circ \varphi} = (U_\alpha)^\varphi$ is a graphon that is weakly isomorphic to U_α. Furthermore, $U_{\alpha \circ \varphi}$ is monotone decreasing, and clearly 0-1 valued. Conversely, every monotone decreasing 0-1 valued graphon is almost everywhere equal to a threshold graphon (see Exercise 14.53).

Threshold graphons form a simple variety. Let $\widehat{C}_4$ denote a signed 4-labeled 4-cycle, with two opposite edges signed "+", the other two signed "−". Then, a kernel $W \in \mathcal{W}$ is a threshold graphon if and only if

(14.8) $$t_{x_1 x_2 x_2 x_4}(\widehat{C}_4, W) = 0$$

almost everywhere. The necessity of this condition is trivial, for the (elementary) proof of the sufficiency, see Exercise 14.55. It follows by Lemma 14.39 that if $W \in \mathcal{W}_0$, then it is enough to assume that $t(\llbracket \widehat{C}_4 \rrbracket, W) = 0$. ♦

EXAMPLE 14.43 (**Excluded induced subgraphs**). From any class of graphs that is characterized by a finite number of excluded induced subgraphs, we get a graphon variety by taking the closure (this is immediate by Proposition 14.21). It seems that the study of this closure often leads to quite interesting questions, which are mostly unexplored.

As an interesting special case, we can consider graphs not containing an induced path on 4 nodes. These graphs are called *complement reducible*, or *cographs*, and have many interesting properties and characterizations into which we don't go here. The closure consists of graphons W satisfying the equation $t_{\text{ind}}(P_4, W) = 0$. While a complete characterization of such graphons is awkward, it turns out that the regular ones among them (those satisfying $t(P_3, W) = t(K_2 K_2, W)$) have a quite pretty characterization (we refer to Lovász and Szegedy [2011] for details). ♦

14.4.2. Stepfunctions and finite rank kernels. Let $\mathcal{S}_k$ denote the set of kernels that are almost everywhere equal to a stepfunction with k steps.

PROPOSITION 14.44. *The set $\mathcal{S}_k$ is a simple kernel variety, defined by an equation $t(f, .) = 0$, where f is a simple quantum graph whose constituents have at most $(k+1)(k+2)$ nodes.*

Proof. It is clear that the set $\mathcal{S}_k$ is closed under weak isomorphism. Every function $U \in \mathcal{S}_k$ satisfies the following equation:

(14.9) $$\prod_{1 \leq i < j \leq k+1} \bigl(U(x_{ij}, x_i) - U(x_{ij}, x_j)\bigr) = 0$$

for almost all choices of the variables x_i ($1 \leq i \leq k+1$) and x_{ij} ($1 \leq i < j \leq k+1$). Indeed, there are always two variables with one index, say x_i and x_j, which belong to the same step, and then the corresponding factor in (14.9) is 0 for any choice of x_{ij}.

Conversely, suppose that $U \in \mathcal{W}$ is not a stepfunction with k steps. Then there is a set of $(k+1)$-tuples $(x_1, \ldots, x_{k+1})$ with positive measure such that no two of the x_i are twins. For every $(k+1)$-tuple in this set, there is a positive measure of choices for x_{ij} such that $U(x_{ij}, x_i) \neq U(x_{ij}, x_j)$. So (14.9) fails to hold on a set of positive measure.

Now (14.9) can be written as $t_x(g, U) = 0$ for an appropriate simple m-labeled quantum graph g, where $m = k + 1 + \binom{k+1}{2} = \binom{k+2}{2}$. By Lemma 14.37, this can be expressed as $t(g, U) = 0$ with a simple unlabeled quantum graph g. This implies that $\mathcal{S}_k$ is a simple kernel variety. Let us note in addition that every constituent of g is bipartite and fully labeled, hence the construction in the proof of Lemma 14.37 is finished in two steps, and it only doubles the number of nodes. □

We define the *rank* of a kernel W as the rank of the corresponding kernel operator T_W. This is usually infinite, but we will be interested in the cases when it is finite. Since every nonzero eigenvalue of T_W has finite multiplicity, we know that a kernel has finite rank if and only if it has a finite number of distinct eigenvalues. Every stepfunction has finite rank. It is easy to see that the sum and product of two kernels with finite rank have finite rank (see Exercise 14.56).

If the rank r of W is finite, then some of the formulas in Section 7.5 become simpler: the spectral decomposition (7.19) of W will be finite (and therefore, it will hold almost everywhere, not just in L_2):

$$(14.10) \qquad W(x, y) = \sum_{k=1}^{r} \lambda_k f_k(x) f_k(y).$$

The expression (7.25) for the density of a graph F in W also becomes finite:

$$(14.11) \qquad t(F, W) = \sum_{\chi: E \to [r]} \prod_{e \in E} \lambda_{\chi(e)} \prod_{v \in V} M_\chi(f).$$

EXAMPLE 14.45 (**Stepfunctions**). Every stepfunction has finite rank. Indeed, if W is a stepfunction with m steps, then every function in the range of T_W is a stepfunction with the same m steps, and so the dimension of this range is at most m. ♦

EXAMPLE 14.46. If $W(x, y)$ is a (symmetric) polynomial in x and y, then W has finite rank. It suffices to prove this for the elementary symmetric polynomials $x + y$ and xy, in which case the corresponding operators have rank 2 and 1, respectively. ♦

PROPOSITION 14.47. *(a) Kernels with at most m different nonzero eigenvalues form a simple variety. (b) Kernels of rank at most m form a simple variety.*

Proof. (a) We have seen in the proof of Theorem 13.47 that a kernel has at most m different nonzero eigenvalues if and only if the determinant (13.18) vanishes. This determinant can be expressed as $t(f, W)$, where f is a quantum graph (in which every constituent is a disjoint union of cycles). Thus W has at most m nonzero eigenvalues if and only if $t(f, W) = 0$. (b) Let $x_1, \ldots x_m, y_1, \ldots, y_m \in [0, 1]$, and consider the matrix $W[x,y] = (W(x_i,y_j))_{i,j=1}^m$. A kernel has rank at most m if and only if for $\det W[x,y] = 0$ for almost all $x, y \in [0,1]^m$. This condition can be transformed into the form of $t(f, W) = 0$ similarly to the proof of Proposition 14.44. □

Example 14.46 shows that not every kernel with finite rank is a stepfunction. The following theorem asserts that, at least from the point of view of varieties, the two classes are not very far.

THEOREM 14.48. *If a kernel variety contains a kernel with finite rank, then it contains a stepfunction.*

Proof. Let the variety $\mathcal{V}$ be defined by the equations
$$t(g_i, X) = 0 \qquad (i = 1, \ldots, m),$$
and let $W \in \mathcal{V}$ have finite rank r. Let $\mathcal{H}$ be the set of all constituents of the g_i. The equations
$$t(F, X) = t(F, W) \qquad (F \in \mathcal{H})$$
define a kernel variety $\mathcal{V}' \subseteq \mathcal{V}$.

Let $f_1, \ldots, f_r$ be the eigenfunctions of T_W. By (14.11), if $(u_1, \ldots, u_r)$ is another set of bounded measurable functions that satisfy

(14.12) $$M_a(u) = M_a(f)$$

for every vector of exponents $a \in [M]^r$ (where $M = \max\{e(F) : F \in \mathcal{H}\}$), then the kernel $U = \sum_{t=1}^r \lambda_t u_t(x) u_t(y)$ satisfies $t(F, U) = t(F, W)$ for all $F \in \mathcal{H}$ and so $U \in \mathcal{V}'$. By Proposition A.25 in the Appendix there is a system of functions u satisfying (14.12) which are stepfunctions, and then U is also a stepfunction. □

We will see that every stepfunction in $\mathcal{W}$ forms a simple variety in itself (Corollary 16.47). Hence the family of stepfunctions in Theorem 14.48 could not be replaced by any other family of finite rank kernels (e.g., polynomials).

EXERCISE 14.49. Show how the facts that regular graphons form simple varieties and 0-1 valued kernels form a variety (Examples 14.32, 14.34 and 14.36) follow immediately from the unlabeling method.

EXERCISE 14.50. Prove that the union and intersection of two graphon varieties are graphon varieties.

EXERCISE 14.51. Graphons with values in a fixed finite set $S \subseteq \mathbb{R}$ form a variety, but this variety is not simple unless $|S| = 1$.

EXERCISE 14.52. Show that $\mathcal{W}_0$ is not a variety in $\mathcal{W}$.

EXERCISE 14.53. Prove that every monotone decreasing 0-1 valued graphon is almost everywhere equal to a threshold graphon.

EXERCISE 14.54. We say that a kernel W on $[0,1]$ is an *equivalence graphon* if there is a partition $[0,1] = \bigcup_{i \in I} S_1$ into measurable parts such that for almost all pairs $x, y \in [0,1]$, $W(x, y) = \mathbb{1}(i = j)$ if $x \in S_i$ and $y \in S_j$. Prove that equivalence graphons (up to isomorphism modulo a 0 set) form a simple variety in $\mathcal{W}_0$.

EXERCISE 14.55. Assume that a graphon W satisfies (14.8). Prove the following consequences: (a) W is 0-1 valued almost everywhere. (b) Defining $N(x) = \{y \in [0,1] : W(x, y) = 1\}$, either $\lambda(N(x) \setminus N(y)) = 0$ or $\lambda(N(y) \setminus N(x)) = 0$ for all $x, y \in [0,1]$. (c) Defining $\alpha(x) = \frac{1}{2}(1 + \lambda\{z \in [0,1] : \lambda(N(z) \setminus N(x)) = 0\} - \lambda(N(x)))$, we have $\alpha \geq 0$; (d) $W(x, y) = \mathbb{1}(\alpha(x) + \alpha(y) \leq 1)$ almost everywhere.

EXERCISE 14.56. Let U and W be two kernels with finite rank. Prove that the kernels $U+W$, UW, and $U \otimes W$ have finite rank. If U has finite rank and W is arbitrary, then the kernel $U \circ W + W \circ U$ has finite rank.

14.5. Random graphons

Going on with the study of the graphon space, we would like to introduce probability measures on it.

It is not clear that random graphons can be defined at all: in general, one cannot define the notion of a random measurable function on an uncountable set in a useful way. However, the situation is better with unlabeled graphons (equivalence classes of graphons under weak isomorphism): they form a compact metric space, and the σ-algebra $\mathcal{B}$ of Borel subsets in this metric space has good properties (see e.g. Exercise 14.63). A probability distribution on $\mathcal{B}$ will be called a *random graphon model*.

EXAMPLE 14.57 (**Limits of growing graph sequences**). Let $(G_1, G_2, \dots)$ be a sequence of graphs growing under some randomized rule, such that it is convergent with probability 1. Sometimes the limit graphon is uniquely determined: Among our examples of convergent graph sequences, growing uniform attachment graphs (Example 11.39), growing ranked attachment graphs (Exercise 11.47) and growing prefix attachment graphs (Example 11.41) had this property. On the other hand, a growing preferential attachment graph sequence (Example 11.44) converges to a proper distribution on graphons (this distribution is supported on rather simple graphons in this case, namely constant functions where the constant is chosen from an appropriate distribution). ♦

EXAMPLE 14.58 (**The ultimate random graph**). When one learns about random graphs, it is a very counterintuitive fact that they look so much alike: you expect to see a great variety in their edge-densities, triangle-densities, etc., to call them "random" in the everyday sense of word. Well, here is a construction that satisfies this (it is not clear though whether it can be used in actual constructions or proofs, due to the non-constructive steps in the definition).

We start with constructing a random graphon model. Let $(F_1, F_2, \dots)$ be any ordering of all connected simple graphs, up to isomorphism. (For example, we could order them by increasing number of nodes, and within this, randomly). We are going to specify the values of the densities $t(F_i, W)$ one by one as follows. Suppose that we have fixed $a_1 = t(F_1, W), \dots, a_m = t(F_m, W)$, so that the set $\mathcal{W}_{a_1 \dots a_m}$ of graphons satisfying all these conditions is nonempty. Since the functionals $t(F, .)$ are continuous on $(\widetilde{\mathcal{W}_0}, \delta_\square)$, this set is closed, and hence compact. Consider the set

$$I_{a_1 \dots a_m} = \{t(F_{m+1}, W) : W \in \mathcal{W}_{a_1 \dots a_m}\}.$$

This is a compact subset of $[0,1]$, and hence measurable. If it has positive measure, we choose from it uniformly a random real number for a_{m+1}. If it is of measure 0, then we choose its minimal element as a_{m+1}.

When we are done with our infinitely many choices, we have a system of equations $t(F_i, W) = a_i$, $i = 1, 2, \dots$. By construction, any finite number of these are satisfiable, so by the compactness of the graphon space, all of them are simultaneously satisfiable. Furthermore, since the subgraph densities determine the graphon up to weak isomorphism, we get a unique unlabeled random graphon **W**.

This defines a probability distribution on graphons. Now if we want a large simple "ultimately random" graph, we construct $\mathbb{G}(n, \mathbf{W})$.

This is of course a toy example at the moment. It would be interesting to see if this or perhaps other, similarly constructed "ultimately random" graphs have any interesting applications. ♦

14.5.1. Equivalent descriptions of a random graphon. The following theorem (combining results from Diaconis and Janson [2008] and Lovász and Szegedy [2012a]) characterizes random graphon models. Another way of looking at it is that it shows that for each of the objects in Theorem 11.52 we can relax the conditions to get another important family of cryptomorphic structures. For example, to relax the condition of multiplicativity (for a normalized parameter), we consider isolate-indifferent paremeters (invariant under deletion of isolated nodes). A motivation for the introduction of random graphon models is that they give a representation of graph parameters that are more general than those represented by graphons.

THEOREM 14.59. *The following are equivalent (cryptomorphic):*

(a) an isolate-indifferent, normalized simple graph parameter with nonnegative upper Möbius inverse;

(b) a consistent random graph model;

(c) a countable random graph model;

(d) a random graphon model.

Proof. The proof is similar to the proof of Theorem 11.52. The construction (a)→(b)→(c) is essentially the same as (b)→(c)→(d) there. The remaining two steps need some modification.

(c)→(d). Let $\mathbf{G}$ be a random countable graph from a consistent countable random graph model. Let $\mathbf{G}_n$ be the finite graph spanned by the first n nodes of $\mathbf{G}$. By Proposition 11.14, the graph sequence $(\mathbf{G}_n)$ is almost certainly convergent. Thus by Theorem 11.52, it tends to a limit graphon $\mathbf{W}$ with probability 1.

This way we have described a method to generate a random graphon $\mathbf{W}$. It is not hard to check that the distribution of $\mathbf{W}$ is a probability measure on the Borel sets of $(\widetilde{\mathcal{W}}_0, \delta_\square)$. For every graph F, this random graphon satisfies

$$t(F, \mathbf{W}) = \lim_{n\to\infty} t(F, \mathbf{G}_n) = \lim_{n\to\infty} t_{\text{inj}}(F, \mathbf{G}_n).$$

By the consistency of $\mathbf{G}$, the expectation of $t_{\text{inj}}(F, \mathbf{G}_n)$ is independent of n for $n \geq k = |V(F)|$, and so

$$\mathsf{E}\big(t(F, \mathbf{W})\big) = \lim_{n\to\infty} \mathsf{E}\big(t_{\text{inj}}(F, \mathbf{G}_n)\big) = \mathsf{E}\big(t_{\text{inj}}(F, \mathbf{G}_k)\big) = \mathsf{P}(F \subseteq \mathbf{G}_k).$$

(d)→(a). Let $\mathbf{W}$ be a random graphon from any random graphon model. This defines a graph parameter f by

$$f(F) = \mathsf{E}\big(t(F, \mathbf{W})\big).$$

For every fixed graphon W, the graph parameter $f(\cdot) = t(\cdot, W)$ is normalized, isolate-indifferent (since it is multiplicative), and has nonnegative Möbius inverse (by Theorem 11.52). Trivially, these properties are inherited by the expectation. □

14.5.2. Equivalent properties of the descriptions. Graph parameters as well as random graph models in Theorems 11.52 and 14.59 have a number of alternative characterizations. The next Propositions give some of these.

PROPOSITION 14.60. *Let f be an isolate-indifferent, normalized simple graph parameter. Then the following are equivalent:*

(a) *f is reflection positive;*

(b) *f is flatly reflection positive (i.e., connection matrices for fully labeled graphs are semidefinite);*

(c) *f is nonnegative on signed graphs;*

(d) *f has nonnegative Möbius inverse;*

(e) *$f = \mathsf{E}(t(.,\mathbf{W}))$, where $\mathbf{W}$ is a random graphon from a random graphon model;*

(f) *f is in the convex hull of the functions $t(.,W)$, where W is a graphon.*

Proof. (a) The implications $(a) \Rightarrow (b)$, $(c) \Rightarrow (d)$ and $(e) \Rightarrow (f)$ are trivial.

$(b) \Rightarrow (c)$: The Lindström–Wilf Formula (see (A.1) in the Appendix) gives the following diagonalization of $M^{\text{flat}}(f,k)$: Let $\mathcal{F}'_k$ denote the set of fully labeled simple graphs on $[k]$. Let Z denote the $\mathcal{F}'_k \times \mathcal{F}'_k$ matrix defined by $Z_{F_1,F_2} = \mathbb{1}_{F_1 \subseteq F_2}$. Let D be the diagonal matrix with $D_{F,F} = f^\uparrow(F)$. Then $M^{\text{flat}}(f,k) = Z^\mathsf{T} D Z$. This implies that $M^{\text{flat}}(f,k)$ is positive semidefinite if and only if $f^\uparrow(F) \geq 0$ for all graphs with k nodes.

$(d) \Rightarrow (c)$: Let $F = (V, E_1, E_2)$ be a signed graph. Then an easy computation shows that

$$f(F) = \sum_{\substack{E_1 \subseteq E(F') \subseteq E_1 \cup E_2 \\ V(F')=V}} (-1)^{\mathsf{e}(F')-|E_1|} f(F') = \sum_{\substack{E_1 \subseteq E(F'), E_2 \cap E(F')=\emptyset \\ V(F')=V}} f^\uparrow(F') \geq 0.$$

$(c) \Rightarrow (e)$: Let f be an isolate-indifferent, normalized graph parameter with nonnegative Möbius inverse. By Theorem 14.59, it defines a random graphon $\mathbf{W}$ such that $f = \mathsf{E}(t(\cdot, \mathbf{W}))$.

$(f) \Rightarrow (a)$: Every function $t(.,W)$ is reflection positive, and this is clearly inherited by their convex hull. $\square$

Among graph parameters described above, the multiplicative ones have different characterizations as well.

PROPOSITION 14.61. *Let f be a graph parameter satisfying the conditions of Proposition 14.60. Then the following are equivalent:*

(c) *f is multiplicative;*

(b) *$f = t(.,W)$, where W is a graphon;*

(c) *f is the limit of homomorphism density functions $t(.,G)$, where G is a simple graph.*

Proof. Clearly both (b) and (c) imply (a). Conversely, if f is multiplicative, then by Theorem 11.52, there is graphon W such that $f = \mathsf{E}(t(.,W))$. By Corollary 11.15, the function $t(.,W)$ is the limit of homomorphism density functions for every W. $\square$

The following proposition describes a connection between graph-theoretic and group-theoretic properties of countable random graph models, and indicate a connection with ergodic theory. Recall from Section 11.2.2 that a countable random graph model is given by a probability measure on the sigma algebra $\mathcal{A}$ on the set of graphs on $\mathbb{N}^*$, invariant under permutations of $\mathbb{N}^*$ (for short, an invariant measure). We say that an invariant measure μ is *ergodic*, if no subset $T \in \mathcal{A}$ with $0 < \mu(T) < 1$ is invariant under permutations of the nodes.

We note that invariant measures on $\mathcal{A}$ form a convex set K, and μ is ergodic if and only if it is an extreme point of K. Indeed, if we can write $\mu = \frac{1}{2}(\mu_1 + \mu_2)$, where μ_1, μ_2 are different invariant measures, then the positive support of the signed measure $\mu_1 - \mu_2$ is an invariant set in $\mathcal{A}$ with positive measure. Conversely, if $T \in \mathcal{A}$ is an invariant set with $0 < \mu(T) < 1$, then μ is a convex combination of the invariant measures $\mu_1(X) = \mu(X \cap T)/\mu(T)$ and $\mu_2 = \mu(X \setminus T)/(1 - \mu(T))$.

PROPOSITION 14.62. *A countable random graph model is local if and only if it is ergodic.*

Proof. Let μ be an invariant measure on $\mathcal{A}$. First, suppose that μ is ergodic, then μ is an extreme point of K. By Theorem 14.59, there is a probability distribution π on graphons representing μ in the following sense: if $\mathbf{W}$ is a random graphon drawn from this distribution, then $\mathbb{G}(\mathbb{N}^*, \mathbf{W})$ has distribution μ. It follows that μ is a convex combination of distributions $\mathbb{G}(\mathbb{N}^*, W)$ where W is in the support of π. Since μ is an extreme point of K, the distribution π must be concentrated on a single graphon W_0. Thus μ is the distribution of $\mathbb{G}(\mathbb{N}^*, W_0)$, and Theorem 11.52 implies that μ is local.

Conversely, if μ is not ergodic, then we can write it as $\mu = \frac{1}{2}(\mu_1 + \mu_2)$, where $\mu_1, \mu_2 \in K$ and $\mu_1 \neq \mu_2$. Let $\mathbb{G}_1$ and $\mathbb{G}_2$ be random countable graphs from the distributions μ_1 and μ_2, respectively, and let $\mathbb{G}$ be $\mathbb{G}_1$ with probability $1/2$ and $\mathbb{G}_2$ with probability $1/2$. Let $S, T \subseteq \mathbb{N}^*$ be finite sets with $|T| = |S|$ and $T \cap S = \emptyset$, and consider a labeled graph F on $|S|$ nodes such that $\mathsf{P}(\mathbb{G}_1[S] = F) \neq \mathsf{P}(\mathbb{G}_2[S] = F)$. Set $a_1 = \mathsf{P}(\mathbb{G}_1[S] = F) = \mathsf{P}(\mathbb{G}_1[T] = F)$ (by invariance, these two probabilities are equal), and define a_2 analogously.

Thus we have

$$\mathsf{P}(\mathbb{G}[S] = F, \mathbb{G}[T] = F) - \mathsf{P}(\mathbb{G}[S] = F)\mathsf{P}(\mathbb{G}[T] = F)$$
$$= \frac{1}{2}\Big(\mathsf{P}(\mathbb{G}_1[S] = F, \mathbb{G}_1[T] = F) + \mathsf{P}(\mathbb{G}_2[S] = F, \mathbb{G}_2[T] = F)\Big)$$
$$- \frac{1}{4}\Big(\mathsf{P}(\mathbb{G}_1[S] = F) + \mathsf{P}(\mathbb{G}_2[S] = F)\Big)\Big(\mathsf{P}(\mathbb{G}_1[T] = F) + \mathsf{P}(\mathbb{G}_2[T] = F)\Big)$$
$$= \frac{1}{2}(a_1^2 + a_2^2) - \frac{1}{4}(a_1 + a_2)^2 = \frac{1}{4}(a_1 - a_2)^2 > 0.$$

This shows that the events $\mathbb{G}[S] = F$ and $\mathbb{G}[T] = F$ are not independent, and hence μ is not local. $\square$

EXERCISE 14.63. Prove that the sigma-algebra of Borel sets in the metric space $(\widetilde{\mathcal{W}}_0, \delta_\square)$ is generated by the "semivarieties" $\mathcal{S}(F, a) = \{W \in \widetilde{\mathcal{W}}_0 : t(F, W) \geq a\}$, where F is a simple graph and a is a rational number.

14.6. Exponential random graph models

In this last section of our study of the graphon space, we sketch some results in the theory of random graphs where viewing the graphon space as a single compact metric space is crucial. Chatterjee and Varadhan [2011] applied the theory of graph limits to the theory of large deviations for Erdős–Rényi random graphs. This was extended by Chatterjee and Diaconis [2012] to more general distributions on graphs, which they call *exponential random graph models*. We summarize their ideas without going into the details of the proofs.

Let f be a bounded graph parameter such that for every convergent graph sequence (G_n), the numerical sequence $f(G_n)$ is convergent. Such parameters are called *estimable*. The canonical examples of such parameters are subgraph densities $t(F.,)$, but there are many others. We will return to them in Section 15.1 to study their estimation through sampling and other characterizations. Right now we only need the fact that every such parameter can be extended to the graphon space $\widetilde{\mathcal{W}}_0$ so that if $G_n \to W$ then $f(G_n) \to f(W)$, and the extension is continuous in the distance $\delta_\square$ (in particular, the extension is invariant under weak isomorphism). These facts are immediate consequences of the definition.

Suppose that we want to understand the structure of a random graph, but under the condition that $f(G)$ is small. For example, Chatterjee and Varadhan were interested in random graphs $\mathbb{G}(n, 1/2)$ in which the triangle density is much less than $1/8$ (the expectation). To this end, we consider a weighting of all simple graphs on n nodes by $e^{-f(G)n^2}$; this will emphasize those graphs for which $f(G)$ is small. The factor n^2 in the exponent is needed to make the logarithms of the weights to have the same order of magnitude as the logarithm of the total number of simple graphs on $[n]$ (which is just $\binom{n}{2}$, if you take binary logarithm). We introduce the probability distribution φ_n on $\mathcal{F}_n^{\mathrm{simp}}$ by

$$\varphi_n(G) = \frac{e^{-f(G)n^2}}{\sum_{G' \in \mathcal{F}_n^{\mathrm{simp}}} e^{-f(G')n^2}}.$$

Let ψ_n denote the normalizing factor in the denominator. It looks quite hairy, but Chatterjee and Diaconis derived an asymptotic formula for it. To state their result, we need some notation. For $W \in \mathcal{W}_0$, consider the entropy-like functional

$$I(W) = \frac{1}{2} \int_{[0,1]^2} W(x,y) \log W(x,y) + \big(1 - W(x,y)\big) \log\big(1 - W(x,y)\big).$$

Chatterjee and Varadhan proved that this functional is invariant under weak isomorphism and lower semicontinuous on the space $(\widetilde{\mathcal{W}}_0, \delta_\square)$ (this fact is quite similar to Lemma 14.16).

The formula of Chatterjee and Diaconis can be stated as follows:

THEOREM 14.64. *If f is an estimable graph parameter, then*

$$\lim_{n \to \infty} \psi_n = \sup_{W \in \mathcal{W}_0} \big(f(W) - I(W)\big). \quad \square$$

Using this, they prove the following result about the behaviour of a random graph drawn from the distribution φ_n. Since $f(W) - I(W)$ is upper semicontinuous on the compact space $(\widetilde{\mathcal{W}}_0, \delta_\square)$, the supremum in the above formula is in fact a maximum, and it is attained on a compact set $K_f \subseteq \mathcal{W}_0$.

THEOREM 14.65. *Let f be an estimable graph parameter, and let $\mathbf{G}_n$ be a random graph from the distribution φ_n. Then for every $\eta > 0$ there are $C, \varepsilon > 0$ such that*

$$\mathsf{P}(\delta_\square(W_{\mathbf{G}_n}, K_f) > \eta) \leq Ce^{-\varepsilon n^2}. \quad \square$$

This implies that if we choose a random $\mathbf{G}_n$ from φ_n for every n, then $\delta_\square(W_{\mathbf{G}_n}, K_f) \to 0$ with probability 1. If K_f consists of a single graphon W_0 (which is in a sense the "generic" case), then $\mathbf{G}_n \to W_0$ with probability 1.

Theorems 14.64 and 14.65 provide a framework for analyzing the behavior of exponential random graph models. This is by no means easy, and the results are interesting. Most work has been done for the parameters of the form $f(G) = \beta_1 t(K_2, G) + \beta_2 t(K_3, G)$ (extending, in a sense, our discussions in Sections 2.1.1 and 16.3.2). We refer to Chatterjee and Diaconis [2012], Aristoff and Radin [2012], and Radin and Yin [2012] for recent results.

CHAPTER 15

Algorithms for large graphs and graphons

We have seen in the Introduction that different kinds of algorithmic questions can be asked for a very large graph: we may want to estimate a parameter, test a property, or compute (in some sense) an additional structure for the graph. We sneak in a fourth one, making a distinction between "property distinction" and "property testing". We will see that the theory of graph limits and other methods developed in this book provide valuable tools for the theoretical understanding of all these types of algorithmic problems.

15.1. Parameter estimation

We want to determine some parameter of a very large graph G. Of course, we'll not be able to determine the exact value of this parameter; the best we can hope for is that if we take a sufficiently large sample, we can find the approximate value of the parameter with high probability.

To be precise, a graph parameter f is *estimable*, if for every $\varepsilon > 0$ there is a positive integer k such that if G is a graph with at least k nodes and we select a random k-set X of nodes of G, then from the subgraph $G[X]$ induced by them we can compute an estimate $g(G[X])$ of f such that

(15.1) $$\mathsf{P}(|f(G) - g(G[X])| > \varepsilon) < \varepsilon.$$

We call the parameter g a *test parameter for f*. However, we don't really need this notion: we can always use $g = f$ (cf. Goldreich and Trevisan [2003]). Indeed, (15.1) implies that $\mathsf{P}(|f(G[X]) - g(G[X])| > \varepsilon) < \varepsilon$, and so

$$\mathsf{P}(|f(G) - f(G[X])| > 2\varepsilon) \leq \mathsf{P}(|f(G) - g(G[X])| > \varepsilon) \\ + \mathsf{P}(|g(G[X]) - f(G[X])| > \varepsilon) < \varepsilon + \varepsilon = 2\varepsilon,$$

so we can choose the threshold k belonging to $\varepsilon/2$ in the original definition to get the condition obtained by replacing g by f.

It is easy to see that estimability is equivalent to saying that for every convergent graph sequence (G_n), the sequence of numbers $\big(f(G_n)\big)$ is convergent. (So graph parameters of the form $t(F,.)$ are estimable by the definition of convergence.) Using this, for any estimable parameter f we can define a functional $\widehat{f}$ on $\mathcal{W}_0$, where $\widehat{f}(W)$ is the limit of $f(G_n)$ for any sequence of simple graphs $G_n \to W$. It is also immediate that this functional $\widehat{f}$ is continuous on $(\widetilde{\mathcal{W}}_0, \delta_\square)$. The functional $\widehat{f}$ does not determine the graph parameter f: defining $f_0(G) = \widehat{f}(W_G)$ we get a graph parameter with $\widehat{f_0} = \widehat{f}$, but f could be any parameter of the form $f_0 + h$, where $h(G) \to 0$ if $\mathsf{v}(G) \to \infty$.

All this is, however, more-or-less just a reformulation of the definition. Borgs, Chayes, Lovász, Sós and Vesztergombi [2008] gave a number of more useful conditions characterizing testability of a graph parameter. We formulate one, which is perhaps easiest to verify for concrete parameters.

THEOREM 15.1. *A graph parameter f is estimable if and only if the following three conditions hold:*

(i) *If G_n and G'_n are simple graphs on the same node set ($n = 1, 2, \ldots$) and $d_\square(G_n, G'_n) \to 0$, then $f(G_n) - f(G'_n) \to 0$.*

(ii) *For every simple graph G, $f(G(m))$ has a limit as $m \to \infty$ (recall that $G(m)$ denotes the graph obtained from G by blowing up each node into m twins).*

(iii) *$f(GK_1) - f(G) \to 0$ if $\mathsf{v}(G) \to \infty$ (recall that GK_1 is obtained from G by adding a single isolated node).*

Note that all three conditions are special cases of the statement that

(iv) *if $|V(G_n)|, |V(G'_n)| \to \infty$ and $\delta_\square(G_n, G'_n) \to 0$, then $f(G_n) - f(G'_n) \to 0$.*

This condition is also necessary, so it is equivalent to its own three special cases (i)–(iii) in the Theorem.

Proof. The necessity of condition (iv) (which implies (i)–(iii)) is easy: Suppose that there are two sequences of graphs G_n such that $|V(G_n)|, |V(G'_n)| \to \infty$ and $\delta_\square(G_n, G'_n) \to 0$, but $f(G_n) - f(G'_n) \not\to 0$. By selecting a subsequence, we may assume that $|f(G_n) - f(G'_n)| > \varepsilon$ for all n for some $\varepsilon > 0$. Going to a further subsequence, we may assume that the sequences $(G_1, G_2, \ldots)$ and $(G'_1, G'_2, \ldots)$ are convergent. But then $\delta_\square(G_n, G'_n) \to 0$ implies that the interlaced graph sequence $(G_1, G'_1, G_2, G'_2, \ldots)$ is convergent as well. However, the numerical sequence $(f(G_1), f(G'_1), f(G_2), f(G'_2), \ldots)$ is not convergent, a contradiction.

To prove the sufficiency of (i)–(iii), we start with proving the following stronger form of (i):

(i') *If G_n and G'_n are simple graphs with the same number of nodes ($n = 1, 2, \ldots$) and $\delta_\square(G_n, G'_n) \to 0$, then $f(G_n) - f(G'_n) \to 0$.*

This follows by Theorem 9.29, which implies that one can overlay the graphs G_n and G'_n so that $d_\square(G_n, G'_n) \to 0$.

Consider a convergent graph sequence $(G_1, G_2, \ldots)$, we prove that the sequence $(f(G_1), f(G_2), \ldots)$ is convergent. Let $\varepsilon > 0$. Using (i'), we can choose an $\varepsilon_1 > 0$ so that if $\delta_\square(G, G') \leq \varepsilon_1$, then $|f(G) - f(G')| \leq \varepsilon$. Since the graph sequence is convergent, we can choose and fix an integer $n \geq 1$ so that $\delta_\square(G_n, G_m) \leq \varepsilon_1/2$ for $m \geq n$. By (ii), the sequence $\bigl(f(G_n(p)) : p = 1, 2, \ldots\bigr)$ is convergent. Let a be its limit, then we can choose a threshold $p_0 \geq 1$ so that $\bigl|f(G_n(p)) - a\bigr| \leq \varepsilon$ for every integer $p \geq p_0$. We may assume that $p_0 \geq 4/\varepsilon_1$. Finally, based on (iii), we can chose a threshold $q \geq 1$ such that $|f(GK_1) - f(G)| \leq \varepsilon/\mathsf{v}(G)$ whenever $\mathsf{v}(G) \geq q$.

Now consider a member G_m of the sequence for which $m \geq n$ and $\mathsf{v}(G_m) \geq \max(q, p_0 \mathsf{v}(G_n), 4\mathsf{v}(G_n)/\varepsilon_1)$. We can write $\mathsf{v}(G_m) = p\mathsf{v}(G_n) + r$, where $p \geq p_0$ and $0 \leq r < \mathsf{v}(G_n)$. Then G_m and $G' = G_n(p)K_1^r$ have the same number of nodes. Furthermore,

$$\delta_\square(G_m, G') \leq \delta_\square\bigl(G_m, G_n(p)\bigr) + \delta_\square\bigl(G_n(p), G'\bigr)$$
$$\leq \delta_\square(G_m, G_n) + \frac{2r}{p\mathsf{v}(G_n)} \leq \varepsilon_1.$$

Hence $|f(G_m) - f(G')| \leq \varepsilon$, and so
$$|f(G_m) - a| \leq |f(G_m) - f(G')| + |f(G') - f(G_n(p))| + |f(G_n(p)) - a|$$
$$\leq \varepsilon + r\frac{\varepsilon}{\mathsf{v}(G_n)} + \varepsilon < 3\varepsilon.$$

So for any two indices m_1 and m_2 that are large enough, we have $|f(G_{m_1}) - f(G_{m_2})| < 6\varepsilon$, which proves that f is estimable. □

EXAMPLE 15.2 (**Maximum cut**). As a basic example, consider the density of maximum cuts (recall Example 5.18). One of the first substantial results on property testing (Goldreich, Goldwasser and Ron [1998], Arora, Karger and Karpinski [1995]) is that this parameter is estimable. In the introduction we gave an argument (which can be made precise using high concentration results like Azuma's inequality) that if S is a sufficiently large random subset of nodes of G, then $\mathsf{maxcut}(G[S]) \geq \mathsf{maxcut}(G) - \varepsilon$: a large cut in G, when restricted to S, gives a large cut in $G[S]$. It is harder, and in fact quite surprising, that if most subgraphs $G[S]$ have a large cut, then so does G. This follows from Theorem 15.1 above, since conditions (i)–(iii) are easily verified for $f = \mathsf{maxcut}$. ◆

EXAMPLE 15.3 (**Free energy**). The "free energy" is a statistical physical quantity. Recall the definition of the energy of a map $\sigma : V(G) \to [q]$ (a configuration) from the introduction:
$$(15.2) \qquad H(\sigma) = -\sum_{uv \in E(G)} J_{\varphi(u),\varphi(v)},$$

and also the partition function
$$(15.3) \qquad Z = \sum_{\sigma:V(G)\to[q]} e^{-H(\sigma)/T},$$

where T is the temperature (for simplicity, we don't consider an external field). The *mean field partition function* of G can be obtained (formally) by considering a very high temperature:
$$(15.4) \qquad Z_{\mathrm{mean}} = \sum_{\sigma:V(G)\to[q]} e^{-H(\sigma)/\mathsf{v}(G)}.$$

The *free energy* is defined by
$$(15.5) \qquad \mathbf{F}(G,H) = -\frac{\ln Z(G,H)}{\mathsf{v}(G)}.$$

It would exceed the framework of this book to explain the physics behind these names; let us just treat them as graph parameters related to homomorphism numbers. Note that the normalization is different from (2.15) in the exponent and therefore we only divide by $\mathsf{v}(G)$ (as opposed to (5.33), for example). For more about this connection, we refer to Borgs, Chayes, Lovász, Sós and Vesztergombi [2012].

The free energy (for a fixed weighted graph H) is a more complicated example of a estimable parameter, which illustrates the power of Theorem 15.1. It is difficult to verify directly either the definition, or say condition (iv). The theorem splits this task into three: condition (i) is easy by the definition of $d_\square(G,G')$; (ii) is an exercise in classical combinatorics, in which we have to count mappings that split the twin classes in given proportions; finally, (iii) is trivial. ◆

EXERCISE 15.4. Show that neither one of the three conditions in Theorem 15.1 can be dropped.

EXERCISE 15.5. Use Theorem 15.1 to prove that the i-th largest eigenvalue of a graph is an estimable parameter for every fixed $i \geq 1$. Give a new proof of Theorem 11.54 based on this argument.

EXERCISE 15.6. Fix a graphon W, then $\mathsf{cut}(W, F)$, as a function of F, defines a simple graph parameter. Prove that it is estimable.

15.2. Distinguishing graph properties

An algorithmic task of different nature is to test whether a given graph G has a certain property (e.g., is it connected, bipartite, or perfect). Before (e.g. in Section 4.3) we considered graph properties as 0-1 valued graph parameters. In this case, this would not be a useful approach, at least not if we wanted to reduce the question to parameter estimation: getting a 0-1 valued parameter with less than 50 percent error is tantamount to getting it exactly, which is clearly too much to require for very large graphs. Therefore we have to modify the definition to allow an error (which is then equal to 1, so very large in our setting), but with small probability.

We start with a version of the problem that turns out simpler: how to distinguish two properties? Deciding whether a graph has a given property $\mathcal{P}$ will be then treated as distinguishing graph property $\mathcal{P}$ from the set of graphs that are far from having this property.

Let $\mathcal{P}_1$ and $\mathcal{P}_2$ be two graph properties that are exclusive (i.e., $\mathcal{P}_1 \cap \mathcal{P}_2 = \emptyset$). We want to design a method that, given an arbitrarily large graph G, looks at a sample of some fixed size k, and guesses whether the graph has property $\mathcal{P}_1$ or $\mathcal{P}_2$. If the graph has property $\mathcal{P}_1$ or property $\mathcal{P}_2$, we would like to guess right with rather high probability, say at least $2/3$. If the graph does not have either one of the properties, then we don't care what the guess was.

The guess will be based on a third graph property $\mathcal{Q}$, which we call the *test property*. If the sample $\mathbb{G}(k, G)$ has property $\mathcal{Q}$, we guess that G has property $\mathcal{P}_1$; else, we guess that G has property $\mathcal{P}_2$. So the precise definition is the following: we call properties $\mathcal{P}_1$ and $\mathcal{P}_2$ *distinguishable by sampling*, if there exists a positive integer k and a test property $\mathcal{Q}$ such that for every graph G with at least k nodes

$$\mathsf{P}\big(\mathbb{G}(k, G) \in \mathcal{Q}\big) \begin{cases} \geq \dfrac{2}{3}, & \text{if } G \in \mathcal{P}_1, \\ \leq \dfrac{1}{3}, & \text{if } G \in \mathcal{P}_2. \end{cases}$$

The following lemma shows that the numbers $1/3$ and $2/3$ in this definition are arbitrary; we could replace them with any two numbers a and b with $0 < a < b < 1$. Furthermore, we can replace the number k by every sufficiently large integer.

LEMMA 15.7. *Let $\mathcal{P}_1$ and $\mathcal{P}_2$ be two graph properties with $\mathcal{P}_1 \cap \mathcal{P}_2 = \emptyset$. Let $0 < a < b < 1$ and $0 < c < d < 1$. Suppose that there is a positive integer k and a test property $\mathcal{Q}$ such that for every graph G with at least k nodes*

$$\mathsf{P}\big(\mathbb{G}(k, G) \in \mathcal{Q}\big) \begin{cases} \geq b & \text{if } G \in \mathcal{P}_1, \\ \leq a & \text{if } G \in \mathcal{P}_2. \end{cases}$$

Then for every positive integer k' that is large enough there is a test property $\mathcal{Q}'$ such that for every graph G with at least k' nodes

$$\mathsf{P}\bigl(\mathbb{G}(k',G)\in\mathcal{Q}'\bigr)\begin{cases}\geq d & \text{if } G\in\mathcal{P}_1,\\ \leq c & \text{if } G\in\mathcal{P}_2.\end{cases}$$

Proof. Let $k' > k$. For every graph F on k' nodes, let

$$f(F) = \mathsf{P}(\mathbb{G}(k,F)\in\mathcal{Q}),$$

and define the property $\mathcal{Q}'$ as the set of graphs F on k' nodes such that $f(F) \geq (a+b)/2$.

Let $G \in \mathcal{P}_1$, $\mathsf{v}(G) \geq k'$. Since $\mathbb{G}\bigl(k,\mathbb{G}(k',G)\bigr)$ is a random k-node subgraph of G, we have

$$f_0 = \mathsf{E}\bigl(f(\mathbb{G}(k',G))\bigr) = \mathsf{P}(\mathbb{G}(k,G)\in\mathcal{Q}) \geq b.$$

Furthermore, the graph parameter f has the property that if we change edges in F incident with a given node v, then the validity of the event $\mathbb{G}(k,F) \in \mathcal{Q}$ changes only if the random k-subset contains v, which happens with probability k/k'. So the value $f(F)$ changes by at most k/k'. We can apply the Sample Concentration Theorem 10.2 to the parameter $(k'/k)f$, and get that

$$\mathsf{P}\bigl(\mathbb{G}(k',G)\notin\mathcal{Q}'\bigr) = \mathsf{P}\Bigl(f(\mathbb{G}(k',G)) \leq \frac{a+b}{2}\Bigr) \leq \mathsf{P}\Bigl(f(\mathbb{G}(k',G)) \leq f_0 - \frac{b-a}{2}\Bigr) \leq e^{-t},$$

where $t = (b-a)^2 k'/(8k^2)$. Choosing k' large enough, this will be less than $1-d$, proving that $\mathcal{Q}'$ and k' satisfy the first condition in the lemma. The second condition follows similarly. $\square$

Using our Sampling Lemmas, we can give the following characterization of distinguishable properties.

THEOREM 15.8. *For two graph properties $\mathcal{P}_1$ and $\mathcal{P}_2$, the following are equivalent:*

(a) *$\mathcal{P}_1$ and $\mathcal{P}_2$ are distinguishable by sampling;*

(b) *there exists a positive integer k such that for any $G_i \in \mathcal{P}_i$ with $\mathsf{v}(G_i) \geq k$, we have $\delta_\square(G_1, G_2) \geq 1/k$;*

(c) *there exists a positive integer k such that for any $G_i \in \mathcal{P}_i$ with $\mathsf{v}(G_i) \geq k$, we have*

$$d_{\mathrm{var}}\bigl(\mathbb{G}(k,G_1), \mathbb{G}(k,G_2)\bigr) \geq \frac{1}{3}.$$

Note that (b) could be phrased as $\overline{\mathcal{P}_1} \cap \overline{\mathcal{P}_2} = \emptyset$.

Proof. (a)$\Rightarrow$(c): Let $\mathcal{P}_1$ and $\mathcal{P}_2$ be distinguishable with sample size k and test property $\mathcal{Q}$. Then for any two graphs $G_1, G_2 \in \mathcal{P}_i$, we have

$$\mathsf{P}\bigl(\mathbb{G}(k,G_1)\in\mathcal{Q}\bigr) - \mathsf{P}\bigl(\mathbb{G}(k,G_2)\in\mathcal{Q}\bigr) \geq \frac{1}{3}.$$

On the other hand,
$$\mathsf{P}\big(\mathbb{G}(k,G_1) \in \mathcal{Q}\big) - \mathsf{P}\big(\mathbb{G}(k,G_2) \in \mathcal{Q}\big)$$
$$= \sum_{F \in \mathcal{Q}} \Big(\mathsf{P}\big(\mathbb{G}(k,G_1) = F\big) - \mathsf{P}\big(\mathbb{G}(k,G_2) = F\big)\Big)$$
$$\leq \sum_F \big|\mathsf{P}\big(\mathbb{G}(k,G_1) = F\big) - \mathsf{P}\big(\mathbb{G}(k,G_2) = F\big)\big|_+$$
$$= d_{\mathrm{var}}\big(\mathbb{G}(k,G_1), \mathbb{G}(k,G_2)\big).$$

This proves (c).

(c)⇒(b): This follows immediately from the Counting Lemma 10.22.

(b)⇒(a): Let $\delta_\square(\overline{\mathcal{P}_1}, \overline{\mathcal{P}_2}) = c > 0$. Let k be large enough, and define the test property $\mathcal{Q}$ by

(15.6) $$\mathcal{Q} = \Big\{F :\ \mathsf{v}(F) = k,\ \delta_\square(F, \mathcal{P}_1) \leq \frac{c}{2}\Big\}.$$

To see that this is a valid test property, consider any graph $G \in \mathcal{P}_1$ with $\mathsf{v}(G) \geq k$. By the Second Sampling Lemma 10.15, we have with probability at least $2/3$ that

$$\delta_\square\big(G, \mathbb{G}(k,G)\big) \leq \frac{20}{\sqrt{\log k}} < \frac{c}{2},$$

and if this happens, then $\mathbb{G}(k,G) \in \mathcal{Q}$. The condition on graphs in $\mathcal{P}_2$ follows similarly. □

As a consequence of this proof, we can make the following observation: While the test property $\mathcal{Q}$ looks like the key to this testing method, it can in fact be chosen in a very specific way, in the form (15.6).

We can also talk about distinguishing two graphon properties by testing. We assume that we can get information about a given graphon W by generating a W-random graph $\mathbb{G}(k,W)$. All the above can be repeated in this model, and we will not go through the details.

15.3. Property testing

Testing for a single property is a more complicated business than distinguishing two properties. There is a large literature on this subject; here we restrict our attention to aspects that are related to graph limit theory, based on Lovász and Szegedy [2010a]. We start with a discussion of testing for a graphon property (which is easily handled using the results about distinguishing two properties), and then add the necessary work to apply this to testing graph properties.

15.3.1. Testable graphon properties.
We call a graphon property $\mathcal{R}$ *testable* if it is closed in the $\delta_\square$ metric, and there is a graph property $\mathcal{R}'$ (called the *test property* for $\mathcal{R}$), such that

(a) for every graphon $W \in \mathcal{R}$ and every $k \geq 1$, we have $\mathbb{G}(k,W) \in \mathcal{R}'$, with probability at least $2/3$, and

(b) for every $\varepsilon > 0$ there is a $k_\varepsilon \geq 1$ such that for every graphon W with $d_1(W, \mathcal{R}) > \varepsilon$ and every $k \geq k_\varepsilon$ we have $\mathbb{G}(k,W) \notin \mathcal{R}'$ with probability at least $2/3$.

The definition above is clearly one-sided, there is no symmetry between the property and its complement. Several other versions could be defined, but we'll restrict our attention to this one. We could require (a) only for $k \geq k_0$, but we can put all smaller graphs into $\mathcal{R}'$ for free, so this would not give anything different. On the other hand, we need a threshold in (b) to depend on ε: for a fixed k, if $d_1(W, \mathcal{R})$ is very small, then there is a graphon $U \in \mathcal{R}$ such that $d_1(W, U)$ is very small, and the distributions of $\mathbb{G}(k, U)$ and $\mathbb{G}(k, W)$ are almost the same, so no test property $\mathcal{R}'$ can distinguish them.

EXAMPLE 15.9 (**Complete or edgeless**). Let $\mathcal{R}$ be the graphon property satisfied by the identically-0 and identically-1 graphons. Then the graph property "complete or edgeless" is a good test property for $\mathcal{R}$, so $\mathcal{R}$ is testable. ♦

EXAMPLE 15.10 (**Constant graphon**). Let $\mathcal{R}$ be the graphon property satisfied by the identically-1/2 graphon U. We show that this property is not testable. Consider a random graph $G_n = \mathbb{G}(n, 1/2)$; then $\|W_{G_n} - U\|_\square \to 0$ with probability 1. Fix a sequence for which this happens. The distribution of $\mathbb{G}(k, W_{G_n})$ tends to the distribution of $\mathbb{G}(k, 1/2)$ for every fixed k, so for every possible test property $\mathcal{R}'$, if $\mathbb{G}(k, 1/2) \in \mathcal{R}'$ with probability at least $2/3$, then $\mathbb{G}(k, W_{G_n}) \in \mathcal{R}'$ with probability at least $1/2$ (for every k, if n is large enough). On the other hand, $d_1(W_{G_n}, \mathcal{R}) = 1/2$ for every n, and hence we should have $\mathbb{G}(k, W_{G_n}) \notin \mathcal{R}'$ with probability at least $2/3$ if k is large enough.

We note that the complementary property $\mathcal{W}_0 \setminus \mathcal{R}$ is testable: since $d_1(W, \mathcal{R}^c) = 0$ for every graphon W, the identically true property as a good test property. ♦

Similarly as for property distinguishing, one may feel that the tricky choice of the test property $\mathcal{R}'$ is crucial. but in fact, once a property is testable, we can use a very simple test property:

PROPOSITION 15.11. *Let $\mathcal{R}$ be a testable graphon property. Then*

$$\mathcal{R}' = \left\{F : \mathsf{v}(F) = 1 \text{ or } \delta_\square(W_F, \mathcal{R}) \leq \frac{20}{\sqrt{\log \mathsf{v}(F)}}\right\}$$

is a valid test property for $\mathcal{R}$.

Proof. First, suppose that $W \in \mathcal{R}$, and let $k \geq 2$. By the Second Sampling Lemma 10.16, we have

$$\delta_\square\big(W, \mathbb{G}(k, W)\big) \leq \frac{20}{\sqrt{\log k}}$$

with probability at least $2/3$. Thus $\mathbb{G}(k, W) \in \mathcal{R}'$ with probability larger than $2/3$.

Second, let $\mathcal{R}''$ be any valid test property for $\mathcal{R}$. By definition, for every $\varepsilon > 0$ there is a $k \geq 1$ (depending on ε, $\mathcal{R}$ and $\mathcal{R}''$) such that whenever $d_1(W, \mathcal{R}) > \varepsilon$ for some graphon W, then $\mathsf{P}(\mathbb{G}(k, W) \in \mathcal{R}'') \leq 1/3$. Let $U \in \mathcal{R}$, then $\mathsf{P}(\mathbb{G}(k, U) \in \mathcal{R}'') \geq 2/3$. This implies that the variation distance of the distributions of $\mathbb{G}(k, W)$ and $\mathbb{G}(k, U)$ is at least $1/3$. By Corollary 10.25, this implies that $\delta_\square(W, U) \geq \frac{1}{3}2^{-k^2}$. This holds for all $U \in \mathcal{R}$, so $\delta_\square(W, \mathcal{R}) \geq \frac{1}{3}2^{-k^2}$. Let n be large enough (depending on k), and consider the W-random graph $\mathbb{G}(n, W)$, and the corresponding graphon $\mathbb{W}_n = W_{\mathbb{G}(n,W)}$. Then with high probability

$$\delta_\square\big(\mathbb{W}_n, \mathcal{R}\big) \geq \frac{1}{3}2^{-k^2} - \delta_\square\big(\mathbb{W}_n, W\big) \geq \frac{1}{3}2^{-k^2} - \frac{20}{\sqrt{\log n}} > \frac{20}{\sqrt{\log n}}.$$

So $\mathbb{G}(n,W) \notin \mathcal{R}'$ with high probability, and this shows that $\mathcal{R}'$ is a valid test property. $\square$

It is not always easy to decide about a graphon property whether it is testable, and we will have to develop some theory to prove properties of further, more interesting examples. We start with showing that testability and distinguishability are closely related. Let $\mathcal{R}_\varepsilon^c$ denote the set of graphons W such that $d_1(W, \mathcal{R}) \geq \varepsilon$.

PROPOSITION 15.12. *A graphon property $\mathcal{R}$ is testable if and only if it is distinguishable by sampling from the property $\mathcal{R}_\varepsilon^c$ for every $\varepsilon > 0$.* $\square$

Proof. The "only if" part is straightforward to check. To verify the "if" part, suppose that $\mathcal{R}$ and $\mathcal{R}_\varepsilon^c$ are distinguishable by sampling for every $\varepsilon > 0$. Using Lemma 15.7, this means that for every $\varepsilon > 0$ there is a positive integer k_ε and a test property $\mathcal{Q}_\varepsilon$ such that for every $k \geq k_\varepsilon$, we have

$$\mathsf{P}\big(\mathbb{G}(k,W) \in \mathcal{Q}_\varepsilon\big) \begin{cases} \geq \dfrac{2}{3}, & \text{if } G \in \mathcal{R}, \\ \leq \dfrac{1}{3}, & \text{if } G \in \mathcal{R}_\varepsilon^c. \end{cases}$$

We may assume that k_ε increases if ε decreases.

One difficulty is that we need to define a single test property $\mathcal{Q}$, not the family $\{\mathcal{Q}_\varepsilon\}$. But this is easy. Let $F \in \mathcal{Q}$ if and only if $k_{1/m} \leq \mathsf{v}(F) < k_{1/(m+1)}$ and $F \in \mathcal{Q}_{1/m}$ for some positive integer m. Then $\mathcal{Q}$ works for every $\varepsilon > 0$.

The second difficulty is that we want $\mathsf{P}(\mathbb{G}(k,W) \in \mathcal{Q})$ for all k, not just for $k \geq k_\varepsilon$. But this property holds for $k \geq k_1$ by definition, so all we have to do is to include all graphs with fewer that k_1 nodes in $\mathcal{Q}$. $\square$

The following corollary to Theorem 15.8 and Proposition 15.12 provides an analytic characterization of testable graphon properties. Recall that the distances d_1 and $d_\square$ are related trivially by $d_\square \leq d_1$. Testability of a property means, in a sense, an inverse relation:

COROLLARY 15.13. *A closed graphon property $\mathcal{R}$ is testable if and only if $\delta_\square(\mathcal{R}, \mathcal{R}_\varepsilon^c) > 0$ for every $\varepsilon > 0$.*

The condition given in this corollary can be rephrased in various ways, for example: *For every sequence of graphons (W_n) such that $d_\square(W_n, \mathcal{R}) \to 0$, we have $d_1(W_n, \mathcal{R}) \to 0$.* We show that this condition is equivalent to a seemingly weaker condition. (Recall the definition of flexing from Section 14.3.3.)

LEMMA 15.14. *A closed graphon property $\mathcal{R}$ is testable if and only if for every $U \in \mathcal{R}$ and every sequence of graphons (W_n) such that $W_n \to U$ and every W_n is a flexing of U, we have $d_1(W_n, \mathcal{R}) \to 0$.*

Proof. Let (W_n) be a sequence of graphs such that $d_\square(W_n, \mathcal{R}) \to 0$. If $d_1(W_n, \mathcal{R}) \not\to 0$, then we may take a subsequence for which $\liminf d_1(W_n, \mathcal{R}) > 0$, and then a further subsequence such that $\delta_\square(W_n, U) \to 0$ for some $U \in \mathcal{R}$. Define

$$W_n'(x,y) = \begin{cases} U(x,y) & \text{if } U(x,y) \in \{0,1\}, \\ W_n(x,y) & \text{otherwise.} \end{cases}$$

Then W_n' is a flexing of U. Furthermore, we have

$$\|W_n - W_n'\|_1 = \int_{U=0} W_n + \int_{U=1} (1-W_n) \to 0$$

by Lemma 8.22. This implies that $\delta_\square(W_n, W_n') \to 0$, and hence $\delta_\square(W_n', U) \to 0$. By the hypothesis of the lemma implies that $d_1(W_n', \mathcal{R}) \to 0$. Hence

$$d_1(W_n, \mathcal{R}) \leq d_1(W_n', \mathcal{R}) + \|W_n - W_n'\|_1 \to 0. \qquad \square$$

Since the condition of the last lemma is trivially fulfilled if $\mathcal{R}$ is flexible, we get a useful corollary:

COROLLARY 15.15. *Every closed flexible graphon property is testable. In particular, the closure of every hereditary property is testable.*

The following result can be viewed as the graphon analogue of the theorem of Fischer and Newman [2005] (from which the finite theorem can be derived).

THEOREM 15.16. *A closed graphon property $\mathcal{R}$ is testable if and only if the functional $d_1(., \mathcal{R})$ is continuous in the cut norm.*

Proof. If $d_1(., \mathcal{R})$ is continuous, then Corollary 15.13 implies that $\mathcal{R}$ is testable.

Suppose that $\mathcal{R}$ is testable. The functional $d_1(., \mathcal{R})$ is lower semicontinuous in the cut norm by Lemma 14.15. To prove upper semicontinuity, let $W, W_n \in \mathcal{W}_0$ and let $\|W_n - W\|_\square \to 0$. We claim that $\limsup_n d_1(W_n, \mathcal{R}) \leq d_1(W, \mathcal{R})$.

Let $\varepsilon > 0$, and let $U \in \mathcal{R}$ be such that $\|W - U\|_1 \leq d_1(W, \mathcal{R}) + \varepsilon$. By Proposition 8.25, there is a sequence of graphons U_n such that $\|U_n - U\|_\square \to 0$ and $\|U_n - W_n\|_1 \to \|U - W\|_1$. By Corollary 15.13 (in the form in the remark after its statement), it follows that $d_1(U_n, \mathcal{R}) \to 0$, and so

$$d_1(W_n, \mathcal{R}) \leq \|W_n - U_n\|_1 + d_1(U_n, \mathcal{R}) \to \|U - W\|_1.$$

Hence

$$\limsup_{n \to \infty} d_1(W_n, \mathcal{R}) \leq \|U - W\|_1 \leq d_1(W, \mathcal{R}) + \varepsilon.$$

Since $\varepsilon > 0$ is arbitrary, this implies that $d_1(., \mathcal{R})$ is upper semicontinuous. $\square$

EXAMPLE 15.17 (**Neighborhood of a property**). Let $\mathcal{S} \subseteq \mathcal{W}_0$ be an arbitrary graphon property and let $a > 0$ be an arbitrary number. Then the property $\mathcal{R} = \{U \in \mathcal{W}_0 : \delta_\square(U, \mathcal{S}) \leq a\}$ is testable.

To show this, we use Corollary 15.13. For $\varepsilon > 0$ define $\varepsilon' = a\varepsilon/2$. Let $W \in B_\square(\mathcal{R}, \varepsilon')$. Then $W \in B_\square(\mathcal{S}, a+\varepsilon')$, and so there is a $U \in \mathcal{S}$ such that $\|U - W\|_\square \leq a + 2\varepsilon'$. Consider $Y = (1-\varepsilon)W + \varepsilon U$. Then $\|Y - U\|_\square = \|(1-\varepsilon)(U-W)\|_\square \leq (1-\varepsilon)(a+2\varepsilon') < a$, so $Y \in \mathcal{R}$. Furthermore, $\|W - Y\|_1 = \|\varepsilon(W-U)\|_1 \leq \varepsilon$, and so $W \in B_1(\mathcal{R}, \varepsilon)$. Since W was an arbitrary element of $B_\square(\mathcal{R}, \varepsilon')$, this implies that $\delta_\square(\mathcal{R}, \mathcal{R}_\varepsilon^c) > \varepsilon'$. ♦

EXAMPLE 15.18 (**Subgraph density**). For every fixed graph F and $0 < c < 1$, the property $\mathcal{R}$ of a graphon W that $t(F, W) = c$ is testable. Let us verify that for every $\varepsilon > 0$ there is an $\varepsilon' > 0$ such that $d_1(W, \mathcal{R}) \geq \varepsilon$ implies that $d_\square(W, \mathcal{R}) \geq \varepsilon'$. Assume that $d_1(W, \mathcal{R}) \geq \varepsilon$, then $t(F, W) \neq c$; let (say) $t(F, W) > c$. The graphons $U_s = (1-s)W$, $0 \leq s \leq \varepsilon$, are all in $B_1(W, \varepsilon)$, and hence not in $\mathcal{R}$. It follows that $t(F, U_s) > c$ for all $0 \leq s \leq \varepsilon$. Since $t(F, U_\varepsilon) = (1-\varepsilon)^{\mathsf{e}(F)} t(F, W)$, this implies that $t(F, W) > (1-\varepsilon)^{-\mathsf{e}(F)} c$. Thus for every $U \in \mathcal{R}$ we have $t(F, W) - t(F, U) \geq$

$((1-\varepsilon)^{-\mathsf{e}(F)} - 1)c$. By the Counting Lemma 10.23 this implies that $\delta_\square(U,W) \geq ((1-\varepsilon)^{-\mathsf{e}(F)} - 1)c/\mathsf{e}(F)$. Choosing the right hand side in this inequality as ε', we get that $\delta_\square(W,\mathcal{R}) \geq \varepsilon'$, which we wanted to verify.

Fixing two subgraph densities, however, may yield a non-testable property: for example, $t(K_2, W) = 1/2$ and $t(C_4, W) = 1/16$ imply that $W \equiv 1/2$ (see Section 1.4.2), and we have seen that this graphon property is not testable. ♦

15.3.2. Testable graph properties. We call a graph property $\mathcal{P}$ *testable*, if for every $\varepsilon > 0$, graphs in $\mathcal{P}$ can be distinguished from graphs farther than ε from $\mathcal{P}$ in the edit distance. To be more precise, recall that $d_1(F,G)$ is defined for two graphs on the same node set, and it denotes their normalized edit distance. So for a graph G on n nodes, $d_1(G,\mathcal{P})$ is the minimum number of edges to be changed in order to get a graph with property $\mathcal{P}$, divided by n^2. If there is no graph in $\mathcal{P}$ on n nodes, then we define $d_1(F,\mathcal{P}) = 1$. Let $\mathcal{P}_\varepsilon^c$ denote the set of simple graphs F such that $d_1(F,\mathcal{P}) \geq \varepsilon$; then we want to distinguish $\mathcal{P}$ from $\mathcal{P}_\varepsilon^c$ by sampling.

This notion of testability is usually called *oblivious testing*, which refers to the fact that no information about the size of G is assumed.

Using our analytic language, we can give several reformulations of the definition of testability of a graph property, which are often more convenient to use.

(T1) if (G_n) is a sequence of graphs such that $\mathsf{v}(G_n) \to \infty$ and $\delta_\square(G_n, \mathcal{P}) \to 0$, then $d_1(G_n, \mathcal{P}) \to 0$;

(T2) for every $\varepsilon > 0$ there is an $\varepsilon' > 0$ such that if G and G' are simple graphs such that $\mathsf{v}(G'), \mathsf{v}(G') \geq 1/\varepsilon'$, $G \in \mathcal{P}$ and $\delta_\square(G, G') < \varepsilon'$, then $d_1(G', \mathcal{P}) < \varepsilon$;

(T3) $\overline{\mathcal{P}} \cap \overline{\mathcal{P}_\varepsilon^c} = \emptyset$ for every $\varepsilon > 0$.

The equivalence of these with testability follows by Theorem 15.8.

From this characterization of testability it follows that if $\mathcal{P}$ is a testable property such that infinitely many graphs have property $\mathcal{P}$, then for every n that is large enough, it contains a graph on n nodes. Indeed, suppose that for infinitely many n, $\mathcal{P}$ contains a graph G_n on n nodes but none on $n+1$ nodes. We may assume that $G_n \to W \in \mathcal{W}_0$. Then $G_n K_1 \in \mathcal{P}_{1/2}^c$ and $G_n K_1 \to W$, so $W \in \overline{\mathcal{P}} \cap \overline{\mathcal{P}_\varepsilon^c}$, a contradiction.

It is surprising that this rather restrictive definition allows many testable graph properties: for example, bipartiteness, triangle-freeness, every property definable by a first order formula (Alon, Fischer, Krivelevich and Szegedy [2000]). Let us begin with some simple examples.

EXAMPLE 15.19 (**Nonempty**). Let $\mathcal{P}$ be the graph property that "G has at least one edge". This is testable with the identically true test property. ♦

This example sounds like playing in a trivial way with the definition. The following examples are more substantial.

EXAMPLE 15.20 (**Large clique**). Let $\mathcal{P}$ be the graph property $\omega(G) \geq \mathsf{v}(G)/2$. Then $\mathcal{P}$ is testable. This can be verified using (T2): we show that for every $\varepsilon > 0$ there is an $\varepsilon' > 0$ such that $\delta_\square(G, \mathcal{P}) \leq \varepsilon'$ implies that $d_1(G, \mathcal{P}) \leq \varepsilon$. We show that $\varepsilon' = \exp(-10000/\varepsilon^6)$ does the job.

Indeed, if $\delta_\square(G, \mathcal{P}) \leq \varepsilon'$, then there is a graph $H \in \mathcal{P}$ such that $\delta_\square(G, H) \leq 2\varepsilon'$. Let $V(H) = [q]$ and $V(G) = [p]$, then $\delta_\square\big(G(q), H(p)\big) \leq 2\varepsilon'$, and since $v\big(G(q)\big) = v\big(H(p)\big) = pq$, Theorem 9.29 implies that $\widehat{\delta}_\square\big(G(q), H(p)\big) \leq 45/\sqrt{-\log(2\varepsilon')} <$

$\varepsilon^3/2$, which means that $G(q)$ and $H(p)$ can be overlaid so that $d_\square\big(G(q), H(p)\big) \leq \varepsilon^3/2$. Since $H \in \mathcal{P}$, it contains a complete graph of size at least $\mathsf{v}(H)/2$, and so $H(p)$ contains a complete graph K of size at least $pq/2$. Now the definition of the cut distance implies that $G(q)[V(K)]$ is almost complete: it misses at most $\varepsilon^3(pq)^2/2$ edges. Let q_i denote the number of nodes in K that come from node i of G, so that $0 \leq q_i \leq q$ and $\sum_i q_i = pq$. For simplicity of presentation, assume that $p = 2r$ is even and $\varepsilon \leq 1/2$. Also assume that $q_1 \geq q_2 \geq \cdots \geq q_p$.

Let k denote the number of q_i with $i \leq r$ and $q_i \leq \varepsilon q$. We claim that $k \leq 2r\varepsilon$. We may assume that $k > 0$, then $q_{r+1}, \ldots, q_p < \varepsilon q$, and so $(r-k)q + (r+k)\varepsilon q \geq \sum_i q_i = rq$. This implies the bound on k. Let $G_1 = G[1, \ldots, r]$, then

$$\frac{1}{2}\varepsilon^3 p^2 q^2 \geq \sum_{ij \in E(\overline{G_1})} q_i q_j \geq (\mathsf{e}(\overline{G_1}) - kr)(\varepsilon q)^2,$$

whence

$$\mathsf{e}(\overline{G_1}) \leq kr + \frac{\varepsilon}{2}p^2 \leq \frac{\varepsilon}{2}p^2 + \frac{\varepsilon}{2}p^2 < \varepsilon p^2.$$

So adding at most εp^2 edges to G, we can create a complete subgraph with $p/2$ nodes, showing that $d_1(G, \mathcal{P}) \leq \varepsilon$. ♦

EXAMPLE 15.21 (**Triangle-free**). Let $\mathcal{P}$ be the property of being triangle-free. Then $\mathcal{P}$ is a valid test property for itself. It is trivial that if G is triangle-free, then any sample $\mathbb{G}(k, G)$ is also triangle-free. The other condition is, however, far from being trivial. If $\mathbb{G}(k, G)$ is triangle-free with probability at least $2/3$, and k is large enough, then G has very few triangles. Hence by the Removal Lemma 11.64, we get that we can change (in this case, delete) a small number of edges so that we get rid of all triangles. In fact, the Removal Lemma is equivalent to the testability of triangle-freeness.

Theorem 15.24 below will give a general sufficient condition for testability, which will imply the Removal Lemma. ♦

There is a tight connection between testability of graph properties and the testability of their closures. To formulate it, we need a further definition. A graph property $\mathcal{P}$ is *robust*, if for every $\varepsilon > 0$ there is an $\varepsilon_0 > 0$ such that if G is a graph with $\mathsf{v}(G) \geq 1/\varepsilon_0$ and $d_1(W_G, \overline{\mathcal{P}}) \leq \varepsilon_0$, then $d_1(G, \mathcal{P}) \leq \varepsilon$. Another way of stating this is that if (G_n) is a sequence of graphs such that $d_1(W_{G_n}, \overline{\mathcal{P}}) \to 0$, then $d_1(G_n, \mathcal{P}) \to 0$. (For a more combinatorial formulation of this property, see Exercise 15.30.)

THEOREM 15.22. (a) *A graphon property is testable if and only if it is the closure of a testable graph property.*

(b) *A graph property is testable if and only if it is robust and its closure is testable.*

It is not the first time in this book that results about graphons are nice and easy, but to describe the connection between the notions for graphons and the corresponding notions for graphs is the hard part. This is true in this case too, and we will omit some details of the rather long proof of this theorem; see Lovász and Szegedy [2010a] for these details.

Before going into the proof, let us look at an example.

EXAMPLE 15.23 (**Alternating**). A graph property with a testable closure need not be itself testable. Let $\mathcal{P}$ be the graph property that the graph is complete if the number of nodes is even, but edgeless if the number of nodes is odd. Clearly, this is not testable. The closure of $\mathcal{P}$ is the graphon property valid for $W \equiv 0$ and $W \equiv 1$, which is testable (Example 15.9). ◆

Proof. We start with proving that the closure of a testable graph property is testable. It suffices to prove that if (W_n) is a sequence of graphons such that $d_\square(W_n, \overline{\mathcal{P}}) \to 0$, then $d_1(W_n, \overline{\mathcal{P}}) \to 0$. We may assume that the sequence W_n is convergent, so $W_n \to U$ for some $U \in \mathcal{W}_0$ (in the $\delta_\square$ distance). Clearly $U \in \overline{\mathcal{P}}$, so by the definition of closure, there are graphs $H_n \in \mathcal{P}$ such that $H_n \to U$.

Fix any $\varepsilon > 0$. By Theorem 15.16, there is an $\varepsilon' > 0$ such that if $|V(G)|, |V(H)|$ are large enough, $H \in \mathcal{P}$, and $\delta_\square(G, H) < \varepsilon'$, then $d_1(G, \mathcal{P}) < \varepsilon$. Furthermore, there is an $n_\varepsilon \geq 1$ such that if $n \geq n_\varepsilon$, then $\delta_\square(W_{H_n}, U), \delta_\square(W_n, U) \leq \varepsilon'/3$.

Fix any $n \geq n_\varepsilon$, and let $G_{n,m}$ ($m = 1, 2, \ldots$) be a sequence of graphs such that $G_{n,m} \to W_n$ as $m \to \infty$. Then, provided m is large enough,

$$\delta_\square(H_n, G_{n,m}) \leq \delta_\square(W_{H_n}, U) + \delta_\square(U, W_n) + \delta_\square(W_n, W_{G_{n,m}}) < \varepsilon'.$$

Hence by the choice of ε', we have $d_1(G_{n,m}, \mathcal{P}) \leq \varepsilon$. This means that there are graphs $J_{n,m} \in \mathcal{P}$ with $V(J_{n,m}) = V(G_{n,m})$ such that $d_1(G_{n,m}, J_{n,m}) \leq \varepsilon$. By choosing a subsequence, we can assume that $J_{n,m} \to U_n$ as $m \to \infty$ for some $U_n \in \overline{\mathcal{P}}$. Applying Lemma 14.16 we obtain that

$$d_1(W_n, \overline{\mathcal{P}}) \leq \delta_1(W_n, U_n) \leq \liminf_{m \to \infty} \delta_1(W_{G_{n,m}}, W_{J_{n,m}}) \leq \liminf_{m \to \infty} d_1(G_{n,m}, J_{n,m}) \leq \varepsilon.$$

The proof of the converse in (a) (which we will not use in this book) is omitted.

Next we show that every testable graph property $\mathcal{P}$ is robust. Let (G_n) be a sequence of graphs such that $d_1(W_{G_n}, \overline{\mathcal{P}}) \to 0$, then $d_\square(W_{G_n}, \overline{\mathcal{P}}) \to 0$. This implies that there are graphons $U_n \in \overline{\mathcal{P}}$ such that $d_\square(W_{G_n}, U_n) \to 0$. By the definition of $\overline{\mathcal{P}}$, this implies that there are simple graphs $H_n \in \mathcal{P}$ such that $\delta_\square(G_n, H_n) \to 0$. This means that $\delta_\square(G_n, \mathcal{P}) \to 0$. Since $\mathcal{P}$ is testable, this implies that $d_1(G_n, \mathcal{P}) \to 0$.

Finally, we show that if a graph property $\mathcal{P}$ is robust and its closure is testable, then $\mathcal{P}$ is testable. Let (G_n) be a sequence of graphs such that $d_\square(G_n, \mathcal{P}) \to 0$. Then $d_\square(W_{G_n}, \overline{\mathcal{P}}) \to 0$. Since $\overline{\mathcal{P}}$ is testable, this implies that $d_1(W_{G_n}, \overline{\mathcal{P}}) \to 0$. By robustness, we get that $d_1(G_n, \mathcal{P}) \to 0$, which proves that $\mathcal{P}$ is testable. □

15.3.3. Hereditary and testable properties. A surprisingly general sufficient property for testability was found by Alon and Shapira [2008].

THEOREM 15.24 (**Alon–Shapira**). *Every hereditary graph property is testable.*

We have seen that the Removal Lemma is perhaps the simplest nontrivial special case of this theorem. The analytic proof of the Removal Lemma we described in Section 11.8 can be extended; we don't go into the details of this, instead we state and prove a more general result, which gives a full characterization of testable graph properties in terms of being "almost hereditary".

Testable properties are not necessarily hereditary, as Examples 15.19 and 15.20 show, so this condition is not necessary and sufficient. But there is a weaker version of heredity that can be used to characterize testability: we consider only those induced subgraphs that are close to the original graph in the cut-distance. (Recall that by the Second Sampling Lemma 10.15, almost all sufficiently large subgraphs have this property.) Clearly this theorem implies Theorem 15.24. The proof will be quite involved, using much of the material developed earlier in this chapter.

THEOREM 15.25. *A graph property $\mathcal{P}$ is testable if and only if for every $\varepsilon > 0$ there is an $\varepsilon' > 0$ such that if $H \in \mathcal{P}$ and G is an induced subgraph of H with $\mathsf{v}(G) \geq 1/\varepsilon'$ and $\delta_\square(G, H) < \varepsilon'$, then $d_1(G, \mathcal{P}) < \varepsilon$.*

Another way to state the condition is that if $H_n \in \mathcal{P}$ ($n = 1, 2, \ldots$) is a sequence of simple graphs, G_n is an induced subgraph of H_n, and $\delta_\square(G_n, H_n) \to 0$, then $d_1(G_n, \mathcal{P}) \to 0$. Informally, induced subgraphs inherit the property of the big guy, but they have to pay an inheritance tax; the tax is however small if the descendants are also big and they are close to the big guy.

Proof. The "only if" part is trivial, since the condition is a special case of the reformulation (T2) of testability. By theorem 15.22, it suffices to prove that $\overline{\mathcal{P}}$ is testable and $\mathcal{P}$ is robust.

We start with proving a graphon version of the condition in the theorem.

CLAIM 15.26. *Let $U \in \overline{\mathcal{P}}$ and let (G_n) be a sequence of simple graphs with $G_n \to U$. Also assume that $t_{\mathsf{ind}}(G_n, U) > 0$. Then $d_1(G_n, \mathcal{P}) \to 0$.*

Since $U \in \overline{\mathcal{P}}$, there is a sequence of simple graphs $H_m \in \mathcal{P}$ ($m = 1, 2, \ldots$) such that $H_m \to U$. Condition $t_{\mathsf{ind}}(G_n, U) > 0$ implies that $t_{\mathsf{ind}}(G_n, H_m) > 0$ for every n if m is large enough. Furthermore, both $G_n \to U$ and $H_m \to U$, and hence $\delta_\square(G_n, H_m) \to 0$ if $n, m \to \infty$. So if n is large enough, we can select an $m(n)$ such that G_n is an induced subgraph of $H_{m(n)}$ and $\delta_\square(G_n, H_{m(n)}) \to 0$. The condition in the Theorem implies the claim.

To prove that $\overline{\mathcal{P}}$ is testable, we use Lemma 15.14. So let us consider a graphon $U \in \overline{\mathcal{P}}$ and a sequence of graphons $W_n \to U$ where every W_n is a flexing of U. We want to prove that $d_1(W_n, \overline{\mathcal{P}}) \to 0$. For each n, we choose a simple graph G_n such that

(15.7) $\qquad\qquad\qquad \mathsf{v}(G_n) \geq n,$

(15.8) $\qquad\qquad\qquad t_{\mathsf{ind}}(G_n, W_n) > 0,$

(15.9) $\qquad\qquad\qquad \delta_\square(G_n, U) \leq \delta_\square(W_n, U) + \dfrac{1}{n},$

(15.10) $\qquad\qquad\qquad d_1(G_n, \mathcal{P}) \geq d_1(W_n, \overline{\mathcal{P}}) - \dfrac{1}{n}.$

This is not difficult: $G_n = G_{nk} = \mathbb{G}(k, W_n)$ will satisfy these conditions with high probability if k is sufficiently large. Indeed, (15.7) and (15.8) are essentially trivial, and (15.9) follows by Lemma 10.16. To verify (15.10), we select a graph $H_{nk} \in \mathcal{P}$ with $V(H_{nk}) = [k]$ such that

$$d_1(G_{nk}, \mathcal{P}) = d_1(G_{nk}, H_{nk}) \geq \delta_1(W_{G_{nk}}, W_{H_{nk}}).$$

Let $k \to \infty$, then $W_{G_{nk}} \to W_n$ in the $\delta_\square$-distance with probability 1, and (by selecting an appropriate subsequence) $W_{H_{nk}} \to U_n \in \overline{\mathcal{P}}$. By Lemma 14.16, we get that

$$\liminf_{n \to \infty} \delta_1(W_{G_{nk}}, W_{H_{nk}}) \geq \delta_1(W_n, U_n) \geq \delta_1(W_n, \overline{\mathcal{P}}),$$

which proves that (15.10) is satisfied if k is large enough.

Claim 15.26 implies that $d_1(G_n, \mathcal{P}) \to 0$. Indeed, condition (15.8) implies that $t_{\mathsf{ind}}(G_n, U) > 0$ (here we use that W_n is a flexing of U). Furthermore, $G_n \to U$ by (15.9), so Claim 15.26 applies.

From here, the testability of $\overline{\mathcal{P}}$ follows easily:

$$d_1(W_n, \overline{\mathcal{P}}) \leq \delta_1(W_n, \overline{\mathcal{P}}) \leq \delta_1(G_n, \mathcal{P}) + \frac{1}{n} \to 0.$$

Our second task is to prove that $\mathcal{P}$ is robust: if (G_n) is a sequence of simple graphs such that $d_1(W_{G_n}, \overline{\mathcal{P}}) \to 0$, then $d_1(G_n, \mathcal{P}) \to 0$. Let $W_n \in \overline{\mathcal{P}}$ be such that $\|W_{G_n} - W_n\|_1 \to 0$. By selecting an appropriate subsequence, we may assume that $\delta_\square(W_n, U) \to 0$, for some graphon U. Clearly $U \in \overline{\mathcal{P}}$ and $G_n \to U$.

Consider the random graph $G'_n = \mathbb{G}'(\mathsf{v}(G_n)), U)$. We have $t_{\mathsf{ind}}(G'_n, U) > 0$ with probability 1, and by Lemma 10.18, with probability tending to 1, $G'_n \to U$. Furthermore, an easy computation (cf. Exercise 10.14) gives that $\mathsf{E}\big(d_1(G_n, G'_n)\big) = \mathsf{E}(\|W_{G_n} - W_{G'_n}\|_1) = \|W_{G_n} - W_n\|_1 \to 0$, and so with high probability, we have $d_1(G_n, G'_n) \to 0$. By Claim 15.26, this implies that $d_1(G'_n, \mathcal{P}) \to 0$. Hence $d_1(G_n, \mathcal{P}) \leq d_1(G_n, G'_n) + d_1(G'_n, \mathcal{P}) \to 0$, which proves that $\mathcal{P}$ is robust. □

Other characterizations of testable graph properties are known. Alon, Fischer, Newman and Shapira [2006] characterized testable graph properties in terms of Szemerédi partitions (we refer to their paper for the formulation). Fischer and Newman [2005] connected testability to estimability. We already stated a version of this result for graphons (Theorem 15.16), from which it can be derived (we don't go into the details):

THEOREM 15.27. *A graph property is testable if and only if the normalized edit distance from the property is an estimable parameter.* □

EXERCISE 15.28. Prove that the graph property $\omega(G) \geq \mathsf{v}(G)/2$ satisfies the condition given in Theorem 15.25.

EXERCISE 15.29. Prove the following analogue of Proposition 15.11 for finite graphs: If $\mathcal{P}$ is a testable graph property, then

$$\mathcal{P}' = \left\{F : \mathsf{v}(F) = 1 \text{ or } \delta_\square(F, \mathcal{P}) \leq \frac{20}{\sqrt{\log \mathsf{v}(F)}}\right\}$$

is a valid test property for $\mathcal{P}$.

EXERCISE 15.30. Prove that graph property $\mathcal{P}$ is robust if and only if for every $\varepsilon > 0$ there is an $\varepsilon_0 > 0$ such that if G is a graph with $\mathsf{v}(G) \geq 1/\varepsilon_0$ and G has infinitely many near-blowups G' with $d_1(G', \mathcal{P}) \leq \varepsilon_0$, then $d_1(G, \mathcal{P}) \leq \varepsilon$.

15.4. Computable structures

15.4.1. Similarity distance and representative sets.
Let us recall from the Introduction that if we want to run algorithms on a very large graph G, it is very useful to define and also compute a "representative set" of nodes, a (fairly large, but bounded size) subset $R \subseteq V(G)$ such that every node is "similar" to one of the nodes in R. We want to define a distance that reflects for two nodes how "similar" their positions in the graph are.

We define the similarity distance of two nodes $s, t \in V(G)$ as follows: for any node w, compute

$$a(s, t; w) = \frac{1}{n}\big||N(w) \cap N(s)| - |N(w) \cap N(t)|\big|,$$

and let
$$(15.11) \qquad d_{\text{sim}}(s,t) = \frac{1}{n} \sum_{w \in V(G)} a(s,t;w).$$

We can think of $a(s,t;w)$ as a measure of how different s and t are from the point of view of w; then $d_{\text{sim}}(s,t)$ is an average measure of this difference. Of course, w could be more myopic and not look for neighbors of s and t among its own neighbors, but look only for s and t; then $d_{\text{sim}}(s,t)$ would measure the size of the symmetric difference of the neighborhoods of s and t. As explained in the Introduction, this is a perfectly reasonable definition, but it would not measure what we want. The node w could also look for second or third neighbors of s and t, but this would not give anything more useful than this definition, at least for dense graphs.

There are many ways to rephrase this definition. We can pick three random nodes $\mathbf{w}, \mathbf{v}, \mathbf{u} \in V(G)$, and define
$$(15.12) \qquad d_{\text{sim}}(s,t) = \mathsf{E}_\mathbf{w} \big| \mathsf{E}_\mathbf{v}(a_{s\mathbf{v}} a_{\mathbf{vw}}) - \mathsf{E}_\mathbf{u}(a_{t\mathbf{u}} a_{\mathbf{uw}}) \big|,$$
where (a_{ij}) is the adjacency matrix of G. We could use $\mathbf{v} = \mathbf{u}$ here, I just used different variables to make the correspondence with the definition clearer. We can also notice that $d_{\text{sim}}(s,t)$ is the L_1 distance of rows s and t of the square of the adjacency matrix, normalized by n^2. Finally, the similarity distance is quite closely related to the distance $\bar{r}_{W_G}$, discussed in Section 13.4: $d_{\text{sim}}(s,t) = \bar{r}_{W_G}(x,y)$, where x and y are arbitrary points of the intervals representing s and t in W_G.

There is an easy algorithm to compute (approximately) the similarity distance of two nodes.

ALGORITHM 15.31.
Input: A graph G given by a sampling oracle, two nodes $s, t \in V$, and an error bound $\varepsilon > 0$.
Output: A number $\mathbf{D}(s,t) \geq 0$ such that with probability at least $1 - \varepsilon$,
$$\mathbf{D}(s,t) - \varepsilon \leq d_{\text{sim}}(s,t) \leq \mathbf{D}(s,t) + \varepsilon.$$

The algorithm is based on (15.12). Select a random node $\mathbf{w}$ and fix it temporarily. Select $O(1/\varepsilon^2)$ random nodes $\mathbf{v}$ and compute the average of $a_{s\mathbf{v}} a_{\mathbf{vw}}$, to get a number that is within an additive error of $\varepsilon/4$ to $\mathsf{E}_\mathbf{v}(a_{s\mathbf{v}} a_{\mathbf{vw}})$. Estimate $\mathsf{E}_\mathbf{v}(a_{t\mathbf{v}} a_{\mathbf{vw}})$ similarly. This gives an estimate for $|\mathsf{E}_\mathbf{v}(a_{\mathbf{uw}}(a_{s\mathbf{v}} - a_{t\mathbf{v}}))|$ with error at most $\varepsilon/2$ with high probability. Repeat this $O(1/\varepsilon^2)$ times and take the average to get $\mathbf{D}(s,t)$.

Next, we specialize Theorem 13.31 to graphs.

THEOREM 15.32. *Let $G = (V, E)$ be a graph.*

(a) *If $\mathcal{P} = \{S_1, \ldots, S_k\}$ is a partition of $V(G)$ such that $d_\square(G, G_\mathcal{P}) = \varepsilon$, then we can select a node $v_i \in S_i$ from each partition class such that the average d_{sim}-distance from $S = \{v_1, \ldots, v_k\}$ is at most 4ε.*

(b) *If $S \subseteq V$ is a subset such that the average d_{sim}-distance from $S = \{v_1, \ldots, v_k\}$ is ε, then the Voronoi cells of S form a partition $\mathcal{P}$ such that $d_\square(G, G_\mathcal{P}) \leq 8\sqrt{\varepsilon}$.* □

We define a *representative set* with error $\varepsilon > 0$ as a subset $R \subseteq V(G)$ such that any two elements of R are at a (similarity) distance at least $\varepsilon/2$, and the average distance of nodes from R is at most 2ε. (The first condition is not crucial for the applications we want to give, but it guarantees that the set is chosen economically.) Theorem 13.31 implies that such a set R exists with $|R| \leq 2^{32/\varepsilon^2}$. Furthermore, such a set can be constructed in our model.

ALGORITHM 15.33.
Input: A graph G given by a sampling oracle, and an error bound ε.

Output: A random set $\mathbf{R} \subseteq V(G)$ such that $|\mathbf{R}| \leq (64/\varepsilon^2)^{1028/\varepsilon^2}$, and with probability at least $1 - \varepsilon$, $\mathbf{R}$ is a representative set with error ε.

The set R is grown step by step, starting with the empty set. At each step, a new uniform random node $\mathbf{w}$ of G is generated, and the approximate distances $\mathbf{D}(\mathbf{w}, v)$ are computed for all $v \in \mathbf{R}$ with error less than $\varepsilon/4$ with high probability. If all of these are larger than $3\varepsilon/4$, then $\mathbf{w}$ is added to $\mathbf{R}$. Else, $\mathbf{w}$ is discarded and a new random node is generated. If $\mathbf{R}$ is not increased in $k = \lceil 2000 \frac{1}{\varepsilon} \log \frac{1}{\varepsilon} \rceil$ steps, the algorithm halts.

We have to make sure that we don't make the mistake of stopping too early. It is clear that as long as the average distance from $\mathbf{R}$ is larger than 2ε, then the probability that a sample has distance at least ε is at least ε, and so the probability that in k iterations we don't pick a node whose distance from R is less than ε is less than $e^{-k\varepsilon}$. If we find a good node $\mathbf{u}$, then with high probability the approximate distance satisfies $\mathbf{D}(\mathbf{u}, \mathbf{R}) > 3\varepsilon/4$, and so we add $\mathbf{u}$ to $\mathbf{R}$. Hence the probability that we stop prematurely is less than $e^{-k\varepsilon}\mathsf{E}(|\mathbf{R}|) \leq \varepsilon$.

The size of $\mathbf{R}$ can be bounded using Proposition 13.32, which gives the bound on the output size. We can say more. Suppose that there exists a representative set R with error ε. Then only a fraction of $2\sqrt{\varepsilon}$ nodes of G (call these nodes "remote") are at a distance more than $\sqrt{\varepsilon}$ from R. Let us run the above algorithm with ε replaced by $2\sqrt{\varepsilon}$, to get a representative set R' with error $q\sqrt{\varepsilon}$. The set R' will contain at most one non-remote node from every Voronoi cell of R. We have little control over how many remote nodes we selected, but we can post-process the result. The $\sqrt{\varepsilon}$-balls around the non-remote nodes in R' cover all the non-remote nodes, so only leave out a fraction of $2\sqrt{\varepsilon}$ of all nodes. By sampling and brute force, we can select the smallest subset $R'' \subseteq R'$ with this property. This way we have constructed a representative set R'' with $|R''| \leq |R|$ and error at most $3\sqrt{\varepsilon}$.

As a special case, if there is a representative set whose size is polynomially bounded in the error ε, our algorithm will find one with a somewhat worse polynomial bound.

REMARK 15.34. One could try to work with a stronger notion: define a *strong representative set* with distance $\varepsilon > 0$ as a subset $R \subseteq V(G)$ such that any two elements of R are at a (similarity) distance at least ε, and any other node of G is at a distance at most ε from R.

It is trivial that every graph contains a strong representative set: just take a maximal set of nodes any two of which are at least ε apart. Furthermore, Proposition 13.32 shows that the size of such a set can be bounded by a function of ε. There are, however, several problems with the idea of computing and using it. First, in our very large graph model, the similarity distance cannot be computed exactly; second (and more importantly) the graph can have a tiny remote part which no sampling will discover but a representative of which should be included in the strong representative set.

15.4.2. Computing regularity partitions.

As an easy application of Theorem 15.32, we give an algorithm to compute a weak Szemerédi partition in a huge graph. Our goal is to illustrate how an algorithm works in the pure sampling model, as well as in what form the result can be returned. (This way of presenting the output of an algorithm for a large graph was proposed by Frieze and Kannan [1999]). A polynomial time algorithm to compute a regularity partition in the traditional setting of graph algorithms was given by Alon, Duke, Lefmann, Rödl and Yuster [1994].

Algorithm 15.33 enables us to encode a partition of $V(G)$ as a subset $R \subseteq V(G)$: for each $r \in R$, we can define the partition class V_r as the Voronoi cell of r. Ties will be broken arbitrarily, and nodes to which there are several "almost closest" nodes may be misclassified, but this is the best one can hope for. To formalize,

ALGORITHM 15.35.
Input: A graph G given by a sampling oracle, a subset $R \subseteq V(G)$, a node $u \in V$, and an error bound $\varepsilon > 0$.
Output: A (random) node $\mathbf{v} \in R$ almost closest to v in the sense that with probability at least $1 - \varepsilon$, $d_{\text{sim}}(u, \mathbf{v}) \leq (1 + \varepsilon) d_{\text{sim}}(u, R)$.

This algorithm uses Algorithm 15.31 to compute (approximately) the distances $d_{\text{sim}}(u, r)$, $r \in R$, and returns the node $r \in R$ that it finds closest to u. In other words, we compute the Voronoi cells of the set R.

Theorem 15.32 says in this context that the partition determined by Algorithms 15.33 and 15.35 satisfies $d_\square(G, G_\mathcal{P}) \leq \varepsilon$ with high probability. We omit the details of the error analysis. So we get a weak regularity partition. It is not known whether stronger regularity partitions can be computed in the sampling model.

15.4.3. Computing a maximum cut.

The algorithm to approximately compute the maximum cut is similar.

ALGORITHM 15.36.
Input: A graph G given by a sampling oracle, a subset $R \subseteq V(G)$, and an error bound $\varepsilon > 0$.
Output: A partition $R = R_1 \cup R_2$.

For the partition $\mathcal{P}$ implicitly determined above, we can also compute the edge densities between the partition classes, which we use to weight the edges of the complete graph on R, so that we get a weighted graph H. We find the maximum cut in H by brute force, to get a partition $R = R_1 \cup R_2$. This gives an implicit definition of a cut in G, where a node u is put on the left side of the cut iff $\mathbf{D}(u, R_1) < Db(u, R_2)$ for the approximate distances computed by Algorithm 15.31.

REMARK 15.37. 1. A notion closely related to testability and to computing a structure, called "local reparability", was introduced by Austin and Tao [2010]. We only give an informal definition. We start with a graph property $\mathcal{P}$ and a graph for which almost all induced subgraphs of size N have property $\mathcal{P}$; we say that G has the property N-*locally*. Now we want to repair G to have property $\mathcal{P}$ itself, by changing a small fraction of the edges. So far, this is essentially the same as testability, but we have to do the repair by a local algorithm as follows. We select a random sample $A \subseteq V(G)$ of size k (where $k \leq N$ is chosen appropriately), and for every pair of nodes $u, v \in V(G)$, we compute whether they should be adjacent, knowing only the induced subgraph $G[A \cup \{u,v\}]$. We could decide simply the adjacency of them in G, but then we would not do any repair. Taking the subgraph $G[A]$ and its connections to u and v also into account, our algorithm will define in a modified graph G'. This graph G' should have property $\mathcal{P}$, and its edit distance from G should be arbitrarily small if N and k are large enough.

It may or may not be possible to do so. We say that $\mathcal{P}$ is *locally reparable* if it is always possible. Austin and Tao prove, among others, that every hereditary property is reparable. For the exact definitions, formulation, proofs, generalizations to hypergraphs and other results we refer to the paper.

2. There is a natural nondeterministic version of testability, introduced by Lovász and Vesztergombi [2012]. A property of finite graphs is called *nondeterministically testable* if it has a "certificate" in the form of a coloring of the nodes and edges with a bounded number of colors, adding new edges with other colors, and orienting the edges, such that once the certificate is specified, its correctness can be verified by random local testing. Here are a few examples of properties that are nondeterministically testable in a natural way: "the graph is 3-colorable;" "the graph contains a clique on half of its nodes;" "the graph is transitively orientable"; "one can add at most $\mathsf{v}(G)^2/100$ new edges to make the graph perfect."

Using the theory of graph limits, it is proved that every nondeterministically testable property is deterministically testable. In a way, this means that $P = NP$ in the world of property testing for dense graphs. (Many, but not all, of the properties described above are also covered by Theorem 15.24.)

We will see that for bounded-degree graphs, the analogous statement does not hold. In fact, the study of nondeterministic certificates will lead to a new interesting notion of convergence (Section 19.2).

CHAPTER 16

Extremal theory of dense graphs

Extremal graph theory was one of the motivating fields for graph limit theory, as described in the Introduction. It is also one of the most fertile fields of applications of graph limits. In this chapter we give an exposition of some of the main directions. We start with two sections developing some technical tools, reflection positivity and variational calculus. Then we discuss extremal problems for complete graphs and some other specific problems. We re-prove some classical general results in extremal graph theory, and finally, we treat some very general questions (formulated in the introduction) about decidability of extremal graph problems and the possible structure of extremal graphs.

16.1. Nonnegativity of quantum graphs and reflection positivity

We are interested in linear inequalities between the densities of some subgraphs. (Why just linear? We have seen in the Introduction that, using the multiplicativity of subgraph densities, algebraic inequalities between them can be replaced by equivalent linear inequalities. To be sure, there are nontrivial non-algebraic inequalities that hold between subgraph densities; see Exercise 16.21. But their theory is virtually completely unexplored.)

As we have seen in the Introduction, many results in extremal graph theory can be stated in this form. We have also seen an example of a proof that relied on simple computations with quantum graphs. To make the problem and the methods precise, for a set $\mathcal{U} \subseteq \mathcal{W}$ and for a quantum graph x we write $x \geq 0$ (for $\mathcal{U}$) if $t(x, W) \geq 0$ for every graphon $W \in \mathcal{U}$. Most of the time (but not always) we will be concerned with $\mathcal{U} = \mathcal{W}_0$, and then we will suppress the "(for $\mathcal{U}$)" part of the notation. The condition $x \geq 0$ is equivalent to requiring that $t(x, G) \geq 0$ for every simple graph G.

Reflection positivity of the subgraph densities (Proposition 7.1) implies many inequalities of this type. (In a sense it implies all, as we will see in Theorem 16.41.) For every $k \geq 0$, every graphon W, every set of simple k-labeled graphs $\{F_1, \ldots, F_m\}$, and every vector $a \in \mathbb{R}^m$, we have

$$\sum_{i,j=1}^{m} a_i a_j t(\llbracket F_i F_j \rrbracket, W) \geq 0,$$

or, using our notation,

$$\sum_{i,j=1}^{m} a_i a_j \llbracket F_i F_j \rrbracket \geq 0.$$

This is equivalent to

(16.1) $$\left[\!\left[\Big(\sum_{i,j=1}^m a_i F_i\Big)^2\right]\!\right] \geq 0.$$

So we get the fact, used implicitly in the Introduction (Section 2.1.3), that unlabeling the square of a k-labeled quantum graph, we get a nonnegative quantum graph. Let us also recall the trivial fact that adding or deleting isolated nodes to a graph F does not change the homomorphism densities $t(F,.)$. We call a quantum graph g a *square-sum* if there are k-labeled quantum graphs $y_1, \ldots, y_m$ for some k such that g can be obtained from $\sum_i y_i^2$ by unlabeling and adding or deleting isolated nodes. We have just shown that every square-sum satisfies $g \geq 0$.

Another important property of semidefinite matrices is that their determinant is nonnegative: for any set $\{F_1, \ldots, F_m\}$ of graphs and any graphon W, we have

$$\begin{vmatrix} t([\![F_1 F_1]\!], W) & \cdots & t([\![F_1 F_m]\!], W) \\ \vdots & & \vdots \\ t([\![F_m F_1]\!], W) & \cdots & t([\![F_m F_m]\!], W) \end{vmatrix} \geq 0.$$

Expanding this determinant, we get a polynomial inequality, which can be turned into a linear inequality involving subgraph densities, using multiplicativity of the parameter $t(., W)$. This can be expressed as nonnegativity of a quantum graph:

(16.2) $$\begin{vmatrix} [\![F_1 F_1]\!] & \cdots & [\![F_1 F_m]\!] \\ \vdots & & \vdots \\ [\![F_m F_1]\!] & \cdots & [\![F_m F_m]\!] \end{vmatrix} \geq 0.$$

This is still a rather complicated inequality, but one special case will be useful:

(16.3) $$[\![F_1 F_1]\!][\![F_2 F_2]\!] \geq [\![F_1 F_2]\!]^2.$$

Another consequence of reflection positivity is the following: let $(a_{ij})_{i,j=1}^m$ be a symmetric positive semidefinite matrix, then

(16.4) $$\sum_{i,j=1}^m a_{ij} [\![F_i F_j]\!] \geq 0.$$

We can add two relations related to subgraphs. First, adding an isolated node to a graph F does not change its density in any graph or graphon, and so

$$FK_1 - F \geq 0, \quad \text{but also} \quad F - FK_1 \geq 0.$$

Furthermore, if F' is a subgraph of F, then $t(F', W) \geq t(F, W)$ for every graphon W, and hence

(16.5) $$F' - F \geq 0.$$

EXERCISE 16.1. Show that inequalities (16.2) (16.4) and (16.5) can be derived from the inequalities (16.1).

EXERCISE 16.2. Prove the following "supermodularity" inequality: if F_1 and F_2 are two simple graphs on the same node set, then $F_1 \cup F_2 + F_1 \cap F_2 \geq F_1 + F_2$.

16.2. Variational calculus of graphons

One advantage of working in the large space of graphons instead of the much simpler (discrete) space of graphs is that we can do continuous deformations of a graphon. This is useful when trying to optimize some functional, or when studying conditions that uniquely determine graphons. The following considerations are conceptually not too difficult, but technically more involved.

To express the variation of graphon functionals, we need some notation. For every multigraph $F = (V, E)$ and node $i \in V$, let F^i denote the 1-labeled quantum graph obtained by labeling i by 1. For every edge $ij \in E$, let F^{ij} denote the 2-labeled quantum graph obtained from F by deleting the edge ij, and labeling i by 1 and j by 2. The 2-labeled quantum graph F_{ij} is constructed similarly, but the edge ij is not deleted. So $F_{ij} = K_2^{\bullet\bullet} F^{ij}$. Let $F^\dagger = \sum_{i \in V} F^i$ and $F^\ddagger = \frac{1}{2} \sum_{i,j: ij \in E} F^{ij}$ (each edge contributes two terms, since its endpoints can be labeled in two ways; this is why it will be convenient to divide by 2). Similarly, let $F^\flat = \frac{1}{2} \sum_{i,j: ij \in E} F_{ij}$. We extend the operators $F \mapsto F^\dagger$, $F \mapsto F^\ddagger$ and $F \mapsto F^\flat$ linearly to all quantum graphs. We have $x^\flat = K_2^{\bullet\bullet} x^\ddagger$.

EXAMPLE 16.3. Clearly $C_n^\ddagger = nP_n^{\bullet\bullet}$, where $P_n^{\bullet\bullet}$ denotes the path on n nodes with its endpoints labeled. So $t_{xy}(C_n^\ddagger, W) = nt_{xy}(P_n^{\bullet\bullet}, W) = nW^{\circ(n-1)}$. ♦

The Edge Reconstruction Conjecture 5.31 says in this language that if F and G are simple graphs that are large enough (both have at least four non-isolated nodes), then $[\![F^\ddagger]\!] = [\![G^\ddagger]\!]$ implies that $F \cong G$.

We study two kinds of variations of a kernel W: in the more general version, we change the values of W at every node. However, it is often easier to construct variations in which the measure on $[0,1]$ is rescaled. This simpler kind variation will have the advantage that if we start with a graphon, then we don't have to worry about the values of W running out of the interval $[0,1]$. We start with describing the variation of the measure.

Consider a family $\alpha_s : [0,1] \to \mathbb{R}_+$ ($s \in [0,1]$) of weight functions such that $\int_0^1 \alpha_s(x)\,dx = 1$ for every s. Every such function defines a probability measure μ_s on $[0,1]$ by

$$\mu_s(A) = \int_A \alpha_s(x)\,dx.$$

Every kernel W on $[0,1]^2$ gives rise to a family of kernels W_s, where $W_s = ([0,1], \mu_s, W)$. For every finite graph F, we have

$$t(F, W_s) = \int_{[0,1]^{V(F)}} \prod_{ij \in E(F)} W(x_i, x_j) \prod_{i \in V(F)} d\mu_s(x_i)$$

$$= \int_{[0,1]^{V(F)}} \prod_{ij \in E(F)} W(x_i, x_j) \prod_{i \in V(F)} \alpha_s(x_i) \prod_{i \in V(F)} dx_i.$$

We say that the family (α_s) has *uniformly bounded derivative*, if for every $x \in [0,1]$ the derivative $\dot{\alpha}_s(x) = \frac{d}{ds}\alpha_s(x)$ exists, and there is a constant $M > 0$ such that $|\dot{\alpha}_s(x)| \leq M$ for all x and s. If α_s is a family of weight functions with uniformly bounded derivative, then by elementary analysis it follows that the function

$t(F, W_s)$ is differentiable as a function of s, and

(16.6) $$\frac{d}{ds} t(F, W_s) = \langle \dot{\alpha}_s, t_x(F^\dagger, W_s) \rangle.$$

In the second type of variation, we consider a family $U_s \in \mathcal{W}$ ($0 \leq s \leq 1$) of kernels. We say that the family U_s has *uniformly bounded derivative*, if for every $x, y \in [0, 1]$ the derivative $\dot{U}_s(x, y) = \frac{d}{ds} U_s(x, y)$ exists, and there is a constant $M > 0$ such that $|\dot{U}_s(x)| \leq M$ for all x and s. If U_s is a uniformly bounded family of kernels with uniformly bounded derivative, then the function $t(F, W_s)$ is differentiable as a function of s, and

(16.7) $$\frac{d}{ds} t(F, U_s) = \langle \dot{U}_s, t_{xy}(F^\ddagger, U_s) \rangle.$$

As an application of these formulas, we derive versions of the Kuhn–Tucker conditions in optimization for a graphon minimizing a smooth function of a finite number of homomorphism densities (rephrasing conditions of Razborov [2007, 2008] in our language). For a functional ω on $\mathcal{W}_0$, we say that a graphon W is a *local minimizer*, if there is an $\varepsilon > 0$ such that $\omega(U) \geq \omega(W)$ for every graphon U with $\|U - W\|_1 < \varepsilon$. (We could define this notion with respect to other norms, but this will be the version we use.)

LEMMA 16.4. *Let g be a simple quantum graph, and suppose that W is a local minimizer of $t(g, W)$ over $W \in \mathcal{W}_0$.*

(a) *For almost all $x \in [0, 1]$, $t_x(g^\dagger, W) = t([\![g^\dagger]\!], W)$.*

(b) *For almost all $x, y \in [0, 1]$,*

$$t_{xy}(g^\ddagger, W) \begin{cases} = 0 & \text{if } 0 < W(x, y) < 1, \\ \geq 0, & \text{if } W(x, y) = 0, \\ \leq 0, & \text{if } W(x, y) = 1. \end{cases}$$

(c) *For almost all $x, y \in [0, 1]$, $t_{xy}(g^\flat, W) \leq 0$.*

We will prove a more general lemma.

LEMMA 16.5. *Let $\Phi : \mathbb{R}^m \to \mathbb{R}$ be a differentiable function, and let $\Phi_i = \frac{\partial}{\partial x_i} \Phi$. Let $F_1, \ldots, F_m$, simple graphs, and suppose that W is a local minimizer of $\Phi(t(F_1, W), \ldots, t(F_m, W))$ over $W \in \mathcal{W}_0$. Set $a_i = \Phi_i(t(F_1, W), \ldots, t(F_m, W))$.*

(a) *For almost all $x \in [0, 1]$,*

$$\sum_{i=1}^m a_i \big(t_x(F_i^\dagger, W) - \mathsf{v}(F_i) t(F_i, W) \big) = 0.$$

(b) *For almost all $x, y \in [0, 1]$,*

$$\sum_{i=1}^m a_i t_{xy}(F_i^\ddagger, W) \begin{cases} = 0 & \text{if } 0 < W(x, y) < 1, \\ \geq 0, & \text{if } W(x, y) = 0, \\ \leq 0, & \text{if } W(x, y) = 1. \end{cases}$$

(c) *For almost all $x, y \in [0, 1]$,*

$$\sum_{i=1}^m a_i t_{xy}(F_i^\flat, W) \leq 0.$$

Proof. (a) Let $\varphi : [0,1] \to [-1,1]$ be a measurable function such that $\int \varphi = 0$. For $s \in [-1,1]$, we re-weight the points of $[0,1]$ by $\alpha_s(x) = 1 + s\varphi(x)$ to get the graphon W_s. Using (16.6), we get

$$\frac{d}{ds}\Phi\big(t(F_1, W_s),\ldots, t(F_m, W_s)\big)\bigg|_{s=0} = \sum_{i=1}^{m} a_i \int_0^1 \varphi(x) t_x(F_i^\dagger, W)\, dx$$

$$= \int_0^1 \varphi(x) \sum_{i=1}^{m} a_i t_x(F_i^\dagger, W)\, dx.$$

Since W is a local minimizer, and $\|W_s - W\|_1$ is arbitrarily small if s is small (see Exercise 13.7), this derivative must be 0 for all φ. This implies that $\sum_i a_i t_x(F_i^\dagger, W)$ is a constant function of x almost everywhere. Integrating over x, we recover the value of the constant, which proves (a).

(b) Let $U \in \mathcal{W}_1$ be a function such that $U(x,y) \geq 0$ if $W(x,y) = 0$ and $U(x,y) \leq 0$ if $W(x,y) = 1$. Then for every $s \geq 0$ in a small neighborhood of 0, $W + sU \in \mathcal{W}_0$. Using (16.7),

$$\frac{d}{ds}\Phi\big(t(F_1, W+sU),\ldots, t(F_m, W+sU)\big)\bigg|_{s=0}$$

$$= \int_{[0,1]^2} U(x,y) \sum_{i=1}^{m} a_i' t_{xy}(F_i^\ddagger, W)\, dx\, dy.$$

Since W is a local minimizer, we must have

$$\int_{[0,1]^2} U(x,y) \sum_{i=1}^{m} a_i' t_{xy}(F_i^\ddagger, W)\, dx\, dy \geq 0.$$

This must hold for all functions $U \in \mathcal{W}_1$ such that $U(x,y) \geq 0$ if $W(x,y) = 0$ and $U(x,y) \leq 0$ if $W(x,y) = 1$, which implies (b).

(c) This follows from (b) by multiplying by $W(x,y)$. $\square$

EXERCISE 16.6. For $W \in \mathcal{W}_0$, define $\delta(W) = \min_{x \in [0,1]} t_x(K_2^\bullet, W)$ and $\Delta(W) = \max_{x \in [0,1]} t_x(K_2^\bullet, W)$ (minimum degree and maximum degree). Prove that for any tree T, $\delta(W)^{v(T)-1} \leq t(T, W) \leq \Delta(W)^{v(T)-1}$.

EXERCISE 16.7. For a simple graph F, prove that
$\sup\{t(K_2, W) : W \in \mathcal{W}_0, t(F, W) = 0\} = \sup\{\delta(W) : W \in \mathcal{W}_0, t(F, W) = 0\}$.

16.3. Densities of complete graphs

The problem of describing relationships between densities of complete graphs in a graph G has received special attention. In this section we survey results about these questions in the framework of our book.

16.3.1. Linear inequalities. We start with a result that characterizes linear inequalities between complete subgraph densities. This was proved in a special case by Bollobás [1976], but, as observed by Schelp and Thomason [1998], his proof extends to the proof of the general theorem. Schelp and Thomason give further applications of the method (see Exercise 16.16).

THEOREM 16.8. *Let g be a quantum graph whose constituents are complete graphs. Then $g \geq 0$ if and only if $t(g, K_n) \geq 0$ for every $n \geq 1$.*

Note that the case $n = 1$ is included, which means that $t(g, 0) \geq 0$; in other words, the coefficient of K_0 in g is nonnegative. Since K_n tends to the all-one function as $n \to \infty$, the condition implies that $t(g, 1) \geq 0$. Stating the result more directly: an inequality of the form

$$(16.8) \qquad \sum_{i=1}^{m} a_i t(K_i, G) \geq 0$$

holds for every graph G if and only if

$$(16.9) \qquad \sum_{i=1}^{m} a_i \frac{(n)_i}{n^i} \geq 0$$

holds for every integer $n \geq 1$.

Proof. The "only if" direction is trivial. To prove the "if" direction, suppose that $t(g, K_n) \geq 0$ for all n; we want to prove that $t(g, W) \geq 0$ for every $W \geq 0$. It suffices to prove this for any dense set of graphons W, and we choose the set graphons W_H, where H is a node-weighted simple graph (all edgeweights are 0 or 1). Let $V(H) = [q]$, and let $\alpha_1, \ldots, \alpha_q \geq 0$ be the nodeweights (we allow 0 nodeweights for this argument). We may assume that $\sum_i \alpha_i = 1$. Supposing that there is an H with $t(g, H) < 0$, choose one with minimum number of nodes, and choose the nodeweights so as to minimize $t(g, H)$. Then all the nodeweights must be positive, since a node with weight 0 could be deleted without changing any subgraph density, contradicting the minimality of q. Clearly $t(g, H)$ is a polynomial in the nodeweights α_i. Furthermore, the assumptions that the constituents of g are complete and H has no loops imply that every homomorphism contributing to $t(g, H)$ is injective, and so $t(g, H)$ is multilinear.

Next we prove that H must be complete. Indeed, if (say) nodes $1, 2 \in V(H)$ are nonadjacent, then $t(g, H)$ has no term containing the product $\alpha_1 \alpha_2$, i.e., fixing the remaining variables, $t(g, H)$ is a linear function of α_1 and α_2. Since only the sum of α_1 and α_2 is fixed, we can shift them keeping the sum fixed and not increasing the value of $t(g, H)$ until one of them becomes 0. This is a contradiction, since we know that all weights must be positive.

To show that all weights are equal, let us push the argument above a bit further. Fixing all variables but α_1 and α_2, we can write $t(g, H) = a + b_1 \alpha_1 + b_2 \alpha_2 + c \alpha_1 \alpha_2$. Since H is complete, we know that $t(g, H)$ is a symmetric multilinear polynomial in $\alpha_1, \ldots, \alpha_q$, and so $b_1 = b_2$. Since $\alpha_1 + \alpha_2$ is fixed, we get $t(g, H) = a' + c \alpha_1 \alpha_2$, where a' does not depend on α_1 or α_2. If $c \geq 0$, then this is minimized when $\alpha_1 = 0$ or $\alpha_2 = 0$, which is a contradiction as above. Hence we must have $c < 0$, and in this case $t(g, H)$ is minimized when $\alpha_1 = \alpha_2$. Since this holds for any two variables, all the α_i are equal.

But this means that $t(g, H) = t(g, K_q)$, which is impossible since $t(g, K_q) \geq 0$ by hypothesis. This completes the proof. $\square$

As a corollary (which is in fact equivalent to the theorem) we get the following. Fix an integer $m \geq 1$, and associate with every graphon W the vector $\mathbf{t}_W = \big(t(K_2, W), \ldots, t(K_m, W)\big)$. Let T_m denote the set of the vectors $\mathbf{t}_W$. It follows from Theorem 11.21 and Corollary 11.15 that T_m is the closure of the points $\mathbf{t}_G$, where G is a simple graph (we write $\mathbf{t}_G$ for $\mathbf{t}_{W_G}$).

COROLLARY 16.9. *The extreme points of the convex hull of T_m are the vectors $\mathbf{t}_{K_n}$ ($n = 1, 2, \ldots$) and $(1, \ldots, 1)$.*

The following corollary is interesting to state in view of the undecidability result of Hatami and Norine [2011] already mentioned in the Introduction (which will be proved as Theorem 16.34 a little later). It is easy to design an algorithm to check whether (16.9) holds for every n, and hence:

COROLLARY 16.10. *For quantum graphs g with rational coefficients whose constituents are complete graphs, the property $g \geq 0$ is algorithmically decidable.*

As a further corollary, we derive Turán's Theorem for graphons.

COROLLARY 16.11. *For every $r \geq 2$, we have*
$$\max\{t(K_2, W) : W \in \widetilde{\mathcal{W}}_0, \ t(K_r, W) = 0\} = 1 - \frac{1}{r-1},$$
and the unique optimizer is $W = W_{K_{r-1}}$.

One could prove this result along the lines of several well-known proofs of Turán's Theorem; we could also prove a generalization of Goodman's inequality 2.2. Specializing the proof of Theorem 16.8 above just to this case, we get the proof by "symmetrization", due to Zykov [1949].

Proof. Let us prove the inequality
$$(16.10) \qquad r^r t(K_r, W) - (r-1)t(K_2, W) + r - 2 \geq 0.$$
By Theorem 16.8, it suffices to verify this inequality when $W = W_{K_n}$ for some $n \geq 1$. This is straightforward, and we also see that equality holds for $n = r - 1$ only. Corollary 16.9 implies that equality holds in (16.10) only if $W = W_{K_{r-1}}$. In the special case with $t(K_r, W) = 0$, we get Corollary 16.11. $\square$

16.3.2. Edges vs. triangles. In the introduction (Section 2.1.1) we mentioned several results about the number of triangles in a graph, if the number of edges is known: Goodman's bound and its improvements, and the Kruskal–Katona Theorem. In this Section we describe the exact relationship between the edge density and triangle density in a graph, i.e., we describe the set $D_{2,3}$; for convenience, we recall how it looks (Figure 16.1; also recall that the figure is distorted to be able to see some features better).

As a special case of Corollary 16.9, we get the result of Bollobás [1976] mentioned in the introduction:

COROLLARY 16.12. *The set $D_{2,3}$ is contained in the convex hull of the points $(1, 1)$ and*
$$\mathbf{t}_n = \big(t(K_2, K_n), t(K_3, K_n)\big) = \left(\frac{n-1}{n}, \frac{(n-1)(n-2)}{n^2}\right) \qquad (n = 1, 2, \ldots).$$

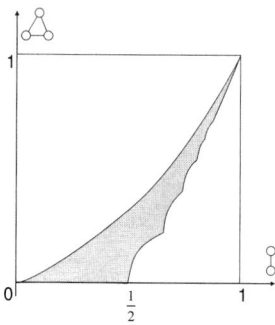

FIGURE 16.1

However, a quick look at Figure 2.1 shows that Corollary 16.12 does not tell the whole story: between any two special points (including the endpoints of the upper boundary), the domain $D_{2,3}$ is bounded by a curve that appears to be concave. It turns out that these curves are indeed concave, which can be proved by the same kind of argument as used in the proof of Theorem 16.8 (see Exercise 16.18). The formula for these curves (cubic equations) is more difficult to obtain, and this will be our main concern in the rest of this section.

The Kruskal–Katona bound. Let us start with the curve bounding the domain T from above, which is not hard to determine: its equation is $y = x^{2/3}$. As we mentioned in the introduction, this follows from (a very special case of) the Kruskal–Katona Theorem in extremal hypergraph theory. Here we give a short direct proof using the formalism of graph algebras. Applying (16.3) with $F_1 = P_3^{\bullet\bullet}$ and $F_2 = P_2^{\bullet\bullet}$, we get

$$\triangle\triangle = [\![(\triangleleft\!\bullet)(\;\!\bullet\;\!)]\!]^2 \leq [\![\triangleleft\!\bullet^2]\!][\![\;\!\bullet\;\!^2]\!] = \bowtie\;\!\bullet \leq \bowtie\bowtie$$

(the last step uses the trivial monotonicity (16.5)). This shows that $t(K_3, W) \leq t(K_2, W)^{3/2}$ for every graphon W, what we wanted to prove.

We also want to prove that this upper bound on the triangle density is sharp. For $n \geq 1$, let G consist of a complete graph on k nodes and $n - k$ isolated nodes. Then $t(K_2, G) = (k)_2/n^2$ and $t(K_3, G) = (k)_3/n^3$. Clearly, points of the form $((k)_2/n^2, (k)_3/n^3)$ get arbitrarily close to any point on the curve $y = x^{2/3}$.

Razborov's Theorem. To determine the lower bounding curve of $D_{2,3}$ is much harder (Razborov [2008]); even the result is somewhat lengthy to state. Perhaps the best way to remember it is to describe a family of extremal graphons (which are all node-weighted complete graphs).

THEOREM 16.13. *For all $0 \leq d \leq 1$, the minimum of $t(K_3, W)$ subject to $W \in \mathcal{W}_0$ and $t(K_2, W) = d$, is attained by the stepfunction $W = W_H$, where H is a weighted complete graph on $k = \lceil \frac{1}{1-d} \rceil$ nodes with edgeweights 1 and appropriate nodeweights: $k-1$ of the nodeweights are equal and the last one is at most as large as these.*

One indication of the difficulty of the proof is that the extremal graphon is not unique, except for the special values $d = 1 - 1/k$. Let us consider the interval I representing the smallest weighted node and the interval J representing any other node. Restricted to $I \cup J$, the graphon is bipartite and hence triangle-free. If we replace the function W_H on $(I \cup J) \times (I \cup J)$ by any other triangle-free function with the same integral, then neither the edge density nor the triangle density changes, but we get a different extremal graphon.

The nodeweights can be determined by simple computation. With a convenient parametrization suggested by Nikiforov [2011], they can be written as $(1+u)/k, \ldots, (1+u)/k, (1-(k-1)u)/k$. The edge density in the extremal graph is

$$(16.11) \qquad t(K_2, H) = d = \frac{k-1}{k}(1-u^2) = \frac{k-1}{k}(1+u)(1-u),$$

and the triangle density is

$$(16.12) \quad t(K_3, H) = \frac{(k-1)(k-2)}{k^2}(1-3u^2-2u^3) = \frac{(k-1)(k-2)}{k^2}(1+u)^2(1-2u).$$

This gives a parametric equation for the cubic curve bordering the domain in Figure 2.1 in the interval $\left[\frac{k-2}{k-1}, \frac{k-1}{k}\right]$. We can solve (16.11) for u as a function of d, and then substitute this in (16.12) to get an explicit expression for $t(K_3, H)$ as a function of d (this hairy formula is not the way to understand or remember the result; but we need it in the proof):

$$(16.13) \qquad t(K_3, H) = f(d) = \frac{(k-1)(k-2)}{k^2}(1+u)^2(1-2u)$$

$$= \frac{(k-1)(k-2)}{k^2}\left(1+\sqrt{1-\frac{kd}{k-1}}\right)^2\left(1-2\sqrt{1-\frac{kd}{k-1}}\right).$$

(where $k = \lceil \frac{1}{1-d} \rceil$). This function is rather complicated, and I would not even bother to write it out, except that we need its explicit form in the proof below. Perhaps the following form says more:

$$(16.14) \qquad \left(1 - \frac{3kd}{2(k-1)} + \frac{k^2 f(d)}{2(k-1)(k-2)}\right)^2 = \left(1 - \frac{kd}{k-1}\right)^3.$$

This shows that after an appropriate affine transformation, every concave piece of the boundary of the region $D_{2,3}$ looks alike (including the curve bounding the region from above). Perhaps this is trying to tell us something—I don't know.

These considerations allow us to reformulate Theorem 16.13 in a more direct form:

THEOREM 16.14. *If G is a graph with $t(K_2, G) = d$, then $t(K_3, G) \geq f(d)$.*

The original proof of this theorem uses Razborov's flag algebra technique, which is basically equivalent to the methods developed in this book. Since then, the result has been extended by Nikiforov [2011] to the number of K_4's and by Reiher [2012] to all complete graphs. We describe the proof of Razborov's Theorem in our language. (Reiher's proof can be viewed as a generalization of this argument to all complete graphs; the generalization is highly nontrivial.)

Proof. Let $W \in \mathcal{W}_0$ minimize $t(K_3, W) - f\big(t(K_2, W)\big)$ subject to $\frac{k-2}{k-1} \leq t(K_2, W) \leq \frac{k-1}{k}$, and suppose (by way of contradiction) that the minimum value is negative. Set $d = t(K_2, W)$ and

$$(16.15) \qquad \lambda = f'(d) = \frac{3(k-2)}{k}\left(1+\sqrt{1-\frac{kd}{k-1}}\right) = \frac{3(k-2)}{k}(1+u).$$

(The representation in terms of the parameter u will be useful if you want to follow some of the computations below). Since the objective function is 0 at the endpoints of the interval, we must have $\frac{k-2}{k-1} < d < \frac{k-1}{k}$, and so W is a local minimizer in $\mathcal{W}_0$.

Let us simplify notation by writing $g_1 \equiv g_2$ (for W) and $g_1 \leq g_2$ (for W) for two quantum graphs g_1 and g_2 if $t(g_1, W) = t(g_2, W)$ and $t(g_1, W) \leq t(g_2, W)$, respectively. We invoke the formulas obtained by variational calculus on graphons, and get by Lemma 16.5(a) that

(16.16) $\qquad 3\, [\text{path}_\bullet] - 2\lambda\, [\text{edge}_\bullet] = 3\, [\text{path}] - 2\lambda\, [\text{edge}]$ (for W)

(if a graph pictogram has a black node, this means that the equation or inequality holds for every choice of the image of the black node). Multiplying (16.16) by the edge with one node labeled, and then unlabeling, we get

(16.17) $\qquad 3\, [\text{4-graph}_1] - 2\lambda\, [\text{3-graph}_1] = 3\, [\text{4-graph}_2] - 2\lambda\, [\text{3-graph}_2]$ (for W).

By Lemma 16.5(c),

(16.18) $\qquad 3\, [\text{path}_\bullet] - \lambda\, [\text{edge}_\bullet] \leq 0$ (for W).

Multiplying with the signed 3-node path with both edges negative and with both endpoints labeled, we get

(16.19) $\qquad 3\, [\text{4-graph}] \leq \lambda\, [\text{3-graph}]$ (for W),

which can be written as

$$3\, [\text{G}_1] - 6\, [\text{G}_2] + 3\, [\text{G}_3] \leq \lambda\, [\text{H}_1] - 2\lambda\, [\text{H}_2] + \lambda\, [\text{H}_3] \text{ (for } W\text{)},$$

or, simplifying and leaving just the 4-node graphs on the left,

(16.20) $\qquad 3\, [\text{G}_1] - 6\, [\text{G}_2] \leq (\lambda - 3)\, [\text{H}_1] + \lambda\, [\text{H}_2] - 2\lambda\, [\text{H}_3]$ (for W).

We get a third inequality from inclusion-exclusion:

$$[\text{A}] = [\text{B}] - 3\, [\text{C}] + 3\, [\text{D}] - [\text{E}],$$

whence

(16.21) $\qquad 3\, [\text{X}_1] - 3\, [\text{X}_2] = [\text{Y}_1] - [\text{Y}_2] - [\text{Y}_3] \leq [\text{Z}_1] - [\text{Z}_2]$ (for W).

Adding (16.17), (16.20) and (16.21), we get

$$0 \leq (\lambda - 2)\, [\text{a}] + \lambda\, [\text{b}] + 3\, [\text{c}] - 2\lambda\, [\text{d}] - [\text{e}] \text{ (for } W\text{)}.$$

We can replace every $[\text{edge}]$ by d, to get

(16.22) $\qquad (\lambda + 3d - 2)\, [\text{tri}] \geq \lambda(2d^2 - d) + [K_4]$ (for W).

For $k = 3$ (i.e., $\frac{1}{2} < d < \frac{2}{3}$), we can just ignore the K_4 term (it is nonnegative), and hence (upon verifying that $\lambda + 3d - 2 > 0$)

$$t(K_3, W) \geq \frac{\lambda(2d^2 - d)}{\lambda + 3d - 2} = \frac{f'(d)(2d^2 - d)}{f'(d) + 3d - 2} = f(d)$$

(where the last equality is easy to check).

However, if $k \geq 4$, then we need a nontrivial lower bound for $t(K_4, W)$ (note that the extremal graph H contains many K_4-s). Let

(16.23) $\qquad \mu = 2\lambda d - 3t(K_3, W) \quad \text{and} \quad w(z) = t_z(K_2^\bullet, W) = \int_0^1 W(z, y)\, dy.$

Then we can write (16.16) as

(16.24) $$3t_z(K_3^\bullet, W) = 2\lambda w(z) - \mu.$$

The numbers λ and μ will play an important role in our computations, so let us have a closer look at them. Recall that $\lambda = f'(d)$ is given by (16.15), which yields the bounds

(16.25) $$\frac{3(k-2)}{k} \leq \lambda \leq \frac{3(k-2)}{k-1}.$$

For μ we don't have an explicit formula in terms of d alone, but we can get rather tight bounds: from Goodman's Theorem and the indirect hypothesis we get $d(2d-1) \leq t(K_3, W) < f(d)$, which in turn implies

$$2df'(d) - 3f(d) < \mu \leq 2df'(d) - 3d(2d-1).$$

From here, it takes straightforward computation to verify that

(16.26) $$\frac{k-1}{k-2} < \frac{3\mu}{\lambda^2} \leq \frac{k-2}{k-3}$$

We see that $\mu > 0$. Furthermore, inequality 16.18 implies that

(16.27) $$3t_z(K_3^\bullet, W) \leq \lambda w(z),$$

for almost all $z \in [0, 1]$, and then (16.27) and (16.24) imply that

$$\frac{\mu}{2\lambda} \leq w(z) \leq \frac{\mu}{\lambda}.$$

For any point $z \in [0, 1]$, the density of K_4-s containing it is just the density of triangles in its neighborhood. To be precise, for any point $z \in [0, 1]$, we define a new graphon W_z on $[0, 1]$, by keeping the same W but adding a nodeweight function $W(z, .)/w(z)$. Then

$$t(K_2, W_z) = \frac{t_z(K_3^\bullet, W)}{w(z)^2} = \frac{2\lambda w(z) - \mu}{3w(z)^2}.$$

The right hand side is a monotone increasing function of $w(z)$ for $w(z) \in [\frac{\mu}{2\lambda}, \frac{\mu}{\lambda}]$, and hence

$$t(K_2, W_z) = \frac{2\lambda w(z) - \mu}{3w(z)^2} \leq \frac{\lambda^2}{3\mu} \leq \frac{k-2}{k-1}.$$

So by induction on k, we know that $t(K_3, W_z) \geq f(t(K_2, W_z))$, and

$$t_z(K_4^\bullet, W) = t(K_3, W_z)w(z)^3 \geq f\left(\frac{2\lambda w(z) - \mu}{3w(z)^2}\right)w(z)^3.$$

Hence

$$t(K_4, W) \geq \int_0^1 f\left(\frac{2\lambda w(z) - \mu}{3w(z)^2}\right)w(z)^3 \, dz.$$

The ugly integrand is hopeless to evaluate directly; the trick is to find a lower bound that is linear in $w(z)$ and approximates it well enough. Set $g(w) = f\left(\frac{2\lambda w - \mu}{3w^2}\right)w^3$. Let $w_0 \in [\frac{\mu}{2\lambda}, \frac{\mu}{\lambda}]$ be chosen so that

(16.28) $$\frac{2\lambda w_0 - \mu}{3w_0^2} = \frac{k-3}{k-2}.$$

This is possible, since the left side, as a function of w_0, ranges from 0 to $\lambda^2/(3\mu)$ on this interval, and the target value $(k-3)/(k-2)$ is in this range by (16.26). Then

$$g(w_0) = f\left(\frac{k-3}{k-2}\right)w_0^3 = \frac{(k-3)(k-4)}{(k-2)^2}w_0^3.$$

The linear function we will use goes through the point $(w_0, g(w_0))$ and has appropriate slope:

CLAIM 16.15. *Let λ and μ be real numbers satisfying (16.25) and (16.26), and let w_0 satisfy (16.28). Then for every $w \in [\frac{\mu}{2\lambda}, \frac{\mu}{\lambda}]$, we have*

$$g(w) - g(w_0) \geq \frac{1}{3}(2\lambda^2 - 3\mu)(w - w_0).$$

All functions in this claim are explicit, which makes it a (hard and tedious) exercise in first year calculus; we do not reproduce the details of its proof. Using this claim, we have

$$t(K_4, W) \geq \int_0^1 g(w(z))\,dz \geq g(w_0) - \frac{1}{3}(2\lambda^2 - 3\mu)w_0 + \frac{1}{3}(2\lambda^2 - 3\mu)\int_0^1 w(z)\,dz$$

$$= \frac{(k-3)(k-4)}{(k-2)^2}w_0^3 + \frac{1}{3}(2\lambda^2 - 3\mu)(d - w_0).$$

Hence, returning to (16.22),

(16.29) $\quad (\lambda + 3d - 2)t(K_3, W) \geq \lambda(2d^2 - d) + \dfrac{(k-3)(k-4)}{(k-2)^2}w_0^3 + \dfrac{1}{3}(2\lambda^2 - 3\mu)(d - w_0).$

This is another messy formula, but we can express the variables in terms of u and $y = w_0/(1+u)$ (the latter is, of course, chosen with hindsight): we already have expressions for λ and d; we have $w_0 = y(1+u)$, and then μ can be expressed using (16.28). With these substitutions, the difference of the two sides looks like

(16.30) $\quad (1+u)^2\left(y - \dfrac{k-2}{k}\right)$

$$\times \left(\frac{k-1}{k}(1-u) - \left(\frac{k-3}{k-2}u + \frac{3k-7}{k-2}\right)y + \frac{2(k-1)(k-3)}{k(k-2)}(1+u)y^2\right).$$

We know that $0 < u < \frac{1}{k-1}$, and (16.28) and (16.26) imply that $\frac{k-2}{k} \leq y \leq \frac{(k-2)^2}{k(k-3)}$. Then we face another exercise in calculus, to show that in this range (16.30) is negative (we don't describe the details). This contradicts (16.29), and completes the proof. $\square$

It would of course be important to find a more "conceptual" proof of Theorem 16.13. As a couple of examples of the kind of general question that arises, is an algebraic inequality between densities of complete graphs decidable? Does such an inequality hold true if it holds true for all node-weighted complete graphs?

EXERCISE 16.16. Let g be a quantum graph such that every constituent with negative coefficient is complete. Prove that $t_{\text{inj}}(g, W) \geq 0$ for every graphon W if and only if $t_{\text{inj}}(g, K_n) \geq 0$ for all $n \geq 1$ (Schelp and Thomason [1998]).

EXERCISE 16.17. Prove that for quantum graphs g with rational coefficients whose constituents are complete graphs, the property $g \geq 0$ is in P. (The input length is the total number of digits in the numerators and denominators of the coefficients, and complete k-graphs ($0 \leq k \leq m$) contribute 1 even it their coefficient is 0.)

EXERCISE 16.18. Adopt the proof of Theorem 16.8 to prove that the boundary of the domain $D_{2,3}$ (Fig. 2.1) is concave between any two special points $\mathbf{t}_n$ and $\mathbf{t}_{n+1}$.

EXERCISE 16.19. (a) Let K'_r denote the graph obtained by deleting an edge from K_r. Prove that
$$t(K'_{r+1}, G) \geq \frac{t(K_r, G)^2}{t(K_{r-1}, G)}.$$

(b) Prove that
$$t(K_{r+1}, G) - t(K_r, G) \leq r\bigl(t(K_{r+1}, G) - t(K'_{r+1}, G)\bigr).$$

(c) Prove that
$$r\frac{t(K_r, G)}{t(K_{r-1}, G)} \leq (r-1)\frac{t(K_{r+1}, G)}{t(K_r, G)} + 1.$$

(d) Prove the following result of Moon and Moser [1962]: If N_r denotes the number of complete r-graphs in G, then
$$\frac{N_{r+1}}{N_r} \geq \frac{1}{r^2 - 1}\left(r^2 \frac{N_r}{N_{r-1}} - N_1\right).$$

EXERCISE 16.20. Prove the following generalization of Goodman's Theorem (2.2): if $d = t(K_2, W)$ is the edge density of a graph G, then
$$t(K_r, G) \geq d(2d - 1)(3d - 2) \cdots \bigl((r-1)d - (r-2)\bigr).$$

EXERCISE 16.21. Prove that
$$t(K'_4, G) \geq t(K_3, G)^2 \log^*\bigl(1/t(K_3, G)\bigr).$$

[T. Tao; hint: use the Removal Lemma.]

16.4. The classical theory of extremal graphs

Following the exposition of this topic in the introduction, let us state now in more general and more precise terms the extremal graph results of Erdős and Stone [1946], Erdős and Simonovits [1966], and Simonovits [1968]. In this more general setting, we exclude several graphs $L_1, \ldots, L_k$ as subgraphs of a simple graph G, and we want to determine the maximum number of edges of G, given the number of nodes n. In our formalism, we want to solve

(16.31) $\max\{t(K_2, G) : t_{\text{inj}}(L_1, G) = \cdots = t_{\text{inj}}(L_k, G) = 0\}.$

Turán's Theorem (Corollary 16.11) is a special case when $k = 1$ and $L_1 = K_r$.

The key results are summed up in the following theorem.

THEOREM 16.22. *Let $L_1, \ldots, L_k$ be simple graphs and let $r = \min_i \chi(L_i)$. Suppose that a simple graph G does not contain any L_i as a subgraph. Then*

(16.32) $t(K_2, G) \leq 1 - \dfrac{1}{r-1} + o(1) \qquad (\mathsf{v}(G) \to \infty).$

Asymptotic equality holds $G = T(n, r-1)$ is the Turán graph on $n = \mathsf{v}(G)$ nodes with $r - 1$ color classes. Furthermore, this extremal graph is stable in the following sense: For every $\varepsilon > 0$ there is an $\varepsilon' > 0$ (depending on $L_1, \ldots, L_k$ and ε, but not on G) such that if $t(K_2, G) \geq 1 - 1/(r-1) - \varepsilon'$, then $\widehat{\delta}_1(G_n, T(n, r-1)) \leq \varepsilon$.

Theorem 16.22 can be translated to the language of graphons, and proved quite easily using our general results on graph limits. As in almost all of our applications of graph limit theory, the original treatment has the advantage that it provides explicit bounds for the $o(1)$ term as well as the dependence of ε' on ε in the theorem above.

THEOREM 16.23. *Let $L_1, \ldots, L_k$ be simple graphs and let $r = \min_i \chi(L_i)$. Then*

$$\max\{t(K_2, W) : W \in \mathcal{W}_0, \ t(L_1, W) = \cdots = t(L_k, W) = 0\} = 1 - \frac{1}{r-1},$$

and the unique optimizer (up to weak isomorphism) is $W = W_{K_{r-1}}$.

Proof. Let, say $\chi(L_1) = r$. Then $t(L_1, K_r) > 0$, and hence it follows easily that $t(K_r, W) = 0$ (Exercises 7.6, 13.25). Application of Turán's Theorem for graphons (Corollary 16.11) completes the proof. □

It is not quite trivial, but not very hard either, to derive the classical results mentioned above from Theorem 16.23. Let us illustrate this by the derivation of stability. If stability fails, then there exists a sequence G_n of simple graphs such that $t_{\mathsf{inj}}(L_1, G_n) = \cdots = t_{\mathsf{inj}}(L_k, G_n) = 0$ and $t(K_2, G_n) \to 1 - 1/(r-1)$, but $\widehat{\delta}_1(G_n, T(n, r-1)) \not\to 0$. By Theorem 9.30, this implies that $\delta_1(G_n, T(n, r-1)) \not\to 0$. In graphon language, this means that $\delta_1(W_{G_n}, W_{K_{r-1}}) \not\to 0$. By choosing a subsequence and than a subsequence of that, we may assume that $\delta_1(W_{G_n}, W_{K_{r-1}}) > a$ for some $a > 0$ for all n, and $G_n \to U$ for some graphon U. Then $t(L_1, U) = \cdots = t(L_k, U) = 0$ and $t(K_2, U) = 1 - 1/(r-1)$, so by Theorem 16.23, U is weakly isomorphic to $W_{K_{r-1}}$. So $G_n \to W_{K_{r-1}}$. Since $W_{K_{r-1}}$ is 0-1 valued, it follows by Proposition 8.24 that $\delta_1(W_{G_n}, W_{K_{r-1}}) \to 0$, a contradiction.

16.5. Local vs. global optima

One advantage of embedding the set of graphs to the large space of graphons is that for every optimization problem, we can define local optima, and study then with the methods of analysis. We describe three problems where this treatment gives interesting results.

In our first example, every local optimum is also global by convexity, and this fact can be exploited to get a short proof. In the second example, we don't know whether local and global optima are the same; we can determine the local ones, and perhaps these are also global. In the third example, local and global optima are quite different, and the consideration of local optima leads to interesting results.

16.5.1. The distance from a hereditary graph property. A surprisingly general result in extremal graph theory is the theorem of Alon and Stav [2008], proving that for every hereditary property, a random graph with appropriate density is asymptotically the farthest from the property in edit distance.

THEOREM 16.24 **(Alon and Stav)**. *For every hereditary graph property $\mathcal{P}$ there is a number p, $0 \leq p \leq 1$, such that for every graph G with $\mathsf{v}(G) = n$,*

$$d_1(G, \mathcal{P}) \leq \mathsf{E}\big(d_1(\mathbb{G}(n, p), \mathcal{P})\big) + o(1) \qquad (n \to \infty).$$

The following theorem of Lovász and Szegedy [2010a] states a graphon version of this fact. (Recall that, by Proposition 14.25, the closure of a hereditary graph property is flexible. We omit the details of the derivation of Theorem 16.24 from Theorem 16.25.)

THEOREM 16.25. *If $\mathcal{R}$ is a flexible graphon property, then the maximum d_1-distance from $\mathcal{R}$ is attained by a constant function.*

Proof. This proof illustrates the power of extending graph problems to a continuum. By Proposition 14.26, the set $\mathcal{W}_0 \setminus \mathcal{R}$ is convex. Hence it follows that the d_1 distance from $\mathcal{R}$ is a concave function on $\mathcal{W}_0 \setminus \mathcal{R}$. We also know by Proposition 15.15 and Theorem 15.16(b) that $d_1(.,\mathcal{R})$ is a continuous function on $(\widetilde{\mathcal{W}_0}, \delta_\square)$, and hence it assumes its maximum. Let M be the set of maximizing graphons in $\mathcal{W}_0$; this is a convex, closed subset of $\mathcal{W}_0$. Since $\mathcal{W}_0 \setminus \mathcal{R}$ is invariant under the group of invertible measure preserving transformations of $[0, 1]$, so is M, and hence M is also compact in the pseudometric $\delta_\square$.

This implies in many ways that M contains a constant function; here is a fast argument. Let $W \in M$ have minimum L_2-norm (such a graphon exists, since by Lemma 14.15 the L_2-norm is lower semicontinuous with respect to the cut norm). For every measure preserving transformation φ, we have $(W + W^\varphi)/2 \in M$. Furthermore,

$$\left\| \frac{W + W^\varphi}{2} \right\|_2 \leq \frac{1}{2}(\|W\|_2 + \|W_\varphi\|_2) = \|W\|_2.$$

By the choice of W, we must have equality here, which implies that W and W^φ are proportional, but since they have the same integral, they must be equal almost everywhere. Since this holds for any φ, W must be constant almost everywhere. $\square$

16.5.2. The Sidorenko Conjecture. Sidorenko [1991, 1993] conjectured that the inequality

(16.33) $$t(F, W) \geq t(K_2, W)^{e(F)}.$$

holds for all bipartite graphs F and all $W \in \mathcal{W}$, $W \geq 0$. Several special cases of this inequality were mentioned in the Introduction, Section 2.1.2. Sidorenko in fact formulated this not only for graphs but for graphons, being perhaps the first to use the integral expression for $t(.,.)$ as a generalization of subgraph counting. (The conjecture extends to non-symmetric functions W, but we restrict our attention to the symmetric case here.) A closely related conjecture in extremal graph theory was raised earlier by Simonovits [1984]. In spite of its very simple form and a lot of effort, this conjecture is unproven in general.

It is easy to see that every graph satisfying Sidorenko's Conjecture must be bipartite. Indeed, if $W = W_{K_2}$, then the right side of (16.33) is positive, but the left side is positive only if F is bipartite.

We can view this as an extremal problem in two ways: (1) for every nonnegative $W \in \mathcal{W}$, matchings minimize $t(F, W)$ among all bipartite graphs with a given number of edges; (2) for every bipartite graph F, constant functions W minimize $t(F, W)$ among all nonnegative kernels W with a given integral. Since both sides of (16.33) are homogeneous in W of the same degree, we can scale W and assume that $t(K_2, W) = 1$. Then we want to conclude that $t(F, W) \geq 1$ for every bipartite graph F.

There are partial results in the direction of the conjecture. Sidorenko proved it for a fairly large class of graphs, including trees, complete bipartite graphs, and all bipartite graphs with at most 4 nodes in one of the color classes. After a long period of little progress, several new (but unfortunately still partial) results were obtained recently. Each of these is in one way or other related to the material in this book, so we discuss them in some detail.

We have defined weakly norming graphs in Section 14.1. Hatami [2010] gives a proof of the following (easy) fact, attributing it to B. Szegedy : *If a bipartite graph*

F is weakly norming, then it satisfies the Sidorenko conjecture. Combined with the result of Hatami (Proposition 14.2) that all cubes are weakly norming, it follows that all cubes satisfy Sidorenko's conjecture.

In another direction, Conlon, Fox and Sudakov [2010] proved that the conjecture is satisfied by every bipartite graph that contains a node connected to all nodes on the other side. Their proof uses a sophisticated probabilistic argument. Li and Szegedy [2012] give a shorter analytic proof, which extends to a larger class of graphs. Szegedy [unpublished] uses entropy arguments to prove the conjecture for an even larger class, which includes of all previously settled special cases. The smallest graph for which the conjecture is not known is the Möbius ladder of length 5 (equivalently, a 10-cycle with the longest diagonals added).

16.5.3. Local Sidorenko Conjecture. We can ask for conditions on W, rather than on F, that suffice to prove the conjectured inequality (16.33). Lovász [2011] proved that for every bipartite graph F, the constant 1 kernel minimizes $t(F, W)$ at least locally:

THEOREM 16.26. *Let F be a bipartite graph with m edges. Let W be a kernel with $\int W = 1$ and $\|W - 1\|_\infty \leq 1/(4m)$. Then $t(F, W) \geq 1$.* □

The proof of this theorem is not given here; instead, we prove the following easier related result. Let us say that a graph F has the *local Sidorenko property* if for every kernel $W \geq 0$ there is an $\varepsilon_W > 0$ such that for every $0 \leq \varepsilon \leq \varepsilon_W$, we have $t(F, 1 + \varepsilon(W - 1)) \geq 1$. (For a graph satisfying the Sidorenko conjecture, we have $\varepsilon_W = 1$ for every $W \geq 0$.) With this weaker notion, a graph does not have to be bipartite to satisfy it. In fact, we have the following characterization of these graphs:

PROPOSITION 16.27. *A graph has the local Sidorenko property if and only if either it is a forest or its girth is even.*

In particular, every bipartite graph has the local Sidorenko property.

Proof. Let us start with the "if" part. Replacing W by $1 + (W-1)/\|W-1\|_\infty$, we may assume that $0 \leq W \leq 2$. Let $U = W - 1$, then $U \in \mathcal{W}_1$. The homomorphism density $t(F, W) = t(F, 1 + U)$ can be expanded in terms of the subgraphs of F, and so what we want to prove is

$$\sum_{F' \subseteq F} t(F', \varepsilon U) \geq 1$$

for a sufficiently small $\varepsilon > 0$. (Let us agree that, for the rest of this section, $F' \subseteq F$ means that F' is a subgraph of F without isolated nodes.) The term with $F' = \emptyset$ is 1, and so (pulling out the ε factors) we want to prove that

(16.34) $$\sum_{\emptyset \neq F' \subseteq F} t(F', U) \varepsilon^{e(F')} \geq 0.$$

It follows from the definition of U that $t(K_2, U) = 0$, and so every term in (16.34) where F' is a matching cancels. If F itself is a matching, we have nothing to prove. Otherwise, the next smallest term is $t(P_3, U)$, which is nonnegative, since $P_3 = [\![(K_2^\bullet)^2]\!]$ is a square. If $t(P_3, U) > 0$, then for every sufficiently small $\varepsilon > 0$ it dominates the sum (16.34), and we are done. So suppose that $t(P_3, U) = 0$, then

$t_x(K_2^\bullet, U) = 0$ for almost all x. This implies that $t(F', U) = 0$ whenever U has a node of degree 1. In particular, if F is a forest, we are done.

Suppose that F is not a forest, then the nonzero term in 16.34 with the smallest number of edges is $t(C_{2r}, U)$, where C_{2r} is the shortest cycle in F. Since $t(C_{2r}, U)^{1/(2r)}$ is a norm by Proposition 14.2, this term is nonzero if $U \neq 0$, and so for a sufficiently small $\varepsilon > 0$, it dominates the remaining terms.

To prove the "only if" part, suppose that the girth of F is odd. Let U be the graphon defined by the matrix $\begin{pmatrix} -1 & 1 \\ 1 & -1 \end{pmatrix}$. Then $t_x(K_2^\bullet, U) = 0$ for every x, and hence all those terms in (16.34) are 0 in which F has a node with degree 1. So the nonzero terms with the smallest exponent of ε correspond to the shortest (odd) cycles. Trivially $t(F', W) = -1$ for such a term, and so for a sufficiently small ε the whole expression (16.34) will be negative. □

The proof of Theorem 16.26 expands the idea of the proof above, but one has to do much more careful estimations. Many steps in the proof can be viewed as using a version of the calculus "for $\mathcal{W}_0$" developed before, but this time "for $\mathcal{W}_1$". Some steps are described in Exercises 16.29-16.31 to illustrate this potentially useful technique.

16.5.4. Common graphs. The following inequality is closely related to Goodman's Theorem (2.2), and it can be proved along the same lines:

$$(16.35) \qquad t(K_3, G) + t(K_3, \overline{G}) \geq \frac{1}{4},$$

and equality holds asymptotically if G is a random graph with edge density $1/2$. Erdős conjectured that a similar inequality will hold for K_4 in place of K_3, but this was disproved by Thomason [1998]. More generally, one can ask which graphs F satisfy

$$t_{\mathsf{inj}}(F, G) + t_{\mathsf{inj}}(F, \overline{G}) \geq (1 + o(1))2^{1-\mathsf{e}(F)},$$

where the $o(1)$ refers to $\mathsf{v}(G) \to \infty$. Going to the limit, we get a formulation free of remainder terms: Which simple graphs F satisfy

$$(16.36) \qquad t(F, W) + t(F, 1 - W) \geq 2^{1-\mathsf{e}(F)} = 2t(F, \frac{1}{2})$$

for every graphon W? Such graphs F are called *common graphs*. So the triangle is common, but K_4 is not. Are there any other common graphs?

Sidorenko [1996] studied graphs with this and other "convexity" properties. Let F be a graph satisfying Sidorenko's conjecture. Then

$$t(F, W) + t(F, 1 - W) \geq t(K_2, W)^{\mathsf{e}(F)} + t(K_2, 1 - W)^{\mathsf{e}(F)}$$
$$\geq 2\left(\frac{t(K_2, W) + t(K_2, 1 - W)}{2}\right)^{\mathsf{e}(F)} = 2^{1-\mathsf{e}(F)},$$

so F is common. Sidorenko's conjecture would imply that all bipartite graphs are common, and all bipartite graphs mentioned above for which Sidorenko's conjecture is verified are common. Among non-bipartite graphs, not many common graphs are known. Jagger, Štovíček and Thomason [1996] showed that no graph containing K_4 is common.

Franek and Rödl [1992] showed that if we delete an edge from K_4, the obtained graph K_4' is common. Recently Hatami, Hladky, Král, Norine and Razborov [2011] proved that the 5-wheel is common, using computers to find appropriate nonnegative expressions in the flag algebra. We cannot reproduce their proof here; instead, let us give the proof of the fact that K_4' is common, which should give a feeling for this technique. We do our computations in the graph algebra, instead of the flag algebra.

We start with rewriting (16.36) as follows. Let $U = 2W - 1$, then substituting U in (16.36) and multiplying by $2^{\mathsf{e}(F)}$, we get

$$(16.37) \qquad t(F, 1+U) + t(F, 1-U) \geq 2,$$

which should hold for every $U \in \mathcal{W}_1$. In other words, F is common iff the left side is minimized by $U = 0$.

The subgraph densities on the left side of (16.37) can be expanded as before, and we get

$$t(F, 1+U) + t(F, 1-U) = 2 \sum_{\substack{F' \subseteq F \\ \mathsf{e}(F') \text{ even}}} t(F', U).$$

The term with $\mathsf{e}(F') = 0$ gives 2, the value on the right side of (16.37), so F is common if and only if

$$(16.38) \qquad \sum_{\substack{F' \subseteq F \\ \mathsf{e}(F') > 0, \text{ even}}} t(F', U) \geq 0$$

for every $U \in \mathcal{W}_1$. Note that inequality (16.38) has to be true for all $U \in \mathcal{W}_1$, not just for $U \in \mathcal{W}_0$, so the fact that all terms on the left have nonnegative coefficient does not make this relation trivial.

As an example, in the case $F = K_4'$ we get from (16.38) that its commonness follows if we can show that

$$(16.39) \qquad 2\,\vcenter{\hbox{⚬⚬}} + 8\,\vcenter{\hbox{⚬⚬}} + \vcenter{\hbox{⚬⚬}} + 4\,\vcenter{\hbox{⚬⚬}} \geq 0 \text{ (for } \mathcal{W}_1\text{)}.$$

Here we can write the left side as

$$\left[\!\left[2\,\vcenter{\hbox{⚬}}^2 + 4\,\vcenter{\hbox{⚬}}^2 + \left(\vcenter{\hbox{⚬}} + 2\,\vcenter{\hbox{⚬}}\right)^2 \right]\!\right] + 4(\vcenter{\hbox{⚬}} - \vcenter{\hbox{⚬}}).$$

It is easy to see (see Exercise 16.30) that the last term is nonnegative. The other terms are squares, which proves (16.39).

Locally common graphs. We say that a graph F is *locally common*, if for every $U \in \mathcal{W}_1$ there is a $0 < \varepsilon_U \leq 1$ such that if $0 < \varepsilon < \varepsilon_U$, then $t(F, 1+\varepsilon U) + t(F, 1-\varepsilon U) \geq 2$.

Franek and Rödl [1992] proved that K_4 is locally common. In fact, the following more general result holds, and can be proved along the lines of the proof of Proposition 16.27, using formula 16.38.

PROPOSITION 16.28. *Let G be a graph in which the subgraph with the minimum number of edges such that all degrees are at least 2 and the number of edges is even is an even cycle. Then G is locally common.* □

In particular, every bipartite graph is locally common (this follows by Proposition 16.27 as well), and so is every simple graph containing a 4-cycle. Combining with the theorem of Jagger, Štovíček and Thomason [1996] mentioned above, it follows that every graph that contains a K_4 is not common but locally common. Not all graphs are locally common (see Exercise 16.33).

EXERCISE 16.29. Prove that $C_2 \geq C_4 \geq C_6 \geq \ldots$ (for $\mathcal{W}_1$).

EXERCISE 16.30. Prove that $[\![F_1^2 F_2]\!] \leq [\![F_1^2]\!]$ (for $\mathcal{W}_1$) for any two k-labeled multigraphs F_1 and F_2.

EXERCISE 16.31. Suppose that a bipartite graph F contains a 4-cycle. Prove that $F \leq C_4$ (for $\mathcal{W}_1$). More generally, if F is not a forest, then $F \leq C_{2r}$ (for $\mathcal{W}_1$), where C_{2r} is the shortest cycle in F.

EXERCISE 16.32. Prove that triangles are common (16.35).

EXERCISE 16.33. Prove that $[\![C_7^\bullet C_{11}^\bullet]\!]$ is not locally common.

16.6. Deciding inequalities between subgraph densities

Now we turn to more general questions in extremal graph theory, as we already indicated in the Introduction.

16.6.1. Undecidability of density inequalities. In analogy with Artin's Theorem for real polynomials (see Appendix A.7), we may try to represent quantum graphs g with $g \geq 0$ as sums of squares or, more generally, as quotients of sums of squares: if y and z are square-sums and $y = zg$, then $g \geq 0$. The following theorem by Hatami and Norine [2011] tells us once and forever that no such "certificate" of nonnegativity can be given.

THEOREM 16.34. *It is algorithmically undecidable whether an inequality $g \geq 0$ holds (where g is a quantum graph with rational coefficients).*

It is a related, but easier, fact that the following is algorithmically undecidable: given simple graphs $F_1, \ldots, F_m$ and rational coefficients $a_1 \ldots, a_m$, decide whether $\sum_i a_i \hom(F, G_i) \geq 0$ holds for every simple graph G. This follows from a result of Ioannidis and Ramakrishnan [1995], proved in a completely different setting of databases; for the adaptation to the graph case, see Exercise 16.43, based on the lecture notes of Kopparty [2011].

The proof will consist of a reduction of the problem of deciding whether a given polynomial $p \in \mathbb{Z}[x_1, \ldots, x_k]$ is nonnegative for every $x_1, \ldots, x_k \in \mathbb{N}$, to the problem of deciding whether $x \geq 0$ for a quantum graph g with rational coefficients. Since the latter problem is undecidable by the Theorem of Matiyasevich (see Section A.7 in the Appendix), this will imply that so is the former. To this end, we need a version of Problem A.33, which asks for nonnegativity on another set of numbers, namely on the set

$$A = \left\{ 1 - \frac{1}{n} : n = 1, 2, \ldots \right\}.$$

The variables will be represented by edge densities and triangle densities in appropriate graphs, and the fact that Goodman's bound is only attained for edge densities in A (Corollary 16.12) will then be used to force the edge-densities to be of this form.

Proof of Theorem 16.34. We start with some algebraic reductions.

CLAIM 16.35. *It is algorithmically undecidable whether an inequality $p \geq 0$ ($p \in \mathbb{Z}[x_1, \ldots, x_k]$) holds on A^k.*

Indeed, one can reduce deciding if $p \geq 0$ on $\mathbb{N}^k$ to deciding whether $p \geq 0$ on A^k by a straightforward change of variables.

The connection to graph theory is established by the following reduction. Consider the set $D_{2,3}$ of all pairs $(t(K_2, W), t(K_3, W))$ where W is a graphon (Figures

2.1 and 16.1). Let $D \subseteq D_{2,3}$ consist of the points (x, y) where $x \in A$ and $y = 2x^2 - x$ (i.e., pairs $\bigl(t(K_2, K_n), t(K_3, K_n)\bigr)$ together with $(0,0)$ and $(1,1)$).

For a polynomial $p \in \mathbb{Z}[x_1, \ldots, x_k]$, construct the polynomial in $2k$ variables:

$$p^*(x_1, y_1, \ldots, x_k, y_k) = \prod_{i=1}^{k}(1-x_i)^2 p(x_1, \ldots, x_k) + \sum_{i=1}^{k} M(y_i - 2x_i^2 + x_i),$$

where $M = 2\max\{\|(\mathrm{grad}\, p)(x)\|_\infty : x \in [0,1]^k\}$. We also construct a polynomial in $3k$ variables:

$$p^{**}(u_1, v_1, w_1, \ldots, u_k, v_k, w_k) = (u_1 \ldots u_k)^N p^*\Bigl(\frac{v_1}{u_1^2}, \frac{w_1}{u_1^3}, \ldots, \frac{v_k}{u_k^2}, \frac{w_k}{u_k^3}\Bigr),$$

(where $N = 5\deg(p^*)$ is large enough to cancel all denominators).

CLAIM 16.36. *The following are equivalent:* (i) $p \geq 0$ *on* A^k; (ii) $p^* \geq 0$ *on* $D_{2,3}^k$; (iii) $p^* \geq 0$ *on* D^k.

Trivially, (ii)⇒(iii)⇒(i). To prove that (i)⇒(ii), assume that $p \geq 0$ on A^k, and let $(x_i, y_i) \in D_{2,3}$. We want to prove that $p^*(x_1, y_1, \ldots, x_k, y_k) \geq 0$. We start with bounding the first term in the definition of p^*. Let $z_i \in A$ be closest to x_i. Then $p(z_1, \ldots, z_k) \geq 0$, and so

$$p(x_1, \ldots, x_k) \geq p(x_1, \ldots, x_k) - p(z_1, \ldots, z_k) = (x-z) \cdot (\mathrm{grad}\, p)(\xi),$$

where $\xi \in [0,1]^k$. By the definition of M, we get that

$$(16.40) \quad \prod_{i=1}^{k}(1-x_i)^2 p(x_1, \ldots, x_n) \geq -\prod_{i=1}^{k}(1-x_i)^2 \frac{M}{2} \sum_{i=1}^{k}|x_i - z_i|$$

$$\geq -\frac{M}{2}\sum_{i=1}^{k}(1-x_i)^2 |x_i - z_i|.$$

We show that each term is compensated for by the corresponding term in the other part of p^*, i.e.,

$$(16.41) \quad \frac{1}{2}(1-x_i)^2 |x_i - z_i| \leq y_i - 2x_i^2 + x_i.$$

Let us assume e.g. that $x_i \leq z_i$ (the other case is similar). Let $w_i \in A$ be the closest point to x_i with $w_i < x_i$.

By Corollary 16.12, y_i is above the chord between z_i and w_i of the parabola $2x^2 + x$. On the other hand, $2x_i^2 + x_i$ is below the chord between z_i and $(z_i + w_i)/2$. The slope of the first chord is $2z_i + 2w_i - 1$; the slope of the second, $3z_i + w_i - 1$. The difference in slopes is $z_i - w_i$, and so $y_i - 2x_i^2 + x_i \geq (z_i - w_i)(z_i - x_i)$. Simple computation shows that

$$z_i - w_i = \frac{(1-w_i)^2}{2-w_i} \geq \frac{1}{2}(1-x_i)^2.$$

This proves (16.41) and thereby also Claim 16.36.

As a preparation for the rest of the proof, we need to construct some special graphs. We fix a simple graph F with node set $[k]$ that has no automorphisms. For any set of positive integers $n_1, \ldots, n_k$, let $F(n_1, \ldots, n_k)$ denote the (unlabeled) graph obtained from F by replacing every node i by a set of n_i twins. We will call these n_i nodes the *clones* of i.

As a related construction, we define F_{ir} ($i \in [k]$, $r \geq 0$) by adding r new twins of node i, making them mutually adjacent and adjacent to i, and labeling the original nodes $1, \ldots, k$. The node i and the new nodes will be called the *clones* of i. So $F_{1,r}$ is a k-labeled version of $F(r+1, 1, \ldots, 1)$.

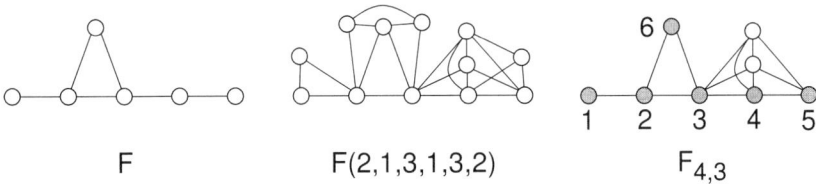

FIGURE 16.2. Auxiliary graphs for the proof of theorem 16.34.

As a further step, we add all missing edges to F_{ir} with negative sign, to get the k-labeled signed graph $\widehat{F}_{ir}$ (as we have seen, this can be considered as a k-labeled quantum graph). Recall that $\widehat{F}$ is defined analogously.

CLAIM 16.37. *Every homomorphism of $\widehat{F}$ into any graph G is an induced embedding.*

Indeed, every homomorphism preserves both edges and non-edges by the definition of signed graphs. Suppose that two nodes $u, v \in V(F)$ are mapped onto the same node of G. Then every further node of F must be connected to u and v in the same way, and so interchanging u and v is an automorphism of F, which contradicts the choice of F.

CLAIM 16.38. *For every homomorphism of $\widehat{F}$ into any of the special graphs $F(n_1, \ldots, n_k)$, each node $i \in V(F)$ is mapped onto a clone of i.*

The proof is similar to the previous one. We already know that the map is injective. For $u \in V(F)$, let $\sigma(u)$ be defined as the node of F whose clones in $F(n_1, \ldots, n_k)$ contain the image of u. No two nodes $u, v \in V(F)$ have $\sigma(u) = \sigma(v)$: similarly as before, interchanging two such nodes would be an automorphism of F. Hence σ is an automorphism of F, and hence σ must be the identity. This proves the Claim.

Our next observation is that homomorphism densities from the signed graphs F_{ir} into any simple graph G can be expressed quite simply. Let G be any simple graph, and let $\varphi : [k] \to V(G)$, and let $S = \varphi([k])$ be its range. Let $U_{\varphi,i}$ be the set of nodes in $V(G) \setminus S$ which are connected to $\varphi(i)$ and all the neighbors of $\varphi(i)$ in S, but to no other node in S.

We claim that

$$(16.42) \quad \mathsf{hom}_\varphi(\widehat{F}_{ir}, G) = \begin{cases} \mathsf{hom}(K_r, G[U_{\varphi,i}]) & \text{if } \varphi \text{ is an induced embedding of } F, \\ 0 & \text{otherwise.} \end{cases}$$

Assume first that φ is an induced embedding of F into G. It is clear that if ψ is any homomorphism of $\widehat{F}_{ir}$ into G extending φ, then all the clones of i in $\widehat{F}_{ir}$ must be mapped onto nodes in $U_{\varphi,i}$. Since these twins form a complete graph K_r, the number of ways to map these twins into $G[U_{\varphi,i}]$ homomorphically is $\mathsf{hom}(K_r, G[U_{\varphi,i}])$, and every such map, together with φ, forms a homomorphism of $\widehat{F}_{ir}$ into G. Claim

16.37 implies that $\hom_\varphi(\widehat{F}_{ir}) = 0$ if φ is not an induced embedding of F. Hence (16.42) follows.

Normalizing the homomorphism numbers in (16.42), we get that if φ is an induced embedding, then

$$(16.43) \qquad t_\varphi(\widehat{F}_{ir}, G) = \frac{|U_{\varphi,i}|^r}{\mathsf{v}(G)^r} t(K_r, G[U_{\varphi,i}]).$$

We want to reduce the problem of deciding whether $p \geq 0$ on A^k (for a given polynomial $p \in \mathbb{Z}[x_1, \ldots, x_k]$) to deciding whether $g \geq 0$ for a quantum graph g. Given p, we construct the polynomials $p^* \in \mathbb{Z}[x_1, y_1, \ldots, x_k, y_k]$ and $p^{**} \in \mathbb{Z}[u_1, v_1, w_1, \ldots, u_k, v_k, w_k]$ as above, and define the k-labeled quantum graph

$$g = p^{**}(\widehat{F}_{11}, \widehat{F}_{12}, \widehat{F}_{13}, \ldots, \widehat{F}_{k1}, \widehat{F}_{k2}, \widehat{F}_{k3}).$$

The key step in the proof is the following:

CLAIM 16.39. *We have $[\![g]\!] \geq 0$ if any only if $p \geq 0$ on A^k.*

To start with the "if" direction, assume that $p \geq 0$ on A^k. Then $p^* \geq 0$ on $D_{2,3}^k$ by Claim 16.36. We want to prove that $t([\![g]\!], G) \geq 0$ for every graph G. This follows if we show that $t_\varphi(g, G) \geq 0$ for every map $\varphi : [k] \to V(G)$. To simplify notation, set $t_{ir} = t_\varphi(\widehat{F}_{ir}, G)$. Then

$$t_\varphi(g, G) = p^{**}(t_{11}, t_{12}, t_{13}, \ldots, t_{k1}, t_{k2}, t_{k3})$$
$$= (t_{11} \ldots t_{k1})^N p^* \left(\frac{t_{12}}{t_{11}^2}, \frac{t_{13}}{t_{11}^3}, \ldots, \frac{t_{k2}}{t_{k1}^2}, \frac{t_{k3}}{t_{k1}^3}\right).$$

By (16.43), we have $(t_{12}/t_{11}^2, t_{13}/t_{11}^3) = \big(t(K_2, G[U_{\varphi,i}]), t(K_3, G[U_{\varphi,i}])\big) \in D_{2,3}$, and hence it follows that $t_\varphi(g, G) \geq 0$.

To prove the converse, assume that $[\![g]\!] \geq 0$. By Claim 16.36, it suffices to prove that $p^* \geq 0$ on D^k. Let $x_i = (n_i - 1)/n_i \in A$ and $y_i = 2x_i^2 - x_i = (n_i^2 - 3n_i + 2)/n_i^2$, and consider the graph $G = F(n_1+1, \ldots, n_k+1)$. By our assumption, $t([\![g]\!], G) \geq 0$. We can write

$$t([\![g]\!], G) = \sum_{\varphi:[k] \to V(G)} \frac{1}{\mathsf{v}(G)^k} t_\varphi(g, G).$$

In every constituent of g, the labeled nodes induce a copy of $\widehat{F}$, which implies by (16.42) that only those terms where φ is an induced embedding are nonzero. By Claim 16.38, such mappings φ map every node of F onto a clone of it, and hence $t_\varphi(g, G)$ is the same for every induced embedding. This implies that $t_\varphi(g, G) \geq 0$ for any such map φ. Let us fix an induced embedding φ of F into G.

Now we can apply (16.43): the set $U_{\varphi,i}$ induces a complete graph with n_i nodes, and hence

$$t_\varphi(\widehat{F}_{ir}, G) = \frac{|n_i|^r}{n^r} t(K_r, K_{n_i}) = \frac{(n_i)_r}{n^r}.$$

This implies that

$$t_\varphi(g, G) = p^{**}\left(\frac{n_1}{n}, \frac{(n_1)_2}{n^2}, \frac{(n_1)_3}{n^3}, \ldots, \frac{n_k}{n}, \frac{(n_k)_2}{n^2}, \frac{(n_k)_3}{n^3}\right)$$
$$= \left(\frac{n_1 \ldots n_k}{n^k}\right)^N p^*(x_1, y_1, \ldots, x_k, y_k).$$

Since the left side is nonnegative, this implies that $p^*(x_1, y_1, \ldots, x_k, y_k) \geq 0$ as claimed.

This proves Claim 16.39, and together with Claim 16.35, it completes the proof of the theorem. □

As mentioned in the introduction, an inequality $g \geq 0$, where g is a quantum graph, is decidable with an arbitrarily small error:

PROPOSITION 16.40. *There is an algorithm that, given a quantum graph g with rational coefficients and an error bound $\varepsilon > 0$, decides either that $g \not\geq 0$ or that $g + \varepsilon K_1 \geq 0$ (if both inequalities are true, then it may return either answer).*

Proof. This will follow from Theorem 16.41 in the next section, but let us describe a simple direct proof suggested by Pikhurko. Let $g = \sum_F a_F F$ be the given quantum graph, let $a = \sum_F |a_F| \mathsf{e}(F)$, $\varepsilon_1 = \varepsilon/a$. By Corollary 9.25, there is an integer $k \geq 1$ such that all simple graphs with k nodes form an ε_1-net in $(\mathcal{W}_0, \delta_\square)$. Let us check the inequality $\sum_F a_F t(F, G) \geq 0$ for all simple graphs G with at most k nodes. If we find a graph that violates it, we know that $g \not\geq 0$. Else, let G be any simple graph. By the definition of k, there is a simple graph G' on k nodes such that $\delta_\square(G, G') \leq \varepsilon_1$, and hence by the Counting Lemma 10.22, we have

$$\sum_F a_F t(F,G) \geq \sum_F a_F(t(F,G') - \mathsf{e}(F)\varepsilon_1) \geq \sum_F a_F t(F,G') - \varepsilon \geq -\varepsilon = -\varepsilon t(K_1, G'),$$

so we can conclude that $g + \varepsilon K_1 \geq 0$. □

16.6.2. Positivstellensatz for graphs. Is there a quantum graph $g \geq 0$ which is not a square sum? Hatami and Norine [2011] constructed such a quantum graph. In fact, the existence of such a quantum graph follows from Theorem 16.34, stating that it is algorithmically undecidable whether a quantum graph with rational coefficients is nonnegative. To see this, consider two Turing machines, both working on an input which is a quantum graph g with rational coefficients. We may assume that the constituents of g have no isolated nodes. One of them will look for a graph G with $t(g, G) < 0$; the other, for a representation of g as a square-sum. If for every input one of them halts, then we know whether or not $g \geq 0$. So there must be an input g on which both Turing machines run forever; then we have $g \geq 0$, and g is not a square sum.

(To be precise, we must add that if g is a square-sum, then it is a square-sum where the coefficients in the quantum graphs y_i in the definition are algebraic real numbers; then there are only a countable number of possibilities, and the second Turing machine can check them in an appropriate order. One needs a method to check that given k-labeled quantum graphs $y_i, \ldots, y_k$ with algebraic coefficients, whether g is obtained from $\sum_i y_i^2$ by unlabeling and deleting isolated nodes. Such an algorithm follows from Tarski's Theorem on the decidability of the first order theory of real numbers.)

But not all is lost: the following weaker result was proved by Lovász and Szegedy [2012a].

THEOREM 16.41. *Let f be a quantum graph. Then $f \geq 0$ if and only if for every $\varepsilon > 0$ there is a square-sum g such that $\|f - g\|_1 < \varepsilon$.*

An analogous theorem for nonnegative polynomials was proved by Lasserre [2007].

Proof. The "if" part is trivial. The idea of the proof of the "only if" part is the following. Consider the (unlabeled) quantum graph $g = \sum_F a_F F$ (where only a finite number of the a_F are nonzero). We may assume that no graph F with $a_F > 0$ contains an isolated node, since removing isolated nodes does not change $t(g, W)$. The condition $g \geq 0$ means that $h(g) \geq 0$ for every graph parameter h of the form $h = t(., W)$ with $W \in \mathcal{W}_0$. This constraint is linear, so we can equivalently require the inequality for every graph parameter of the form $h = \mathsf{E}(t(.,\mathbf{W}))$, where the expectation is over some probability distribution on graphons (see Section 14.5). By Proposition 14.60, this is equivalent to requiring that h is normalized, isolate-indifferent and reflection positive. We can forget about the normalization, since the condition $\sum_F a_F h(F) \geq 0$ is homogeneous. So the question is: Does the inequality

$$\sum_F a_F h(F) \geq 0 \tag{16.44}$$

hold for every isolate-indifferent and reflection positive graph parameter h?

This problem can be rephrased in terms of the connection matrix $X = M(h, \mathbb{N})$, whose entries we consider as unknowns. These unknowns are not all different: if (for this proof only) F' denotes the graph obtained from F by removing its isolated nodes, then we have $X_{F_1,F_2} = X_{G_1,G_2}$ whenever $[\![F_1 F_2]\!]' \cong [\![G_1 G_2]\!]'$. The reflection positivity conditions mean that $X \succeq 0$. The question is: Do these constraints imply the inequality

$$\sum_F a_F X_{F,K_0} \geq 0 \tag{16.45}$$

(In this last sum, all the graphs F are unlabeled.)

This is just a feasibility problem in semidefinite programming —apart from the "minor" problem that the unknowns form an infinite matrix, and have to satisfy an infinite number of constraints. We will have to cut the program to finite size, which will bring in the error in the theorem.

But let us ignore the problems with infinities, and apply the Semidefinite Farkas Lemma: We have to assign, as a Lagrange multiplier, a matrix $Y \succeq 0$, which has to satisfy

$$\sum_{F_1,F_2:\, [\![F_1 F_2]\!]' \cong F} Y_{F_1,F_2} = a_F \tag{16.46}$$

for every $F \in \mathcal{F}^{\mathrm{simp}}$ (where the summation extends over all partially labeled simple graphs F_1 and F_2). We can rewrite this as

$$g = \sum_F a_F F = \sum_{F_1,F_2} Y_{F_1,F_2} [\![F_1 F_2]\!]'$$

Let us write $Y = ZZ^\mathsf{T}$ with some matrix Z; this takes care of the semidefiniteness condition (remember, we are ignoring the problem that these matrices are infinite). Then

$$g = \sum_{F_1,F_2} \sum_m Z_{F_1,m} Z_{F_2,m} [\![F_1 F_2]\!]' = \sum_m [\![\Big(\sum_F Z_{m,F} F\Big)^2]\!].$$

showing that g is a square-sum.

Now we have to make this argument precise. Let $\mathcal{F}'_k$ denote the set of fully labeled graphs on $[k]$. Let $\mathcal{M}$ denote the linear space of all symmetric matrices indexed by partially labeled simple graphs, let $\mathcal{P}$ be the subset of $\mathcal{M}$ consisting of positive semidefinite matrices, and let $\mathcal{L}$ denote the subspace of matrices satisfying $X_{F_1,F_2} = X_{G_1,G_2}$ whenever $[\![F_1 F_2]\!]' \cong [\![G_1 G_2]\!]'$. Clearly, $\mathcal{P}$ is a convex cone. Let

Φ_k denote the operator mapping a matrix in $\mathcal{M}$ to its restriction to $\mathcal{F}'_k \times \mathcal{F}'_k$ (this is a finite matrix!). Then $\mathcal{M}_k = \Phi_k \mathcal{M}$ is the space of all symmetric $\mathcal{F}'_k \times \mathcal{F}'_k$ matrices, and $\mathcal{P}_k = \Phi_k \mathcal{P}$ is the positive semidefinite cone in $\mathcal{M}_k$. It is also clear that $\mathcal{L}_k = \Phi_k \mathcal{L}$ consists of those matrices $X \in \mathcal{M}_k$ for which $X_{F_1,F_2} = X_{G_1,G_2}$ whenever $[\![F_1 F_2]\!]' \cong [\![G_1 G_2]\!]'$. Clearly,

(16.47) $$\Phi_k(\mathcal{P} \cap \mathcal{L}) \subseteq \mathcal{P}_k \cap \mathcal{L}_k,$$

but equality may not hold in general.

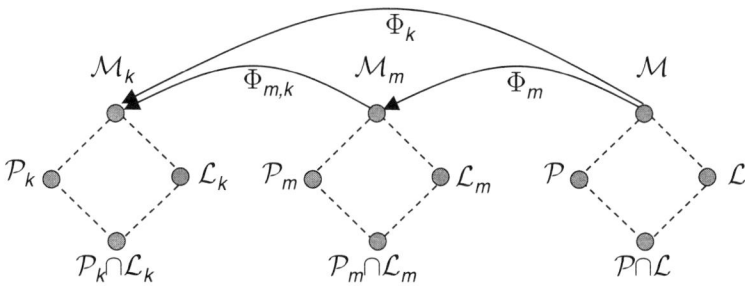

FIGURE 16.3. Spaces and cones in the proof of Theorem 16.41.

We note that the entries of every matrix $X \in \mathcal{P}_k \cap \mathcal{L}_k$ are in the interval $[0, X_{\emptyset,\emptyset}]$. Indeed, for any $F \in \mathcal{F}'_k$ and the fully labeled edgeless graph $U_k \in \mathcal{F}'_k$, the condition $X \in \mathcal{L}_k$ implies that $X_{U_k,F} = X_{F,F} = X_{F,U_k}$, and so $X \in \mathcal{P}_k$ implies that $0 \leq X_{F,F} \leq X_{\emptyset,\emptyset}$. Since $X_{F,G} = X_{FG,FG}$, the claim follows.

For $k \leq m$, we embed $\mathcal{F}'_k$ into $\mathcal{F}'_m$, by adding $m - k$ isolated nodes labeled $k+1, \ldots, m$. The corresponding operator restricting $\mathcal{F}'_m \times \mathcal{F}'_m$ matrices to $\mathcal{F}'_k \times \mathcal{F}'_k$ will be denoted by $\Phi_{m,k}$.

We claim that the following weak converse of (16.47) holds:

(16.48) $$\Phi_k(\mathcal{P} \cap \mathcal{L}) = \bigcap_{m \geq k} \Phi_{m,k}(\mathcal{P}_m \cap \mathcal{L}_m).$$

Indeed, let A be a matrix that is contained in the right hand side. Then for every $m \geq k$ we have a matrix $B_m \in \mathcal{P}_m \cap \mathcal{L}_m$ such that A is a restriction of B_m. Now let $m \to \infty$; by selecting a subsequence, we may assume that all entries of B_m tend to a limit. This limit defines a graph parameter f, which is normalized, isolate-indifferent and flatly reflection positive. By Proposition 14.60, f is reflection positive, and so the matrix $M(f)$ is in $\mathcal{P} \cap \mathcal{L}$ and $\Phi_k M(f) = A$.

We may assume that $|V(F)| = k$ whenever $a_F \neq 0$. Let $A \in \mathcal{M}_k$ denote the matrix

$$A_{FG} = \begin{cases} a_F, & \text{if } F = G, \\ 0, & \text{otherwise.} \end{cases}$$

Then $g \geq 0$ means that $A \cdot Z \geq 0$ for all $Z \in \Phi_k(\mathcal{P} \cap \mathcal{L})$ (where the inner product $A \cdot Z$ of two matrices is defined as $\sum_{i,j} A_{ij} Z_{ij}$). In other words, A is in the dual cone of $\Phi_k(\mathcal{P} \cap \mathcal{L})$. From (16.48) it follows that there are diagonal matrices $A_m \in \mathcal{M}_k$ such that $A_m \to A$ and $A_m \cdot Y \geq 0$ for all $Y \in \Phi_{m,k}(\mathcal{P}_m \cap \mathcal{L}_m)$. In other words, $A_m \cdot \Phi_{m,k} Z \geq 0$ for all $Z \in \mathcal{P}_m \cap \mathcal{L}_m$, which can also be written as $\Phi^*_{m,k} A_m \cdot Z \geq 0$, where $\Phi^*_{m,k} : \mathcal{M}_k \to \mathcal{M}_m$ is the adjoint of the linear map $\Phi_{m,k} : \mathcal{M}_m \to \mathcal{M}_k$. (This adjoint acts by adding 0-s.) So $\Phi^*_{m,k} A_m$ is in the polar cone of $\mathcal{P}_m \cap \mathcal{L}_m$, which is $\mathcal{P}_m^* + \mathcal{L}_m^*$. The positive semidefinite cone is self-polar. The linear space $\mathcal{L}_m^*$ consists of those matrices $B \in \mathcal{M}_m$ for which $\sum_{F_1,F_2} B_{F_1,F_2} = 0$, where the summation extends over all pairs $F_1, F_2 \in \mathcal{F}'M$ for which $F_1 F_2 \simeq F_0$, for every fixed graph F_0. Thus we have $\Phi_{m,k}^* A_m = P + L$, where P is positive semidefinite and $L \in \mathcal{L}_m^*$. Since P is positive semidefinite, we can write it as $P = \sum_{k=1}^N v_k v_k^{\mathsf{T}}$, where $v_k \in \mathbb{R}^{\mathcal{F}'_m}$. We can write this as

$$\sum_{\substack{F_1, F_2 \\ F_1 F_2 \simeq F_0}} \sum_{k=0}^N v_{k,F_1} v_{k,F_2} = \begin{cases} (A_m)_{F_0,F_0}, & \text{if } F_1 F_2 \simeq F_0 \in \mathcal{F}'_k, \\ 0, & \text{otherwise.} \end{cases}$$

In other words,

$$\sum_{k=1}^N \left(\sum_F v_{k,F} F\right)^2 = \sum_{F_0} (A_m)_{F_0,F_0} F_0.$$

So the quantum graph on the right side is a sum of squares. Furthermore, if $m \to \infty$, then $A_m \to A$ and so

$$\sum_{F_0} (A_m)_{F_0,F_0} F_0 \to \sum_{F_0} A_{F_0,F_0} F_0 = f. \qquad \square$$

In view of the usefulness of extending graphs to graphons, it seems natural to define graph algebras of infinite formal linear combinations of graphs with appropriate convergence properties; in other words, of graph parameters. It has not been worked out, however, what the structure of the resulting algebra is, and how it is related to graphons.

Proposition 14.61 suggests that it should be enough to use fully labeled quantum graphs when approximating a nonnegative quantum graph as a square sum. This is indeed true (see Exercise 16.44), but the approximation is very inefficient, as shown by the following example.

EXAMPLE 16.42. Consider the quantum graph

$$g = \text{\raisebox{-2pt}{⋀}} - \text{\raisebox{-2pt}{∷}} \geq 0.$$

If we allow unlabeled nodes, then g can be represented (up to labels and isolated nodes) as a square sum (in fact, as a single square):

$$\left(\text{\raisebox{-2pt}{⋮}} - \text{\raisebox{-2pt}{⋮•}}\right)^2.$$

But g cannot be represented as a square sum of fully labeled graphs. To see this, let S_k denote the set of quantum graphs obtained by deleting isolates and unlabeling square sums $\sum_i y_i^2$, where every constituent of y_i is a fully labeled graph on k nodes. Consider the graph parameter

$$f(G) = \begin{cases} 3\binom{k}{4} & \text{if } \mathsf{e}(G) = 0, \\ \binom{k-2}{2} & \text{if } \mathsf{e}(G) = 1, \\ 1 & \text{if } E(G) \text{ consists of two disjoint edges,} \\ 0 & \text{otherwise.} \end{cases}$$

Then $f(g) < 0$. On the other hand, we claim that $f(y^2) \geq 0$ if every constituent of y is a fully labeled graph on k nodes. This means that the flat connection matrix $M^{\text{flat}}(f, k)$ is positive semidefinite. By the Lindström–Wilf Formula (A.1), this is equivalent to saying that the upper Möbius inverse $f^\uparrow$ (see (4.1)) is nonnegative, which is easy to check. Since f is isolate-indifferent, this proves that f is nonnegative on every quantum graph in S_k. It follows that $g \notin S_k$ for any k.

But g can be approximated by members of S_k for large k. To construct such an approximation, let H_{ij} ($1 \leq i < j \leq k$) consist of $V(G) = [k]$ and a single edge connecting i to j. Expanding and unlabeling the quantum graph

$$h_k^2 = \left(\frac{1}{k-1} \sum_{2 \leq i \leq k} H_{1i} - \frac{1}{\binom{k-1}{2}} \sum_{2 \leq i < j \leq k} H_{ij} \right)^2,$$

we get

$$(\text{\raisebox{-2pt}{$\wedge$}} - \text{\raisebox{-2pt}{$\circ\circ$}}) + \frac{1}{(k-1)(k-2)} \left(k\,\text{\raisebox{-2pt}{$\mid$}} - (k-6)\,\text{\raisebox{-2pt}{$\wedge$}} + (4k-10)\,\text{\raisebox{-2pt}{$\circ\circ$}} \right).$$

The last term tends to 0 as $k \to \infty$, so g is arbitrarily well approximated by h_k^2. ♦

Razborov and his collaborators found several concrete applications of the method of semidefinite programming in extremal graph theory. Let us mention one. Paul Erdős [1984] conjectured in 1984 that the number of pentagons in a triangle-free graph with a given number of nodes is maximized by blow-ups of a pentagon. In our language, if G is a triangle-free simple graph, then

(16.49) $$t(C_5, G) \leq t(C_5, C_5) = \frac{2}{625}.$$

In spite of its naturality and simple form, this conjecture remained unproven until recently, when Hatami, Hladky, Kral, Norine and Razborov [2011] and independently Grzesik [2011] found a proof, using flag algebras and computer-assisted solutions to semidefinite programs.

> EXERCISE 16.43. Prove that the problem whether $\sum_i a_i \text{hom}(F, G_i) \geq 0$ holds for every simple graph G (for given simple graphs $F_1, \ldots, F_m$ and integer coefficients $a_1 \ldots, a_m$) is undecidable. [Hint: Use the result of Exercise 5.50 and Matiyasevich's Theorem.]
>
> EXERCISE 16.44. Use the method of Example 16.42 to show that if a quantum graph g can be written as a square sum (unlabeled and isolates removed), then it can be approximated up to an arbitrarily small error by square-sums of quantum graphs with fully labeled constituents.

16.7. Which graphs are extremal?

Is there a "template" describing the structure of extremal graphs for an extremal graph problem? For "classical" extremal problems, discussed in Section 16.4, this has been answered by the Erdős–Simonovits–Stone theory of extremal graphs (at least in the dense case). As we have seen, in an asymptotic sense (for large graphs) the only extremal graphs are the Turán graphs. In the language of graphons, every optimization problem in the sense of Theorem 16.23 has an optimum solution of the form W_{K_q} for some $q \geq 2$, and the solution is unique.

More general stepfunctions (with positive integral values) have been found to serve as optimum solutions of generalizations of these problems to directed graphs and multigraphs, in the work of Brown, Erdős and Simonovits [1973, 1978].

In this discussion we stay with simple graphs, but we consider a much more general type of graph theoretic extremal problem:

(16.50)
$$\text{maximize } t(f, W)$$
$$\text{subject to } t(g_1, W) = a_1$$
$$\vdots$$
$$t(g_k, W) = a_k$$

where $f, g_1, \ldots, g_k$ are given simple quantum graphs. Most of the graphon versions of extremal problems discussed so far fit this scheme. Is there a special family of graphons such that every extremal graph problem has a solution from this family? We define an interesting class of graphons that are all needed, and we conjecture that they are also sufficient.

16.7.1. Finitely forcible graphons. To motivate this definition, recall the result of Chung, Graham and Wilson from Section 1.4.2 that a sequence of graphs $(G_n, n = 1, 2, \ldots)$ with the property that $t(K_2, G_n) \to p$ and $t(C_4, G_n) \to p^4$ for some $0 < p < 1$ is quasirandom, so it is convergent and its limit is the all-p function. In other words, $t(F, G_n) \to t(F, L_p)$ for every simple graph F, where L_p is the weighted graph with one node with a loop of value p.

Are there other convergent graph sequences that can be characterized by the convergence of a finite number of homomorphism densities $t(F, G_n)$? Are perhaps all convergent graph sequences like this?

We can translate this question to the language of graphons, which allows a more precise statement: Which graphons W have a finite "forcing" set of simple graphs $F_1, \ldots, F_m$ such that if any other graphon U has the property that $t(F_i, U) = t(F_i, W)$ ($i = 1, \ldots, m$), then U and W are weakly isomorphic, i.e., $t(F, U) = t(F, W)$ for every simple graph F? Such a graphon will be called *finitely forcible*. In other words, a finitely forcible graphon forms a single-element variety in the graphon space. The theorem of Chung, Graham and Wilson quoted above shows that every constant function is finitely forcible.

Coming down to the earth, this notion has the following version for finite graphs: a convergent simple graph sequence (G_n) is *finitely forcible*, if there is a finite set of simple graphs $F_1, \ldots F_m$ such that whenever H_n is another graph sequence such that $\lim_n t(F_i, G_n) = \lim_n t(F_i, H_n)$ for $i = 1, \ldots, m$, then $\lim_n t(F, G_n) = \lim_n t(F, H_n)$ for every simple graph F. So it follows from the characterization of quasirandom graph sequences by Chung, Graham and Wilson [1989] in Section 1.4.2 that every quasirandom graph sequence is finitely forcible.

We will see that "most" graphons (in a Baire category sense) are not finitely forcible. On the other hand, there are interesting families of finitely forcible graphons. In the next section, we describe some such families. See Lovász and Szegedy [2011] for more constructions of finitely forcible graphons.

Trivially, every finitely forcible graphon is the unique solution of an extremal problem of type (16.50). Perhaps the converse is also true:

CONJECTURE 16.45. *Every extremal problem has a finitely forcible optimum.* In other words, if a finite set of constraints of the form $t(F_i, W) = a_i$ is satisfied by some graphon, then we can add a finite number of further constraints of this type to make the solution unique (up to weak isomorphism).

16.7.2. Many finitely forcible graphons.
In this section, we describe some reasonably general constructions for finitely forcible graphons.

Stepfunctions. Almost all classical extremal problems have a solution whose "template" is a stepfunction. It was shown by Lovász and Sós [2008] that every stepfunction is finitely forcible, which means that every stepfunction is the template of an appropriate extremal graph problem. (It was conjectured that these are the only ones, but as we shall see there are many more.) We give a somewhat stronger result with a better bound on the size of the forcing graphs F.

THEOREM 16.46. *Let H be a weighted graph with nonnegative nodeweights and with edgeweights from $[0,1]$, and let U be a kernel. If $t(F,U) = t(F,W_H)$ for all graphs F with $\mathsf{v}(F) \leq 4(2\mathsf{v}(H)+3)^8$, then U is weakly isomorphic to W_H.*

Proof. By Proposition 14.44 we know that stepfunctions with $q = \mathsf{v}(H)$ steps form a variety: there is a simple quantum graph g such that $t(g, W) = 0$ if and only if $W \in \mathcal{S}_q$, where every constituent of g has $(q+2)(q+1)$ nodes. So $t(g, W_H) = t(g, H) = 0$, and hence $t(g, U) = 0$, which implies that U is a stepfunction with q steps. So we may assume that there is a weighted graph H' with q nodes such that $U = W_{H'}$ almost everywhere. It follows that for every simple graph F with at most $4(2q+3)^8$ nodes, we have $t(F, H') = t(F, U) = t(F, H)$. By Theorem 5.33, this implies that $H \cong H'$, and so $U = W_H$ almost everywhere. □

COROLLARY 16.47. *Every stepfunction is finitely forcible.*

Theorem 16.46 implies for finite graphs:

COROLLARY 16.48. *If H is weighted graph with edgeweights from $[0,1]$, and (G_n) is a graph sequence such that $t(F, G_n) \to t(F, H)$ for every simple graph F with $\mathsf{v}(F) \leq 4(2\mathsf{v}(H)+3)^8$, then $t(F, G_n) \to t(F, H)$ for every simple graph F (in other words, $G_n \to W_H$).*

Threshold graphons. The simple threshold graphon $\mathbb{1}(x+y \leq 1)$ introduced in Example 11.36 is perhaps the simplest finitely forcible graphon that is not a stepfunction. We in fact have a more general class. Let $p(x,y)$ be a symmetric real polynomial. We call the graphon $U(x,y) = \mathbb{1}(p(x,y) > 0)$ the *p-threshold graphon*.

PROPOSITION 16.49. *If p is monotone decreasing on $[0,1]^2$, then the p-threshold graphon U is finitely forcible among kernels.*

The proof will show that U is determined by the equations

$$(16.51) \qquad t_{x_1 x_2 x_3 x_4}(\widehat{C}_4, W) = 0$$

and

$$(16.52) \qquad t(K_{a,b}, W) = t(K_{a,b}, U) \qquad (1 \leq a, b \leq 2\deg(p)+4)$$

(here $\widehat{C}_4$ is the signed 4-cycle defined in Example 14.42). It can be conjectured that the monotonicity condition is not needed.

Proof. The condition on the monotonicity of p implies that $W = U$ satisfies (16.51). It is trivial that equations (16.52) are satisfied by $W = U$.

Let $W \in \mathcal{W}$ be any graphon satisfying (16.51)-(16.52). As discussed in Example 14.42, we may assume that W is a threshold graphon, i.e., 0-1 valued and monotone decreasing. Let $S_W = \{(x,y) : W(x,y) = 1\}$. We have

$$t(K_{a,b}, W) = \int_{[0,1]^a} \int_{[0,1]^b} \prod_{i=1}^{a} \prod_{j=1}^{b} W(x_i, y_j) \, dy \, dx.$$

Split this integral according to which x_i and which y_j is the largest. Restricting the integral to, say, the domain where x_1 and y_1 are the largest, we have that whenever $W(x_1, y_1) = 1$ then also $W(x_i, y_j) = 1$ for all i and j, and hence

$$\int_{x_1 \in [0,1]} \int_{x_2, \ldots, x_a \leq x_1} \int_{y_1 \in [0,1]} \int_{y_2, \ldots, y_b \leq y_1} \prod_{i=1}^{a} \prod_{j=1}^{b} W(x_i, y_j) \, dy \, dx$$

$$= \int_{x_1 \in [0,1]} \int_{y_1 \in [0,1]} W(x_1, y_1) x_1^{a-1} y_1^{b-1} \, dy_1 \, dx_1 = \int_{(x,y) \in S_W} x^{a-1} y^{b-1} \, dy \, dx.$$

Since there are a choices for the largest x_i and b choices for the largest y_j, this implies that

$$(16.53) \qquad t(K_{a,b}, W) = ab \int_{(x,y) \in S_W} x^{a-1} y^{b-1} \, dy \, dx.$$

Let us approximate W by a threshold graphon V for which the boundary curve ∂S_V is smooth (except for the endpoints of the intersection of S_V with boundary of the square), and $\varepsilon = \lambda(S_W \triangle S_V)$ is very small. Let ds be the arc length of the boundary curve ∂S_V, and let $n(x,y) = (n_1(x,y), n_2(x,y))$ denote the outward normal of ∂S_V at point (x,y). By the Gauss–Ostrogradsky Theorem, we can rewrite (16.53) as an integral along the boundary:

$$t(K_{a,b}, V) = b \int_{\partial S_V} x^a y^{b-1} n_1(x,y) \, ds.$$

Interchanging the roles of x and y, and adding, we get

$$(16.54) \quad \int_{\partial S_V} x^a y^b \big(n_1(x,y) + n_2(x,y)\big) \, ds = \frac{1}{a+1} t(K_{a+1,b}, V) + \frac{1}{b+1} t(K_{a,b+1}, V).$$

Consider the following integral:

$$I(V) = \int_{\partial S_V} x(1-x) y(1-y) \, p(x,y)^2 \big(n_1(x,y) + n_2(x,y)\big) \, ds.$$

By (16.54), this can be expressed as a linear combination

$$I(V) = \sum_{a,b=1}^{2 \deg(p) + 4} c_{ab} t(K_{a,b}, V),$$

where the coefficients depend only on a, b and p. Furthermore, we have

$$I(V) = \sum_{a,b=1}^{2\deg(p)+4} c_{ab} t(K_{a,b}, V) \approx \sum_{a,b=1}^{2\deg(p)+4} c_{ab} t(K_{a,b}, W)$$

$$= \sum_{a,b=1}^{2\deg(p)+4} c_{ab} t(K_{a,b}, U) = I(U) = 0,$$

where the error at the $\approx$ sign is bounded by $C\varepsilon$ for some C that depends only on a, b and p.

On the other hand, the integrand in $I(V)$ is nonnegative everywhere. Letting $\varepsilon \to 0$, we see that the $x(1-x)y(1-y)\,p(x,y)^2 = 0$ on $\partial(S_W)$. Since p is monotone decreasing, this implies that $\partial(S_W) = \partial(S_U)$, and hence $W = U$ except perhaps on the boundary (which is of measure 0). $\square$

Complement reducible graphs. We have seen that Complement reducible graphs form a variety (Example 14.43). Many of them are finitely forcible, and they are of a rather different nature, sort of fractal-like. Here is a specific example.

PROPOSITION 16.50. *For $x, y \in [0,1]$, let $U(x,y) = 1$, if the first bit where the binary expansions of x and y differ is at an odd position, and let $U(x,y) = 0$ otherwise. Then U is finitely forcible.* $\square$

Our examples of finitely forcible graphons have had finite range (the function W assumed only a finite set of values). We refer to the paper of Lovász and Szegedy [2011] for the proof of Proposition 16.50 and for further constructions of finitely forcible graphons, where the range of the function W contains an interval.

16.7.3. Not too many finitely forcible graphons. Are there any graphons that are not finitely forcible? A natural extension of the class of stepfunctions is the class of kernels with finite rank. However, we don't get any new finitely forcible graphons in this class. In fact, Theorem 14.48 implies:

COROLLARY 16.51. *Every finitely forcible kernel with finite rank is a stepfunction.*

In view of Proposition 16.49, the following further corollary may be surprising:

COROLLARY 16.52. *Assume that $W \in \mathcal{W}_0$ can be expressed as a non-constant polynomial in x and y. Then W is not finitely forcible.*

We want to derive more general necessary conditions for being finitely forcible. We start with a rather strong property of finitely forcible functions. Recalling the definition of the 2-labeled graph $F^\ddagger$ from Section 16.2, let $\mathcal{L}(W)$ be the linear space generated by all 2-variable functions $t_{xy}(F^\ddagger, W)$, where F ranges over all simple graphs.

LEMMA 16.53. *Suppose that $W \in \mathcal{W}$ is forced (in $\mathcal{W}$) by the simple graphs $F_1, \ldots, F_m$. Then either the functions $t_{xy}(F_1^\ddagger, W), \ldots, t_{xy}(F_m^\ddagger, W)$ are linearly dependent, or they generate $\mathcal{L}(W)$ (or both).*

Proof. Suppose not, then there is a simple graph F_{m+1} such that the functions $t_{xy}(F_i^\ddagger, W)$ ($i = 1, \ldots, m+1$) are linearly independent. For $U \in \mathcal{W}$, set

$h_k(U; x, y) = t_{xy}(F_k^\ddagger, U)$. So $h_k(U) \in \mathcal{W}$, and $h_k : \mathcal{W} \to \mathcal{W}$ is a (non-linear) operator. Let $\Phi(U) \in \mathcal{W}$ denote the component of $h_{m+1}(U)$ orthogonal to the subspace of $\mathcal{W}$ generated by $h_1(U), \ldots, h_m(U)$.

We need the following technical claim.

CLAIM 16.54. *There is an open L_∞-ball $\mathcal{U}$ in $\mathcal{W}$ centered at the function W such that $\Phi \colon \mathcal{W} \to \mathcal{W}$ is Lipschitz (in the L_∞ norm) on $\mathcal{U}$.*

The operators h_k are L_∞-Lipschitz on $\mathcal{W}$. This follows from the inequality

$$\|t_{xy}(F, U) - t_{xy}(F, U')\|_\infty \leq e(F) \|U - U'\|_\infty,$$

which is a simple exercise. Define the (numerical) matrix

$$B(U) = \begin{pmatrix} \langle h_1(U), h_1(U) \rangle & \ldots & \langle h_1(U), h_m(U) \rangle \\ \vdots & & \vdots \\ \langle h_m(U), h_1(U) \rangle & \ldots & \langle h_m(U), h_m(U) \rangle \end{pmatrix}$$

and the matrices $B_i(U)$ obtained from B by replacing the i-th column by $\langle h_1(U), h_{m+1}(U) \rangle, \ldots, \langle h_m(U), h_{m+1}(U) \rangle$. By elementary linear algebra, we have

$$\Phi(U) = h_{m+1}(U) + \sum_{i=1}^m \frac{\det(B_i(U))}{\det(B(U))} h_i(U).$$

The denominator is bounded away from 0 in a neighborhood $\mathcal{U}$ of W, and all the other functions are Lipschitz in this neighborhood, proving that Φ is Lipschitz. This completes the proof of the Claim.

By classical results on differential equations in Banach spaces (see e.g. Zeidler [1985]), there exists a $b > 0$ and a differentiable family $\{U_s : s \in [-b, b]\}$ of functions in $\mathcal{U}$ satisfying the differential equation

$$\dot{U}_s = \Phi(U_s), \quad U_0 = W.$$

By (16.7), we have

$$\frac{d}{ds} t(F_i, U_s) = \langle \Phi(U_s), t_2(F_i^\ddagger, U_s) \rangle = 0$$

for $i = 1, \ldots, m$, and hence $t(F_i, U_s) = t(F_i, U_0) = t(F_i, W)$ for all $s \in [0, c]$. Since the graphs F_i force W, it follows that the U_s are weakly isomorphic to W, and so $t(F_{m+1}, U_s) = t(F_{m+1}, W)$. But then $\langle \Phi(W), t_2(F^\ddagger, W) \rangle = \frac{d}{ds} t(F, U_s)\big|_{s=0} = 0$, and so $\langle \Phi(W), \Phi(W) \rangle = \langle \Phi(W), t(F_{m+1}^\ddagger, W) \rangle = 0$, which is a contradiction, since $\Phi(W) \neq 0$. □

A corollary of the previous theorem is that every finitely forcible kernel satisfies a "nontrivial" relation of the form $t_{xy}(f, W) = 0$. To specify what we mean by "nontrivial", let us say that a connected component of a partially labeled graph F is a *floating component*, if it contains no labeled nodes.

COROLLARY 16.55. *If the kernel $W \in \mathcal{W}$ is finitely forcible, then there is a nonzero simple 2-labeled quantum graph g with nonadjacent labeled nodes and no floating components such that $t_{xy}(g, W) = 0$ almost everywhere.*

Proof. The linear dependence of the functions $t_{xy}(F^\ddagger, W)$ gives a simple 2-labeled quantum graph with nonadjacent labeled nodes of the form $f = \sum_i \alpha_i F_i^\ddagger$ that satisfies $t_{xy}(f, W) = 0$ almost everywhere. To see that $f \neq 0$, it suffices to note that from any constituent of $F_i^\ddagger$ we can recover F_i by connecting its labeled nodes, so no cancellation will occur. □

As an application of Corollary 16.55, we prove:

THEOREM 16.56. *The set of finitely forcible graphons is of first category in $(\widetilde{\mathcal{W}}_0, \delta_\square)$.*

Proof. For a fixed set $\{F_1, \ldots, F_k\}$ of simple 2-labeled graphs with no floating components, let $\mathcal{T}(F_1, \ldots, F_k)$ denote the set of graphons W for which there is a nonzero quantum graph of the form $f = \sum_{i=1}^{k} a_i F_i$ satisfying $t_{xy}(f, W) = 0$ for all $x, y \in [0, 1]$. Corollary 16.55 implies that every finitely forcible graphon belongs to one of the sets $\mathcal{T}(F_1, \ldots, F_k)$, so it suffices to prove that these sets are nowhere dense. We do so in two steps.

CLAIM 16.57. *Let $F_1, \ldots, F_k$ be simple 2-labeled graphs with no floating components, and let $W \in \mathcal{W}_0$. Then every neighborhood of W contains a graphon W' such that $t_{xy}(F_1, W'), \ldots, t_{xy}(F_k, W')$ are linearly independent.*

It is easy to see that there is a simple 2-labeled graph G such that $[\![GF_1]\!], \ldots, [\![GF_k]\!]$ are mutually non-isomorphic (a large complete graph $K_n^{\bullet\bullet}$, with an edge incident with the node labeled 1 removed, suffices). Proposition 5.44 implies that there are graphons $U_1, \ldots, U_k$ such that the matrix $\bigl(t([\![GF_i]\!], U_j)\bigr)_{i,j=1}^{k}$ is nonsingular. For $0 < \varepsilon < 1/k$, define

$$W^\varepsilon = (1 - k\varepsilon)W \oplus (\varepsilon)U_1 \oplus \cdots \oplus (\varepsilon)U_k$$

(so the components of W^ε are $W, U_1, \ldots, U_k$, scaled by $1 - k\varepsilon, \varepsilon, \ldots, \varepsilon$).

First we show that $W^\varepsilon \to W$ in $(\widetilde{\mathcal{W}}_0, \delta_\square)$ if $\varepsilon \to 0$. Indeed, for every connected simple graph F, we have

$$t(F, W^\varepsilon) = (1 - k\varepsilon)^{\mathsf{v}(F)} t(F, W) + \varepsilon^{\mathsf{v}(F)}\bigl(t(F, U_1) + \cdots + t(F, U_k)\bigr),$$

and hence $t(F, W^\varepsilon) \to t(F, W)$ as $\varepsilon \to 0$.

Next, we show that $t_{xy}(F_1, W^\varepsilon), \ldots, t_{xy}(F_k, W^\varepsilon)$ are linearly independent for all $\varepsilon > 0$. If not, then there are real numbers a_i such that

$$\sum_{i=1}^{k} a_i t_{xy}(F_i, W^\varepsilon) = 0$$

for all $x, y \in [0, 1]$. Suppose $1 - k\varepsilon + (j-1)\varepsilon \leq x, y \leq 1 - k\varepsilon + j\varepsilon$, then every choice of the variables for which one of the unlabeled nodes has value outside the interval $[1 - k\varepsilon + (j-1)\varepsilon, 1 - k\varepsilon + j\varepsilon]$ contributes 0 to $t_{xy}(F_i, W^\varepsilon)$. Hence

$$\sum_{i=1}^{k} a_i \varepsilon^{\mathsf{v}(F_i)} t_{xy}(F_i, U_j) = 0 \qquad (j = 1, \ldots, k)$$

for all $x, y \in [0, 1]$. Multiplying by $t_{xy}(G, U_j)$ and integrating, we get

$$\sum_{i=1}^{k} a_i \varepsilon^{\mathsf{v}(F_i)} t([\![GF_i]\!], U_j) = 0 \qquad (j = 1, \ldots, k).$$

But this contradicts the nonsingularity of the matrix $\bigl(t([\![GF_i]\!], U_j)\bigr)$, and proves the claim.

CLAIM 16.58. *If $F_1, \ldots, F_k$ are simple 2-labeled graphs with no floating components, then $\mathcal{T}(F_1, \ldots, F_k)$ is nowhere dense in $(\widetilde{\mathcal{W}}_0, \delta_\square)$.*

Indeed, Claim 16.57 implies that every nonempty open set in $(\widetilde{\mathcal{W}}_0, \delta_\square)$ contains a graphon U such that $t_{xy}(F_1, U), \ldots, t_{xy}(F_k, U)$ are linearly independent. Then their Gram determinant $\left|t(\llbracket F_i F_j \rrbracket, U)\right|_{i,j=1}^k$ is positive. But this determinant is a continuous in U, and so there is a neighborhood of $\mathcal{U}$ in which it does not vanish, and hence $t_{xy}(F_1, U'), \ldots, t_{xy}(F_k, U')$ are linearly independent in this neighborhood. This proves the claim and thereby the theorem. $\square$

16.7.4. Infinitesimally finitely forcible graphons. Lemma 16.53 suggests the following notion. We say that W is *infinitesimally finitely forcible* if $\mathcal{L}(W)$ has finite dimension. To explain the name, suppose that the functions $t_{xy}(F_1^\ddagger, W), \ldots, t_{xy}(F_k^\ddagger, W)$ generate $\mathcal{L}(W)$. Informally, this means that every infinitesimal change in W that preserves $t(F_1, W), \ldots, t(F_k, W)$, also preserves $t(F, W)$ for every F.

The following observation, contrasted with Corollary 16.51, shows that infinitesimal finite forcibility and finite forcibility behave quite differently:

LEMMA 16.59. *Every infinitesimally finitely forcible kernel has finite rank.*

Proof. If $\mathcal{L}(W)$ has finite dimension d, then consider the functions

$$t_{xy}(C_k^\ddagger, W) = k t_{xy}(P_k^{\bullet\bullet}, W) = kW^{\circ k} \in \mathcal{L}(W), \qquad k = 3, \ldots, d+3.$$

These are linearly dependent, and so W satisfies a polynomial equation as an operator. This means that it has a finite number of different nonzero eigenvalues. Since every nonzero eigenvalue has finite multiplicity, W has finite rank. $\square$

The following corollary shows that the (false) conjecture mentioned above that only stepfunctions are finitely forcible is true in a weaker sense.

COROLLARY 16.60. *Graphons that are both finitely forcible and infinitesimally finitely forcible are exactly the stepfunctions.*

This corollary implies that our examples of finitely forcible non-step-functions (e.g., the simple threshold graphon) are finitely forcible but not infinitesimally finitely forcible. We don't know any examples for the converse.

Proof. If a graphon is both finitely forcible and infinitesimally finitely forcible, then it has finite rank by Lemma 16.59, and so it is a stepfunction by Corollary 16.51.

Conversely, we know by Theorem 16.46 that every stepfunction W is finitely forcible. Since every function $t_{xy}(F^\ddagger, W)$ is itself a stepfunction with the same steps, it follows that $\mathcal{L}(W)$ is finite dimensional, so W is infinitesimally finitely forcible. $\square$

Summarizing Lemma 16.53 and Corollary 16.60, we get the following.

COROLLARY 16.61. *Suppose that $W \in \mathcal{W}$ is forced (in $\mathcal{W}$) by the simple graphs $F_1, \ldots, F_m$. Then either W is a stepfunction, or the functions $t_{xy}(F_i^\ddagger, W)$ ($i = 1, \ldots, m$) are linearly dependent.* $\square$

REMARK 16.62. 1. All the examples of finitely forcible graphons discussed above, indeed all the examples we know of, have dimension at most 1 (in the sense of the topology of the graphon discussed in Section 13.4). Most likely this is just due to the lack of more involved constructions; but it is not too far fetched to ask: *Does every finitely forcible graphon have finite dimension?* Together with Proposition 13.34 this would imply that finitely forcible graphons have polynomial-size weak regularity partitions. Together with Conjecture 16.45 and the properties of finitely forcible graphons proved above, this would provide nontrivial "templates" for extremal graphs, and possibly provide some help in finding the extremal graphs for specific extremal graph problems by imposing limitations on them.

2. We have seen a number of "finiteness" conditions on a graphon W: (a) W is a stepfunction; (b) W has finite rank; (c) W is finitely forcible; (d) W is infinitesimally finitely forcible; (e) the graph parameter $t(.,W)$ has finite connection rank, or equivalently, the corresponding gluing algebras $\mathcal{Q}_k/W$ have finite dimension; (f) the spaces (J, r_W) and/or $(\overline{J}, \overline{r}_W)$ are finite dimensional. We could add further such conditions, like (g) the algebras $\mathcal{Q}_k/W$ are finitely generated (this is true not only for stepfunctions, but also for a simple threshold function, for example). Several implications between these finiteness properties have been proved in this book, but several others are only conjectured.

EXERCISE 16.63. Prove that the simple threshold graphon (Example 11.36) is forced by the conditions (16.51) and $t(P_3, W) - t(K_2, W) + 1/6 = 0$.

EXERCISE 16.64. Show that for every kernel W there is a nonzero simple 2-labeled quantum graph g with nonadjacent labeled nodes (which may have floating components) such that $t_{xy}(g, W) = 0$.

EXERCISE 16.65. Which implications among the finiteness conditions (a)-(g) in Remark 16.62 are proved in this book? Which others are trivial/easy/possible?

CHAPTER 17

Multigraphs and decorated graphs

Limit objects can be defined for multigraphs, directed graphs, colored graphs, hypergraphs etc. In many cases, like directed graphs without parallel edges, or graphs with nodes colored with a fixed number of colors, this can be done along the same lines as for simple graphs.

Turning to multigraphs, even the definition of homomorphisms is not unique, as we have discussed in Chapter 5. In one version, a homomorphism $F \to G$ is a map $V(F) \to V(G)$ where the image of any edge has at least as large multiplicity as the edge itself (node-homomorphism); in another version, to specify a homomorphism between multigraphs, we have to tell the image of every node as well as the image of every edge (node-and-edge homomorphism). We also mentioned homomorphisms that preserve edge-multiplicities (induced homomorphisms). But this is not the main complication. To illuminate the content of this chapter, let us discuss informally convergence of multigraphs. We get to the most general question in several steps. We want to define convergence of a multigraph sequence $(G_1, G_2, \ldots)$ in terms of the convergence of the homomorphism densities $t(F, G_n)$ for every F, and want to construct a limit object that appropriately reflects the limiting values.

(1) In the previous chapters, this program was carried out in detail (maybe even in more detail than you wished to see) in the case when the graphs G_n as well as the graphs F were simple.

(2) Suppose that the graphs G_n are multigraphs, but we care about the densities of simple graphs F only. In this case, node-homomorphisms mean nothing new, but node-and-edge homomorphisms do. Let us assume for the time being that the edge multiplicities in the graphs G_n remain uniformly bounded by a fixed constant d. This case is quite easy, and it has been settled (even in greater generality) by Borgs, Chayes, Lovász, Sós and Vesztergombi [2008]: the limit object can be described by a kernel with values in $[0, d]$, and the proofs are rather straightforward generalizations of the proofs from case (1).

(3) Let the graphs G_n be multigraphs with bounded edge multiplicities as before, but we want the limit object to correctly reflect densities of multigraphs F. This case is more interesting. It turns out that whether we consider node-homomorphisms or node-and-edge homomorphisms does not matter much (this is not obvious at the first sight). Nor do the numerical values of the edge multiplicities: we can think of them just as decorations of the edges from the set $K = \{0, 1, \ldots, d\}$, and the only relevant property of this set is that it is finite. Here comes the first surprise: the limit object can again be defined as a function on $[0, 1]^2$, but its values are not numbers, but probability distributions on K (in other words, d-tuples of numbers). The second surprise is that one can generalize the results to decorations from a set K that is any compact Hausdorff space. Once the right statement of

the results is found, the proofs can be obtained by essentially the same techniques as before. These results of B. Szegedy and the author will be discussed in Section 17.1.

(4) Let us backtrack and generalize in another direction: we allow unlimited edge multiplicities for the graphs G_n, but are only interested in densities of simple graphs F. The limit object is, not too surprisingly, an unbounded kernel. But the treatment becomes more technical; one needs appropriate bounds on the growth of edge multiplicities, and even then, one has to modify the definition of the cut norm and strengthen the Regularity Lemma to get the proofs. Some preliminary results of L. Szakács and the author [unpublished] are described in the internet notes [Notes].

(5) Finally, if we have sequences of graphs with unbounded edge-multiplicities and we want a limit object that correctly reflects densities of multigraphs, then we have to combine the ideas of questions (3) and (4). Here the cases of node-homomorphism densities and node-and-edge homomorphism densities diverge: there will be graph sequences that are convergent in the node-homomorphism sense but not in the node-and-edge homomorphism sense. Kolossváry and Ráth [2011] showed how to assign limit objects if we work with node-homomorphisms; these results can also be derived from the results mentioned in point (3) above, by compactifying the set of integers. The limit object is a function defined on $[0,1]^2$, whose values are probability distributions on $\mathbb{N}$. One expects that under appropriate bounds on the growth of edge multiplicities, these limit objects will also be valid for the node-and-edge homomorphism densities. However, as far as I know, no details have been worked out here.

17.1. Compact decorated graphs

17.1.1. Sampling and homomorphism numbers. We often encounter graphs with a special decoration: most often we color nodes or edges with a finite number of colors, but in some cases the objects used for decoration are more complicated, like kernels in $\mathcal{W}$. Multiplicities of edges can be thought of as decorations from $\mathbb{N}$, and nodeweights and edgeweights can be thought of as decorations from $\mathbb{R}$. In this section we sketch how to extend the results about convergence and limit of simple graphs to the more general setting when we decorate every edge ij of a simple graph G by an element β_{ij}^G of an arbitrary, but fixed compact Hausdorff space K. Most of this is based on the work of Lovász and Szegedy [2012b].

It will be convenient to assume that K contains a special element called 0, where an edge decorated with 0 means that it is missing (one can always add an element to K to play this role). This way we may assume that the underlying simple graph is K_n°, the complete graph on $[n]$ with a loop edge on every node. We denote by $\mathcal{F}_n(K)$ the set of all K-decorated graphs on $[n]$, and by $\mathcal{F}(K)$, the set of all K-decorated graphs.

If K is finite, then so is $\mathcal{F}_n(K)$. If K is endowed with a topology, then $\mathcal{F}_n(K)$ is a topological space, endowed with the product topology of a finite number of copies of K. Compactness of K implies that $\mathcal{F}_n(K)$ is compact. We can identify graphs in $\mathcal{F}_n(K)$ that are isomorphic (in the obvious sense: the nodes can be permuted so that we get the same decoration for every edge); we get a topological space $\mathcal{F}_n(K)/S_n$, which is also compact.

To define subgraph sampling for K-decorated graphs is straightforward: For $G \in \mathcal{F}(K)$ and $k \in [\mathsf{v}(G)]$, let $\mathbb{G}(k, G)$ denote the K-decorated graph obtained by selecting a random ordered subset $(v_1, v_2, \ldots, v_k)$ of $V(G)$ uniformly, and decorating the edge ij of K_n° by β_{v_i,v_j}^G. While $\mathbb{G}(k, G)$ comes with labeled nodes, it is clear that this graph with any other labeling of its nodes arises with the same probability.

Here comes the first (little) surprise: to define homomorphism numbers and homomorphism densities for K-decorated graphs is not straightforward. There is no natural way to define $\mathsf{hom}(F, G)$ for two K-decorated graphs F and G. What we can do is the following. Let $\mathcal{C}$ denote the space of continuous real valued functions on K. For every map $\varphi \colon V(F) \to V(G)$, where F is a $\mathcal{C}$-decorated graph and G is a K-decorated graph, we define a real value

$$\mathsf{hom}_\varphi(F, G) = \prod_{1 \le i < j \le k} \beta_{ij}^F(\beta_{\varphi(i)\varphi(j)}^G).$$

(Recall that β_{ij}^F is a function on K while $\beta_{\varphi(i)\varphi(j)}^G$ is an element of K, so $\beta_{ij}^F(\beta_{\varphi(i)\varphi(j)}^G)$ is well defined.) The *homomorphism number* $\mathsf{hom}(F, G)$ is defined, as earlier, by

$$\mathsf{hom}(F, G) = \sum_{\varphi \colon V(F) \to V(G)} \mathsf{hom}_\varphi(F, G),$$

and we define $\mathsf{inj}(F, G)$, as before, by summing over injective maps. We also define the *homomorphism density* by

$$t(F, G) = \frac{\mathsf{hom}(F, G)}{\mathsf{v}(G)^{\mathsf{v}(F)}}.$$

These subgraph densities have some new features relative to those used so far. First of all, they are not necessarily in $[0, 1]$. Second, while for simple graphs homomorphism numbers and sample distributions were easily expressed in terms of each other (recall Proposition 5.5), in this more general setting the situation is different. For a fixed (large) K-decorated graph G, sampling from G assigns probabilities to K-decorated graphs, while the homomorphism numbers into G assign real numbers to $\mathcal{C}$-decorated graphs.

We are going to characterize convergence of a graph sequence in terms of homomorphism numbers from $\mathcal{C}$-decorated graphs. This seems to be quite wasteful, since in the case of simple graphs, only a countable number of convergence conditions had to be assumed. But we can restrict ourselves to graphs decorated by elements from an appropriate subset of $\mathcal{C}$. We say that a set $\mathcal{B} \subseteq \mathcal{C}$ is a *generating system* if the linear space generated by the elements of $\mathcal{B}$ is dense in $\mathcal{C}$ in the L_∞ norm. If $\mathcal{C}$ is finite dimensional, then it is the most economical to choose a basis of $\mathcal{C}$ for $\mathcal{B}$. It turns out that the choice of the family $\mathcal{B}$ has combinatorial significance, as the following examples show.

EXAMPLE 17.1 (**Simple graphs**). Let K be the discrete space with two elements called "edge" and "non-edge" or shortly 1 and 0. The set $\mathcal{C}$ consists of all maps $\{0, 1\} \to \mathbb{R}$, i.e., of all pairs $(f(0), f(1))$ of real numbers. A natural generating subset (in fact, a basis) in $\mathcal{C}$ consists of the pairs $f_0 = (1, 1)$ and $f_1 = (0, 1)$. Sampling, convergence, and homomorphism densities correspond to these notions introduced for simple graphs.

One may, however, take another basis in $\mathcal{C}$, namely the pair $g_0 = (0, 1)$ and $g_1 = (1, 0)$. Then again $\mathcal{B}$-decorated graphs can be thought of as simple graphs, and $\hom(F, G)$ counts the number of maps that preserve both adjacency and non-adjacency. ♦

EXAMPLE 17.2 (**Colored graphs**). Let K be a finite set of "colors" with the discrete topology. Continuous functions on K can be thought of as vectors in $\mathbb{R}^K$. The standard basis $\mathcal{B}$ in this space corresponds to elements of K, and so $\mathcal{B}$-decorated graphs are just the same as K-decorated graphs. The homomorphism density $t(F, G)$ is the probability that a random map $V(F) \to V(G)$ preserves edge colors. ♦

EXAMPLE 17.3 (**Multigraphs**). Let G be a multigraph with edge multiplicities at most d. Then G can be thought of as a K-decorated graph, where $K = \{0, 1, \ldots, d\}$. However, there are several meaningful ways of picking a basis in $\mathcal{C}$, giving rise different notions of homomorphisms.

- Taking the standard basis in $\mathcal{C} = \mathbb{R}^K$ means that we think of the edge multiplicities just as different labels (colors). The graph F will be decorated with edge multiplicities too, and then a homomorphism must preserve edge multiplicities. This is equivalent to Example 17.2; it can be thought of as the induced version of homomorphisms between multigraphs.
- Take the basis $\mathcal{B} = \{(1, 0, 0, \ldots, 0), (1, 1, 0, \ldots, 0), \ldots, (1, 1, 1, \ldots, 1)\}$ in $\mathcal{C}$. Again, we can think of a $\mathcal{B}$-decorated graph as a multigraph with edge multiplicities at most d. Then a map $\varphi : V(F) \to V(G)$ counts as a homomorphism if and only if the multiplicity of each target edge $\varphi(i)\varphi(j) \in E(G)$ is at least as large as the multiplicity of $ij \in E(F)$; in other words, it counts node-homomorphisms.
- Take the functions $\mathcal{B} = \{1, x, \ldots, x^d\}$ in $\mathcal{C}$. Again, we can think of a $\mathcal{B}$-decorated graph as a multigraph with edge multiplicities at most d, where an edge decorated by x^i is represented by i parallel edges. With this choice, $\hom(F, G)$ is the number of node-and-edge homomorphisms of F into G as multigraphs. ♦

EXAMPLE 17.4 (**Weighted graphs**). Let $\mathcal{K} \subseteq \mathbb{R}$ be a bounded closed interval. Let $\mathcal{B}$ be the collection of functions $x \mapsto x^j$ for $j = 0, 1, 2, \ldots$ on $\mathcal{K}$; then $\mathcal{B}$ is a generating system. It is natural to consider a $\mathcal{B}$-decorated graph F as a multigraph, and then $\hom(F, G)$ is just our usual homomorphism into a weighted graph. Note that Example 17.3 for the third choice of the basis of $\mathcal{C}$ is a special case. ♦

Much of the theory of graph homomorphisms can be built up for compact decorated graphs without much difficulty, but with some care. For example, the relationships between $\hom$ and $\inj$ (equations (5.16) and (5.24)) can be extended for any $\mathcal{C}$-labeled graph F and $\mathcal{K}$-labeled graph G:

$$(17.1) \qquad \hom(F, G) = \sum_P \inj(F/P, G),$$

and

$$(17.2) \qquad \inj(F, G) = \sum_P \mu_P \hom(F/P, G),$$

where we have to re-define F/P for a partition $P = V_1, \ldots, V_q$ of $V(F)$: the nodes are the partition classes, and an edge (V_i, V_j) is decorated by $\prod_{u \in V_i, v \in V_j} \beta^F_{uv}$.

(It may seem strange at the first sight that we decorate the edges of F/P by the product of their inverse images, and not by the sum, say. Looking at Example 17.1 gives an explanation: a missing edge is decorated by the function $(1, 1)$, an edge present, by the function $(0,1)$, so the product is $(0,1)$ if and only if at least one edge is present.)

From the results in Part 2, we state a generalization of one result (the first statement of Theorem 5.29), which we will need later.

THEOREM 17.5. *Two K-decorated graphs $G_1, G_2 \in \mathcal{F}(K)$ are isomorphic if and only if $\hom(F, G_1) = \hom(F, G_2)$ for every $\mathcal{B}$-decorated graph F with $\mathsf{v}(F) \leq \max(\mathsf{v}(G_1), \mathsf{v}(G_2))$.*

Proof. The "only if" part is obvious. For the converse, suppose that $\hom(F, G_1) = \hom(F, G_2)$ for every $\mathcal{B}$-decorated graph F. In particular, this holds for $F = K_1$, which implies that $\mathsf{v}(G_1) = \mathsf{v}(G_2) = n$ (say). Furthermore,

- $\hom(F, G_1) = \hom(F, G_2)$ for every F with $\mathsf{v}(F) \leq n$ whose edges are decorated by linear combinations of functions from $\mathcal{B}$. Indeed, expanding the product in the definition of the homomorphism number, we see that $\hom(F, G_1)$ can be written as a linear combination of values $\hom(F', G_1)$, where every F' is $\mathcal{B}$-decorated. We get a similar expression for $\hom(F, G_2)$ in terms of the values $\hom(F', G_2)$. Since $\hom(F', G_1) = \hom(F', G_2)$ for all these graphs F' by hypothesis, it follows that $\hom(F, G_1) = \hom(F, G_2)$.

- $\hom(F, G_1) = \hom(F, G_2)$ for every $\mathcal{C}$-decorated F with $\mathsf{v}(F) \leq n$. This follows since linear combinations of functions in $\mathcal{B}$ are dense in $\mathcal{C}$.

- $\mathsf{inj}(F, G_1) = \mathsf{inj}(F, G_2)$ for every $\mathcal{C}$-decorated F with $\mathsf{v}(F) \leq n$. This follows from (17.2).

Now let S be the set of all elements of K occurring as edge-decorations in G_1 or G_2. Let $F = K_n^\circ$, and let us decorate the edges of F with functions $f_e \in \mathcal{C}$ such that the values $f_e(s)$ ($e \in E(F)$, $s \in S$) are algebraically independent transcendentals (such functions clearly exist). In the equation $\mathsf{inj}(F, G_1) = \mathsf{inj}(F, G_2)$, every term is a product of these transcendentals, so for the equation to hold, we need that every term on the left cancels a term on the right, and vice versa. But if the term corresponding to an (injective) map $\varphi: V(F) \to V(G_1)$ cancels the term corresponding to $\psi: V(F) \to V(G_2)$, then $\varphi^{-1} \circ \psi$ is an isomorphism between G_1 and G_2. $\square$

17.1.2. Convergence. The definition of convergence in terms of sampling is rather straightforward, after we remark that the samples $\mathbb{G}(k, G)$ define a distribution on a compact space $\mathfrak{F}_k(K)$ (and not on a finite space as before), and probability distributions on a compact space behave nicely (see Appendix A.3.3). We say that a sequence $(G_1, G_2, \ldots)$ of K-decorated graphs is *convergent*, if $\mathsf{v}(G_n) \to \infty$, and for every $k \geq 1$, the samples $(\mathbb{G}(k, G_n) : n = 1, 2, \ldots)$ are weakly convergent in distribution. In other words, for every continuous function $f: \mathcal{F}_k(K) \to \mathbb{R}$, the limit $\lim_{n \to \infty} \mathsf{E}(f(\mathbb{G}(k, G_n)))$ exists. We note that it is enough to require this for continuous functions $f: \mathcal{F}_k(K)/S_k \to \mathbb{R}$ (where S_k is the symmetric group on $[k]$), or in other words, for continuous functions $f: \mathcal{F}_k(K) \to \mathbb{R}$ that are invariant under relabeling of the nodes. Indeed, the distribution of $\mathbb{G}(k, G_n)$ is invariant under relabeling, so if we define

$$\overline{f}(G) = \frac{1}{k!} \sum_{\pi \in S_k} f(G^\pi)$$

(where G^π denotes the graph obtained by permuting the labels in G according to π), then $\mathsf{E}\bigl(f(\mathbb{G}(k,G_n))\bigr) = \mathsf{E}\bigl(\overline{f}(\mathbb{G}(k,G_n))\bigr)$. If the values on the right converge, then so do the values on the left.

The next theorem shows that our somewhat complicated definition of homomorphism functions is right for the notion of convergence.

THEOREM 17.6 (**Equivalence of convergence notions**). *Let $(G_1, G_2, \dots)$ be a sequence of K-decorated graphs with $\mathsf{v}(G_n) \to \infty$, and let $\mathcal{B} \subseteq \mathcal{C}$ be a generating system. Then the following are equivalent:*

(a) *$(G_1, G_2, \dots)$ is convergent;*

(b) *for every $\mathcal{C}$-decorated graph F, the numerical sequence $t(F, G_n)$ is convergent;*

(c) *for every $\mathcal{B}$-decorated graph F, the numerical sequence $t(F, G_n)$ is convergent.*

An important consequence of this theorem is that if we define convergence of the sequence (G_n) in terms of the convergence of $t(F, G_n)$ for all $\mathcal{B}$-decorated graph, then the definition is independent of the choice of $\mathcal{B}$. In particular, convergence of a multigraph sequence is the same whether we use induced homomorphisms, node-homomorphisms, or node-and-edge homomorphisms in the definition.

Proof. (a)⇒(b)⇒(c): From the identity

$$\text{(17.3)} \qquad t_{\mathsf{inj}}(F, G_n) = \mathsf{E}\Bigl(t_{\mathsf{inj}}\bigl(F, \mathbb{G}(\mathsf{v}(F), G_n)\bigr)\Bigr)$$

we see that $t_{\mathsf{inj}}(F, G_n)$ is convergent. Just as for simple graphs, the convergence of the sequence $t_{\mathsf{inj}}(F, G_n)$ is equivalent to convergence of $t(F, G_n)$. Trivially, (c) also follows.

(c)⇒(a): Let f be any continuous function on $\mathcal{F}_k(K)/S_k$. We claim that F can be approximated uniformly in the L_∞ norm by functions of the type $\mathsf{hom}(h, .)$, where h is a $\mathcal{B}$-decorated quantum graph. (The value $\mathsf{hom}(h, G)$ is invariant under relabeling of the nodes of G, and hence $\mathsf{hom}(h, .)$ can be considered as a function on $\mathcal{F}_k(K)/S_k$.) We will use the Stone–Weierstrass Theorem. The set of functions $\mathsf{hom}(h, .)$ is closed under linear combination and multiplication, and contains the constant functions. Furthermore, it separates elements of $\mathcal{F}_k(K)/S_k$, i.e., for two non-isomorphic decorated graphs $G_1, G_2 \in \mathcal{F}_k(K)$ there is a $\mathcal{B}$-decorated quantum graph h such that $\mathsf{hom}(h, G_1) \neq \mathsf{hom}(h, G_2)$. This is just the content of Theorem 17.5.

It follows that if (G_n) is a sequence of decorated graphs that satisfies (c), then for every continuous function $\mathcal{F}_k(K)/S_k \to \mathbb{R}$, the sequence $f(G_n)$ is convergent. Hence (G_n) is convergent. □

17.1.3. Limit objects. Somewhat surprisingly, limit objects of K-decorated graph sequences are not 2-variable functions with values in K, but 2-variable functions with values that are probability measures on K. (This may be the precept of this whole section!) Let us denote by $\mathcal{W}(K)$ the set of functions $\omega : [0,1]^2 \to \mathcal{P}(K)$ that are measurable and symmetric ($\omega(x,y) = \omega(y,x)$ for every $(x,y) \in [0,1]^2$). Elements of $\mathcal{W}(K)$ will be called K-*graphons*.

Note that these K-graphons are different from the random graphons introduced in Section 14.5. In a random graphon $\mathbf{W}$, the values $\mathbf{W}(x,y)$ are also random variables, but for different points (x,y) they are correlated in specific ways (like in Example 11.44, where we obtained a distribution on constant functions). On the other hand, in a K-graphon we think of the probability distribution $\omega(x,y)$ as independent for different x and y. Technically, this cannot be defined (we don't have uncountably many independent random variables), but this will be how we treat them as soon as we can restrict our attention to a finite or countable set of pairs (x,y).

The above definition of K-graphons through measure-valued functions is rather abstract; we can give a more down-to-earth definition as follows. Let ω be a K-graphon and let $f \in \mathcal{C}$. Define $W_f : [0,1]^2 \mapsto \mathbb{R}$ by $W_f(x,y) = \int_K f\, d\omega(x,y)$ (recall that $\omega(x,y)$ is a probability measure on the Borel sets of K for every x and y). For every fixed $f \in \mathcal{C}$, the function W_f is a kernel.

Conversely, if we specify a measurable function $W : \mathcal{C} \times [0,1]^2 \to \mathbb{R}$ such that $W_f(x,y)$ is linear in f, $W_{f \cong 1}(x,y) = 1$ and $W_f(x,y) > 0$ for every function $f > 0$ for all $(x,y) \in [0,1]$, then there is a K-graphon represented by $W_f(x,y)$. Indeed, for every fixed $(x,y) \in [0,1]^2$, the functional $f \mapsto W_f(x,y)$ is a linear functional on $\mathcal{C}$ that is positive on positive functions. The Riesz representation theorem implies that there is a probability measure $\omega(x,y)$ such that $W_f(x,y) = \int_K f\, d\omega(x,y)$, and it is not hard to check that $\omega(x,y)$ defines a K-graphon.

It is enough to know the values $W_f(x,y)$ for $f \in \mathcal{B}$, where $\mathcal{B}$ is a generating system; this determines the values $W_f(x,y)$ for all $f \in \mathcal{C}$, and through this, the probability distributions $\omega(x,y)$. We call the system of functions $(W_f : f \in \mathcal{B})$ the $\mathcal{B}$-*moment representation* of ω. (The name refers to the fact that for various natural choices of K and $\mathcal{B}$, the numbers $t(F, W_f)$ $(F \in \mathcal{B})$ behave similarly to the moments of a single-variable function. See Example 17.3).

The construction assigning a graphon to every simple graph extends in a straightforward manner: every K-decorated graph G gives rise to a K-graphon ω_G as follows. Let $V(G) = [n]$. We split the unit interval into n intervals $J_1, \ldots, J_n$ of length $1/n$, and let $\omega_G(x,y) = \beta_{ij}^G$ for $x \in J_i$, $y \in J_j$ (here we identify the element $\beta_{ij}^G \in K$ with the probability distribution concentrated on β_{ij}^G).

It is also straightforward to extend homomorphism densities. For every K-graphon ω and $\mathcal{C}$-decorated graph F on $V(F) = [k]$, we introduce the homomorphism density $t(F,\omega)$ by

$$t(F,\omega) = \int_{[0,1]^k} \prod_{1 \leq i < j \leq k} W_{\beta_{i,j}^F}(x_i, x_j)\, dx_1 \ldots dx_k.$$

It is easy to see that for every K-decorated graph G and $\mathcal{C}$-decorated graph F,

$$t(F, \omega_G) = t(F, G).$$

Note that if F is $\mathcal{B}$-decorated for some generating system $\mathcal{B} \subseteq \mathcal{C}$, then $t(F, \omega)$ is expressed in terms of the $\mathcal{B}$-moment representation of ω.

EXAMPLE 17.7. It may be worthwhile to revisit our examples from Section 17.1.1.

Simple graphs could be thought of as K-decorated graphs with $K = \{0,1\}$. Every probability distribution on K can be represented by a number between 0 and 1, which is the probability of being adjacent (i.e., the probability of the element $1 \in K$). So a K-graphon is described by a symmetric measurable function $W : [0,1]^2 \mapsto [0,1]$, i.e., by a graphon.

For edge-colored graphs, K is a finite set of "colors". Probability distributions on K can be described by the probabilities of its points. So a K-graphon is represented by $k = |K|$ measurable functions $W_i : [0,1]^2 \mapsto [0,1]$ with $\sum_i W_i(x,y) = 1$.

For multigraphs with bounded edge multiplicities, K-graphons have the same description as above, only the "colors" correspond to "multiplicities" now. It is remarkable that this notion (and the notion of convergent sequence) does not depend on which definition of homomorphisms we adopt.

Finally, in the case of weighted graphs with edge weights from $K = [0,1]$ (for example), a K-graphon has values that are probability distributions on $[0,1]$. These are uniquely determined by their moments, so we can describe a K-graphon as a sequence of measurable functions $W_n : [0,1]^2 \mapsto [0,1]$ ($n = 0, 1, \dots$) such that $(W_n(x,y) : n = 0, 1, \dots)$ is the sequence of moments of a random variable with values in $[0,1]$ for all $x, y \in [0,1]$. ♦

The limit of a convergent sequence of K-decorated graphs can be represented by a K-graphon:

THEOREM 17.8. *Let $(G_1, G_2, \dots)$ be a convergent sequence of K-decorated graphs. Then there is a K-graphon ω such that $t(F, G_n) \to t(F, \omega)$ as $n \to \infty$ for every $\mathcal{C}$-decorated graph F.*

In Chapter 11, we based the proof of the analogous theorem (Theorem 11.21) on the compactness of the graphon space. One can extend that proof; the difficulty is that it is awkward to define the cut distance and prove its basic properties. One can bypass the cut distance and work directly with homomorphism densities. For the details of the proof, we refer to the paper of Lovász and Szegedy [2012b].

It is worth pointing out that while we can select different generating sets in $\mathcal{C}$, and we get seemingly very different sequences of homomorphism densities, the limit object is independent of which generating system we use.

We do get compactness of the K-graphon space, without the metrization given by the cut distance, by an easy observation. Just as for simple graphs and graphons, not only sequences of graphs but also sequences of K-graphons have convergent subsequences, which can be proved along the same lines:

THEOREM 17.9. *Let $\mathcal{B}$ be a countable generating set and let $W_1, W_2, \dots$ be a sequence of K-graphons. Then we can select an infinite subsequence for which $(t(F, W_n))$ is a convergent sequence for every $\mathcal{B}$-decorated graph F.* □

17.1.4. An old debt. It is time to sketch the proof of Theorem 5.61 characterizing multiplicative and reflection positive graph parameters with finite connection rank $r(f, 2)$. (We do use a fair bit of the material in previous chapters.)

Proof. The necessity of the conditions is easy, along the lines of the proofs of Propositions 5.64 and 7.1.

The sufficiency takes more work, but the proof can be put together from arguments that we have established before. Let f be a multiplicative, reflection positive multigraph parameter for which $r(f, 2)$ is finite. Replacing f by $f(G)/(f(K_1)^{\mathsf{v}(G)} f(K_2)^{\mathsf{e}(G)})$, we may assume that $f(K_1) = f(K_2) = 1$. We establish several representations of f.

1. First, we represent f as an expectation of homomorphism-like quantities. For every $n \geq 1$, there is a distribution on $[-1, 1]$-weighted graphs on $[n]$ such

that if $\mathbf{H}_n$ is chosen from this distribution, then $f(G) = \mathsf{E}\big(\mathsf{inj}(G, \mathbf{H}_n)\big)$ for every multigraph G on $[n]$. This can be proved using Proposition A.24 in the Appendix.

2. If the weighted graph Z_n on $[n]$ is chosen from the distribution in Step 1, then $f(G) = \lim_{n\to\infty} t(G, Z_{n^2})$ with probability 1 for every multigraph G. This can be proved by modifying the proof of Lemma 11.8 appropriately (in fact, it suffices to compute the second moments only).

3. There is a $[-1, 1]$-graphon W (in the sense of Section 17.1.3) such that $f = t(., W)$. This can be deduced from Theorem 17.8.

4. If $W = (W_0, W_1, \dots)$ be the moment sequence representation of W, then every W_n is a stepfunction with the same steps. This is proved using the assumption that $r(f, 2)$ is finite in a way similar to the proof of Theorem 13.47.

Knowing this, it is easy to construct the randomly weighted graph H. Its nodes correspond to the steps $S_1, \dots, S_q$ of the W_n, and the measures of the steps also give their nodeweights. The random variable $W(x, y)$ ($x \in S_i, y \in S_j$) gives the decoration of the edge ij. It is not hard to check that the construction gives the right graph parameter. □

17.2. Multigraphs with unbounded edge multiplicities

An obvious drawback of our compactly decorated graph model is that it does not include multigraphs with unbounded multiplicities. We can think of these as $\mathbb{N}$-decorated graphs, but this is not a compact space in the discrete topology. One way out is to compactify this space; the simplest compactification consists of adding a single new element ∞, whose neighborhoods are complements of finite subsets of $\mathbb{N}$. Let $\overline{\mathbb{N}}$ denote this compactification of $\mathbb{N}$.

Continuous functions $\overline{\mathbb{N}} \to \mathbb{R}$ are convergent sequences of real numbers, where the value on ∞ is the limit of the sequence. Let $\mathcal{C} = \mathcal{C}(\overline{\mathbb{N}})$ be the set of these functions. Similarly as in Example 17.3 above, we can choose $\mathcal{B}_0 = \{f_0, f_1, \dots, f_\infty\}$, where $f_i = \mathbb{1}_{\{i\}}$ for finite i, and $f_\infty \equiv 1$; or we can choose $\mathcal{B}_1 = \{g_0, g_1, \dots, g_\infty\}$, where $g_i = \mathbb{1}_{\{1,\dots,i\}}$ for finite i, and $g_\infty \equiv 1$. In the first case, $\mathsf{hom}(F, G)$ counts induced homomorphisms, while in the second, it counts node-homomorphisms. However, taking the functions x^i for $x > 1$ is not allowed here, since these values don't form a convergent sequence. So counting node-and-edge homomorphisms between multigraphs with unbounded edge multiplicities does not fit in this model.

We know that convergence of a sequence does not depend on the choice of the generating system, so we can characterize it either through induced homomorphisms or node-homomorphisms. It is easy to see that sequence of multigraphs $(G_n : n = 1, 2, \dots)$ is convergent if and only if for every $k \geq 0$, truncating the edge multiplicities at k gives a convergent sequence (now graphs with bounded edge multiplicities).

This is a reasonable definition, but if we want to describe convergence of node-and-edge homomorphism densities, or convergence of a sequence of weighted graphs with unbounded edge-weights, then we have to work more, as the following example shows.

EXAMPLE 17.10. Let G_n be the multigraph on $[n]$ where the multiplicity of the edge connecting 1 to 2^k is 4^k for $1 \leq k \leq \log n$, and all other edge multiplicities are 1. This graph sequence is convergent in the compactification sense. However, the edge densities $t(K_2, G_n)$ do not form a convergent sequence: we have $t(K_2, G_n) \sim 11/3$ if n is a power of 2, but $t(K_2, G_n) \sim 5/3$ if n is one less. ♦

Lovász and Szakács [unpublished] gave a different definition of convergence for multigraph sequences with unbounded edge multiplicities and constructed an appropriate limit object such that the densities of simple graphs converge to a limit for every convergent multigraph sequence; see Lovász [Notes].

Part 4

Limits of bounded degree graphs

CHAPTER 18

Graphings

This next part of the book treats convergence and limit objects of bounded degree graphs. We fix a positive integer D, and consider graphs with all degrees bounded by D. Unless explicitly said otherwise, this degree bound will be tacitly assumed.

In this chapter we introduce infinite graphs that generalize finite bounded degree graphs. Their main role will be to serve as limit objects for sequences of bounded degree graphs, analogous to the role of graphons in the previous part. Graphons (symmetric functions in two variables) are very common objects and of course they have been studied for many reasons since the dawn of analysis. Graphings are less common; however, they are interesting on their own right, and in fact, they too have been studied in other contexts, mainly in connection with group theory.

The situation will be more complex than in the dense case, and there will be no single "true" limit object. But the connection between these objects is quite interesting. As a further warning, it is not known whether the objects to be discussed in this chapter are all limit objects of sequences of finite graphs. This makes it even more justified to treat them separately from convergent finite graph sequences.

In this part we will consider finite graphs, countably infinite graphs, and even larger graphs (typically of continuum cardinality). To keep notation in check, we will denote finite graphs by $F, F', G, G', G_1 \ldots$, and families of finite graphs by calligraphic letters. In particular, we denote by $\mathcal{G}$ the family of all finite graphs (with all degrees bounded by D). We denote countable graphs by $H, H', H_1, \ldots$, and their families by Gothic letters. In particular, $\mathfrak{G}$ denotes the family of all countable graphs (with all degrees bounded by D). Graphs with larger cardinality will be denoted by boldface letters like $\mathbf{G}, \mathbf{G}', \ldots$; we will not talk about families of them.

18.1. Borel graphs

We start with a quite general notion. Let $(\Omega, \mathcal{B})$ be a Borel sigma-algebra; then $(\Omega, \mathcal{B})$ is separating and generated by a countable family $\mathcal{J} = \{J_1, J_2, \ldots\}$ of subsets of Ω. It will be convenient to assume that the generator set $\mathcal{J}$ is closed under complementation and finite intersections, which implies that it is a Boolean algebra (not a sigma-algebra!). We call the sets in $\mathcal{B}$ and also in $\mathcal{B} \times \mathcal{B}$ etc. *Borel sets*. (The reader who likes more concrete structures can think of this as the sigma-algebra of Borel sets in $[0, 1]$, with $\mathcal{J}$ consisting of finite unions of open intervals with rational endpoints.)

Let $\mathbf{G}$ be a graph with node set $V(\mathbf{G}) = \Omega$. We call $\mathbf{G}$ a *Borel graph*, if its edge-set is a Borel set in $\mathcal{B} \times \mathcal{B}$. For example, the complete graph on Ω is Borel, but we will be interested in graphs with all degrees bounded by D, and will tacitly assume this condition for these infinite graphs too.

EXAMPLE 18.1. For a fixed $a \in (0,1)$, we define a graph $\mathbf{P}_a$ on $[0,1]$ by connecting two points x and y if $|x-y| = a$. This defines a Borel graph. The graph structure of this Borel graph is quite simple: it is the union of finite paths. If $a > 1/2$, then it is just a matching together with isolated nodes. Of course, the Borel structure adds additional structure.

We can make the example more interesting, if we wrap the interval $[0,1]$ around, and consider the graph $\mathbf{C}_a$ on $[0,1)$ in which a node x is connected to $x+a \pmod 1$ and $x-a \pmod 1$. If a is irrational, we get a graph that consists of two-way infinite paths; if a is rational the graph will consist of cycles. ♦

The following lemma is very useful and it also motivates some of the definitions in the sequel.

LEMMA 18.2. *A graph $\mathbf{G}$ on a Borel space $(\Omega, \mathcal{B})$ is a Borel graph if and only if for every Borel set $B \in \mathcal{B}$, the neighborhood $N_{\mathbf{G}}(B)$ is Borel.*

Proof. Suppose that $\mathbf{G}$ is a Borel graph, and let $B \in \mathcal{B}$. Then $B' = E(G) \cap (B \times \Omega)$ is a Borel set. Furthermore, if we project B' to the second coordinate, then the inverse image of any point is finite. A classical theorem of Lusin [1930] implies that the projection is also Borel; but this projection is just $N_{\mathbf{G}}(B)$.

Conversely, assume that $\mathbf{G}$ has the property that the neighborhood of any Borel set is also Borel. Let $\mathcal{P}_i$ ($i = 1, 2, \dots$) range over all partitions of Ω into a finite number of sets in $\mathcal{J}$. We claim that

$$(18.1) \qquad E(\mathbf{G}) = \bigcap_i \bigcup_{J \in \mathcal{P}_i} (J \times N_{\mathbf{G}}(J));$$

this will prove that $\mathbf{G}$ is Borel.

First, let $(x,y) \in E(\mathbf{G})$. If $x \in J \in \mathcal{P}_i$, then $y \in N_{\mathbf{G}}(J)$ and so $(x,y) \in J \times N_{\mathbf{G}}(J)$. Hence it follows that (x,y) is contained in the right hand side of (18.1).

Second, let (x,y) be a pair contained in the right hand side of (18.1), then for every $i \geq 1$ there is a set $J \in \mathcal{P}_i$ for which $(x,y) \in J \times N_{\mathbf{G}}(J)$. Hence $y \in N_{\mathbf{G}}(J)$, and so there is a point $z_i \in J$ such that $y \in N_{\mathbf{G}}(z_i)$, which means that $z_i \in N_{\mathbf{G}}(y)$. But $N_{\mathbf{G}}(y)$ is a finite set, and so this can hold for all sets $J \in \mathcal{P}_i$, $J \ni x$ only if $x \in N_{\mathbf{G}}(y)$. This shows that $(x,y) \in E(\mathbf{G})$. □

There is a rather rich theory of Borel graphs (see e.g. Kechris and Miller [2004]). We state and prove only a few results that we need. The following theorem, which extends Brooks' Theorem from finite graphs to Borel graphs, was proved by Kechris, Solecki and Todorcevic [1999]. We say that a coloring of the nodes of a Borel graph is a *Borel coloring*, if nodes with any given color form a Borel set.

THEOREM 18.3. *Every Borel graph has a Borel coloring with $D+1$ colors.*

Proof. We start with constructing a countable Borel coloring. Consider the countable Boolean algebra $\mathcal{J} = \{J_1, J_2, \dots\}$ generating Borel sets. For every node $v \in \Omega$ there is a set J_i such that $N(v) \subseteq J_i$ but $v \notin J_i$. Let $\psi(v)$ denote the smallest index i for which J_i has this property. Then trivially the sets $\psi^{-1}(i)$ ($i = 1, 2, \dots$) are disjoint. Two adjacent nodes u and v cannot have $\psi(u) = \psi(v)$, so the sets $\psi^{-1}(i)$ ($i=1,2,\dots$) consist of nonadjacent points. It is easy to check (using Lemma 18.2) that these sets are Borel sets.

Next, for each $i=1,2,\dots$, we recolor each node in $\psi^{-1}(i)$ with the least color that does not occur among its neighbors. We do this in order, so the nodes in $\psi^{-1}(2)$ get recolored before the nodes in $\psi^{-1}(3)$ etc. Using Lemma 18.2, it is easy to prove that those points that switch their color to a given j form a stable Borel set. Hence the final coloring is Borel. It is trivial that every new color is bounded by $D+1$. □

Now we turn to coloring the edges. Shannon's Theorem asserts that the edges of a multigraph with maximum degree D can be colored by at most $3D/2$ colors. For simple graphs, Vizing's Theorem gives the better bound of $D+1$. For Borel graphs, only a weaker result is known (which is quite trivial in the finite case):

THEOREM 18.4. *Every Borel graph has a Borel edge coloring with $2D-1$ colors.*

Proof. We define a new Borel graph, the "line-graph" $L(\mathbf{G})$ of $\mathbf{G}$, defined on the set $E(\mathbf{G})$ (which, as a subset of $\Omega \times \Omega$, is equipped with a sigma-algebra of Borel sets), where two edges of $\mathbf{G}$ are adjacent if and only if they share a common endpoint. It is straightforward to see that this graph $L(\mathbf{G})$ is Borel, and has degrees at most $2D-2$. Applying Theorem 18.3 to $L(\mathbf{G})$, we get the Theorem. □

If we have a Borel graph, then various basic constructions lead to Borel sets and functions. We state and prove this fact for the degree function, but many other similar assertions can be proved (see the Exercises below). For every set $A \subseteq \Omega$ and $x \in \Omega$, let $\deg_A^{\mathbf{G}}(x)$ denote the number of neighbors of x in A. We suppress the superscript $\mathbf{G}$ if there is only one graph around.

LEMMA 18.5. *Let $\mathbf{G}$ be a Borel graph. Then for every Borel set $A \subseteq V(\mathbf{G})$, $\deg_A(x)$ is a Borel function of x.*

Proof. Let $\mathcal{P}_i$ ($i=1,2,\dots$) range over all partitions of $V(\mathbf{G})$ into D sets from the generator set $\mathcal{J}$. Then
$$\deg_A(x) = \max_{i \in \mathbb{N}} \sum_{J \in \mathcal{P}_i} \mathbb{1}_{N_{\mathbf{G}}(A \cap J)}(x),$$
showing that this function is Borel. □

Let G be a graph (of any cardinality). We define the *local distance* of two nodes $u,v \in V(G)$ by
$$d^\circ(u,v) = \inf\{2^{-r} : B_{G,r}(u) \cong B_{G,r}(v)\}.$$
This turns $V(G)$ into a semimetric space. Unfortunately, two different nodes may be at distance 0: this happens exactly when there is an automorphism of G moving one onto the other. (This is the first occurrence of the "curse of symmetry" in this part; it has caused difficulties in Chapter 6, and it will haunt us when constructing graphings or designing local algorithms for large graphs.) We call the topology defined by this semimetric the *local topology*. This defines a *local topology* on $V(G) \times V(G)$.

Assuming that there are no points at distance 0, local distance defines an *ultrametric space* (i.e., the triangle inequality holds in a very strong sense: $d^\circ(x,y) \leq \max\{d^\circ(x,z), d^\circ(z,y)\}$). This implies (or it is easy to see directly) that the set of nodes whose r-neighborhood has a fixed isomorphism type is both closed and open, and such sets form an open basis. The space is totally disconnected.

PROPOSITION 18.6. *Let G be a bounded degree graph (of any cardinality), and suppose that G has no automorphism. Then $E(G)$ is closed in the local topology, and hence G is a Borel graph with respect to the Borel space defined by the local topology.*

Proof. Let $xy \notin E(G)$, and let $y_1, \ldots, y_d$ be the neighbors of x. Since G has no automorphism, there is an $r \geq 1$ such that $B_{G,r}(y) \not\cong B_{G,r}(y_1), \ldots, B_{G,r}(y_d)$. We claim that if $u, v \in V(G)$ such that $d^\circ(u,x) < 2^{-r}$ and $d^\circ(v,y) < 2^{-r}$, then $uv \notin E(G)$.

Assume that $uv \in E(G)$. By the definition of the distance function, $d^\circ(u,x) < 2^{-r}$ implies that $B_{G,r+1}(x) \cong B_{G,r+1}(u)$. Let, say, y_1 correspond to v under this isomorphism, then $B_{G,r}(y_1) \cong B_{G,r}(v)$. But $d^\circ(u,x) < 2^{-r}$ implies that $B_{G,r}(v) \cong B_{G,r}(y)$, so $B_{G,r}(y_1) \cong B_{G,r}(y)$, a contradiction. □

What to do if G has automorphisms? One possibility is to decorate the nodes from some set K of "colors", in order to break all automorphisms. A similar construction will be described in Section 18.3.4, and here we don't go into the details.

EXERCISE 18.7. Let **G** be a Borel graph, and let us add all edges that connect nodes at distance 2. Prove that the resulting graph $\mathbf{G}^2$ is Borel.

EXERCISE 18.8. Let **G** be a Borel graph. Prove that for every 1-labeled simple graph F, the quantity $\hom_u(F, \mathbf{G})$ is well-defined, and it is a Borel function of $u \in V(\mathbf{G})$.

EXERCISE 18.9. Let **G** be a Borel graph and let V_k denote the set of nodes with degree k. Prove that V_k is a Borel set.

EXERCISE 18.10. Let **G** be a Borel graph and let V_k denote the union of its finite components with k nodes. Prove that V_k is a Borel set.

EXERCISE 18.11. Prove that every Borel graph has a maximal stable set of nodes that is Borel.

EXERCISE 18.12. Prove that if a graph with bounded degree has no automorphism, then its cardinality is at most continuum.

18.2. Measure preserving graphs

Now we come to the definition of graphs that will serve as limit objects for convergent sequences of bounded degree graphs. We endow the sigma-algebra $(\Omega, \mathcal{B})$ with a probability measure λ. We say that a graph **G** with node set Ω is *measure preserving*, or a *graphing*, if it is Borel and for any two measurable sets A and B, we have

$$(18.2) \qquad \int_A \deg_B(x)\, d\lambda(x) = \int_B \deg_A(x)\, d\lambda(x).$$

In other words, "counting" the edges between A and B from A, we get the same as counting them from B. To be precise, a graphing is a quadruple $\mathbf{G} = (\Omega, \mathcal{B}, \lambda, E)$, where $\Omega = V(G)$ is a set, $\mathcal{B} = \mathcal{B}(\mathbf{G})$ is a Borel sigma-algebra on Ω, $\lambda = \lambda_\mathbf{G}$ is a probability measure on $\mathcal{B}$, and $E = E(\mathbf{G}) \in \mathcal{B} \times \mathcal{B}$ is a Borel set satisfying the measure preserving condition (18.2).

REMARK 18.13 (**On terminology**). The name "graphing" was introduced by Adams [1990], and it refers to the representation of the classes of an equivalence relation as the connected components of a Borel graph. It seems, however, that the usage is shifting to the one above. Since these objects are analogous to our "graphons" in the dense case (whose name comes from the contraction of graph-function), I like the parallel "graphon–graphing", and will adopt the above meaning.

Besides the probability measure λ on the points, there are two (related) measures that often play a role. The integral measure of the degree function is often called the *volume*:

$$(18.3) \qquad \operatorname{Vol}(A) = \int_A \deg(x)\, d\lambda(x).$$

The volume of the whole underlying set is the average degree:

$$(18.4) \qquad d_0 = \operatorname{Vol}(\Omega) = \int_\Omega \deg(x)\, d\lambda(x).$$

We can normalize the volume to get a probability measure $\lambda^*(A) = \operatorname{Vol}(A)/\operatorname{Vol}(\Omega)$. We call the distribution λ^* the *stationary distribution* of $\mathbf{G}$; the name refers to the random walk on $\mathbf{G}$.

We can define a finite measure $\eta = \eta_\mathbf{G}$ on $(\Omega \times \Omega, \mathcal{B} \times \mathcal{B})$ by

$$(18.5) \qquad \eta(A \times B) = \int_A \deg_B(x)\, d\lambda(x)$$

for product sets $(A, B \in \mathcal{B})$. It is not hard to see that Caratheodory's Theorem applies and we can extend η to the sigma-algebra $\mathcal{B} \times \mathcal{B}$. If we want a probability measure on the edges, we can normalize by the average degree: the measure η/d_0 can be considered as the uniform probability measure on $E(\mathbf{G})$. Equation (18.2) implies that η is invariant under interchanging the coordinates. Both marginals of η give the volume measure.

LEMMA 18.14. *The measure η is concentrated on $E(\mathbf{G})$.*

Proof. Let $\mathcal{J} = \{J_1, J_2, \dots\}$. We claim that

$$(18.6) \qquad E(\mathbf{G}) = (\Omega \times \Omega) \setminus \bigcup_{i=1}^\infty \bigl(J_i \times (\Omega \setminus N_\mathbf{G}(J_i))\bigr).$$

It is clear that $E(\mathbf{G})$ is contained in the right hand side. Conversely, if $(x,y) \notin \bigcup_i \bigl(J_i \times (\Omega \setminus N_\mathbf{G}(J_i))\bigr)$, then for each i for which $J_i \ni x$, we have $y \in N_\mathbf{G}(J_i)$. So there is a $z_i \in J_i$ adjacent to y. Since y has finite degree, this can hold for each J_i only if x is adjacent to y. This proves (18.6).

Since $\eta\bigl(J_i \times (\Omega \setminus N_\mathbf{G}(J_i))\bigr) = 0$ by the definition of η, equation (18.6) implies that $\eta(\Omega \times \Omega \setminus E(\mathbf{G})) = 0$. $\square$

Assuming that the average degree is positive, one way to generate a random edge from the distribution η/d_0 is to select a point x from the distribution λ^*, and then select an edge incident with x uniformly at random. Conversely, selecting a random edge from the distribution η/d_0, and then selecting randomly one of its endpoints, we get a point from the distribution λ^*. To describe the connection between λ and λ^* in this language, we can generate a point from λ^* by generating a random point x according to λ, and keeping it with probability $\deg(x)/D$ (else, rejecting it and generating a new one). If there are no isolated nodes, we can generate a point from λ by generating a random point x according to λ^*, and keeping it with probability $1/\deg(x)$.

EXAMPLE 18.15. If $D = 1$, then every graphing $\mathbf{G}$ is the graph of an involution $\varphi : S \to S$ for some set $S \subseteq V(\mathbf{G})$. Since $S = N_{\mathbf{G}}(V(\mathbf{G}))$, it is measurable. Furthermore, for any measurable $A \subseteq S$ we have

$$\lambda(\varphi^{-1}(A)) = \int_{\varphi^{-1}(A)} \deg_A(x)\,d\lambda(x) = \int_A \deg_{\varphi^{-1}(A)}(x)\,d\lambda(x) = \lambda(A),$$

and so φ is a measure preserving map. ◆

EXAMPLE 18.16 (**Graphings from graphs**). For any finite graph F, we define a graphing $\mathbf{G}_F$ as follows. Let $V(F) = [n]$, and let us split the unit interval $[0,1)$ into n intervals $J_i = [(i-1)/n, i/n)$. For every edge $ij \in E(F)$ with $i < j$, let us connect every point $x \in J_i$ to $x + (j-i)/n \in J_j$. It is not hard to verify that the resulting graph $\mathbf{G}_F$ is measure preserving. Every connected component of $\mathbf{G}_F$ is isomorphic to F. See Figure 18.1(a) for the graphing on $[0,1]$ representing of the pentagon. The picture is similar to the pixel picture of the graphon associated with a simple graph, except that instead of a black square, we have a white square with a diagonal. ◆

EXAMPLE 18.17 (**Cyclic graphing**). Consider the graph $\mathbf{C}_a$ introduced in Example 18.1. Endowing it with the uniform measure on $[0,1]$ turns it into a graphing. If a is rational, then every connected component of $\mathbf{C}_a$ is a cycle; else, every connected component is a 2-way infinite path. In this latter case, we call $\mathbf{C}_a$ an *irrational cyclic graphing* (Figure 18.1(b)). ◆

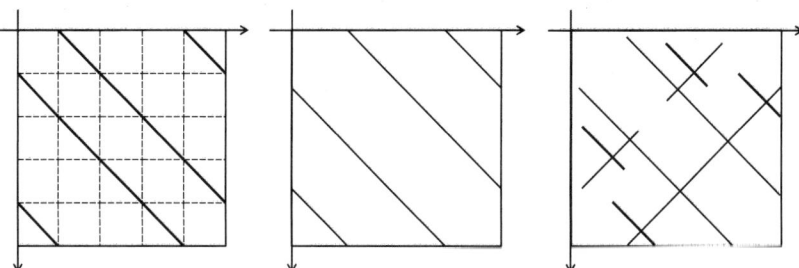

FIGURE 18.1. Three graphings: (a) the pentagon as a graphing, (b) a cyclic graphing, (c) the union of a symmetric finite set of segments with slopes ± 1; every such picture defines a graphing. The slopes of ± 1 correspond the measure preservation.

REMARK 18.18. It may be useful to allow graphs measurable with respect to the completion of $\mathcal{B}$ instead of $\mathcal{B}$ with respect to the probability measure λ (in the case of the interval $[0,1]$, this means allowing Lebesgue measurable sets instead of Borel sets). We call a graph *Lebesgue measurable* if for every set $A \in \mathcal{A}$, its neighborhood $N_{\mathbf{G}}(A) \in \mathcal{A}$. The correspondence between graphings and Lebesgue measurable graphs is described in Exercises 18.30–18.32.

18.2.1. Verifying measure preservation. Suppose that we have a measurable graph $\mathbf{G}$ with a probability measure λ on the node set. How to verify that this graph is measure preserving? Let us describe some methods to do so.

Edge measure. The simplest method, which often works well, is to specify a measure η on the edge set satisfying (18.5). To be more precise, we consider a probability space $(\Omega, \mathcal{B}, \lambda)$ and a Borel graph $\mathbf{G}$ on it. Suppose that there exists a finite measure η on the Borel sets in $\Omega \times \Omega$, which is invariant under interchanging the coordinates, concentrated on E, and its marginal is the volume measure Vol. This trivially implies that (18.2) holds.

Borel subgraphs. Every subgraph of a graphing that is in a sense explicitly definable is itself a graphing: there is no constructive way to violate (18.2). The following lemma makes this precise.

LEMMA 18.19. *Let $\mathbf{G} = (\Omega, \mathcal{B}, \lambda, E)$ be a graphing, and let $L \subseteq E$ be a symmetric Borel set. Then $\mathbf{G}' = (\Omega, \mathcal{B}, \lambda, L)$ is a graphing.*

Proof. Let $A, B \subseteq \Omega$ be Borel sets. We want to show that

$$(18.7) \qquad \int_A d_B^{\mathbf{G}'}(x)\, d\lambda(x) = \int_B d_A^{\mathbf{G}'}(x)\, d\lambda(x).$$

First we prove that this equation holds when $L = E \cap \big[(S \times T) \cup (T \times S)\big]$ with two disjoint Borel sets S, T. Indeed, for any two Borel sets A and B,

$$\int_A d_B^{\mathbf{G}'}(x)\, d\lambda(x) = \int_{A \cap S} d_{B \cap T}^{\mathbf{G}}(x)\, d\lambda(x) + \int_{A \cap T} d_{B \cap S}^{\mathbf{G}}(x)\, d\lambda(x)$$

$$= \int_{B \cap T} d_{A \cap S}^{\mathbf{G}}(x)\, d\lambda(x) + \int_{B \cap S} d_{A \cap T}^{\mathbf{G}}(x)\, d\lambda(x) = \int_B d_A^{\mathbf{G}'}(x)\, d\lambda(x).$$

A similar computation shows that (18.7) holds if $L = E \cap (S \times S)$ for any Borel set S.

To prove the lemma in general, we use induction on the degree bound D. For $D = 1$ the assertion is trivial.

Let $\varepsilon > 0$, let $\mathcal{J} = \{J_1, J_2, \ldots\}$ be a countable generator set of $\mathcal{B}$, and let $\mathcal{P}_n$ be a partition of $V(\mathbf{G})$ into the atoms generated by $J_1, \ldots, J_n$. Let X_n be the set of points with degree D all whose neighbors belong to the same class of $\mathcal{P}_n$. Since any two points are separated by $\mathcal{P}_n$ if n is large enough, we have $\cap_n X_n = \emptyset$, and hence $\lambda(X_n) \leq \varepsilon$ if n is large enough. Let us fix such an n.

For $S \in \mathcal{P}_n$, let $S' = S \setminus X_n$. For every $S, T \in \mathcal{P}_n$, the graph $\mathbf{G}(S, T)$ obtained by restricting the edge set of $\mathbf{G}$ to $(S' \times T') \cup (T' \times S')$ is measure preserving by the special case proved above. In $\mathbf{G}(S, T)$, each point has degree at most $D - 1$, by the definition of S' and T'. Hence by the induction hypothesis, restricting the edge set to $E(\mathbf{G}(S, T)) \cap L$ we get a graphing $\mathbf{G}'(S, T)$, which means that

$$\int_A d_B^{\mathbf{G}'(S,T)}\, d\lambda(x) = \int_B d_A^{\mathbf{G}'(S,T)}(x)\, d\lambda(x).$$

Since the graphings $\mathbf{G}'(S, T)$ are edge-disjoint, it follows that $\mathbf{G}_1 = \cup_{S,T} \mathbf{G}'(S, T)$ is measure preserving. We get $\mathbf{G}_1$ from $\mathbf{G}'$ by deleting all edges incident with X_n, and hence for any two measurable sets A and B, we have

$$\int_A d_B^{\mathbf{G}'}(x)\, d\lambda(x) = \int_A d_B^{\mathbf{G}_1}(x)\, d\lambda(x) + \int_A d_{B \cap X_n}^{\mathbf{G}'}(x)\, d\lambda(x) + \int_{A \cap X_n} d_{B \setminus X_n}^{\mathbf{G}'}(x)\, d\lambda(x).$$

Here

$$\int_A d^{\mathbf{G}'}_{B\cap X_n}(x)\,d\lambda(x) \leq \int_A d^{\mathbf{G}}_{B\cap X_n}(x)\,d\lambda(x) = \int_{B\cap X_n} d^{\mathbf{G}}_A(x)\,d\lambda(x) \leq D\lambda(X_n) \leq D\varepsilon,$$

and

$$\int_{A\cap X_n} d^{\mathbf{G}'}_{B\setminus X_n}(x)\,d\lambda(x) \leq D\lambda(X_n) \leq D\varepsilon.$$

Hence

$$\left| \int_A d^{\mathbf{G}'}_B(x)\,d\lambda(x) - \int_B d^{\mathbf{G}'}_A(x)\,d\lambda(x) \right| \leq \left| \int_A d^{\mathbf{G}_1}_B(x)\,d\lambda(x) - \int_B d^{\mathbf{G}_1}_A(x)\,d\lambda(x) \right| + 2D\varepsilon$$

$$= 2D\varepsilon.$$

Since ε was arbitrarily small, this proves (18.7). $\square$

COROLLARY 18.20. *The intersection and union of two graphings on the same probability space are graphings.*

Proof. Let $\mathbf{G}_1$ and $\mathbf{G}_2$ be the two graphings, then consider $\mathbf{G}_1 \cap \mathbf{G}_2$ (we keep the underlying point set and do the set operation on the edge set). This is a Borel subgraph of $\mathbf{G}_1$, and hence it is a graphing.

The assertion about the union is trivial if the graphings are edge-disjoint. In the general case, consider the graphs $\mathbf{G}_1 \setminus \mathbf{G}_2$, $\mathbf{G}_2 \setminus \mathbf{G}_1$ and $\mathbf{G}_1 \cap \mathbf{G}_2$. These three graphs are Borel subgraphs of one of the graphings G_1 and G_2, and hence they are graphings. But then so is their union, which is just $\mathbf{G}_1 \cup \mathbf{G}_2$. $\square$

Measure preserving families. Another way to "certify" the measure preservation condition is to use the simpler notion of invertible measure preserving maps.

Let $A_1, \ldots, A_k, B_1, \ldots, B_k$ be measurable subsets of a Borel space $(\Omega, \mathcal{B})$, and let $\varphi_i : A_i \to B_i$ be invertible measure preserving maps. The tuple $H = (\varphi_1, \ldots, \varphi_k)$ will be called a *measure preserving family* (see Gaboriau [2002], Kechris and Miller [2004]). From every measure preserving family H we get a directed multigraph $\overrightarrow{\mathbf{G}}$ on Ω by connecting $x, y \in \Omega$ for every i such that $y = \varphi_i(x)$. The edges of this digraph are colored with k colors in such a way that each color-class defines a measure preserving bijection between two measurable subsets of Ω.

Forgetting the orientation and the edge-colors of this digraph, we get a graph $\mathbf{G}$ with degrees bounded by $2k$, which we call the *support graph* of the measure preserving family.

It is more natural perhaps to assume that the maps $\varphi_1, \ldots, \varphi_k$ are involutions, in which case we get an undirected graph right away. We say that the measure preserving family is *involutive*. A little advantage of working with involutions is that we could extend the maps φ_i to measure preserving involutions $\Omega \to \Omega$, and would not have to worry about domains A_i and ranges B_i.

A graphing with its edges colored and oriented so that each color defines an invertible measure preserving map is equivalent to a measure preserving family. Conversely, every graphing can be "certified" by an appropriate measure preserving family.

THEOREM 18.21. *A graph supporting a measure preserving family is a graphing. Conversely, for every graphing* **G** *there is an involutive measure preserving family with at most $2D - 1$ parts supported by* **G**.

This measure preserving family is not unique in general.

Proof. If the family consists of a single measure preserving map φ, then it is easy to check that its support graph is measure preserving. For a general measure preserving family, this follows by Corollary 18.20.

To prove the converse, Theorem 18.4 implies that the edges of **G** can be split into at most $2D-1$ Borel sets that are matchings. By Lemma 18.19, the involutions that these matchings define are measure preserving. □

While it is nicer to work with measure preserving involutions, the generality allowed in the first statement of the previous theorem has some merits. It is often easier to construct a family of non-involutory measure preserving maps to support the graph. Furthermore, while the minimum number of maps in a measure preserving family with given support graph, and the minimum number of maps in an involutive family can be mutually bounded by factors of 2, the former may be smaller. Both of these merits are illustrated by the following example.

EXAMPLE 18.22. Consider an irrational cyclic graphing $\mathbf{C}_a$ as in Example 18.17; we may assume without loss of generality that $0 < a < 1/2$. This graphing is 2-regular, and as a graph it consists of disjoint 2-way infinite paths.

We claim that $\mathbf{C}_a$ cannot be represented by two involutions. Suppose that the involutions φ_1 and φ_2 define $\mathbf{C}_a$. Then each point $x \in V(\mathbf{C}_a)$ is matched with $x - a \pmod{1}$ by φ_1 and to $x + a \pmod{1}$ by φ_2, or the other way around. Let A_1 denote the set of points matched the first way, and A_2, the rest. Then trivially A_1 and A_2 are Borel sets, $A_1 \cup A_2 = V(\mathbf{C}_a)$, and $A_2 = A_1 + a \pmod{1}$. But this is impossible by basic results in ergodic theory, since the map $x \mapsto x + a$ is ergodic.

On the other hand, we can represent $\mathbf{C}_a$ by three involutions: One matches points x and $x + a$ if $0 < x \leq 1 - a$ and $x \in (2ka, (2k+1)a]$ for some $k \in \mathbb{N}$; the other matches points x and $x + a$ if $0 < x \leq 1 - a$ and $x \in \big((2k+1)a, (2k+2)a\big]$ for some $k \in \mathbb{N}$; the third matches points x and $x + a - 1$ if $1 - a < x \leq 1$. ◆

EXAMPLE 18.23 (**Squaring the circle**). Answering a problem of Tarski, Laczkovich [1990] proved that a circular disc D can be partitioned into a finite number of sets, and these can be translated so that they form a partition of a square S with the same area. (It is not known whether this can be achieved by measurable pieces.) This result gives rise to interesting graphings. Let $X_1, X_2, \ldots, X_m$ be the pieces of D, and let $v_1, \ldots, v_m$ be the translation vectors (so that $X_1 + v_1, \ldots, X_m + v_m$ form a partition of S). We may assume that D and S are disjoint, and $\lambda(D) = \lambda(S) = 1/2$. We define a bipartite graph **G** on $D \cup S$ by connecting $x \in D$ to $y \in S$ iff $y - x \in \{v_1, \ldots, v_m\}$. Clearly **G** is a Borel graph, and every point has degree at most m. Furthermore, those edges that are defined by the same vector v_i define a measure preserving map between $D \cap (S - v - i)$ and $(D + v_i) \cap S$. Hence by Theorem 18.21, **G** is a graphing.

The theorem of Laczkovich is equivalent to saying that the vectors $v_1, \ldots, v_m$ can be chosen so that the resulting graphing **G** has a perfect matching. It is not known whether (for an appropriately rich family of translations) it has a Borel perfect matching. (Exercise 18.29 shows that a graphing can have a perfect matching, but no Borel perfect matching.) ♦

EXERCISE 18.24. Let **G** be a graphing, and let us add all edges that connect nodes at distance 2. Prove that the resulting graph $\mathbf{G}^2$ is a graphing.

EXERCISE 18.25. Let **G** be a graphing in which every connected component has at most k nodes. Let $S \subseteq V(\mathbf{G})$ be a measurable set that intersects every connected component. Prove that $\lambda(S) \geq 1/k$.

EXERCISE 18.26. Let **G** be a graphing on $[0,1]$, let $E' \subseteq E(\mathbf{G})$ be a symmetric Borel set, and $E'' = E(\mathbf{G}) \backslash E'$. Consider the graphings $\mathbf{G}'$ and $\mathbf{G}''$ on $[0,1]$ defined by the edge sets E' and E'' (cf. Lemma 18.19). Prove that $\eta_\mathbf{G} = \eta_{\mathbf{G}'} + \eta_{\mathbf{G}''}$.

EXERCISE 18.27. Let **G** be a graphing, and let $S \subseteq E(\mathbf{G})$ be a (not necessarily symmetric) Borel set. For $x \in V(\mathbf{G})$, let $d_S^+(x)$ denote the number of pairs $(x,y) \in S$, and let $d_S^-(x)$ denote the number of pairs $(y,x) \in S$. Prove that $\int_{V(G)} d_S^+\, d\lambda = \int_{V(G)} d_S^-\, d\lambda$.

EXERCISE 18.28. Let G_i ($i=1,2$) be a graphing on $[0,1]$. Define the categorical product $G_1 \times G_2$ (as a graph on $[0,1] \times [0,1]$), and prove that it is a graphing (cf. Aldous and Lyons [2007]).

EXERCISE 18.29. Let $\mathbf{C}_a$ be an irrational cyclic graphing. (a) Show that it contains a perfect matching, but no Borel measurable perfect matching (Laczkovich [1988]). (b) Show that if M is any Borel measurable matching in $\mathbf{C}_a$, then there is an augmenting path: a path of odd length such that its endpoints are not covered by M, but every second edge on the path belongs to M (Elek and Lippner [2010]).

EXERCISE 18.30. Prove that for every Lebesgue measurable graph **G** there is a set $T \subseteq V(\mathbf{G})$ with measure 1 such that the subgraph $\mathbf{G}[T]$ obtained by restricting the edge set to $T \times T$ is Borel.

EXERCISE 18.31. Show by an example that there is a Borel graph on $[0,1]$ that is not Lebesgue measurable.

EXERCISE 18.32. (a) Let **G** be a Borel graph such that $\lambda(N_\mathbf{G}(A)) = 0$ for all $A \subseteq V(\mathbf{G})$ with $\lambda(A) = 0$. Prove that **G** is Lebesgue measurable. (b) Prove that every graphing is Lebesgue measurable.

18.3. Random rooted graphs

The construction of Benjamini and Schramm [2001] for limit objects of bounded degree graph sequences is different from graphings, but closely related to them. (It is interesting to note that the relationship of these objects with convergent graph sequences is not completely known, but their relationship with each other is quite well understood.) They will be helpful throughout, in particular in extending the notion of weak isomorphism to graphings and characterizing it.

Let us pick a random point x of a graphing **G**. (When talking about a random point of a graphing $\mathbf{G} = (\Omega, \mathcal{B}, E, \lambda)$, we mean that it is selected according to the probability measure λ.) Consider the connected component $\mathbf{G}_x$ containing it: This is a countable graph with bounded degree, and it has a special "root", namely the node x. So we have generated a random rooted connected graph with bounded degrees. We start with making precise sense of this, by constructing an interesting Borel graph, the "graph of graphs".

18.3.1. The graph of graphs.

Let $\mathfrak{G}^\bullet$ denote the set of connected countable graphs (with all degrees bounded by D) that also have a specified node called their *root*. We denote the root of a graph $H \in \mathfrak{G}^\bullet$ by $\operatorname{root}(H)$. (We could consider these graphs as 1-labeled, but in this context calling the single labeled node the "root" is common.) Sometimes we will write we also write $H = (H', v)$, where $v = \operatorname{root}(H)$, and $H' = [\![H]\!]$ is the unrooted graph underlying H. For every rooted graph H, we denote by $\deg(H)$ the degree of its root. The set of finite graphs in $\mathfrak{G}^\bullet$ will be denoted by $\mathcal{G}^\bullet$.

We consider two graphs in $\mathfrak{G}^\bullet$ the same if there is an isomorphism between them that preserves the root. Let $\mathfrak{B}_r \subseteq \mathcal{G}^\bullet$ denote the set of r-balls, i.e., the set of finite rooted graphs in which every node is at a distance at most r from the root. (Since we keep the degree bound D fixed, the set $\mathfrak{B}_r$ is finite.) For a rooted countable graph $(H, v) \in \mathfrak{G}^\bullet$, let $B_{H,r} = B_{H,r}(v) \in \mathfrak{B}_r$ denote the neighborhood of the root with radius r. For every r-ball F, let $\mathfrak{G}^\bullet_F$ denote the set of "extensions" of F, i.e., the set of those graphs $H \in \mathfrak{G}^\bullet$ for which $B_{H,r} \cong F$ (as rooted graphs).

With all this notation, we can define something more interesting. First we define a graph $\mathbf{H}$ on the set $\mathfrak{G}^\bullet$. Let $(H, v) \in \mathfrak{G}^\bullet$. For every edge $e = vv' \in E(H)$, connect (H, v) by an edge to the rooted graph $(H, v') \in \mathfrak{G}^\bullet$. So every edge of H incident with v gives rise to an edge of $\mathbf{H}$ incident with (H, v). In particular, all degrees in $\mathbf{H}$ are bounded by D. We call $\mathbf{H}$ the "Graph of Graphs".

The r-neighborhood of a rooted graph H in $\mathbf{H}$ is almost the same as the r-neighborhood of the root in H. To be precise, if $[\![H]\!]$ has no automorphism, then $B_{\mathbf{H},r}(H) \cong B_{H,r}(\operatorname{root}(H))$. The image of $v \in V(H)$ under this isomorphism is obtained by moving the root of H to v. However, if there is an automorphism of H moving $\operatorname{root}(H)$ to v, then the "curse of symmetry" strikes again, and this map is not one-to-one.

We endow the set $\mathfrak{G}^\bullet$ with a metric: For two graphs $H_1, H_2 \in \mathfrak{G}^\bullet$, define their *ball distance* by

$$d^\bullet(H_1, H_2) = \inf\{2^{-r} : B_{H_1,r} \cong B_{H_2,r}\}.$$

(This is reminiscent of the semimetric defining the local topology of a graph, but it is defined on a different set.) This turns $\mathfrak{G}^\bullet$ into a metric space. It is easy to see (Exercise 18.43) that the sets $\mathfrak{G}^\bullet_F$ are both closed and open, they form an open basis, and the space $(\mathfrak{G}^\bullet, d^\bullet)$ is compact and totally disconnected. The sigma-algebra of Borel sets of $(\mathfrak{G}^\bullet, d^\bullet)$ will be denoted by $\mathfrak{A}$.

As usual, every subset of $\mathfrak{G}^\bullet$ you will ever need, and every function $\mathfrak{G}^\bullet \to \mathbb{R}$ you will ever define, will be Borel. The graph $\mathbf{H}$ is Borel with respect to the sigma-algebra $\mathfrak{A}$. This follows by the same argument as Proposition 18.6.

18.3.2. Invariant measures.

You may have noticed that we have not defined any measure on the set $\mathfrak{G}^\bullet$. We will in fact consider many probability measures on it; these measures will carry the real information. Let σ be any probability measure on $(\mathfrak{G}^\bullet, \mathfrak{A})$. It is easy to see that the degree $\deg$ of the root is a measurable function on $\mathfrak{G}$. Nodes with different degrees cause some complication here, and it will be best to introduce right away another probability measure on $\mathfrak{G}^\bullet$: we define

$$\sigma^*(A) = \int_A \deg\, d\sigma \Big/ \int_{\mathfrak{G}^\bullet} \deg\, d\sigma.$$

Clearly these integrals are finite (at most D). If the denominator is 0, then σ is concentrated on the graph consisting of a single node (the only connected graph with average degree 0). In this trivial case, we set $\sigma^* = \sigma$.

Next, we introduce a very important condition on the distribution, which expresses that all possible roots of a graph are taken into account judiciously. (The meaning of this condition will be clearer when we get to limits of graph sequences.) Select a rooted graph H according to the distribution σ^* and then select a uniform random edge e from the root. We consider e as oriented away from the root. This way we get a probability distribution $\sigma^{\rightarrow}$ on the set $\mathfrak{G}^{\rightarrow}$ of graphs in $\mathfrak{G}^{\bullet}$ with an oriented edge (the "root edge") from the root also specified. We say that σ is *involution invariant* (another name commonly used is *unimodular*) if the map $\mathfrak{G}^{\rightarrow} \to \mathfrak{G}^{\rightarrow}$ obtained by reversing the orientation of the root edge is measure preserving with respect to $\sigma^{\rightarrow}$. By an *involution invariant random graph* we mean a random rooted connected graph drawn from an involution invariant probability measure on $\mathfrak{G}^{\bullet}$.

EXAMPLE 18.33. Let $G \in \mathcal{G}$ be a connected finite graph. Selecting a root from $V(G)$ uniformly at random defines a probability distribution σ_G on $\mathfrak{G}^{\bullet}$ (concentrated on rooted copies of G). If we select the root v with probability proportional to the degree of v, and a root edge e incident with v uniformly, then simple computation shows that the edge is uniformly distributed among all oriented edges, and so the distribution σ_G is involution invariant. ♦

EXAMPLE 18.34 (**Path**). Let P denote the two-way infinite path with any node chosen as a root. The distribution on $\mathfrak{G}^{\bullet}$ concentrated on P is involution invariant, since selecting any root edge we still get a distribution concentrated on a single graph, so reversing the edge preserves this distribution. ♦

EXAMPLE 18.35 (**Triangular Ribbon**). Let P be the 2-way infinite path and let R be the "ribbon" obtained from P by connecting every pair of nodes at distance 2 (Figure 18.2(a)). If we specify any node as its root, we get a connected countable 4-regular rooted graph $R^{\bullet}$. The distribution on $\mathfrak{G}^{\bullet}$ (where $D = 4$) concentrated on $R^{\bullet}$ is involution invariant. To see this, note that if we select an oriented edge as a root, we get only two edge-rooted graphs H' and H'' (up to isomorphism): either an edge of P is selected, or an edge not on P. Furthermore, reversing the edge yields an isomorphic edge-rooted graph, so the distribution on $\{H', H''\}$ remains involution invariant. ♦

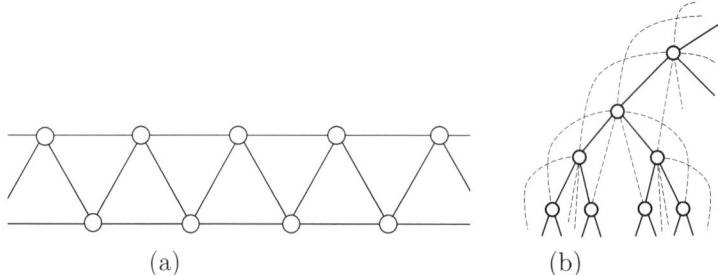

FIGURE 18.2. (a) The triangular ribbon. (b) The grandmother graph. Note that the neighborhood of a node reveals the orientation of the tree.

EXAMPLE 18.36 (**Grandmother graph**). Let T be a two-way infinite binary tree, and let us connect every node to its grandparent (Figure 18.2(b)). The resulting 8-regular connected graph H has a node-transitive automorphism group, and hence if specify any node as its root, we get isomorphic rooted graphs. The distribution on $\mathfrak{G}^\bullet$ (where $D = 8$) concentrated on H is, however, *not* involution invariant.

To see this, we again determine the possible edge-rooted graphs that we obtain from H. We get 4 types: an edge of T oriented "up", an edge of T oriented "down", an edge not in T oriented "up", and an edge not in T oriented "down". It is not hard to check that these are non-isomorphic, and the probabilities they are obtained with are (in the above order) $2/8, 1/8, 4/8, 1/8$. Reversing the root edge interchanges the first two and the last two probabilities, so the distribution is not involution invariant. ♦

18.3.3. Graphings and random rooted graphs. Let us recall the simple construction at the beginning of this section, providing a link between graphings and involution invariant distributions. Let $\mathbf{G}$ be a graphing and choose a random point $x \in V(\mathbf{G})$. The connected component $\mathbf{G}_x$ of $\mathbf{G}$ containing x, with a root x, is a graph in $\mathfrak{G}^\bullet$, which we call a *random rooted component* of $\mathbf{G}$. The map $x \mapsto \mathbf{G}_x$, which we will call the *component map*, is measurable as a map $(V(\mathbf{G}), \mathcal{A}) \to (\mathfrak{G}^\bullet, \mathfrak{A})$, and thus every graphing $\mathbf{G}$ defines a probability distribution $\sigma = \sigma_\mathbf{G}$ on $(\mathfrak{G}^\bullet, \mathfrak{A})$. Selecting x from the distribution λ^*, the graph $\mathbf{G}_x$ will be a random rooted connected graph from the distribution σ^*. Selecting an edge of $\mathbf{G}_x$ incident with x, we get an edge of $\mathbf{G}$ from the probability distribution $\eta_\mathbf{G}/d_0$, together with an orientation. Since $\eta_\mathbf{G}$ is symmetric, shifting the root to the other endpoint does not change the distribution. Hence σ is involution invariant.

So every graphing gives rise to a (well-defined) involution invariant random graph; we also say that the graphing *represents* this distribution. The following converse to this statement was known in various related contexts for some time; for written versions, see Aldous and Lyons [2007] and Elek [2007a].

THEOREM 18.37. *Every involution invariant probability distribution on $\mathfrak{G}^\bullet$ can be represented by a graphing.*

Here we don't claim uniqueness any more. This will be quite relevant a little later! Before proving this theorem, let us consider a couple of examples.

EXAMPLE 18.38 (**Grid**). Consider the involution invariant random graph concentrated on the infinite planar grid (with any root). We can construct a graphing representing this by taking two irrational reals α and β that are independent over the rationals, and connecting every $x \in [0,1)$ to $x + \alpha \pmod 1$, $x - \alpha \pmod 1$, $x + \beta \pmod 1$ and $x - \beta \pmod 1$. There are many other constructions, for example, we could take the unit square $[0,1)^2$ as the underlying probability space, and connect (x, y) to $(x \pm \alpha \pmod 1, y \pm \beta \pmod 1)$. Every connected component of this graphing will be an infinite grid. ♦

EXAMPLE 18.39. Consider the involution invariant random graph that is concentrated on the D-regular tree with a root (which is unique up to isomorphism). How to represent this by a graphing?

For $D = 2$, an irrational cyclic graphing is a graphing representation. Here is another one: let us randomly 2-color the path with colors 0 and 1. The sequence of colors to the right from the root (including the root) can be thought of as a number $x \in [0, 1]$. (Let us ignore the ambiguity that one can write rational numbers whose denominator is a power of two in two different ways; this involves a set of measure 0 anyway.) Similarly, the sequence of colors to the left of the root (this time excluding the root) gives a number $y \in [0, 1]$. So every point of the unit square corresponds to a 2-colored 2-way infinite path (with a root), and this correspondence is bijective. To shift the root to the right by one step corresponds to replacing x by $2x \pmod 1$ and y by $y/2$ if $x < 1/2$ and by $y/2+1/2$ if $x \geq 1/2$. (This map, as a transformation of $\{0,1\}^{\mathbb{Z}}$, is called a Bernoulli shift. In its other incarnation as a transformation of the unit square, it is sometimes called the *dough folding map*.) The graphing will be defined on $[0, 1]^2$ (with the Lebesgue measure), and every point (x, y) will be connected to its image and to its inverse image under the dough folding map.

For $D > 2$ a geometric construction of a representing graphing is more complicated. We can start from the fact that a D-regular tree is the Cayley graph of a group freely generated by D involutions. This group can be represented, for example, by reflections in D generic hyperplanes through the origin in D-space. If we take the surface of the unit sphere in $\mathbb{R}^D$ with the uniform probability distribution, and connect every point to its images and inverse images under these reflections, we get a graphing representing the infinite D-regular tree. (Points on the D hyperplanes in which we reflect will have lower degree than D; but we can delete these points and all their images under the group, which is a set of measure 0, and then we get a graphing in which every connected component is a D-regular tree.) ♦

As a preparation for the proof of Theorem 18.37, we describe a rather simple construction of a measurable graph from an involution invariant random graph (which, unfortunately, does not always represent the right measure). Consider the "graph of graphs" **H** constructed in Section 18.3.1. We have seen that **H** is a Borel graph. Unfortunately, this Borel graph with the measure σ does not represent the involution invariant distribution σ in general. For example, in Example 18.34 the graph **H** will have a single node (the rest is of measure 0), and this is too simple to represent anything nontrivial. Exercise 18.47 describes an even worse example, showing that an involution invariant measure does not necessarily turn the graph **H** into a graphing.

The problem is clearly caused by the symmetries in the graph σ is concentrated on. This motivates the following lemma.

LEMMA 18.40. *Let σ be an involution invariant distribution on $\mathfrak{G}^{\bullet}$ such that almost all graphs from σ have no automorphism (as unrooted graphs). Then $(\mathbf{H}, \sigma)$ is a graphing that represents σ.*

Proof. First, we prove that $(\mathbf{H}, \sigma)$ is a graphing. Choose a rooted graph $(H, v) \in \mathfrak{G}^{\bullet}$ from the distribution σ^*, and a random neighbor u of v (uniformly from the neighbors). By the assumption that almost surely H has no nontrivial automorphism, the graph (H, u) is almost surely different from (H, v). The pair $(H, u)(H, v)$ is an edge of **H**, and selecting another neighbor of v, we would get a different edge of **H**. This describes a way to generate a random edge of **H** (with an orientation). It follows from the involution invariance of σ that this distribution on edges of **H** is invariant under flipping the orientation. Since the marginal of η is σ^*, this implies that $(\mathbf{H}, \sigma)$ is measure preserving.

Let (H, v) be any rooted graph from σ with no automorphism. We want to argue that the connected component of **H** containing (H, v) as a root is isomorphic to (H, v). Indeed, assigning the role of the root to different nodes of H gives non-isomorphic rooted graphs, and so we get an injection of $V(H) \to \mathfrak{G}^\bullet$. From the definition of adjacency in **H**, this embedding preserves adjacency and non-adjacency, and the range is a connected component of **H**. This proves that $(\mathbf{H}, \sigma)$ represents σ. $\square$

18.3.4. The Bernoulli Graphing. To prove Theorem 18.37, we have to break the symmetries of the graphs from σ. For this, we generalize the "graph of graphs" construction. Let $\mathfrak{G}^+$ denote the set of triples (H, v, α), where $(H, v) \in \mathfrak{G}^\bullet$, and $\alpha : V(H) \to [0, 1]$ is a weighting of the nodes of H. Two such rooted, weighted graphs are considered the same, if there is an isomorphism between the graphs that preserves the root and preserves the weights. Let $\mathcal{A}^+$ be the sigma-algebra on $\mathfrak{G}^+$ generated by the following cylinder sets: for an $r \geq 0$, we fix the isomorphism type of the ball $B \in \mathbf{B}_r$ with radius r about the root, and also for every node in B, we specify a Borel set in $[0, 1]$ from which the weight is to be chosen. (The choice of the interval $[0, 1]$ to use for weighting is arbitrary; we could have decorated the nodes by the points of any other Borel probability space. In fact, other decoration will be needed later.) It is easy to see that $(\mathfrak{G}^+, \mathcal{A}^+)$ is a Borel sigma-algebra.

We can define a graph on $\mathfrak{G}^+$, the *Graph of Weighted Graphs* $\mathbf{H}^+$, as follows: we connect two nodes (G, α) and (G', α') by an edge if G' arises from G by shifting the root to one of its neighbors (while keeping all the nodeweights); in other words, if there is an isomorphism ι from G to G' (as unrooted graphs) such that $\alpha'\big(\iota(u)\big) = \alpha(u)$ for every $u \in V(G)$, and $\iota\big(\text{root}(G)\big)$ is a neighbor of $\text{root}(G')$.

Given a probability distribution σ on $(\mathfrak{G}^\bullet, \mathcal{A})$, we can define a probability distribution σ^+ on $(\mathfrak{G}^+, \mathcal{A}^+)$ as follows: Select a random graph $H \in \mathfrak{G}^\bullet$ from the distribution σ, and assign independent, uniform random weights from $[0, 1]$ to the nodes.

LEMMA 18.41. *If σ is an involution invariant probability distribution on $(\mathfrak{G}^\bullet, \mathcal{A})$, then $\mathbf{B}_\sigma = (\mathfrak{G}^+, \mathcal{A}^+, \sigma^+)$ is a graphing, and it represents σ.*

This construction associates a graphing with every involution-invariant distribution, which we call the *Bernoulli graphing* representing σ. (The name refers to its close relationship with the Bernoulli shift in Example 18.39.) This lemma also provides the proof of Theorem 18.37.

Proof. The proof is essentially the same as the proof of Lemma 18.40, since assigning the random weights to the nodes of a graph G chosen from σ almost surely destroys all automorphisms. $\square$

EXERCISE 18.42. Prove that for $r > 2$, an r-ball has at most D^r nodes, and the number of non-isomorphic r-balls is bounded by D^{D^r}.

EXERCISE 18.43. Prove that the sets $\mathfrak{G}_F^\bullet$ are closed and open in the metric space $(\mathfrak{G}^\bullet, d^\bullet)$, they form an open basis, and the space is homeomorphic to a Cantor set.

EXERCISE 18.44. Show that a function $f : \mathfrak{G}^\bullet \to \mathbb{R}$ is continuous if and only if for every finite rooted graph $F \in \mathfrak{B}_r$ there is an $\varepsilon > 0$ such that for all graphs $H \in \mathfrak{G}^\bullet$ with $B_{H,r} \cong F$, we have $|f(G) - f(F)| < \varepsilon$.

EXERCISE 18.45. Prove that the following functions, defined for $H \in \mathfrak{G}^\bullet$, are Borel:
(a) $\mathbb{1}(H \cong H_0)$, where $H_0 \in \mathfrak{G}$ and isomorphism is meant as isomorphism of unlabeled graphs; (b) $\omega(H)$; (c) $\chi(H)$; (d) $f(H) = \limsup_{n\to\infty} e(B_r(G))/v(B_r(G))$.

EXERCISE 18.46. Let $H \in \mathfrak{G}^\bullet$, and consider the probability distribution on $\mathfrak{G}^\bullet$ concentrated on H. Prove that this distribution is involution invariant if and only if the automorphism group of H is transitive on the nodes, and for every oriented edge $\vec{e}$, the orbit of $\vec{e}$ (as a directed graph) has equal indegrees and outdegrees.

EXERCISE 18.47. Let G be a countably infinite graph consisting of a two-way infinite path with two nodes of degree 1 hanging from every node of the path. Let G_1 and G_2 be the two rooted graphs obtained from G by selecting a node of degree 4 and a node of degree 1 as its root, respectively. Let π be the probability distribution on $\mathfrak{G}$ in which $\pi(G_1) = 1/3$ and $\pi(G_2) = 2/3$. Show that π is involution invariant, but $(\mathbf{H}, \pi)$ is not measure preserving.

18.4. Subgraph densities in graphings

Our next goal is to generalize our central notion, graph homomorphism, to involution-invariant random graphs and to graphings. Following our general framework, we consider homomorphisms from a small graph into (say) an involution-invariant random graph, as well as homomorphisms from an infinite graph into small graphs. In both cases, there will be some nontrivial preparation that is needed, including proving some results that are important on their own right.

In this section we address the easier task, defining homomorphism densities in involution-invariant random graphs and graphings. This takes some preparation, discussing an important consequence of involution invariance.

18.4.1. Mass Transport Principle. Let us consider the set $\mathfrak{G}^{\bullet\bullet}$ of 2-labeled connected countable graphs (again, graphs that are isomorphic as 2-labeled graphs are identified). We can endow this set with a compact topology just like we did for $\mathfrak{G}^\bullet$, and then Borel functions are defined.

The following very useful characterization of involution invariance was proved by Aldous and Lyons [2007] (it was in fact this form how Benjamini and Schramm first defined involution-invariant measures).

PROPOSITION 18.48 (**Mass Transport Principle**). *Let σ be a probability distribution on $\mathfrak{G}^\bullet$. Then σ is involution invariant if and only if for every Borel function $f: \mathfrak{G}^{\bullet\bullet} \to \mathbb{R}_+$ the following identity holds:*

$$\mathsf{E}\Big(\sum_u f(H, v, u)\Big) = \mathsf{E}\Big(\sum_u f(H, u, v)\Big), \tag{18.8}$$

where $(H, v) \in \mathfrak{G}^\bullet$ is randomly chosen from the distribution σ.

Equation (18.8) allows that both sides be infinite. One can generalize the principle to functions without the nonnegativity condition, by applying it to separately to the negative and positive parts (however, one has to make sure that no infinite expectations occur). The name refers to the following interpretation: if we transport $f(H, u, v)$ amount of mass from node u to node v in the countable graph H (where this amount depends only on the isomorphism type of (H, u, v), and it depends on it in a Borel measurable way), then on the average, no node gains or loses. This is trivial for finite graphs, but it does not automatically hold for countable graphs, since there are distributions on $\mathfrak{G}^\bullet$ that are not involution invariant.

One can formulate a related identity for graphings, which shows that the Mass Transport principle is in a sense a form of Fubini's Theorem. To illustrate that we can vary the conditions, let us say that a function $f: S \times S \to \mathbb{R}$ (where S is a set of any cardinality) is *locally finite*, if the sums $\sum_{x \in S} f(x,y)$ and $\sum_{y \in S} f(x,y)$ are absolutely convergent (this includes that they have a countable number of nonzero terms).

PROPOSITION 18.49. *Let $\mathbf{G}$ be a graphing, and let $f: V(\mathbf{G}) \times V(\mathbf{G}) \to \mathbb{R}$ be a locally finite Borel function. Assume that $f(x,y) = 0$ unless $y \in V(\mathbf{G}_x)$. Then*

$$\int_{V(\mathbf{G})} \sum_y f(x,y)\,dx = \int_{V(\mathbf{G})} \sum_x f(x,y)\,dy$$

If f is the indicator function of edges between two Borel sets A and B, then this identity gives the basic measure preservation identity 18.2. The Mass Transport Principle can be used to prove properties of "typical" graphs from an involution invariant distribution; see Exercises 18.52 and 18.53.

We describe the proof of the graphing version; Proposition 18.48 can be proved along the same lines.

Proof. It suffices to prove this identity for nonnegative Borel functions, since we can write a general f as the difference of two such functions, which will also be locally finite. It suffices to prove it for bounded Borel functions, since we can obtain an unbounded nonnegative f as the limit of an increasing sequence of bounded Borel functions. By scaling, we may assume that the range of f is contained in $[0,1]$. We may assume that there is an $r \in \mathbb{N}$ such that $f(x,y) = 0$ unless $y \in B_{\mathbf{G},r}(x)$, since we can obtain f as the limit of an increasing sequence of such functions. Finally, it suffices to consider 0-1 valued Borel functions, since we can write f as

$$f(x,y) = \int_0^1 \mathbb{1}(f(x,y) \geq t)\,dt,$$

and here the function $\mathbb{1}(f(x,y) \geq t)$ is a 0-1 valued Borel function for every t.

A 0-1 valued Borel function corresponds to a Borel subset $S \subseteq V(G) \times V(G)$. Consider the graphing $\mathbf{G}^r$ obtained from $\mathbf{G}$ by connecting any two nodes at distance at most r. (This is indeed a graphing by Exercise 18.7.) The set S is a Borel subset of $E(\mathbf{G}^r)$, and hence by Exercise 18.27 we have $\int d_S^+\,d\lambda = \int d_S^-\,d\lambda$. But this is just the identity to be proved. □

18.4.2. Homomorphism frequencies. Recall that in a finite graph G, $t^*(F,G)$ can be interpreted as the expectation of $\hom_u(F',G)$, where F' is obtained from F by labeling one of its nodes, and u is a random node of G. This can be generalized to homomorphisms into an involution-invariant random graph. Indeed, let σ be an involution-invariant distribution, and let (H,v) denote a random rooted graph from σ. Then $\hom_v(F',H)$ is a bounded nonnegative integer, and since it depends only on a bounded neighborhood of the root v, it is a Borel function of (H,v). So $t^*(F',\sigma) = \mathsf{E}\bigl(\hom_v(F',H)\bigr)$ is well defined. Based on the finite case, we expect that $t^*(F',\sigma)$ is independent of the node labeled in F, and so we can define $t^*(F,\sigma) = t^*(F',\sigma)$. This is correct, but not obvious.

PROPOSITION 18.50. *Let F' and F'' be two 1-labeled graphs obtained from the same unlabeled connected graph F. Let σ be an involution-invariant distribution, then $t^*(F',\sigma) = t^*(F'',\sigma)$.*

Proof. Let F^* be the 2-labeled graph obtained by labeling both nodes that are labeled in F' or F''. Then for every rooted graph (H, u) generated according to σ, we have $\hom_u(F', H) = \sum_v \hom_{uv}(F^*, H)$ and $\hom_u(F'', H) = \sum_v \hom_{vu}(F^*, H)$. Applying the Mass Transport Principle to the function $f(H, u, v) = \hom_{uv}(F^*, H)$, we get the assertion. $\square$

If we want to define $t(F, \mathbf{G})$ for a graphing $\mathbf{G}$, it suffices to note that the graphing defines a unique involution invariant distribution σ, and so we can define $t(F, \mathbf{G}) = t(F, \sigma)$. Explicitly, $\hom_u(F', \mathbf{G})$ is a bounded measurable function of $u \in V(\mathbf{G})$, so the definition $t^*(F, \mathbf{G}) = \int \hom_u(F', \mathbf{G}) \, d\lambda(u)$ makes sense. By the argument above, it follows that this value is independent of the choice of the labeled node in F.

EXERCISE 18.51. Show that the distribution concentrated on the Grandmother Graph (Example 18.36) violates the Mass Transport Principle.

EXERCISE 18.52. Suppose that an involution-invariant random rooted graph is almost always infinite. Prove that the expected degree of the root is at least 2 (Aldous and Lyons [2007]).

EXERCISE 18.53. If $G \in \mathfrak{G}^\bullet$ is a random graph from an involution invariant distribution, then with probability 1, G has 0, 1, 2 or infinitely many ends. (An end of a graph is defined as an equivalence class of one-way infinite paths, where two paths are equivalent if they cannot be separated by a finite set of nodes.)

18.5. Local equivalence

Two graphings $\mathbf{G}_1$ and $\mathbf{G}_2$ are *locally equivalent*, if they have the same subgraph densities: $t^*(F, \mathbf{G}_1) = t^*(F, \mathbf{G}_2)$ for every connected simple graph F. We can formulate this notion in terms of the sample distributions. Recall that if G is a finite graph, then $\rho_{G,r}(B)$ ($B \in \mathfrak{B}_r$) is the probability that the r-neighborhood of a random node of G is isomorphic to the r-ball B. This definition extends to graphings verbatim. Two graphings $\mathbf{G}_1$ and $\mathbf{G}_2$ are locally equivalent if and only if the neighborhood distributions $\rho_{\mathbf{G}_i, r}$ are the same for every radius r. This is also equivalent to saying that they represent the same involution invariant distribution on the "graph of graphs" $\mathfrak{G}^\bullet$.

Our goal is to characterize weakly equivalent pairs of graphings. To this end, we have to introduce some special maps that certify weak equivalence.

Let $\mathbf{G}_1$ and $\mathbf{G}_2$ be two graphings. We call a measure preserving map $\varphi : V(\mathbf{G}_1) \to V(\mathbf{G}_2)$ a *local isomorphism*, if its restriction to almost every connected component of $\mathbf{G}_1$ is an isomorphism with one of the connected components of $\mathbf{G}_2$. To be more precise, if x is a random point of $V(\mathbf{G}_1)$, then $\varphi(x)$ is a random point of $V(\mathbf{G}_2)$, and $(\mathbf{G}_1)_x \cong (\mathbf{G}_2)_{\varphi(x)}$ with probability 1 (as rooted graphs). So the involution invariant distributions defined by $\mathbf{G}_1$ and $\mathbf{G}_2$ are the same. Note, however, that a local isomorphism may not be invertible.

EXAMPLE 18.54 (**Cycles vs. bicycles**). Consider the cyclic graphing $\mathbf{C}_a$ for an irrational real number a, and let $\mathbf{C}'_a$ be defined by connecting every $x \in [0, 1]$ to $x \pm (a/2) \mod 1/2$ if $x < 1/2$, and to $1/2 + (x \pm (a/2) \mod 1/2)$ if $x \geq 1/2$. Informally, $\mathbf{C}'_a$ consists of two disjoint copies of $\mathbf{C}_a$, shrunk by a factor of 2.

We claim that the map $\varphi : x \mapsto 2x \mod 1$ is a weak isomorphism from $\mathbf{C}'_a$ to $\mathbf{C}_a$. It is easy to see that this map is measure preserving. Furthermore, any connected component G of $\mathbf{C}'_a$ lies either entirely in $[0, 1/2)$ or entirely in $[1/2, 1)$, and therefore φ, restricted to G, is an isomorphism with an appropriate component of $\mathbf{G}_a$. ♦

EXAMPLE 18.55 (**Grid II**). Let $\mathbf{G}_1$ and $\mathbf{G}_2$ be the two graphings defined in Example 18.38 representing the infinite grid, where $\mathbf{G}_1$ is defined on $[0,1)$ and $\mathbf{G}_2$, on $[0,1)^2$. Then the map $(x,y) \mapsto x+y \pmod 1$ defines a local isomorphism from $\mathbf{G}_2$ to $\mathbf{G}_1$: it is trivially measure preserving, and it is an isomorphism when restricted to any connected component of $\mathbf{G}_2$. ♦

The relation "$\mathbf{G}_1$ has a local isomorphism into $\mathbf{G}_2$" is transitive (since local isomorphisms can be composed), but not symmetric (since local isomorphism may not be invertible). To make it symmetric, we define (temporarily, as we shall see) two graphings to be *bi-locally isomorphic*, if there is a third graphing that has local isomorphisms into both. This is now a symmetric relation, but don't we lose transitivity? No, we don't; the next lemma will be a main step in proving this.

LEMMA 18.56. *Let $\mathbf{G}_1$ and $\mathbf{G}_2$ be two graphings that both have a local isomorphism into a third graphing $\mathbf{G}_0$. Then they are bi-locally isomorphic.*

Proof. Let $\varphi_i : \mathbf{G}_i \to \mathbf{G}_0$ be a local isomorphism. Consider the set
$$\Omega = \{(x_1, x_2) : x_i \in V(\mathbf{G}_i),\ \varphi_1(x_1) = \varphi_2(x_2)\},$$
We have $\Omega \in \mathcal{A}(\mathbf{G}_1) \times \mathcal{A}(\mathbf{G}_2)$, which follows from the easy-to-check formula
$$(18.9) \qquad (V(\mathbf{G}_1) \times V(\mathbf{G}_2)) \setminus \Omega = \bigcup_{i=1}^{\infty} \varphi_1^{-1}(J_i) \times \varphi_2^{-1}(V(\mathbf{G}_0) \setminus J_i),$$
where $\{J_1, J_2, \dots\}$ is a countable generating Boolean algebra for $\mathcal{A}(\mathbf{G}_0)$. Let $\mathcal{A}$ denote the sigma-algebra obtained by restricting $\mathcal{A}(\mathbf{G}_1) \times \mathcal{A}(\mathbf{G}_2)$ to Ω.

Next, we define an appropriate probability measure on $(\Omega, \mathcal{A})$, or more conveniently phrased, a coupling measure on $V(\mathbf{G}_1) \times V(\mathbf{G}_2)$ concentrated on Ω. Proposition A.7 in the Appendix implies that such a measure exists. We denote by λ its restriction to $(\Omega, \mathcal{A})$.

Finally, we have to define an appropriate graph on the probability space $(\Omega, \mathcal{A}, \lambda)$. We first define it on the product $V(\mathbf{G}_1) \times V(\mathbf{G}_2)$ as the categorical product $\mathbf{G}_1 \times \mathbf{G}_2$, and then take the induced subgraph on Ω. (The graph $\mathbf{G}_1 \times \mathbf{G}_2$ has degrees bounded by D^2, but its restriction to Ω has degrees at most D, which will follow from the proof below, but can be checked directly.) The projection map from $V(\mathbf{G}_1) \times V(\mathbf{G}_2)$ onto $V(\mathbf{G}_i)$ can be restricted to Ω, to get a map $\psi_i : \Omega \to V(\mathbf{G}_i)$. We claim that ψ_i is local isomorphism.

First, we show that ψ_i is measure preserving. Indeed, for any set $A_1 \in \mathcal{A}(\mathbf{G}_1)$, we have $\psi_i^{-1}(A_1) = (A_1 \times V(\mathbf{G}_2)) \cap \Omega$, and hence
$$\lambda(\psi_i^{-1}(A_1)) = \lambda((A_1 \times V(\mathbf{G}_2)) \cap \Omega) = \lambda(A_1 \times V(\mathbf{G}_2))$$
$$= \lambda(\varphi_1(A) \cap \varphi_2(V(\mathbf{G}_2))) = \lambda_1(A_1).$$
(where λ_i is the probability measure of the graphing $\mathbf{G}_i$).

Let $(x_1, x_2) \in \Omega$ be a random point from the distribution λ, and let H denote its connected component. Let $H_i = (\mathbf{G}_i)_{x_i}$ denote the connected component of $\mathbf{G}_i$ containing x_i, and let H_0 be the connected component of $\mathbf{G}_0$ containing $\varphi_1(x_1) = \varphi_2(x_2)$. With probability 1, these three graphs are isomorphic, and the maps φ_i give isomorphisms $H_1 \cong H_0 \cong H_2$. For every node $v \in V(H_0)$, let $\alpha_i(v) \in V(H_i)$

be the node with $\varphi_i(\alpha_i(v)) = v$. (Note that $\varphi_i^{-1}(v)$ is not necessarily a singleton, but exactly one of its elements belongs to H_i.)

Let $\alpha(v) = (\alpha_1(v), \alpha_2(v)) \in \Omega$, then $\alpha(u)$ and $\alpha(v)$ are adjacent in $\mathbf{G}$ if and only if u and v are adjacent in H_0, by the definition of the product graph. So α is an embedding of H_0 into H as an induced subgraph. We want to argue that $\alpha(H_0) = H$. Indeed, if not, then there is a node $\alpha(u)$ that is connected by an edge of H to a node $(w_1, w_2) \in V(H) \setminus \alpha(V(H_0))$. Now $\alpha_1(u)$ is connected to w_1 in H_1 by the definition of the product graph, and hence $z = \varphi_1(w_1)$ is connected to u in H_0, and so it is a node in H_0. Similarly, $\varphi_2(w_2)$ is in H_0. Furthermore, $(w_1, w_2) \in V(H) \subseteq \Omega$, and hence $\varphi_1(w_1) = \varphi_2(w_2) = z$. But then $(w_1, w_2) = \alpha(z)$, a contradiction.

It follows that with probability 1, $\mathbf{G}_{(x_1,x_2)} \cong (\mathbf{G}_i)_{x_i}$, and ψ_i provides this isomorphism, which proves that ψ_i is a local isomorphism. Thus $\mathbf{G}_1$ and $\mathbf{G}_2$ are bi-locally isomorphic. □

COROLLARY 18.57. *Bi-local isomorphism is a transitive relation.*

Proof. Figure 18.3 tells the whole story: composing two bi-local isomorphisms, the middle part can be "flipped up" by Lemma 18.56 to get a single bi-local isomorphism. □

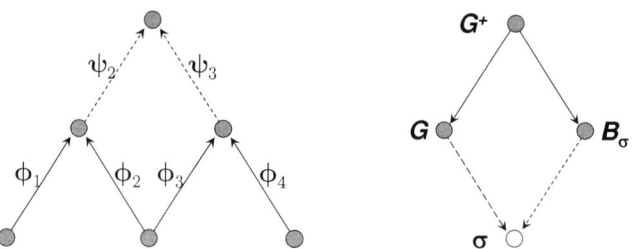

FIGURE 18.3. (a) Bi-local isomorphism is transitive. (b) The relationship between graphings, Bernoulli graphings and Bernoulli lifts.

We need a construction, introduced by Hatami, Lovász and Szegedy [2012], which is similar to the Bernoulli graphing defined in Section 18.3.3. For every graphing $\mathbf{G}$, we define the graphing $\mathbf{G}^+$, which we call the *Bernoulli lift* of $\mathbf{G}$. The points of $\mathbf{G}^+$ will be pairs (x, ξ), where $x \in V(\mathbf{G})$ and $\xi : V(\mathbf{G}_x) \to [0, 1]$. We connect (x, ξ) to (y, ζ) if y is a neighbor of x and $\xi = \zeta$ (note that if y is a neighbor of x, then $\mathbf{G}_x = \mathbf{G}_y$, so ξ and ζ are weightings of the same graph). Let Ω be the set of such pairs. We define a sigma-algebra $\mathcal{A}$ on Ω generated by the sets $\varphi^{-1}(A)$ ($A \in \mathcal{A}(\mathbf{G})$) and $\psi^{-1}(A)$ ($A \in \mathcal{B}$). To define a measure on $(\Omega, \mathcal{A})$, it is perhaps easiest to describe how a random element is generated: we pick a random point x of $\mathbf{G}$, and then assign independent random weights $\xi(u)$ to the nodes u of $(\mathbf{G})_x$. It is easy to see that $\mathbf{G}^+$ is a graphing.

There is a natural map $\varphi : V(\mathbf{G}^+) \to V(\mathbf{G})$, which simply forgets the weighting. There is also a natural map $\psi : V(\mathbf{G}^+) \to V(\mathbf{B}_\sigma)$, where σ is a involution-invariant distribution represented by $\mathbf{G}$, which forgets all about $\mathbf{G}$ except the distribution σ: this is defined by $\psi(x, \xi) = (\mathbf{G}_x, \xi)$. The following lemma is straightforward to verify.

LEMMA 18.58. *The maps $\varphi : V(\mathbf{G}^+) \to V(\mathbf{G})$ and $\psi : V(\mathbf{G}^+) \to V(\mathbf{B}_\sigma)$ defined above are local isomorphisms.*

Now we are able to prove the main result in this section.

THEOREM 18.59. *Two graphings are locally equivalent if and only if they are bi-locally isomorphic.*

Proof. The "if" part is trivial by the discussion above. To prove the "only if" part, let $\mathbf{G}_1$ and $\mathbf{G}_2$ be two locally equivalent graphings, we want to prove that they are bi-locally isomorphic. They define the same involution invariant distribution σ on $\mathfrak{G}$, and so they are both locally equivalent to the Bernoulli graphing $\mathbf{B}_\sigma$. Lemma 18.58 implies that they are both bi-locally isomorphic to $\mathbf{B}_\sigma$. Corollary 18.57 implies that they are bi-locally isomorphic. $\square$

EXERCISE 18.60. Let $0 \leq a \leq 1$ be an irrational number, and define a graphing $\mathbf{C}_a''$ (related to the graphings $\mathbf{C}_a$ and $\mathbf{C}_a'$ in Example 18.54): we connect every $x \in [0,1]$ to $1/2 + (x \pm (a/2) \mod 1/2)$ if $x < 1/2$, and to $x \pm (a/2) \mod 1/2$ if $x \geq 1/2$. Informally, we consider two circles of circumference $1/2$, and connect every point on one to the two points $a/2$ away from the corresponding point on the other circle. Prove that $\mathbf{C}_a''$ is locally equivalent to $\mathbf{C}_a$, and construct a local isomorphism $\mathbf{C}_a'' \to \mathbf{C}_a$.

18.6. Graphings and groups

Let $(\Omega, \varphi_1, \ldots, \varphi_m)$ be a measure preserving family, where $\varphi_i : \Omega \to \Omega$ are measure preserving maps defined on the whole Ω. The maps φ_i generate a group Γ of measure preserving maps.

Conversely, let Γ be a finitely generated group, with generators $g_1, \ldots, g_m$. Let us assume, for simplicity, that together with each g_i, its inverse g_i^{-1} is also among the generators. Let H be the Cayley graph of Γ: $V(H) = \Gamma$, and for every $x \in \Gamma$ and $1 \leq i \leq m$, we connect x to xg_i by an edge. We will get every edge in both directions, so we may consider G as an undirected graph.

Consider the random rooted graph model which is concentrated on G with root 1 (the identity element of Γ). This is involution invariant (see Exercise 18.46), so it defines a graphing $\mathbf{G}$ (Theorem 18.21). In fact, every oriented edge of H is marked by a generator of the group, and this marking is inherited by $\mathbf{G}$, and we can use this marking to construct the graphing in Theorem 18.21.

This correspondence between finitely generated groups, graphings and measure preserving families explains the interest of group theorists in the limit theory of bounded degree graphs. We do not elaborate this quite broad and very active area in this book; see e.g. Kechris and Miller [2004].

CHAPTER 19

Convergence of bounded degree graphs

Convergence of a graph sequence with bounded degree was perhaps the first which was formally defined (Benjamini and Schramm [2001]), but it is a more complex notion than convergence in the dense case. There are more than one non-equivalent reasonable definitions, which capture different aspects of the notion that graphs in a sequence are becoming "more and more similar" to each other. We treat two such notions in this Chapter.

19.1. Local convergence and limit

19.1.1. Distances. Just like in the dense case, we need to introduce some notions of a distance between two bounded degree graphs before starting the treatment of convergence. We don't have a good analogue of the cut distance, and therefore we will have to do with the sampling distance. This is a simpler notion, but of course less powerful, since knowing that two graphs are close in sampling distance does not translate into information about their global structure.

Recall that we have defined (in the introduction, informally) the *sampling distance* of two graphs $F, F' \in \mathcal{G}$. To make the definition precise, we start with the *sampling distance of depth* r, which is just the variational distance of neighborhood distributions, and we simply sum these with convenient (but ad hoc) weights:

$$(19.1) \qquad \delta_\odot^r(F, F') = d_{\text{var}}(\rho_{F,r}, \rho_{F',r}), \quad \delta_\odot(F, F') = \sum_{r=0}^{\infty} \frac{1}{2^r} \delta_\odot^r(F, F').$$

Note that in the second expression, the term with $r = 0$ is 0.

Lacking a good analogue of the cut distance, this sampling distance will be our main tool when comparing graphs. Since we can sample from a graphing just as well as we can sample from a graph, this distance is defined for two graphings, and also for a graphing and a graph. Since the sample distributions $\rho_{F,r}$ and $\rho_{\mathbf{G}_F, r}$ are the same (where $\mathbf{G}_F$ is the graphing on $[0, 1]$ representing the finite graph F), we have

$$(19.2) \qquad \delta_\odot^r(F, F') = \delta_\odot^r(\mathbf{G}_F, \mathbf{G}_{F'}), \quad \text{and} \quad \delta_\odot(F, F') = \delta_\odot(\mathbf{G}_F, \mathbf{G}_{F'}).$$

Using the trivial inequality $\delta_\odot^r(G, G') \leq \delta_\odot^{r+1}(G, G')$, we get that for every $r \geq 1$,

$$(19.3) \qquad \frac{1}{2^r} \delta_\odot^r(G, G') \leq \delta_\odot(G, G') \leq \frac{1}{2^r} + \delta_\odot^r(G, G').$$

An easy consequence of inequalities 19.3 is that if we want to estimate $\delta_\odot(G, G')$ of two finite graphs (which by definition depends on infinitely many radii r) with an error less than $\varepsilon > 0$, then we can do the following. We choose a positive integer $k > \log(3/\varepsilon)$, and then we take sufficiently many samples from both G and G' so that the empirical distributions φ_r and φ'_r of r-balls in these samples satisfy

$d_{\text{var}}(\rho_{G,r}, \varphi_r) \leq \varepsilon/6$ and $d_{\text{var}}(\rho_{G',r}, \varphi'_r) \leq \varepsilon/6$ with high probability. We claim that $A = \sum_{r=0}^{k} 2^{-r} d_{\text{var}}(\varphi_r, \varphi'_r)$ is a good estimate of $\delta_\odot(G, G')$. Indeed, with high probability,

$$|\delta_\odot(G, G') - A| \leq \sum_{r=0}^{k} \frac{1}{2^r} d_{\text{var}}(\rho_{G,r}, \varphi_r) + \sum_{r=0}^{k} \frac{1}{2^r} d_{\text{var}}(\rho_{G',r}, \varphi'_r)$$
$$+ \sum_{r=k+1}^{\infty} \frac{1}{2^r} d_{\text{var}}(\rho_{G,r}, \rho_{G',r}).$$

Here the first term is bounded by $(1+1/2+\cdots+1/2^k)\varepsilon/6 < \varepsilon/3$, and similar bound applies for the second term. The last term is bounded by $1/2^{k+1} + 1/2^{k+2} + \cdots = 1/2^k \leq \varepsilon/3$. So the total error is less than ε.

Often we have to compare two graphings $\mathbf{G}_1, \mathbf{G}_2$ that are defined on the same Borel graph $\mathbf{G}$, and only differ in the invariant distributions π_1, π_2 on them. In this case the sampling distance can be bounded by the variational distance of π_1 and π_2; it is easy to see that for every $r \geq 1$, we have

(19.4) $\quad \delta_\odot^r(\mathbf{G}_1, \mathbf{G}_2) \leq d_{\text{var}}(\pi_1, \pi_2), \quad \text{and} \quad \delta_\odot(\mathbf{G}_1, \mathbf{G}_2) \leq d_{\text{var}}(\pi_1, \pi_2).$

We will also need the edit distance of graphs/graphings on the same node set. For two graphs $G, G' \in \mathcal{G}$ with $V(G) = V(G') = [n]$, this is defined as

$$d_1(G, G') = \frac{1}{n}|E(G) \triangle E(G')|.$$

The difference from the dense case is in the normalization. (We will not need the "best overlay" version δ_1.) To extend the edit distance to two graphings $\mathbf{G}, \mathbf{G}'$ with $V(G) = V(G') = [0,1]$, there is a little subtlety. To "count" the edges to be edited, we use the edge measure defined by (18.5); but these two graphings have different edge measures, which edge measure to use? After a little thought, the solution is natural:

$$d_1(\mathbf{G}, \mathbf{G}') = \eta_{\mathbf{G}}(E(\mathbf{G}) \setminus E(\mathbf{G}')) + \eta_{\mathbf{G}'}(E(\mathbf{G}') \setminus E(\mathbf{G})).$$

We note that $\eta_{\mathbf{G}}(E(\mathbf{G}) \cap E(\mathbf{G}')) = \eta_{\mathbf{G}'}(E(\mathbf{G}') \cap E(\mathbf{G}))$ (Exercise 18.26).

An easy inequality between the edit distance and sampling distances is stated in the following proposition.

PROPOSITION 19.1. *For any two graphings* $\mathbf{G}$ *and* $\mathbf{G}'$ *on the same underlying probability space and* $r \in \mathbb{N}$, *we have*

$$\delta_\odot^r(\mathbf{G}, \mathbf{G}') \leq 2D^r d_1(\mathbf{G}, \mathbf{G}'),$$

and

$$\delta_\odot(\mathbf{G}, \mathbf{G}') \leq 3 d_1(\mathbf{G}, \mathbf{G}')^{1/\log(2D)}.$$

In particular, these bounds hold for finite graphs.

Proof. Let $S = E(\mathbf{G}) \setminus E(\mathbf{G}')$ and $S' = E(\mathbf{G}') \setminus E(\mathbf{G})$.

CLAIM 19.2. *Let* $\mathbf{x}$ *be a random point in* $\mathbf{G}$, *then the probability that the ball* $B_{\mathbf{G},r}(\mathbf{x})$ *contains any edge in* S *is bounded by* $2D^r \eta_{\mathbf{G}}(S)$.

The number of points x for which $B_{\mathbf{G},r}(x)$ contains a given edge is bounded by $2D^r$ (this follows by an elementary computation). In the finite case, this implies the Claim by an easy double counting. For graphings, this double counting can be justified using the Mass Transport Principle for graphings, Proposition 18.49. For two nodes of $\mathbf{G}$, let $f(x,y) = \deg_S(y)\mathbb{1}(x \in B_{\mathbf{G},r}(y))$ (where $\deg_S(y)$ denotes the number of edges in S incident with y). Let $\mathbf{x}$ be a random point of $V(\mathbf{G})$. Then

$$\lambda\{x: \ E(B_{\mathbf{G},r}(x)) \cap S \neq 0\} \leq \mathsf{E}(|B_{\mathbf{G},r}(\mathbf{x}) \cap S|) \leq \mathsf{E}\Big(\sum_{y \in B_{\mathbf{G},r}(\mathbf{x})} \deg_S(y)\Big)$$

$$= \mathsf{E}\Big(\sum_y f(\mathbf{x},y)\Big) = \mathsf{E}\Big(\sum_y f(y,\mathbf{x})\Big) \leq 2D^r \mathsf{E}(\deg_S(\mathbf{x})) = 2D^r \eta(S).$$

This implies the Claim.

Applying the inequality in Claim 19.2 to S' as well, we see that with probability at least $1 - 2D^r\eta(S) - 2D^r\eta'(S')$, the ball $B_{\mathbf{G},r}(\mathbf{x})$ contains no edge in S and the ball $B_{\mathbf{G}',r}(\mathbf{x})$ contains no edge of S'. In this case these two balls are isomorphic, proving that

$$\delta_\odot^r(\mathbf{G},\mathbf{G}') = d_{\mathrm{var}}(B_{\mathbf{G},r}(\mathbf{x}), B_{\mathbf{G}',r}(\mathbf{x}')) \leq 2D^r\eta(S) + 2D^r\eta'(S') = 2D^r d_1(\mathbf{G},\mathbf{G}').$$

This proves the first inequality. The second inequality follows from the first by using (19.3) with $r = \lceil -\log(2d_1(\mathbf{G},\mathbf{G}'))/\log(2D)\rceil$. □

19.1.2. Locally convergent sequences. A sequence of graphs G_n with $\mathsf{v}(G_n) \to \infty$ is *locally convergent* if the r-neighborhood densities $\rho_{G_n}(F)$ converge for every r and every r-ball F. Similarly as for the subgraph sampling in the dense case, there are equivalent frequency type parameters whose convergence could be used instead of the neighborhood densities: we could stipulate the convergence of $t^*(F, G_n)$ for every connected graph F, as proved in Proposition 5.6, and also the convergence of $t^*_{\mathrm{inj}}(F, G_n)$ or $t^*_{\mathrm{ind}}(F, G_n)$ for every connected graph F. All these versions would lead to the same convergent sequences.

We can also describe convergent sequence of graphs as Cauchy sequences in the sampling distance. From (19.3) it is easy to check that a graph sequence (G_n) with bounded degrees and with $\mathsf{v}(G_n) \to \infty$ is convergent if and only if it is Cauchy in the sampling distance. Of course, this is essentially just a reformulation of the definition, and not a structural characterization of convergence as Theorem 11.3 was in the dense case.

For every G_n and every positive integer r, neighborhood sampling provides a probability distribution $\rho_{G_n,r}$ on the set $\mathfrak{B}_r$ of r-balls. By the definition of convergence, for every fixed r, this distribution tends to a limit distribution σ_r. The sequence of these distributions has some special properties. First of all, it is *consistent*, in the sense that selecting a random r-ball from σ_r, and deleting from it the nodes at distance r from the root, we get an $(r-1)$-ball from distribution σ_{r-1}.

There is another, more subtle consistency property, which is a finite version of involution invariance for a distribution on rooted countable graphs. Note that an r-ball contains other $(r-1)$-balls, centered at the neighbors of the original root, and these from these balls we should also be able to recover σ_{r-1}. Since there are several of these in any given r-ball, we have to be a bit careful with the counting. As done before, we bias the distribution by the degree of the root: for $F \in \mathfrak{B}_r$, define

$$\sigma_r^*(F) = \frac{\deg(F)\sigma_r(F)}{\sum_{H \in \mathfrak{B}_r} \deg(H)\sigma_r^*(H)}.$$

Select a random r-ball F from σ_r^*, and a random edge uv from the root u of F. We can create two random $(r-1)$-balls with a root edge: one, we delete from F the nodes at distance more than $r-1$ from the root u; two, we delete all the nodes at distance more than $r-1$ from v, and consider v the root and vu the root edge. If we get the same distribution on $(r-1)$-balls with a root edge with both construction, and this holds for every $r \geq 1$, we say that the sequence $(\sigma_1, \sigma_2, \dots)$ is *involution invariant*. To sum up, every convergent graph sequence gives rise to an involution invariant and consistent probability measure on $\mathfrak{B}_r$.

We have defined involution invariance for measures on the "graph of graphs", and of course the two notions are closely related. From every probability distribution σ on $(\mathfrak{G}^\bullet, \mathfrak{A})$, we get a probability distribution σ_r on $\mathfrak{B}_r$ by selecting a random countable graph from σ and taking the r-ball about its root. It is trivial that this sequence $(\sigma_1, \sigma_2, \dots)$ is consistent.

Conversely, from every consistent sequence $(\sigma_1, \sigma_2, \dots)$ we get a distribution σ on $(\mathfrak{G}^\bullet, \mathfrak{A})$, by defining $\sigma(\mathfrak{G}^\bullet_F) = \sigma_r(F)$ for every r-ball F. It is also straightforward to check that $(\sigma_1, \sigma_2, \dots)$ is involution invariant if and only if σ is.

So there is a bijective correspondence between consistent involution invariant sequences $(\sigma_1, \sigma_2, \dots)$, where σ_r is a distribution on $\mathfrak{B}_r$, and involution invariant probability distributions on $(\mathfrak{G}^\bullet, \mathfrak{A})$. Through this correspondence, every locally convergent graph sequence gives rise to an involution invariant distribution σ on the sigma-algebra $(\mathfrak{G}^\bullet, \mathcal{A})$. This is the *Benjamini–Schramm limit* or *local limit* of the sequence.

By Theorem 18.37, it follows that there is a graphing $\mathbf{G}$ such that $\rho_{G_n, r} \to \rho_{\mathbf{G}, r}$ for every $r \geq 1$. We write $G_n \to \mathbf{G}$, and say that this graphing "represents" the limit; but one should be careful not to call it "the" limit; all locally equivalent graphings represent the same limit object.

EXAMPLE 19.3 (**Cycles III**). Consider the sequence of cycles (C_n). It is easy to see that the Benjamini–Schramm limit is the involution invariant distribution concentrated on the two-way infinite path (with any node specified as the root). The graphing $\mathbf{C}_a$ constructed in Example 18.1 represents the limit of this sequence for any irrational number a. All connected components of this graphing are two-way infinite paths, so generating a random point $x \in [0,1]$, its connected component $(\mathbf{C}_a)_x$ has the Benjamini-Schramm limit distribution.

Every graphing locally equivalent to $\mathbf{C}_a$ (i.e., in which almost all connected components are two-way infinite paths) provides a representation of the limit object. Example 18.54 shows two different graphings representing this limit. ♦

EXAMPLE 19.4 (**Grids**). Let G_n be the $n \times n$ grid in the plane. The r-neighborhood of a node v is a $(2r+1) \times (2r+1)$ grid (rooted in the middle), provided v is farther than $r-1$ from the boundary. This holds for $(n-2r)^2$ of the nodes, which means almost all nodes if $n \to \infty$. So in the weak limit, every r-neighborhood is a $(2r+1) \times (2r+1)$ grid. Hence the Benjamini–Schramm limit of this sequence is concentrated on the infinite square grid (with a root). We have seen (Example 18.38) how to represent this involution invariant distribution as a graphing. ♦

EXAMPLE 19.5 (**Penrose tilings**). This is a more elaborate example, but interesting in many respects. We can tile the plane with the two rhomboids of the left side of Figure 19.1. This is no big deal, if we can use them periodically (for example, as in the middle of Figure 19.1); but we put decorations on the edges, and impose the restriction that these decorations must match along every common edge

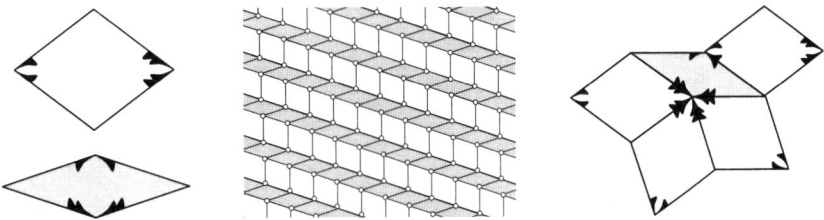

FIGURE 19.1. The Penrose rhomboids, an illegal tiling, and how they should be attached.

(as on the right side of Figure 19.1); in particular, we are not allowed to combine two of the same kind into a single parallelogram. It turns out that you can tile the whole plane this way (in fact, in continuum many ways), but there is no periodic tiling. Figure 19.2 shows the graph obtained from a Penrose tiling of the plane. There is a related (in fact, equivalent) version, in which we use two deltoids instead of two rhomboids; such a tiling is also shown in Figure 19.2. A deltoid tiling can be obtained from a rhomboid tiling by cutting up the rhomboids into a few pieces and recombining these to form deltoids. (To figure out the details is left as a challenge to the reader.)

One of the interesting (and nontrivial) features of such tilings is that every one of them contains each of the two rhomboids with the same frequency. Similar property holds for every configuration of rhomboids: if a finite configuration F of tiles can be completed to a tiling at all, then this configuration occurs in every Penrose tiling with the same frequency. To be precise, if we take a $K \times K$ square about the origin in the plane, and count how many copies of F it contains, then this number, divided by K^2, tends to a limit if $K \to \infty$. Moreover, this limit is independent of the Penrose tiling that we are studying.

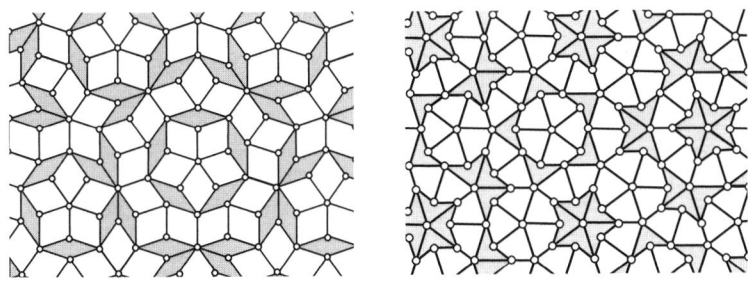

FIGURE 19.2. A piece of a Penrose rhomboid tiling and of a deltoid tiling.

We are not going to dive into the fascinating theory of Penrose tilings, but point out that their basic properties can be translated into graph limits. Let G_n be the graph obtained by restricting the graph of a Penrose rhomboid tiling to the $n \times n$ square about the origin. The above properties of the Penrose tiling imply that this sequence is convergent, and in fact it remains convergent if we interlace it with a sequence obtained from a different Penrose tiling. In other words, these finite pieces of any Penrose tiling converge to the same limit. The Benjamini–Schramm limit will be not the original Penrose tiling, but a probability distribution on all

Penrose tilings. (This illuminates that in Example 19.4 of grids we end up with a single limiting grid only because grids are periodic.)

A graphing representation of the limit of Penrose rhomboid tilings can be described based on their characterization by de Bruijn [1981] (Figure 19.3); this was pointed out by M. Bárász. ♦

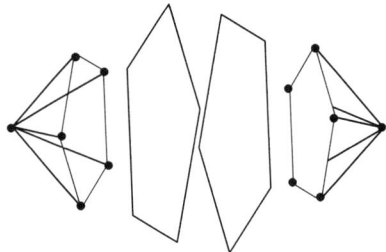

FIGURE 19.3. A graphing describing the limit of Penrose rhomboid tilings. The underlying set is the union of parallel slices of a rhombic icosahedron through its vertices. The edges of the graphing are all translates of the edges of the polytope that connect two points on these planes.

EXAMPLE 19.6 (**Large girth graphs**). Let G_n be a sequence of D-regular graphs whose girth (length of the shortest cycle) tends to infinity (it is well known that such graph sequences exist). For every $r \geq 0$ and sufficiently large n, the r-neighborhood $B_{G_n,r}(v)$ of any node v is a rooted tree $T_{r,D}$ of depth r, in which all the nodes closer to the root than r have degree D. So the limiting sequence of distributions is concentrated on these trees $T_{r,D}$. The Benjamini–Schramm limit of this sequence is concentrated on a single countable graph, namely the D-regular tree (it does not matter where we put the root). We have seen (Example 18.39) how to construct a graphing representation of this involution invariant distribution. ♦

EXAMPLE 19.7 (**Various random D-regular graphs**). Let $\mathbb{G} = \mathbb{G}(n,D)$ denote a random D-regular multigraph. This notion itself is a bit tricky. We can of course define it as the uniform distribution over all D-regular graphs on n nodes; but this definition is quite difficult to handle. A more useful definition is called the *configuration model*. We start with a set S of nD nodes, partitioned into n sets $S_1, \ldots, S_n$ of size D. We take a random perfect matching on S (we better assume that nD is even), and then identify every S_i into a single node labeled i. This way we obtain a random D-regular multigraph. If we want a random simple graph, we reject it if we get a graph with multiple edges, and try again.

It is easy to compute that the expected number of loops, as well as the number of multiple edges in $\mathbb{G}(n,D)$ is bounded by a function of D. More generally, the expected number of k-cycles is bounded by $(D-1)^k/2^k + o(1)$ (when $n \to \infty$). Hence it follows that for every fixed r and k, almost all nodes will be farther than r from any cycle of length k or less. In other words, almost all r-neighborhoods will be D-ary trees of depth r. So this sequence is locally convergent with probability 1, and its local limit is the same infinite D-regular tree as in the previous example.

Finally, random bipartite D-regular graphs will be interesting for us. These can be generated just like above, except that we assume that $n = 2m$ is even, and we use a random perfect matching between $S_1 \cup \cdots \cup S_m$ and $S_{m+1} \cup \cdots \cup S_n$. By computations similar to the above, we can see that random D-regular bipartite graphs tend to the same local limit as random D-regular graphs, the D-regular rooted tree. ♦

19.1.3. Which distributions are limits? A big difference from the dense case is that there is no easy way to construct a sequence of finite graphs that converges to a given graphing (or involution invariant distribution). In fact, we don't know whether all involution invariant distributions arise as limit objects:

CONJECTURE 19.8 (**Aldous–Lyons** [2007]). *Every involution invariant distribution on $(\mathfrak{G}^\bullet, \mathcal{A})$ is the limit of a locally convergent bounded-degree graph sequence.*

Since every involution invariant distribution can be represented by a graphing (Theorem 18.37), this is equivalent to asking whether every graphing is the local limit of a locally convergent sequence of bounded-degree graphs. This conjecture, which is a central unsolved problem in the limit theory of bounded-degree graphs, generalizes a long-standing open problem about sofic groups. It is known in some special cases: when the distribution is concentrated on trees (Bowen [2004], Elek [2010b]; see Exercise 19.12), and also when the graphing is "hyperfinite" (to be discussed in Section 21.1).

The following is an interesting reformulation of this conjecture. Let $A_r \subseteq \mathbb{R}^{\mathfrak{B}_r}$ denote the set of all probability distributions $\rho_{G,r}$, where G ranges through all finite graphs. Let $A'_r \subseteq \mathbb{R}^{\mathfrak{B}_r}$ denote the set of probability distributions $\rho_{\mathbf{G},r}$, where $\mathbf{G}$ ranges through all graphings. Equivalently, A'_r consists of probability distributions on $\mathfrak{B}_r$ induced by an involution invariant probability distribution on $\mathfrak{G}^\bullet$. Clearly $A_r \subseteq A'_r$.

PROPOSITION 19.9. (a) *The closure $\overline{A}_r$ of A_r is a compact convex set.* (a) *A'_r is a compact convex set.*

While most of the time the limit theory of graphs with bounded degree is more complicated than the dense theory, Proposition 19.9 represents an opposite case: in the dense case, even the set $D_{2,3}$ discussed in Section 16.3.2 was non-convex with a complicated structure.

Proof. (a) Let G_1 and G_2 be two finite graphs, and consider the graph $G = G_1^{\mathsf{v}(G_2)} G_2^{\mathsf{v}(G_1)}$ consisting of $\mathsf{v}(G_2)$ copies of G_1 and $\mathsf{v}(G_1)$ copies of G_2. Then

$$\rho_{G,r}(B) = \frac{1}{2}\bigl(\rho_{G_1,r}(B) + \rho_{G_2,r}(B)\bigr)$$

for every r-ball B. This implies that $\overline{A}_r$ is convex. Since it is a bounded closed set in a finite dimensional space, it is compact.

(b) The fact that A'_r is closed follows from general considerations: the set M of involution-invariant measures, as a subset of the set of all probability measures on the compact metric space $\mathfrak{G}^\bullet$, is closed in the weak topology, and so it is compact. Using that each of the cylinders $\mathfrak{G}^\bullet_F$ is open-closed, the projection of M onto $\mathbb{R}^{\mathfrak{B}_r}$ is continuous, and hence the image, which is just A'_r, is compact. The convexity of A'_r follows by a construction similar to that in (a). □

The Aldous–Lyons Conjecture is equivalent to saying that $\overline{A}_r = A'_r$ for every r. So if the conjecture fails to hold, then there is an $r \in \mathbb{N}$ and a linear inequality on $\mathbb{R}^{\mathfrak{B}_r}$ that is valid for $\overline{A}_r$ but not for A'_r. This would be a linear inequality between r-neighborhood densities that holds for every finite graph, but fails to hold for all graphings, a "positive" consequence of a "negative" fact.

There is a finite version of the Aldous–Lyons conjecture, which was raised by this author at a conference, and was proved, at least in a non-effective sense, quickly by Alon [unpublished]:

PROPOSITION 19.10. *For every $\varepsilon > 0$ there is a positive integer n such that for every graph $G \in \mathcal{G}$ there is a graph $G' \in \mathcal{G}$ such that $\mathsf{v}(G') \leq n$ and $\delta_\odot(G, G') \leq \varepsilon$.*

Proof. Let $r = \lceil \log(2/\varepsilon) \rceil$, and let $G_1, \ldots, G_m$ be any maximal family of graphs in $\mathcal{G}$ such that $\delta^r_\odot(G_i, G_j) > \varepsilon/2$ for all $1 \leq i < j \leq m$. Such a family is finite, since every graph is represented by a point in A_r, which is a bounded set in a finite dimensional space, and these points are at least $\varepsilon/2$ apart in the total variation distance. It follows that $n = \max_i \mathsf{v}(G_i)$ is finite. By the maximality of the family, for every graph G there is an $i \leq m$ such that $\delta^r_\odot(G, G_i) \leq \varepsilon/2$. We have $\mathsf{v}(G_i) \leq n$, and by (19.3)

$$\delta_\odot(G, G_i) \leq \frac{1}{2^r} + d^r_\odot(G, G_i) \leq \frac{1}{2^r} + \frac{\varepsilon}{2} \leq \varepsilon. \qquad \square$$

Unfortunately, no effective bound on n follows from the proof (one can easily get an explicit bound on m, the number of graphs in the representative family, but not on the size of these graphs). It would be very interesting to give any explicit bound (as a function of D and ε), or to give an algorithm to construct H from G. Ideally, one would like to design an algorithm that would work locally, in the sampling framework, similarly to the algorithm in Section 15.4.2 in the dense case.

Proposition 19.10 is related to the Aldous–Lyons Conjecture 19.8. Indeed, the Aldous–Lyons Conjecture implies that for any graphing $\mathbf{G}$ there is a finite graph G whose neighborhood distribution is arbitrarily close; Proposition 19.10 says that for any finite graph G there is a finite graph H of bounded size whose neighborhood distribution is arbitrarily close. Suppose that we have a constructive way of finding, for an arbitrarily large graph G with bounded degree, a graph H of size bounded by a function of r and ε that approximates the distribution of r-neighborhoods in G with error ε. With luck, the same construction could also work with a graphing in place of G, proving the Aldous–Lyons Conjecture.

One route to disproving the Aldous–Lyons Conjecture could be to explicitly find the sets $\overline{A}_r$ and A'_r for some r, and see that they are disjoint. Since the dimension of $\overline{A}_r$ grows very fast with r, it seems useful to consider even simpler questions. Instead of looking at A_r and A'_r, we could fix a finite set $\{F_1, \ldots, F_m\}$ of simple graphs, assign the vector $(t^*(F_1, G), \ldots, t^*(F_m, G))$ to every graph $G \in \mathcal{G}$, and consider the set $T(F_1, \ldots, F_m)$ of all such vectors. We define the set $T'(F_1, \ldots, F_m)$ analogously, replacing graphs by graphings. By the same argument as above, the sets $\overline{T}(F_1, \ldots, F_m)$ and $T(F_1, \ldots, F_m)$ are convex. The Aldous–Lyons Conjecture is equivalent to saying that $\overline{T}(F_1, \ldots, F_m) = T(F_1, \ldots, F_m)$ for every $F_1, \ldots, F_m$.

This leads us to the problem, very interesting on its own right, to determine the sets $\overline{T}(F_1, \ldots, F_m)$ and $T(F_1, \ldots, F_m)$, and more generally, to extremal problems for bounded degree graphs. This should be the title of a chapter, but very little has been done in this direction. There are, of course, many results in extremal graph theory that concern graphs with bounded degree; but the limit theory of bounded degree graphs has not been applied to extremal graph theory in a sense in which the limit theory of dense graphs has been. One notable exception is the result of Harangi [2012], who determined the sets $\overline{T}(K_3, K_4)$ and $T(K_3, K_4)$ for D-regular graphs. He found the same answer in both cases (so this did not give a counterexample to the conjecture).

19.1.4. On colored graphs. It will be useful at various points to extend our constructions and results to colored graphs, where the nodes are colored with b node-colors and the edges are colored with c edge-colors (where b and c are fixed positive integers). Colored graphings can be defined analogously, where every node set and edge set with a given color is Borel. Colored graphs and graphings can be used to express some properties and additional structures which we want to pass to the limit. For example, we can express measure preserving families (used in Section 18.2.1 to certify that a measurable graph is measure preserving) by edge-coloring.

We could have formulated all our arguments in the previous chapter and this one in the more general context of colored graphs and graphings. Alternatively, we could repeat these arguments now for colored graphs. We will do neither; we point out in a few sentences how these generalizations would work, and leave it to the interested reader to think through that the arguments can be extended to colored graphs.

Sampling from a colored graph results in a distribution of colored balls, and since there is only a finite number of them, all the arguments above remain valid. We can extend the notion of convergence to colored graphs of a fixed type (b, c), i.e., to graphs that are node-colored with b colors and edge-colored with c colors. The sampling process returns a *colored r-ball*, which is node-colored with b colors, edge-colored with c colors, and has a specified root. As before, we denote by $\rho_{G,r}$ the probability distribution on colored r-balls about a random node (where the type (b, c) is understood). We say that the colored graph sequence is *locally convergent*, if the sequence $(\rho_{G_n,r}(F) : n = 1, 2, \dots)$ converges for every r and every colored r-ball F.

We can define the "graph of colored graphs": its nodes will be all connected colored rooted countable graphs (with the same degree bound as always). Adjacency is defined as before; we color the node (H, v) with the color of v, and we color the edge $(H, v)(H, u)$ with the color of the edge vu.

Every convergent colored graph sequence has a limit object in the form of an involution invariant probability distribution on the "graph of colored graphs", which in turn can be represented by a colored graphing.

One could go a step further, and decorate every node and/or every edge by an element of a fixed compact Hausdorff space K. (For the dense case, a similar extension was treated in Section 17.1.) One could extend the notions of involution invariant distributions and measure preserving graphs to this case, but it would take more effort, and would have fewer applications. One example of an application would be the assignment of weights α to the nodes of graphs in $\mathfrak{G}^\bullet$ in the proof of Theorem 18.37, which could be phrased as using a node-decoration from the compact space $[0, 1]$. We will use this more general construction in the next section, where colored graphs will play and important role.

EXERCISE 19.11. Let $G \in \mathcal{G}$ and let $S \subseteq E(G)$, $|S| = \varepsilon \mathsf{v}(G)$. Prove that $\delta_\odot(G, G \setminus S) \leq 4\varepsilon^{1/\log D}$.

EXERCISE 19.12. Prove that if σ is an involution-invariant distribution such that a rooted graph chosen from σ is almost always a tree, then σ is the local limit of a finite graph sequence (Elek [2010b]).

EXERCISE 19.13. Prove that merging two node colors or two edge colors, every convergent colored graph sequence remains convergent.

19.2. Local-global convergence

Are the notion of convergence and the limit object constructed above informative enough? The limit graphon of a dense sequence of graphs contains very much information about the asymptotic properties of the sequence. This is not quite so for the bounded degree case, unfortunately. Let us illustrate this by a simple example.

EXAMPLE 19.14. Let G_n be a sequence of random 3-regular bipartite graphs. Let H_n consist of two disjoint copies of G_n. The Benjamini–Schramm limit of both sequences is a distribution concentrated on a single 3-regular rooted tree T_3.

This limit graphing is not uniquely determined. Among others, we have Bernoulli graphing $\mathbf{T}_3$ associated with T_3, but one could take the disjoint union $\mathbf{T}_3^2$ of two such graphings (with the node measure scaled down by 2). It seems that $\mathbf{T}_3$ represents the limit of G_n "better", while $\mathbf{T}_3^2$ represents the union the limit of H_n "better". As another example, if we consider the free group F_3 with three generators acting without fixed points on a probability space, then the corresponding graphing (obtained by connecting every point to its images under any of the generators) represents the Benjamini–Schramm limit. One feels that the limit of the sequence (G_n) is 'better" represented if the action of the free group is ergodic, while for the limit of H_n, the space should be split into two invariant subsets of measure $1/2$. ♦

This example suggests that in the limit object, the underlying σ-algebra carries combinatorial information. This is in stark contrast with the dense case (cf. Remark 10.1 and the discussion in that section).

In this section we define a notion of convergence for graphs with bounded degree that is stronger than the local convergence (Hatami, Lovász and Szegedy [2012]). Among others, if a sequence of graphs is convergent in this stronger sense, then we can read off from the limit whether the graphs are expanders (up to a non-expanding part of negligible size).

19.2.1. Nondeterministic sampling distance.
First, we define a new version of the sampling distance. Let $G_1, G_2 \in \mathcal{G}$, then their *non-deterministic sampling distance of depth r for k colors* is defined as the least $c > 0$ with the following property: for every k-coloring α_1 of $V(G_1)$ there exists a k-coloring α_2 of $V(G_2)$ such that $\delta_\odot^r\big((G_1, \alpha_1), (G_2, \alpha_2)\big) \leq c$, and vice versa. (The sampling distance of (G_1, α_1) and (G_2, α_2) means their sampling distance as colored graphs.) We denote this distance by $\delta_\odot^{(r,k)}(G_1, G_2)$. We then take, similarly as before,

$$\delta_\odot^{\mathrm{nd}}(G_1, G_2) = \sum_{k=0}^{\infty} \sum_{r=0}^{\infty} \frac{1}{2^{r+k}} \delta_\odot^{(r,k)}(G_1, G_2).$$

It is easy to see that these formulas define a metric on finite graphs.

We can define the non-deterministic sampling distance of two graphings $\mathbf{G}_1$ and $\mathbf{G}_2$ similarly, except that we only allow Borel k-colorings, and have to use infimum instead of a minimum:

$$\delta_\odot^{(r,k)}(\mathbf{G}_1, \mathbf{G}_2) = \inf\{c : \forall \alpha_1 \exists \alpha_2 \delta_\odot^r((\mathbf{G}_1, \alpha_1), (\mathbf{G}_2, \alpha_2)) \leq c, \text{ and} \tag{19.5}$$
$$\forall \alpha_2 \exists \alpha_1 \delta_\odot^r((\mathbf{G}_1, \alpha_1), (\mathbf{G}_2, \alpha_2)) \leq c\}.$$

The quantity $\delta_\odot^{\mathrm{nd}}(\mathbf{G}, \mathbf{G}')$ is defined from this just like in the case of graphs.

We say that two graphings $\mathbf{G}$ and $\mathbf{G}'$ are *locally-globally equivalent*, if $\delta_\odot^{\mathrm{nd}}(\mathbf{G}, \mathbf{G}') = 0$. A sequence of graphs (G_n) is *locally-globally convergent* if it is a Cauchy sequence in the nondeterministic distance, i.e., $\delta_\odot^{\mathrm{nd}}(G_n, G_m) \to 0$ as $n, m \to \infty$. We say that its *local-global limit* is graphing $\mathbf{G}$, if $\delta_\odot^{\mathrm{nd}}(G_n, \mathbf{G}) \to 0$. It is clear that we could replace d^{nd} by $d^{k,r}$ in these definitions, and require the conditions for all $k, r \geq 1$.

We have defined nondeterministic distance and local-global equivalence in terms of coloring the nodes. We could allow coloring of the edges as well without changing the notion of equivalence. Let me elaborate this for local-global equivalence.

PROPOSITION 19.15. *Suppose that two graphings $\mathbf{G}$ and $\mathbf{G}'$ are locally-globally equivalent. Then for any $\varepsilon > 0$, $k \geq 1$, and Borel k-edge-coloring α and Borel k-point-coloring β of $\mathbf{G}$, there exists a Borel k-edge-coloring α' and Borel k-point-coloring β' of $\mathbf{G}'$ such that $\delta_\odot((\mathbf{G}, \alpha, \beta), (\mathbf{G}', \alpha', \beta')) \leq \varepsilon$.*

Proof. We want to encode the edge-coloring into a node-coloring. The first trick is to construct (independently of the coloring α) another Borel edge-coloring γ with $2D-1$ colors such that no two adjacent edges have the same γ-color (Theorem 18.4). Using this, we define a point-coloring ζ: we color a point x with the pair $(\beta(x), \sigma(x))$, where $\sigma(x)$ is the set of all pairs $(\alpha(x,y), \gamma(x,y))$, where $(x,y) \in E(\mathbf{G})$. This point-coloring uses a finite set $K = [k] \times 2^{[k] \times [2D-1]}$ of colors. From this point-coloring, we can recover the original coloring easily: for every point x, $\beta(x)$ is the first element of $\zeta(x)$, and for every edge (x,y), $\sigma(x) \cap \sigma(y)$ has a unique element (a, b), and the original color of (x,y) was a. (The only role of the coloring γ was to make sure that this common element of $\sigma(x)$ and $\sigma(y)$ is unique.)

By the definition of local-global equivalence, there is a K-coloring ζ' of the points of $\mathbf{G}'$ such that $\delta_\odot((\mathbf{G}, \zeta), (\mathbf{G}', \zeta')) \leq \varepsilon$. We define an edge-coloring α' and a point-coloring β' of $\mathbf{G}'$ as follows. Since $\zeta(x) \in K$, we can write $\zeta(x) = (b(x), s(x))$, where $b(x) \in [k]$ and $s(X) \subseteq [k] \times [2D-1]$. Let $\beta'(x) = b(x)$; let $\alpha'(x,y)$ be the smallest $a \in [k]$ for which there is a $g \in [2D-1]$ such that $(a, g) \in s(x) \cap s(y)$; if no such g exists, then let $\alpha'(x, y) = 1$ (this will happen only for a small set of edges).

Now comes the key observation: Whenever the r-neighborhoods of point x in $(\mathbf{G}, \zeta)$ and of point y in $(\mathbf{G}', \zeta')$ are isomorphic, then the r-neighborhoods of x in $(\mathbf{G}, \alpha, \beta)$ and of x' in $(\mathbf{G}', \alpha', \beta')$ are also isomorphic. The rules for obtaining α and β from ζ and α' and β' from ζ' work the same way in both graphs. Hence $\delta_\odot((\mathbf{G}, \alpha, \beta), (\mathbf{G}', \alpha', \beta')) \leq \delta_\odot((\mathbf{G}, \zeta), (\mathbf{G}', \zeta')) \leq \varepsilon$. □

19.2.2. Graphings as local-global limits. We have seen that limits of locally convergent graph sequences can be described as involution-invariant distributions, and this representation of the limit is unique. We could also represent the limit by a graphing, but this was not unique, which means that graphings are more complicated objects than necessary. Why bother with graphings at all, why not use involution invariant distributions only? One justification for considering graphings is the following result of Hatami, Lovász and Szegedy [2012].

THEOREM 19.16. *For every locally-globally convergent sequence of finite graphs there is a graphing that is its local-global limit.*

For the proof, we need a couple of lemmas about k-colorings. First, let us discuss continuous k-colorings of a graphing. For this to make sense, we have to fix a topology on $V(\mathbf{G})$. Of course, we should not use the standard topology of (say) $[0,1]$: this would not admit nontrivial k-colorings. But if we use the local topology as defined in Section 18.1, we get interesting continuous colorings. Recall that this topology can be defined by a metric, where two points are 2^{-r}-close if their r-neighborhoods are isomorphic. This topology is totally disconnected, so there will be nontrivial continuous k-colorings. A k-coloring will be continuous if an only if the color of a node can be determined just from the isomorphism type of its r-neighborhood for some finite r.

We know by Lusin's Theorem that every Borel function on a compact probability space can be approximated by a continuous function γ in the sense that the set $\{\alpha \neq \gamma\}$ has arbitrarily small measure. Here, in general, γ will not be a k-coloring (if the underlying space of $\mathbf{G}$ is the unit interval, for example, then its range will be an interval, not $[k]$). However, with an appropriate topology on $V(\mathbf{G})$, the approximating function can be chosen to be a coloring itself.

LEMMA 19.17. *Let K be a compact metric space that is totally disconnected, let π be a probability measure on K, and let $\alpha: K \to [k]$ be a Borel k-coloring of K. Then for every $\varepsilon > 0$ there is a continuous k-coloring $\delta: K \to [k]$ such that $\pi\{\alpha \neq \delta\} \leq \varepsilon$.*

Proof. By Lusin's Theorem, there is a continuous function β on K such that $T = \{x \in K : \alpha = \beta\}$ has measure at least $1 - \varepsilon$. Open-closed sets in K separate any two points, hence by the Stone–Weierstrass Theorem, there is a stepfunction γ whose steps are open-closed (i.e., γ is continuous), and $|\beta(x) - \gamma(x)| < 1/3$ for every x. If a step S of γ contains a point $y \in T$, then we fix one such point, and define $\delta(x) = \alpha(y) = \beta(y)$ for all $x \in S$; else, we define $\delta(x) = 1$. This way we get a continuous k-coloring δ.

Let $x \in T$ and let S be the step of γ containing x. Then $\alpha(x) = \beta(x)$, and so, for the point $y \in S \cap T$ used in the definition of $\delta(x)$ (which may or may not be x), we have

$$|\alpha(x) - \delta(x)| \leq |\beta(x) - \gamma(x)| + |\gamma(x) - \delta(x)| = |\beta(x) - \gamma(x)| + |\gamma(y) - \beta(y)| \leq \frac{2}{3}.$$

Since $\alpha(x)$ and $\delta(x)$ are integers, this implies that $\alpha(x) = \delta(x)$ for $x \in T$. $\square$

The second lemma we need is similar to Proposition 19.10. It shows that a uniformly bounded number of k-colorings can approximate all k-colorings (in the sense of neighborhood statistics) of an arbitrarily large graph.

LEMMA 19.18. *For every $k, r \geq 1$ and $\varepsilon > 0$ there is an integer $M = M(k, r, \varepsilon) \geq 1$ such that every graph $G \in \mathcal{G}$ has M k-colorings $\alpha_1, \ldots, \alpha_M$ such that for every k-coloring β of G there is an i ($1 \leq i \leq M$) with*

$$\delta_\odot^r\big((G, \beta), (G, \alpha_i)\big) \leq \varepsilon.$$

Of course, M depends on D too, but this is tacitly assumed to be a constant in all our discussions in this part.

Proof. Let $\{F_1, \ldots, F_M\}$ be any maximal family of k-colored graphs in $\mathcal{G}$ such that $\delta^r_\odot(F_i, F_j) > \varepsilon/2$ for all $1 \leq i < j \leq M$. Since the distributions $\rho_{G,r}$ on k-colored r-balls B belong to a bounded set in a finite dimensional space, such a family is finite.

Let G be any finite graph. For every $i \leq m$, select a k-coloring α_i of G such that $\delta^r_\odot\big((G, \alpha_i), F_i\big) \leq \varepsilon/2$, if such a k-coloring exists (call such an i *relevant*); else, let α_i be an arbitrary k-coloring of G. We claim that the k-colorings α_i constructed this way have the required property.

For every k-coloring β of G, there is an $i \leq M$ such that $\delta^r_\odot\big((G, \beta), F_i\big) \leq \varepsilon/2$, by the maximality of the family $\{F_1, \ldots, F_M\}$. Then clearly this i is relevant, and so for the corresponding α_i, we have

$$\delta^r_\odot\big((G, \beta), (G, \alpha_i)\big) \leq \delta^r_\odot\big((G, \beta), F_i\big) + \delta^r_\odot\big(F_i, (G, \alpha_i)\big) \leq \frac{\varepsilon}{2} + \frac{\varepsilon}{2} = \varepsilon. \qquad \square$$

Proof of Theorem 19.16. We apply Lemma 19.17 with $\varepsilon = 2^{-r}$, and denote $M(k, r, 2^{-r})$ by $M(k, r)$. We fix a set of $M(k, r)$ k-colorings as in Lemma 19.17 for every graph $G \in \mathcal{G}$, and call them its *representative k-colorings*.

Consider the product space $K = \prod_{k,r=1}^{\infty} [k]^{M(k,r)}$; this is compact and totally disconnected. We start with constructing a decoration $\chi = \chi_G : V(G) \to K$ for every $G \in \mathcal{G}$. Given a node $v \in V(G)$, we consider the representative k-colorings $\alpha_1, \ldots, \alpha_{M(k,r)}$ of G, and concatenate the sequences $(\alpha_1(v), \ldots, \alpha_{M(k,r)}(v))$ for $k, r = 1, 2, \ldots$ to get $\chi(v)$.

Using the decoration χ_G and the projection map $\varphi_{k,r} : K \to [k]^{M(k,r)}$, we can manufacture many k-colorings of G as $\beta = \psi \circ \varphi_{k,r} \circ \chi$, where $\psi : [k]^{M(k,r)} \to [k]$ is any map. We call these k-colorings "special". It follows from the construction of χ that the representative k-colorings of G are special. Hence for every graph G, every $k, r \geq 1$, and every k-coloring α of $V(G)$, there is a special k-coloring β close to α, in the sense that $\delta_\odot\big((G, \alpha), (G, \beta)\big) \leq 2^{-r}$.

The graphing $\mathbf{H}^K$ we construct is similar to the "Graph of Weighted Graphs" $\mathbf{H}^+$ introduced in Section 18.3.3, but instead $[0, 1]$, we use weights from K. We construct probability measures on $\mathbf{H}^K$ to get representations of finite graphs and then, representations of the limit. With the decoration χ_G, and any choice of a root $v \in V(G)$, the triple (G, v, χ_G) is a point of $\mathbf{H}^K$. The map $\tau_G : v \mapsto (G, v, \chi_G)$ defines an embedding $G \to \mathbf{H}^K$ onto a connected component of $\mathbf{H}^K$ (the fact that this map is injective is clear, since for any two nodes $u, v \in V(G)$ one of the k-colorings in Lemma 19.18 must distinguish them once r is large enough). Let ζ_G be the uniform distribution on $\tau_G(V(G))$. Since G is finite, this distribution is involution-invariant on $\mathbf{H}^K$.

Let (G_n) be a locally-globally convergent graph sequence. By Prokhorov's Theorem (see Appendix A.3.3), we can replace our graph sequence by a subsequence such that the distributions ζ_{G_n} converge weakly to a distribution ζ on $\mathbf{H}^K$. Since every ζ_{G_n} is involution-invariant, so is ζ, and hence $\mathbf{G} = (\mathbf{H}^K, \zeta)$ is a graphing.

We claim that $G_n \to \mathbf{G}$ in the local-global sense. To prove this convergence, we need the following auxiliary fact.

CLAIM 19.19. *Let β be a continuous k-coloring of $\mathbf{G}$, and let $\beta_n = \beta \circ \tau_{G_n}$ be the k-coloring it induces on G_n. Then $\delta_\odot\big((G_n, \beta_n), (\mathbf{G}, \beta)\big) \to 0$ as $n \to \infty$.*

To prove this, fix $r \geq 1$, and express the frequency of a k-colored r-ball B_0 in $(\mathbf{G}, \beta)$ as an integral:

$$\rho_{\mathbf{G},\beta,r}(B_0) = \int_{\mathbf{H}^K} \mathbb{1}\big(B_{\mathbf{G},\beta,r}(x) \cong B_0\big)\, d\zeta(x).$$

By the definition of β_n, we have a similar expression for the frequency of B_0 in (G_n, β_n):

$$\rho_{G_n,\beta_n,r}(B_0) = \int_{\mathbf{H}^K} \mathbb{1}\big(B_{\mathbf{G},\beta,r}(x) \cong B_0\big)\, d\zeta_{G_n}(x).$$

The main observation we need is that the integrand is continuous. Indeed, suppose that $H_n \to H$ in the topology of $\mathbf{H}^K$, where $H_n, H \in \mathbf{H}^K$ are rooted K-decorated countable graphs. Then for a sufficiently large n, the balls $B_{H_n,r}$ and $B_{H,r}$ are isomorphic, and moreover, there is an isomorphism $\sigma_n : B_{H,r} \to B_{H_n,r}$ such that $\chi_{H_n}(\sigma_n(x)) \to \chi_H(x)$ for every $x \in V(B_{H,r})$. This means that $\chi_H(x)$ and $\chi_{H_n}(\sigma_n(x))$ agree in more and more coordinates as n grows, which implies that $\beta(\sigma_n(x)) \to \beta(x)$, since β is continuous. Since β has finite range, this implies that $\beta(\sigma_n(x)) = \beta(x)$ if n is large enough. But then $\mathbb{1}\big(B_{\mathbf{G},\beta,r}(H_n) \cong B_0\big) = \mathbb{1}\big(B_{\mathbf{G},\beta,r}(H) \cong B_0\big)$ if n is large enough, which proves that the integrand is continuous.

Hence it follows by the weak convergence $\zeta_{G_n} \to \zeta$ that

$$\int_{\mathbf{H}^K} \mathbb{1}(B_{\mathbf{G},\beta,r} \cong B_0)\, d\zeta_{G_n} \longrightarrow \int_{\mathbf{H}^K} \mathbb{1}(B_{\mathbf{G},\beta,r} \cong B_0)\, d\zeta,$$

which proves that $\rho_{G_n,\beta_n,r}(B_0) \to \rho_{\mathbf{G},\beta,r}(B_0)$ for every k-colored r-ball B_0. This proves the claim.

Let us return to the proof of the local-global convergence $G_n \to \mathbf{G}$. By the definition of the nondeterministic sampling distance, we have to verify two things for every $r, k \geq 1$: every k-coloring of G_n can be "matched" by a k-coloring of $\mathbf{G}$ so that the distributions of r-neighborhoods are close, and vice versa. Let $\varepsilon > 0$; we may assume that $\varepsilon \geq 2^{-r}$, since larger neighborhoods are more difficult to match.

First, let α be a Borel k-coloring of $\mathbf{G}$. Then by Lemma 19.17, there is another Borel k-coloring β such that β is continuous in the topology of $\mathbf{H}^K$ and $\alpha = \beta$ on a set of measure at least $1 - \varepsilon(2D)^{-r}$. Then $\delta_\odot^r\big((\mathbf{G}, \alpha), (\mathbf{G}, \beta)\big) \leq \varepsilon/2$ by (19.4). For every n, the k-coloring β gives a k-coloring β_n of the nodes of G_n, under the embedding τ_{G_n}. By Claim 19.19, we have $\delta_\odot\big((G_n, \beta_n), (\mathbf{G}, \beta)\big) \leq \varepsilon/2$ if n is large enough. This implies that $\delta_\odot^r\big((G_n, \beta_n), (\mathbf{G}, \alpha)\big) \leq \varepsilon$.

Second, let n be large enough so that for all $m \geq n$, we have $\delta_\odot^{(r,k)}(G_n, G_m) \leq \varepsilon/3$, and let α_n be a k-coloring of G_n. Then for every $m \geq n$ there is a k-coloring α_m of G_m such that $\delta_\odot^r\big((G_n, \alpha_n), (G_m, \alpha_m)\big) \leq \varepsilon/3$. Furthermore, there is a special k-coloring $\beta_m = \psi_m \circ \varphi_{k,r} \circ \chi_{G_m}$ of G_m (with an appropriate $\psi_m : [k]^{M(k,r)} \to [k]$) such that $\delta_\odot^r\big((G_m, \alpha_m), (G_m, \beta_m)\big) \leq \varepsilon/3$. It follows that $\delta_\odot^r\big((G_n, \alpha_n), (G_m, \beta_m)\big) \leq 2\varepsilon/3$. We can select an infinite subsequence such that $\psi_m = \psi$ is independent of m, so that $\beta_m(v)$ depends only on the decoration $\chi_{G_m}(v)$ of the node $v \in V(G_m)$. We can use the same map ψ to get an k-coloring of $\mathbf{G}$: we color every $x \in \mathbf{H}^K$ with $\beta(x) = \psi\big(\varphi_r(\chi(x))\big)$, where $\chi(x)$ is the decoration of $\mathrm{root}(x)$. This coloring is continuous, and on $\tau_{G_m}(v)$ it coincides with $\beta_m(v)$. Claim (19.19) implies that $\delta_\odot^r\big((G_m,\beta_m),(\mathbf{G},\beta)\big) \to 0$. Hence $\delta_\odot^r\big((G_n,\alpha_n),(\mathbf{G},\beta)\big) \leq \varepsilon$ if n is large enough. □

EXERCISE 19.20. Let F_1 and F_2 be two finite graphs and let $\mathbf{G}_{F_1}$ and $\mathbf{G}_{F_2}$ denote the associated graphings. Prove that $\delta_\odot^{\mathrm{nd}}(F_1,F_2) = \delta_\odot^{\mathrm{nd}}(\mathbf{G}_{F_1},\mathbf{G}_{F_2})$.

EXERCISE 19.21. For an (uncolored) graph G, let $\mathcal{Q}_{r,k}(G)$ denote the set of all neighborhood distributions $\rho_{G^*,r}$, where G^* is a k-colored version of G. Prove that
$$\delta_\odot^{(r,k)}(G,G') = d_{\mathrm{var}}^{\mathrm{Haus}}\big(\mathcal{Q}_{r,k}(G),\mathcal{Q}_{r,k}(G')\big).$$

EXERCISE 19.22. (a) Let (G_n) be a locally-globally convergent graph sequence. Prove that the numerical sequences $\alpha(G_n)/\mathsf{v}(G_n)$ and $\mathsf{Maxcut}(G_n)/\mathsf{v}(G_n)$ are convergent. (b) Show by an example that this does not hold for every locally convergent sequence.

CHAPTER 20

Right convergence of bounded degree graphs

20.1. Random homomorphisms to the right

Homomorphisms from a large bounded-degree graph G into a small weighted graph H are the bread and butter of statistical physics, as we have illustrated in the Introduction (Chapter 2.2). What happens if we go to the limit with G through bounded degree graphs? Does it make sense to talk about a random homomorphism from a countable graph into a weighted graph H? Or from a graphing? It is natural that statistical physicists have worked out a theory that is able to answer these questions. In this section we reproduce some of these results. We will need these, among others, in the next section, where we discuss right-convergence.

To start with a trivial example, let G be a countably infinite graph with bounded degree, and let H be the looped complete graph K_q°. Then we can map the nodes of G independently, which defines a perfectly fine probability distribution on maps $V(G) \to V(H)$. Unfortunately, if we delete any edge from K_q°, then the probability that a random map $V(G) \to [q]$ remains a homomorphism is 0. So we could not define a random homomorphism $G \to H$ by taking a random map $V(G) \to V(H)$ and condition on its being a homomorphism. It turns out that random homomorphisms from countable graphs into weighted graphs can be defined in some cases: when the maximum degree of G is small and the edgeweights of H are close to 1. (We will not attempt to define a random homomorphism from a graphing.)

It turns out that the construction for a random homomorphism $G \to H$ for infinite graphs G is made possible by another important phenomenon, this time for finite graphs. In its simplest version, let u and v be two nodes of G that are far from each other, and consider a random homomorphism $G \to H$. Are the images of u and v essentially independently distributed? This is not always so; for example, if G is a connected bipartite graph, then there are two homomorphisms $G \to K_2$, and the image of one node determines the images of all of the others. We will start with showing that under similar conditions as we mentioned above (the maximum degree of G is small and the edgeweights of H are close to 1), the images of distant nodes will be essentially independent. This important result, called the Dobrushin Uniqueness Theorem, will be stated and proved first. There is of course a lot more in the literature about this theorem and its applications, see e.g. Georgii [1988] or Simon [1993]. (We have to postpone the explanation of the word "uniqueness" to the end of the next section.)

Since we have graphons at hand, we can replace H by a graphon W and get more general results with only a small amount of additional hassle.

20.1.1. Homomorphisms and Markov chains. We start with defining random homomorphisms from a finite bounded degree into a weighted graph and into graphon. Let $G = (V, E)$ be a simple graph and let H be a weighted graph with nonnegative edgeweights.

We have considered random maps $\varphi : V(H) \to V$ where the probability of φ was proportional to α_φ. It is also quite natural to bias these with the product of the edgeweights. In other words, let the probability of φ be

$$\pi_{G,H}(\varphi) = \alpha_\varphi \frac{\hom_\varphi(G, H)}{\hom(G, H)}. \tag{20.1}$$

In the special case when H is a looped-simple unweighted graph, this is the uniform distribution on the set $\mathrm{Hom}(G, H)$.

EXAMPLE 20.1 (**Ising model**). Recall the example from the Introduction (Section 2.2). There is a very large graph G (most often, a grid) whose nodes are the atoms and whose edges are bonds between these atoms. There is a small graph H, whose nodes represent the possible states of an atom. (In the case of the Ising model, H has two nodes only, representing the spins "UP" and "DOWN".) The nodeweights $\alpha_i = e^{-h_i}$ represent the influence of an external field on an atom in state i, and the edgeweights $\beta_{ij} = e^{-J_{ij}}$ represent the interaction energy between two adjacent atoms in states i and j (we ignore the dependence on the temperature for this discussion). A possible configuration is a map $\sigma : V(G) \to V(H)$, and its energy is

$$H(\sigma) = -\sum_u h_{\sigma(u)} - \sum_{uv \in E(G)} J_{\sigma(u),\sigma(v)}.$$

In the introduction we were focusing on the partition function Z of the system, which turned out to be $\hom(G, H)$. But the exponential of the energy

$$e^{-H(\sigma)} = \alpha_\sigma \hom_\sigma(G, H)$$

is also very important, because the system will be in state σ with probability

$$\frac{e^{-H(\sigma)}}{Z} = \frac{\alpha_\sigma \hom_\sigma(G, H)}{\hom(G, H)} = \pi_{G,H}(\sigma).$$

So this distribution on homomorphisms introduced above expresses the fundamental physical state of a material. ◆

It is not hard to generalize these constructions to the case when we replace H by a graphon $W \neq 0$ on $[0, 1]$ (and of course the formulas become simpler). We will call maps $V \to [0, 1]$, "weightings". For a measurable set $X \subseteq [0, 1]^V$, we define

$$\pi_{G,W}(X) = \frac{\int_X t_x(G, W)\, dx}{t(G, W)}. \tag{20.2}$$

We call a random map drawn from the distribution $\pi_{G,W}$ a *random W-weighting of G*. We could of course replace W by a weakly isomorphic graphon on some other probability space, and this could be more natural in some cases (think of the generalization of the Ising model, where the spins can be arbitrarily unit vectors in $\mathbb{R}^3$).

Let $S \subseteq V$, $X \subseteq [0, 1]^S$ and suppose that we fix a partial weighting $y \in [0, 1]^{V \setminus S}$. We can define a kind of conditional distribution

$$\pi_y(X) = \frac{\int_X t_{y,x}(G, W)\, dx}{t_y(G, W)} \tag{20.3}$$

(The condition $x|_S = y$ may have probability 0, but the formula works.) In the special case when $S = V \setminus \{v\}$, the distribution π_y can be identified with a distribution on $[0, 1]$, which we denote by $\pi_{y,v}$. It will be important to notice that in this case the distribution π_y is determined by the restriction of y to $N_G(v)$.

Is there a more tangible way of defining this distribution? A general technique of generating random elements of complicated distributions and studying their properties is to construct a Markov chain with the given stationary distribution. In this case, there is a rather simple Markov chain $\mathcal{M}$ on weightings in $[0, 1]^V$ with this property. (In the special case when $W = W_{K_q}$, this will specialize to the "heat-bath" chain, or "Glauber dynamics", on q-colorings of G.) One step of this Markov chain is described as follows: Given a weighting x, we select a uniform random node $\mathbf{v} \in V$ (which we call the *pivot node*) and reweight it from the distribution $\pi_{x,\mathbf{v}}$. All other nodeweights remain unchanged. It is not hard to check that $\pi_{G,W}$ is a stationary distribution of this Markov chain.

Let us fix a set $U \subseteq V$ and its complement $Z = V \setminus U$. We can modify the Markov chain $\mathcal{M}$ by selecting the pivot node $\mathbf{v}$ from Z only. This modified Markov chain preserves the weighting of U; if we restrict it to the extensions of a partial weighting $a \in [0, 1]^U$, then we get a Markov chain $\mathcal{M}_a$, whose stationary distribution is π_a.

Next, we define a Markov chain $\mathcal{M}_2$ on pairs $(x, y) \in [0, 1]^V \times [0, 1]^V$. Given (x, y), we generate a random pivot node $\mathbf{v} \in Z$ and modify both x and y according to $\mathcal{M}$, separately but not independently: using the same pivot node, we generate a random weight $\mathbf{x}$ from the distribution $\pi_{x,\mathbf{v}}$, and a random weight $\mathbf{y}$ from the distribution $\pi_{y,\mathbf{v}}$, and couple $\mathbf{x}$ and $\mathbf{y}$ optimally, so that $\mathsf{P}(\mathbf{x} \neq \mathbf{y}) = d_{\mathrm{tv}}(\pi_{x,\mathbf{v}}, \pi_{y,\mathbf{v}})$. We change the weight of $\mathbf{v}$ in x to $\mathbf{x}$, and in y to $\mathbf{y}$. Note that for fixed $a, b \in [0, 1]^U$, the set of pairs of weightings (x, y) with $x|_U = a$ and $y|_U = b$ is invariant. Let $\mathcal{M}_{a,b}$ denote the Markov chain restricted to such pairs.

The stationary distribution of this Markov chain is difficult to construct directly, but at least it exists:

LEMMA 20.2. *The Markov chain $\mathcal{M}_{a,b}$ has a stationary distribution with marginals π_a and π_b.*

This is trivial if $\mathcal{M}_{a,b}$ has a finite number of states (which happens if W is a stepfunction, i.e., we are studying homomorphisms into a finite weighted graph). For the general case, the proof follows by more advanced arguments in probability theory, and is not given here (see Lovász [Notes]).

These Markov chains (especially the simplest chain $\mathcal{M}$) are quite important in simulations in statistical physics and also in theoretical studies. A lot of work has been done on their mixing times and other properties. For us, however, the main consequence of their introduction will be the existence of the stationary distribution of $\mathcal{M}_{a,b}$.

20.1.2. Correlation decay. Our next goal is to state and prove the fact, mentioned above, that $\pi_{G,W}$ has no long-rage interaction: under appropriate conditions, the weights of two distant nodes in a random W-weighting from $\pi_{G,W}$ are essentially independent. We start with an easy observation (the verification is left to the reader as an exercise).

PROPOSITION 20.3. *Let $G = (V, E)$ be a simple graph, and let W be a graphon of rank 1. Then $\pi_{G,W}$ is a product measure on $[0, 1]^V$. In other words, if $\mathbf{x}$ is a* random W-weighting of G, then the weights $\mathbf{x}_u$ ($u \in V(G)$) are independent (as random variables).

The Dobrushin Uniqueness Theorem, in its combinatorial form, will imply that if the adjacency matrix of H is close to a rank-1 matrix, then there is almost no correlation between $\mathbf{x}_u$ and $\mathbf{x}_v$ (where $\mathbf{x}$ is a random W-weighting of G), provided the degrees of G are small and the distance of u and v is large. In fact, the theorem is stronger: it implies that conditioning on the weights of *all* nodes far from a given set $U \subseteq V$ has very little influence on the weighting of U.

To state this result, we need the following parameter of graphons. Let $r \leq D$, and consider the star S_{r+1} on $\{0, 1 \ldots r\}$ with center 0. Let us define the *Dobrushin value* of W as

(20.4) $$\mathsf{dob}(W) = \sup_{r \leq D} \sup_{x,y} d_{\mathrm{tv}}(\pi_x, \pi_y),$$

where $x, y \in [0,1]^{[r]}$ range through weightings of the leaves of S_{r+1} that differ only for a single node $u \in [r]$. If $\mathsf{dob}(W)$ is small, then changing the weight of a neighbor of a node has little influence on the weight of the node. Changing the weight of neighbors one by one, we get by induction that for any graph $G \in \mathcal{G}$, node $v \in V(G)$ and $x, y \in [0,1]^V$, we have

(20.5) $$d_{\mathrm{tv}}(\pi_{x,v}, \pi_{y,v}) \leq \mathsf{dob}(W) \big|\{u \in N(v) : x_u \neq y_u\}\big|.$$

THEOREM 20.4. *Let $G = (V, E)$ be a (finite) graph with all degrees bounded by D, and let W be a graphon. Then for any partition $V = Z \cup U$ and any two maps $a, b \in [0,1]^U$, the distributions π_a and π_b have a coupling κ such that for every node $v \in Z$ and every pair $(\mathbf{x}, \mathbf{y})$ of random W-weightings from the distribution κ, we have*

$$\mathsf{P}(\mathbf{x}_v \neq \mathbf{y}_v) \leq (\mathsf{dob}(W) D)^{d(v,U)},$$

where $d(v, U)$ denotes the distance of v from U in G.

What is important in this theorem is that it gives an exponentially decaying correlation between the weight of v and the weights of nodes far away, provided $\mathsf{dob}(W) < 1/D$.

Proof. We assume that $\mathsf{dob}(W) < 1/D$ (else, there is nothing to prove). Let κ be the stationary distribution of the Markov chain $\mathcal{M}_{a,b}$ with marginals π_a and π_b. So κ is a coupling of these distributions.

Let $x, y \in [0,1]^V$, and let (x', y') be obtained from (x, y) by making one step of $\mathcal{M}_{a,b}$, using a random pivot node $\mathbf{v} \in Z$. Let $n = |Z|$. Then for any node $w \in Z$,

(20.6) $$\mathsf{P}(x'_w \neq y'_w) = \frac{n-1}{n} \mathsf{P}(x'_w \neq y'_w \mid \mathbf{v} \neq w) + \frac{1}{n} \mathsf{P}(x'_w \neq y'_w \mid \mathbf{v} = w).$$

Here $\mathsf{P}(x'_w \neq y'_w \mid \mathbf{v} \neq w) = \mathbb{1}(x_w \neq y_w)$ (since nothing changes at w under this condition), and

$$\mathsf{P}(x'_w \neq y'_w \mid \mathbf{v} = w) = \mathsf{P}(\mathbf{i} \neq \mathbf{j} \mid \mathbf{v} = w) = d_{\mathrm{tv}}(\pi_{x,w}, \pi_{y,w}).$$

Substituting in (20.6), we get

(20.7) $$\mathsf{P}(x'_w \neq y'_w) = \frac{n-1}{n} \mathbb{1}(x_w \neq y_w) + \frac{1}{n} d_{\mathrm{tv}}(\pi_{x,w}, \pi_{y,w}).$$

Now let $(\mathbf{x},\mathbf{y})$ be a random pair from κ, and average (20.7) over $\mathbf{x}$ and $\mathbf{y}$, to get

$$\text{(20.8)} \qquad \mathsf{P}(\mathbf{x}'_w \neq \mathbf{y}'_w) = \frac{n-1}{n}\mathsf{P}(\mathbf{x}_w \neq \mathbf{y}_w) + \frac{1}{n}\mathsf{E}\big(d_{\mathrm{tv}}(\pi_{\mathbf{x},w}, \pi_{\mathbf{y},w})\big).$$

By the definition of stationary distribution, $(\mathbf{x}',\mathbf{y}')$ has the same distribution as $(\mathbf{x},\mathbf{y})$, and hence $\mathsf{P}(\mathbf{x}'_w \neq \mathbf{y}'_w) = \mathsf{P}(\mathbf{x}_w \neq \mathbf{y}_w)$. Substituting in (20.8), we get

$$\text{(20.9)} \qquad \mathsf{P}(\mathbf{x}_w \neq \mathbf{y}_w) = \mathsf{E}\big(d_{\mathrm{tv}}(\pi_{\mathbf{x},w}, \pi_{\mathbf{y},w})\big).$$

So far, we have not used the Dobrushin parameter $\mathsf{dob}(W)$. By (20.5), we get

$$\text{(20.10)} \qquad \mathsf{E}\big(d_{\mathrm{tv}}(\pi_{\mathbf{x},w}, \pi_{\mathbf{y},w})\big) \leq \mathsf{dob}(W) \sum_{u \in N(w)} \mathsf{P}(\mathbf{x}_u \neq \mathbf{y}_u).$$

Define $f(u) = \mathsf{P}(\mathbf{x}_u \neq \mathbf{y}_u)$. We have $f(u) \in \{0,1\}$ if $u \in U$, and (20.9) and (20.10) imply that

$$\text{(20.11)} \qquad f(w) \leq \mathsf{dob}(W) \sum_{u \in N(w)} f(u)$$

holds for all $w \in Z$. Inequality (20.11) says that the function f is strictly subharmonic at the nodes of Z. It is easy to derive from this fact an estimate on f. Let us start a random walk $(v^0 = v, v^1, \dots)$ on G from $v \in Z$, and let T be the (random) time when this random walk hits U (if the connected component of v does not intersect U, then $f = 0$ on this connected component and the conclusion below is trivial). Consider the random variables $X_t = f(v^t)(\mathsf{dob}(W)D)^t$. It follows from (20.11) that these form a submartingale, and hence by the Martingale Stopping Theorem A.11, we get

$$f(v) = X^0 \leq \mathsf{E}(X^T) = \mathsf{E}\big((\mathsf{dob}(W)D)^T f(v^T)\big) \leq \mathsf{E}\big((\mathsf{dob}(W)D)^T\big).$$

Since trivially $T \geq d(v,U)$, this completes the proof. $\square$

It is important that the coupling κ constructed above is independent of the node v. This means that if we want to estimate the probability that $\mathbf{x}|_S \neq \mathbf{y}|_S$ for some subset $S \subseteq Z$, then we get the same coupling distribution κ, and so we can use the union bound:

COROLLARY 20.5. *Under the conditions of Theorem 20.4, every $S \subseteq Z$ satisfies*

$$\mathsf{P}(\mathbf{x}|_S \neq \mathbf{y}|_S) \leq (\mathsf{dob}(W)D)^{d(S,U)}|S|.$$

Let us formulate some other consequences. First, consider proper q-colorings of G, i.e., homomorphisms $G \to K_q$. For S_{r+1} in the definition of the Dobrushin parameter, let φ and ψ be two q-colorings of the leaves that differ at node 1 only. Then $\pi_{\varphi,0}$ is the uniform distribution on the set $[q]\setminus\varphi([r])$, and $\pi_{\psi,0}$ has an analogous description. These sets have Hamming distance at most $2D$, and hence their total variation distance is at most $1/(q-D)$. So $\mathsf{dob}(W_{K_q}) < 1/D$ is satisfied if $q > 2D$, and we get:

COROLLARY 20.6. *Let $G = (V,E)$ be a graph with all degrees bounded by D, and let $q > 2D$. Then for any $U \subseteq V$, any two proper q-colorings α and β of $G[U]$, and any $v \in V \setminus U$, the random extensions $\boldsymbol{\varphi}$ and $\boldsymbol{\psi}$ of α and β to proper q-colorings of G satisfy*

$$d_{\mathrm{tv}}(\boldsymbol{\varphi}(v), \boldsymbol{\psi}(v)) \leq \Big(\frac{D}{q-D}\Big)^{d(v,Z)}.$$

We can generalize this corollary to homomorphisms into any looped-simple graph H, assuming the maximum degree $\Delta(\overline{H})$ of its complement $\overline{H}$ (among looped-simple graphs) is small:

COROLLARY 20.7. *Let $G = (V, E)$ be a simple graph with all degrees bounded by D, and let H be a looped-simple graph with $2D\Delta(\overline{H}) < \mathsf{v}(H)$. Then for any subset $U \subseteq V$, any two homomorphisms $\alpha, \beta : G[U] \to H$, and any $v \in V \setminus U$, the uniform random extensions $\boldsymbol{\varphi}$ and $\boldsymbol{\psi}$ of α and β to homomorphisms $G \to H$, restricted to the node v, satisfy*

$$d_{\mathrm{tv}}(\boldsymbol{\varphi}(v), \boldsymbol{\psi}(v)) \leq \Big(\frac{D\Delta(\overline{H})}{\mathsf{v}(H) - D\Delta(\overline{H})}\Big)^{d(v,Z)}.$$

20.1.3. The Dobrushin value. Which graphons W have small Dobrushin value? This property is related to the approximability of W by rank-1 graphons (see Exercises 20.10 and 20.11, and also Proposition 20.3), but for us, the case when W is close to the special rank-1 function 1 will be important. For a graphon W, define

$$(20.12) \qquad \overline{\Delta}(W) = \sup_{x \in [0,1]} \int_0^1 W(x,y)\, dy.$$

(the "maximum degree"; note that $\delta(W_H) = \Delta(H)/\mathsf{v}(H)$ if H is a looped-simple graph). The quantity provides a useful upper bound on the Dobrushin value.

LEMMA 20.8. *Every graphon W satisfies*

$$\mathsf{dob}(W) \leq \frac{\overline{\Delta}(W)}{1 - D\overline{\Delta}(W)}.$$

In particular, the Dobrushin condition is satisfied if $\overline{\Delta}(W) < 1/(2D)$.

Proof. The proof is just computation (although a bit tedious). Let $z, w \in [0,1]^r$ be two weightings of the leaves of S_{r+1} that differ only at node 1. Let

$$g(x) = \prod_{i=2}^r W(x, z_i) = \prod_{i=2}^r W(x, w_i) \quad \text{and} \quad s(x) = \int_0^1 g(y) W(x,y)\, dy.$$

The density functions of the distributions $\pi_{z,0}$ and $\pi_{w,0}$ are $g(x)W(x,z_1)/s(z_1)$ and $g(x)W(x,w_1)/s(w_1)$, respectively, and hence

$$(20.13) \qquad d_{\mathrm{tv}}(\pi_z, \pi_w) = \frac{1}{2}\int_0^1 g(x) \Big| \frac{W(x,z_1)}{s(z_1)} - \frac{W(x,w_1)}{s(w_1)} \Big|\, dx.$$

We may assume without loss of generality that $s(z_1) \geq s(w_1)$, then

$$\begin{aligned}
d_{\text{tv}}(\pi_z, \pi_w) &= \frac{1}{2}\int_0^1 g(x)\left|\frac{W(x,z_1)}{s(z_1)} - \frac{W(x,w_1)}{s(w_1)}\right|dx \\
&= \frac{1}{2}\int_0^1 g(x)\left|\frac{W(x,z_1)}{s(z_1)} - \frac{W(x,z_1)}{s(w_1)} + \frac{W(x,z_1) - W(x,w_1)}{s(w_1)}\right|dx \\
&\leq \frac{1}{2}\int_0^1 g(x)\left(\left|\frac{W(x,z_1)}{s(z_1)} - \frac{W(x,z_1)}{s(w_1)}\right| + \left|\frac{W(x,z_1) - W(x,w_1)}{s(w_1)}\right|\right)dx \\
&= \frac{1}{2}\int_0^1 g(x)\left(\frac{W(x,z_1)}{s(w_1)} - \frac{W(x,z_1)}{s(z_1)} + \left|\frac{W(x,z_1) - W(x,w_1)}{s(w_1)}\right|\right)dx \\
&= \frac{1}{2}\int_0^1 g(x)\left(\frac{W(x,z_1)}{s(w_1)} - \frac{W(x,w_1)}{s(w_1)} + \frac{|W(x,z_1) - W(x,w_1)|}{s(w_1)}\right)dx \\
&= \frac{1}{2}\int_0^1 g(x)\frac{W(x,z_1) - W(x,w_1) + |W(x,w_1) - W(x,z_1)|}{s(w_1)}dx
\end{aligned}$$

(we have used here that $\int g(x)W(x,z_1)/s(z_1) = 1 = \int g(x)W(x,w_1)/s(w_1)$). It is easy to check that $W(x,z_1) - W(x,w_1) + |W(x,w_1) - W(x,z_1)| \leq 2 - 2W(x,w_1)$, and using the trivial fact that $g(x) \leq 1$, we get

$$d_{\text{tv}}(\pi_x, \pi_y) \leq \int_0^1 g(x)\frac{1 - W(x,w_1)}{s(w_1)}dx \leq \frac{1}{s(w_1)}\int_0^1 (1 - W(x,w_1))\,dx \leq \frac{\overline{\Delta}(W)}{s(w_1)}.$$

To estimate the denominator, note that

$$g(x)W(x,w_1) = \prod_{i=1}^r W(x,w_i) \geq 1 - \sum_{i=1}^r (1 - W(x,w_i)),$$

and so

$$s(w_1) = \int_0^1 g(x)W(x,w_1)\,dx \geq \int_0^1 1 - \sum_{i=1}^r (1 - W(x,z_i))\,dx \geq 1 - D\overline{\Delta}(W).$$

This proves the lemma. □

EXERCISE 20.9. For a random 3-coloring of the path P_n, determine the correlation between the colors of the endnodes. Does it decay exponentially?

EXERCISE 20.10. Prove that for every graphon W there is a kernel of rank 1 such that $\|U - W\|_1 \leq \mathsf{dob}(W)$.

EXERCISE 20.11. Let $U > 0$ be a kernel of rank 1, and let W be an arbitrary kernel. Prove that $\mathsf{dob}(W) \leq 4\|(W/U) - 1\|_\infty$.

20.1.4. Random homomorphisms from infinite graphs into graphons.
Our goal is to define a random homomorphism $G \to W$, where G is a countable graph with degrees bounded by D, and W is graphon.

First, some technicalities: the W-weightings of $V(G)$ form the product space $[0,1]^{V(G)}$, which we endow with the product sigma-algebra $\mathcal{A}$. A random homomorphism will be defined by a probability measure on $\mathcal{A}$. To specify such a measure, it suffices to specify its values on cylinder sets obtained by restricting the weight of a finite number of nodes of G to given Borel sets. In other words, we can specify a distribution π on W-weightings of G by specifying the distribution of its restriction $\pi|_S$ to every finite set $S \subseteq V(G)$. Of course, these restrictions must satisfy appropriate consistency conditions: if $S \subseteq T$, then $(\pi_T)|_S = \pi_S$. Once we have a family

(π_S) of distributions satisfying these consistency relations, then the Extension Theorem of Kolmogorov gives us a probability distribution π on all W-weightings such that $\pi|_S = \pi_S$ for all finite sets $S \subseteq V(G)$.

So far, this is quite simple. There are many ways to specify such a family (π_S) of distributions. However, we would like other conditions to be satisfied. Let us formulate two:

- **Markov property.** If G_1 and G_2 are two finite k-labeled graphs, φ is a random W-weighting of G_1G_2, and we condition on $\varphi|_{[k]}$, then $\varphi|_{V(G_1)}$ and $\varphi|_{V(G_2)}$ become independent. This is just another way to express the product formula (5.53). This property can be generalized to infinite graphs. Of course, we have to exercise some care, since G_1 and G_2 may be infinite. Let $S \subseteq V(G)$ be a finite set and suppose that $G \setminus S$ is the disjoint union of two graphs G_1 and G_2. Let z be a W-weighting of S, and let $\mathbf{x}$ denote a random W-weighting of $V \setminus S$ from the distribution obtained by conditioning on $\mathbf{x}|_S = \alpha$. We require that the random weightings $\mathbf{x}|_{V(G_1)}$ and $\mathbf{x}|_{V(G_2)}$ be independent. We say that the distribution of $\mathbf{x}$ has the *Markov property*, if this condition holds for every finite subset $S \subseteq V(G)$ and every W-weighting z of S.

- **Locality.** For a finite set $S \subseteq V(G)$, we would like to get a good idea of the distribution π_S by looking at a sufficiently large neighborhood of S. Let $B(S,r) = \{v \in V(G) : d(v,S) \leq r\}$ be the r-neighborhood of S, and let $\mathbf{x}_r$ denote a random W-weighting of $G[B(S,r)]$. Then we want that $\mathbf{x}_r|_S \to \mathbf{x}|_S$ in distribution as $r \to \infty$. We call the distribution of $\mathbf{x}$ *local* if this holds.

These conditions are not too strong, as the following classical theorem shows (see [1988] and [1993] for slightly different statements of this fact).

THEOREM 20.12. *Let G be a countable graph with degrees bounded by D, and let W be a graphon such that $\mathsf{dob}(W) < 1/D$. Then there is a unique local probability distribution $\pi_{G,W}$ on W-weightings of G with the Markov property.*

Proof. Let $S \subseteq V(G)$ be a finite set, and let $\mathbf{x}_r$ be a random W-weighting of $G[B(S,r)]$.

CLAIM 20.13. *The distribution of $\mathbf{x}_r|_S$ tends to a limit as $r \to \infty$.*

We show that these distributions form a Cauchy-sequence in the total variation distance. Let $\varepsilon > 0$. Since $\mathsf{dob}(W) < 1/D$, we can choose r large enough so that $(D\mathsf{dob}(W))^r \leq \varepsilon/|S|$. Let $m, n > r$, we claim that the distributions of $\mathbf{x}_m|_S$ and $\mathbf{x}_n|_S$ are ε-close in total variation distance. Let $\mathbf{z}_n$ be the restriction of $\mathbf{x}_n$ to $B(S,n) \setminus B(S,r)$, and let $\mathbf{x}'_n$ be the random weighting of $G[B(S,n)]$, obtained by conditioning on $\mathbf{z}_n$. By the Markov property (we are using it for finite graphs here!), $\mathbf{x}'_n$ has the same distribution as $\mathbf{x}_n$. We define $\mathbf{z}_m$ and $\mathbf{x}'_m$ analogously.

Now we fix any two weightings z_n of $B(S,n) \setminus B(S,r)$ and z_m of $B(S,m) \setminus B(S,r)$, and let $\mathbf{y}_n$ and $\mathbf{y}_m$ be obtained by conditioning $\mathbf{x}_n$ and $\mathbf{x}_m$ on these partial weightings. By Theorem 20.4, $\mathbf{y}_n$ and $\mathbf{y}_m$ can be coupled so that $\mathsf{P}(\mathbf{y}_n(v) \neq \mathbf{y}_m(v)) \leq \varepsilon/|S|$ for every $v \in S$. This implies that

$$d_{\mathrm{tv}}(\mathbf{y}_n|_S, \mathbf{y}_m|_S) \leq \varepsilon.$$

Since this holds for fixed $\mathbf{z}_n$ and $\mathbf{z}_m$, it also holds if they are random restrictions of $\mathbf{x}_n$ and $\mathbf{x}_m$, so it holds for $\mathbf{x}'_n$ and $\mathbf{x}'_m$. Since these weightings have the same distribution as $\mathbf{x}_n$ and $\mathbf{x}_m$, the claim follows.

Now we are able to define the distribution on W-weightings. For a finite set $S \subseteq V(G)$, let π_S be the limit of the distributions of $\mathbf{x}_r|_S$ as $r \to \infty$. It is easy to check (using similar arguments as in the proof of Claim 20.13 above), that the family (π_S) of distributions is consistent, and the distribution $\pi_{G,W}$ they define has the Markov property. Uniqueness follows immediately from locality. $\square$

A probability distribution on W-weightings of G is called a *Gibbs state* if it is invariant under the Markov chain $\mathcal{M}$ of local re-weightings (as used in the proof of theorem 20.4 in the finite case). It can be proved that under the condition that $\mathsf{dob}(W) < 1/(2D)$, the Gibbs state is unique.

REMARK 20.14. In a sense, the construction of a random homomorphism can be extended to graphings. The method is similar to the Bernoulli lift of a graphing (Section 18.5). Given a graphing $\mathbf{G}$ and a graphon W on $[0,1]$ such that $\mathsf{dob}(W) < 1/(2D)$, we define a graphing $\mathbf{G}[W]$ on the Graph of Weighted Graphs $\mathbf{H}^+$. To describe the probability distribution on $\mathfrak{G}^+$, we generate a random element from it as follows: pick a point $x \in V(\mathbf{G})$ and generate a random W-weighting of $\mathbf{G}_x$ as described above. If $W \equiv 1$, we get the Bernoulli lift.

We cannot randomly map all points of a graphing into $[0,1]$ in any reasonable way; this is impossible even if the graphing has no edges. But if we select any countable subset, this can be mapped, and the graphing $\mathbf{G}[W]$ contains the necessary information. I don't know of any applications of this construction, but I like the fact that our two basic limit objects, graphings and graphons, can be combined this way.

20.2. Convergence from the right

While the theory of convergent sequences of bounded degree graphs lacks some of the key facts and constructions that apply in the dense case (most notably a good notion of distance), it is nicer in at least one respect: convergence of a graph sequence can be characterized by convergence of (appropriately normalized) homomorphism numbers into certain fixed graphs so we don't have to switch to maximization of multicuts as in the dense case. This result is due to Borgs, Chayes, Kahn and Lovász [2012]. We note that the necessity of the right-convergence condition follows only for target graphs that satisfy the Dobrushin condition (but under this condition, it follows more generally for homomorphism numbers into graphons).

To state this result, let us define, for a simple graph G with bounded degrees and graphon W, the (sparse, normalized) *homomorphism entropy*

$$\mathsf{ent}^*(G, W) = \frac{\log t(G, W)}{\mathsf{v}(G)},$$

In the case when $W = W_H$ for some weighted graph H on q nodes, we write $\mathsf{ent}^*(G, H)$. In this special case, we could replace $t(G, H)$ by $\mathsf{hom}(G, H)$ in this definition: this would mean simply adding $\log \alpha_H$ to the value, so it is a matter of taste which version one uses in the definition.

To see the meaning of $\mathsf{ent}^*(G, W)$, consider the case when $W = W_H$ for some simple graph H. Then $\log \mathsf{hom}(G, H)/\mathsf{v}(G)$ expresses the freedom (entropy) we have in choosing the image of a node $v \in V(G)$ in a homomorphism $G \to H$, and $\mathsf{ent}^*(G, H)$ (which is always nonpositive) expresses the loss of entropy per node due to taking homomorphisms instead of all maps.

The main result in this section is the following.

THEOREM 20.15. *For any sequence (G_n) of graphs in $\mathcal{G}$, the following are equivalent:*

(i) *(G_n) is locally convergent;*

(ii) *for every graphon W with $\mathsf{dob}(W) \leq 1/D$, the sequence $\mathsf{ent}^*(G_n, W)$ is convergent;*

(iii) *there is an $\varepsilon > 0$ such that for every looped-simple graph H with $\Delta(\overline{H}) \leq \varepsilon \mathsf{v}(H)$ the sequence $\mathsf{ent}^*(G_n, H)$ is convergent.*

The equivalence of conditions (ii) and (iii) is analogous to the equivalence of conditions (ii) and (iii) in Theorem 12.20, and similarly as there, we could replace them by any condition "inbetween", like weighted graphs satisfying the Dobrushin condition.

In the special case when $H = K_q$, we have $\Delta(\overline{K_q}) = 1$, and $\hom(G, K_q)$ is the number of q-colorings of G. So it follows that if (G_n) is convergent and $q > 2D$, then the number of q-colorings grows as $c^{\mathsf{v}(G_n)}$ for some $c > 1$. It is easy to see that some condition on q is needed: for example, if G_n is the n-cycle and $q = 2$, then $\mathsf{ent}^*(G_n, K_2)$ oscillates between $-\infty$ and ≈ 0 as a function of n.

Lemma 20.8 says that $\overline{\Delta}(W) < 1/(2D)$ is sufficient for (ii) to apply. This condition could not be relaxed by more than a constant factor, as the following example shows.

EXAMPLE 20.16. Let $\mathbb{G}_n$ be a random D-regular graph on $2n$ nodes, and $\mathbb{G}'_n$ be a random bipartite D-regular graph on $2n$ nodes. The interlaced sequence $(\mathbb{G}_1, \mathbb{G}'_1, \mathbb{G}_2, \mathbb{G}'_2, \dots)$ is locally convergent with high probability (almost all r-neighborhoods are D-regular trees if r is fixed and n is large enough). Let H be obtained from K_2° by weighting the non-loop edge by 1 and the loops by 2^c. Inequality (5.33) can be generalized to give the bounds

$$c \, \mathsf{maxcut}'(G) \leq \mathsf{ent}^*(G, H) \leq c \, \mathsf{maxcut}'(G) + 1.$$

(Here $\mathsf{maxcut}'(G) = \mathsf{Maxcut}(G)/\mathsf{v}(G)$ is normalized differently from the normalization in (5.33).) The maximum cut in $\mathbb{G}'_n$ has $Dn/2$ edges, but the maximum cut in $\mathbb{G}_n$ has at most $Dn/3$ edges with high probability (see Bertoni, Campadelli, Posenato [1997] for a sharp estimate). Hence

$$\mathsf{ent}^*(\mathbb{G}'_n, H) = \frac{cD}{2}, \quad \text{but} \quad \mathsf{ent}^*(\mathbb{G}_n, H) \leq 1 + \frac{cD}{3}$$

If $cD/2 - cD/3 = cD/6 > 1$, then the sequence $(\mathsf{ent}^*(\mathbb{G}_1), \mathsf{ent}^*(\mathbb{G}'_1), \mathsf{ent}^*(\mathbb{G}_2),$ $\mathsf{ent}^*(\mathbb{G}'_2), \dots)$ cannot be convergent with high probability. So assuming $\overline{\Delta}(W) \leq 7/D$ would not be enough in (ii). ◆

While Theorem 20.15 sounds similar to the results in Chapter 12 (in particular Theorem 12.20), it is both more and less than that theorem. We get a characterization of convergence in terms of left and right homomorphisms, but no analogue of the characterization as a Cauchy sequence in the cut metric. Also, convergence is not established for all soft-core graphs H, just for those close to a complete graph. On the other hand, the proof below says more, since it provides explicit formulas relating left and right homomorphism numbers. Furthermore, homomorphism densities into graphons are considered, not just weighted graphs; recall that the corresponding extension of Theorem 12.20 to graphons if false (Remark 12.22).

Supposing that we have a convergent sequence (G_n) tending to an involution-invariant distribution σ (or to a graphing $\mathbf{G}$), what is the limit of the homomorphism entropies? The answer is not trivial, since there is no good way to "count" homomorphisms of an infinite graph (or of a graphing) into a weighted graph H. Even for the number of q-colorings ($q > 2D$), and for a sequence of D-regular graphs with girth tending to ∞, where the Benjamini–Schramm limit is a single (infinite) D-regular rooted tree, the limiting value is nontrivial to determine. A natural guess would be that starting at the root of the infinite tree, and working our way out, we have q choices for the color of the root and $q-1$ choices for every other node, which suggests an entropy of $\mathsf{ent}^*(G_n, K_q) \to \log(1 - 1/q)$. But this is not the right answer, which was determined by Bandyopadhyay and Gamarnik [2008]:

$$(20.14) \qquad \mathsf{ent}^*(G_n, K_q) \to \frac{D}{2} \log\left(1 - \frac{1}{q}\right).$$

To motivate the following description of the limiting homomorphism entropy in the general case, consider a finite graph G. For any ordering $(v_1, \ldots, v_n)$ of its nodes, we consider the graphs $G_i = G[v_1, \ldots, v_i]$. Then

$$(20.15) \qquad \mathsf{ent}^*(G, W) = \frac{1}{n} \sum_{i=1}^{n} \log \frac{t(G_i, W)}{t(G_{i-1}, W)}.$$

If $W = W_H$ for some looped-simple graph H, then the fraction inside the logarithm is the conditional probability that a random map $V(G_i) \to V(H)$ is a homomorphism, given that its restriction to G_{i-1} is a homomorphism. In general, it can be expressed as

$$\frac{t(G_i, W)}{t(G_{i-1}, W)} = \mathsf{E}_\mathbf{y} \mathsf{E}_\mathbf{x} \left(\prod_{u \in N(v)} W(\mathbf{x}, \mathbf{y}_u) \right),$$

where $\mathbf{x}$ is a uniform random number in $[0, 1]$, and $\mathbf{y}$ is a random W-weighting of G_{i-1}.

If we try to extend this to infinite graphs, the formula makes sense, but as Example 20.16 shows, it may give the wrong result. The trick is to average over all orderings of $V(G)$. We generate an ordering by a random map $\tau : V(G) \to [0, 1]$. We denote the set of nodes $u \in V(G)$ with $\tau(u) < \tau(v)$ by $V_\tau(v)$, and set $N_\tau(v) = N(v) \cap V_\tau(v)$. We can view (20.15) as another averaging (over a random node of G). Thus we get

$$\mathsf{ent}^*(G, W) = \mathsf{E}_v \mathsf{E}_\tau \log \mathsf{E}_\mathbf{y} \mathsf{E}_\mathbf{x} \left(\prod_{u \in N_\tau(v)} W(\mathbf{x}, \mathbf{y}_u) \right),$$

where $\mathbf{x}$ is a uniform random number in $[0, 1]$, where $\mathbf{y}$ is a random W-weighting of $G[V_\tau(v)]$.

Now this formula extends to involution-invariant distributions σ. Instead of a random node v, we consider a random rooted graph (G, v) from σ. Instead of random bijection $V(G) \to [n]$, we consider a random map $\tau : V(G) \to [0, 1]$. Assuming G satisfies the Dobrushin condition, so does $G[V_\tau(v)]$, and so the random W-weighting $\mathbf{y}$ is well defined. So we can define

$$(20.16) \qquad \mathsf{ent}^*(\sigma, W) = \mathsf{E}_{(G,v)} \mathsf{E}_\tau \log \mathsf{E}_\mathbf{y} \mathsf{E}_\mathbf{x} \left(\prod_{u \in N_\tau(v)} W(\mathbf{x}, \mathbf{y}_u) \right).$$

Then we have the following supplement to Theorem 20.15.

SUPPLEMENT 20.17. *Let (G_n) be a locally convergent sequence of graphs with degrees at most D, and let σ be the involution-invariant distribution representing its limit. Let W be a graphon with $\mathsf{dob}(W) < 1/D$. Then*
$$\mathsf{ent}^*(G_n, W) \to \mathsf{ent}^*(\sigma, W).$$

Let us illustrate that the rather hairy formula (20.16) does allow us to determine the limiting values of homomorphism entropies, at least in a simple case.

EXAMPLE 20.18. Suppose that G_n is D-regular and the girth of G_n tends to infinity. Let $H = K_q$, so that $\hom(G, H) = \mathsf{chr}(G, q)$. Then G_n tends to the involution-invariant distribution concentrated on the infinite D-regular tree (at least we don't have to take expectation over this). Specializing (20.16), we get
$$\mathsf{ent}^*(\sigma, W) = \mathsf{E}_\tau \log \mathsf{E}_\mathbf{y} \mathsf{E}_\mathbf{x} \Big(\prod_{u \in N_\tau(v)} \mathbb{1}(\mathbf{x} \neq \mathbf{y}_u) \Big).$$

Here $\mathbf{y}$ is a random coloring with colors from $[q]$, and $\mathbf{x}$ is a random color. Whatever $V_\tau(v)$ is, $\mathbf{y}$ assigns uniform and independent colors to the nodes in $N_\tau(v)$, since our graph is a tree. Hence for every $\mathbf{x}$,
$$\mathsf{E}_\mathbf{y}\Big(\prod_{u \in N_\tau(v)} \mathbb{1}(\mathbf{x} \neq \mathbf{y}_u) \Big) = \Big(\frac{q-1}{q}\Big)^{|N_\tau(v)|},$$
and hence
$$\mathsf{ent}^*(\sigma, W) = \mathsf{E}_\tau \log \Big(\Big(\frac{q-1}{q}\Big)^{|N_\tau(v)|} \Big) = \mathsf{E}_\tau(N_\tau(v)) \log\Big(1 - \frac{1}{q}\Big) = \frac{D}{2} \log\Big(1 - \frac{1}{q}\Big).$$
So we get the theorem of Bandyopadhyay and Gamarnik (20.14). ♦

Proof of Theorem 20.15. (i)⇒(ii) Let $G = (V, E)$ be a simple graph with degrees bounded by D. We may assume that $\alpha_H = 1$. We use the formula (20.15) derived above, and concentrate on the innermost expression
$$\mathsf{s}(v, \tau, y) = \mathsf{E}_\mathbf{x}\Big(\prod_{u \in N_\tau(v)} W(\mathbf{x}, y_u) \Big).$$
The Dobrushin Uniqueness Theorem 20.4 implies that we don't change the expression by much if we restrict everything to the r-neighborhood $N_r(v)$. To be precise, let $c = D\mathsf{dob}(H) < 1$, and define $G^r = G[N_r(v)]$, $V^r_\tau(v) = N_r(v) \cap V_\tau(v)$, and let s^r denote the function s defined for the graph G^r. Let $\mathbf{z}$ be a random W-weighting of $G[V^r_\tau(v)]$, then Theorem 20.4 implies that the distributions of $\mathbf{y}$ and $\mathbf{z}$, when restricted to v and its neighbors, are closer that $(D+1)c^{r-1}$ is total variation distance. This implies that
$$\big|\mathsf{E}_\mathbf{z} s^r(v, \tau, \mathbf{z}) - \mathsf{E}_\mathbf{y} \mathsf{s}(v, \tau, \mathbf{y})\big| \leq (D+1)c^{r-1},$$
and hence
(20.17) $$\big|\mathsf{ent}^*(G, H) - \mathsf{E}_\mathbf{v} F_r(\mathbf{v})\big| \leq (D+1)c^{r-1},$$
where
(20.18) $$F_r(v) = \mathsf{E}_\tau \log\big(\mathsf{E}_\mathbf{z} s^r(v, \tau, \mathbf{z})\big).$$
(We can take expectation over the same τ, since it induces a uniform random permutation of $V(G^r)$ as well as of $V(G)$.)

Let us note that in (20.18) $F_r(v)$ depends only on the r-ball $B = N_r(v)$, and we can denote it by $F(B)$. This allows us to express $\mathsf{E}_\mathbf{v} F_r(\mathbf{v})$ in terms of the distribution $\sigma_{G,r}$ of r-neighborhoods in G. Thus (20.17) implies

$$(20.19) \qquad \left| \mathsf{ent}^*(G,H) - \sum_{B \in \mathfrak{B}_r} \sigma_{G,r}(B) F(B) \right| \leq (D+1) c^{r-1}.$$

Now, let (G_n) be a locally convergent sequence tending to an involution-invariant distribution σ. Then (20.19) implies that

$$\limsup_n \mathsf{ent}^*(G_n, H) \leq \sum_{B \in \mathfrak{B}_r} \sigma_r(B) F(B) + (D+1) c^{r-1},$$

and hence

$$\limsup_n \mathsf{ent}^*(G_n, H) \leq \liminf_r \sum_{B \in \mathfrak{B}_r} \sigma_r(B) F(B).$$

A similar argument proves that $\liminf_n \geq \limsup_r$, which implies that both limits exist.

(ii)$\Rightarrow$(iii) is trivial.

(iii)$\Rightarrow$(i) We switch to the natural logarithm, since we are going to use analytic formulas (this only means that all formulas are multiplied by $\ln 2$). We express the logarithm of $t(G,H)$ as

$$(20.20) \qquad \ln t(G,H) = \sum_{S \leq V(G)} \ell(G[S], H),$$

where by Möbius inversion,

$$(20.21) \qquad \ell(G,H) = \sum_{S \subseteq V(G)} (-1)^{|V(G)|-|S|} \ln t(G[S], H).$$

Using that $\ln t(., H)$ is an additive graph parameter for any fixed H, it is easy to see that $\ell(F, H) = 0$ unless F is a connected graph together with isolated nodes (cf. Exercise 4.2). The term corresponding to the edgeless graph is 0, and so we can modify (20.20) so that the summation runs over connected induced subgraphs of G. Collecting terms with isomorphic graphs, we get

$$(20.22) \qquad \mathsf{ent}^*(G,H) = \sum_F \frac{\mathsf{ind}(F,G)}{\mathsf{v}(G)} \cdot \frac{\ell(F,H)}{\mathsf{aut}(F)},$$

where the summation ranges over all isomorphism types of connected graphs F; but of course, only a finite number of terms are non-zero for any fixed G.

So we can express the homomorphism entropies $\mathsf{ent}^*(G_n, H)$ as linear combinations of the induced subgraph densities $\mathsf{ind}(F, G_n)/\mathsf{v}(G_n)$. This suggests a heuristic for the proof: We show that the system of equations (20.22) can be inverted, to express the induced subgraph densities as linear combinations of the homomorphism entropies. It follows then that if the homomorphism entropy into any given graph converges to some value, then so does the frequency of each induced subgraph.

This heuristic is of course very naive: (20.22) is an infinite system of equations, and so to do anything with it we need tail bounds; furthermore, the coefficient $\ell(F, H)$ is defined by the hairy formula (20.21), which has all the unpleasant features one can think of: it has an exponential number of terms, these terms alternate in sign, and the terms themselves are logarithms of simpler functions.

The identities developed in Section 5.3.1 come to rescue. We can get rid of the logarithms using Corollary 5.22. Substituting the formula for $\ln t(G,H)$ in the definition of $\ell(G,H)$, we get a lot of cancellation, which leads to the formula

$$(20.23) \qquad \ell(F,H) = \sum_{m=1}^{\infty} \frac{(-1)^m}{m!} \sum_{\substack{J_1,\ldots,J_m \in \mathrm{Conn}(F) \\ \cup_i V(J_i) = V(F)}} (-1)^{\sum_i \mathsf{e}(J_i)}$$

$$\times \mathsf{cri}\big(L(J_1,\ldots,J_m)\big) \prod_{r=1}^{k} t(J_r, \overline{H}).$$

(It is not clear at this point that this is any better than (20.21), but be patient.)

Next we turn to inverting the expression (20.22). Let $m \geq 1$ and let $\{F_1,\ldots,F_N\}$ be the set of all connected simple graphs with $2 \leq \mathsf{v}(F_i) \leq m$. Let $q > m/\varepsilon$, add $q - \mathsf{v}(F_i) \geq 1$ new isolated nodes to F_i, and take the complement to get a looped-simple graph H_i on $[q]$ with loops added at all nodes. We weight each node of H_i by $1/q$. Every node in H_i has degree at least $q - m$, so $\Delta(\overline{H}_i) \leq \varepsilon q$.

Consider any graph G with all degrees at most D. We write (20.22) in the form

$$(20.24) \qquad \mathsf{ent}^*(G,H_j) = \sum_{i=1}^{N} \frac{\mathsf{ind}(F_i,G)}{\mathsf{v}(G)} \cdot \frac{\ell(F_i,H_j)}{\mathsf{aut}(F_i)} + R(G,H_j),$$

where

$$(20.25) \qquad R(G,H_j) = \sum_{\mathsf{v}(F)>m} \frac{\mathsf{ind}(F,G)\ell(F,H_j)}{\mathsf{aut}(F)\mathsf{v}(G)}$$

is a remainder term.

We can view (20.24) as a system of N equations in the N unknowns $x_i = \mathsf{inj}(F_i,G)/\mathsf{v}(G)$. Let $A = \big(\ell(F_i,H_j)/\mathsf{aut}(F_i)\big)_{i,j=1}^{N}$ be the matrix of this system, and let $s, R \in \mathbb{R}^N$ be defined by $s_j = \mathsf{ent}^*(G,H_j)$ and $R_j = R(G,H_j)$, then we have $A^\mathsf{T} x = s - R$. Assuming that A is invertible (which we will prove momentarily), let $B = (A^\mathsf{T})^{-1}$. Then the system can be solved: $x = B(s-R)$, or

$$(20.26) \qquad \frac{\mathsf{ind}(F_i,G)}{\mathsf{v}(G)} = \sum_{j=1}^{N} B_{ij} \mathsf{ent}^*(G,H_j) + r_i(G),$$

where

$$r_i = r_i(G) = \sum_{j=1}^{N} B_{ij} R(G,H_j)$$

is a remainder term.

We have to show that the matrix A is invertible (at least if q is large enough) and estimate the remainder terms. We use (20.23):

$$(20.27) \qquad \ell(F,H_i) = \sum_{k=1}^{\infty} \frac{(-1)^k}{k!} \sum_{\substack{J_1,\ldots,J_k \in \mathrm{Conn}(F) \\ \cup V(J_i) = V(F)}} (-1)^{\sum_i \mathsf{e}(J_i)}$$

$$\times \mathsf{cri}\big(L(J_1,\ldots,J_k)\big) \prod_{r=1}^{k} t(J_r, \overline{H}_i).$$

By the construction of H_i, we have $t(J_r, \overline{H_i}) = q^{-\mathsf{v}(J_r)}\mathsf{hom}(J_r, F_i)$, and so

$$\prod_{r=1}^{k} t(J_r, \overline{H_i}) = q^{-\sum_r \mathsf{v}(J_r)} \prod_{r=1}^{k} t(J_r, F_i).$$

Note that for a nonzero term the exponent of q is less than $-\mathsf{v}(F)$ except for $k = 1$ and $V(J_1) = V(F)$, and that the last product does not depend on q. Hence for any simple graph F,

(20.28) $$\ell(F, H_i) = q^{-\mathsf{v}(F)} \sum_{J \in \mathrm{Csp}(F)} (-1)^{\mathsf{e}(J)-1} t(J, F_i) + O(q^{-\mathsf{v}(F)-1}).$$

(Here and in what follows, the constants implied in the big-O notation may depend on m, but not on q and G). By Proposition 5.43, the matrix $M = \bigl(t(F_i, F_j)\bigr)_{i,j=1}^{N}$ is nonsingular. Let L be the $N \times N$ matrix with entries $L_{ij} = \mathbb{1}(F_i \in \mathrm{Csp}(F_j))$, and let P and Q denote the diagonal matrices with entries $P_{ii} = (-1)^{\mathsf{e}(F_i)-1}$ and $Q_{ii} = q^{\mathsf{v}(F_i)}\mathsf{aut}(F_i)$, respectively. Clearly L, P and Q are nonsingular. By (20.28), we have

$$QA^\mathsf{T} = L^\mathsf{T} PM + O(q^{-1}),$$

which implies that A is nonsingular if q is large enough. Furthermore,

$$B_{ij} = q^{\mathsf{v}(F_i)}\mathsf{aut}(F_i)((M^\mathsf{T} PL)^{-1})_{ij} + O(q^{\mathsf{v}(F_j)-1}),$$

and so

(20.29) $$|B_{ij}| = O(q^{\mathsf{v}(F_i)}) = O(q^m).$$

Using this, the remainder terms can be estimated as follows:

$$|R(G, H_j)| \leq \sum_{r=m+1}^{\infty} \sum_{\mathsf{v}(F)=r} \frac{\mathsf{ind}(F, G)}{\mathsf{aut}(F)\mathsf{v}(G)} |\ell(F, H_j)|$$

$$= \sum_{r=m+1}^{\infty} \sum_{\mathsf{v}(F)=r} \frac{\mathsf{ind}(F, G)}{\mathsf{aut}(F)\mathsf{v}(G)} O(q^{-r})$$

(20.30) $$= \sum_{r=m+1}^{\infty} 2^{Dr} O(q^{-r}) = O(q^{-m-1}).$$

and

(20.31) $$r_i(G) = \sum_{j=1}^{N} B_{ji} R(G, H_j) = O(q^m) O(q^{-m-1}) = O(q^{-1}).$$

So we have proved that in (20.26), for fixed m, the error term r_i tends to 0 as $q \to \infty$.

The rest of the proof is standard analysis: Assume that $\mathsf{ent}^*(G_n, H) \to S_j$ ($n \to \infty$) for every looped-simple graph H with $\Delta(\overline{H}) \leq \varepsilon$. Consider any simple graph F_i on m nodes. Equation (20.26) implies that

(20.32) $$\left| \frac{\mathsf{ind}(F_i, G_n)}{\mathsf{v}(G_n)} - \sum_{j=1}^{N} B_{ji} S_j \right| \leq \sum_{j=1}^{N} |B_{ji}| |\mathsf{ent}^*(G_n, H_j) - S_j| + |r_i(G_n)|.$$

Let $\delta > 0$ be given, and choose q large enough so that $|r_i(G_n)| \leq \delta/2$ for every n (recall that the big-O in (20.31) does not depend on G). Since $\mathsf{ent}^*(G_n, H_j) \to S_j$, the first term on the right side of (20.32) is at most $\delta/2$ if n is large enough. It follows that $\mathsf{ind}(F, G_n)/\mathsf{v}(G_n)$ is a Cauchy sequence, which means that the sequence (G_n) is locally convergent. $\square$

The proof of the Supplement is based on similar arguments and not given here in detail. The proof method used above for (iii)$\Rightarrow$(i) can also be used to prove a somewhat weaker version of (ii), replacing the Dobrushin condition $\mathsf{dob}(W) < 1/D$ by $8D\Delta(\overline{W}) < 1$. In fact, the expression (20.22) yields itself more directly to a proof of (i)$\Rightarrow$(ii) than to a proof of (b): naively, if the frequency of any induced subgraph converges to some value, then so do the homomorphism entropies. The main issue is to obtain good tail bounds, which can be done similarly as in the proof above, as long as we are satisfied with proving the convergence for very small $\Delta(W)$; but if we want a bound that is sharp up to a constant, then we need more technical computations. We refer to the paper of Borgs, Chayes, Kahn and Lovász [2012] for these details.

REMARK 20.19. It is a natural question to ask which sequences of bounded degree graphs are right-convergent in the sense that their homomorphism entropies converge for all soft-core target graphs. Gamarnik [2012] studies this problem for sparse random graphs, but the general question is unsettled. It is also natural to ask whether local-global convergence can be characterized by any right-convergence condition.

EXERCISE 20.20. Let G and G' be two graphs on the same set of nodes $[n]$, such that $|E(G)\triangle E(G')| \leq \varepsilon n$. Prove that $\delta_\odot^{\mathrm{nd}}(G_1, G_2) \leq 2\varepsilon$.

EXERCISE 20.21. Let H be a weighted graph with positive edgeweights and (G_n), a bounded degree graph sequence for which the sequence $(\mathsf{ent}^*(G_n, H))$ is convergent. Let G'_n be obtained from G_n by deleting $o(v(G_n))$ nodes and edges. Prove that $(\mathsf{ent}^*(G'_n, H))$ is convergent.

EXERCISE 20.22. Let H be a weighted graph with at least one positive edgeweight. Prove that the sequence $\mathsf{ent}^*(P_n\square P_m, H)$ is convergent as $n, m \to \infty$, and the same holds for the sequence $\mathsf{ent}^*(C_n\square P_m, H)$, provided n is restricted to even numbers.

EXERCISE 20.23. Let H be a weighted graph whose edges with positive weight form a connected and nonbipartite graph. Prove that the sequence $\mathsf{ent}^*(C_n\square P_m, H)$ is convergent as $n, m \to \infty$.

EXERCISE 20.24. Let σ be an involution-invariant measure. Show how to express $\mathsf{s}(\sigma)$ in terms of the associated Bernoulli graphing.

CHAPTER 21

On the structure of graphings

21.1. Hyperfiniteness

A notion related to Følner sequences in the theory of amenable groups is "hyperfiniteness" for general graph families with bounded degree, which can be extended to graphings in a natural way. This notion was introduced (in different settings) by Kechris and Miller [2004], Elek [2007b] and Schramm [2008]. Hyperfiniteness of a graph family has a number of important consequences, like testability of many graph properties. Quoting an informal remark by Elek, hyperfinite bounded-degree graph families and graphings behave as nicely as dense graph sequences and graphons do.

21.1.1. Hyperfinite graph families. A graph $G \in \mathcal{G}$ is called (ε, k)-*hyperfinite* ($\varepsilon \in (0,1), k \in \mathbb{N}$), if we can delete $\varepsilon \mathsf{v}(G)$ edges and get a graph in which every connected component has at most k nodes. Let $\mathcal{H} \subseteq \mathcal{G}$ be any family of finite graphs (recall that all degrees are bounded by D). We say that $\mathcal{H}$ is *hyperfinite*, if for every $\varepsilon > 0$ there is a $k = k(\varepsilon) > 0$ such that every $G \in \mathcal{H}$ is (ε, k)-hyperfinite.

We could talk about deleting nodes instead of edges. Indeed, deleting one endnode of every edge in S results in even smaller components. Conversely, if deleting a set T of nodes results in a graph with small components, then deleting the set S of edges incident with any node in T leaves small components, and $|S| \leq D|T|$.

We will see that hyperfinite families are very well-behaved, often as well as dense graphs for analogous questions. How special are they? The examples below show that several important graph families are hyperfinite. (In fact, one has to work to construct a family that is not hyperfinite.) It is also likely that many large real-life networks can be thought of as hyperfinite, showing the potential applicability of the theory of hyperfinite families.

EXAMPLE 21.1 (**Trees**). The family of trees in $\mathcal{G}$ is hyperfinite. Indeed, select an endpoint r as the root in a tree T and fix an integer $k \geq 1$. It is easy to see that if $\mathsf{v}(T) > k$, then there is always an edge such that the branch rooted at it has at least $k/(D-1)$ but at most k nodes. If we delete recursively such edges, then the number of edges deleted is at most $(D-1)(\mathsf{v}(T)-1)/k$, and every connected component of the remaining forest has at most k nodes. So our tree is $((D-1)/k, k)$-hyperfinite.
♦

EXAMPLE 21.2 (**Grids**). From an $n \times m$ grid G, delete the edges inside every M-th vertical and horizontal ribbon of squares (starting from the top and from the left, say). The number of edges deleted is at most $m((n-1)/M) + n((m-1)/M) < 2\mathsf{v}(G)/M$, and every connected component of the remaining graph has at most M^2 nodes. So this grid is $(M^2, 2/M)$ hyperfinite.
♦

EXAMPLE 21.3 (**Planar graphs**). More generally, the family of planar graphs with degree bounded by D is hyperfinite. Indeed, let G be such a graph on n nodes. The Lipton–Tarjan Planar Separation Theorem [1979] says that G has a set S of at most $3\sqrt{n}$ nodes such that every connected component of $G - S$ has at most $2n/3$ nodes. We repeat this with every connected component of the remaining graph until all components have at most K nodes.

The estimation of the number of deleted nodes is somewhat tricky and it is left to the reader as Exercise 21.21. ◆

EXAMPLE 21.4 (**Random regular graphs**). Let us generate a random D-regular graph $\mathbb{G}_n$ on n nodes, for every even n, by choosing one of the D-regular graphs uniformly at random. The family of graphs obtained is *not* hyperfinite with probability 1 if $D \geq 3$. The following heuristic argument to prove this can be made precise rather easily. If $\mathbb{G}_n$ is (ε, k)-hyperfinite, then $V(G)$ can be split into two sets of size between $n/2 - k$ and $n/2 + k$ in such a way that the number of edges between the two classes is at most εn. On the other hand, let $\{S_1, S_2\}$ be a partition of $[n]$ into two classes of size about $n/2$, and let Z denote the number of edges in $\mathbb{G}_n$ connecting the two classes. The expected number of such edges is about $Dn/4$. Furthermore, Z is highly concentrated around its mean, and so the probability that $Z \leq \varepsilon n$ is $o(2^{-n})$ if ε is small enough. There are fewer than 2^n such partitions of $[n]$, so with high probability all of these have more than εn edges connecting the two classes. ◆

EXAMPLE 21.5 (**Expanders**). We call a family $\mathcal{E} \subseteq \mathcal{G}$ of graphs an *expander family* if there is a $c > 0$ such that for every graph $G \in \mathcal{E}$ and every $S \subseteq V(G)$ with $|S| \leq \mathsf{v}(G)/2$, we have $e_G(S, V(G) \setminus S) \geq c|S|$. An infinite family of expander graphs is not hyperfinite. Indeed, $T \subseteq E(G)$ and $G - T$ has components $G_1, \ldots, G_r$, and all of these have fewer than $\mathsf{v}(G)/2$ nodes, then

$$|T| = \frac{1}{2} \sum_{i=1}^{r} e_G\big(V(G_i), V(G) \setminus V(G_i)\big) \geq \sum_{i=1}^{r} c\mathsf{v}(G_i) = c\mathsf{v}(G).$$

It can be shown that the family of random D-regular graphs in Example 21.4 is an expander family with probability 1.

The following explicit construction for an expander graphing was given (in a different context) by Margulis. [1973] Consider the space $\mathbb{R}^2/\mathbb{Z}^2$, a.k.a. the torus. Let us connect every point (x, y) to the points $(x \pm y, y)$ and $(x, y \pm x)$ (additions modulo 1; we can leave out the axes if we don't want loops). This graph is the support of the measure preserving family consisting of the two maps $(x, y) \mapsto (x + y, y)$ and $(x, y) \mapsto (x, x + y)$, and hence it is a graphing. Furthermore, this graphing is an expander, and hence not hyperfinite. This is not easy to prove; for a proof based on Fourier analysis, see Gabber and Galil [1981]. ◆

EXAMPLE 21.6. A special case of a hyperfinite family is a family of graphs with subexponential growth, familiar from group theory. To be precise, for a function $f : \mathbb{N} \to \mathbb{N}$ we say that a family $\mathcal{H}$ of graphs has *f-bounded growth*, if for any graph $G \in \mathcal{H}$, any $v \in V(G)$ and any $m \in \mathbb{N}$, the number of nodes in the m-neighborhood of v is at most $f(m)$. We say that $\mathcal{H}$ has *subexponential growth*, if it has f-bounded growth for some function f such that $\big(\ln f(m)\big)/m \to 0$ $(m \to \infty)$. It was asked by Elek and proved by Fox and Pach [unpublished] that this property implies hyperfiniteness (Exercise 21.22). ◆

The following important class of hyperfinite families, generalizing Example 21.3 was found by Benjamini, Schramm and Shapira [2010].

PROPOSITION 21.7. *Every minor-closed family of graphs in $\mathcal{G}$ that does not contain all graphs is hyperfinite.*

Proof (sketch). Alon, Seymour and Thomas [1990] proved that the Planar Separator Theorem extends to every minor-closed property not containing all graphs. The same argument as described in Example 21.3 can be carried through to give hyperfiniteness. □

21.1.2. Hyperfinite graphings. Hyperfiniteness can be generalized to graphings; in fact, we don't have to talk about classes of graphings, the notion makes sense for a simple graphing, and leads to some nontrivial questions.

A graphing $\mathbf{G}$ is (ε, k)-*hyperfinite* ($\varepsilon \in (0, 1), k \in \mathbb{N}$), if there is a Borel set $S \subseteq E(\mathbf{G})$ with $\eta(S) \leq \varepsilon$ such that every connected component of $\mathbf{G} - S$ has at most k nodes. A graphing $\mathbf{G}$ is *hyperfinite*, if for every $\varepsilon > 0$ there is a positive integer k such that $\mathbf{G}$ is (ε, k)-hyperfinite.

We could relax this definition and ask for a set $S \subseteq E(\mathbf{G})$ with $\eta(S) \leq \varepsilon$ such that every connected component of $\mathbf{G} - S$ is finite. But this would not change the notion of hyperfiniteness. Indeed, suppose that $\mathbf{G}$ satisfies the relaxed condition; we show that it satisfies this stronger condition as well. We choose a set S' for $\varepsilon/2$ in place of ε. Let $V_m \subset V(\mathbf{G}) \setminus S'$ be the set of points contained in some connected component of $\mathbf{G} - S'$ with m nodes. Then V_m is measurable (Exercise 18.10), and we have $\sum_m \eta(E(\mathbf{G}[V_m])) = \eta(E(\mathbf{G})) \leq D$. It follows that there is a positive integer K such that $\sum_{m \geq K} \eta(E(\mathbf{G}[V_m])) \leq \varepsilon/2$, and so the set $S = S' \cup \bigcup_{m \geq K} E(\mathbf{G}[V_m])$ is a set of measure at most ε such that every connected component of $\mathbf{G} - S$ has fewer than K nodes.

Similarly as for graphs, we could define the same notion by the existence of a Borel set of points $S \subseteq V(\mathbf{G})$ such that $\lambda(S) \leq \varepsilon$ and every connected component of $\mathbf{G} - S$ is finite.

For a graphing, (ε, k)-hyperfiniteness can be expressed in a local-global way, through a red-blue coloring of the edges such that the η-measure of red edges is less than ε, and every connected k-node subgraph contains a red edge. Using this, it is easy to verify the following fact.

PROPOSITION 21.8. *Let $\mathbf{G}$ and $\mathbf{G}'$ be locally-globally equivalent graphings. If $\mathbf{G}$ is (ε, k)-hyperfinite, then $\mathbf{G}'$ is (ε', k)-hyperfinite for every $\varepsilon' > \varepsilon$.*

This proposition does not remain true for locally equivalent graphings (see Example 21.12 below). The following important and surprisingly non-trivial theorem, which is a version of a result of Schramm (see Theorem 21.13 below) shows that hyperfiniteness (without parameters) is preserved by local equivalence.

THEOREM 21.9. *Let $\mathbf{G}$ and $\mathbf{G}'$ be locally equivalent graphings. If $\mathbf{G}$ is hyperfinite, then so is $\mathbf{G}'$.*

The proof is best understood if we introduce a fractional version of hyperfiniteness. The motivation comes from combinatorial optimization. Let $\mathbf{G}$ be a graphing, and let $\mathcal{R}$ denote the set of subsets $Y \subseteq V = V(\mathbf{G})$ that induce a connected subgraph of $\mathbf{G}$ and have at most k elements. This can be considered as a subset of $U = V \cup V^2 \cup \cdots \cup V^k$, and it is a Borel set if we endow U with the natural sigma-algebra it inherits from V. Let $S \subseteq E(\mathbf{G})$ be a Borel set such that every connected component of $\mathbf{G} \setminus S$ has at most k nodes. For $x \in V$, let C_x denote the node set of the connected component of $\mathbf{G} \setminus S$ containing x. The sets C_x partition V. Let $\mathbf{x}$ be a random point of $\mathbf{G}$, then $C_\mathbf{x}$ is a random member of $\mathcal{R}$, which has the following two properties:

(1) If we select $C_\mathbf{x}$ first, and then select a uniform random point $\mathbf{y} \in C_\mathbf{x}$ (note that $C_\mathbf{x}$ is finite!), then $\mathbf{y}$ is distributed according to λ.

(2) If $\partial(X)$ denotes the number of edges of $\mathbf{G}$ connecting X to $V(\mathbf{G}) \setminus X$ (where X is a finite subset of $V(\mathbf{G})$), then

$$\mathsf{E}\Big(\frac{\partial(C_\mathbf{x})}{|C_\mathbf{x}|}\Big) = \eta(S) \leq \varepsilon.$$

Both of these properties can be easily verified using the Mass Transport Principle (similarly to the proof of Proposition 18.50).

This motivates the next definition: we call a probability distribution τ on $\mathcal{R}$ a *fractional partition* (into parts in $\mathcal{R}$), if selecting $\mathbf{Y} \in \mathcal{R}$ according to τ, and then a point $\mathbf{y} \in \mathbf{Y}$ uniformly, we get a point distributed according to λ; and we define the *boundary value* of τ as

$$\partial(\tau) = \mathsf{E}\Big(\frac{\partial(\mathbf{Y})}{|\mathbf{Y}|}\Big).$$

We say that $\mathbf{G}$ is *fractionally (ε, k)-hyperfinite*, if there is a fractional partition τ such that $\partial(\tau) \leq \varepsilon$. It follows from the discussion above that every (ε, k)-hyperfinite graphing is fractionally (ε, k)-hyperfinite. The converse is not true (cf. Example 21.12), but we have the following weak converse:

LEMMA 21.10. *If a graphing is fractionally (ε, k)-hyperfinite, then it is $(\varepsilon \log(8D/\varepsilon), k)$-hyperfinite.*

Proof. We use the Greedy Algorithm to construct a partition from a fractional partition τ that establishes that $\mathbf{G}$ is fractionally (ε, k)-hyperfinite. Similar algorithms are well known in combinatorial optimization, but here we have to be careful, since we are going to construct an uncountable family of sets, and have to make sure that the partition we obtain has the property that the set of edges connecting different classes is Borel (and of course has small measure). We do our construction in $r = \lfloor \log(2D/\varepsilon) \rfloor$ phases.

We start with $\mathcal{R}_0 = U_0 = \emptyset$. In the j-th phase, let $U_{j,0} = U_{j-1}$ be the union of previously selected sets. Let $\mathcal{R}_{j,1}$ be the set of sets $Y \in \mathcal{R}$ such that $\partial(Y) < \varepsilon 2^{j-1}|Y \setminus U_{j,0}|$. Let $\mathcal{Q}_{j,1}$ be a maximal Borel set of sets $\mathcal{R}_{j,1}$ such that the sets $Y \setminus U_{j,0}$ are disjoint. Such a set exists by the following construction. Let $\mathbf{H}_0$ be the intersection graph of $\mathcal{R}_{j,0}$. It is easy to see that $\mathbf{H}_0$ is a Borel graph with bounded degree, and so it contains a maximal stable set that is Borel (this is implicit in the proof of Theorem 18.3, see Exercise 18.11). Let $U_{j,1}$ be the union of $U_{j,0}$ and the sets in $\mathcal{Q}_{j,1}$.

The phase is not over; we select a maximal Borel family $\mathcal{Q}_{j,2}$ of sets $Y \in \mathcal{R}$ such that the sets $Y \setminus U_{j,1}$ are disjoint and $\partial(Y) < \varepsilon 2^{j-1}|Y \setminus U_{j,1}|$. We let $U_{j,2}$ be the union of $U_{j,1}$ and the sets in $\mathcal{Q}_{j,2}$. We repeat this $k+1$ times, to finish the j-th phase (after a while, we may not be adding anything). Let $\mathcal{Q}_j$ be the family of sets $Y \in \mathcal{R}$ selected in the j-th phase, and let $U_j = U_{j,k}$ be their union.

We repeat this for $j = 1, \ldots, r$. Let $\mathcal{Q} = \mathcal{Q}_1 \cup \cdots \cup \mathcal{Q}_r$ be the set of all sets $Y \in \mathcal{R}$ selected. For every $Y \in \mathcal{Q}_j$, let $Y^0 = Y \setminus U_{j-1}$ (this is the set of nodes first covered by Y). Let T_0 be the set of all edges incident with any node of $V \setminus U_r$, and let T_1 denote the set of all edges connecting any set $Y \in \mathcal{Q}$ to its complement. Clearly every connected component of $\mathbf{G} - (T_0 \cup T_1)$ has at most k nodes.

Next we show that

(21.1) $$\partial(Y) \geq \varepsilon 2^{j-1}|Y \setminus U_j|$$

for every $Y \in \mathcal{R}$ (selected or not) and $1 \leq j \leq r$. Suppose (by way of contradiction) that $\partial(Y) < \varepsilon 2^{j-1}|Y \setminus U_j|$. Then $Y \not\subseteq U_j$, and hence it was not selected in the j-th phase or before. But Y was eligible for selection throughout the j-th phase, and if it was not selected, then (by the maximality of the family selected) Y must contain a point of $U_{j,i} \setminus U_{j,i-1}$ for $i = 1, \ldots, k+1$, which is impossible since $|Y| \leq k$. This proves (21.1).

We want to bound the measure of $T_0 \cup T_1$. We start with T_0. Select a random set $\mathbf{Y} \in \mathcal{R}$ from the distribution τ, and a random point $\mathbf{y} \in Y$. Then $\mathbf{y}$ is distributed according to λ, and hence by (21.1) and the definition of the fractional partition τ we have

(21.2) $$\lambda(V \setminus U_j) = \mathsf{P}(\mathbf{y} \notin U_j) = \mathsf{E}\left(\frac{|\mathbf{Y} \setminus U_j|}{|\mathbf{Y}|}\right) \leq \mathsf{E}\left(\frac{\partial(\mathbf{Y})}{\varepsilon 2^{j-1}|\mathbf{Y}|}\right) \leq 2^{1-j}$$

for every $1 \leq j \leq r$. In particular, we have $\lambda(V \setminus U_r) \leq 2^{1-r} \leq 2\varepsilon/D$ by the choice of r, and hence

$$\eta(T_0) \leq \int_{V \setminus U_r} \deg(x)\, dx \leq D\lambda(V \setminus U_r) \leq 2\varepsilon.$$

Turning to T_1, we select a random point $\mathbf{y}$ of $\mathbf{G}$ again, and consider the set $\mathbf{Y} \in \mathcal{Q}$ that is the first set added containing $\mathbf{y}$. Since sets added at the same time are disjoint, this is well-defined, unless $\mathbf{y} \notin U_r$, in which case we take $\mathbf{Y} = \{\mathbf{y}\}$. We consider $\mathbf{Y}$ a random set, from some distribution
$alpha$). We can generate a random $\mathbf{y}$ by generating $\mathbf{Y}$ first according to α, and then selecting a uniform random element of $\mathbf{Y}^0$. Counting every edge in T_1 with its endpoint that was selected first (breaking ties arbitrarily), we have

$$\eta(T_1) \leq \mathsf{E}\left(\frac{\partial(\mathbf{Y})}{|\mathbf{Y}^0|}\right) \leq \varepsilon \sum_{j=1}^{r} 2^{j-1}\bigl(\lambda(U_j) - \lambda(U_{j-1})\bigr)$$

$$= \varepsilon \sum_{j=1}^{r} 2^{j-1}\bigl(\lambda(V \setminus U_{j-1}) - \lambda(V \setminus U_j)\bigr).$$

Doing partial summation and using (21.2) again,

$$\eta(T_1) \leq \varepsilon \sum_{j=1}^{r} 2^{j-1} \lambda(V \setminus U_j) \leq \varepsilon r \leq \varepsilon \log(2D/\varepsilon).$$

Hence
$$\eta(T_0 \cup T_1) \leq 2\varepsilon + \varepsilon \log(2D/\varepsilon) = \varepsilon \log(8D/\varepsilon). \qquad \square$$

Theorem 21.9 follows from our characterization of local equivalence (Theorem 18.59), Lemma 21.10 and the following rather simple couple of facts.

PROPOSITION 21.11. *Let $\varphi : \mathbf{G}_1 \to \mathbf{G}_2$ be a local isomorphism between graphings $\mathbf{G}_1$ and $\mathbf{G}_2$.*

(a) If $\mathbf{G}_2$ is (ε, k)-hyperfinite, then so is $\mathbf{G}_1$.

(b) If $\mathbf{G}_1$ is fractionally (ε, k)-hyperfinite, then so is $\mathbf{G}_2$.

Proof. (a) Let $S \subseteq E(\mathbf{G}_2)$ be a Borel set such that every connected component of $\mathbf{G}_2 \setminus S$ is has at most k nodes. Then $S' = \varphi^{-1}(S)$ is a Borel set in $E(\mathbf{G}_1)$ with $\eta_1(S') = \eta_2(S)$. We claim that almost every connected component of $\mathbf{G}_1 \setminus S'$ has at most k nodes. Indeed, let X be the union of components of $\mathbf{G}_1 \setminus S'$ with more than k nodes, then for almost all $x \in X$, φ is an isomorphism on $(\mathbf{G}_1)_x$, and hence $\varphi((\mathbf{G}_1)_x)$ is a connected subgraph of $\mathbf{G}_2 \setminus S$, which is a contradiction unless X has measure 0.

(b) Let τ_1 be a probability distribution on the set $\mathcal{R}_1$ of the connected induced subgraphs of $\mathbf{G}_1$ with at most k nodes that is a fractional partition with $\partial(\tau) \le \varepsilon$. Select $\mathbf{Y} \in \mathcal{R}_1$ randomly according to τ_1; then $\mathbf{Z} = \varphi(\mathbf{Y})$ is a random connected induced subgraph of $\mathbf{G}_2$ with at most k nodes. Let τ_2 denote the distribution of $\mathbf{Z}$.

We claim that τ_2 is a fractional partition. Indeed, we can generate a uniform random point of $\mathbf{Z}$ by generating a uniform random point $\mathbf{y} \in \mathbf{Y}$, and taking $\mathbf{z} = \varphi(\mathbf{y})$. (We use here that φ is a local isomorphism, and so it yields and isomorphism between $\mathbf{Y}$ and $\mathbf{Z}$.) Since $\mathbf{y}$ is distributed according to $\lambda_{\mathbf{G}_1}$, it follows that $\mathbf{z}$ is distributed according to $\lambda_{\mathbf{G}_2}$.

By a similar argument, we have

$$\mathsf{E}\left(\frac{\partial(\mathbf{Z})}{|\mathbf{Z}|}\right) = \mathsf{E}\left(\frac{\partial(\mathbf{Y})}{|\mathbf{Y}|}\right) \le \varepsilon.$$

This proves (b). $\square$

We note that the stronger assertion, namely that (ε, k)-hyperfiniteness of a graphing would be invariant under local equivalence is false. As a simple example, consider the two graphings $\mathbf{C}_a$ and $\mathbf{C}_a''$ (Exercise 18.60). These are locally equivalent, but $\mathbf{C}_a''$ contains a perfect matching that is a Borel set (the set of edges of the form $(x, 1/2 + (x - a/2 \mod 1/2))$, where $x \le 1/2$), and hence it is $(1/2, 2)$-hyperfinite. No such perfect matching exists in $\mathbf{C}_a$ (indeed, such a perfect matching would give a measure-preserving involution by Lemma 18.19, and its complement would be another one, contradicting the argument in Example 18.22). Hence $\mathbf{C}_a$ is not $(1/2, 2)$-hyperfinite.

In this example, (ε, k)-hyperfiniteness is almost preserved, in the sense that $\mathbf{C}_a$ is not $(1/2, 2)$-hyperfinite, but $(1/2 + \varepsilon, 2)$-hyperfinite for every $\varepsilon > 0$. We will see that this is a general phenomenon among hyperfinite graphings (Corollary 21.18), but not among all graphings, as the following example shows.

EXAMPLE 21.12. For a 3-regular graph G, define a new graph $G^\triangle$ as follows. We replace every node by a triangle (call these principal triangles), and then replace every old edge by a copy of K_4^-, with the two nodes of degree 2 identified with the endpoints of the edge (Figure 21.1). The graph $G^\triangle$ has $6n$ nodes and $21n/2$ edges.

If G is bipartite, then $G^\triangle$ is $(3/4, 3)$-hyperfinite. Indeed, in this case $V(G^\triangle)$ can be covered by disjoint triangles: we select all principal triangles corresponding to nodes in one color class, and all non-principal triangles disjoint from them. The number of remaining edges is $|E(G^\triangle)| - |V(G^\triangle)| = 9n/2 = (3/4)|V(G^\triangle)|$.

Now let G be nonbipartite, and let S be a minimum set of edges such that every connected component of $G^\triangle \setminus S$ has at most 3 nodes. Let A be the set of nodes in G for which the corresponding principal triangle is a component of $G^\triangle \setminus S$. It is

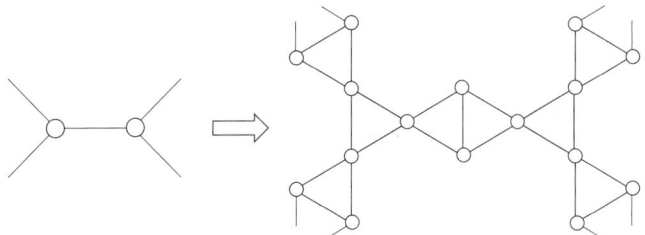

FIGURE 21.1. $(3/4, 3)$-hyperfiniteness is not even approximately preserved by local equivalence

easy to check that for every edge of G induced by A or induced by $V(G) \setminus A$, there is at least one node of degree at most one among the corresponding four nodes of $G^{\triangle} \setminus S$. The remaining nodes of $G^{\triangle} \setminus S$ have degree at most 2. Hence

$$|E(G^{\triangle}) \setminus S| \leq \frac{1}{2}\bigl(|E(G)| - e_G(A, V \setminus A) + 2(|V(G^{\triangle})| - |E(G)| + e_G(A, V \setminus A))\bigr)$$
$$\leq \frac{9n}{2} + \mathsf{Maxcut}(G),$$

and so

$$|S| \geq 6n - \mathsf{Maxcut}(G) > \frac{9n}{2},$$

since G is nonbipartite. So $G^{\triangle}$ is $(3/4, 3)$-hyperfinite. For the appropriate choice of G, we can prove more: let $\mathbb{G}_n$ be a random D-regular graph and $\mathbb{G}'_n$, a random D-regular bipartite graph on n nodes, then $(\mathbb{G}'_n)^{\triangle}$ is $(3/4, 3)$-hyperfinite. On the other hand, $\mathsf{Maxcut}(\mathbb{G}_n) < 1.41n$ with high probability (McKay [1982], see also Hladky [2006]), and we get that $(\mathbb{G}_n)^{\triangle}$ is not even $(4/5, 3)$-hyperfinite.

Let $\mathbf{G}$ and $\mathbf{G}'$ be local-global limit graphings of the sequences $(\mathbb{G}_n)^{\triangle}$ and $(\mathbb{G}'_n)^{\triangle}$, respectively (or of appropriate subsequences), then $\mathbf{G}$ and $\mathbf{G}'$ are locally equivalent, $\mathbf{G}$ is $(3/4, 3)$-hyperfinite, but $\mathbf{G}'$ is not even $(4/5, 3)$-hyperfinite. ♦

Our argument proving Theorem 21.9, which was motivated by an argument of Schramm [2008], says that if a graphing is (ε, k)-hyperfinite, then every graphing locally equivalent to it is (ε', k)-hyperfinite with a somewhat larger ε' than ε (namely, $\varepsilon' = O(\varepsilon \log(1/\varepsilon))$). We could have based another proof on the graph partitioning algorithm of Hassidim, Kelner, Nguyen, and Onak; [2009] this would show that if a graphing is (ε, k)-hyperfinite, then every graphing locally equivalent to it is (ε, k')-hyperfinite with some larger k'.

The following theorem was proved (in a different formulation, using involution invariant random rooted graph models) by Schramm [2008].

THEOREM 21.13. *Let (G_n) be a sequence of graphs in $\mathcal{G}$, converging to a graphing $\mathbf{G}$. Then $\mathbf{G}$ is hyperfinite if and only if the family $\{G_n : n = 1, 2, \ldots\}$ is hyperfinite.*

Proof. The "if" part is not hard. Suppose that (G_n) is hyperfinite. Let $\varepsilon > 0$, we want to show that $\mathbf{G}$ is hyperfinite. Let $\varepsilon > 0$, and let $k \geq 1$ be chosen so that for every n that is large enough there is a set $S_n \subseteq V(G_n)$ with $|S_n| \leq \varepsilon \mathsf{v}(G_n)$ such that every connected component of $G_n - S_n$ has at most k nodes. Consider the pairs (G_n, S_n) as graphs with their nodes 2-colored, and choose a subsequence that is convergent as a sequence of colored graphs. The limit can be represented by a colored graphing $(\mathbf{G}', S)$. It follows from the definition of convergence that $|S_n|/\mathsf{v}(G_n) \to \lambda(S)$ (where λ is the node measure in $(\mathbf{G}', S)$), and also that almost all connected components of $\mathbf{G}' - S$ have at most k nodes. Hence the uncolored graphing $\mathbf{G}'$ is hyperfinite. Since $\mathbf{G}'$ and $\mathbf{G}$ are locally equivalent, it follows by Theorem 21.9 that $\mathbf{G}$ is hyperfinite.

To prove the "only if" part, we invoke Theorem 19.16. By selecting an appropriate subsequence, we may assume that the sequence (G_n) is locally-globally convergent, and so it has a limit graphing $\mathbf{G}'$ such that $\delta_\odot^{\mathrm{nd}}(G_n, \mathbf{G}') \to 0$ for every $k \geq 0$. Clearly $\mathbf{G}'$ and $\mathbf{G}$ are locally equivalent, and hence $\mathbf{G}'$ is hyperfinite by Theorem 21.9. This means that for every $\varepsilon > 0$ there is an $m \geq 1$ such that $V(\mathbf{G}')$ has a Borel 2-coloring with read and blue such that every connected m-node subgraph contains a red point, and $\lambda'\{\text{red points}\} \leq \varepsilon$. These properties can be read off from the 1-balls and the m-balls, respectively. It follows by the assumption that $G_n \to \mathbf{G}'$ in the local-global sense that for a large enough n, G_n has a 2-coloring such that the set R_n of red nodes satisfies $|R_n| \leq 2\varepsilon\mathsf{v}(G_n)$, and the number of m-neighborhoods that contain a connected blue subgraph with $m+1$ nodes is at most $\varepsilon\mathsf{v}(G_n)$. Adding the roots of these m-neighborhoods to R_n, we get a set $R'_n \subseteq V(G_n)$ with $|R'_n| \leq 3\varepsilon n$ such that every connected component of $G_n - R_n$ has at most m nodes. □

We state another result, in a sense dual to Theorem 21.13:

THEOREM 21.14. *A graphing is hyperfinite if and only if it is the limit of a hyperfinite graph sequence.*

Proof. In view of Theorem 21.13, it suffices to prove that every hyperfinite graphing is the limit of a locally convergent graph sequence. (So the Aldous–Lyons conjecture holds for hyperfinite graphings.) Let $\mathbf{G}$ be a hyperfinite graphing, and let $\varepsilon > 0$. Let S be a subset of edges with $\eta(S) = \varepsilon$ such that every connected component of $\mathbf{G} - S$ is finite. Proposition 19.1 implies that

(21.3) $$\delta_\odot(\mathbf{G}, \mathbf{G} - S) \leq 4\varepsilon^{1/\log(2D)}.$$

For every graph $F \in \mathcal{G}$, let a_F be the measure of points in $\mathbf{G} - S$ whose connected component is isomorphic to F. Since $\sum_F a_F = 1$, we can choose a finite set $\mathcal{H}$ of graphs such that $\sum_{F \notin \mathcal{H}} a_F \leq \varepsilon/D$. Let $n > (D/\varepsilon) \sum_{F \in \mathcal{H}} \mathsf{v}(F)$, and $n_F = \lfloor a_F n/\mathsf{v}(F) \rfloor$ (so that the rationals $n_F \mathsf{v}(F)/n$ approximate the real numbers a_F with common denominator). For every $F \notin \mathcal{H}$, let us delete the edges of all connected components of $\mathbf{G} - S$ isomorphic to F. For every $F \in \mathcal{H}$, let us delete the edges of a set of connected components of $\mathbf{G} - S$ isomorphic to F so that the remaining connected components cover a set of measure $n_F \mathsf{v}(F)/n$; it is not hard to see that this can be done so that a Borel graph remains. The measure of the set T of deleted edges can be bounded as follows:

$$\eta(T) \leq \sum_{F \notin \mathcal{H}} \frac{D}{2} a_F + \sum_{F \in \mathcal{H}} \frac{D}{2}\left(a_F - \frac{n_F \mathsf{v}(F)}{n}\right) \leq \frac{\varepsilon}{2} + \frac{D}{2} \sum_{F \in \mathcal{H}} \frac{\mathsf{v}(F)}{n} \leq \varepsilon.$$

Hence it follows just like above that

(21.4) $$\delta_\odot(\mathbf{G} - S, \mathbf{G} - S - T) \leq 4\varepsilon^{1/\log(2D)}.$$

Let G be a graph on n nodes consisting of n_F copies of each F, together with sufficiently many isolated nodes. Then $\mathbf{G} - S - T$ and G have the same connected components, with the same frequencies; hence $\delta_\odot(\mathbf{G}-S-T,G)=0$, and so by (21.3) and (21.4),

$$\delta_\odot(\mathbf{G},G) \leq \delta_\odot(\mathbf{G},\mathbf{G}-S) + \delta_\odot(\mathbf{G}-S,\mathbf{G}-S-T) \leq 8\varepsilon^{1/\log(2D)}.$$

So $\mathbf{G}$ can be approximated arbitrarily well in the $\delta_\odot$ distance by finite graphs. □

Kaimanovich [1997] proved the following characterization of hyperfinite graphings, which we quote without proof. (For a proof, see also Elek [2012a].)

THEOREM 21.15. *A graphing $\mathbf{G}$ is not hyperfinite if and only if it has a subgraphing $\mathbf{F}$ such that $\eta_\mathbf{G}(E(\mathbf{F})) > 0$ and there is an $\varepsilon > 0$ such that $\partial_\mathbf{F}(Y) \geq \varepsilon \mathsf{v}(Y)$ for every finite connected subgraph Y of $\mathbf{F}$.*

We conclude our discussion of hyperfiniteness with a result of Hatami, Lovász and Szegedy [2012] and Elek [2012a], showing that in the hyperfinite world, local and local-global are equivalent.

THEOREM 21.16. *Any two locally equivalent hyperfinite atom-free graphings are locally-globally equivalent.*

As a preparation for the proof, we prove a somewhat stronger statement in a special case.

LEMMA 21.17. *Let $\mathbf{G}$ and $\mathbf{G}'$ be two locally equivalent graphings such that all components of them are finite and have at most k nodes ($k \geq 1$). Then for every Borel m-coloring β of $\mathbf{G}'$ there is a Borel m-coloring γ of $\mathbf{G}$ such that $(\mathbf{G},\gamma)$ and $(\mathbf{G},\beta)$ are locally equivalent as colored graphings.*

Proof. By Theorem 18.59, we may assume that there is a local isomorphism $\varphi: \mathbf{G}' \to \mathbf{G}$ or a local isomorphism $\varphi: \mathbf{G} \to \mathbf{G}'$. The second alternative is trivial, so we assume the first.

Applying Theorem 18.3 to the graphing obtained by filling up every connected component of $\mathbf{G}$ to a complete graph (which results in a Borel graph, cf. Exercise 18.7), we get that there is a Borel k-coloring $\alpha: V(\mathbf{G}) \to [k]$ such that any two nodes in the same component have different colors.

For every isomorphism type F of k-colored graphs with at most k nodes, let U_F be the union of all connected components C of $\mathbf{G}$ for which $C \cong F$. It is easy to see that the U_F are Borel sets.

Next, we pull back the coloring to $\mathbf{G}'$: let $U'_F = \varphi^{-1}(U_F)$ and $\alpha' = \varphi \circ \alpha$. It follows from the definition of local isomorphism that every connected component of $\mathbf{G}'[U'_F]$ is isomorphic to F as a colored graph (with colors according to α').

Consider the given m-coloring β of $\mathbf{G}'$. On every connected component C of U'_F, the colors according to α' are all different, and hence β can be represented as $\beta = \alpha' \circ f_C$ with an appropriate map $f_C: [k] \to [m]$. For every isomorphism class F of k-colored graphs with at most k nodes and every map $f: [k] \to [m]$, let $U'_{F,f}$ be the union of all connected components C for which $C \cong F$ and $f_C = f$. This partitions every set U'_F into m^k sets $U_{F,f}$, which are Borel (as it is easy to see).

The images $\varphi(U'_{F,f})$ are not necessarily Borel sets, but we can partition every set U_F into m^k Borel sets $U_{F,f}$ so that $\lambda_\mathbf{G}(U_{F,f}) = \lambda_{\mathbf{G}'}(U'_{F,f})$. Let us color $x \in U_{F,f}$ with color $f(\alpha(x))$, to get an m-coloring γ. Then the whole component of a point $x \in U_{F,f}$ is colored by γ the same way as the component of any $y \in U'_{F,f}$. This proves that γ satisfies the conclusion of the Lemma. □

Proof of Theorem 21.16. As above, we may assume that there is a local isomorphism $\varphi : \mathbf{G}' \to \mathbf{G}$. We want to prove that for every m-coloring β of $\mathbf{G}'$ there is an m-coloring of $\mathbf{G}$ for which the sampling distance of these colored graphings is arbitrarily small; and vice versa. In fact, the "vice versa" part is trivial (we can just pull back the k-coloring of $\mathbf{G}$ by φ). The first assertion, however, takes more work.

Let $\varepsilon > 0$. By hyperfiniteness, there is a set $S \subseteq E(\mathbf{G})$ and a $k \geq 1$ such that $\eta_{\mathbf{G}}(S) \leq \varepsilon'$, where $\varepsilon' = \frac{1}{4}\varepsilon^{\log(2D+2)}$, and every connected component of $\mathbf{G} \setminus S$ has at most k nodes. Let $S' = \varphi^{-1}(S)$, then $\eta_{\mathbf{G}'}(S') = \eta_{\mathbf{G}}(S) \leq \varepsilon'$ and φ is a local isomorphism from $\mathbf{G}' \setminus S'$ to $\mathbf{G} \setminus S$.

Since β is a Borel m-coloring of $\mathbf{G}'$, Lemma 21.17 implies that there is a Borel m-coloring γ of $\mathbf{G} \setminus S$ such that $(\mathbf{G} \setminus S, \gamma)$ and $(\mathbf{G}' \setminus S', \beta)$ are locally equivalent as colored graphings. By Proposition 19.1, we have

$$\delta_{\odot}((\mathbf{G}, \gamma), (\mathbf{G} \setminus S, \gamma)) \leq 2d_1((\mathbf{G}, \gamma), (\mathbf{G} \setminus S, \gamma))^{1/\log(2D+2)} \leq \frac{\varepsilon}{2},$$

and similar inequality holds for $\mathbf{G}'$. Hence

$$\delta_{\odot}((\mathbf{G}, \gamma), (\mathbf{G}', \beta)) \leq \delta_{\odot}((\mathbf{G}, ,\gamma), (\mathbf{G} \setminus S, \gamma)) + \delta_{\odot}((\mathbf{G} \setminus S, \gamma), (\mathbf{G}' \setminus S', \beta))$$
$$+ \delta_{\odot}((\mathbf{G}' \setminus S', \beta), (\mathbf{G}', \beta)) \leq \frac{\varepsilon}{2} + 0 + \frac{\varepsilon}{2} = \varepsilon. \qquad \square$$

This theorem has some interesting consequences. We have seen (Example 21.12) that a graphing may be (ε, k) hyperfinite and a locally equivalent graphing may not be $(16\varepsilon/15, k)$ hyperfinite. However, if the graphing is hyperfinite, this cannot occur:

COROLLARY 21.18. *Let $\mathbf{G}$ and $\mathbf{G}'$ be two hyperfinite locally equivalent graphings, and assume that $\mathbf{G}$ is (ε, k)-hyperfinite. Then $\mathbf{G}'$ is (ε', k)-hyperfinite for every $\varepsilon' > \varepsilon$.*

The second corollary shows that the two notions of convergence discussed in sections 19.1 and 19.2 are equivalent for hyperfinite graph sequences.

COROLLARY 21.19. *Every locally convergent hyperfinite graph sequence is locally-globally convergent.*

Proof. Let (G_n) be a locally convergent hyperfinite sequence, then it converges locally to a hyperfinite graphing $\mathbf{G}$ by Theorem 21.13. If (G_n) does not converge locally-globally to $\mathbf{G}$, then it has a locally-globally convergent subsequence whose limit graphing $\mathbf{G}'$ is not locally-globally equivalent to $\mathbf{G}$. Since $\mathbf{G}$ and $\mathbf{G}'$ are locally equivalent, this contradicts Theorem 21.16. $\qquad \square$

EXERCISE 21.20. Let $G \in \mathcal{G}$ be an (ε, k)-hyperfinite graph, and let $0 \leq \delta \leq \varepsilon$. Prove that there exists an $(\varepsilon - \delta, k)$ hyperfinite graph G' for which $\delta_{\odot}(G, G') \leq 4\delta^{1/\log D}$. State and prove the analogous assertion for graphings.

EXERCISE 21.21. Prove that for every planar graph G on n nodes and every integer $K \leq n$, one can delete at most $60n/\sqrt{K} - 30\sqrt{n}$ nodes from G so that every connected component of the remaining graph has at most K nodes. (The strange formula is given as help, to facilitate induction.)

EXERCISE 21.22. Prove that every family of graphs in $\mathcal{G}$ with subexponential growth is hyperfinite.

EXERCISE 21.23. Formulate and prove a version of Corollary 21.18 for finite graphs.

21.2. Homogeneous decomposition

Hyperfinite graphs can be decomposed into bounded size graphs by deleting a small fraction of the edges. How far can we simplify a general bounded degree graph by deleting a small fraction of the edges? It was proved recently by Angel and Szegedy [unpublished], and independently by Elek and Lippner [2011], that every graph with degrees bounded by D can be decomposed into a bounded number of "homogeneous" parts by deleting small number of edges.

To be precise, let us call a subset $U \subseteq V(G)$ an (ε, δ)-*island*, if $e_G(U, V(G)\setminus U) \leq \delta \mathsf{v}(G)$, $|U| \geq \varepsilon \mathsf{v}(G)$, and $\delta_\odot(G[U], G) \geq \varepsilon$. We say that a graph $G \in \mathcal{G}$ (ε, δ)-*homogeneous*, if it contains no (ε, δ)-island. Clearly, an (ε, δ)-island is also an (ε', δ')-island if $\varepsilon' \leq \varepsilon$ and $\delta' \geq \delta$. Hence an (ε, δ)-homogeneous graph is also (ε', δ')-homogeneous, if $\varepsilon' \leq \varepsilon$ and $\delta' \geq \delta$.

EXAMPLE 21.24. An $m \times m$ grid $G = P_m \square P_m$ is (ε, δ)-homogeneous if $\delta \leq \varepsilon^2/18$ and m is large enough. Indeed, suppose that $U \subseteq V(G)$ is an (ε, δ)-island. Consider any $r \geq 0$. We claim that most nodes of $G[U]$ are "orderly" in the sense that they have the same r-neighborhood as a node in an infinite grid. Indeed, if $v \in U$ is not orderly, then either it is closer to the boundary than r, or the r-ball around it contains one of the edges leaving U. It is easy to check that any edge leaving U can be counted at most $2r^2$ times, hence the number of "disorderly" nodes is at most $4rm + 2r^2 \delta m^2 < 3r^2 \delta m^2$ (if m is large enough). The proportion of "disorderly" nodes in the whole grid G is even smaller, and hence

$$\delta_\odot^r(G, G[U]) < \frac{3r^2 \delta m^2}{\varepsilon m^2} = \frac{3r^2 \delta}{\varepsilon}.$$

Summing,

$$\delta_\odot(G, G[U]) < \sum_{r=0}^{\infty} 2^{-r} \frac{3r^2 \delta}{\varepsilon} = \frac{18\delta}{\varepsilon} \leq \varepsilon.$$

This shows that U is not an (ε, δ)-island. ♦

THEOREM 21.25. *For every $\varepsilon > 0$ there is a $\delta > 0$ such that from every graph $G \in \mathcal{G}$ we can delete $\varepsilon \mathsf{v}(G)$ edges in such a way that every component of the remaining graph is (ε, δ)-homogeneous.*

The dependence of δ on ε is explicit, and only moderately bad: choose $r = 1 + \lceil \log(1/\varepsilon) \rceil$ (so that $2^r \approx \varepsilon/2$), let $b = |\mathfrak{B}_r| \leq D^{D^r}$ be the number of r-balls (Exercise 18.42), and define $\delta = \varepsilon^5/(4D^r b) = 2^{-1/\varepsilon^{O(\log D)}}$.

Proof. The proof follows the argument of Angel and Szegedy, which is reminiscent of the proof of the Regularity Lemma. By (19.3), an (ε, δ)-island U will satisfy $d_{\mathrm{var}}(\rho_{G[U],r}, \rho_{G,r}) \geq \varepsilon/2$.

For a graph $G \in \mathcal{G}$ with connected components $G_1, \ldots, G_k$, we define

$$f(G) = \sum_{i=1}^{k} \frac{\mathsf{v}(G_i)}{\mathsf{v}(G)} \sum_{B \in \mathfrak{B}_r} \rho_{G_i, r}(B)^2.$$

Trivially, $0 \leq f(G) \leq 1$.

Let, say, $G_1, \ldots, G_m$ be those components of G that are not (ε, δ)-homogeneous, and suppose that $\sum_{i=1}^{m} \mathsf{v}(G_i) = p > (\varepsilon/2D)n$. Let $V_i' \subseteq V(G_i)$ be an (ε, δ)-island, and let $V_i'' = V(G_i) \setminus V_i'$. Let C_i be the set of edges connecting V_i' and V_i'', then $|C_i| \leq \delta |V_i'|$. Finally, let G' be obtained from G by removing the edges in the sets

C_i. We want to show that if many of the parts are not ε-homogeneous, then $f(G')$ is substantially larger than $f(G)$. To keep the notation in check, set $G'_i = G[V'_i]$, $G''_i = G[V''_i]$, $n = \mathsf{v}(G)$, $n_i = \mathsf{v}(G_i)$, $n'_i = \mathsf{v}(G'_i)$ etc. Since the radius r is fixed, we don't have to show it in notation, and write $\rho_i = \rho_{G_i,r}$, $\rho'_i = \rho_{G'_i,r}$ etc.

Fix any $i \in [k]$ and any $B \in \mathfrak{B}_r$, and consider the difference of their contributions to $f(G')$ and $f(G)$:

(21.5)
$$\frac{n'_i}{n}\rho'_i(B)^2 + \frac{n''_i}{n}\rho''_i(B)^2 - \frac{n_i}{n}\rho_i(B)^2$$
$$= \frac{n'_i}{n}\big(\rho'_i(B) - \rho_i(B)\big)^2 + \frac{n''_i}{n}\big(\rho''_i(B) - \rho_i(B)\big)^2$$
$$+ \frac{2}{n}\rho_i(B)\big(n'_i\rho'_i(B) + n''_i\rho''_i(B) - n_i\rho_i(B)\big).$$

Here the first term will provide the gain, the second is nonnegative, while the third is an error term. To estimate the "gain" term, first we sum over all balls B:

$$\sum_B \big(\rho'_i(B) - \rho_i(B)\big)^2 \geq \frac{1}{b}\Big(\sum_B |\rho'_i(B) - \rho_i(B)|\Big)^2 = \frac{4}{b}d_{\text{var}}(\rho'_i, \rho_i)^2 \geq \frac{\varepsilon^2}{b}.$$

Summing over i and using that $n'_i \geq \varepsilon n_i$ by the definition of an island, we get

$$\sum_{i,B} \frac{n'_i}{n}\big(\rho'_i(B) - \rho_i(B)\big)^2 \geq \sum_i \frac{n_i}{n}\varepsilon\frac{\varepsilon^2}{b} \geq \frac{\varepsilon^4}{Db}.$$

To estimate the error term, we argue that the quantity $|n'_i\rho'_i(B) + n''_i\rho''_i(B) - n_i\rho_i(B)|$ is the increase or decrease in the number of neighborhoods isomorphic to B when the edges in C_i are deleted; since deletion of an edge can change at most $2D^r$ balls with radius r, we have

$$\sum_B |n'_i\rho'_i(B) + n''_i\rho''_i(B) - n_i\rho_i(B)| \leq 2D^r|C_i|,$$

and so the total contribution of the error term is at most

$$\frac{2D^r}{n}\sum_i |C_i| \leq 2D^r\delta < \frac{\varepsilon^4}{2Db}.$$

It follows that the value of $f(G)$ increases by at least $\varepsilon^4/(2Db)$. The number of edges deleted is at most $\sum_i |C_i| \leq \delta n$. We can repeat this until the number of nodes in non-(ε, δ)-homogeneous components drops below $\varepsilon n/D$. This happens after at most $2Db/\varepsilon^4$ repetitions (since $f(G) \leq 1$), and when we get stuck, the number of deleted edges is at most $(2Db\delta/\varepsilon^4)n < (\varepsilon/2)n$, and the number of nodes in those components that are not (ε, δ)-homogeneous is less than $(\varepsilon/D)n$. Deleting all edges in these components means the deletion of no more than $(\varepsilon/2)n$ further edges. This turns these remaining components into isolated nodes, which count as (ε, δ)-homogeneous components. This proves the theorem. $\square$

21.2.1. The quest for a regularity lemma.
Is there a good analogue of the Regularity Lemma for bounded degree graphs? The Regularity Lemma, as discussed in Chapter 9, does not say anything about non-dense graphs.

What do we expect from such a lemma? If we think about the many uses of the dense Regularity Lemma, there is no single answer to this question.

- It gives a partition of the nodes such that most bipartite graphs between different classes are homogeneous (random-like). Several extensions of the Regularity Lemma to sparse graphs in this sense are known (see e.g. Kohayakawa [1997], Gerke and Steger [2005], Scott [2011]), but they are more-or-less meaningless, or very weak, for graphs that have bounded degree.

- It gives a decomposition of the graph into simpler, homogeneous subgraphs. Theorem 21.25 describes such a decomposition. However, this result is clearly not the ultimate word: the (ε, δ)-homogeneous pieces it produces can still have a very complicated structure.

- It implies that an arbitrarily large (simple, dense) graph can be "scaled down" to a graph whose size depends on the error bound only, and which is almost indistinguishable from the original by sampling. Proposition 19.10 shows that such a "downscaling" is also valid for bounded degree graphs; Unfortunately, it is non-effective, and provides no algorithm for the construction of the smaller graph.

- It provides an approximate code for the graph, which has bounded size (depending on the error we allow), from which basic parameters of the graph can be reconstructed, and from which graphs can be generated on an arbitrary number of nodes that are almost indistinguishable from the original graph by sampling. In this sense, a Regularity Lemma may exist, and should be very useful once we learn how to work with it. While not quite satisfactory, I feel that the results mentioned above justify cautious optimism.

CHAPTER 22

Algorithms for bounded degree graphs

The algorithmic theory of large graphs with bounded degree is quite extensive. Similarly as in the case of dense graphs, we can formulate the problems of parameter estimation, property distinction, property testing, and computing a structure.

However, it seems that the theory in the bounded degree case is lacking the same sort of general treatment as dense graphs had, in the form of useful general conditions for parameter estimations (like Theorem 15.1), treatment of property distinction in the limit space (Section 15.3), and the use of regularity partitions and representative sets in the design of algorithms (Section 15.4). The most important tools that are missing are analogues of the Regularity Lemma and of the cut distance.

Our discussions in this chapter, accordingly, will be more an illustration of several interesting and nontrivial results than a development of a unifying theory. But even so, graph limit theory provides a useful point of view for these results.

22.1. Estimable parameters

We call a graph parameter defined on bounded degree graphs *estimable*, if it is bounded, and for every $\varepsilon > 0$ there is a positive integer k and an "estimator" function $g : (\mathfrak{B}_k)^k \to \mathbb{R}$ such that for every graph $G \in \mathcal{G}$ and uniform, independently chosen random nodes $v_1, \ldots, v_k \in V(G)$, we have

$$(22.1) \qquad \mathsf{P}\big(|f(G) - g(B_{G,k}(v_1), \ldots, B_{G,k}(v_k))| > \varepsilon\big) \leq \varepsilon.$$

In other words, g estimates f from a sample chosen according to the rules of sampling from a bounded degree graph. (For convenience, we use the same ε to bound the error in the function value and the probability of a large error; also the same k for the number of samples and the radius of balls we explore around the sampling points. In specific algorithms, one may want to distinguish these values, but this would not alter the notion of estimability.)

In the dense case, we did not need a separate estimator function g; we could use $g = f$. This is not the case here.

EXAMPLE 22.1. Let $f(G)$ be the fraction of nodes of G of degree 1. If G is a 3-regular graph with large girth, then every sample $B_{G,k}(v)$ is a tree with more than half of its nodes of degree 1; but G itself has no nodes of degree 1. ♦

It is also easy to see that it would not be enough to use just one sample ball.

EXAMPLE 22.2. Let G be a 2-regular graph on n nodes, G', a 3-regular graph on n nodes, and GG', their disjoint union. Let the parameter to estimate be the average degree. In a single sample you see only nodes of degree 2 or nodes of degree 3, no matter how far you explore the graph. No matter how large neighborhoods you take, and what function of them you compute, this single bit of information (degree 2 or degree 3) will not distinguish three possibilities (G, G' and GG'). ♦

On the other hand, some other facts extend from the dense case with more or less difficulty. The following theorem of Elek [2010a] connects parameter estimation with convergence (recall that the analogous result for dense graphs was trivial).

THEOREM 22.3. *A bounded graph parameter f is estimable if and only if for every locally convergent graph sequence (G_n), the sequence of numbers $(f(G_n))$ is convergent.*

Proof. The "only if" part is easy: from similar graphs we get similar samples and so we compute similar estimates. Let us make this precise. Suppose that f is estimable, and let (G_n) be a locally convergent graph sequence. Let $0 < \varepsilon < 1/8$, we want to show that $|f(G_n) - f(G_m)| < \varepsilon$ if n, m are large enough. By the definition of estimability, we have a positive integer k and an estimator function $g : (\mathfrak{B}_k)^k \to \mathbb{R}$ such that (22.1) holds. If n, m are large enough, then $\delta_\odot(G_n, G_m) \leq 1/(4k2^k)$, and hence $d_{\mathrm{var}}(\rho_{k,G_n}, \rho_{k,G_m}) \leq 1/(4k)$. This means that we can couple a random node $\mathbf{v} \in V(G_n)$ with a random node $\mathbf{u} \in V(G_m)$ so that $B_{G_n,k}(\mathbf{v}) \cong B_{G_m,k}(\mathbf{u})$ with probability at least $1 - 1/(4k)$. If we sample k independent nodes $\mathbf{v}_1, \ldots, \mathbf{v}_k$ from G_n and k independent nodes $\mathbf{u}_1, \ldots, \mathbf{u}_k$ from G_m, then with probability more than $3/4$, we have $B_{G_m,k}(\mathbf{u}_1) \cong B_{G_n,k}(\mathbf{v}_1), \ldots, B_{G_m,k}(\mathbf{u}_k) \cong B_{G_n,k}(\mathbf{v}_k)$. With positive probability, we have simultaneously $B_{G_m,k}(\mathbf{u}_1) \cong B_{G_n,k}(\mathbf{v}_1), \ldots, B_{G_m,k}(\mathbf{u}_k) \cong B_{G_n,k}(\mathbf{v}_k)$, $|f(G_n) - g(B_{G_n,k}(\mathbf{v}_1), \ldots, B_{G_n,k}(\mathbf{v}_k))| \leq \varepsilon$ and $|f(G_m) - g(B_{G_m,k}(\mathbf{u}_1), \ldots, B_{G_m,k}(\mathbf{u}_k))| \leq \varepsilon$. But in this case we have $|f(G_n) - f(G_m)| \leq 2\varepsilon$, which we wanted to prove.

The converse is a bit trickier. Suppose that $(f(G_n))$ is convergent for every locally convergent graph sequence (G_n). Given $\varepsilon > 0$, we want to find a suitable positive integer k and construct an estimator $g : (\mathfrak{B}_k)^k \to \mathbb{R}$. The condition on f implies that for every $\varepsilon > 0$ there is an $\varepsilon' > 0$ such that if $\delta_\odot(G, G') \leq \varepsilon'$ then $|f(G) - f(G')| \leq \varepsilon$. Let r be chosen so that $2^{1-r} < \varepsilon'$, and let $k > 2r/(\varepsilon \varepsilon')$.

The estimator we construct will only depend on the r-balls around the roots of the k-balls. So we will construct a function $g : (\mathfrak{B}_r)^k \to \mathbb{R}$. For every sequence $\mathbf{b} = (B_1, \ldots, B_k) \in (\mathfrak{B}_r)^k$, let $\rho_\mathbf{b}$ denote the distribution of a randomly chosen element of the sequence. We define the estimator as follows:

$$g(\mathbf{b}) = \begin{cases} f(G) & \text{where } G \text{ is any graph with } d_{\mathrm{var}}(\rho_{G,r}, \rho_\mathbf{b}) \leq \varepsilon'/4, \\ & \text{if such a graph exists,} \\ 0 & \text{otherwise.} \end{cases}$$

To show that this is a good estimator, let $G \in \mathcal{G}$ be any graph, and let $\mathbf{v}_1, \ldots, \mathbf{v}_k \in V(G)$ be uniformly chosen random nodes, and let $\mathbf{b} = (B_{G,r}(\mathbf{v}_1), \ldots, B_{G,r}(\mathbf{v}_k))$. By the choice of k, elementary probability theory gives that with probability at least $1 - \varepsilon$, we have $d_{\mathrm{var}}(\rho_\mathbf{b}, \rho_{G,r}) \leq \varepsilon'/4$. If this happens, then in the definition of $g(\mathbf{b})$ the first alternative applies, and so $g(\mathbf{b}) = f(G')$ for some graph G' that satisfies $d_{\mathrm{var}}(\rho_{G',r}, \rho_\mathbf{b}) \leq \varepsilon'/4$. This implies that $d_{\mathrm{var}}(\rho_{G',r}, \rho_{G,r}) \leq \varepsilon'/2$. Then we have by (19.3)

$$\delta_\odot(G, G') \leq \frac{1}{2^r} + \frac{\varepsilon'}{2} \leq \varepsilon'.$$

By the definition of ε', this implies that $|f(G) - f(G')| \leq \varepsilon$. □

COROLLARY 22.4. *For every estimable graph parameter f there exists a graphing parameter $\widehat{f}$ that is continuous in the $\delta_\odot$ distance such that $f(G_n) \to \widehat{f}(\mathbf{G})$ whenever $G_n \to \mathbf{G}$.*

Notice that continuity in the $\delta_\odot$ distance implies invariance under local equivalence.

Proof. It is easy to see (using Theorem 22.3) that the parameter $\widehat{f}$ is uniquely determined for graphings that represent limits of convergent graph sequences, and it is continuous in the $\delta_\odot$ distance. However, we don't know if all graphings are like that (cf. Conjecture 19.8). To complete the proof, we can use Tietze's Extension Theorem to extend the definition of $\widehat{f}$ to all graphings. □

This possible non-uniqueness of the extension may be connected with the fact that it is typically not easy to see the "meaning" of the extension of quite natural graph parameters (cf. Supplement 20.17).

Our discussion in Section 20.2 shows that parameters of the type $G \mapsto \text{ent}^*(G, H) = \log t(G, H)/\mathsf{v}(G)$ are estimable provided the weighted graph H is sufficiently dense. In the next two sections we describe a couple of further interesting examples of estimable graph parameters.

Not all natural parameters of graphs are estimable (see Examples 20.16 and 22.5). However, it was shown by Elek [2010a] that if we restrict ourselves to testing properties on hyperfinite graphs, then many of these become testable. The method is similar to property testing for hyperfinite graphs, which will be discussed later.

EXAMPLE 22.5 (**Independence ratio**). Recall that $\alpha(G)$ denotes the maximum size of a stable set in graph G. The independence ratio $\alpha(G)/\mathsf{v}(G)$ is not estimable. Let $\mathbb{G}_n$ be a random D-regular graph on $2n$ nodes, and $\mathbb{G}'_n$ be a random bipartite D-regular graph on $2n$ nodes. It is clear that $\alpha(\mathbb{G}'_n) = n$. In contract, $\alpha(\mathbb{G}_n) \leq (1 - 2c_D)n$ with high probability, where $c_D > 0$ depends only (Bollobás [1980]). The interlaced sequence $(\mathbb{G}_1, \mathbb{G}'_1, \mathbb{G}_2, \mathbb{G}'_2, \dots)$ is locally convergent (as discussed in Example 19.7), but the independence ratios oscillate between $1/2$ and something less than $\frac{1}{2} - c_D$, so they don't converge.

However, it we restrict ourselves to the sequence $\mathbb{G}_n$, then the independence ratios form a convergent sequence; this is a recent highly nontrivial result of Bayati, Gamarnik and Tetali [2011]. ♦

22.1.1. Number of spanning trees.
Lyons [2005] proved that the number of spanning trees $\text{tree}(G)$, suitably normalized, is an estimable parameter of bounded degree graphs. He in fact proved a more general result, allowing the degrees to be unbounded, as long as the average degree remains bounded and the degrees don't vary too much; we treat the bounded case only, and refer for the exact statement of the more general result to the paper.

Let G be a connected graph with n nodes and m edges, whose degrees are bounded by D as always in this part of the book. It is easy to see that $\text{tree}(G) \leq D^{\mathsf{v}(G)}$; a bit sharper,
$$\text{tree}(G) \leq \prod_{v \in V(G)} \deg(v),$$
whence
$$\frac{1}{n} \log \text{tree}(G) \leq \frac{1}{n} \sum_{v \in V(G)} \log \deg(v).$$

The right hand side is clearly bounded and estimable, which reassures us that we have the right normalization.

THEOREM 22.6. *The graph parameter* $(\log \mathsf{tree}(G))/\mathsf{v}(G)$ *is estimable for connected bounded degree graphs.*

Proof. Let G be a connected graph with all degrees bounded by D. It will be convenient to choose D generously so that all degrees are in fact at most $D/2$. We add $D - \deg(v)$ loops to each node v, to make the graph regular (here, a loop adds only 1 to the degree) and also to make sure that its adjacency matrix A is positive semidefinite. This does not change the number of spanning trees, and from the samples of the original graph the samples of this augmented graph are easily generated just by adding loops.

We start with developing formulas for $\mathsf{tree}(G)$ and its logarithm. Here we face an embarrassment of riches: there are many formulas for $\mathsf{tree}(G)$ in the literature, and possibly others would also work. We use (5.41), which we write as

$$(22.2) \quad \frac{1}{n} \log \mathsf{tree}(G) = \frac{n-1}{n} \log D - \frac{\log n}{n} - (\log e) \sum_{r=1}^{\infty} \frac{1}{r} \left(\frac{t^*(C_r, G)}{D^r} - \frac{1}{n} \right).$$

For every fixed r, the quantity $t^*(C_r, G) - 1/n$ is estimable. Since the other terms in (22.2) are trivially estimable, we are almost done. But the problem is that we have an infinite sum, and we need a convergent majorant. (This is where it becomes important that we have subtracted $1/(nr)$ in every term!)

LEMMA 22.7. *For any $r \geq 0$ and $v \in V(G)$,*

$$\frac{1}{n} \leq \frac{\mathsf{hom}_v(C_r^\bullet, G)}{D^r} \leq \frac{1}{n} + \frac{2D^{1/3}}{(r+1)^{1/3}}.$$

(It may help with the digestion of this formula that $D^r = \mathsf{hom}_v(P_r^\bullet, G)$, and so the ratio in the middle expresses the probability that a random walk started at v returns to v after r steps. Since the endpoint of a random walk becomes more and more independent of the starting point, this probability tends to $1/n$ by elementary properties of random walks. The main point is that the upper bound gives a uniform bound on the rate of this convergence.)

Averaging over all nodes v, the lemma implies that

$$(22.3) \quad 0 \leq \frac{t^*(C_r, G)}{D^r} - \frac{1}{n} \leq \frac{2D^{1/3}}{(r+1)^{1/3}}.$$

This gives a convergent majorant, independent of G, for the infinite sum in (22.2), which proves that $(1/n) \log \mathsf{tree}(G)$ is estimable. $\square$

Proof of Lemma 22.7. Let $P = (1/D)A$ (this is the transition matrix of the random walk on G), and $\mathbf{y}_r = P^r \mathbb{1}_v$ (this is the distribution of a random walk after r steps). Clearly $t_v^*(C_r^\bullet, G)/D^r = \mathbb{1}_v^\mathsf{T} P^r \mathbb{1}_v = \mathbb{1}_v \mathbf{y}_r$. Since P is positive semidefinite, we see from here that the values $\mathbf{y}_r(v)$ are monotone decreasing, and since $P^r \to \frac{1}{n} J$ (where J is the all-1 matrix), $\mathbf{y}_r(v) \to \frac{1}{n}$ as $r \to \infty$. This implies the lower bound in the lemma.

To get the upper bound, we note that

$$(22.4) \quad \sum_{t=0}^{\infty} \mathbf{y}_t^\mathsf{T}(I - P)\mathbf{y}_t + \sum_{t=0}^{\infty} \mathbf{y}_t^\mathsf{T}(P - P^2)\mathbf{y}_t = 1 - \frac{1}{n}$$

Indeed, the matrices $I - P$ and $P - P^2$ are positive semidefinite, hence all terms here are nonnegative. Furthermore, $P\mathbf{y}_t = \mathbf{y}_{t+1}$, so if we stop the sums at m steps, then the middle terms telescope out, and we are left with $\mathbf{y}_0^\mathsf{T}\mathbf{y}_0 - \mathbf{y}_{m+1}^\mathsf{T}\mathbf{y}_{m+1}$, where $\mathbf{y}_0^\mathsf{T}\mathbf{y}_0 = 1$ and $\mathbf{y}_{m+1}^\mathsf{T}\mathbf{y}_{m+1} \to 1/n$.

From (22.4) it follows that there is a $t \leq r$ such that $\mathbf{y}_t^\mathsf{T}(I - P)\mathbf{y}_t \leq 1/(r+1)$. Let $x = \frac{1}{2}(\mathbf{y}_t(v) + \frac{1}{n})$, and let u be the closest node to v with $\mathbf{y}_t(u) \leq x$. Consider a shortest path $v_0 v_1 \ldots v_k$, where $v_0 = v$ and $v_k = u$. Since $\mathbf{y}_t(v_0), \ldots, \mathbf{y}_z(v_{k-1}) \geq x$, we must have $k \leq 1/x$. On the other hand,

$$\begin{aligned}
\bigl(\mathbf{y}_t(v) - x\bigr)^2 &\leq \bigl(\mathbf{y}_t(v_0) - \mathbf{y}_t(v_k)\bigr)^2 \\
&= \bigl((\mathbf{y}_t(v_0) - \mathbf{y}_t(v_1)) + \cdots + (\mathbf{y}_t(v_{k-1}) - \mathbf{y}_t(v_k))\bigr)^2 \\
&\leq k\bigl((\mathbf{y}_t(v_0) - \mathbf{y}_t(v_1))^2 + \cdots + (\mathbf{y}_t(v_{k-1}) - \mathbf{y}_t(v_k))^2\bigr) \\
&\leq Dk\mathbf{y}_t^\mathsf{T}(I - P)\mathbf{y}_t \leq \frac{Dk}{r+1} \leq \frac{D}{x(r+1)}.
\end{aligned}$$

Hence

$$(22.5) \qquad (\mathbf{y}_t(v) - x)^2 x \leq \frac{D}{r+1}.$$

Substituting the definition of x, we get

$$\left(\mathbf{y}_t(v) - \frac{1}{n}\right)^3 \leq 8(\mathbf{y}_t(v) - x)^2 x \leq \frac{8D}{r+1}.$$

Since we know that $\mathbf{y}_r(v) \leq \mathbf{y}_t(v)$, this proves the lemma.

We note that Lyons gets a better estimate, with $1/2$ in the exponent of $r+1$ rather than $1/3$, but the simpler bound above was good enough for our purposes. □

Once we know that the graph parameter $(\log \mathsf{tree}(G))/\mathsf{v}(G)$ is estimable, we also know that if G_n is a locally convergent graph sequence, then $(\log \mathsf{tree}(G_n))/\mathsf{v}(G_n)$ tends to a limit. From the proof, it is not difficult to figure out what the limiting graphing parameter (or involution-invariant-distribution-parameter) is. We formulate the answer for a graphing, but it is easy to translate this to the Benjamini–Schramm model. Given a graphing $\mathbf{G}$ and a number D such that $D/2$ is an upper bound on the degrees, pick a random node x, and start a random walk from x, where you have to add $D - \deg(y)$ loops to node y as you go along. Let X_r be the indicator that the random walk returns to x after r steps (not necessarily the first time). With this notation, we have

$$(22.6) \qquad \frac{\log \mathsf{tree}(G_n)}{\mathsf{v}(G_n)} \longrightarrow \log D - \sum_{r=1}^{\infty} \frac{1}{r}\mathsf{E}(X_r).$$

The expression on the right describes the limit as a function of the limiting graphing. (Note that its value may be $-\infty$.)

EXERCISE 22.8. Suppose that we want to estimate the variance of the degrees, i.e., $\sum_{v \in V(G)}(\deg(v) - d_0)^2/\mathsf{v}(G)$, where d_0 is the average degree. (a) Show that this parameter is estimable. (b) Prove that we cannot estimate it using an estimator of the form $g(B_{G,k}(v_1), \ldots, B_{G,k}(v_k)) = (h(B_{G,k}(v_1)) + \cdots + h(B_{G,k}(v_k)))/k$ with any function $h: \mathfrak{B}_k \to \mathbb{R}$.

22.2. Testable properties

22.2.1. Distinguishing properties. Similarly as in the dense case (Section 15.2), our next step is discuss the problem of distinguishing two disjoint graph properties. The setup is very similar to the dense case. Let $\mathcal{P}_1, \mathcal{P}_2 \subseteq \mathcal{G}$ be two graph properties with $\mathcal{P}_1 \cap \mathcal{P}_2 = \emptyset$. We call properties $\mathcal{P}_1$ and $\mathcal{P}_2$ *distinguishable by sampling*, if there exist positive integers r and k, and a property $\mathcal{Q}$ of k-tuples of r-balls, such that for every graph $G \in \mathcal{G}$ and random nodes $v_1, \ldots, v_k$,

$$\mathsf{P}\big((B_{G,r}(v_1), \ldots, B_{G,r}(v_k)) \in \mathcal{Q}\big) \begin{cases} \geq \dfrac{2}{3}, & \text{if } G \in \mathcal{P}_1, \\ \leq \dfrac{1}{3}, & \text{if } G \in \mathcal{P}_2. \end{cases}$$

The following analogue of Theorem 15.8 is due to Benjamini, Schramm and Shapira [2010].

THEOREM 22.9. *Two graph properties* $\mathcal{P}_1, \mathcal{P}_2 \subseteq \mathcal{G}$ *are distinguishable by sampling if and only if* $\delta_\odot(\mathcal{P}_1, \mathcal{P}_2) > 0$.

Proof. The necessity of the condition is easy to prove, along the same lines as it was done for dense graphs in the proof of Theorem 15.8; we don't go into the details. The sufficiency is done differently, and unfortunately it is not constructive.

Suppose that $\delta_\odot(G_1, G_2) = \varepsilon > 0$. Similarly as in the proof of Proposition 19.10, we can select a finite set $\mathcal{Q}_1 \subseteq \mathcal{P}_1$ of graphs such that for every graph $G \in \mathcal{P}_1$ there is a graph $H \in \mathcal{Q}_1$ such that $\delta_\odot(G, H) \leq \varepsilon/4$. Now if we want to decide whether $G \in \mathcal{P}_1$ or $G \in \mathcal{P}_2$, then we compute $\delta_\odot(G, H)$ with error less than $\varepsilon/4$ for all $H \in \mathcal{Q}_1$. (This can be done with error probability less than $1/3$ by taking a large enough sample of large enough balls.) If there is an H for which we find that $\delta_\odot(G, H) \leq \varepsilon/2$, we conclude that $G \in \mathcal{P}_1$. If no such H exists, we conclude that $G \in \mathcal{P}_2$. It is straightforward to check that if $G \in \mathcal{P}_i$, then the answer will be correct with probability larger than $2/3$. □

22.2.2. Testable properties and their closures. Now we come to testing a single property $\mathcal{P}$. Just as in the dense case, we either want to conclude that a given graph G does not have the property, or that one can change a small number of adjacencies so that the property is restored. To be more precise, let $\overline{\mathcal{P}}_\varepsilon = \{G \in \mathcal{G} : d_1(G, \mathcal{P}) > \varepsilon\}$. We say that a property $\mathcal{P}$ of bounded degree graphs is *testable* if for every $\varepsilon > 0$ there are integers $r = r(\varepsilon) \geq 1$ and $k = k(\varepsilon)$ such that sampling k neighborhoods of radius r from a graph $G \in \mathcal{G}$, we can compute "YES" or "NO" so that:

(a) if $G \in \mathcal{P}$, then the answer is "YES" with probability at least $2/3$;

(b) if $G \notin \overline{\mathcal{P}}_\varepsilon$, then the answer is "NO" with probability at least $2/3$.

EXAMPLE 22.10 (**Forests**). Let us look at a simple example that illustrates some of the difficulties in designing algorithms for property testing for graphs with bounded degree, even for monotone properties. Suppose that we want to test whether a graph G is a forest. Our first thought might be to test whether a random ball contains a cycle. Certainly, if it does, then the graph is not a forest. But drawing a conclusion in the other direction is not justified: if the graph G has large girth, then every ball will be tree, and G would be very far from being a forest. This shows that (unlike in the dense case) $\mathcal{P}$ is not a good test property for itself. If in addition to this we estimate the average degree and eliminate small components, we can design a test for being a forest (Goldreich and Ron [2008]). To fill in the details makes an interesting exercise. ♦

We can use limit objects to give the following condition for testability (the proof is immediate).

PROPOSITION 22.11. *A graph property $\mathcal{P}$ is not testable if and only if there exists an $\varepsilon > 0$ and two convergent sequences of graphs (G_n) and (H_n) with $G_n \in \mathcal{P}$ and $d_1(H_n, \mathcal{P}) > \varepsilon$ that have a common local limit.*

22.2.3. Hyperfinite properties. Hyperfiniteness is particularly important in property testing. Just as in the dense case, property testing is about the interplay between the sampling distance and the edit distance, and these two distances are intimately related for hyperfinite graphs.

Benjamini, Schramm and Shapira show that hyperfiniteness is in a sense testable. This does not make sense as said, since hyperfiniteness is a property of a family of graphs, not of a single graph. But if we quantify hyperfiniteness, then we can turn it into a meaningful statement.

PROPOSITION 22.12. *For every ε there is an ε' such that for any positive integer k, the properties $\mathcal{P}_1 = \{(\varepsilon', k)\text{-hyperfinite}\}$ and $\mathcal{P}_1 = \{\text{not } (\varepsilon, k)\text{-hyperfinite}\}$ are distinguishable.*

Proof. Suppose that this is false, then by Theorem 22.9 there exist an $\varepsilon > 0$, a sequence $\varepsilon_n \to 0$, graphs $G_n, G'_n \in \mathcal{G}$ and positive integers k_n such that G_n is (ε_n, k_n)-hyperfinite, G'_n is not (ε, k_n)-hyperfinite, and $\delta_\odot(G_n, G'_n) \to 0$. Then the sequence (G_n) is hyperfinite. Let us select a convergent subsequence, then the limit graphing of this subsequence is hyperfinite by Theorem 21.13. But the sequence (G'_n) has the same limit, so it must be hyperfinite, by the same theorem. This implies that there is a positive integer k such that all members of the sequence are (ε, k)-hyperfinite. Since (G'_n) is not (ε, k_n)-hyperfinite, it follows that $k_n < k$ for all n. This implies that almost all connected components of $\mathbf{G}$ have at most k elements, but then it follows from $G_n \to \mathbf{G}$ that all but a $o(1)$ fraction of connected components of G'_n have at most k elements, which implies that (G'_n) is an (ε, k)-hyperfinite sequence. □

Benjamini, Schramm and Shapira [2010] proved an important analogue of Theorem 15.24: every minor-closed property of bounded degree graphs is testable. As noted by Elek, the theorem can be extended to any monotone hyperfinite graph property.

THEOREM 22.13. *Every monotone hyperfinite property of graphs with bounded degree is testable.*

The property of being a forest is certainly minor-closed, so the example discussed at the end of the introduction above is a special case. As another special case, planarity of bounded degree graphs is testable.

Proof. Let $\mathcal{P}$ be a monotone hyperfinite graph property, and suppose that it is not testable. Then there exist an $\varepsilon > 0$ and two sequences of graphs (G_n) and (F_n) such that $G_n \in \mathcal{P}$, $d_1(F_n, \mathcal{P}) > \varepsilon$ and $\delta_\odot(G_n, F_n) \to 0$. We may assume that both sequences are locally convergent, and so they have a common weak limit graphing

G. Since $\mathcal{P}$ is hyperfinite, so is the sequence (G_n). Theorem 21.13 implies that **G** is hyperfinite, and applying this theorem again, we get that (F_n) is hyperfinite. By Theorem 21.19, the interlaced sequence $G_1, F_1, G_2, F_2, \ldots$ is locally-globally convergent, and hence we may assume that $G_n, F_n \to \mathbf{G}$ in the local-global sense.

Since **G** is hyperfinite, it has a Borel 2-coloring $\alpha : V(\mathbf{G}) \to [2]$ such that $\lambda(\alpha^{-1}(1)) \leq \varepsilon'$ and there is a $k \in \mathbb{N}$ such that every connected subgraph with $k+1$ nodes contains a point v with $\alpha(v) = 1$. By local-global convergence, if n is large enough, then G_n has a 2-coloring $\alpha_n : V(G_n) \to [2]$ such that $\delta_\odot^k((\mathbf{G}, \alpha), (G_n, \alpha_n)) \leq \varepsilon'/k$. In particular, $|\alpha_n^{-1}(1)| \leq 2\varepsilon' \mathsf{v}(G_n)$ and the union of connected subgraphs with $k+1$ nodes that contain no node with color 1 has at most $\varepsilon' \mathsf{v}(G_n)$ nodes. The graph F_n has a 2-coloring β_n with similar properties. It follows that $\delta_\odot^k((G_n, \alpha_n), (F_n, \beta_n)) \leq 2\varepsilon'$.

Let G'_n be obtained from G_n by deleting all edges incident with any node of S_n as well as all edges in connected components of $G_n[T_n]$ that have more than k nodes. This way we delete at most $3\varepsilon' D \mathsf{v}(G_n)$ edges. Furthermore, every connected component of G'_n has at most k nodes. We define F'_n analogously.

It is important that whether or not an edge is deleted is determined locally, which implies that whenever $v \in V(G_n)$ and $u \in V(F_n)$ satisfied $B_{G_n, \alpha_n, k}(v) \cong B_{F_n, \beta_n, k}(u)$, then also $B_{G'_n, \alpha_n, k}(v) \cong B_{F'_n, \beta_n, k}(u)$, which means simply that the connected component of G'_n containing v is isomorphic to the connected component of F'_n containing u. Hence $\delta_\odot^k(G'_n, F'_n) \leq \delta_\odot^k((G_n, \alpha_n), (F_n, \beta_n)) \leq 2\varepsilon'$. Let $\mathcal{Y}_k$ denote the set of connected graphs with at most k nodes (up to isomorphism), let a_Y denote the number of connected components of G'_n isomorphic to $Y \in \mathcal{Y}_k$, and let b_Y be defined analogously. In these terms, we have

$$\sum_{Y \in \mathcal{Y}_k} \left| \frac{a_Y \mathsf{v}(Y)}{\mathsf{v}(G_n)} - \frac{b_Y \mathsf{v}(Y)}{\mathsf{v}(F_n)} \right| \leq 4\varepsilon'.$$

We may assume that $\mathsf{v}(G_n) \geq \mathsf{v}(F_n)$. Let $c_Y = \min(b_Y, \lfloor a_Y \mathsf{v}(F_n)/\mathsf{v}(G_n) \rfloor)$. Let us keep c_Y copies of every $Y \in \mathcal{Y}_k$ in F'_n and delete the edges of the rest, to get a graph F''_n. The number of edges to delete is bounded by $2\varepsilon' D \mathsf{v}(F_n) + Dk|\mathcal{Y}_k| < (\varepsilon/2) \mathsf{v}(F_n)$ if n is large enough, and so $d_1(F_n, F''_n) \leq 3\varepsilon' D + \varepsilon/2 < \varepsilon$. Furthermore, F''_n is isomorphic to a subgraph of G_n, and hence $F''_n \in \mathcal{P}$ by monotonicity. This implies that $d_1(F_n, \mathcal{P}) < \varepsilon$, a contradiction. $\square$

COROLLARY 22.14. *Every minor-closed property of graphs with bounded degree is testable.*

Monotonicity of the property $\mathcal{P}$ was used in the proof above only in a somewhat annoying technical way, and one would like to extend the argument to all hyperfinite properties $\mathcal{P}$. One must be careful though: the property that "G is a planar graph with an even number of nodes" is hyperfinite, but not testable (a large grid with an even number of nodes cannot be distinguished from a large grid with an odd number of nodes by neighborhood sampling). But the method works with a little twist: suppose that two graphs F and G have the same number of nodes, and we know that $G \in \mathcal{P}$, and the sampling distance of G and F is small; then it follows that the edit distance of F from $\mathcal{P}$ is small. This implies that $\mathcal{P}$ is testable in a non-uniform sense. For an exact formulation and details, see Newman and Sohler [2011].

EXERCISE 22.15. Let G and G' be (ε, k)-hyperfinite graphs with the same number of nodes n. Prove that they can be overlayed so that

$$\frac{1}{n}|E(G) \triangle E(G')| \leq 2(1 + D^k)\varepsilon + D\delta_\odot^k(G, G').$$

22.3. Computable structures

Suppose that we want to compute some structure on a very large graph with bounded degree: say, a maximum matching, a maximum flow, a spanning tree, a maximum cut, a 3-coloring. Even if we can compute (approximately) the appropriate number, what does it mean to compute this object? Similarly as in the dense case, the answer is not obvious.

One possible answer is similar to what we did in the dense case. We offer a service: if somebody comes with a question about a particular node v ("How is this node matched in your maximum matching?"), we can answer this question just by inspecting a bounded neighborhood of v (determine the mate u of v in the matching, or conclude that v is unmatched). These answers must be consistent, so for example in the case of matchings, if somebody comes with the request concerning the node u, we must match it with v. Furthermore, the proportion of unmatched nodes should be at most ε higher than for the true maximum matching.

In the bounded-degree world, however, there is another, equivalent model, that is perhaps easier to understand and analyze. Let us place an "agent" on every node of the graph G. These agents are allowed to communicate with each other, but only with their neighbors along the edges and for a bounded time. At the end, they have to decide whether they would be matched with any neighbor at all, and if so, to which of their neighbors. This model is called *distributed computing*.

The two models are essentially equivalent.

• Suppose that the agents can compute something; then in the "service" model, if somebody comes with a question about a node v, we just look up what our agent responsible for v has computed. All the information the agent has collected from the neighbors can be gathered by exploring a bounded neighborhood of the node.

• Suppose that we can provide the service correctly. Then we can instruct each agent to do the computation we would have done if the node they are responsible for were queried. Of course, the agent has to gather all the information about the neighborhood we would have explored, but this can be done by communicating with his/her neighbors. Of course, we also have to instruct them to provide and communicate the information that is needed for this. All of this takes a bounded number of bits. We assume here that the agents can generate a name for themselves that identifies them, at least locally. We don't go into other details of this model, like whether the communication between agents is synchronized (sending a bit on every tick of the clock).

We will in fact use a kind of hybrid description of the algorithms, where the agents are allowed to explore their neighborhood to a bounded depth (including the colors and weights of the other agents in this neighborhood, which we have to use in some cases). Using results of Nguyen and Onak [2008] and Csóka [2012a] we illustrate the power and some of the subtleties of these models.

Symmetry. There is a fundamental difficulty with distributed algorithms: symmetry. Suppose that we want to construct a matching in a very long cycle. All our agents see the same neighborhood of any given radius, so they will all compute the same answer: that they want to remain unmatched (which gives an empty matching, very far from being optimal). Symmetry does not allow them to give any other answer, at least deterministically.

We can break the symmetry and find a matching close to the optimum, if we allow the agents to flip coins. We can consider the coinflips generated by any agent as a real number between 0 and 1, and call this the *local random seed* of the agent. This takes an infinite number of coinflips, but only a finite and bounded number of them has to be generated during the run of the algorithm.

Preprocessing. In our examples for computing a structure in the dense case, preprocessing (computing a representative set) played a large role. In the bounded degree case, there is less room for preprocessing. We can think of two kinds of preprocessing:

• We can do some preliminary computation (perhaps randomized) independently of the graph, and inform the agents about the result. If we are lazy, we can just let the agents do this computation for themselves. The only information they need for this is the random seed we use during the computation. So it suffices to generate a random number in $[0,1]$ and tell it to all the agents. We call this number the *global random seed*. (Note: they could generate a random number themselves, but this would not be the same for all agents!)

• We can do preliminary computation using information about the graph. This could be based on the distribution of r-balls in G for some fixed r (which is the realistic possibility for us to obtain information about the graph), but perhaps we have some other information about the graph (like somebody tells us that it is connected). Again, we can let the agents work, just have to pass on to them the information about the graph they need. In the strongest form, we let the agents know what the graph is (up to isomorphism).

The task. Assume that our agents have to compute a decoration $f : V(G) \to C$, where C is a finite set. Not all decorations will be feasible, but we assume that the feasibility criterion is local, i.e., there is an $r \in \mathbb{N}$ and a set of feasible C-decorated r-neighborhoods such that a decoration is feasible if and only if every r-neighborhood is feasible.

The goal is to find an "optimal" decoration. The decoration is evaluated locally in the following sense: we associate a value $\omega(B) \in [0,1]$ to every C-decorated r-ball $F \in \mathcal{F}$, and we want to minimize the average value of r-balls. Setting $\omega(v) = \omega(B_{G,r}(v), f\mid_{B_{G,r}(v)})$, the *cost* of the decoration is defined by

$$w(f) = \frac{1}{\mathsf{v}(G)} \sum_{v \in V(G)} \omega(v).$$

The agents want to compute a decoration f for which $w(f)$ is as small as possible.

EXAMPLE 22.16 (**Proper coloring**). Suppose that we want to compute a proper k-coloring of G. Then we choose $C = [K]$ for some very large K, the feasibility criterion is that the coloring should be proper (clearly this can be verified from the 1-neighborhoods), and we evaluate the coloring by imposing a penalty of 1 on every node with color larger than k. ♦

EXAMPLE 22.17 (**Maximum matching**). Suppose that we want to compute a maximum matching in G. Then we can take $C = [D+1]$. Decoration with i,

$i \leq D$, means that the node is matched with its i-th largest neighbor (in the order of their local seeds); decoration with $D+1$ means that the node is unmatched. The feasibility criterion is clearly local. We impose a penalty of 1 for decoration by $D+1$. ♦

EXAMPLE 22.18 (**Max-flow-min-cut**). Suppose that we are given a 3-coloring of the nodes of a graph G by red, white and green. We consider all edges to have capacity 1, and would like to find a maximum flow from the red nodes to the green nodes. This means a decoration of every node v by a rational vector $f(v) = (f_1, \ldots, f_D)$, where f_i is the flow it is sending to its i-th highest weighted neighbor (this can be negative or positive). The sum of entries of $f(v)$ is the *gain* $\gamma(v)$ of the node. Feasibility means that the flow on any edge uv, indicated in the decoration of u, is the negative of the flow on this edge indicated in the decoration of v; furthermore, the gain is 0 at every white node, nonnegative at every red node, and nonpositive at every green node. The objective function is the sum of $\gamma(v)$ over the red nodes.

Computing the minimum cut fits in the framework quite easily too: We decorate every node by either "LEFT" or "RIGHT". Feasibility means that all red nodes are decorated by "LEFT" and all green nodes are decorated by "RIGHT". The objective value is half of the average number of neighbors of a node on the other side. ♦

The computational model. We compare four settings, getting increasingly more powerful:

(A) The agents don't get any preprocessing information, and have to work deterministically;

(B) The agents don't get any preprocessing information, but have access to their own random number generator;

(C) The agents have access to their own random number generator, and in addition they get the same global seed chosen uniformly from $[0, 1]$;

(D) The agents have access to their own random number generator, to the public random number $\mathbf{g}_0$ as in (B), and in addition they know the graph up to isomorphism (but they don't know at which node of the graph they sit).

Note that in models (A)-(C), the agents can see their own r-neighborhood, and can hear about other r-neighborhoods at a bounded distance from them, but they will not be able to learn global statistics. The agents themselves will not know, for example, the average degree of the graph. In model (D), one may be concerned how an arbitrarily large graph can be communicated to our agents; instead, we could say that this model allows any kind of information (any number of graph parameters) to be passed to the agents: the most natural would be neighborhood statistics, but the model allows non-testable graph parameters and properties like the chromatic number or connectivity to be used.

We will see that there are nontrivial algorithms in the weakest model (A); that (B) is strictly stronger than (A), and (C) is strictly stronger than (B); but every problem solvable in (D) is also solvable in (C) with an arbitrarily small increase in the cost.

22.3.1. Matchings. We start with describing an algorithm in model (B), designed by Nguyen and Onak, [2008] to find an (almost) maximum matching.

ALGORITHM 22.19.

Input: A graph G with maximum degree D and no isolated nodes in the agent model, and an error bound ε.

Output: A random matching M such that with probability at least $1 - \varepsilon$, $|M| \geq (1 - \varepsilon)\nu(G)$.

As in most matching algorithms, we start with the empty matching, and augment it using *augmenting paths*: these are paths that start and end at unmatched nodes, and every second edge of them belongs to the current matching M. Augmenting along such a path interchanging the matching edges and non-matching edges) increases the size of the matching by 1. Of course, in our setting we will have to augment simultaneously along many disjoint augmenting paths, to make measurable progress. We will augment along augmenting paths of length at most $k = \lceil 3/\varepsilon \rceil$, which we call *short augmenting paths*.

This is again done in rounds. It will be convenient to assume that in each round, a new local seed is generated for every agent v. (They could get this from a single random real number in $[0, 1]$, by using all the bits in even position in the first round, half of the remaining bits in the second, etc.) This way the rounds will be independent of each other in the probabilistic sense. Augmentation along many disjoint augmenting paths will be carried out simultaneously by our agents. It is clear that agents looking at their neighborhoods with radius k will discover all short augmenting paths. The problem is that there will be conflicts: these short augmenting paths are not disjoint. To this end, we define when a path is better than another, and we will augment only along those paths that are better than any path intersecting them.

To be precise, we define that path P is *better* than path Q if walking along both paths, starting from their endnodes with higher local seed, the first node that is different has higher local seed in P than in Q. We will augment along paths that are better than any path intersecting them; we call such a path *locally best*. If we allow agents to explore their neighborhoods with radius $2k$, then every locally best short augmenting path will be discovered by at least one agent, who will carry out the augmentation (i.e., send a message to the agents along the path how their mates are to be changed). Several agents may do so for a given path, but there will be no conflict between their messages.

The above is repeated $q = 4D^{2k}\lceil\log(1/\varepsilon)\rceil$ times, then we stop and output the current matching.

The idea in the analysis is that in a particular phase, we either find many good short paths, and hence make substantial progress, or the number of all short augmenting paths is small, in which case we have an almost maximum matching.

Let as call a node *eligible* (at a certain phase) if at least one short augmenting path starts at it (such a node is of course unmatched). Let M_i be matching after the i-th round, and let X_i denote the number of eligible nodes. Let M' be a maximum matching, and consider the set $M_i \cup M'$. This set of edges consists of the common edges of M_i and M', and cycles and paths whose edges alternate between M and M'. Every cycle contains the same number of edges from M_i and M'. Paths that contain more edges from M' than from M_i have to end with edges in M' at both ends, and so they are augmenting paths. Thus the number of augmenting paths is at least $|M'| - |M_i| = \nu(G) - |M_i|$. The number of augmenting paths among these that have length more than k is less than $2|M'|/k$, so there are at least

$(1-2/k)\nu(G) - |M_i|$ short augmenting paths, and $X_i \geq \big(2 - (4/k)\big)\nu(G) - 2|M_i|$ eligible nodes.

Let u be an eligible node after phase i. All the short augmenting paths intersecting any of the short augmenting paths starting at u stay within $B_{G,2k}(u)$. Since $|B_{G,2k}(u)| \leq D^{2k}$, there is a chance of at least $p = 1/D^{2k}$ that u is the node with highest local seed among them. Then the best path starting at u will be augmented upon, and hence u has a chance of at least p to become matched in that round. This means that

$$\mathsf{E}\big(|M_{i+1}| \,\big|\, M_i\big) \geq |M_i| + \frac{1}{2}pX_i \geq |M_i| + p\Big(\frac{k-2}{k}\nu(G) - |M_i|\Big)$$

(here expectation is taken over random choices in the $(i+1)$-st round), which we can write as

$$\mathsf{E}\Big(\frac{k-2}{k}\nu(G) - |M_{i+1}| \,\Big|\, M_i\Big) \leq (1-p)\Big(\frac{k-2}{k}\nu(G) - |M_i|\Big).$$

Taking expectation over M_i, we get

$$\mathsf{E}\Big(\frac{k-2}{k}\nu(G) - |M_{i+1}|\Big) \leq (1-p)\mathsf{E}\Big(\frac{k-2}{k}\nu(G) - |M_i|\Big).$$

Hence

$$\mathsf{E}\Big(\frac{k-2}{k}\nu(G) - |M_q|\Big) \leq (1-p)^q \mathsf{E}\Big(\frac{k-2}{k}\nu(G) - |M_0|\Big) = (1-p)^q \frac{k-2}{k}\nu(G).$$

By Markov's Inequality, this implies that

$$\mathsf{P}\Big(\frac{k-2}{k}\nu(G) - |M_q| > \frac{k-2}{k^2}\nu(G)\Big) \leq k(1-p)^q \leq ke^{-pq} \leq \varepsilon.$$

So with probability at least $1-\varepsilon$, we have

$$|M_q| \geq \frac{k-2}{k}\nu(G) - \frac{k-2}{k^2}\nu(G) = \frac{(k-1)(k-2)}{k^2}\nu(G) \geq (1-\varepsilon)\nu(G).$$

This proves that the algorithm works as claimed.

We have seen that without local seeds, there is no way to approximately compute a maximum matching. So the matching problem can be solved in model (B) but not in (A).

22.3.2. Maximum flow: an algorithm in the weakest model. An algorithm based on similar ideas was developed by Csóka [2012a] to find an almost maximum flow (Example 22.18). This algorithm too looks for short augmenting paths. We do not describe the details here, but point out one interesting feature.

Using the random local seeds, the agents compute a flow that is almost optimal, in a way similar to the maximum matching algorithm described above. A different choice of seeds would give a different flow. But the expected flow values (expectation taken over all random seeds) also give a valid, almost maximum flow, which is independent of any random seeds. This expectation can be computed locally, from the neighborhoods with radius $2r$. Thus to compute the maximum flow, we don't need the random seeds after all: deterministic agents can compute it. (This does not contradict our arguments about the curse of symmetry, because matchings are not invariant under all automorphism of the given graph, but there is a "canonical" maximum flow that is invariant under automorphisms preserving the sets of sources and sinks: the average of all maximum flows.)

Csóka also describes an algorithm in model (C) to find an almost minimum cut, and proves that for this, a public random number is needed; in other words, the problem cannot be solved in model (B).

22.3.3. Knowing the graph does not help. Our next goal is to show that model (D), which seems to be a lot stronger than (C), is in fact equivalent to it (up to an arbitrarily small increase of the cost; Csóka [2012a]).

THEOREM 22.20. *Suppose that there is an algorithm in model* (D) *by which the agents compute, for every graph G, a feasible decoration f with cost $c(G)$. Then for every $\varepsilon > 0$ there is an algorithm in model* (C) *by which the agents compute, for every graph G, a feasible decoration f with cost at most $c(G) + \varepsilon$.*

Proof. Recall that A_r denotes the set of all probability distributions $\rho_{G,r}$, where G ranges through finite graphs. By Proposition 19.9, its closure $\overline{A}_r$ is convex.

Let us fix a large graph G. We may assume that $c(G)$ is the optimal cost of a decoration the agents can compute in model (D). If the agents mistakenly believe that they are working on the graph F, then they will compute (in model (D)) a decoration f_F of G (which is random, as a function of the public random seed and the private random seeds). The expectation of $\omega(f_F)$ will depend on the true distribution $\rho_{G,r}$. Setting $\omega_F(v) = \omega(B_{G,r}(v), f_F\,|_{B_{G,r}(v)})$, we get by the linearity of expectation,

$$\mathsf{E}\big(w(f_F)\big) = \frac{1}{\mathsf{v}(G)} \sum_{v \in V(G)} \mathsf{E}\big(\omega_F(v)\big).$$

The last expectation depends only on F and on $B_{G,2r}(v)$ (since the distribution of $f_F(u)$ depends only on the r-neighborhood of u, and to compute the distribution of $\omega_F(v)$ it suffices to know the joint distribution of the decorations $f_F(u)$ in the r-neighborhood of v). For $B \in \mathfrak{B}_{2r}$, let $a(F, B) = \mathsf{E}\big(\omega_F(\mathrm{root}(B))\big)$, then

$$\mathsf{E}\big(w(f_F)\big) = \sum_{B \in \mathfrak{B}_{2r}} \rho_{G,2r}(B) a(F, B) = L_F(\rho_{G,2r}),$$

where $L_F: A_{2r} \to \mathbb{R}$ is a homogeneous linear function. Clearly $L_F(\rho_{G,2r}) \geq c(G)$ for any F, and our assumption about the quality of the output of (D) implies that $L_G(\rho_{G,2r}) = c(G)$.

Next, we define a function $u: \overline{A}_{2r} \to \mathbb{R}$ by

$$u(\rho) = \liminf c(G_n),$$

where the limes inferior is to be taken over all sequences (G_n) for which $\rho_{G_n,2r} \to \rho$. A similar argument as in the proof of Proposition 19.9 shows that the function u is convex. For any graph G, considering the special sequence $(G^n : n = 1, 2, \dots)$, we get

(22.7) $$u(\rho_{G,2r}) \leq c(G).$$

We claim that for every $\rho \in A_{2r}$ there is a graph F such that

(22.8) $$L_F(\rho) < u(\rho) + \varepsilon.$$

Indeed, let (G_n) be a sequence of graphs such that $\rho_n = \rho_{G_n,2r} \to \rho$ and $c(G_n) \to u(\rho)$. Then

$$L_{G_n}(\rho) = L_{G_n}(\rho_n) + o(1) = c(G_n) + o(1) = u(\rho) + o(1).$$

Thus $F = G_n$ can be chosen in (22.8) for a sufficiently large n.

Next, we engage in a little convex geometry. Consider the set $K \subseteq \mathbb{R}^{\mathfrak{B}_r} \times \mathbb{R}$, defined by $K = \{(\rho, y) : y \geq u(\rho)\}$. It is clear that this set is convex. For every graph F, we consider the halfspace $H_F = \{(\rho, y) : y \leq L_F(\rho) - \varepsilon\}$. We claim that

$$K \cap \bigcap_F H_F = \emptyset.$$

Indeed, suppose that (ρ, y) is a point contained in the left side, then $(\rho, y) \in K$, and (22.8) implies that there is an $F \in \mathcal{G}$ such that $y \geq u(\rho) > L_F(\rho) - \varepsilon$. On the other hand, $(\rho, y) \in H_F$ implies that $y \leq L_F(\rho) - \varepsilon$, a contradiction.

Hence by Helly's Theorem, there is a finite set of graphs $F_1, \ldots, F_m \in A_{2r}$, where $m \leq |\mathfrak{B}_r| + 1$, such that $K \cap H = \emptyset$, where $H = \bigcap_{i \leq m} H_{F_i}$. Since K and H are convex, there is a halfspace defined by a linear inequality $y - L(\rho) \leq b$ containing H but disjoint from K. This means two things:

(a) The inequality $y \geq u(\rho)$ ($\rho \in \overline{A}_{2r}$) implies that $y - L(\rho) > b$. This last condition means that $u(\rho) > L(\rho) + b$ for every $\rho \in \overline{A}_{2r}$.

(b) The linear inequalities $y - L_{F_i}(\rho) \leq \varepsilon$ imply the inequality $y - L(\rho) \leq b$. By the Farkas Lemma, there are nonnegative numbers α_i such that $\sum_i \alpha_i = 1$, $\sum_i \alpha_i L_{F_i} = L$, and $b \geq -\varepsilon$. The numbers α_i form a probability distribution α on $[m]$.

Now we can give the following instruction to our agents: Use the even bits of the public random number $\mathbf{g}_0$ to pick an $\mathbf{i} \in [m]$ from the distribution α. (All agents will pick the same $\mathbf{i}$.) Then pretend that you are working on the graph $F_\mathbf{i}$, and compute the decoration $f_{F_\mathbf{i}}$ according to algorithm (C) (using the remaining bits of $\mathbf{g}_0$ as the public random number). Then (using (22.7) in the last step) the agents achieve a cost that is almost as good as the cost they could achieve knowing the graph:

$$\mathsf{E}\big(w(f_{F_\mathbf{i}})\big) = \sum_j \alpha_j \mathsf{E}\big(w(f_{F_j})\big) = \sum_j \alpha_j L_{F_j}(\rho_{G,2r}) = L(\rho_{G,2r})$$
$$\leq u(\rho_{G,2r}) - b \leq u(\rho_{G,2r}) + \varepsilon \leq c(G) + \varepsilon. \qquad \square$$

22.3.4. Computable structures and Borel sets. We conclude this chapter with sketching a connection between algorithmic problems and measure theory. Elek and Lippner [2010] give another algorithm for computing an almost maximum matching in a large bounded degree graph. Their approach is based on the connections with Borel graphs, which were discussed in Section 18.1. Instead of describing a second matching algorithm, we only illustrate the idea on a simpler example, by showing how the proof that every Borel graph with degrees at most D has a Borel coloring with $D+1$ colors (Theorem 18.3) can be turned into an algorithm.

In the proof of Theorem 18.3, we start with constructing a countable Borel coloring. This part of the argument can be translated easily. We have to select an explicit countable basis for the Borel sets in $[0,1)$; for example, we can choose intervals of the form $[a/b, (a+1)/b)$, where $0 \leq a < b$ are integers. We have to assign a positive integer index to each of these intervals, say $(2a+1)2^b$. Then every agent picks the interval with smallest index that contains his local seed but not the local seed of any of his neighbors. Now the agents have indices (they can forget the seeds from now on). Trivially, adjacent agents have different indices.

Next, every agent whose index is smaller than the indices of his neighbors changes his index to 1, and labels himself FINISHED. (In the proof of Theorem

18.3, only those with index 2 not adjacent to any node with index 1 did so in the first round; but it is easy to see that eventually all nodes with a locally minimal index will change to 1). Next, all those agents whose index is smaller than the indices of all their unfinished neighbors change their indices to the smallest possible, etc.

At this point comes an important difference: for Borel coloring, we could repeat this infinitely many times, but here we have a time bound. Those nodes that managed to change their indices are now properly colored with $D+1$ colors; however, there will be some who are stuck with their large original indices. Down-to-earth work starts here to show that their number is a small fraction of $v(G)$. We don't go into the details (see Exercises 22.22, 22.23).

> EXERCISE 22.21. Describe how Algorithm 22.19 can be simplified in two simpler versions of the problem: (a) we only want to find a maximal (non-extendable) matching (of course, with an error); (b) somebody marks a matching for us, and we have to test whether it is maximum (again, with some error).
>
> EXERCISE 22.22. Prove that any constant time distributed algorithm that constructs a legitimate coloring of a cycle will use, with high probability, more than 100 colors.
>
> EXERCISE 22.23. Prove that for every $\varepsilon > 0$ there is a $k \geq 1$ such that the algorithm (with k rounds) as described above will produce a coloring in which fewer than $\varepsilon v(G)$ nodes have color larger than D.

Part 5

Extensions: a brief survey

CHAPTER 23

Other combinatorial structures

The ideas of characterizing homomorphism functions, connection ranks, regularity lemmas and limit objects have been extended to several combinatorial structures besides graphs. Some of these extensions are rather involved and deep, like the limit theory of hypergraphs; others can be described as "analogous" (at least after finding the right definitions). Without attempting to be complete, we survey several of these extensions.

23.1. Sparse (but not very sparse) graphs

The obvious big gap in our treatment of limits of growing graph sequences is any sequence of graphs with density tending to 0, but maximum degree tending to infinity. Some interesting examples are the point-line incidence graphs of finite projective planes (about $n^{3/2}$ edges, if n is the number of nodes), and d-cubes ($n \log n$ edges).

Some work has been done. We have mentioned extensions of the Regularity Lemma to sparser graphs by Kohayakawa [1997], Gerke and Steger [2005], and Scott [2011]). While the case of bounded degree graphs is open, these results are highly nontrivial for sparse (but not very sparse) graphs, and have important applications. They are very likely to play an important role in the limit theory of such graphs. In a substantial paper, Bollobás and Riordan [2009] investigate many of the techniques discussed in this book and elsewhere, mostly from the point of view of extending them from the case of dense graphs to sparser classes.

Lyons [2005] extended the convergence theory of bounded degree graphs to graph sequences with bounded average degree, under a condition called tightness (this guarantees that the sequence of sample distributions has a limit distribution). The following example shows that some condition like this is necessary: the average degree of a subdivision G' of any graph G (dense or not) is bounded by 4. Clearly, to properly describe the limit of the graph sequence (G'_n), the description must contain essentially the same information as the limit of the sequence (G_n). So limits of graphs with bounded average degree are as complex as limits of any graph sequence (dense or sparse).

Graphons and graphings generalize dense graphs and bounded degree graphs, respectively, and they can be considered as the two extremes as far as edge density goes. One common feature is that we can do a random walk on each of them. More precisely, there is a Markov chain on a graphon, as well as on a graphing, and we are going to show that this Markov chain contains all the necessary information about these objects.

Let $W: \Omega^2 \to [0,1]$ be a graphon with density $\omega = t(K_2, W)$. We can define a Markov chain on Ω by

$$P_u(A) = \frac{1}{d_W(u)} \int_A W(u,v)\, dv$$

(this is defined for almost all u). This Markov chain has a stationary distribution, defined by

$$\pi(A) = \frac{1}{\omega} \int_{A \times [0,1]} W(x,y)\, dx\, dy.$$

It is also easy to check that this Markov chain is reversible. The step distribution of this Markov chain is proportional to the integral measure of W. Note that the Markov chain does not change if we scale W, so we have to remember the "density" ω if we want to preserve all information about W. But the Markov chain together with the density does determine the graphon.

Next, consider a graphing $\mathbf{G}$ on Ω. We can define a Markov chain by

$$P_u(A) = \frac{\deg_A(u)}{\deg(u)}.$$

Then the random walk defined by this chain is just the random walk on this graph in the usual sense. The measure preservation condition (18.2) says that this Markov chain is reversible. A stationary measure of this random walk is λ^* (as defined in Section 18.2), its step distribution is η/ω. So the step distribution of the Markov chain is the same as the probability measure on the edges of a graphing. The graphing is determined by this Markov chain. It is a fascinating open problem whether Markov chains can be used to define convergence and limit objects for graph sequences that are neither dense nor of bounded degree.

23.2. Edge-coloring models

23.2.1. Edge-connection matrices. We consider multigraphs with loops. It will be useful to allow a single edge with no endpoints; we call this graph the *circle*, and denote it by ◯.

We can define edge-connection matrices that are analogous to the connection matrices defined before: Instead of gluing graphs together along nodes, we glue them together along edges. To be precise, we define a *k-broken graph* as a k-labeled graph in which the labeled nodes have degree one. (It is best to think of the labeled nodes as not nodes of the graph at all, rather, as points where the k edges sticking out of the rest of the graph are broken off.) We allow that both ends of an edge be broken off.

For two k-broken graphs G_1 and G_2, we define $G_1 * G_2$ by gluing together the corresponding broken ends of G_1 and G_2. These ends are not nodes of the resulting graph any more, so $G_1 * G_2$ is different from the graph $G_1 G_2$ we would obtain by gluing together G_1 and G_2 as k-labeled graphs. We can glue together two copies of an edge with both ends broken off; the result is the circle ◯. One very important difference is that while $G_1 G_2$ is k-labeled, $G_1 * G_2$ has no broken edges any more, and so it is not k-broken but 0-broken. This fact leads to considerable difficulties in the treatment of edge models.

For every graph parameter f and integer $k \geq 0$, we define the *edge-connection matrix* $M'(f,k)$ as follows. The rows and columns are indexed by isomorphism types of k-broken graphs. The entry in the intersection of the row corresponding to G_1 and the column corresponding to G_2 is $f(G_1 * G_2)$. Note that for $k = 0$, we have $M(f,0) = M'(f,0)$, but for other values of k, connection and edge-connection matrices are different.

Let G be a finite graph. An *edge-coloring model* is determined by a mapping $h : \mathbb{N}^q \to \mathbb{R}$, where q is positive integer. We call h the *node evaluation function*. Here we think of $[q]$ as the set of possible edge colors; for any coloring of the edges and $d \in \mathbb{N}^q$, we think of $h(d)$ as the "value" of a node incident with d_c edges with the color c ($c \in [q]$). In statistical physics this is called a *vertex model*: the edges can be in one of several states, which are represented by the color; an edge-coloring represents a state of the system, and (assuming that $h > 0$) $\ln h(d)$ is the contribution of a node (incident with d_c edges with color c) to the energy of the state.

There are many interesting and important questions to be investigated in connection with edge-coloring models; we will only consider what in statistical physics terms would be called its "partition function". To be more precise, for an edge-coloring $\varphi : E(G) \to [q]$ and node v, let $\deg_c(\varphi, v)$ denote the number of edges e incident with node v with color $\varphi(e) = c$. So the vector $\deg(\varphi, v) \in \mathbb{N}^q$ is the "local view" of node v. The *edge-coloring function* of the model is defined by

$$\mathsf{col}(G, h) = \sum_{\varphi:\ E(G) \to [q]} \prod_{v \in V(G)} h\big(\deg(\varphi, v)\big).$$

Recall that we allow the graph $\bigcirc$ consisting of a single edge with no endpoints; by definition, $\mathsf{col}(\bigcirc, h) = q$. We also allow that $q = 0$, in which case $\mathsf{col}(G, h) = 1$ if G has no edges, and $\mathsf{col}(G, h) = 0$ otherwise. We could of course allow complex valued node evaluation functions, in which case the value of the edge-coloring function can be complex.

EXAMPLE 23.1 (**Number of perfect matchings**). The number of perfect matchings can be defined by coloring the edges by two colors, say black and white, and requiring that the number of black edges incident with a given node be exactly one. This means that this number is $\mathsf{col}(., h)$, where $h : \mathbb{N}^2 \to \mathbb{R}$ is defined by $h(d_1, d_2) = \mathbb{1}(d_1 = 1)$. The number of all matchings could be expressed similarly.
♦

EXAMPLE 23.2 (**Number of 3-edge-colorings**). This number is $\mathsf{col}(., h)$, where $h : \mathbb{N}^3 \to \mathbb{R}$ is defined by $h(d_1, d_2, d_3) = \mathbb{1}(d_1, d_2, d_3 \leq 1)$. ♦

EXAMPLE 23.3 (**Spectral decomposition of a graphon**). Recall the definition (7.18) and expression (7.25) for $t(F, W)$ in terms of the spectrum of T_W. We can consider χ as a coloring of $E(F)$ with colors $1, 2, \ldots$. Then $M_\chi(v)$ depends only on the numbers of edges with different colors, and so we can write $M_\chi(v) = h\big(\deg(\chi, v)\big)$, and we get

$$t(F, W) = \mathsf{col}(G, \lambda, h) = \sum_{\chi:\ E(G) \to [q]} \prod_{e \in E(F)} \lambda_{\chi(e)} \prod_{v \in V(G)} h\big(\deg(\chi, v)\big).$$

However, this is not a proper edge-coloring model, since the value of the circle, which is the number of colors, is infinite in general. ♦

The following facts about the edge-connection matrices of edge-coloring functions are easy to prove along the same lines as Proposition 5.64:

PROPOSITION 23.4. *For every edge-coloring model $h : \mathbb{N}^q \to \mathbb{R}$, the graph parameter $\mathsf{col}(., h)$ is multiplicative, its edge-connection matrices $M'(f, k)$ are positive semidefinite, and $\mathrm{rk}(M'(f, k)) \leq q^k$.* □

B. Szegedy [2007] showed that the first two of these properties suffice to give a characterization of edge-coloring functions.

THEOREM 23.5. *A graph parameter f can be represented as an edge-coloring function if and only if it is multiplicative and $M'(f, k)$ is positive semidefinite for all $k \geq 0$.* □

The proof of this theorem is quite involved and not reproduced here; it is based on ideas similar to those used in Section 6.6 to prove Theorem 5.57, but using quite a bit more involved tools: The use of the Nullstellensatz and simple semidefiniteness arguments must be replaced by real versions of the Nullstellensatz (Positivstellensatz), and the simple symmetry arguments must be replaced by deeper results from the representation theory of algebras. (As a historical comment, the proof of Theorem 23.5 came first, and Schrijver's proof of Theorem 5.57 was motivated by this.)

Draisma, Gijswijt, Lovász, Regts and Schrijver [2012] give characterizations of complex valued edge-coloring functions. Let us state without proof a result of Schrijver [2012], which shows that a condition on the growth of the edge-connection rank (along with minor other constraints) can characterize complex edge-coloring models.

THEOREM 23.6. *A complex valued graph parameter f is an edge-coloring function of a complex model if and only if it is multiplicative, $f(\bigcirc)$ is real, and $\mathrm{rk}(M'(f, k)) \leq f(\bigcirc)^k$ for every k.* □

23.2.2. Tensor algebras. There is a surprisingly close connection between edge-coloring models and rather general multilinear algebra. Given an edge-coloring model (with color set $[q]$), we can think of the nodes as little gadgets with wires (or legs) sticking out corresponding to the edges incident with it. If we assign colors to the wires, the gadget outputs a number (this could be real or complex). The graph parameter defined by this edge-coloring model is the expectation of the product of these numbers, one for each node, where the edge-coloring is random.

Formulating the question like this, we see two restrictions that look artificial:

— There is only one gadget for each degree. Why not have several?

— We have assumed that the output of a gadget depends only on the number of legs with each color; in other words, it is invariant under the permutation of the legs. Why not drop this condition?

If we relax these conditions, then every gadget would be described by a real array $(H_{i_1,\ldots,i_d} : i_r \in [q])$, where d is the number of legs. In other words, the gadget is described by a tensor with d slots over $\mathbb{R}^q$. Furthermore, we have to indicate for each node v which gadget is sitting there, and how its legs correspond to the edges incident with v.

In more mathematical terms, we have a graph where a tensor is associated with every node, and an index associated with every edge, so that the slots (indices) of the tensor correspond to the edges incident with the node. Let us call such a graph a *tensor network*. The corresponding graph parameter is evaluated by taking the

23.2. EDGE-COLORING MODELS

product of these tensors, and then summing over all choices of the indices. Note that every index occurs twice, so we could call this "tracing out" every index.

These tensor networks play an important role in several areas of physics, but we can't go into this topic in this book.

This setup allows for a more general construction. If we have a tensor network with k broken edges, then the value associated with the graph will depend on the color of these edges, in other words, it will be described by an array $(A_{i_1,\ldots,i_k} : i_r \in [q])$. So the graph with k broken edges can be considered as a gadget itself.

We can break down the procedure of assembling a tensor network from the gadgets (with or without broken edges) into two very simple steps:

(a) We can take the disjoint union of two gadgets; if the gadgets have k and l legs, respectively, the union has $k + l$ legs. In terms of multilinear algebra, this means to form the tensor product of two tensors.

(b) We can fuse two legs of a gadget. If $(A_{i_1,\ldots,i_k} : i_r \in [q])$ is the tensor describing the gadget, and (say) we fuse legs $k-1$ and k, then we get the tensor

$$B_{i_1,\ldots,i_{k-2}} = \sum_{j \in [q]} A_{i_1,\ldots,i_{k-2},j,j}.$$

In multilinear algebra slang, we trace out the last two indices.

It is easy to see that with these operations, we can construct every tensor network with or without broken edges, and we get the corresponding tensor.

Supposing that we have a starting kit of gadgets, we can look at the set of all tensors that can be realized by assembling tensor graphs with broken edges from these gadgets. In the spirit of linear algebra, we take all linear combinations of the obtained tensors with the same number of slots. Every tensor obtained this way will be called an *assembled tensor*.

It is clear from (a) and (b) above that the set of assembled tensors has the following structure: For every k, there is a linear space $\mathcal{T}_k$ of tensors over $\mathbb{R}^q$ with k slots. For every $A \in \mathcal{T}_k$ and $B \in \mathcal{T}_l$, the tensor product $A \otimes B \in \mathcal{T}_{k+l}$. For every $A \in \mathcal{T}_k$, and any two indices in A, tracing out these two indices results in a tensor in $\mathcal{T}_{k-2}$. We call such a set of tensors a *traced tensor algebra*.

Conversely, every traced tensor algebra arises as the set of assembled tensors: for every number k of slots, we select a basis of the space $\mathcal{T}_k$, and use the resulting set of tensors as the starting kit.

It is quite fruitful to use this connection; one can obtain results that are new both for graphs and for tensor algebras. We describe one important result with combinatorial connections.

Given a starting kit K, how can we decide about a tensor whether it can be assembled from this kit? In other words, is it contained in the traced tensor algebra generated by K? A beautiful answer to this question was found by Schrijver [2008a], which we describe in graph-theoretic terms (the proof uses the representation theory of algebras, and we do not give it here; cf. also Schrijver [2008b, 2009]).

Recall that we work over a fixed vector space $\mathbb{R}^q$. Every $q \times q$ real matrix A is a gadget in itself, with two legs. If it is symmetric, then the legs are interchangeable, but in general we have to talk about a "left leg" (corresponding to the row index) and a "right leg" (corresponding to the column index). Connecting the gadgets for matrices A and B in series gives a gadget representing the matrix AB.

Orthogonal matrices will play a special role. One observation is that if A is orthogonal, then $AA^\mathsf{T} = I$ (the identity matrix), and so if we have a gadget graph with no broken edges, and replace any edge by a path of length 3 with A and A^T sitting on it:

If we replace every edge by this path of length 3, then the value of the graph does not change. However, we can group together every original gadget B with the orthogonal matrices next to, to get a gadget B^A, which—in multilinear algebra terms—is obtained from B by applying the linear transformation A to every slot. If we replace every gadget B in the kit by B^A, then the value of the tensor network does not change.

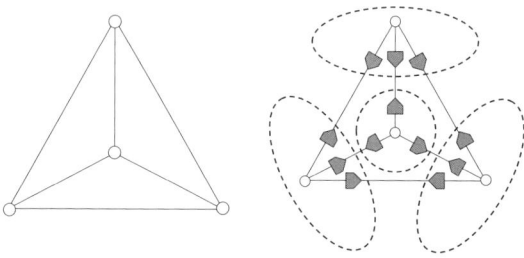

FIGURE 23.1. Replacing every edge by a path with the same orthogonal transformation at both inner nodes (just facing the opposite direction), and regrouping does not change the value.

Now consider a tensor network with broken edges. If we replace every tensor B in the kit by B^A, then the matrices A and A^T along the unbroken edges still cancel each other, but on the broken edges, one copy still remains. In other words, if we apply the same orthogonal transformation to every slot of every tensor in the kit, then the tensor defined by a tensor network with broken edges undergoes the same transformation.

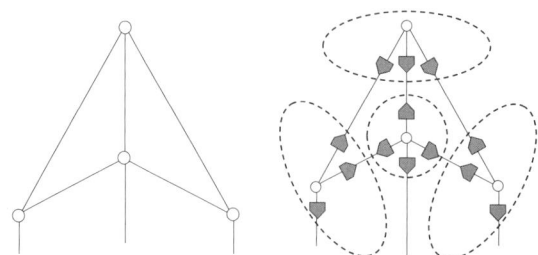

FIGURE 23.2. Applying the same orthogonal transformation to all slots of all tensors in the kit results in applying the same orthogonal transformation to the slots of the assembled tensor.

In particular, if all tensors in the kit have the property that a particular orthogonal transformation applied to all their slots leaves them invariant, then the same holds for every assembled tensor. The theorem of Schrijver [2008a] asserts that this is the only obstruction to assembling a given tensor.

THEOREM 23.7. *Let $\mathcal{T}$ be a traced tensor algebra generated by a set $\mathcal{S}$ of tensors, including the identity tensor $\mathbb{1}(i = j)$ $(i, j \in [q])$. Then a tensor T is in $\mathcal{T}$ if and only if it is invariant under every orthogonal transformation that leaves every tensor in $\mathcal{S}$ invariant.* □

The special case when the generating tensors are symmetric describes edge-coloring models. This can be viewed as an analogue of Theorem 6.38, with the role of the edges and nodes interchanged. Regts [2012] showed how Theorem 23.7 yields an exact formula for the edge-connection rank of edge-coloring models.

EXAMPLE 23.8 (**Number of perfect matchings revisited**). The tensor model for this graph parameter is a bit more complicated than in Example 23.1. We have 2 edge colors (which will be convenient to call 0 and 1), so we work over $\mathbb{R}^2$; but we need to specify a tensor for every degree d, expressing that exactly one edge is black:
$$T_{i_1,\ldots,i_d} = \mathbb{1}(i_1 + \cdots + i_d = 1).$$
It is easy to see that no orthogonal transformation, applied to all slots, leaves this tensor invariant, so it follows from Theorem 23.7 that every tensor can be assembled from this kit. (We note that the tensor is invariant under permuting the slots; however, this symmetry is not preserved under composition of tensor networks.) ♦

EXAMPLE 23.9 (**Number of 3-edge-colorings revisited**). To construct a tensor model for the number of 3-edge-colorings, we work over $\mathbb{R}^3$. We again need to specify a tensor for every degree expressing that the edges have different colors:
$$T_{i_1,\ldots,i_d} = \mathbb{1}(i_1, \ldots, i_d \text{ are different})$$
(for $d > 3$, we get the 0 tensor). Permuting the colors (i.e., the coordinates in the underlying vector space $\mathbb{R}^3$) leaves this tensor invariant, and these are the only orthogonal transformations of $\mathbb{R}^3$ with this property. Theorem 23.7 implies that a tensor is invariant under the permutations of the coordinates of $\mathbb{R}^3$ if and only if it can be assembled from this kit. ♦

23.3. Hypergraphs

When talking about generalizing results on graphs, the first class of structures that comes to mind is hypergraphs (at least to a combinatorialist). So it is perhaps surprising that to extend the main concepts and methods developed in this book (quasirandomness, limit objects, Regularity Lemma, and Counting Lemma) to hypergraphs is highly nontrivial. Even the "right" formulation of the Regularity Lemma took a long time to find, and in the end both the Regularity Lemma and the limit object turned out quite different from what one would expect as a naive generalization. Nevertheless, the issue is essentially solved now, thanks to the work of Chung, Elek, Graham, Gowers, Rödl, Schacht, Skokan, Szegedy, Tao and others. A full account of this work would go way beyond the possibilities of this book, but we will give a glimpse of the results.

By an *r-uniform hypergraph*, or briefly *r-graph*, we mean a pair $H = (V, E)$, where $V = V(H)$ is a finite set and $E = E(H) \subseteq \binom{V}{r}$ is a collection of r-element subsets. The elements of V are called *nodes*, the elements of E are called *edges*. So 2-graphs are equivalent to simple graphs. We can define the homomorphism number $\hom(G, H)$ of an r-graph G into an r-graph H in the natural way, as the number of maps $\varphi : V(G) \to V(H)$ for which $\varphi(A) \in E(H)$ for every $A \in E(G)$. The *homomorphism density* of G in H is defined as one expects, by the formula

$$t(G, H) = \frac{\hom(G, H)}{|V(G)|^{|V(H)|}}$$

Quasirandomness can be defined by generalizing the condition on the density of quadrilaterals. We need to define a couple of special hypergraph classes. Let K_n^r denote the complete r-uniform hypergraph on $[n]$ (i.e., $E(K_n^r) = \binom{[n]}{r}$). Let L_k^r be the "complete r-partite hypergraph" defined on the node set $V_1 \cup \cdots \cup V_r$, where the V_i are disjoint k-sets, and the edges are all r-sets containing exactly one element from each V_i. Clearly $t(K_r^r, H) = t(L_1^r, H)$ is the edge density of H. It is not hard to prove that $t(L_k^r, H) \geq t(K_r^r, H)^{k^r}$ for every H (this generalizes inequality 2.9 from the Introduction). We define the *quasirandomness* of H as the difference $\mathsf{qr}(H) = t(L_2^r, H) - t(K_r^r, H)^{2^r}$.

A sequence (H_n) of hypergraphs is called *quasirandom* with density p if $t(K_r^r, H_n) \to p$ and $\mathsf{qr}(H_n) \to 0$, or equivalently, $t(L_2^r, H_n) \to p^{2^r}$. It was proved by Chung and Graham [1989] that this implies that $t(G, H_n) \to p^{e(G)}$ for every r-graph G, so the equivalence of conditions (QR2) and (QR3) for quasirandomness in the Introduction (Section 1.4.2) generalizes nicely.

As a first warning that not everything extends in a straightforward way, let us try to generalize (QR5). A first guess would be to consider disjoint sets $X_1, \ldots, X_r \subseteq V$, and then stipulate that the number of edges with one endpoint in each of them is $p|X_1|\ldots|X_r| + o(n^r)$. (For simplicity of presentation, we assume that $v(H_n) = n$.) This property is indeed valid for every quasirandom sequence, but it is strictly weaker than quasirandomness. It is not well-defined what the "right" generalization is; we state one below, which is a version of a generalization found by Gowers. Several other equivalent conditions are given by Kohayakawa, Rödl and Skokan [2002].

PROPOSITION 23.10. *A sequence (H_n) of hypergraphs is quasirandom with density p if and only if for every $(r-1)$-graph G_n on $V(H_n)$, the number of edges of H_n that induce a complete subhypergraph in G_n is $t(K_r^{r-1}, G_n)t(K_r^r, H_n)\binom{n}{r} + o(n^r)$.* □

In the case of simple graphs ($r = 2$), let H_n be a simple graph with edge density p. The 1-graph G_n means simply a subset of $V(H_n)$, and K_2^1 is just a 2-element set. So the condition says that the number of edges of the graph H_n induced by the set G_n is asymptotically

$$t(K_2^1, G_n)t(K_2^2, H_n)\binom{n}{2} = \Big(\frac{|G_n|}{n}\Big)^2 \frac{2\mathsf{e}(H_n)}{n^2}\binom{n}{2} \sim p\binom{|G_n|}{2},$$

and so we get condition (Q4). For general r, the condition can be rephrased as follows: for a random r-set $X \subseteq V$, the events that X is complete in G_n and X is an edge in H_n are asymptotically independent.

The last remark takes us to another complication.

EXAMPLE 23.11. Let $\mathbb{G}(n, 1/2)$ be a random graph and let $\mathbb{T}_n$ denote the 3-graph formed by the triangles in $\mathbb{G}(n, 1/2)$. Then $\mathbb{T}_n$ is a 3-graph with density $1/8$, which is random in some sense, but it is very different from the random 3-graph $\mathbb{H}_n$

on $[n]$ obtained by selecting every edge independently with probability $1/8$. In fact, the sequence $(\mathbb{H}_n)$ is quasirandom with probability 1 (this is not hard to see), while $\mathbb{T}_n$ has a very small intersection with every quasirandom 3-graph by Proposition 23.10. Also, $\mathbb{T}_n$ has some special features, like no 4-set of nodes contains exactly 3 edges of H_n.

On the other hand, $\mathbb{T}_n$ is totally homogeneous. It has no special global structure; more concretely: on any two disjoint k-sets we see independent copies of the same random hypergraph. If we want to generalize the Regularity Lemma, it has to reflect the difference between $\mathbb{T}_n$ and $\mathbb{H}_n$, and similarly for the generalization of the notion of graphons. Which of these sequences should tend to a constant function?
♦

We show how to overcome this difficulty, starting with the construction of the limit object. We say that a sequence of r-graphs (H_n) is *convergent*, if $v(H_n) \to \infty$ and $t(F, H_n)$ has a limit as $n \to \infty$ for every r-graph F. Let $t(F)$ denote this limit. How to represent this limit function, in other words, what is the hypergraph analogue of a graphon? The natural guess would be a symmetric r-variable function $W : [0,1]^r$, which would represent the limit by

$$t(F, W) = \int_{[0,1]^r} \prod_{\{i_1,\ldots,i_r\} \in E(F)} W(x_{i_1}, \ldots, x_{i_r^2}) \, dx.$$

The example of the hypergraphs $\mathbb{H}_n$ and $\mathbb{T}_n$ above show that this cannot be right. The only reasonable candidate for their limit object would be the function $W \equiv 1/8$, which represents correctly the limiting densities for the sequence $\mathbb{H}_n$, but not for the sequence $\mathbb{T}_n$. We could make life even more complicated, and consider the intersection $\mathbb{H}_n \cap \mathbb{T}_n$, which is a random 3-graph with expected density $1/64$, and the limiting densities are even more complicated.

For $r > 3$, one could construct a whole zoo of homogeneous random hypergraphs, generalizing the construction of $\mathbb{H}_n$ and $\mathbb{T}_n$. After several steps of generalization, one arrives at the following: we generate a random coloring of K_n^j for every $0 \leq j \leq r$ (with any number of colors). To decide whether an r-subset $X \subseteq [n]$ should be an edge, we look at the colors of its subsets, and see if this coloring belongs to some prescribed family of colorings of 2^X. (We assume that the prescribed family is invariant under permutations of X.)

While this example warns us of complications, it also suggests a way out: we describe the limit not in the r-dimensional but in the 2^r-dimensional space. In fact, the limit object turns out to be a subset, rather than a function, which is a gain (it is of course very little relative to the increase in the number of coordinates).

Consider the set $[0,1]^{2^{[r]}}$ (so we have a coordinate x_I for every $I \subseteq [r]$; the coordinate for $\emptyset$ will play no role, we can think of it as 0). Let us note that the symmetric group S_r acts on the power set $2^{[r]}$, and hence also on $[0,1]^{2^{[r]}}$. Let $U \subseteq [0,1]^{2^{[r]}}$ be a measurable set that is invariant under the action of S_r. We call such a set a *hypergraphon*.

For every hypergraphon U, we define the density of an r-graph F as follows. We assign independent random variables X_S, uniform in $[0,1]$, to every subset $S \subseteq V(F)$ with $|S| \leq r$. For every edge $A = \{a_1, \ldots, a_r\} \in E(F)$, and every $I \subseteq [r]$, we denote by A_I the subset $\{a_i : i \in I\}$, and we consider the point $X(A) \in [0,1]^{2^{[r]}}$ defined by $(X(A))_I = X_{A_I}$ (this depends on the ordering of A, but

this will not matter thanks to our symmetry assumption about U). Now we define
$$t(F, U) = \mathsf{P}\big(X(A) \in U \text{ for all } A \in E(F)\big).$$

To illuminate the meaning of this formula a little, consider the case $r = 2$. Then we have $U \subseteq [0,1]^3$, where the three coordinates correspond to the sets $\{1\}$, $\{2\}$ and $\{1, 2\}$ (as we remarked above, the empty set plays no role). For a graphon W, we define the set $U_W = \{(x_1, x_2, x_{12}) \in [0,1]^3 : x_{12} \leq W(x_1, x_2)\}$. Then it is easy to see that $t(F, U_W) = t(F, W)$ for any simple graph F.

Elek and Szegedy [2012] prove the following.

THEOREM 23.12. *For every convergent sequence (H_n) of r-graphs there is a hypergraphon U such that $t(F, H_n) \to t(F, U)$ for every r-graph F.* □

The limit graphon is essentially unique up to some "structure preserving transformations", which are more difficult to define than in the case of graphs and we don't go into the details. Elek and Szegedy [2012] give several applications of Theorem 23.12. For a given hypergraphon U, they define U-random hypergraphs and prove that they converge to U. They derive from it the Hypergraph Removal Lemma due to Frankl and Rödl [2002], Gowers [2006], Ishigami [2006], Nagle, Rödl and Schacht [2006] and Tao [2006a]. As a refreshing exception, the statement of this lemma is a straightforward generalization of the Removal Lemma for graphs (Lemma 11.64); the proof of Elek and Szegedy is similar to our second proof in Section 11.8. They also derive the Hypergraph Regularity Lemma using Theorem 23.12, using a stepfunction approximation of hypergraphons.

This brings us to the Hypergraph Regularity Lemma, a very important but also quite complicated statement. There are several essentially equivalent, but not trivially equivalent forms, due to Frankl and Rödl [1992], Gowers [2006, 2007], Rödl and Skokan [2004], Rödl and Schacht [2007a, 2007b]. Proving the appropriate Counting Lemma for these versions is a further difficult issue, and I will not go into it. But I must not leave this topic without stating at least one form, based on the formulation of Elek and Szegedy [2012], which in fact generalizes the strong form of the Regularity Lemma (Lemma 9.5).

We have to define what we mean by "regularizing" a hypergraph. For $\varepsilon, \delta > 0$ and $k \in \mathbb{N}$, we define an (α, β, k)-*regularization* of a r-graph H on $[n]$ as follows. For every $i \in [r]$, we partition the complete hypergraph K_n^i into r-graphs $G_{i,1}, \ldots, G_{i,k}$. Let us think of the edges in $G_{i,j}$ as colored with color j. This defines a partition $\mathcal{P}$ of the edges of K_n^r, where two r-sets are in the same class if the colorings of their subsets are isomorphic. The family $\{G_{i,j} : i \in [r], j \in [k]\}$, together with an r-graph G on $[n]$ will be called an (α, β, k)-regularization of H, if

(a) every r-graph $G_{i,j}$ has quasirandomness at most α, and

(b) G is the union of some of the classes of $\mathcal{P}$, and

(c) $|E(H) \triangle E(G)| \leq \beta \binom{n}{r}$.

Now we can state one version of the Hypergraph Regularity Lemma.

LEMMA 23.13 (**Strong Hypergraph Regularity Lemma**). *For every $r \geq 2$ and every sequence $\epsilon = (\varepsilon_0, \varepsilon_1, \ldots)$ of positive numbers there is a positive integer k_ϵ such that for every r-graph H there is an integer $k \leq k_\epsilon$ such that H has an $(\varepsilon_k, \varepsilon_0, k)$-regularization.*

The main point is that to regularize H, we have to partition not only its node set, but also the set of i-tuples for all $i \leq r$. Just like in the graph case, we could demand that the i-graphs $G_{i,j}$ have almost the same number of edges for every fixed i. Of course, the prize we have to pay for stating a relatively compact version is that it takes more work to apply it; but we don't go in that direction.

The extension of the theory exposed in this book to hypergraphs is not complete, and there is space for a lot of additional work. Just to mention a few loose ends, it seems that no good extension of the distance $\delta_\square$ has been found to hypergraphs (just as in the case of limit objects or the regularity lemma, the first natural guesses are not really useful). Another open question is to extend these results to nonuniform hypergraphs, with unbounded edge-size. The semidefiniteness conditions for homomorphism functions can be extended to hypergraphs (see e.g. Lovász and Schrijver [2008]), but perhaps this is just the first, "naive" extension. One area of applications of these conditions is extremal graph theory. The work of Razborov [2010] shows that generalizations of graph algebras and of the semidefiniteness conditions can be useful in extremal hypergraph theory. However, we have seen that graph algebras can be defined in the setting of gluing along nodes and also along edges, and this indicates that for hypergraphs a more general concept of graph algebras may be useful.

23.4. Categories

The categorial way of looking at mathematical structures is quite prevalent in many branches of mathematics. In graph theory, the use of categories (as a language and also as guide for asking question in a certain way) has been practiced mainly by the Prague school, and has lead to many valuable results; see e.g. the book by Hell and Nešetřil [2004].

One can go a step further and consider categories (with appropriate finiteness assumptions) as objects of combinatorial study on their own right. After all, categories are rather natural generalizations of posets, and there is a huge literature on the combinatorics of posets. However, surprisingly little has happened in the direction of a combinatorial theory of categories; some early work of Isbell [1991], Lovász [1972] and Pultr [1973], and the more recent work of Kimoto [2003a, 2003b] can be cited.

Working with graph homomorphisms, we have found not only that the categorial language suggests very good questions and a very fruitful way of looking at our problems, but also that several of the basic results about graph homomorphism and regularity can be extended to categories in a very natural way. The goal of this section is to describe these generalizations, and thereby encourage a combinatorial study of categories. (Appendix A.8 summarizes some background.)

23.4.1. Cancellation laws. Counting homomorphisms has been a main tool for proving cancellation laws for finite relational structures in Section 5.4, and it is not surprising that these results can be extended to locally finite categories (Lovász [1972], Pultr [1973]). The following two theorems generalize Theorem 5.34, Proposition 5.35(b) and Lemma 5.38 to categories.

THEOREM 23.14. *Let a and b be two objects in a locally finite category such that the direct powers $a^{\times k}$ and $b^{\times k}$ exist and are isomorphic. Then a and b are isomorphic.*

THEOREM 23.15. *Let a, b, c be three objects in a locally finite category $\mathcal{K}$ such that the direct products $a \times c$ and $b \times c$ exist and are isomorphic.*

(a) *If both a and b have at least one morphism into c, then a and b are isomorphic.*

(b) *There exists an isomorphism from $a \times c$ to $b \times c$ that commutes with the projections of $a \times c$ and $b \times c$ to c.* □

So if there is any isomorphism σ in Figure 23.3, then there is one for which the diagram commutes.

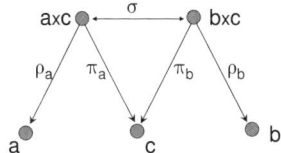

FIGURE 23.3

23.4.2. Connection matrices and algebras of morphisms. For the next theorem, we assume that $\mathcal{K}$ is a locally finite category that has a zero object, a left generator, pushouts, and epi-mono decompositions. Let f be a real valued function defined on the objects, invariant under isomorphism. We say that f is *multiplicative over coproducts*, if $f(a \oplus b) = f(a)f(b)$ for any two objects a and b.

For every object a, we define a (possibly infinite) symmetric matrix $M(f,a)$, whose rows and columns are indexed by morphisms in $\mathcal{K}_a^{\text{in}}$, and whose entry in row α and column β is $f\bigl(t(\alpha \vee \beta)\bigr)$ (since $\alpha \vee \beta$ is determined up to isomorphism, this is well defined). Note that specializing to the category of graph homomorphisms, $M(f,a)$ corresponds to the multiconnection matrix; to get the simple connection matrix, we have to restrict the row and column indices to monomorphisms.

One can extend the characterization of homomorphism functions in Corollary 5.58 to categories (Lovász and Schrijver [2010]); this theorem will also contain the dual characterization Theorem 5.59.

THEOREM 23.16. *Let $\mathcal{K}$ be a locally finite category that has a zero object z, a left generator, pushouts, and epi-mono decompositions. Let f be a function defined on the objects, invariant under isomorphism. Then there is an object b such that $f = |\mathcal{K}(.,b)|$ if and only if the following conditions are fulfilled:*

(F1) $f(z) = 1$,

(F2) *f is multiplicative over coproducts, and*

(F3) *$M(f,a)$ is positive semidefinite for every object a.*

We note that if there is an epimorphism from a to b, then $M(f,b)$ is a submatrix of $M(f,a)$. Thus it would be enough to require the semidefiniteness condition for a left-cofinal subset of elements a.

COROLLARY 23.17. *Conditions (F1)–(F3) of the theorem imply that (a) the values of f are non-negative integers, (b) the rank of $M(f,a)$ is finite for every a.*

Statement (a) of this corollary contrasts it with Theorem 5.54, where (thanks to the weights) the function values can be arbitrary real numbers. An analogue of (b) must be imposed as an additional condition e.g. in the characterization in Theorem 5.54, while in this version it follows from the other assumptions.

The conditions are very similar to those in Theorem 5.54, except that there the graphs cannot have loops and the matrices are indexed by monomorphisms only. As a consequence, the characterization concerns homomorphism numbers into weighted graphs, which has not been extended to categories so far.

The proof of Theorem 23.16 is built on similar ideas as the proof of Theorem 5.54 in Chapter 6, using algebras associated with the category. Since it is instructive how such algebras can be defined, we describe their construction below; for the details of the proof, we refer to the paper of Lovász and Schrijver [2010].

For two objects a and b in a locally finite category $\mathcal{K}$, a formal linear combination (with real coefficients) of morphisms in $\mathcal{K}(a,b)$ will be called a *quantum morphism*. Quantum morphisms between a and b form a finite dimensional linear space $\mathcal{Q}(a,b)$. Let

$$x = \sum_{\varphi \in \mathcal{K}(a,b)} x_\varphi \varphi \in \mathcal{Q}(a,b) \qquad \text{and} \qquad y = \sum_{\psi \in \mathcal{K}(b,c)} y_\psi \psi \in \mathcal{Q}(b,c),$$

then we define

$$xy = \sum_{\substack{\varphi \in \mathcal{K}(a,b) \\ \psi \in \mathcal{K}(b,c)}} x_\varphi y_\psi \varphi\psi \in \mathcal{Q}(a,c).$$

With this definition, quantum morphisms form a category $\mathcal{Q}$ on the same set of objects as $\mathcal{K}$. (Of course, $\mathcal{Q}$ is not locally finite any more, but it is locally finite dimensional.)

We can be more ambitious and take formal linear combinations of morphisms in $\mathcal{K}_a^{\text{out}}$ (for a fixed object a), to get a linear space $\mathcal{Q}_a^{\text{out}}$. This space will be infinite dimensional in general, but it has interesting finite dimensional factors. For each object a, the pushout operation $\wedge$ defines a semigroup on $\mathcal{K}_a^{\text{out}}$. Let $\mathcal{Q}_a^{\text{out}}$ denote its semigroup algebra of all formal finite linear combinations of morphisms in $\mathcal{K}_a^{\text{out}}$. So $\mathcal{Q}_a^{\text{out}} = \bigoplus_b \mathcal{Q}(a,b)$.

Just as in the case of graphs, every function $f : \text{Ob}(\mathcal{K}) \to \mathbb{R}$ defines an inner product on $\mathcal{Q}_a^{\text{out}}$, by $\langle \alpha, \beta \rangle = f\big(h(\alpha \wedge \beta)\big)$. Condition (F3) in Theorem 23.16 implies that this inner product is positive semidefinite. Factoring out its kernel, we get a Frobenius algebra, which is finite dimensional (this takes a separate argument, since unlike in the proof of Theorem 5.54, this is not assumed directly). The proof of Theorem 23.16, just like the proof of Theorem 5.54, is built on studying the idempotent bases in these algebras.

EXAMPLE 23.18 (**Graph algebras**). If the category is the category of graph homomorphisms, and a is the k-labeled graph with k nodes and no edges, then $\mathcal{Q}_a^{\text{out}}$ is the gluing algebra of k-multilabeled graphs. ♦

EXAMPLE 23.19 (**Flag algebras**). Razborov's "flag algebras" [2007] can be defined in our setting as follows. We consider the category of embeddings (injective homomorphisms) between graphs. Fixing a graph F (which Razborov calls a "type"), the morphisms from F correspond to graphs with a specified subgraph isomorphic with F (which Razborov calls a "flag"). The pushout of two such morphisms results in an object obtained by gluing together the two graphs along the image of F, which is exactly how Razborov defines the product in flag algebras. So flag algebras are the algebras $\mathcal{Q}_F^{\text{out}}$ in the category of monomorphisms between graphs. This is a subalgebra of the algebra $\mathcal{Q}_F^{\text{out}}$ defined in terms of all homomorphisms between graphs. ♦

23.4.3. Regularity Lemma for categories.

There are more results on graph homomorphisms that extend quite naturally to the categorial setting. Let us state a generalization of the Regularity Lemma—both in its weak and original form (Lovász [Notes]).

To motivate the definitions below, consider a weighted graph G. This can be viewed as a weighting of the edges of a complete graph, i.e., as a quantum morphism $K_2 \to \tilde{K}_n$, which is symmetric, i.e., it is invariant under swapping the two nodes of K_2. Regularity lemmas try to find a partition (a morphism $K_n^\circ \to \tilde{K}_k$), and weighting of edges of $d = \tilde{K}_k$ (a quantum morphism in $\mathcal{Q}(a,d)$), such that "pulling back" these weights to K_n°, we get a good approximation in the cut norm. How to translate to the categorial language that the cut norm of a weighted graph is small? It means that for every morphism $K_n^\circ \to K_2^\circ$, if we push forward the edgeweights, then the resulting edgeweights of K_2° are all small (this says that versions (a) and (c) of the cut norm in Exercise 8.4 are small, but these are all equivalent up to absolute constant factors). These considerations motivate the following general definitions.

Let $\alpha \in \mathcal{K}(a,b)$ and $\beta \in \mathcal{K}(c,b)$. We define a quantum morphism $\alpha\beta^* \in \mathcal{Q}(a,c)$ by
$$\alpha\beta^* = \sum_{\varphi \in \mathcal{K}(a,c):\, \varphi\beta=\alpha} \varphi.$$
This operation extends linearly to define xy^* for $x \in \mathcal{Q}(a,b)$ and $y \in \mathcal{Q}(c,b)$. It is not hard to check that $x(zy)^* = (xy^*)z^*$, and $\langle x, yz^*\rangle = \langle xz, y\rangle$.

For every quantum morphism $x = \sum_\varphi x_\varphi \varphi \in \mathcal{Q}(a,b)$ and every object c, we define the *c-norm* of x by
$$\|x\|_c = \max_{\beta \in \mathcal{K}(b,c)} \frac{\|x\beta\|_\infty}{|\mathcal{K}(a,b)|}.$$
This norm generalizes the cut norm: if $a = K_2$ and $c = K_2^\circ$, then a symmetric quantum morphism $x \in \mathcal{Q}(a,b)$ is a weighting of the edges of b, and it is not hard to see that $\|x\|_\square/2 \leq \|x\|_c \leq \|x\|_\square$.

Let c^m denote the m-th direct power of the object c. The first inequality in the following lemma generalizes the Frieze–Kannan Weak Regularity Lemma 9.3, while the second implies the Original Regularity Lemma of Szemerédi 9.2.

LEMMA 23.20. *Let $\mathcal{K}$ be a locally finite category having finite direct products. Let a, b and c be three objects in $\mathcal{K}$, and let $m \geq 1$. Then for every $x \in \mathcal{Q}(a,b)$ there exists a morphism $\varphi \in \mathcal{K}(b,c^m)$ and a quantum morphism $y \in \mathcal{Q}(a,c^m)$ such that*
$$\|x - y\varphi^*\|_c \leq \frac{1}{\sqrt{m}}\|x\|_2$$
and
$$\|x - y\varphi^*\|_{c^{2m}} \leq \frac{1}{\sqrt{\log^* m}}\|x\|_2.$$

The Weak Regularity Lemma is obtained, as described above, by taking $a = K_2$ and $c = K_2^\circ$ and applying the first bound. Note that a morphism in $\mathcal{K}(b, c^m)$ corresponds to a partition of $V(G)$ into 2^m classes. The Original Regularity Lemma can be derived from the second bound similarly. Strong versions can be generalized as well, but for the details we refer to Lovász [Notes].

There are many unsolved questions here: can the Counting Lemma be generalized to categories? Do the notions of convergence and limit objects be formulated in an interesting way? Could these results shed new light on hypergraph limits and regularity lemmas? Or perhaps even on sparse regularity lemmas?

EXERCISE 23.21. Let $\mathcal{K}$ be a locally finite category, and let c be an object. Prove that every monomorphism in $\mathcal{K}(c, c)$ is an isomorphism.

EXERCISE 23.22. Let $\mathcal{K}$ be a locally finite category, and let c, d be two objects. Suppose that there are monomorphisms in $\mathcal{K}(c, d)$ and in $\mathcal{K}(d, c)$. Prove that c and d are isomorphic.

EXERCISE 23.23. Let $\mathcal{K}$ be a locally finite category, and let c and d be two objects. For any two morphisms $\alpha \in \mathcal{K}(a, a')$ and $\beta \in \mathcal{K}(b, b')$, let $N_{\alpha,\beta}$ denote the number of 4-tuples of morphisms $(\varphi, \psi, \mu, \nu)$ ($\varphi \in \mathcal{K}(c, a), \psi \in \mathcal{K}(c, b), \mu \in \mathcal{K}(a', d), \nu \in \mathcal{K}(b', d)$) such that $\varphi \alpha \mu = \psi \beta \nu$. Prove that the matrix $N = (N_{\alpha,\beta})$, where α and β range over all morphisms of the category, is positive semidefinite.

EXERCISE 23.24. Let a and b be two objects in a locally finite category. Suppose that the direct powers $a \times a$ and $b \times b$ exist and are isomorphic. Prove that a and b are isomorphic.

EXERCISE 23.25. Let a, b, c, d be four objects in a locally finite category $\mathcal{K}$ such that the direct products $a \times c$, $b \times c$, $a \times d$ and $b \times d$ exist, $a \times c$ and $b \times c$ are isomorphic, and d has at least one morphism into c. Prove that $a \times d$ and $b \times d$ are isomorphic.

23.5. And more...

There are many types of discrete structures for which one can try to define convergence and limit objects for growing sequences. This is typically not straightforward, as one can see from the case of simple graphs with (say) $\Theta(n^{3/2})$ edges. However, this approach has been successful in some cases.

It is a natural question to extend the theory of graph limits to directed graphs. Let us assume that these graphs are simple, so that there are no loops and there is at most one edge between two nodes in a given direction. Diaconis and Janson show that at least some of the theory can be developed based on the theory of exchangeable arrays (see Section 11.3.3). The limit object is a bit more complicated, it can be described by four measurable functions $W_{0,0}, W_{0,1}, W_{1,0}, W_{1,1} : [0,1]^2 \to [0,1]$ such that $W_{0,0}$ and $W_{1,1}$ are symmetric, $W_{0,1}(x, y) = W_{1,0}(y, x)$ and $W_{0,0} + W_{0,1} + W_{1,0} + W_{1,1} = 1$. The function $W_{0,1}(x, y)$ measures the density of edges from an infinitesimal neighborhood of x to an infinitesimal neighborhood of y etc. Some further remarks and observations can be found scattered in papers, but no comprehensive treatment seems to be known. Perhaps most of the extension is rather straightforward (but be warned: the theory of *existence* of homomorphisms between digraphs is much more involved—one can say richer—than for undirected graphs; see Hell and Nešetřil [2004]).

Posets can be considered as special digraphs, but they are sufficiently important in many contexts to warrant a separate treatment. Janson [2011a, 2012] starts a limit theory of posets. The treatment is based on methods similar to the limit theory of dense graphs in this book, but there are some analytic complications and interesting special features, for which we refer to the paper.

Going away from graphs, let us consider the set S_n of permutations of the set $[n]$. Cooper [2004, 2006] defined and characterized quasirandomness for permutations, and proved a regularity lemma for them. Hoppen, Kohayakawa, Moreira, Ráth and Menezes Sampaio [2011, 2011] defined convergent sequences of permutations, and described their limit objects. Given a permutation $\pi \in S_n$ and a subset $A =$

$\{a_1, \ldots, a_k\} \subseteq [n]$, we can define a permutation $\pi[A] \in S_k$ by letting $\pi[A]_i < \pi[A]_j$ iff $\pi_{a_i} < \pi_{a_j}$. For a permutation $\tau \in S_k$, let $\Lambda(\tau, \pi)$ denote the number of sets A with $\pi[A] = \tau$, and define the density of τ in π by $t(\tau, \pi) = \Lambda(\tau, \pi)/\binom{n}{k}$. A sequence of permutations $\pi_1, \pi_2, \ldots$ (on larger and larger sets) is *convergent*, if for every permutation τ, the number $t(\tau, \pi_n)$ tends to a limit as $n \to \infty$. Every convergent permutation sequence has a limit object in the form of a coupling measure on $[0,1]^2$, which is uniquely determined. Král and Pikhurko [2012] have used this machinery of limit objects to prove a conjecture of Graham on permutations.

I have already mentioned the limit theory of metric spaces due to Gromov [1999]. While developed with quite different applications in mind, this turns out to be closely related to our theory of graph limits. Gromov considers metric spaces endowed with a probability measure, and defines distance, convergence and limit notions for them. A simple graph G can be considered as a special case, where the distance of two adjacent nodes is $1/2$, the distance of two nonadjacent nodes is 1, and the probability distribution on the nodes is uniform. Under this correspondence, our notion of graph convergence is a special case of Gromov's "sample convergence" of metric spaces. Vershik [2002, 2004] considers random metric spaces on countable sets, and defines and proves their universality. He also characterizes isomorphism of metric spaces with measures in terms of sampling, analogously to Theorem 13.10. In a recent paper, Elek [2012b] explores this connection and shows how Gromov's notions imply results about graph convergence, and also how results about graph limits inspire answers to some questions about metric spaces. Perhaps Gromov's theory can be applied to graph sequences that are not dense, using the standard distance between nodes in the graph.

One of the earliest limit theories is John von Neumann's theory of continuous geometries. The idea here is that if we look at higher and higher dimensional vector spaces over (say) the real field, then the obvious notion of their limit is the Hilbert space. But, say, we are interested in the behavior of subspaces whose dimension is proportional to the dimension of the whole space. Going to the Hilbert space, this condition becomes meaningless. Neumann constructed a limit object, called a *continuous geometry*, in which the "dimensions" of subspaces are real numbers between 0 and 1. This construction can be extended to certain geometric lattices (Björner and Lovász [1987]), but its connection with the theory in this book has not been explored.

Perhaps most interesting from the point of view of quasirandomness and limits are sequences of integers, due to their role in number theory. (After all, Szemerédi's Regularity Lemma was inspired by his solution of the Erdős–Turán problem on arithmetic progressions in dense sequences of integers.) Often sequences are considered modulo n; this gives a finite group structure to work with, while one does not lose much in generality. Ever since the solution of the Erdős–Turán problem for 3-term arithmetic progressions by Roth [1952], through the general solution by Szemerédi [1975], through the work of Gowers [2001] on "Gowers norms", to the celebrated result of Green and Tao [2008] on arithmetic progressions of primes, a central issue has been to define and measure how random-like a set of integers is. I will not go into this large literature; Tao [2006c] and Kra [2005] give accessible accounts of it. What I want to point out is the exciting asymptotic theory of structures consisting of an abelian group together with a subset of its elements, and more generally, abelian groups with a function defined on them. There has been a lot of parallel developments in this area, most notably the work of Green, Tao and Ziegler [2011] and of Szegedy [2012a]. Not surprisingly, the latter is closer to the point of view taken in this book, and develops a theory of limit objects of functions on abelian groups, which is full of surprises but also with powerful results. (For example, to describe the limits of abelian groups, non-abelian groups are needed!) The theory has connections with number theory, ergodic theory, and higher-order Fourier analysis. This explains why I cannot go into the details, and can only refer to the papers.

APPENDIX A

Appendix

A.1. Möbius functions

Let L be a finite lattice (for us, it will be either the lattice of all subsets, or the lattice of all partitions, of a finite set V). The *Möbius function* of the lattice is a function $\mu : L \times L \to \mathbb{R}$, defined by the equations

$$\mu(x,y) = 0 \quad \text{if } x \not\leq y, \qquad \sum_{x \leq z \leq y} \mu(x,z) = \mathbb{1}(x=y).$$

This is perhaps easier to understand in a matrix algebra setting. Let $\mathcal{M}(L)$ denote the set of $L \times L$ matrices A in which $A_{xy} = 0$ for any two lattice elements $x \not\leq y$. It is easy to see that $\mathcal{M}(L)$ is closed under addition, matrix multiplication and matrix inverse (if an inverse exists), and so it is a matrix algebra. One special matrix of importance is the *zeta matrix* $Z \in \mathcal{M}(L)$ defined by $Z_{xy} = \mathbb{1}(x \leq y)$. Clearly Z is invertible, and $M = Z^{-1}$ is a matrix with integer entries, called the *Möbius matrix*. The entries of M give the Möbius function: $M_{xy} = \mu(x,y)$.

For every function $f : L \to \mathbb{C}$, we define its (upper) summation function $g(x) = \sum_{y \geq x} f(y)$. From g, we can recover function f by the formula $f(x) = \sum_{y \geq x} \mu(x,y)g(y)$. This is again better seen in a matrix form: we consider f and g as vectors in $\mathbb{C}^L$, then $g = Zf$, which is equivalent to $f = Z^{-1}g = Mg$. Of course, we can turn the lattice upside down, and derive similar formulas for the lower summation.

The following simple but very useful matrix identity is due to Lindström [1969] and Wilf [1968]. Let $f : L \to \mathbb{R}$ be any function, A_f be the $L \times L$ matrix with $(A_f)_{xy} = f(x \vee y)$. Then

(A.1) $$A_f = Z\mathrm{diag}(Mf)Z^\mathsf{T}.$$

An important consequence of this identity states that A_f is positive semidefinite if and only if the Möbius inverse of f is nonnegative.

EXAMPLE A.1. If L is the lattice of subsets of a finite set S, then $\mu(X,Y) = (-1)^{|Y \setminus X|}$ for all $X \subseteq Y \subseteq S$. Möbius inversion is equivalent to the inclusion-exclusion formula in this case. ♦

EXAMPLE A.2. Consider the lattice of partitions Π_n of the finite set $[n]$, where the bottom element is the discrete partition P_0 (with n classes), the top element is the indiscrete partition P_1 (with one class), and $P \leq Q$ means that P refines Q. The Möbius function of this lattice is given by the Frucht–Rota–Schützenberger Formula

(A.2) $$\mu_P = \mu(0, P) = (-1)^{n-|P|} \prod_{S \in P} (|S|-1)!$$

where $|P|$ denotes the number of classes in the partition $\mathcal{P}$. (This easily implies a formula for $\mu(Q, P)$, but we won't need it.)

For the partition lattice, we need some simple identities: for every $P \in \Pi_n$,

(A.3) $$\sum_{R \geq P} (x)_{|R|} = x^{|P|}$$

By Möbius inversion,

(A.4) $$\sum_P \mu_P x^{|P|} = (x)_n,$$

and from the Lindström–Wilf Formula,

(A.5) $$\sum_{P,Q} \mu_P \mu_Q x^{|P \vee Q|} = (x)_n.$$

♦

See Van Lint and Wilson [1992] for more on the Möbius function of a lattice.

A.2. The Tutte polynomial

Several important graph invariants can be expressed in terms of the *Tutte polynomial* of the graph $G = (V, E)$ (which may have loops and multiple edges). The quickest way to define this is by the following formula. Let $c(A)$ ($A \subseteq E$) denote the number of connected components of the graph (V, A) (including the isolated nodes); in particular, $c(E)$ is the number of components of G. We define

(A.6) $$\mathsf{tut}(G; x, y) = \sum_{A \subseteq E} (x-1)^{c(A)-c(E)} (y-1)^{c(A)+|A|-\mathsf{v}(G)}.$$

This definition does not in any way indicate the many uses this polynomial has. The recurrence relation

(A.7) $$\mathsf{tut}(G; x, y) = \mathsf{tut}(G - e; x, y) - \mathsf{tut}(G/e; x, y),$$

where $e \in E(G)$ is any edge that is not a cut-edge or a loop, says much more (here G/e denotes the graph obtained from G by contracting e, i.e., deleting one copy of e and identifying its endpoints). If the G has i loops and j cut-edges, and no other edges, then $\mathsf{tut}(G; x, y) = x^i y^j$. The Tutte polynomial is multiplicative over connected components. There are many graph invariants that satisfy recurrence (A.7) (or some very similar recurrence), and these can be expressed as substitutions into the Tutte polynomial (or some slight modification of it).

One often uses the following version of the Tutte polynomial, sometimes called the *cluster expansion polynomial*:

(A.8) $$\mathsf{cep}(G; u, v) = \sum_{A \subseteq E(G)} u^{c(A)} v^{|A|}.$$

This differs from the usual Tutte polynomial $T(x, y)$ on two counts: first, instead of the variables x and y, we use $u = (x-1)(y-1)$ and $v = y - 1$; second, we scale by $u^{c(E)} v^{|V|}$.

The cluster expansion polynomial satisfies the following identities: (a) $\mathsf{cep}(G; u, v) = v\mathsf{cep}(G/e; u, v) + \mathsf{cep}(G - e; u, v)$ for all edges e that are not loops; (b) $\mathsf{cep}(G; u, v) = q\mathsf{cep}(G - i; u, v)$ if i is an isolated node; $\mathsf{cep}(G; u, v) = u^{\mathsf{e}(G)}$ if G is a graph consisting of a single node. These relations determine the value of the polynomial for any substitution. (See e.g. Welsh [1993] for more on the Tutte polynomial.)

Chromatic polynomial. Let $G = (V, E)$ be a multigraph with n nodes. For every nonnegative integer q, we denote by $\mathsf{chr}(G, q)$ the number of q-colorations of G (in the usual sense, where adjacent nodes must be colored differently). Clearly $\mathsf{chr}(G, q)$ does not depend on the multiplicities of edges (as long as these multiplicities are positive), and $\mathsf{chr}(G, q) = 0$ if G has a loop.

Let $\mathsf{chr}_0(G, k)$ denote the number of k-colorations of G in which all colors occur. Then clearly

$$\text{(A.9)} \qquad \mathsf{chr}(G, q) = \sum_{k=0}^{\mathsf{v}(G)} \mathsf{chr}_0(G, k) \binom{q}{k}.$$

This implies that $\mathsf{chr}(G, q)$ is a polynomial in q with leading term q^n and constant term 0, which is called the *chromatic polynomial* of G. One can evaluate this polynomial for non-integral values of q, when it has no direct combinatorial meaning. We define $\mathsf{chr}(K_0, q) = 1$.

It is easy to see that if q is a positive integer, then for every $e \in E(G)$,

$$\text{(A.10)} \qquad \mathsf{chr}(G, q) = \mathsf{chr}(G - e, q) - \mathsf{chr}(G/e, q).$$

Since this equation for polynomials holds for infinitely many values of q, it holds identically. If i is an isolated node of G, then we have $\mathsf{chr}(G, q) = q\mathsf{chr}(G - i; q)$. From these recurrence relations a number of properties of the chromatic polynomial are easily proved, for example, that its coefficients alternate in sign. Most importantly, they imply that the chromatic polynomial is a special substitution of the cluster expansion polynomial: $\mathsf{chr}(G, q) = \mathsf{cep}(G; q, -1)$. From formula (A.8) we get

$$\text{(A.11)} \qquad \mathsf{chr}(G; q) = \sum_{A \subseteq E(G)} (-1)^{|A|} q^{c(A)}.$$

The coefficient of the linear term in the chromatic polynomial is called the *chromatic invariant* of the graph. It will be convenient to consider this quantity with an adjusted sign

$$\mathsf{cri}(G) = \sum_{G'} (-1)^{\mathsf{e}(G') - \mathsf{v}(G) + 1},$$

where G' ranges through all connected spanning subgraphs of G. It follows from (A.10) that if G is a simple graph, then for every $e \in E(G)$,

$$\text{(A.12)} \qquad \mathsf{cri}(G) = \mathsf{cri}(G - e) + \mathsf{cri}(G/e).$$

This implies by induction that $\mathsf{cri}(G) > 0$ if G is connected and $\mathsf{cri}(G) = 0$ if G is disconnected.

Spanning trees. Let $\mathsf{tree}(G)$ denote the number of spanning trees in the graph G. This parameter has played an important role in the development of algebraic graph theory; formulas for its computation go back to the work of Kirchhoff in the mid-19th century. The number of spanning trees satisfies the recurrence relation

$$\text{(A.13)} \qquad \mathsf{tree}(G) = \mathsf{tree}(G - e) + \mathsf{tree}(G/e)$$

for every edge that is not a loop. It is best to define $\mathsf{tree}(K_1) = 1$ and $\mathsf{tree}(K_0) = 0$. One gets by direct substitution in (A.6) that for every connected graph G, $\mathsf{tree}(G) = \mathsf{tut}(G; 1, 1)$.

There are many other expressions in the literature for $\mathsf{tree}(G)$. Perhaps the best known is Kirchhoff's Formula (also called the Matrix Tree Theorem) saying that $\mathsf{tree}(G)$ is equal to any cofactor of the Laplacian $L_G = A_G - D_G$ (here A_G is the adjacency matrix of G and D_G is the diagonal matrix composed of the degrees).

There are many useful inequalities for $\mathsf{tree}(G)$, of which we mention two: the trivial bound

$$\mathsf{tree}(G) \leq \prod_u d_G(u), \tag{A.14}$$

and the relation with the chromatic invariant, which follows easily by induction from the recurrences (A.13) and (A.12):

$$0 \leq \mathsf{chr}(G) \leq \mathsf{tree}(G). \tag{A.15}$$

Nowhere zero flows. Let $\mathsf{flo}(G, q)$ denote the number of nowhere-zero q-flows. To be precise, we fix an orientation of the edges for any graph G, and count maps $f : \vec{E}(G) \to \mathbb{Z}_q$ such that the sum of flow values on edges entering a given node is equal to the sum of flow values on edges leaving the node. This number is given by $|\mathsf{tut}(0, q-1)|$.

A.3. Some background in probability and measure theory

A.3.1. Probability spaces. We have to fix some terminology. A *probability space* is a triple $(\Omega, \mathcal{A}, \pi)$, where $\mathcal{A}$ is a sigma-algebra on the set Ω, and π is a probability measure on π. We say that the space is *separating*, if for any two elements of Ω there is a set in $\mathcal{A}$ containing exactly one of them. The space is *countably generated*, if there is a countable subset $\mathcal{J}$ of $\mathcal{A}$ generating $\mathcal{A}$ (in other words, $\mathcal{A}$ is the smallest sigma-algebra containing $\mathcal{J}$). It is often convenient to assume (which we can do for free), than $\mathcal{J}$ is a set algebra, i.e., it is closed under intersection and complementation. An *atom* of the space is a singleton with positive measure.

Two probability spaces $(\Omega_i, \mathcal{A}_i, \pi_i)$ ($i = 1, 2$) are *isomorphic* if there is an invertible map $\varphi : \Omega_1 \to \Omega_2$ that gives a bijection between $\mathcal{A}_1$ and $\mathcal{A}_2$ and preserves the measure. Two probability spaces $(\Omega_i, \mathcal{A}_i, \pi_i)$ ($i = 1, 2$) are *isomorphic up to nullsets* if one can delete sets $X_i \subseteq \Omega_i$ of measure 0 so that the remaining probability spaces are isomorphic.

From the point of view of basic constructions in probability (independence, expectation and variance of random variables etc.) the underlying probability space does not matter much, at least as long as it is atom-free. But for more advanced technical work, one likes to work with a robust class of them with nice properties. A *Borel sigma-algebra* (also called standard Borel space) is a sigma-algebra isomorphic to the sigma-algebra of Borel subsets of a Borel set in $\mathbb{R}$. It can be shown that this definition would not change if instead of subsets of $\mathbb{R}$ we allowed subsets of $\mathbb{R}^n$, or indeed, of any separable complete metric space. A *Borel probability space* is a probability space defined on a Borel sigma-algebra. Equivalently (this is nontrivial), it is isomorphic up to nullsets to the disjoint union of a closed interval (with the Borel sets and the Lebesgue measure) and a countable set of atoms. Every Borel space is countably generated and separating. Every finite probability space is Borel. A *standard probability space* (with small variations, also called a Lusin, Lebesgue or Rokhlin space) is the completion of a Borel probability space (i.e., we add all subsets of sets of measure 0 to the sigma-algebra). Standard probability spaces have many useful properties, some of which will be mentioned below; in a sense, they behave as you would expect them to behave.

In this sense Borel (or standard) spaces are quite special. On the other hand, they are general enough so that we can restrict our attention to them; this is due to the following fact:

PROPOSITION A.3. *Every probability space on a countably generated separating sigma-algebra can be embedded into a Borel space in the sense that it is isomorphic up to nullsets to the restriction of a Borel space to a subset with outer measure 1.* □

A.3.2. Measure preserving maps. Let $(\Omega_i, \mathcal{A}_i, \pi_i)$ $(i = 1, 2)$ be probability spaces. A map $\varphi : (\Omega_1, \mathcal{A}_1, \pi_1) \to (\Omega_2, \mathcal{A}_2, \pi_2)$ is *measure preserving*, if $\varphi^{-1}(A) \in \mathcal{A}_1$ for every $A \in \mathcal{A}_2$, and $\pi_1(\varphi^{-1}(A)) = \pi_2(A)$. (So the name is a bit misleading, because it is φ^{-1} rather than φ that preserves measure.) A measure preserving map is not necessarily bijective; for example, the map $[0, 1] \to [0, 1]$ defined by $x \mapsto 2x$ mod 1 is measure preserving. We say that a measure preserving map φ is *invertible*, if it is bijective and φ^{-1} is also measure preserving.

If $\varphi : (\Omega_1, \mathcal{A}_1, \pi_1) \to (\Omega_2, \mathcal{A}_2, \pi_2)$ is measure preserving, then for every integrable function $f : (\Omega_2, \mathcal{A}_2, \pi_2) \to \mathbb{R}$ we have

$$(A.16) \qquad \int_{\Omega_1} f(\varphi(x)) \, d\pi_1(x) = \int_{\Omega_2} f(x) d\pi_2(x).$$

Let $\overline{S}_{[0,1]}$ denote the semigroup of measure preserving maps $[0, 1] \to [0, 1]$, and let $S_{[0,1]}$ be the group of invertible measure preserving maps $[0, 1] \to [0, 1]$.

One of the most important properties of standard probability spaces is that under mild conditions, their measure preserving images are also standard.

PROPOSITION A.4. *Let $(\Omega_1, \mathcal{A}_1, \pi_1)$ be a standard probability space and let $(\Omega_2, \mathcal{A}_2, \pi_2)$ be another probability space where $\mathcal{A}_2$ has a countable subset separating any two points of Ω_2. Let $\varphi : \Omega_1 \to \Omega_2$ be a measure preserving map. Then $(\Omega_2, \mathcal{A}_2, \pi_2)$ is standard, and $\Omega_2' = \Omega_2 \setminus \varphi(\Omega_1)$ has measure 0. Furthermore, if φ is bijective, then φ^{-1} is an isomorphism $(\Omega_2', \mathcal{A}_2|_{\Omega_2'}, \pi_2|_{\Omega_2'}) \to (\Omega_1, \mathcal{A}_1, \pi_1)$. In particular, φ^{-1} is also measure preserving.*

REMARK A.5. It is usually a matter of taste or convenience whether we decide to work on a complete space or on a countably generated space. One tends to be sloppy about this, and just say, for example, that the underlying probability space is $[0, 1]$, without specifying whether we mean the sigma-algebra of Borel sets or of Lebesgue measurable sets.

Often, one implicitly assumes that the Borel sigma algebra is defined as the set of Borel sets in a Polish space, and uses topological notions like open sets or continuous functions to define measure theoretic notions. This is sometimes unavoidable (see e.g. the definition of weak convergence below), but the same Borel sigma-algebra can be defined by very different topological spaces, and this is important in some cases even in this book. I will use this topological representation only where it is necessary.

A.3.3. The space of measures. Let T be a topological space, and let $\mathcal{P}(T)$ denote the set of probability measures on the Borel subsets of T. We say that a sequence of measures $\mu_1, \mu_2, \cdots \in \mathcal{P}(T)$ *converges weakly* to a probability measure $\mu \in \mathcal{P}(T)$, if
$$\int_\mathcal{T} f\, d\mu_n \to \int_\mathcal{T} f\, d\mu \quad (n \to \infty)$$
for every continuous bounded function $f: S \to \mathbb{R}$. Most often we need this notion in the case when $\mathcal{T}$ is a compact metric space, so we don't have to assume the boundedness of f. This notion of convergence defines a topology on $\mathcal{P}(T)$, which we call the *topology of weak convergence*.

By Prokhorov's Theorem (see e.g. Billingsley [1999]; this is not the most general form), for a compact metric space K, the space $\mathcal{P}(K)$ is compact in the topology of weak convergence, and also metrizable. (One can describe explicit metrizations, like the Levy-Prokhorov metric, but we don't need them.)

There is an important warning about weak convergence: it is not a purely measure theoretic notion, but topological. In other words, we can have a sequence of measures on a Borel sigma-algebra $(\Omega, \mathcal{B})$ that is weakly convergent if we put one topology on Ω with the given Borel sets, but not convergent if we put another such topology on Ω. Sometimes we play with this, and change the topology (without changing its Borel sets) to suit our needs.

A.3.4. Coupling. A *coupling measure* between two probability spaces $(\Omega_i, \mathcal{A}_i, \pi_i)$ $(i = 1, 2)$ is a probability measure μ on the sigma-algebra $(\Omega_1, \mathcal{A}_1) \times (\Omega_2, \mathcal{A}_2)$ whose marginals are π_1 and π_2, i.e., $\mu(A_1 \times \Omega_2) = \pi_1(A_1)$ and $\mu(\Omega_1 \times A_2) = \pi_2(A_2)$ for all $A_i \in \mathcal{A}_i$. In terms of random variables, a coupling measure is the distribution of a pair (X_1, X_2), where X_i has distribution π_i. The simplest coupling measure is the product measure $\pi_1 \times \pi_2$, corresponding to choosing X_1 and X_2 independently. If $(\Omega_1, \mathcal{A}_1, \pi_1) = (\Omega_2, \mathcal{A}_2, \pi_2)$, then the measure on the diagonal $\{(x,x): x \in \Omega_1\}$ defined by $\mu\{(x,x): x \in A\} = \pi_1(A)$ is another coupling measure.

Suppose that $(\Omega_i, \mathcal{A}_i, \pi_i)$ is the sigma-algebra of Borel sets in a compact metric space K_i. It is easy to see that if we fix the marginal distributions, the set of coupling measures forms a closed (and hence compact) subspace of $\mathcal{P}(K_1 \times K_i)$. This space is in fact much nicer than the space of all measures, as the following proposition shows (for a proof, see [Notes]).

PROPOSITION A.6. *Let K_1, K_2 be compact metric spaces and let $(K_i, \mathcal{B}_i, \lambda_i)$ be probability spaces on their Borel sets. Let $\mu_1, \mu_2, \ldots$ and μ be coupling measures between $(K_1, \mathcal{B}_1, \lambda_1)$ and $(K_2, \mathcal{B}_2, \lambda_2)$. Then the following are equivalent:*

(i) $\mu_n \to \mu$ *weakly;*

(ii) $\mu_n(B_1 \times B_2) \to \mu(B_1 \times B_2)$ *for all sets $B_i \in \mathcal{B}_i$;*

(iii) $\int_{K_1 \times K_2} f\, d\mu_n \to \int_{K_1 \times K_2} f\, d\mu$ *for every function $f: K_1 \times K_2 \to \mathbb{R}$ that is the limit of a uniformly convergent sequence of stepfunctions;*

(iv) *There are measurable functions $f_n, f : [0,1] \to K_1 \times K_2$ such that $\mu_n(X) = \lambda(f_n^{-1}(X))$, $\mu(X) = \lambda(f^{-1}(X))$, and $f_n \to f$ almost everywhere.*

The following construction of coupling measures follows from Proposition 3.8 of Kellerer [1984].

PROPOSITION A.7. *Let* $(\Omega_i, \mathcal{A}_i, \pi_i)$ $(i = 0, 1, 2)$ *be standard probability spaces. Let* $\varphi_i : \Omega_i \to \Omega_0$ $(i = 1, 2)$ *be measure preserving maps. Then there exists a coupling* μ *of* $(\Omega_1, \mathcal{A}_1, \pi_1)$ *and* $(\Omega_2, \mathcal{A}_2, \pi_2)$ *such that* $\mu\{(x_1, x_2) : \varphi_1(x_1) = \varphi_2(x_2)\} = 1$.

A.3.5. Markov chains. Markov chains are very basic material in probability theory, but usually they are defined in a more restrictive setting than what we need, so let us give a brief introduction.

A *Markov chain* is described by a σ-algebra $(\Omega, \mathcal{A})$ (the state space), together with a system of probability distributions $(P_u : u \in \Omega)$ on $(\Omega, \mathcal{A})$ such that $P_u(A)$ is a measurable function of u for each $A \in \mathcal{A}$ (the transition distributions). We call a probability distribution π on $(\Omega, \mathcal{A})$ *stationary*, if

$$\int_A P_u(A) \, d\pi(u) = \pi(A)$$

for all $A \in \mathcal{A}$. If the state space is finite, then the Markov chain always has a stationary distribution. In the general case, this is not always true. One sufficient condition for the existence is that $(\Omega, \mathcal{A})$ is the sigma-algebra of Borel sets in a compact Hausdorff space K, and the map $u \mapsto P_u$ is continuous as a map from K into $\mathcal{P}(K)$ with the weak topology.

The more usual description of a Markov chain as a sequence of random variables is obtained if we also specify a starting distribution σ on $(\Omega, \mathcal{A})$. We start with an $X_0 \in \Omega$ from the distribution σ, and generate X_{n+1} as a random element of Ω from the distribution P_{X_n}. We will call the sequence $(X_0, X_1, X_2, \dots)$ a *random walk* on Ω. If the Markov chain has a stationary distribution π, and X_0 is randomly chosen according to π, then every X_n is also from the stationary distribution, and we call the sequence $(X_0, X_1, X_2, \dots)$ a *stationary walk*.

Every Markov chain defines a probability measure ψ on $\Omega \times \Omega$ by

$$\psi(A \times B) = \int_A P_u(B) \, d\pi(u).$$

We can think of $\psi(A \times B)$ as the frequency with which a stationary walk steps from A to B. We call the measure ψ the *step distribution* of the Markov chain. We say that the Markov chain is *reversible* if $\psi(A \times B) = \psi(B \times A)$ for any two measurable sets A, B.

A.3.6. Martingales. A (finite or infinite) sequence $(X_1, X_2, \dots)$ of real valued random variables is called a *martingale*, if for all $k \geq 0$ we have $\mathsf{E}(|X_k|) < \infty$, and $\mathsf{E}(X_{k+1} \,|\, X_1, \dots, X_k) = X_k$. More generally (and not quite logically), the sequence is called a *supermartingale*, if $\mathsf{E}(X_{k+1} \,|\, X_1, \dots, X_k) \leq X_k$. A *submartingale* is defined analogously.

It is often convenient to define $X_0 = \mathsf{E}(X_1)$ (so this is a random variable that is concentrated on a single value). Clearly all expectations in a martingale are the same: $X_0 = \mathsf{E}(X_1) = \mathsf{E}(X_2) = \dots$ For a supermartingale, the expectations form a non-increasing sequence.

EXAMPLE A.8. Let $Y_1, Y_2, \dots$ be independent random variables such that $\mathsf{E}(Y_k) = 0$. Then the random variables $X_k = Y_1 + \dots + Y_k$ form a martingale.

The condition that $\mathsf{E}(Y_k) = 0$ can of course be arranged, as soon as the expectations exist, by subtracting its expectation from each Y_k, which does not influence the independence of these variables. So the results of martingale theory can be applied to the partial sums of any sequence of independent random variables with finite expectations. ♦

Many applications in combinatorics use martingales through the following construction.

EXAMPLE A.9 (**Doob's Martingale**). Let $(\Omega, \mathcal{A}, \pi)$ be a probability space and let $f : \Omega \to \mathbb{R}$ be an integrable function. Let $Y_1, \ldots, Y_n$ be independent random elements of Ω from the distribution π, and let $X_k = \mathsf{E}(f(Y_1, \ldots, Y_n) \mid Y_1, \ldots, Y_k)$. Then $(X_1, \ldots, X_n)$ is a martingale. ♦

EXAMPLE A.10. Let $f : [0,1] \to \mathbb{R}$ be an integrable function, and let $\mathcal{P}_1, \mathcal{P}_2, \ldots$ be a sequence of partitions of $[0,1]$ into a finite number of measurable parts such that $\mathcal{P}_{n+1}$ is a refinement of $\mathcal{P}_n$. Let $Y \in [0,1]$ be a uniform random point, and consider the sequence $X_k = f_{\mathcal{P}_k}(Y)$. Then $(X_1, X_2, \ldots)$ is a martingale. Instead of $[0,1]$, we could of course consider any probability space, for example, $[0,1]^2$, which shows the connection of martingales with the stepping operator. ♦

There are (at least) three theorems on martingales that are relevant for combinatorial applications; these play an important role in our book as well.

Let $(X_0, X_1, \ldots)$ be a sequence of random variables. A random variable T with nonnegative integral values is called a *stopping time* (for the sequence $(X_0, X_1, \ldots)$), if for every $k \geq 0$, the event $T = k$, conditioned on $X_1, \ldots, X_k$, is independent of the variables $X_{k+1}, X_{k+2} \ldots$ (In computer science, this is often called a *stopping rule*: we decide whether we want to stop after k steps depending on the values of the variables we have seen before, possibly using some new independent coin flips).

The Martingale Stopping Theorem (a.k.a. Optional Stopping Theorem) has many versions, of which we state one:

THEOREM A.11. *Let $(X_1, X_2, \ldots)$ be a supermartingale for which $|X_{m+1} - X_m|$ is bounded (uniformly for all m), and let T be a stopping time for which $\mathsf{E}(T)$ is finite. Then $\mathsf{E}(X_T) \leq X_0$.*

For a martingale, we have equality in the conclusion, and for a submartingale, we have the reverse inequality in the conclusion.

The following fact is called the Martingale Convergence Theorem (again, we don't state it in its most general form).

THEOREM A.12. *Let $(X_1, X_2, \ldots)$ be a martingale such that $\sup_n \mathsf{E}(|X_n|) < \infty$. Then $(X_1, X_2, \ldots)$ is convergent with probability 1.*

Applying this theorem to the martingale in Example A.10, we get that if f is integrable, then the functions $f_{\mathcal{P}_k}$ tend to a limit almost everywhere. This limit may not be the function f itself, but it is equal to f almost everywhere if any two points of $[0,1]$ are separated by one of the partitions $\mathcal{P}_n$ (cf. also Proposition 9.8).

If we want to prove that a random variable is highly concentrated around its average, most of the time we use Azuma's Inequality (or one of its corollaries).

THEOREM A.13. *Let $(X_1, X_2, \ldots)$ be a martingale such that $|X_{m+1} - X_m| \leq 1$ for every $m \geq 0$. Then*

$$\mathsf{P}(X_m > X_0 + \lambda) < e^{-\lambda^2/(2m)}.$$

Applying Azuma's Inequality to the martingale $(-X_1, -X_2, \ldots)$, we can bound the probability that $X_m < X_0 - \lambda$. Applying it to the martingale in Example A.8, we get the following inequality, which (up to minor variations) is called Bernstein's, Chernov's or Hoeffding's:

COROLLARY A.14. *Let $X_1, X_2, \ldots$ be i.i.d. random variables, and assume that $|X_i| \leq 1$. Then*
$$\mathsf{P}\left(\frac{1}{m}\sum_{i=1}^{m} X_m - \mathsf{E}(X_1) > \varepsilon\right) < e^{-\varepsilon^2 m/2}.$$

For us, it will be most convenient to use the following corollary of Azuma's Inequality, obtained by applying it to the martingale in Example A.9:

COROLLARY A.15. *Let $(\Omega, \mathcal{A}, \pi)$ be a probability space, and let $f : \Omega^n \to \mathbb{R}$ be a measurable function such that $|f(x_1, \ldots, x_n) - f(y_1, \ldots, y_n)| \leq 1$ whenever $(x_1, \ldots, x_n)$ and $(y_1, \ldots, y_n)$ differ in one coordinate only. Let $\mathbf{x}$ be a random point of Ω^n (chosen according to the product measure). Then*
$$\mathsf{P}\big(f(\mathbf{x}) - \mathsf{E}(f(\mathbf{x})) > \varepsilon n\big) < e^{-\varepsilon^2 n/2}.$$

There are also reverse martingales. A sequence $(X_1, X_2, \ldots)$ of real valued random variables is called a *reverse martingale*, if for all $k \geq 0$ we have $\mathsf{E}(|X_k|) < \infty$, and $\mathsf{E}(X_k \,|\, X_{k+1}, X_{k+2}, \ldots) = X_{k+1}$. A finite reverse martingale is just a martingale backwards, but infinite reverse martingales are different. While reverse martingales don't seem to be as important as martingales, there is a very important example.

EXAMPLE A.16. Let $(Y_1, Y_2, \ldots)$ be i.i.d. real valued random variables with $\mathsf{E}(|Y_i|) < \infty$, and let $X_k = (Y_1 + \cdots + Y_k)/k$. Then $(X_1, X_2, \ldots)$ is a reverse martingale. So partial sums $Y_1 + \cdots + Y_k$ form a martingale, but dividing by the number of terms, we get a reverse martingale. (The latter is a bit trickier to verify.)
♦

The Martingale Convergence Theorem has an analogue for reverse martingales (which holds under more general conditions and is easier to prove):

THEOREM A.17. *Every reverse martingale is convergent with probability 1.*

Applying this theorem to Example A.16, we can derive the Strong Law of Large Numbers. We refer to the book of Williams [1991] for more on martingales.

A.4. Moments and the moment problem

Throughout this book, we deal with kernels and graphons, which are functions in two variables. We define subgraph densities in them, we consider weak isomorphism and its correspondence with measure preserving changes in the variables, approximate them by stepfunctions, just to name a few analytic techniques with graph-theoretic significance. In this Appendix we summarize some analogous notions and results for functions in a single variable (see Feller [1971], Diaconis and Freedman [2004] for more). Some of these are used in the study of kernels, some others should serve as motivation for the problems and results in the main body of the book.

Let us consider the space $L_\infty[0,1]$ of bounded measurable functions $f : [0,1] \to [0,1]$. For such a function, we consider its moments

$$M_k(f) = \int_0^1 f(x)^k \, dx \qquad (k = 0, 1, 2, \dots).$$

The moment sequence of a function determines it up to a measure preserving transformation:

PROPOSITION A.18. *Two bounded measurable functions $f, g \in L_\infty[0,1]$ have the same moments if and only if there are measure preserving maps $\varphi, \psi \in \overline{S}_{[0,1]}$ such that $f \circ \varphi = g \circ \psi$ almost everywhere. Equivalently, there is a function $h \in L_\infty[0,1]$ and maps $\varphi, \psi \in \overline{S}_{[0,1]}$ such that $f = h \circ \varphi$ and $g = h \circ \psi$.*

What makes this correspondence substantially easier to handle in the one-variable case than in the two-variable case is that in each equivalence class of weak isomorphism there is a special element:

PROPOSITION A.19 (**Monotone Reordering Theorem**). *For every measurable function $f : [0,1] \to \mathbb{R}_+$ there is a monotone decreasing function $h : [0,1] \to \mathbb{R}_+$ and a map $\varphi \in \overline{S}_{[0,1]}$ such that $f = h \circ \varphi$. The function h is uniquely determined up to a set of measure 0.*

Moment sequences can be characterized; this is called the *Hausdorff Moment Problem*. Given a sequence $(a_0, a_1, \dots)$ of nonnegative, we define two infinite matrices $H(a)$ and $M(a)$ by $H(a)_{n,k} = \sum_{j=0}^{k}(-1)^j \binom{k}{j} a_{n+j}$ and $M(a)_{n,k} = a_{n+k}$. Using this notation, moment sequences can be characterized in different ways:

PROPOSITION A.20. *For a sequence (a_n) of nonnegative numbers, the following are equivalent:*

(i) *(a_n) is the moment sequence of a function in $L_\infty[0,1]$;*

(ii) *$a_0 = 1$ and $H(a) \geq 0$ (entry by entry);*

(iii) *$a_0 = 1$ and $M(a)$ is positive semidefinite.*

We call a function $f \in L_\infty[0,1]$ a *stepfunction* if its range is finite. The set $f^{-1}(x)$ for any x in the range is called a *step* of f. Note that its monotone reordering (in the sense of Proposition A.19) is then a stepfunction in the more usual sense, whose steps are intervals.

Moment sequences of stepfunctions can be expressed as finite sums of the form

$$M_k(f) = \sum_{i=1}^{r} \lambda(S_i) f(x_i)^k,$$

where the S_i are the steps and $x_i \in S_i$. Conversely, every exponential sum

$$s(k) = \sum_{i=1}^{k} a_i b_i^k$$

with $a_i > 0$ and $\sum_i a_i = 1$ can be thought of as the moment sequence of a stepfunction. An infinite sum of this type can also be represented as the moment sequence of a function (with countably many "steps"). Proposition A.18 implies that the values $s(k)$ of such an exponential sum uniquely determine the numbers a_i and b_i.

This fact is "self-refining" in the sense that the following seemingly stronger statement easily follows from it.

PROPOSITION A.21. *Let a_i, b_i, c_i, d_i be nonzero real numbers ($i = 1, 2, \ldots$), such that $b_i \neq b_j$ and $d_i \neq d_j$ for $i \neq j$. Assume that there is a $k_0 \geq 0$ such that for all $k \geq k_0$, the sums $\sum_{i=1}^{\infty} a_i b_i^k$ and $\sum_{i=1}^{\infty} c_i d_i^k$ are convergent and equal. Then the two sums are formally equal, i.e., there is a permutation π of $\mathbb{N}$ such that $a_i = c_{\pi(i)}$ and $b_i = d_{\pi(i)}$.*

Returning to stepfunctions with a finite number of steps, we note that they can be characterized in terms of their moment matrices:

PROPOSITION A.22. *A function is a stepfunction if and only if its moment matrix has finite rank. In this case, the rank of the moment matrix is the number of steps.*

Stepfunctions are determined by a finite number of moments, and this fact characterizes them. To be more precise,

PROPOSITION A.23. *(a) Let $f \in L_\infty[0, 1]$ be a stepfunction with m steps, and let $g \in L_\infty[0, 1]$ be another function such that $M_k(f) = M_k(g)$ for $k = 0, \ldots, m$. Then f and g have the same moments.*

(b) For every function $g \in L_\infty[0, 1]$ and every $m \geq 0$ there is a stepfunction $f \in L_\infty[0, 1]$ with m steps so that $M_k(f) = M_k(g)$ for $k = 0, \ldots, m - 1$.

These results can be extended to functions $f : [0, 1] \to [0, 1]^d$ quite easily; we only formulate those that we are using in the book. Such a function is called a *stepfunction* if its range is finite. Moments don't form a sequence, but an array with d indices (a d-array for short). For $a = (a_1, \ldots, a_d) \in \mathbb{N}^d$, the corresponding moment of $f = (f_1, \ldots, f_d)$ is defined by

$$M_a(f) = \int_0^1 f_1(x)^{a_1} \ldots f_d(x)^{a_d}\, dx.$$

For an array $A : \mathbb{N}^d \to \mathbb{R}$, we define its *moment matrix* $M = M(A)$ as the infinite symmetric matrix whose rows and columns are indexed by vectors in $\mathbb{N}^n$, and $M_{u,v} = A_{u+v}$. Semidefiniteness of the moment matrix does not characterize moment sequences if $d \geq 2$, but they do at least when the function values are bounded by 1 (Berg, Christensen and Ressel [1976], Berg and Maserick [1984]):

PROPOSITION A.24. *A d-array A is the moment array of some measurable function $f : [0, 1] \to [-1, 1]^d$ if and only if $A_{0\ldots 0} = 1$, $M(A)$ is positive semidefinite and $|A_v| \leq 1$ for all $v \in \mathbb{N}^d$. Furthermore, f is a stepfunction if and only if $M(A)$ has finite rank, and the rank of $M(A)$ is equal to the number of steps of f.*

Again, stepfunctions are determined by their moments:

PROPOSITION A.25. *(a) Let $f : [0, 1] \to [0, 1]^d$ be a stepfunction with m steps, and let $g : [0, 1] \to [0, 1]^d$ be another function such that $M_a(f) = M_a(g)$ for $a \in \{0, \ldots, m\}^d$. Then there are measure preserving maps $\varphi, \psi \in \overline{S}_{[0,1]}$ such that $f \circ \varphi = g \circ \psi$ almost everywhere. In particular, g is a stepfunction, and $M_a(f) = M_a(g)$ for $a \in \mathbb{N}^d$.*

(b) *For every function* $f : [0,1] \to [0,1]^d$ *and every finite set* $S \subseteq \mathbb{N}^d$, *there is a stepfunction* $g : [0,1] \to [0,1]^d$ *with at most* $|S|+1$ *steps so that* $M_a(f) = M_a(g)$ *for all* $a \in S$.

A next question would be to define moments for functions $f : [0,1]^d \to [0,1]$. It is not enough to use here sequences or arrays. For $d = 2$, the right amount of information is contained in a *graph parameter*, and the subgraph densities $t(F, f)$ show many properties analogous to the classical results described above. Theorem 13.10, Theorem 11.52 together with Proposition 14.61, Theorem 5.54 and Theorem 16.46 are analogues of Theorems A.18, A.20, A.22, and A.23(a). Other results (e.g, the Monotone Reordering Theorem A.19 or Theorem A.23(b) do not seem to generalize to $d = 2$ in any natural way. The case $d \geq 3$ clearly corresponds to hypergraphs, where, as discussed in Chapter 23.3, new difficulties arise, and many of the interesting questions are open.

A.5. Ultraproduct and ultralimit

Ultrafilters. Let $\omega \subseteq 2^{\mathbb{N}}$. We say that ω is an *ultrafilter*, if it is a filter (closed under supersets), it is closed under finite intersections, and for every $X \subseteq \mathbb{N}$, either $X \in \omega$ or $\mathbb{N} \setminus X \in \omega$, but not both. It follows that $\mathbb{N} \in \omega$ and $\emptyset \notin \omega$. (See Bell and Slomson [2006] for more on ultrafilters and other constructions below.)

A trivial example of an ultrafilter is the set of subsets $X \subseteq \mathbb{N}$ containing a given element $n \in \mathbb{N}$; such an ultrafilter is called *principal*. There are non-principle ultrafilters; their existence can be proved using Zorn's Lemma (i.e., the Axiom of Choice). From now on, we fix a non-principal ultrafilter ω (it does not matter which one).

It is sometimes convenient to call the sets in ω *Big*, and the sets in $\mathbb{N} \setminus \omega$, *Small*. (We capitalize to make a distinction from the informal use of these words.) The following properties are not hard to prove:

PROPOSITION A.26. (a) *The union of a finite number of Small sets is Small.*

(b) *The intersection of a finite number of Big sets is Big.*

(c) *Every finite set is Small.*

Ultraproduct of sets. Let $(V_i : i \in \mathbb{N})$ be a sequence of sets. We say that two sequences $(a_i : i \in \mathbb{N})$ and $(b_i : i \in \mathbb{N})$ $(a_i, b_i \in V_i)$ are ω-*equivalent*, if they differ only on a Small set of indices, i.e., if $\{i : a_i = b_i\} \in \omega$. (It is easy to see that this is an equivalence relation.) The *ultraproduct* of the sets V_i (with respect to the ultrafilter ω) is obtained from their cartesian product $\prod_{i \in \mathbb{N}} V_i$ by identifying ω-equivalent sequences. We denote this ultraproduct by $\prod_{\omega} V_i$. Formally, the elements of $\prod_{\omega} V_i$ are ω-equivalence classes of sequences in $\prod_{i \in \mathbb{N}} V_i$; we denote the ω-equivalence class containing a sequence a by $[a]$.

It is not hard to see that the cardinality of the ultraproduct of a sequence of finite non-singleton sets is continuum.

Let $U_i \subseteq V_i$. Consider the set U of sequences $(a_1, a_2, \dots) \in \prod_{i \in \mathbb{N}} V_i$ such that $a_i \in U_i$ for a Large set of indices i. It is clear that if a sequence belongs to U then so does every ω-equivalent sequence; the set of ω-equivalence classes contained in U will be denoted (with a little abuse of notation) by $\prod_{\omega} U_i$.

Ultraproduct of structures. Let $A_i = (V_i, R_{i1}, \dots, R_{ik})$ be relational structures of the same type, where R_{ij} is a relation on V_i with a finite number r_j of variables for $j = 1, \ldots, k$. Their ultraproduct $\prod_\omega A_i$ is defined as the relational structure $(V, R_1, \ldots, R_k)$ of the same type, where $V = \prod_\omega V_i$ and for any r_j sequences $x_i = (x_{i1}, x_{i2}, \ldots)$ $(i = 1, \ldots, r_j)$ we have $([x_1], \ldots, [x_{r_j}]) \in R_j$ if and only if $(x_{i1}, \ldots, x_{ir_j}) \in R_{ij}$ for a large set of indices i. It is easy to see that this definition is correct in the sense that $([x_1], \ldots, [x_{r_j}]) \in R_j$ depends only on the equivalence classes $[x_1], \ldots, [x_{r_j}]$ and not on which representative x_i is chosen from $[x_i]$.

A very important property of ultraproduct of structures is stated in the following theorem:

PROPOSITION A.27 (**Łoś's Theorem**). *If every structure A_i $(i = 1, 2, \ldots)$ satisfies a first order sentence Φ, then their ultraproduct also satisfies Φ.*

As a special case, we can look at a sequence of finite simple graphs $G_i = (V_i, E_i)$, i.e., finite sets V_i with a symmetric irreflexive binary relation E_i. The ultraproduct of them is also a simple graph: the symmetry and irreflexivity of the relation on the ultraproduct is easy to check (or it follows from Łoś's Theorem, since these properties of the relation can be expressed by a first-order sentence: $\forall x \forall y (xy \in E \leftrightarrow yx \in E)$, and $\forall x (xx \notin E)$). If all the graphs have degrees bounded by D, then so does their ultraproduct, since this property can be expressed by a first-order sentence.

Ultralimit of a numerical sequence. As a nice application of an ultrafilter ω we can associate a "limit" to every bounded sequence of numbers. (This is a special construction for a *Banach limit* of bounded sequences.) Let $(a_1, a_2, \ldots)$ $(a_i \in [u, v])$ be a bounded sequence of real numbers. We say that a real number a is the *ultralimit* of the sequence (in notation $\lim_\omega a_i = a$) if for every $\varepsilon > 0$, the set $\{i : |a_i - a| > \varepsilon\}$ is Small. (Note: ordinary convergence to a would require that this set is finite.) It is not hard to prove that every bounded sequence of real numbers has a unique ultralimit. Furthermore, if $\lim_\omega a_i = a$ and $a_i \in [u, v]$ for every i, then $a \in [u, v]$.

Ultraproduct of measures. Let $(V_i, \mathcal{A}_i)$ be a sigma-algebra for $i = 1, 2, \ldots$. The sets of the form $\prod_\omega A_i$ $(A_i \in \mathcal{A}_i)$, considered as subsets of $V = \prod_\omega V_i$, form a Boolean algebra $\mathcal{B}$ (they are closed under finite union, intersection, and complementation). The Boolean algebra $\mathcal{B}$ generates a sigma-algebra on $V = \prod_\omega V_i$, which we denote by $\mathcal{A} = \prod_\omega \mathcal{A}_i$.

Next, suppose that there is a probability measure π_i on $(V_i, \mathcal{A}_i)$; then we define a setfunction on $\mathcal{B}$ by

$$\pi\Big(\prod_\omega A_i\Big) = \lim_\omega \pi_i(A_i).$$

It is not hard to see that π is finitely additive, and a bit harder to see that it is a measure on $\mathcal{B}$, i.e., if $B_1, B_2 \cdots \in \mathcal{B}$ and $\cap_{n=1}^\infty B_n = \emptyset$, then $\lim_n \pi(B_n) = 0$. Trivially $\pi(V) = 1$. It follows by Carathéodory's Measure Extension Theorem (see e.g. Halmos [1950]) that π extends to a probability measure on $\mathcal{A}$ (which we also denote by π). Thus $(V, \mathcal{A}, \pi)$ is a probability space, which we call the *ultraproduct* of the probability spaces $(V_i, \mathcal{A}_i, \pi)$. We write $\pi = \prod_\omega \pi_i$. (This is a special case of a *Loeb space*; see Loeb [1979].)

A.6. Vapnik–Chervonenkis dimension

In probability theory, we often have to prove that out of a large number of "bad" events, with positive probability none happens. The trivial method (which is sufficient surprisingly often) is to use the *union bound*: we can draw this conclusion provided the sum of probabilities of the bad events is less than one. There are of course many less trivial tricks that can be used to improve this method, and one of them is the use of the Vapnik–Chervonenkis dimension. We describe this method only in a simple special case when we need it.

For any set V and family of subsets $\mathcal{H} \subseteq 2^V$, a set $S \subseteq V$ is called *shattered*, if for every $X \subseteq S$ there is a $Y \in \mathcal{H}$ such that $X = Y \cap S$. The *Vapnik–Chervonenkis dimension* or *VC-dimension* $\dim_{\mathrm{VC}}(\mathcal{H})$ of a family of sets is the supremum of cardinalities of shattered sets (Vapnik and Chervonenkis [1971]). For us, this dimension will be always finite.

We recall two basic facts about VC-dimension.

PROPOSITION A.28 **(Sauer–Shelah Lemma)**. *If a family $\mathcal{H}$ of subsets of an m-element set has VC-dimension k, then*

$$|\mathcal{H}| \leq 1 + m + \cdots + \binom{m}{k}.$$

For a family $\mathcal{H}$ of sets, we denote by $\tau(\mathcal{H})$ the minimum cardinality of a set meeting every member of $\mathcal{H}$. The following basic fact about VC-dimension, is based on the results of Vapnik and Chervonenkis [1971]. It was proved by Haussler and Welzl [1987] in a slightly weaker form than stated here, and by Komlós, Pach and Woeginger [1992] in a slightly stronger form.

PROPOSITION A.29. *Let J be a probability space and $\mathcal{H}$, a family of measurable subsets of J such that every $A \in \mathcal{H}$ has measure at least ε. Suppose that $\mathcal{H}$ has finite VC-dimension k. Then $\tau(\mathcal{H}) \leq (8k/\varepsilon)\log(1/\varepsilon)$.*

The following fact follows easily from these basic results (Lovász and Szegedy [2010b]):

PROPOSITION A.30. *Let $\mathcal{H}$ be a family of measurable sets in a probability space with VC-dimension k such that $\pi(A \triangle B) \geq \varepsilon$ for all $A, B \in \mathcal{H}$. Then $|\mathcal{H}| \leq (80k)^k \varepsilon^{-20k}$.*

A.7. Nonnegative polynomials

We say that a real polynomial $p \in \mathbb{R}[x_1, \ldots, x_k]$ is *nonnegative*, in notation $p \geq 0$, if $p(x_1, \ldots, x_k) \geq 0$ for every real $x_1, \ldots, x_k$.

A characterization of nonnegative polynomials is an important problem. A first guess would be that a polynomial is nonnegative if and only if it is a sum of squares of real polynomials. This is indeed true for one and two variables, but fails to hold for three; the polynomial

$$z^6 + x^4 y^2 + x^2 y^4 - 3x^2 y^2 z^2$$

is a counterexample. Hilbert conjectured in 1900 in his famous lecture about 23 mathematical problems (this was number 17), and Artin proved in 1927, that the answer becomes affirmative if we allow rational functions instead of polynomials.

PROPOSITION A.31 **(Artin's Theorem)**. *Every nonnegative polynomial can be written as the sum of squares of rational functions.*

In the same talk, Hilbert suggested (problem number 10) to give an algorithm to solve diophantine equations. It will be useful to contrast this with the problem of solving algebraic equations in reals:

PROBLEM A.32. Given a polynomial $p \in \mathbb{Z}[x_1, \ldots, x_k]$, decide whether or not $p(x_1, \ldots, x_k) \geq 0$ for every real $x_1, \ldots, x_k$.

PROBLEM A.33. Given a polynomial $p \in \mathbb{Z}[x_1, \ldots, x_k]$, decide whether or not $p(x_1, \ldots, x_k) \geq 0$ for every integral $x_1, \ldots, x_k$.

Problem A.32 is decidable; this follows e.g. from the general result of Tarski that the first order theory of real numbers is decidable. Problem A.33 is undecidable; this follows from the solution of Hilbert's Tenth Problem by Matiyasevich [1970].

A.8. Categories

As in other sections of this Appendix, I only summarize some basic notation, definitions and examples that are necessary to understand certain parts of the book. For more definitions and facts in category theory, see e.g. Adámek, Herrlich and Strecker [2006].

A *category* $\mathcal{K}$ consists of a set of *objects* $\mathrm{Ob}(\mathcal{K})$, and, for any two objects $a, b \in \mathrm{Ob}(\mathcal{K})$, a set $\mathcal{K}(a,b)$ of *morphisms*. Two morphisms $\alpha \in \mathcal{K}(a,b)$ and $\beta \in \mathcal{K}(b,c)$ have a product $\alpha\beta \in \mathcal{K}(a,c)$, and this multiplication is associative (whenever defined). For $\alpha \in \mathcal{K}(a,b)$, we set $t(\alpha) = a$ (tail of α) and $h(\alpha) = b$ (head of α). Let $\mathcal{K}_a^{\mathrm{in}}$ [$\mathcal{K}_a^{\mathrm{out}}$] denote the set of morphisms with $h(\alpha) = a$ [$t(\alpha) = a$]. (If you want to think about morphisms as maps, then please note that I am writing the product so that the maps should be applied in the order going from left to right.)

For every object a, we have the identity morphism $\mathrm{id}_a \in \mathcal{K}(a,a)$, which has the property that $\mathrm{id}_a \alpha = \alpha$ for every $\alpha \in \mathcal{K}_a^{\mathrm{out}}$ and $\alpha \mathrm{id}_a = \alpha$ for every $\alpha \in \mathcal{K}_a^{\mathrm{in}}$. An isomorphism between objects a and b is a morphism $\xi \in \mathcal{K}(a,b)$ which has an *inverse* $\zeta \in \mathcal{K}(b,a)$ such that $\xi\zeta = \mathrm{id}_a$ and $\zeta\xi = \mathrm{id}_b$. Two objects are *isomorphic* if there is an isomorphism between them.

For every object a, we introduce an equivalence relation on $\mathcal{K}_a^{\mathrm{in}}$ by $\alpha \simeq \beta$ if and only if $\beta = \alpha\gamma$ for some isomorphism γ. We say that α and β are *left-isomorphic*. It is also clear that if $\alpha_1, \alpha_2 \in \mathcal{K}(a,b)$, $\varphi \in \mathcal{K}(b,c)$, and α_1 and α_2 are left-isomorphic, then so are $\alpha_1\varphi$ and $\alpha_2\varphi$.

We can delete any object from a category and still have a category, so to really work in a category (in particular, to prove the existence of a particular object), we must assume the existence of certain objects and morphisms. The existence of these is usually easily verified when we want to apply our results to specific categories, in particular, to the category of graphs.

Terminal and zero objects. An object is *terminal*, if every object has a unique morphism into it. Any two terminal objects are isomorphic, and we will assume that the terminal object, if it exists, is unique.

Dually, an object is a *zero object*, if it has a unique morphism into any object.

Product and coproduct. A set of morphisms $\pi_i \in \mathcal{K}(c,a_i)$ ($i \in I$) is called a *product*, if for every set of morphisms $\varphi_i \in \mathcal{K}(d,a_i)$ ($i \in I$) there is a unique morphism $\xi \in \mathcal{K}(d,c)$ such that $\varphi_i = \xi\pi_i$ for all $i \in I$. We also say that the object c is the product of objects a_i. It is easy to see that the product is uniquely determined up to isomorphism. For two objects a and b, we denote by $a \times b$ their product. We write $a^{\times k}$ for the k-fold product $a \times \cdots \times a$. We say that the category has products, if every finite set of objects has a product.

Coproducts are defined by turning the arrows around: A set of morphisms $\sigma_i \in \mathcal{K}(a_i, c)$ ($i \in I$) is called a *coproduct*, if for every set of morphisms $\varphi_i \in \mathcal{K}(a_i, d)$ ($i \in I$) there is a unique morphism $\xi \in \mathcal{K}(c, d)$ such that $\varphi_i = \sigma_i \xi$. For two objects a and b, we denote by $a \oplus b$ their coproduct, and by $a^{\oplus k}$, the k-fold coproduct of a with itself.

Pullback and pushout. For two morphisms $\alpha \in \mathcal{K}(a, c)$ and $\beta \in \mathcal{K}(b, c)$, a pair of morphisms $\alpha' \in \mathcal{K}(d, a)$ and $\beta' \in \mathcal{K}(d, b)$ is called a *pullback* of (α, β) if $\alpha'\alpha = \beta'\beta$, and whenever $\xi \in \mathcal{K}(e, a)$ and $\zeta \in \mathcal{K}(e, b)$ are two morphisms such that $\xi\alpha = \zeta\beta$, then there is a unique morphism $\eta \in \mathcal{K}(e, d)$ such that $\eta\alpha' = \xi$ and $\eta\beta' = \zeta$. The morphism $\alpha'\alpha = \beta'\beta$ will be denoted by $\alpha \vee \beta$. The pullback is uniquely determined up to isomorphism: if $\sigma \in \mathcal{K}(d, d_1)$ is an isomorphism, then $(\alpha'\sigma, \beta'\sigma)$ is also a pullback, and we get all pullbacks this way. We say that *the category has pullbacks* if every pair of morphisms into the same object has a pullback.

The *pushout* of two morphisms $\alpha \in \mathcal{K}(c, a)$ and $\beta \in \mathcal{K}(c, b)$ is defined analogously to the pullback, just reversing the arrows: we want a pair of morphisms α' and β' such that $\alpha\alpha' = \beta\beta'$ and for any two morphisms $\xi \in \mathcal{K}(a, e)$ and $\zeta \in \mathcal{K}(b, e)$ such that $\alpha\xi = \beta\zeta$ there exists a unique morphism $\eta \in \mathcal{K}(d, e)$ such that $\alpha'\eta = \xi$ and $\beta'\eta = \zeta$. The morphism $\alpha\alpha' = \beta\beta'$ will be denoted by $\alpha \wedge \beta$.

It is easy to check that for $\alpha_1, \alpha_2, \beta \in \mathcal{K}_a^{\text{out}}$, if $\alpha_1 \simeq \alpha_2$, then $\beta \wedge \alpha_1 \simeq \beta \wedge \alpha_2$. So the operation $\wedge$ is well defined on left-equivalence classes of morphisms, and the object $h(\alpha \wedge \beta)$ is determined up to isomorphism. Similar remarks can be made about pullbacks.

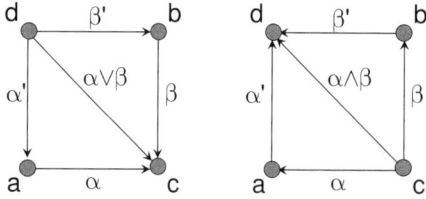

FIGURE A.1. Pushout and pullback

Pushouts generalize coproducts, and pullbacks generalize products: If the category has a zero element z and pushouts, then for any two objects a and b, the pushout of the (unique) morphisms $\alpha \in \mathcal{K}(z, a)$ and $\beta \in \mathcal{K}(z, b)$ points to the coproduct of a and b, and dually, the product can be obtained from the pullback of the unique morphisms into a terminal object.

Generators. An object g is a *right-generator*, if for any two objects a, b and any two different morphisms $\alpha, \beta \in \mathcal{K}(a, b)$, there is a morphism $\eta \in \mathcal{K}(b, g)$ such that $\alpha\eta \neq \beta\eta$. A *left-generator* is defined dually.

Epi-mono decomposition. A morphism $\varphi \in \mathcal{K}(a, b)$ is a *monomorphism*, if for any two morphisms $\alpha, \beta \in \mathcal{K}(c, a)$ such that $\alpha\varphi = \beta\varphi$, it follows that $\alpha = \beta$. An *epimorphism*, is defined dually, reversing the arrows. We say that the category has *epi-mono decompositions*, if every morphism has a decomposition into the product of an epimorphism and a monomorphism.

Monomorphisms of $\mathcal{K}$ are closed under product, and hence they form a category $\mathcal{K}^{\mathrm{mon}}$. The category $\mathcal{K}^{\mathrm{epi}}$ is defined similarly.

Locally finite categories. A category $\mathcal{K}$ is called *locally finite*, if $\mathcal{K}(a,b)$ is finite for all a,b. Categories of homomorphisms between finite objects (graphs, groups etc.) are locally finite. These categories have many pleasant properties, for example, if two objects have monomorphism into each other, then they are isomorphic.

Let us continue with some examples.

EXAMPLE A.34. The category of finite simple graphs with loops (where morphisms are homomorphisms, i.e., adjacency-preserving maps) has all the nice properties defined above.

It is trivial that this category is locally finite and has epi-mono decompositions. The terminal object is the single node with a loop, and the zero object is the empty graph. The looped complete graph on 2 nodes can serve as a right-generator object, and the single node is a left-generator.

To construct the pullback of two homomorphisms $\alpha : a \to c$ and $\beta : b \to c$, take the direct (categorical) product d of the two graphs a and b, together with its projections π_a and π_b onto a and b, respectively, and take the subgraph d' of d induced by those nodes v for which $(\pi_a\alpha)(v) = (\pi_b\beta)(v)$, together with the restrictions of π_a and π_b onto d'. (This essentially the same construction as used in the proof of Lemma 5.38 and in the statement of Theorem 5.59.)

To construct the pushout of two homomorphisms $\alpha : c \to a$ and $\beta : c \to b$, take the disjoint union d of the two graphs a and b, and identify the nodes $\alpha(x)$ and $\beta(x)$ for every node x of c. Note that in the case when c is the edgeless graph on $[k]$, then this is just the product of two k-multilabeled graphs, as defined in Chapter 4.

Coproduct in this category means disjoint union. Product means the direct product of two graphs as defined in Section 3.2. ♦

EXAMPLE A.35. Reversing the arrows in the category of finite simple graphs with loops (Example A.34) gives another category with the above properties, since the collection of these properties is invariant under reversing arrows. ♦

These examples can be extended to simplicial maps between finite simplicial complexes, homomorphisms between directed graphs, hypergraphs, etc.

EXAMPLE A.36 (**Partially ordered sets**). Let $(P, \leq)$ be a partially ordered set. For every pair $x, y \in P$ such that $x \leq y$, we define a unique morphism $\varphi_{x,y}$. There is only one way to define the composition: $\varphi_{x,y}\varphi_{y,z} = \varphi_{x,z}$, which makes sense because of the transitivity of the relation. This category is locally finite, and every morphism is both a monomorphism and an epimorphism, so it has (trivial) epi-mono decompositions.

If the poset is a lattice with lowest element 0 and highest element 1, then 0 is a zero object and 1 is a terminal object. Furthermore, for any two morphisms $\varphi_{c,a}$ and $\varphi_{c,b}$, the morphisms $(\varphi_{a,a\vee b}, \varphi_{b,a\vee b})$ form their pushout. Every element is a left and right generator (in a vacuous way). ♦

Bibliography

M. Abért and T. Hubai: Benjamini-Schramm convergence and the distribution of chromatic roots for sparse graphs, http://arxiv.org/abs/1201.3861

J. Adámek, H. Herrlich and G.E. Strecker: *Abstract and Concrete Categories: The Joy of Cats*, Reprints in Theory and Applications of Categories **17** (2006), 1–507.

S. Adams: Trees and amenable equivalence relations, *Ergodic Theory Dynam. Systems* **10** (1990), 1–14.

R. Ahlswede and G.O.H. Katona: Graphs with maximal number of adjacent pairs of edges, *Acta Math. Hung.* **32** (1978), 97–120.

R. Albert and A.-L. Barabási: Statistical mechanics of complex networks, *Rev. Modern Phys.* **74** (2002), 47–97.

D.J. Aldous: Representations for partially exchangeable arrays of random variables, *J. Multivar. Anal.* **11** (1981), 581–598.

D.J. Aldous: Tree-valued Markov chains and Poisson-Galton-Watson distributions, in: *Microsurveys in Discrete Probability* (D. Aldous and J. Propp, editors), DIMACS Ser. Discrete Math. Theoret. Comput. Sci. **41**, Amer. Math. Soc., Providence, RI. (1998), 1–20.

D. Aldous and R. Lyons: Processes on Unimodular Random Networks, *Electron. J. Probab.* **12**, Paper 54 (2007), 1454–1508.

D.J. Aldous and M. Steele: The Objective Method: Probabilistic Combinatorial Optimization and Local Weak Convergence, in: *Discrete and Combinatorial Probability* (H. Kesten, ed.), Springer (2003) 1–72.

N. Alon (unpublished)

N. Alon, R.A. Duke, H. Lefmann, V. Rödl and R. Yuster: The algorithmic aspects of the regularity lemma, *J. Algorithms* **16** (1994), 80–109.

N. Alon, E. Fischer, M. Krivelevich and M. Szegedy: Efficient testing of large graphs, *Combinatorica* **20** (2000) 451–476.

N. Alon, W. Fernandez de la Vega, R. Kannan and M. Karpinski: Random sampling and approximation of MAX-CSPs, *J. Comput. System Sci.* **67** (2003) 212–243.

N. Alon, E. Fischer, I. Newman and A. Shapira: A Combinatorial Characterization of the Testable Graph Properties: It's All About Regularity, *Proc. 38th ACM Symp. on Theory of Comput.* (2006), 251–260.

N. Alon and A. Naor: Approximating the Cut-Norm via Grothendieck's Inequality SIAM J. Comput. **35** (2006), 787–803.

N. Alon, P.D. Seymour and R. Thomas: A separator theorem for non-planar graphs, *J. Amer. Math. Soc.* **3** (1990), 801–808.

N. Alon and A. Shapira: A Characterization of the (natural) Graph Properties Testable with One-Sided Error, SIAM J. Comput. **37** (2008), 1703–1727.

N. Alon, A. Shapira and U. Stav: Can a Graph Have Distinct Regular Partitions? SIAM J. Discr. Math. **23** (2009), 278–287.

N. Alon and J. Spencer: *The Probabilistic Method*, Wiley–Interscience, 2000.

N. Alon and U. Stav: What is the furthest graph from a hereditary property? *Random Struc. Alg.* **33** (2008), 87–104.

O. Angel and B. Szegedy (unpublished)

D. Aristoff and C. Radin: Emergent structures in large networks, http://arxiv.org/abs/1110.1912

S. Arora, D. Karger and M. Karpinski: Polynomial time approximation schemes for dense instances of NP-hard problems, *Proc. 27th ACM Symp. on Theory of Comput.* (1995), 284–293.

T. Austin: On exchangeable random variables and the statistics of large graphs and hypergraphs, *Probability Surveys* **5** (2008), 80–145.

T. Austin and T. Tao: On the testability and repair of hereditary hypergraph properties, *Random Struc. Alg.* **36** (2010), 373–463.

E. Babson and D. Kozlov: Topological obstructions to graph colorings, *Electr. Res. Announc. AMS* **9** (2003), 61–68.

E. Babson and D.N. Kozlov: Complexes of graph homomorphisms, *Isr. J. Math.* **152** (2006), 285–312.

E. Babson and D.N. Kozlov: Proof of the Lovász conjecture, *Annals of Math.* **165** (2007), 965–1007.

A. Bandyopadhyay and D. Gamarnik: Counting without sampling. Asymptotics of the log-partition function for certain statistical physics models, *Random Struc. Alg.* **33** (2008), 452–479.

A.-L. Barabási: *Linked: The New Science of Networks*, Perseus, Cambridge, MA (2002).

A.-L. Barabási and R. Albert: Emergence of scaling in random networks, *Science* **286** (1999) 509–512.

M. Bayati, D. Gamarnik and P. Tetali: Combinatorial Approach to the Interpolation Method and Scaling Limits in Sparse Random Graphs, *Proc. 42th ACM Symp. on Theory of Comput.* (2010), 105–114.

J.L. Bell and A.B. Slomson: *Models and Ultraproducts: An Introduction*, Dover Publ. (2006).

I. Benjamini, L. Lovász: Global Information from Local Observation, *Proc. 43rd Ann. Symp. on Found. of Comp. Sci.* (2002), 701–710.

I. Benjamini, G. Kozma, L. Lovász, D. Romik and G. Tardos: Waiting for a bat to fly by (in polynomial time), *Combin. Prob. Comput.* **15** (2006), 673–683.

I. Benjamini and O. Schramm: Recurrence of Distributional Limits of Finite Planar Graphs, *Electronic J. Probab.* **6** (2001), paper no. 23, 1–13.

I. Benjamini, O. Schramm and A. Shapira: Every Minor-Closed Property of Sparse Graphs is Testable, *Advances in Math.* **223** (2010), 2200–2218.

C. Berg, J.P.R. Christensen and P. Ressel: Positive definite functions on Abelian semigroups, *Math. Ann.* **223** (1976), 253–272.

C. Berg and P.H. Maserick: Exponentially bounded positive definite functions, *Ill. J. Math.* **28** (1984), 162–179.

A. Bertoni, P. Campadelli and R. Posenato: An upper bound for the maximum cut mean value, *Graph-theoretic concepts in computer science*, Lecture Notes in Comp. Sci., **1335**, Springer, Berlin (1997), 78–84.

P. Billingsley: *Convergence of Probability Measures*, Wiley, New York (1999).

A. Björner and L. Lovász: Pseudomodular lattices and continuous matroids, *Acta Sci. Math. Szeged* **51** (1987), 295-308.

G.R. Blakley and P.A. Roy: A Hölder type inequality for symmetric matrices with nonnegative entries, *Proc. Amer. Math. Soc.* **16** (1965) 1244–1245.

M. Boguñá and R. Pastor-Satorras: Class of correlated random networks with hidden variables, *Physical Review E* **68** (2003), 036112.

B. Bollobás: Relations between sets of complete subgraphs, in: *Combinatorics*, Proc. 5th British Comb. Conf. (ed. C.St.J.A. Nash-Williams, J. Sheehan), Utilitas Math. (1976), 79–84.

B. Bollobás: A probabilistic proof of an asymptotic formula for the number of labelled regular graphs, *Europ. J. Combin.* **1** (1980), 311–316.

B. Bollobás: *Random Graphs*, Second Edition, Cambridge University Press, 2001.

B. Bollobás, C. Borgs, J. Chayes and O. Riordan: Percolation on dense graph sequences, *Ann. Prob.* **38** (2010), 150–183.

B. Bollobás, S. Janson and O. Riordan: The phase transition in inhomogeneous random graphs, *Random Struc. Alg.* **31** (2007) 3–122.

B. Bollobás, S. Janson and O. Riordan: The cut metric, random graphs, and branching processes, *J. Stat. Phys.* **140** (2010) 289–335.

B. Bollobás, S. Janson and O. Riordan: Monotone graph limits and quasimonotone graphs, *Internet Mathematics* **8** (2012), 187–231.

B. Bollobás and V. Nikiforov: An abstract Szemerédi regularity lemma, in: *Building bridges*, Bolyai Soc. Math. Stud. **19**, Springer, Berlin (2008), 219–240.

B. Bollobás and O. Riordan: A Tutte Polynomial for Coloured Graphs, *Combin. Prob. Comput.* **8** (1999), 45–93.

B. Bollobás and O. Riordan: Metrics for sparse graphs, in: *Surveys in combinatorics 2009*, Cambridge Univ. Press, Cambridge (2009) 211–287.

B. Bollobás and O. Riordan: Random graphs and branching processes, in: *Handbook of large-scale random networks*, Bolyai Soc. Math. Stud. **18**, Springer, Berlin (2009), 15–115.

C. Borgs: Absence of Zeros for the Chromatic Polynomial on Bounded Degree Graphs, *Combin. Prob. Comput.* **15** (2006), 63–74.

C. Borgs, J. Chayes, J. Kahn and L. Lovász: Left and right convergence of graphs with bounded degree, http://arxiv.org/abs/1002.0115

C. Borgs, J. Chayes and L. Lovász: Moments of Two-Variable Functions and the Uniqueness of Graph Limits, *Geom. Func. Anal.* **19** (2010), 1597–1619.

C. Borgs, J. Chayes, L. Lovász, V.T. Sós and K. Vesztergombi: Counting graph homomorphisms, in: *Topics in Discrete Mathematics* (ed. M. Klazar, J. Kratochvil, M. Loebl, J. Matoušek, R. Thomas, P. Valtr), Springer (2006), 315–371.

C. Borgs, J.T. Chayes, L. Lovász, V.T. Sós and K. Vesztergombi: Convergent Graph Sequences I: Subgraph frequencies, metric properties, and testing, *Advances in Math.* **219** (2008), 1801–1851.

C. Borgs, J.T. Chayes, L. Lovász, V.T. Sós and K. Vesztergombi: Convergent Graph Sequences II: Multiway Cuts and Statistical Physics, *Annals of Math.* **176** (2012), 151–219.

C. Borgs, J.T. Chayes, L. Lovász, V.T. Sós and K. Vesztergombi: Limits of randomly grown graph sequences, *Europ. J. Combin.* **32** (2011), 985–999.

C. Borgs, J.T. Chayes, L. Lovász, V.T. Sós, B. Szegedy and K. Vesztergombi: Graph Limits and Parameter Testing, STOC38 (2006), 261–270.

L. Bowen: Couplings of uniform spanning forests, *Proc. Amer. Math. Soc.* **132** (2004), 2151–2158.

G. Brightwell and P. Winkler: Graph homomorphisms and long range action, in *Graphs, Morphisms and Statistical Physics*, DIMACS Ser. Disc. Math. Theor. CS, American Mathematical Society (2004), 29–48.

W.G. Brown, P. Erdős and M. Simonovits: Extremal problems for directed graphs, *J. Combin. Theory B* **15** (1973), 77–93.

W.G. Brown, P. Erdős and M. Simonovits: On multigraph extremal problems, *Problèmes combinatoires et théorie des graphes*, Colloq. Internat. CNRS **260** (1978), 63–66.

O.A. Camarena, E. Csóka, T. Hubai, G. Lippner and L. Lovász: Positive graphs, http://arxiv.org/abs/1205.6510

S. Chatterjee and P. Diaconis: Estimating and understanding exponential random graph models, http://arxiv.org/abs/1102.2650

S. Chatterjee and S.R.S Varadhan: The large deviation principle for the Erdős–Rényi random graph, *Europ. J. Combin.* **32** (2011), 1000–1017.

F.R.K. Chung: From Quasirandom graphs to Graph Limits and Graphlets
http://arxiv.org/abs/1203.2269

F.R.K. Chung, R.L. Graham and R.M. Wilson: Quasi-random graphs, *Combinatorica* **9** (1989), 345–362.

F.R.K Chung and R.L. Graham: Quasi-Random Hypergraphs, *Proc. Nat. Acad Sci.* **86** (1989), pp. 8175–8177.

F.R.K Chung and R.L. Graham: Quasi-random subsets of Z_n, *J. Combin. Theory A* **61** (1992), 64–86.

D.L. Cohn: *Measure Theory*, Birkheuser, Boston, 1980.

D. Conlon and J. Fox: Bounds for graph regularity and removal lemmas, http://arxiv.org/abs/1107.4829

D. Conlon, J. Fox and B Sudakov: An approximate version of Sidorenko's conjecture, *Geom. Func. Anal.* **20** (2010), 1354–1366.

J. Cooper: Quasirandom Permutations, *J. Combin. Theory A* **106** (2004), 123–143.

J. Cooper: A permutation regularity lemma, *Electr. J. Combin.* **13** R# 22 (2006).

E. Csóka: Local algorithms with global randomization on sparse graphs (manuscript)

E. Csóka: An undecidability result on limits of sparse graphs, http://arxiv.org/abs/1108.4995

E. Csóka: Maximum flow is approximable by deterministic constant-time algorithm in sparse networks, http://arxiv.org/abs/1005.0513

N.G. de Bruijn: Algebraic theory of Penrose's non-periodic tilings of the plane I-II, *Indag. Math.* **84** (1981) 39–52, 53–66.

P. Diaconis and D. Freedman: On the statistics of vision: the Julesz Conjecture, *J. Math. Psychology* **2** (1981) 112–138.

P. Diaconis and D. Freedman: The Markov Moment Problem and de Finetti's Theorem: Part I, *Math. Zeitschrift* **247** (2004), 183–199.

P. Diaconis, S. Holmes and S. Janson: Threshold graph limits and random threshold graphs, *Internet Math.* **5** (2008), 267–320.

P. Diaconis, S. Holmes and S. Janson: Interval graph limits, http://arxiv.org/abs/1102.2841

P. Diaconis and S. Janson: Graph limits and exchangeable random graphs, *Rendiconti di Matematica* **28** (2008), 33–61.

R.L. Dobrushin: Estimates of semi-invariants for the Ising model at low temperatures, in: *Topics in Statistical and Theoretical Physics* (AMS Translations, Ser. 2) bf 177 (1996), pp. 59–81.

W. Dörfler and W. Imrich: Über das starke Produkt von endlichen Graphen, *Österreich. Akad. Wiss. Math.-Natur. Kl. S.-B. II* **178** (1970), 247–262.

J. Draisma, D.C. Gijswijt, L. Lovász, G. Regts and A. Schrijver: Characterizing partition functions of the vertex model, *J. of Algebra* **350** (2012), 197–206.

G. Elek: On limits of finite graphs, *Combinatorica* **27** (2007), 503–507.

G. Elek: The combinatorial cost, *Enseign. Math.* bf53 (2007), 225–235.

G. Elek: L^2-spectral invariants and convergent sequences of finite graphs, *J. Functional Analysis* **254** (2008), 2667–2689.

G. Elek: Parameter testing with bounded degree graphs of subexponential growth, *Random Struc. Alg.* **37** (2010), 248–270.

G. Elek: On the limit of large girth sequences, *Combinatorica* **30** (2010), 553–563.

G. Elek: Finite graphs and amenability, http://arxiv.org/abs/1204.0449

G. Elek: Samplings and observables. Convergence and limits of metric measure spaces, http://arxiv.org/abs/1205.6936

G. Elek and G. Lippner: Borel oracles. An analytical approach to constant-time algorithms, *Proc. Amer. Math. Soc.* **138** (2010), 2939–2947.

G. Elek and G. Lippner: An analogue of the Szemerédi Regularity Lemma for bounded degree graphs, http://arxiv.org/abs/0809.2879

G. Elek and B. Szegedy: A measure-theory approach to the theory of dense hypergraphs, *Advances in Math.* **231** (2012), 1731–1772.

P. Erdős: On sequences of integers no one of which divides the product of two others and on some related problems, *Mitt. Forsch.-Inst. Math. Mech. Univ. Tomsk* **2** (1938), 74–82;

P. Erdős: On some problems in graph theory, combinatorial analysis and combinatorial number theory, in: *Graph Theory and Combinatorics*, Academic Press, London (1984), 1–17.

P. Erdős, L. Lovász and J. Spencer: Strong independence of graphcopy functions, in: *Graph Theory and Related Topics*, Academic Press (1979), 165-172.

P. Erdős and A. Rényi: On random graphs I, *Publ. Math. Debrecen* **6** (1959), 290–297.

P. Erdős and A. Rényi: On the evolution of random graphs, *MTA Mat. Kut. Int. Közl.* **5** (1960), 17–61.

P. Erdős and M. Simonovits: A limit theorem in graph theory, *Studia Sci. Math. Hungar.* **1** (1966), 51–57.

P. Erdős and A.H. Stone: On the structure of linear graphs, *Bull. Amer. Math. Soc.* **52** (1946), 1087–1091.

W. Feller, *An Introduction to Probability Theory and its Applications*, Second edition, Wiley, New York (1971).

E. Fischer: The art of uninformed decisions: A primer to property testing, The Computational Complexity Column of the *Bulletin of the European Association for Theoretical Computer Science* **75** (2001), 97-126.

E. Fischer and I. Newman: Testing versus Estimation of Graph Properties, *Proc. 37th ACM Symp. on Theory of Comput.* (2005), 138–146.

D.C. Fisher: Lower bounds on the number of triangles in a graph, *J. Graph Theory* **13** (1989), 505–512.

J. Fox: A new proof of the graph removal lemma, *Annals of Math.* **174** (2011), 561–579.

J. Fox and J. Pach (unpublished)

F. Franek and V. Rödl: Ramsey Problem on Multiplicities of Complete Subgraphs in Nearly Quasirandom Graphs, *Graphs and Combin.* **8** (1992), 299–308.

P. Frankl and V. Rödl: The uniformity lemma for hypergraphs, *Graphs and Combin.* **8** (1992), 309–312.

P. Frankl and V. Rödl: Extremal problems on set systems, *Random Struc. Alg.* **20** (2002), 131–164.

M. Freedman, L. Lovász and A. Schrijver: Reflection positivity, rank connectivity, and homomorphisms of graphs, *J. Amer. Math. Soc.* **20** (2007), 37–51.

A. Frieze and R. Kannan: Quick approximation to matrices and applications, *Combinatorica* **19** (1999), 175–220.

D. Gaboriau: Invariants l^2 de relations d'equivalence et de groupes, *Publ. Math. Inst. Hautes. Ètudes Sci.* **95** (2002), 93–150.

O. Gabber and Z. Galil: Explicit Constructions of Linear-Sized Superconcentrators, *J. Comput. Syst. Sci.* **22** (1981), 407–420.

D. Gamarnik: Right-convergence of sparse random graphs, http://arxiv.org/abs/1202.3123

D. Garijo, A. Goodall and J. Nešetřil: Graph homomorphisms, the Tutte polynomial and "q-state Potts uniqueness", *Elect. Notes in Discr. Math.* **34** (2009), 231–236.

D. Garijo, A. Goodall and J. Nešetřil: Contractors for flows, *Electronic Notes in Discr. Math.* **38** (2011), 389–394.

H.-O. Georgii, *Gibbs Measures and Phase Transitions*, de Gruyter, Berlin, 1988.

S. Gerke and A. Steger: The sparse regularity lemma and its applications, *Surveys in Combinatorics* (2005), 227–258.

E.N. Gilbert: Random graphs, *Ann. Math. Stat.* **30** (1959), 1141–1144.

B. Godlin, T. Kotek and J.A. Makowsky: Evaluations of Graph Polynomials, *34th Int. Workshop on Graph-Theoretic Concepts in Comp. Sci.*, Lecture Notes in Comp. Sci. 5344, Springer, Berlin–Heidelberg (2008), 183–194.

C. Godsil and G. Royle: *Algebraic Graph Theory*, Grad. Texts in Math. **207**, Springer, New York (2001).

O. Goldreich (ed): *Property Testing: Current Research and Suerveys*, LNCS 6390, Springer, 2010.

O. Goldreich, S. Goldwasser and D. Ron: Property testing and its connection to learning and approximation, *J. ACM* **45** (1998), 653–750.

O. Goldreich and D. Ron: A sublinear bipartiteness tester for bounded degree graphs, *Combinatorica* **19** (1999), 335–373.

O. Goldreich and D. Ron: Property testing in bounded degree graphs, *Algorithmica* **32** (2008), 302–343.

O. Goldreich and L. Trevisan: Three theorems regarding testing graph properties, *Random Struc. Alg.*, **23** (2003), 23–57.

A.W. Goodman: On sets of aquaintences and strangers at any party, *Amer. Math. Monthly* **66** (1959) 778–783.

W.T. Gowers: Lower bounds of tower type for Szemerédi's Uniformity Lemma, *Geom. Func. Anal.* **7** (1997), 322–337.

W.T. Gowers: A new proof of Szemerédi's theorem, *Geom. Func. Anal.* **11** (2001), 465–588.

W.T. Gowers: Quasirandomness, counting and regularity for 3-uniform hypergraphs, *Combin. Prob. Comput.* **15** (2006), 143–184.

W.T. Gowers: Hypergraph regularity and the multidimensional Szemeredi theorem, *Annals of Math.* **166** (2007), 897–946.

B. Green and T. Tao: The primes contain arbitrarily long arithmetic progressions. *Ann. of Math.* **167** (2008), 481–547.

B. Green, T. Tao and T. Ziegler: An inverse theorem for the Gowers $U^{s+1}[N]$-norm. *Electron. Res. Announc. Math. Sci.* **18** (2011), 69–90.

G.R. Grimmett and D.R. Stirzaker: *Probability and Random Processes*, Oxford University Press (1982).

M. Gromov: *Metric structures for Riemannian and non-Riemannian spaces*, Birkhäuser (1999).

A. Grzesik: On the maximum number of C5's in a triangle-free graph, *J. Combin. Theory B* **102** (2012), 1061–1066.

P. Halmos: *Measure theory*, D. van Nostrand and Co., 1950.

V. Harangi: On the density of triangles and squares in regular finite and unimodular random graphs, http://www.renyi.hu/~harangi/papers/d3d4.pdf

P. de la Harpe and V.F.R. Jones: Graph Invariants Related to Statistical Mechanical Models: Examples and Problems, *J. Combin. Theory B* **57** (1993), 207–227.

A. Hassidim, J.A. Kelner, H.N. Nguyen and K. Onak: Local Graph Partitions for Approximation and Testing, *Proc. 50th Ann. IEEE Symp. on Found. Comp. Science* (2009), 22–31.

H. Hatami: Graph norms and Sidorenko's conjecture, *Israel J. Math.* **175** (2010), 125–150.

H. Hatami, J. Hladky, D. Kral, S. Norine and A. Razborov: On the number of pentagons in triangle-free graphs, http://arxiv.org/abs/1102.1634

H. Hatami, J. Hladky, D. Kral, S. Norine and A. Razborov: Non-three-colorable common graphs exist, http://arxiv.org/abs/1105.0307

H. Hatami, L. Lovász and B. Szegedy: Limits of local-global convergent graph sequences http://arxiv.org/abs/1205.4356

H. Hatami and S. Norine: Undecidability of linear inequalities in graph homomorphism densities, *J. Amer. Math. Soc.* **24** (2011), 547–565.

F. Hausdorff: Summationsmethoden und Momentfolgen I–II, *Math. Z.* **9** (1921), 74–109 and 280–299.

D. Haussler and E. Welzl: ε-nets and simplex range queries, Discr. Comput. Geom. **2** (1987), 127–151.

J. Haviland and A. Thomason: Pseudo-random hypergraphs. Graph theory and combinatorics (Cambridge, 1988). *Discrete Math.* **75** (1989), 255–278.

J. Haviland and A. Thomason: On testing the "pseudo-randomness" of a hypergraph. *Discrete Math.* **103** (1992), 321–327.

P. Hell and J. Nešetřil: On the complexity of H-colouring, *J. Combin. Theory B* **48** (1990), 92–110.

P. Hell and J. Nešetřil: *Graphs and Homomorphisms*, Oxford University Press, 2004.

J. Hladky, D. Kral and S. Norine: Counting flags in triangle-free digraphs, *Electronic Notes in Discr. Math.* **34** (2009), 621–625.

J. Hladky: Bipartite subgraphs in a random cubic graph, Bachelor Thesis, Charles University, 2006.

S. Hoory and N. Linial: A counterexample to a conjecture of Lovász on the χ-coloring complex, *J. Combin. Theory B* **95** (2005), 346–349.

D. Hoover: Relations on Probability Spaces and Arrays of Random Variables, Institute for Advanced Study, Princeton, NJ, 1979.

C. Hoppen, Y. Kohayakawa, C.G. Moreira, B. Rath and R. Menezes Sampaio: Limits of permutation sequences, http://arxiv.org/abs/1103.5844

C. Hoppen, Y. Kohayakawa, C.G. Moreira and R. Menezes Sampaio: Limits of permutation sequences through permutation regularity, http://arxiv.org/abs/1106.1663

Y.E. Ioannidis and R. Ramakrishnan: Containment of conjunctive queries: Beyond relations as sets, *ACM Transactions on Database Systems*, **20** (1995), 288–324.

J. Isbell: Some inequalities in hom sets, *J. Pure Appl. Algebra* **76** (1991), 87–110.

Y. Ishigami: A Simple Regularization of Hypergraphs, http://arxiv.org/abs/math/0612838

C. Jagger, P. Štovíček and A. Thomason: Multiplicities of subgraphs, *Combinatorica* **16** (1996), 123–141.

S. Janson: Connectedness in graph limits, http://arxiv.org/abs/math/0802.3795v2

S. Janson: Graphons, cut norm and distance, couplings and rearrangements, http://arxiv.org/pdf/1009.2376.pdf

S. Janson: Poset limits and exchangeable random posets, *Combinatorica* **31** (2011), 529–563.

S. Janson: Limits of interval orders and semiorders, *J. Combinatorics* **3** (2012), 163–183.

S. Janson: Graph limits and hereditary properties, http://arxiv.org/abs/1102.3571

S. Janson, T. Łuczak and A. Ruczynski: *Random Graphs*, Wiley, 2000.

V. Kaimanovich: Amenability, hyperfiniteness, and isoperimetric inequalities, *C. R. Acad. Sci. Paris Sr. I Math.* **325** (1997), 999–1004.

O. Kallenberg: *Probabilistic Symmetries and Invariance Principles*, Springer, New York, 2005.

Z. Kallus, P. Hága, P. Mátray, G. Vattay and S. Laki: Complex geography of the internet network, *Acta Phys. Pol. B* **52**, 1057–1069.

G.O.H. Katona: Continuous versions of some extremal hypergraph problems, in: *Combinatorics*, (Keszthely, Hungary, 1976), Coll. Math. Soc. J. Bolyai **18** (J. Bolyai Math. Soc., Budapest, 1978), 653–678.

G.O.H. Katona: Continuous versions of some extremal hypergraph problems II, *Acta Math. Hung.* **35** (1980), 67–77.

G.O.H. Katona: Probabilistic inequalities from extremal graph results (a survey), *Ann. Discrete Math.* **28** (1985), 159–170.

A. Kechris and B.D. Miller: *Topics in orbit equivalence theory*, Lecture Notes in Mathematics **1852**. Springer-Verlag, Berlin, 2004.

A. Kechris, S. Solecki and S. Todorcevic: Borel chromatic numbers, *Advances in Math.* **141** (1999) 1–44.

H.G. Kellerer: Duality Theorems for Marginal Problems, *Z. Wahrscheinlichkeitstheorie verw. Gebiete* **67** (1984), 399–432.

K. Kimoto: Laplacians and spectral zeta functions of totally ordered categories, *J. Ramanujan Math. Soc.* **18** (2003), 53–76.

K. Kimoto: Vandermonde-type determinants and Laplacians of categories, preprint No. 2003-24, Graduate School of Mathematics, Kyushu University (2003).

J. Kock: *Frobenius Algebras and 2D Topological Quantum Field Theories*, London Math. Soc. Student Texts, Cambridge University Press (2003).

Y. Kohayakawa: Szemerédi's regularity lemma for sparse graphs, in: *Sel. Papers Conf. Found. of Comp. Math.*, Springer (1997), 216–230.

Y. Kohayakawa, B. Nagle, V. Rödl and M. Schacht: Weak hypergraph regularity and linear hypergraphs, *J. Combin. Theory B* **100** (2010), 151–160.

Y. Kohayakawa and V. Rödl: Szemerédi's regularity lemma and quasi-randomness, in: *Recent Advances in Algorithms and Combinatorics*, CMS Books Math./Ouvrages Math. SMC **11**, Springer, New York (2003), 289–351.

Y. Kohayakawa, V. Rödl and J. Skokan: Hypergraphs, quasi-randomness, and conditions for regularity, *J. Combin. Theory A* **97** (2002), 307–352.

I. Kolossváry and B. Ráth: Multigraph limits and exchangeability, *Acta Math. Hung.* **130** (2011), 1–34.

J. Komlós, J. Pach and G. Woeginger: Almost Tight Bounds for epsilon-Nets, *Discr. Comput. Geom.* **7** (1992), 163–173.

J. Komlós and M. Simonovits: Szemerédi's Regularity Lemma and its applications in graph theory, in: *Combinatorics, Paul P. Erdős is Eighty* (D. Miklós et. al, eds.), Bolyai Society Mathematical Studies **2** (1996), pp. 295–352.

S. Kopparty and B. Rossman: The Homomorphism Domination Exponent, *Europ. J. Combin.* **32** (2011), 1097–1114.

S. Kopparty: Local Structure: Subgraph Counts II, Rutgers University Lecture Notes, http://www.math.rutgers.edu/~sk1233/courses/graphtheory-F11/hom-inequalities.pdf

D. Kozlov: *Combinatorial Algebraic Topology*, Springer, Berlin, 2008.

B. Kra: The Green-Tao Theorem on arithmetic progressions in the primes: an ergodic point of view, *Bull. of the AMS* **43** (2005), 3–23.

D. Král and O. Pikhurko: Quasirandom permutations are characterized by 4-point densities, http://arxiv.org/abs/1205.3074

D. Kunszenti-Kovács (unpublished)

M. Laczkovich: Closed sets without measurable matchings, *Proc. of the Amer. Math. Soc.* **103** (1988), 894–896.

M. Laczkovich: Equidecomposability and discrepancy: a solution to Tarski's circle squaring problem, *J. Reine und Angew. Math.* **404** (1990), 77–117.

M. Laczkovich: Continuous max-flow min-cut theorems, Report 19. Summer Symp. in Real Analysis, *Real Analysis Exchange* **21** (1995–96), 39.

J.B. Lasserre: A sum of squares approximation of nonnegative polynomials, *SIAM Review* **49** (2007), 651–669.

J.L.X. Li and B. Szegedy: On the logarithimic calculus and Sidorenko's conjecture, http://arxiv.org/abs/math/1107.1153v1

B. Lindström: Determinants on semilattices, *Proc. Amer. Math. Soc.* **20** (1969), 207–208.

R. Lipton and R.E. Tarjan: A separator theorem for planar graphs, *SIAM Journal on Applied Mathematics* **36** (1979), 177–189.

P.A. Loeb: An introduction to non-standard analysis and hyperfinite probability theory, *Probabilistic Analysis and Related Topics 2* (A.T. Bharucha-Reid, editor), Academic Press, New York (1979), 105–142.

D. London: Inequalities in quadratic forms, *Duke Mathematical Journal*, **33** (1966), 511–522.

L. Lovász: Operations with structures, *Acta Math. Hung.* **18** (1967), 321–328.

L. Lovász: On the cancellation law among finite relational structures, *Periodica Math. Hung.* **1** (1971), 145–156.

L. Lovász: Direct product in locally finite categories, *Acta Sci. Math. Szeged* **23** (1972), 319–322.

L. Lovász: Kneser's conjecture, chromatic number, and homotopy, *J. Combin. Theory A* **25** (1978), 319-324.

L. Lovász: *Combinatorial Problems and Exercises,* Akadémiai Kiadó - North Holland, Budapest, 1979; reprinted by AMS Chelsea Publishing (2007).

L. Lovász: Connection matrices, in: Combinatorics, Complexity and Chance, A Tribute to Dominic Welsh Oxford Univ. Press (2007), 179–190.

L. Lovász: The rank of connection matrices and the dimension of graph algebras, *Europ. J. Combin.* **27** (2006), 962–970.

L. Lovász: Very large graphs, in: *Current Developments in Mathematics 2008* (eds. D. Jerison, B. Mazur, T. Mrowka, W. Schmid, R. Stanley, and S. T. Yau), International Press, Somerville, MA (2009), 67–128.

L. Lovász: Subgraph densities in signed graphons and the local Sidorenko conjecture, *Electr. J. Combin.* **18** (2011), P127 (21pp).

L. Lovász: Notes on the book: Large networks, graph homomorphisms and graph limits, http://www.cs.elte.hu/~lovasz/book/homnotes.pdf

L. Lovász and A. Schrijver: Graph parameters and semigroup functions, *Europ. J. Combin.* **29** (2008), 987–1002.

L. Lovász and A. Schrijver: Semidefinite functions on categories, *Electr. J. Combin.* **16** (2009), no. 2, Special volume in honor of Anders Björner, Research Paper 14, 16 pp.

L. Lovász and A. Schrijver: Dual graph homomorphism functions, *J. Combin. Theory A* , **117** (2010), 216–222.

L. Lovász and M. Simonovits: On the number of complete subgraphs of a graph, in: *Combinatorics*, Proc. 5th British Comb. Conf. (ed. C.St.J.A.Nash-Williams, J.Sheehan), Utilitas Math. (1976), 439–441.

L. Lovász and M. Simonovits: On the number of complete subgraphs of a graph II, in: *Studies in Pure Math.*, To the memory of P. Turán (ed. P. Erdös), Akadémiai Kiadó (1983), 459-495.

L. Lovász and V.T. Sós: Generalized quasirandom graphs, *J. Combin. Theory B* **98** (2008), 146–163.

L. Lovász and L. Szakács (unpublished)

L. Lovász and B. Szegedy: Limits of dense graph sequences, *J. Combin. Theory B* **96** (2006), 933–957.

L. Lovász and B. Szegedy: Szemerédi's Lemma for the analyst, *Geom. Func. Anal.* **17** (2007), 252–270.

L. Lovász and B. Szegedy: Contractors and connectors in graph algebras, *J. Graph Theory* **60** (2009), 11–31.

L. Lovász and B. Szegedy: Testing properties of graphs and functions, *Israel J. Math.* **178** (2010), 113–156.

L. Lovász and B. Szegedy: Regularity partitions and the topology of graphons, in: *An Irregular Mind, Szemerédi is 70*, J. Bolyai Math. Soc. and Springer-Verlag (2010), 415–446.

L. Lovász and B. Szegedy: Finitely forcible graphons, *J. Combin. Theory B* **101** (2011), 269–301.

L. Lovász and B. Szegedy: Random Graphons and a Weak Positivstellensatz for Graphs, *J. Graph Theory* **70** (2012) 214–225.

L. Lovász and B. Szegedy: Limits of compact decorated graphs http://arxiv.org/abs/1010.5155

L. Lovász and B. Szegedy: The graph theoretic moment problem http://arxiv.org/abs/1010.5159

L. Lovász and K. Vesztergombi: Nondeterministic property testing, http://arxiv.org/abs/1202.5337

N. Lusin: Leçons par les Ensembles Analytiques et leurs Applications, Gauthier–Villars, Paris (1930).

R. Lyons: Asymptotic enumeration of spanning trees, *Combin. Prob. Comput.* **14** (2005) 491–522.

R. Lyons and F. Nazarov: Perfect Matchings as IID Factors on Non-Amenable Groups, *Europ. J. Combin.* **32** (2011), 1115–1125.

J.A. Makowsky: Connection Matrices for MSOL-Definable Structural Invariants, in: *Logic and Its Applications, ICLA 2009* (eds. R. Ramanujam, S. Sarukkai) (2009), 51–64.

W. Mantel: Problem 28, solution by H. Gouwentak, W. Mantel, J. Teixeira de Mattes, F. Schuh and W.A. Wythoff, *Wiskundige Opgaven* **10** (1907), 60–61.

G.A. Margulis: Explicit construction of concentrators, *Probl. Pered. Inform.* **9** (1973), 71–80; English translation: *Probl. Inform. Transmission*, Plenum, New York (1975).

Yu.V. Matiyasevich: The Diophantineness of enumerable sets. Dokl. Akad. Nauk SSSR, 191:279–282, 1970.

J. Matoušek: *Using the Borsuk-Ulam Theorem: Lectures on Topological Methods in Combinatorics and Geometry*, Springer (2003).

B.D. McKay: Maximum Bipartite Subgraphs of Regular Graphs with Large Girth, Proc. 13th SE Conf. on Combin. Graph Theory and Computing, Boca Raton, Florida (1982), 436.

R. McKenzie: Cardinal multiplication of structures with a reflexive relation, *Fund. Math.* **70** (1971), 59–101.

J.W. Moon and L. Moser: On a problem of Turán, *Mat. Kut. Int. Közl.* **7** (1962), 283–286.

H.P. Mulholland and C.A.B. Smith: An inequality arising in genetical theory, *Amer. Math. Monthly* **66** (1959), 673–683.

V. Müller: The edge reconstruction hypothesis is true for graphs with more than $n \cdot \log_2 n$ edges, *J. Combin. Theory B* **22** (1977), 281–283.

B. Nagle, V. Rödl and M. Schacht: The counting lemma for regular k-uniform hypergraphs, *Random Struc. Alg.* **28** (2006), 113–179.

J. Nešetřil and P. Ossona de Mendez: How many F's are there in G? *Europ. J. Combin.* **32** (2011), 1126–1141.

Ilan Newman and Christian Sohler : Every Property of Hyperfinite Graphs is Testable, *Proc. 43th ACM Symp. on Theory of Comput.* (2011) 675–684.

H. N. Nguyen and K. Onak, Constant-time approximation algorithms via local improvements, *Proc. 49th Ann. IEEE Symp. on Found. Comp. Science* (2008), 327–336.

V. Nikiforov: The number of cliques in graphs of given order and size, *Trans. Amer. Math. Soc.* **363** (2011), 1599–1618.

G. Palla, L. Lovász and T. Vicsek, Multifractal network generator, *Proc. NAS* **107** (2010), 7640–7645.

F.V. Petrov, A. Vershik: Uncountable graphs and invariant measures on the set of universal countable graphs, *Random Struc. Alg.* **37** (2010), 389–406.

O. Pikhurko: An analytic approach to stability, Discrete Math, 310 (2010) 2951–2964.

B. Pittel: On a random graph evolving by degrees, *Advances in Math.* **223** (2010), 619–671.

A. Pultr: Isomorphism types of objects in categories determined by numbers of morphisms, *Acta Sci. Math. Szeged* **35** (1973), 155–160.

C. Radin and M. Yin: Phase transitions in exponential random graphs, http://arxiv.org/abs/1108.0649v2

B. Ráth and L. Szakács: Multigraph limit of the dense configuration model and the preferential attachment graph, *Acta Math. Hung.* **136** (2012), 196–221.

A.A. Razborov: Flag Algebras, *J. Symbolic Logic*, **72** (2007), 1239–1282.

A.A. Razborov: On the minimal density of triangles in graphs, *Combin. Prob. Comput.* **17** (2008), 603–618.

A.A. Razborov: On 3-hypergraphs with forbidden 4-vertex configurations, SIAM J. Discr. Math. **24** (2010), 946–963.

G. Regts: The rank of edge connection matrices and the dimension of algebras of invariant tensors, *Europ. J. Combin.* **33** (2012), 1167–1173.

C. Reiher: Minimizing the number of cliques in graphs of given order and edge density (manuscript).

N. Robertson and P. Seymour: Graph Minors. XX. Wagner's conjecture, *J. Combin. Theory B* **92**, 325–357.

V. A. Rohlin: *On the fundamental ideas of measure theory*, Translations Amer. Math. Soc., Series 1, **10** (1962), 1–54. Russian original: *Mat. Sb.* **25** (1949), 107–150.

K. Roth: Sur quelques ensembles d'entiers, C. R. Acad. Sci. Paris **234** (1952), 388–390.

V. Rödl and M. Schacht: Regular partitions of hypergraphs: Regularity Lemmas, *Combin. Prob. Comput.* **16** (2007), 833–885.

V. Rödl and M. Schacht: Regular partitions of hypergraphs: Counting Lemmas, *Combin. Prob. Comput.* **16** (2007), 887–901.

V. Rödl and J. Skokan: Regularity lemma for k-uniform hypergraphs, *Random Struc. Alg.* **25** (2004), 1–42.

R. Rubinfeld and M. Sudan: Robust characterization of polynomials with applications to program testing, SIAM J. Comput. **25** (1996), 252–271.

I.Z. Ruzsa and E. Szemerédi: Triple systems with no six points carrying three triangles, in: *Combinatorics*, Proc. Fifth Hungarian Colloq., Keszthely, Bolyai Society–North Holland (1976), 939–945.

R.H. Schelp and A. Thomason: Remark on the number of complete and empty subgraphs, *Combin. Prob. Comput.* **7** (1998), 217–219.

O. Schramm: Hyperfinite graph limits, *Elect. Res. Announce. Math. Sci. 15* (2008), 17–23.

A. Schrijver: Tensor subalgebras and first fundamental theorems in invariant theory, *J. of Algebra* **319** (2008) 1305–1319.

A. Schrijver: Graph invariants in the edge model, in: *Building Bridges Between Mathematics and Computer Science* (eds. M. Grötschel, G.O.H. Katona), Springer (2008), 487–498.

A. Schrijver: Graph invariants in the spin model, *J. Combin. Theory B* **99** (2009) 502–511.

A. Schrijver: Characterizing partition functions of the vertex model by rank growth (manuscript)

A. Scott and A. Sokal: On Dependency Graphs and the Lattice Gas, *Combin. Prob. Comput.* **15** (2006), 253–279.

A. Scott: Szemerédi's regularity lemma for matrices and sparse graphs, *Combin. Prob. Comput.* **20** (2011), 455–466.

A.F. Sidorenko: Classes of hypergraphs and probabilistic inequalities, *Dokl. Akad. Nauk SSSR* **254** (1980), 540–543.

A.F. Sidorenko: Extremal estimates of probability measures and their combinatorial nature (Russian), *Izv. Acad. Nauk SSSR* **46** (1982), 535–568.

A.F. Sidorenko: Inequalities for functionals generated by bipartite graphs (Russian) *Diskret. Mat.* **3** (1991), 50–65; translation in *Discrete Math. Appl.* **2** (1992), 489–504.

A.F. Sidorenko: A correlation inequality for bipartite graphs, *Graphs and Combin.* **9** (1993), 201–204.

A.F. Sidorenko: Randomness friendly graphs, *Random Struc. Alg.* **8** (1996), 229–241.

B. Simon: *The Statistical Mechanics of Lattice Gasses*, Princeton University Press (1993).

M. Simonovits: A method for solving extremal problems in graph theory, stability problems, in: *Theory of Graphs, Proc. Colloq. Tihany 1966*, Academic Press (1968), 279–319.

M. Simonovits: Extremal graph problems, degenerate extremal problems, and supersaturated graphs, in: *Progress in Graph Theory*, NY Academy Press (1984), 419–437.

M. Simonovits and V.T. Sós: Szemerédi's partition and quasirandomness, *Random Struc. Alg.* **2** (1991), 1–10.

M. Simonovits and V.T. Sós: Hereditarily extended properties, quasi-random graphs and not necessarily induced subgraphs. *Combinatorica* **17** (1997), 577–596.

M. Simonovits and V.T. Sós: Hereditary extended properties, quasi-random graphs and induced subgraphs, *Combin. Prob. Comput.* **12** (2003), 319–344.

Ya.G. Sinai: *Introduction to ergodic theory*, Princeton Univ. Press (1976).

B. Szegedy: Edge coloring models and reflection positivity, *J. Amer. Math. Soc.* **20** (2007), 969–988.

B. Szegedy: Edge coloring models as singular vertex coloring models, in: *Fete of Combinatorics* (eds. G.O.H. Katona, A. Schrijver, T. Szőnyi), Springer (2010), 327–336.

B. Szegedy: Gowers norms, regularization and limits of functions on abelian groups, http://arxiv.org/abs/1010.6211

B. Szegedy (unpublished).

E. Szemerédi: On sets of integers containing no k elements in arithmetic progression", *Acta Arithmetica* **27** (1975) 199–245.

E. Szemerédi: Regular partitions of graphs, *Colloque Inter. CNRS* (J.-C. Bermond, J.-C. Fournier, M. Las Vergnas and D. Sotteau, eds.) (1978) 399–401.

T. Tao: A variant of the hypergraph removal lemma, *J. Combin. Theory A* **113** (2006), 1257–1280.

T.C. Tao: Szemerédi's regularity lemma revisited, *Contrib. Discrete Math.* **1** (2006), 8–28.

T.C. Tao: The dichotomy between structure and randomness, arithmetic progressions, and the primes, in: *Proc. Intern. Congress of Math.* I (Eur. Math. Soc., Zürich, 2006) 581–608.

A. Thomason: Pseudorandom graphs, in: *Random graphs '85* North-Holland Math. Stud. **144**, North-Holland, Amsterdam, 1987, 307–331.

A. Thomason: A disproof of a conjecture of Erdős in Ramsey theory, *J. London Math. Soc.* **39** (1898), 246–255.

P. Turán: Egy gráfelméleti szélsőértékfeladatról. *Mat. Fiz. Lapok* bf 48 (1941), 436–453.

J.H. van Lint and R.M. Wilson: *A Course in Combinatorics*, Cambridge University Press (1992).

V. Vapnik and A. Chervonenkis: On the uniform convergence of relative frequencies of events to their probabilities, *Theor. Prob. Appl.* **16** (1971), 264–280.

A.M. Vershik: Classification of measurable functions of several arguments, and invariantly distributed random matrices, *Funkts. Anal. Prilozh.* **36** (2002), 12–28; English translation: *Funct. Anal. Appl.* **36** (2002), 93–105.

A.M. Vershik: Random metric spaces and universality, *Uspekhi Mat. Nauk* **59** (2004), 65–104; English translation: *Russian Math. Surveys* **59** (2004), 259–295.

D. Welsh: *Complexity: Knots, Colourings and Counting*, London Mathematical Society Lecture Notes **186**. Cambridge Univ. Press, Cambridge, 1993.

D. Welsh and C. Merino: The Potts model and the Tutte polynomial, *Journal of Mathematical Physics*, **41** (2000), 1127–1152.

H. Whitney: The coloring of graphs, *Ann. of Math.* **33** (1932), 688–718.

H.S. Wilf: Hadamard determinants, Möbius functions, and the chromatic number of a graph, *Bull. Amer. Math. Soc.* **74** (1968), 960–964.

D. Williams: *Probability with Martingales*, Cambridge Univ. Press, 1991.

E. Witten: Topological quantum field theory, *Comm. Math. Phys.* **117** (1988), 353–386.

E. Zeidler: *Nonlinear functional analysis and its applications. Part I*, Springer Verlag, New York, 1985.

A.A. Zykov: On Some Properties of Linear Complexes, *Mat. Sbornik* **24** (1949), 163–188; *Amer. Math. Soc. Transl.* **79** (1952), 1–33.

Author Index

Abért, Miklós, xiv
Adámek, Jiři, 447
Adams, Scot, 332
Ahlswede, Rudolph, 28
Albert, Réka, 10
Aldous, David, 23, 180, 184, 338, 341, 344, 346, 357, 358
Alon, Noga, xiv, 8, 128, 143, 151, 154, 160, 230, 272, 274, 276, 279, 294, 358, 385
Angel, Omer, 393
Aristoff, David, 261
Arora, Sanjeev, 19, 265
Artin, Emil, 446
Austin, Timothy, 279

Babson, Eric, 80
Bandyopadhyay, Antar, 377, 378
Barabási, Albert-László, 10
Bayati, Mohsen, 399
Bell, John, 444
Benjamini, Itai, xi, 6, 23, 338, 351, 354, 356, 360, 385, 402, 403
Berg, Christian, 443
Bertoni, Alberto, 376
Billingsley, Patrick, 185, 438
Björner, Anders, 430
Blakley, George, 28
Boguñá, Marian, 157
Bollobás, Béla, 8, 26, 27, 123, 135, 141, 157, 158, 221, 247, 286, 287, 399, 415
Borgs, Christian, xiii, 67, 73, 151, 153, 158, 160, 164, 169, 174, 185, 187, 195, 201, 210, 218, 221, 264, 265, 317, 375, 382
Bowen, Lewis, 357
Brightwell, Graham, 79
Brown, William, 307

Campadelli, Paola, 376
Chatterjee, Sourav, 260
Chayes, Jennifer, xi, xiii, 16, 67, 73, 151, 153, 158, 160, 164, 169, 174, 185, 187, 195, 201, 210, 218, 221, 264, 265, 317, 375, 382
Chervonenkis, Alexey, 446

Christensen, Jens, 443
Chung, Fan, 9, 197, 308, 421, 422
Conlon, David, 142, 143, 296
Cooper, Joshua, 429
Csóka, Endre, xiv, 405, 409, 410

de la Harpe, Pierre, 34, 63
Diaconis, Persi, 8, 14, 157, 178, 180, 184, 193, 218, 221, 253, 257, 260, 441
Dobrushin, Roland, 67, 367, 370, 372, 378
Dörfler, Willibald, 72
Draisma, Jan, 418
Duke, Richard, 279

Elek, Gábor, xiv, 23, 180, 184, 199, 338, 341, 357, 360, 383, 384, 391, 393, 398, 399, 411, 421, 424, 430
Erdős, Paul, xi, 8, 10, 13, 16, 28, 29, 73, 74, 157, 260, 293, 297, 307, 430

Feller, William, 441
Fischer, Eldar, 20, 143, 271, 272, 276
Fisher, David, 27
Fox, Jacob, 142, 143, 198, 296, 384
Franek, Frantisek, 297, 298
Frankl, Peter, 424
Freedman, David, 8, 14, 157, 441
Freedman, Michael, xi, xiii, 47, 64, 75
Frieze, Alan, 127, 142, 145, 146, 230, 279, 428
Frucht, Robert, 433

Gabber, Ofer, 384
Gaboriau, Damien, 336
Galil, Zvi, 384
Gamarnik, David, 377, 378, 382, 399
Garijo, Delia, 97
Gerke, Stefanie, 395, 415
Gijswijt, Dion, 418
Gilbert, Edgar, 8
Godlin, Benny, 46, 51
Goldreich, Oded, xii, 5, 19, 20, 263, 265, 403
Goldwasser, Shafi, xii, 5, 19, 265
Goodall, Andrew, 97

Goodman, Al, 26
Gowers, Timothy, 15, 142, 421, 422, 424, 430
Graham, Ronald, 9, 197, 308, 421, 422
Green, Benjamin, 430
Grimmet, Geoffrey, 191
Gromov, Mikhail, 430
Grothendieck, Alexander, 128
Grzesik, Andrzej, 307

Hága, Péter, 22
Halmos, Paul, 445
Harangi, Viktor, 359
Hassidim, Avinathan, 389
Hatami, Hamed, 30, 239–241, 287, 295–297, 299, 303, 307, 348, 360, 361, 391
Hell, Pavol, 55, 425, 429
Herrlich, Horst, 447
Hilbert, David, 446
Hladky, Jan, 297, 389
Holmes, Susan, 193, 218, 253
Hoory, Shlomo, 80
Hoover, Douglas, 180, 184
Hoppen, Carlos, 429

Imrich, Wilfried, 72
Ioannidis, Yannis, 299
Isbell, John, 425
Ishigami, Yoshiyasu, 424

Jagger, Chris, 297, 298
Janson, Svante, xiv, 8, 122, 123, 131, 157, 158, 178, 180, 184, 193, 218, 221, 247, 248, 253, 257, 429
Jones, Vaughan, 34, 63
Julesz, Bela, 14

Kahn, Jeff, 67, 73, 375, 382
Kaimanovich, Vadim, 391
Kallenberg, Olav, 184, 221
Kallus, Zsófia, 22
Kannan, Ravindran, 127, 142, 145, 146, 160, 230, 279, 428
Karger, David, 19, 265
Karpinski, Marek, 20, 160, 265
Katalin, Vesztergombi, 280
Katona, Gyula O.H., 26–30, 287, 288
Kechris, Alexander, 330, 336, 349, 383
Kellerer, Hans, 438
Kelner, Jonathan, 389
Kimoto, Kazufumi, 425
Kock, Joachim, 84
Kohayakawa, Yoshiharu, 395, 415, 422, 429
Kolmogorov, Andrey, 374
Kolossváry, István, 318
Komlós, János, 446
Kopparty, Swastik, 59, 299
Kotek, Tomer, 46, 51
Kozlov, Dmitri, 80

Kozma, Gadi, 6
Král, Daniel, 297
Kra, Bryna, 430
Krivelevich, Michael, 143, 272
Kruskal, Joseph, 26–28, 30, 287, 288
Kunszenti-Kovács, Dávid, xiv, 239

Laczkovich, Miklós, 138, 337, 338
Laki, Sándor, 22
Lasserre, Jean, 303
Lefmann, Hanno, 279
Li, Xiang, 296
Lindström, Bernt, 433
Linial, Nati, 80
Lippner, Gábor, xiv, 338, 393, 411
Lipton, Richard, 384
Loeb, Peter, 445
London, David, 28
Łoś, Jerzy, 445
Lovász, László, 6, 16, 27, 30, 47, 64, 67–70, 72–78, 80, 98, 101, 125, 139, 141, 143, 149, 151, 153, 157, 158, 160, 164, 166, 167, 169, 174, 178, 180, 185, 187, 195, 201, 210, 218, 221, 226, 232, 245, 253, 257, 264, 265, 268, 280, 294, 296, 303, 308, 309, 311, 317, 318, 324, 348, 360, 361, 375, 382, 391, 418, 425–428, 430, 446
Łuczak, Tomasz, 8
Lusin, Nikolai, 330
Lyons, Russell, xiv, 23, 67, 338, 341, 344, 346, 357, 358, 399, 415

Mátray, Péter, 22
Makowski, Janos, 46, 51
Mantel, W., 25, 26
Margulis, Grigory, 384
Maserick, Peter, 443
Matiyasevich, Yuri, 299, 447
Matoušek, Jiří, 80
McKay, Brandan, 389
McKenzie, Ralph, 72
Menezes Sampaio, Rudini, 429
Merino, Criel, 64
Miller, Benjamin, 330, 336, 349, 383
Moon, John, 293
Moreira, Carlos Gustavo, 429
Moser, Leo, 293
Mulholland, Hugh P., 28
Müller, Vladimir, 69

Nagle, Brendan, 424
Naor, Assaf, 128
Nešetřil, Jarik, xiv, 55, 59, 97, 425, 429
Newman, Ilan, 271, 276, 404
Nguyen, Huy, 24, 389, 405, 407
Nikiforov, Vlado, 27, 141, 289
Norine, Serguey, 30, 287, 297, 299, 303, 307

AUTHOR INDEX

Onak, Krzysztof, 24, 389, 405, 407
Ossona de Mendez, Patrice, 59

Pach, János, 384, 446
Paley, Raymond, 9
Palla, Gergely, 158
Pastor-Satorras, Romualdo, 157
Penrose, Roger, 355
Peres, Yuval, xiv
Petrov, Fedor, 178
Pikhurko, Oleg, xiv, 135, 139, 153, 155
Posenato, Roberto, 376
Pultr, Aleš, 425

Radin, Charles, 261
Ramakrishnan, Raghu, 299
Ráth, Balázs, 318, 429
Razborov, Alexander, xii, 25, 27, 284, 288, 289, 297, 307, 425, 427
Regts, Guus, xiv, 418, 421
Reiher, Christian, 27, 289
Rényi, Alfréd, 8, 10, 157, 260
Ressel, Paul, 443
Riordan, Oliver, 123, 135, 157, 158, 221, 247, 415
Robertson, Neil, 49
Rödl, Vojtech, 279, 297, 298, 421, 422, 424
Romik, Dan, 6
Ron, Dana, xii, 5, 19, 265, 403
Rossman, Benjamin, 59
Rota, Gian-Carlo, 433
Roth, Klaus, 430
Roy, Prabir, 28
Rubinfeld, Ronitt, 19
Ruczinski, Andrzej, 8
Ruzsa, Imre, 198

Sauer, Norbert, 446
Schacht, Matthias, 421, 424
Schelp, Richard, 286, 292
Schramm, Oded, xi, xiv, 23, 338, 351, 354, 356, 360, 383, 385, 389, 402, 403
Schrijver, Alexander, xi–xiii, 53, 64, 75, 76, 108, 174, 185, 418, 419, 421, 425–427
Schützenberger, Marcel-Paul, 433
Scott, Alexander, 67, 395, 415
Seymour, Paul, 49, 385
Shapira, Asaf, 154, 274, 276, 385, 402, 403
Shelah, Saharon, 446
Sidorenko, Alexander, 17, 29, 295, 297
Simonovits, Miklós, xiv, 9, 27, 29, 293, 295, 307
Sinai, Yakov, 121
Skokan, Jozef, 421, 422, 424
Slomson, Alan, 444
Smith, Cedric, 28
Sohler, Christian, 404
Sokal, Alan, 67
Solecki, Slawomir, 330

Sós, Vera T., xi, xiii, xiv, 9, 16, 151, 153, 160, 164, 169, 174, 185, 187, 195, 201, 210, 221, 264, 265, 309, 317
Spencer, Joel, xi, 8, 16, 73, 74
Stav, Uri, 154, 294
Steger, Angelika, 395, 415
Stirzaker, David, 191
Stone, Arthur, 29, 293
Štovíček, Pavel, 297, 298
Strecker, George, 447
Sudakov, Benjamin, 296
Sudan, Madhu, 19
Szakács, László, 318, 326
Szegedy, Balázs, xi, xiii, 16, 30, 77, 78, 98, 108, 125, 139, 141, 143, 149, 157, 166, 167, 178, 180, 184, 199, 226, 232, 234, 245, 253, 257, 268, 294–296, 303, 308, 311, 318, 324, 348, 360, 361, 391, 393, 418, 421, 424, 431, 446
Szegedy, Márió, 143, 272
Szemerédi, Endre, xii, 13, 15, 21, 141, 142, 167, 180, 198, 199, 276, 279, 428, 430

Tao, Terence, 13, 141, 279, 421, 424, 430
Tardos, Gábor, 6
Tarjan, Robert, 384
Tetali, Prasad, 399
Thomas, Robin, 385
Thomason, Andrew, 9, 286, 292, 297, 298
Todorcevic, Stevo, 330
Trevisan, Lucca, 263
Turán, Pál, 13, 25–27, 430
Tutte, William, 48, 64, 434, 453, 462

van Lint, Jakobus, 434
Vapnik, Vladimir, 446
Varadhan, Srinivasa, 260
Vattay, Gábor, 22
Vershik, Anatoly, 430
Vershik, Antoly, 178
Vesztergombi, Katalin, xi, xiii, 16, 151, 153, 160, 164, 169, 174, 185, 187, 195, 201, 202, 210, 221, 264, 265, 317
Vicsek, Tamás, 158
von Neumann, John, 430

Walker, Kevin, xiv
Welsh, Dominic, xiv, 47, 64, 435
Whitney, Hassler, 74
Wilf, Herbert, 433
Williams, David, 441
Wilson, Richard, 9, 197, 308, 434
Winkler, Peter, 79
Witten, Edward, 86
Woeginger, Gerhard, 446

Yin, Mei, 261
Yuster, Raphael, 279

Zeidler, Eberhard, 312
Ziegler, Tamar, 431

Subject Index

adjacency matrix, 39
automorphism, 234
average ε-net, 228
Azuma's Inequality, 163, 265, 440, 441

ball distance, 339
Bernoulli lift, 348
Bernoulli shift, 342
blowup of graph, 40
bond, 38
Borel coloring, 330
Borel graph, 329

Cauchy sequence, 174
chromatic invariant, 435
chromatic polynomial, 47, 435
circle graph, 416
cluster of homomorphisms, 80
component map, 341
concatenation of graphs, 85
conjugate in concatenation algebra, 85
connection matrix, 43
   dual, 76
   flat, 43
connection rank, 44, 107
connector, 94
constituent of quantum graph, 83
continuous geometry, 430
contractor, 94
convergent graph sequence, 173
Counting Lemma, 167
   Inverse, 169
cut
   maximum, 64
cut distance, 12, 128, 132
cut metric, 128
cut norm, 127, 131

distance from a norm, 135
distinguishable by sampling, 266
dough folding map, 342

edge-coloring, 417
edge-coloring function, 417
edge-coloring model, 417

edge-connection matrix, 416
edit distance, 11, 128, 352
epi-mono decomposition, 448
epimorphism, 448
equipartition, 37
equitable partition, 37
equivalence graphon, 255
expansion
   proper, 94

flag algebra, 289, 427
formula
   first order, 50
   monadic second order, 50
   node-monadic second order, 50
free energy, 265
Frobenius algebra, 84
Frobenius identity, 84

grandmother graph, 341
graph
   H-colored
      isomorphic, 71
   ε-homogeneous, 141
   ε-regular, 142
   k-broken, 416
   bi-labeled, 85
   decorated, 120
   directed, 429
   edge-weighted, 38
   eulerian, 63, 89
   expander, 384
   flat, 39
   fully labeled, 39
   gaudy, 51
   H-colored, 71
   hyperfinite, 383
   labeled, 39
   looped-simple, 38
   measure preserving, 332
   multilabeled, 39
   norming, 239
   partially labeled, 39, 85
   planar, 383

470 SUBJECT INDEX

random
  evolving, 8
  involution invariant, 340
  multitype, 8
  randomly weighted, 77
  seminorming, 239
  series-parallel, 88
  signed, 39
  simple, 38
  simply labeled, 39
  W-random, 157
  weakly norming, 239
  weighted, 38
Graph of Graphs, 339
graph parameter, 41
  additive, 41
  contractible, 95
  estimable, 263
  finite rank, 44
  gaudy, 52
  isolate indifferent, 257
  isolate-indifferent, 41
  maxing, 41
  minor-monotone, 42
  multiplicative, 41
  normalized, 41
  reflection positive, 44
    flatly, 44
  simple, 41
graph property, 41
  distinguishable by sampling, 402
  hereditary, 42, 247
  minor-closed, 42
  monotone, 41
  random-free, 247
  robust, 273
  testable, 272, 402
graph sequence
  bounded growth, 384
  hyperfinite, 383
  locally convergent, 16, 353
  quasirandom, 9, 187, 197
    multitype, 10
  subexponential growth, 384
graphing, 23, 332
  bilocally isomorphic, 347
  cyclic, 330
  cylic, 334
  hyperfinite, 385
  local isomorphism, 346
  locally equivalent, 346
  random rooted component, 341
  stationary distribution, 333
graphon, 16, 115, 217
  degree function, 116
  finitely forcible, 308
  infinitesimally finitely forcible, 314
  triangle-free, 247

graphon property, 247
  testable, 268
graphon variety, 250
r-graph, 421
Grothendieck norm, 128
ground state energy, 65
  microcanonical, 204

Hölder property, 240
Hadwiger number, 54
half-graph, 13
Hamilton cycle, 47
Hausdorff Moment Problem, 442
homomorphism, 55
homomorphism density, 6, 58
homomorphism entropy, 202, 204, 375
homomorphism frequency, 59
hypergraph
  r-uniform, 421
  complete, 422
  complete r-partite, 422
hypergraph sequence
  quasirandom, 422

idempotent
  degree, 91
  resolve, 89
  support, 102
idempotent basis, 89
internet, 3
intersection graph, 38
interval graph, 218
isomorphism up to a nullset, 121

kernel, 115, 217
  connected, 122
  direct sum, 122
  pullback, 217
  pure, 222
  regular, 250
  tensor product, 123
  twin-free, 219
  unlabeled, 132
  weakly isomorphic, 121
kernel variety, 250
  simple, 250
K-graphon, 322

left-convergent, 173
left-generator, 448
limit
  dense graph sequence, 180
  weak, 175
line-graph, 38, 331
local distance, 331

Möbius function, 433
Möbius inverse, 42
Möbius matrix, 433

Markov chain, 416, 439
  reversible, 439
  stationary distribution, 439
  step distribution, 439
matching
  perfect, 84
measure
  ergodic, 259
  unimodular, 340
measure preserving, 336
Minkowski dimension, 231
moment matrix, 443
monomorphism, 448
Monotone Reordering Theorem, 253, 442
morphism
  coproduct, 448
  left-isomorphic, 447
  product, 447
  pullback, 448
  pushout, 448
multicut, 64
  restricted, 204
multigraph, 38
multiplicative over coproducts, 426

near-blowup of graph, 40
neighborhood distance, 222
neighborhood sampling, 22
networks, 3
node
  labeled, 28
node cover number, 46
node evaluation function, 417
norm
  invariant, 135

object
  terminal, 447
  zero, 447
oblivious testing, 272
orientation
  eulerian, 49, 97, 119
overlay
  fractional, 129

Pólya urn, 191
Paley graph, 9, 237
parameter estimation, 18
partially ordered set, 429
partition
  fractional, 211, 212
  legitimate, 225
partition function, 265
Penrose tiling, 355
perfect matching, 46, 417
preferential attachment graph
  growing, 191
prefix attachment graph, 188
probability space, 436
  atom, 436
  Borel, 436
  countably generated, 436
  separating, 436
  standard, 437
product of graphs, 40
  gluing, 42
property testing, 5, 19

quantum graph, 83
  k-labeled, 84
  loopless, 84
  simple, 84
quantum morphism, 427
quasirandomness, 422
quotient
  graph
    fractional, 212
quotient graph, 40, 141, 144
  simple, 40
quotient set, 211
  fractional, 212
  restricted, 211

Rado graph, 178
random graph
  ultimate, 256
random graph model, 175
  consistent, 175
  countable, 177
  local, 175
  local countable, 177
random graphon model, 256
random variable
  exchangeable, 184
random walk, 439
rank of kernel, 254
Reconstruction Conjecture, 69
(α, β, k)-regularization, 424
Regularity Lemma, xii, 14, 142, 145, 199
  Strong, 143, 148
  Weak, 142
Removal Lemma, 197, 198, 273
right-generator, 448
root, 339

S-flow, 63
sample concentration, 158, 159
sampling distance, 12, 351
  nondeterministic, 360
Sampling Lemma
  First, 160
  Second, 164, 165
Schatten norm, 135
semidefinite programming, 304
sequence
  well distributed, 185
sequence of distributions
  consistent, 353

involution invariant, 354
similarity distance, 226
square-sum, 282
stability number, 46
stable set polynomial, 65
stationary walk, 439
stepfunction, 115, 442
stepping operator, 144
subgraph sampling, 5
support graph, 336

template graph, 8, 141
tensor network, 418
tensoring method, 240
test parameter, 263
test property, 266, 268
threshold graph
    random, 193
threshold graphon, 187, 253
topology
    local, 331
    weak, 226
tree
    spanning, 48
tree-decomposition, 107
tree-width, 107
Turán graph, 25
Tutte polynomial, 48
twin nodes, 39, 101
twin points, 219
twin reduction, 40

ultrafilter, 444
    principal, 444
ultralimit, 445
ultrametric, 331
ultraproduct, 444
uniform attachment graph, 188
unlabeling, 86

Vapnik–Chervonenkis dimension,
    VC-dimension, 446
variation distance, 12
Voronoi cell, 228

weak convergence, 139

zeta matrix, 433

Notation Index

$\mathbb{1}, \mathbb{1}_A$ indicator function, 37

$\mathfrak{A}$ Borel sets in the graph of graphs, 339
$A \succeq 0$ positive semidefinite, 37
$a \times b$ product in category, 447
$a \oplus b$ coproduct in category, 448
$A \cdot B$ dot product of matrices, 37
$\alpha \vee \beta$ pullback, 448
$\alpha \wedge \beta$ morphisms in pushout, 448
α_φ nodeweight of homomorphism, 56
A_G adjacency matrix, 39
$\alpha(G)$ stability number, 41
$A(G, x)$ characteristic polynomial, 67
α_H nodeweight-sum, 58
$\alpha_i(H)$ nodeweight, 38
$\mathcal{A}^+$ sigma-algebra of weighted rooted graphs, 343
Aut(W) automorphism group, 234
$a_1 W_1 \oplus a_2 W_2 \oplus \dots$ direct sum of kernels, 122

$B_{G,r}(v)$ ball of radius r about v, 38
$B(H, r)$ neighborhood of root, 339
$\beta_{i,j}(H)$ edgeweight, 38
B^m m-bond, 38
$B^{m\bullet\bullet}$ labeled bond, 39
$\mathfrak{B}_r$ r-balls, 339

$\mathbb{C}$ complex numbers, 37
$\mathbf{C}_a$ cyclic graphing, 330
cep$(G; u, v)$ cluster expansion polynomial, 434
chr(G, q) chromatic polynomial, 435
chr$_0(G, k)$ number of colorations, 435
C_n cycle, 38
col(G, h) edge-coloring function, 417
Conn(G) connected subgraphs, 38
cri(G) chromatic invariant, 435
Csp(G) connected spanning subgraphs, 38
$\mathcal{C}(U, H), \mathcal{C}(U, W)$ overlay functional, 205
cut(G, B) maximum multicut, 64

$d_1(G, G')$ edit distance, 128, 352
$d_\square(G, G')$ labeled cut distance, 128, 129

$\widehat{\delta}_\square(G, G'), \delta_\square(G, G')$ unlabeled cut distance, 129
$d_\square(G, G', X)$ overlay cut distance, 129
$\delta_\square(U, W)$ cut distance, 132
$d_\square(U, W)$ cut norm distance, 131
deg, deg$_A$, deg$_A^{\mathbf{G}}$ degree, 331
deg$_c(\varphi, v)$ edge-colored degree, 417
deg(H) degree of root, 339
$d_G(X, Y)$ edge density between sets, 141
$d^{\text{Haus}}(A, B)$ Hausdorff distance, 208
$d^\bullet(H_1, H_2)$ ball distance, 339
dim$_{\text{VC}}(\mathcal{H})$ Vapnik–Chervonenkis dimension, 446
$\delta_\odot^{(r,k)}(G_1, G_2), \delta_\odot^{\text{nd}}(G_1, G_2)$ nondeterministic sampling distances, 360
$\delta_N(U, W), \delta_1(U, W), \delta_2(U, W)$ distances derived from norm, 135
dob(W) Dobrushin value, 370
$\delta_\odot^r(F, F'), \delta_\odot(F, F')$ sampling distance, 351
$\delta_{\text{samp}}(G, G'), \delta_{\text{samp}}(W, W')$ sampling distance, 12, 158
$d_{\text{sim}}(s, t)$ similarity distance, 277
$d^\circ(u, v)$ local distance, 331
$d_{\text{var}}(\alpha, \beta)$ variation distance, 12
$d_W(x)$ normalized degree, 116

E_+, E_- signed edge-sets, 39
$e(G) = |E(G)|$ number of edges, 38
$\eta, \eta_{\mathbf{G}}$ edge measure, 333
$e_G(X, Y)$ set of edges between X and Y, 38
ent(G, H) homomorphism entropy, 202
ent$^*(G, H)$ homomorphism entropy typical, 204
ent$^*(G, W)$ sparse homomorphism entropy, 375
Eul eulerian property, 63
$\overrightarrow{\text{eul}}(F)$ number of eulerian orientations, 49

$f^\uparrow, f^\downarrow, f^\Downarrow$ Möbius inverses, 42
$F_1 F_2$ gluing product, 42
$F^\dagger, F^\ddagger, F^\natural$ labeled quantum graphs from simple graph, 283
$F \circ G$ concatenation of graphs, 85

$\widehat{F}$ signed graph from simple graph, 57
$\mathcal{F}(K), \mathcal{F}_n(K)$ compact decorated graphs, 318
$\mathcal{F}_k^{\text{simp}}$ simple graphs on $[k]$, 38
$F_{k,m}$ bi-labeled graphs, 85, 86
$\mathcal{F}_k^\bullet, \mathcal{F}_S^\bullet$ k-labeled and S-labeled graphs, 39
$\mathcal{F}_k^{\text{mult}}$ multigraphs on $[k]$, 38
$\mathcal{F}_k^{\text{stab}}$ k-labeled graphs with stable labeled set, 95
$\text{flo}(G, q)$ nowhere-zero q-flows, 436
F^* conjugate in concatenation algebra, 85

$G_1 \square G_2$ Cartesian sum, 40
$G_1 \times G_2$ categorical product, 40
$G_1 \boxtimes G_2$ strong product, 40
$G_1 * G_2$ gluing k-broken graphs, 416
$\mathfrak{G}^\rightarrow$ edge-rooted graphs, 340
$\mathfrak{G}^\bullet, \mathcal{G}^\bullet$ rooted graphs, 339
$\mathfrak{G}_F^\bullet$ extensions of F, 339
$\mathbb{G}(H)$ randomized weighted graph, 157
$\mathbb{G}(k, G)$ random induced subgraph, 157
$G^{\times k}$ categorical power, 40
$G(m)$ blow-up of G, 40
$\mathbb{G}(n,p), \mathbb{G}(n,m), \mathbb{G}(n; H)$ random graphs, 8
$\mathbf{G}^+$ Bernoulli lift, 348
$\mathbf{G}^+$ weighted rooted graphs, 343
G_n^{pa} growing preferential attachment graph, 191
G_n^{pfx} prefix attachment graph, 188
$G/\mathcal{P}$ quotient graph, 40, 83, 141, 211
$G_\mathcal{P}$ averaged graph, 141
G/ρ fractional quotient graph, 212
$G[S]$ induced subgraph, 38
G^{simp} simple version of G, 38
$\mathbb{G}(n,W), \mathbb{G}(S,W)$ W-random graphs, 157
G_n^{ua} uniform attachment graph, 188
$[\![G]\!]$ unlabeling, 39
$[\![G]\!]_S$ removing labels not in S, 39
$\mathbf{G}_x$ connected component containing x, 338

$h(\alpha)$ head of morphism, 447
$H(a, B))$ weighted graph with nodeweight vector a and edgeweight matrix B, 38
$\mathbf{H}$ graph of graphs, 339
$\mathbf{H}^+$ graph of weighted graphs, 343
$\hom(F, G)$ homomorphism number, 6, 56, 77, 319
$\text{Hom}(F, G)$ set of homomorphisms, 56
$Hom(F, G)$ graph of homomorphisms, 79
$\mathbf{Hom}(F, G)$ complex of homomorphisms, 80
$\hom_\varphi(F, G)$ weight of homomorphism, 56, 58
$\hom^*(G, H)$ typical homomorphism number, 204
$\hom(F, X), \text{inj}(F, X)$ homomorphism polynomials, 108
$\mathbb{H}(n,W), \mathbb{H}(S,W)$ W-random weighted graphs, 157

$\mathcal{I}(G)$ set of stable sets, 65
$\text{ind}(F, G)$ number of embeddings, 56
$\text{inj}(F, G)$ injective homomorphism number, 56, 57
$I(W)$ entropy functional, 260
$\mathcal{I}(W)$ induced subgraphs of graphon, 247

$[k] = \{1, 2, \ldots, k\}$, 12
$K_1^{\bullet\bullet}$ 2-multilabeled graph on one node, 39
$\mathcal{K}(a, b)$ morphisms, 447
$K_{a,b}^\bullet, K_{a,b}^{\bullet\bullet}$ partially labeled complete bipartite graphs, 39
$\mathcal{K}_a^{\text{in}}, \mathcal{K}_a^{\text{out}}$ morphisms into and out of a, 447
$\chi(G)$ chromatic number, 41
K_n complete graph, 38
K_n° looped complete graph, 38
$K_n^\bullet, K_n^{\bullet\bullet}$ partially labeled complete graphs, 39
K_n^r complete hypergraph, 422

$L(C)$ intersection graph, 38
$L(G)$ line-graph, 38, 331
$\lambda_i(G), \lambda_i'(G), \lambda_i(W), \lambda_i'(W)$ eigenvalues, 195
$\lim_\omega$ ultralimit, 445
$\ln, \log, \log^*$ natural, binary and iterated logarithm, 37
L_n^r complete r-partite hypergraph, 422
$\mathcal{L}(W)$ space of functions t_{xy}, 311

$\text{Maxcut}(G), \text{maxcut}(G)$ maximum cut, 64
$M(f, k), M^{\text{simp}}(f, k), M^{\text{mult}}(f, k), M^{\text{flat}}(f, k), M(f, \mathbb{N})$ connection matrices, 43
$M'(f, k)$ edge-connection matrix, 416
MG Möbius inverse of ZG, 83
$M_{\text{hom}}^A, M_{\text{inj}}^A, M_{\text{surj}}^A, M_{\text{aut}}^A$ matrices of homomorphism numbers, 73
$M_k(f)$ moments of a function, 442
$\mu(x, y), \mu_P$ Möbius function, 433, 434

$\mathbb{N}, \mathbb{N}^*$ nonnegative integers and positive integers, 37
$N(f, k)$ dual connection matrix, 76
$\nu(G)$ matching number, 41
$N_G(v) = N(v)$ neighbors of v, 38
$(n)_k = n(n-1)\ldots(n-k+1)$, 37
$\mathcal{N}_k(f)$ annihilator of quantum graphs, 84
$\nabla(v)$ edges incident with v, 38

$\text{Ob}(\mathcal{K})$ objects of category, 447
$\omega(G)$ size of maximum clique, 41
O_n edgeless graph, 38

$\mathbf{P}_a$ path graphing, 330
$\pi_{G,H}, \pi_{G,W}, \pi_y$ distribution on homomorphisms, 368
$\Pi(\alpha)$ partitions of $[0, 1]$, 205
$\Pi(n), \Pi(n, \alpha)$ partitions of $[n]$, 204
$\Pi^*(n, \alpha)$ fractional partitions, 212

pm(G) number of perfect matchings, 41
P_n path, 38
$P_n^\bullet, P_n^{\bullet\bullet}$ partially labeled paths, 39
$\prod_\omega$ ultraproduct, 444
$\overline{\mathcal{P}}$ closure of graph property, 247
$\mathcal{P}(T)$ probability measures on Borel sets, 438

$\mathcal{Q}_q(G), \mathcal{Q}_\mathbf{a}(G)$ quotient sets, 211
$\mathcal{Q}_q^*(G), \mathcal{Q}_\mathbf{a}^*(G)$ fractional quotient sets, 212
$\mathcal{Q}_k$ k-labeled quantum graphs, 84
$\mathcal{Q}_k/f, \mathcal{Q}_{k,k}/f$ factor algebras, 84
$\mathcal{Q}_k^{\text{stab}}$ k-labeled quantum graphs with stable labeled set, 95
qr(H) quasirandomness, 422

$\mathbb{R}, \mathbb{R}_+$ reals and nonnegative reals, 37
$\mathcal{R}_\varepsilon^c$ graphons far from $\mathcal{R}$, 270
$r(f,k)$ connection rank function, 44
$\rho_{G,r}$ neighborhood sample distribution, 22
root(H) root of graph, 339
$\overline{r}_W(a,b)$ similarity distance, 226
$r_W(x,y)$ neighborhood distance, 222

$\overline{S}_{[0,1]}, S_{[0,1]}$ measure preserving maps, 437
$\sigma_{G,k}$ subgraph sample distribution, 5
S_n star, 38
σ^+ distribution of weighted graphs, 343
Spec(W) spectrum, 124
$\sigma^*(A)$ probability measure scaled by degrees, 339
stab(G) number of stable sets, 46
stab(G,x) stable set polynomial, 65
surj(F,G) surjective homomorphism number, 56

$t(\alpha)$ tail of morphism, 447
$t(F,G), t_{\text{inj}}(F,G), t_{\text{ind}}(F,G)$ homomorphism densities, 58, 319
$t(F,W), t_{\text{ind}}(F,W)$ homomorphism densities in graphons, 116, 117
$t(F,w)$ decorated homomorphism density, 120, 323
$\tau(G)$ node cover number, 46
$\tau(\mathcal{H})$ node-cover of hypergraph, 446
$\mathcal{T}_k$ tensors with k slots, 419
$T(n,r)$ Turán graph, 25
tree(G) number of spanning trees, 435
$t^*(F,G), t_{\text{inj}}^*(F,G), t_{\text{ind}}^*(F,G)$ homomorphism frequencies, 59
tut($G;x,y$) Tutte polynomial, 434
T_W kernel operator, 124
$t_x(F,W)$ labeled homomorphism density, 117

$UW, U \circ W, U \otimes W$ products of kernels, 123
$U[X]$ submatrix of kernel, 160

v(G) = $|V(G)|$ number of nodes, 38

$\mathcal{W}, \mathcal{W}_0, \mathcal{W}_1$ spaces of kernels and graphons, 115
$\widetilde{\mathcal{W}}, \widetilde{\mathcal{W}}_0, \widetilde{\mathcal{W}}_1$ unlabeled kernels, 132
$W^\varphi(x,y)$ variable transformation, 121
W_H graphon from graph, 115
$\mathcal{W}(K)$ K-graphons, 322
$W^n, W^{\circ n}, W^{\otimes n}$ powers of kernels, 123
$W/\mathcal{P}$ quotient graph, 144
$W_\mathcal{P}(x,y)$ stepping operator, 144

$x \geq 0$ (for $\mathcal{U}$) nonnegativity of quantum graphs, 281
$x \equiv y \pmod{f}$ congruence of quantum graphs, 84

$\mathbb{Z}, \mathbb{Z}_q$ integers and integers modulo q, 37
ZG sum of quotients, 83

Published Titles in This Series

60 **László Lovász,** Large Networks and Graph Limits, 2012
58 **Freydoon Shahidi,** Eisenstein Series and Automorphic L-Functions, 2010
57 **John Friedlander and Henryk Iwaniec,** Opera de Cribro, 2010
56 **Richard Elman, Nikita Karpenko, and Alexander Merkurjev,** The Algebraic and Geometric Theory of Quadratic Forms, 2008
55 **Alain Connes and Matilde Marcolli,** Noncommutative Geometry, Quantum Fields and Motives, 2008
54 **Barry Simon,** Orthogonal Polynomials on the Unit Circle: Part 1: Classical Theory; Part 2: Spectral Theory, 2005
53 **Henryk Iwaniec and Emmanuel Kowalski,** Analytic Number Theory, 2004
52 **Dusa McDuff and Dietmar Salamon,** J-holomorphic Curves and Symplectic Topology, Second Edition, 2012
51 **Alexander Beilinson and Vladimir Drinfeld,** Chiral Algebras, 2004
50 **E. B. Dynkin,** Diffusions, Superdiffusions and Partial Differential Equations, 2002
49 **Vladimir V. Chepyzhov and Mark I. Vishik,** Attractors for Equations of Mathematical Physics, 2002
48 **Yoav Benyamini and Joram Lindenstrauss,** Geometric Nonlinear Functional Analysis, 2000
47 **Yuri I. Manin,** Frobenius Manifolds, Quantum Cohomology, and Moduli Spaces, 1999
46 **J. Bourgain,** Global Solutions of Nonlinear Schrödinger Equations, 1999
45 **Nicholas M. Katz and Peter Sarnak,** Random Matrices, Frobenius Eigenvalues, and Monodromy, 1999
44 **Max-Albert Knus, Alexander Merkurjev, Markus Rost, and Jean-Pierre Tignol,** The Book of Involutions, 1998
43 **Luis A. Caffarelli and Xavier Cabré,** Fully Nonlinear Elliptic Equations, 1995
42 **Victor W. Guillemin and Shlomo Sternberg,** Variations on a Theme by Kepler, 1991
40 **R. H. Bing,** The Geometric Topology of 3-Manifolds, 1983
39 **Nathan Jacobson,** Structure and Representations of Jordan Algebras, 1968
38 **O. Ore,** Theory of Graphs, 1962
37 **N. Jacobson,** Structure of Rings, 1956
36 **Walter Helbig Gottschalk and Gustav Arnold Hedlund,** Topological Dynamics, 1955
34 **J. L. Walsh,** The Location of Critical Points of Analytic and Harmonic Functions, 1950
33 **J. F. Ritt,** Differential Algebra, 1950
32 **R. L. Wilder,** Topology of Manifolds, 1949
31 **E. Hille and R. S. Phillips,** Functional Analysis and Semi-groups, 1996
30 **Tibor Radó,** Length and Area, 1948
29 **A. Weil,** Foundations of Algebraic Geometry, 1946
28 **G. T. Whyburn,** Analytic Topology, 1942
27 **S. Lefschetz,** Algebraic Topology, 1942
26 **N. Levinson,** Gap and Density Theorems, 1940
25 **Garrett Birkhoff,** Lattice Theory, 1940
24 **A. A. Albert,** Structure of Algebras, 1939
23 **G. Szego,** Orthogonal Polynomials, 1939
22 **Charles N. Moore,** Summable Series and Convergence Factors, 1938
21 **Joseph Miller Thomas,** Differential Systems, 1937
20 **J. L. Walsh,** Interpolation and Approximation by Rational Functions in the Complex Domain, 1935
19 **N. Wiener and R. C. Paley,** Fourier Transforms in the Complex Domain, 1934
18 **M. Morse,** The Calculus of Variations in the Large, 1934
17 **J. H. M. Wedderburn,** Lectures on Matrices, 1934

PUBLISHED TITLES IN THIS SERIES

16 **Gilbert Ames Bliss,** Algebraic Functions, 1933
15 **M. H. Stone,** Linear Transformations in Hilbert Space and Their Applications to Analysis, 1932
14 **Joseph Fels Ritt,** Differential Equations from the Algebraic Standpoint, 1932
13 **R. L. Moore,** Foundations of Point Set Theory, 1932
12 **Solomon Lefschetz,** Topology, 1930
11 **Dunham Jackson,** The Theory of Approximation, 1930
10 **Arthur B. Coble,** Algebraic Geometry and Theta Functions, 1929
 9 **George D. Birkhoff,** Dynamical Systems, 1927
 8 **L. P. Eisenhart,** Non-Riemannian Geometry, 1927
 7 **Eric T. Bell,** Algebraic Arithmetic, 1927
 6 **Griffith Conrad Evans,** The Logarithmic Potential: Discontinuous Dirichlet and Neumann Problems, 1927
 5 **Griffith Conrad Evans and Oswald Veblen,** The Cambridge Colloquium, 1918
 4 **Leonard Eugene Dickson and William Fogg Osgood,** The Madison Colloquium, 1914
 3 **Gilbert Ames Bliss and Edward Kasner,** The Princeton Colloquium, 1913
 2 **Max Mason, Eliakim Hastings Moore, and Ernest Julius Wilczynski,** The New Haven Colloquium, 1910
 1 **Frederick Shenstone Woods, Henry Seely White, and Edward Burr Van Vleck,** The Boston Colloquium, 1905